Molecular Neurobiology
of the Mammalian Brain

Molecular Neurobiology of the Mammalian Brain

Patrick L. McGeer
The University of British Columbia
Vancouver, British Columbia, Canada

Sir John C. Eccles
Abteilung Neurobiologie
Max-Planck-Institut für Biophysikalische Chemie
Göttingen, West Germany

and

Edith G. McGeer
The University of British Columbia
Vancouver, British Columbia, Canada

PLENUM PRESS · NEW YORK AND LONDON

Library of Congress Cataloging in Publication Data

McGeer, Patrick L
  Molecular neurobiology of the mammalian brain.

  Includes index.
  1. Brain. 2. Molecular biology. 3. Neurobiology. 4. Mammals — Physiology. I. Eccles, John Carew, Sir, joint author. II. McGeer, Edith G., joint author. III. Title. [DNLM: 1. Brain. 2. Molecular biology. WL300.3 M145m]
QP376.M14   599'.01'88   78-14812
ISBN 0-306-40073-1 (pbk.)

First paperback printing — November 1978
Second paperback printing — December 1979

Preface

The human brain is the inner universe through which all external events are perceived. That fact alone should ensure that neuroscience will eventually receive top priority in the list of human endeavors. The brain represents the pinnacle of sophistication in the realm of living systems. Yet it is an imperfect organ, whose failures in disease processes lead to the occupation of more than half of all hospital beds and whose variable performance in the healthy state contributes in undetermined degree to the world's social problems. Every significant advance in our understanding of the brain has yielded enormous practical dividends. There is every reason to believe the future holds even greater promise.

It can be said that brain research took root near the end of the last century when Ramón y Cajal proved beyond doubt that the neuron is the basic functioning unit of the brain and Sherrington revealed its method of transmitting impulses. But it is only in the past two decades that neuroscience has been established as a recognized discipline where the anatomical, physiological, and chemical aspects of neuronal function are treated in a unified fashion. It can be anticipated that this logical advance will allow brain research to reach new levels of sophistication. Already it has resulted in the establishment of graduate programs at dozens of universities, and the founding of numerous journals devoted to reports of interdisciplinary research on the brain.

The activity has not been matched by a comparable appearance of basic texts which present, with balanced emphasis, the traditional disciplines which make up neuroscience. This book is one attempt to fill the gap.

The intention is not to present in detail any aspect of neuroscience. Rather, it is to select for the student certain basic information from each of the subdisciplines in order to build an integrated picture of brain function. We anticipate as well that the book will have value to the specialist, not so much in his own particular realm but in those areas with which he may not be familiar. Thus, it is our hope that the neurophysiologist and neuroanatomist will gain insight into neurochemical and neuropharmacological operations, while the neuropharmacologist and neurochemist will gain some insight into neuro-physiological operations.

The book is developed in three parts. The first deals with the architecture of the neurons and the ways in which neurons signal to each other. The second deals with specific chemical neurotransmitters and the effects of drugs which interact with them. The third deals with the integrative aspects of brain function, drawing heavily on the basic information in the first two parts.

We have concentrated on mammalian brains because, in the neurochemical chapters of Part II in particular, the great bulk of the investigations have been on mammalian brains. The analytical neurobiology of Chapters 1, 2, 3, and 12 is, however, dependent to a considerable extent on the fundamental experiments that have been carried out on nervous systems of lower verte-brates or even of the invertebrate *Loligo*.

To satisfy the needs of economy of space and clarity of presentation, we have avoided detailed documentation and have not attempted to credit every investigator who has made an outstanding contribution to his specialty. Refer-ences, often to review articles, have been quoted by the authors' names in the text and then collected at the end of the book. While these may not present a proper balance of prominent authors, they should provide a satisfactory guide to the reader who wishes to pursue any subject in more depth.

ACKNOWLEDGMENT. A review book of this nature is dependent on good illustra-tions and we are grateful to the publishers and scientists who generously granted permission to reproduce figures from articles by the following authors: K. Akert, G. I. Allen, J. Altman, P. Andersen, M. H. Aprison, H. Asanuma, R. P. Barber, T. V. P. Bliss, L. L. Butcher, R. E. Burke, E. DeRobertis, B. Droz, R. M. Eccles, E. Evarts, P. Fatt, C. A. Fox, K. Fuxe, N. Geschwind, B. Granit, E. G. Gray, P. Greengard, J. Hamori, H. K. Hartline, B. K. Hartman, T. Hattori, E. Henneman, A. L. Hodgkin, T. Hokfelt, M. Ito, M. Jacobson, J. K. S. Jansen, B. Katz, R. D. Keynes, H. H. Kornhuber, S. W. Kuffler, M. Kuno, B. Libet, T. Lømo, A. Lundberg, G. S. Lynch, U. J. Mahan, R. Marsh-banks, P. B. C. Matthews, B. Milner, V. B. Mountcastle, N. H. Neff, R. A. Nicoll, S. Ochs, O. Oscarsson, T. Oshima, D. W. Pfaff, R. Poritsky, G. Raisman, P. Rakic, W. Rall, C. E. Ribak, E. Roberts, A. Rose, P. H. Schiller, G. E. Schneider, T. A. Sears, R. W. Sperry, J. Storm-Mathisen, J. Szentágonthai, T. Takahashi, A. Takeuchi, C. A. Terzuolo, W. T. Thach, C. B. Trevarthen, N. Tsukahara, U. Ungerstedt, Å B. Vallbo, J. R. Walters, S. J. Watson, and J. N. Wiesel. We owe the drawing for Figure 13.10 to P. Scheid and many of the other illustrations are the work of Miss Marguerite Drummond. We would like to thank them as well as Mrs. Elizabeth Lyson and Mrs. Wendy Merlak who spent endless hours typing the manuscript.

Contents

Part II
Specific Neuronal Participants and Their Physiological Actions 141

Chapter 14
Basic Behavioral Patterns 465

List of Figures

Chapter 3
Chemical Synaptic Transmission

Chapter 9
Serotonin Neurons

Chapter 10
The Promising Peptides

Chapter 11
The Putative Transmitters

Chapter 14
Basic Behavioral Patterns

Chapter 15
The Neuronal Mechanisms Involved in Learning and Memory

Chapter 16
Perception, Speech, and Consciousness

I

Architecture and Operation of the Nervous System

It is essential that all investigators of the nervous system understand the principles of operation of the neural machinery that they are investigating. Usually this research is being carried out on some small component that is isolated for study by highly sophisticated techniques, for example, in neurochemistry, neuroimmunology, or electron microscopy. The study may even be at a more elemental level with such techniques as homogenization of some brain constituents such as cerebral nuclei and the isolation from fractions of the homogenates of interesting molecules such as, for example, enzymes or other proteins and putative transmitters or related molecules. The aim of this kind of investigation is to construct metabolic systems that are vitally concerned in chemical synaptic transmission. This important field of inquiry will be dealt with extensively in Part II of this book. In order to achieve neurobiological meaning, however, the detailed accounts of neurochemical and neuropharmacological studies must be related to the structure and operation of the neural machinery of the brain.

In Part I this information is presented in four chapters. It will not be a bare factual treatment. Rather, the attempt is made to outline the important historical events and to show the growth of understanding that occurs with a progressively changing problem situation. Too often it is erroneously believed that science consists of the discovery and presentation of facts. On the contrary, it is concerned with the attempt to understand the phenomena of nature by the development of hypotheses and their rigorous testing by experiments. We hope that this book is an exemplar of this philosophy of science, which makes science an exciting adventure both in creative imagination and in experimental investigation, two aspects of science in vital interaction.

The first chapter gives an account of the structure of the components of the nervous system: neurons, synapses, nerve fibers, glia. Special treatment will be given to organelles, both in the perinuclear region and in the presynap-

1

tic terminals, because of their significance in relation to the theme of Part II. The second chapter concentrates on the beautiful and efficient signaling system of nerve impulses. In the third chapter, synaptic transmission is studied in two locations of peripheral nervous systems because at these favorable sites the principles of operation have been investigated with great precision. When established in this way, it was possible to show that similar principles obtained with transmission at some synapses of the central nervous system that were amenable to comparable investigations. For example, the quantal composition demonstrated for the end-plate potential could be recognized with central synapses (cf. Figure 4.3, Section 4.2). The fourth chapter comprises a wide range of investigations on excitatory and inhibitory synapses, and finishes with a description of some simpler neuronal pathways. This will serve as an introduction to the much more complex situations in Part III.

1

The Fine Structure of the Mammalian Brain

1.1 Introduction

The brains of six mammals drawn on the same scale in Figure 1.1 illustrate the evolutionary development of the brain from the primitive marsupial up to man, where the cerebrum and cerebellum are dominant. But size alone does not account for the preeminent performance of the human brain. This organ, weighing about 1.5 kg, is the most highly organized and the most completely organized matter known in the universe. At our present level of understanding, the brains of higher mammals are not greatly inferior in microstructure and in most aspects of operational performance. The brains of elephants and whales are many times larger—up to five times for a whale—and even the brain of the dolphin is a little larger than the human brain. Yet there is something very special about the human brain. It has a performance in relationship to culture, to consciousness, to language, and to memory, that uniquely distinguishes it from even the most highly developed brains of other animals. That is a problem that we shall discuss in the last chapter. We shall see there that it is beyond our comprehension how these subtle properties of the conscious self came to be associated with a material structure, the human brain that owes its origin to the biological process of evolution. We can state with complete assurance that for each of us, our brain forms the material basis of our experiences and memories, our imaginations, our dreams. Furthermore, it is through our brain that each of us can plan and carry out actions and so achieve expression in the world. We are able to do this because our streams of conceptual thinking can somehow or other activate neuronal changes that eventually result in all the complex movements that give expression. For each of us, our brain is the material basis of our personal identity—distinguished by our selfhood and our character. In summary, it gives for each of us the essential "me." Yet when all this is said we are still only

3

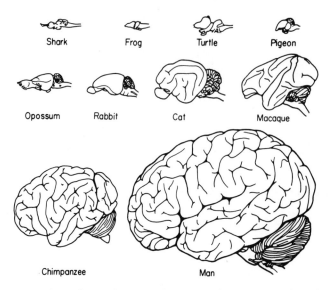

Shark   Frog   Turtle   Pigeon

Opossum   Rabbit   Cat   Macaque

Chimpanzee   Man

Figure 1.1 Brains of vertebrates drawn on the same scale. (Courtesy of Professor J. Jansen.)

at the beginning of comprehending the mystery of what we are. This fundamental philosophical problem is still far beyond our understanding, though in
the last chapter we will suggest that remarkable progress is being made in this,
the greatest of all problems that confront us. It is essentially the problem
formulated long, long ago and defined in the question of Plotinus: "What
am I?"

Meanwhile, we will be engaged in the task of trying to understand the
nervous system as neurobiologists. It is convenient at this stage to consider the
brain as a machine, but it is a special kind of machine of a far higher order of
complexity in performance than any machine designed by man, even the most
complex of computers. Please be warned against the claims of those computer
devotees who seek to alarm us by their arrogant assertions that computers will
soon outsmart man in all that matters. This is science fiction, and these devotees are the modern variants of the idol makers of other superstitious ages;
and like them they seek power through the fostering of idolatry. But we
hasten to add that more than in any other science or technology we shall need
advanced computer instrumentation in our scientific investigations on the
brain.

It will be our task in the first chapter to describe the essential structure of
the brain—the components from which it is built and how they are related to
each other. In the second, third, and fourth chapters we will give an account
of the mode of operation of the simplest components. Each of these two
modes of investigation, the morphological or structural and the physiological,
is complementary to the other and together they form the basis of a further
phase of our inquiry which concerns the linkage of individual components
into the simplest levels or organization. We are beginning to understand the

simpler patterns of neuronal organization and the way they work. We are still at a very early stage of our attempt to understand the brain, however, which may well be the last of all the frontiers of knowledge that man can attempt to penetrate and encompass. We predict that it will occupy hundreds of years into the future. We will never run out of problems on this greatest of all problems confronting man because the problems multiply far faster than we solve them! Vigorous and exciting new disciplines emerge in such fields as neurochemistry, molecular neurobiology, neurogenetics, neuropharmacology, neuromathematics, and neurocommunications.

The cellular components of the brain fall into two major categories: neurons or nerve cells, and neuroglia. The neurons subserve all the specific properties of the brain such as impulse transmission, which operates on the input side in perception and on the output side in movement, and which in all of its patterned diversity underlies the analytic and synthetic properties of the brain. Neuroglia cells are ubiquitous throughout the brain. The word "glia" is derived from "glue" because neuroglia were originally thought of as merely the supporting structures of the brain, but now they are known to have more important and subtle functions, as will be described after the section on neurons.

1.2 The Neuron

1.2.1 General Morphology

Figure 1.2A is a low magnification of a histological section of a small segment of the human visual cortex (Sholl 1956). Each of these densely packed dots is a neuron or a glia cell, and these are the individual components of which the whole brain is built. You have to imagine that the human cerebrum, shown in the drawing of Figure 1.1, has about 10,000 million of those individual neurons; in Figure 1.2A they are shown tightly packed together, but actually the density of the packing is much greater. In Figure 1.2A only the bodies of the neurons are stained. You do not see all the interlacing branching structures stemming from each one of these cell bodies, as shown in Figure 1.2B for a few nerve cells. In this Golgi-stained section of the human cerebral cortex (Ramón y Cajal 1911) at about two times higher magnification you see one of those neurons or nerve cells (D) well displayed right in the center. Growing out upward and sideways from the cell body there are several branching dendrites, as they are called, and projecting straight downward is the fine axon, a nerve fiber. It gives a few branches before leaving the cortex, as also do the axons of cells B, C, and E. Other nerve cells or neurons, A, F, J, and K, in various shapes and sizes, can be seen with their dendrites in different directions and with their axons branching and terminating in that section of the cerebral cortex.

The clarity of Figure 1.2B depends on the fact that in this thin section only about 2% of the cells were stained by this special Golgi technique. But for

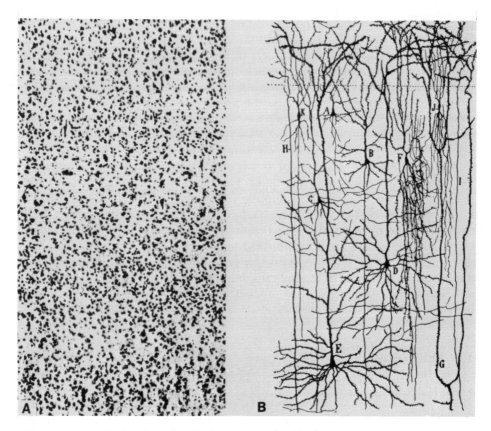

Figure 1.2 Magnified sections of cerebral cortex. A. All cell bodies are stained. (From Sholl 1956.) B. A composite drawing of a Golgi preparation by Ramón y Cajal.

this lucky accident of random selection by the stain, which is still not understood, the composite of nerve cells seen in Figure 1.2A with their dendrites and axons would be solid black, impenetrable to discrimination. This packing in the cerebral cortex is extremely dense and intricate, but we shall see that, despite this congestion, the neuronal components of the cortex (Figure 15.16, Section 15.7) or other regions of the brain are selectively connected to form organized patterns.

We owe particularly to Ramón y Cajal (1895, 1909, 1934), the great Spanish neuroanatomist, the concept that the nervous system is made up of neurons which are isolated cells, not joined together in some syncytium, but each one independently living its own biological life. This concept is called the *neuron theory*. It is already evident from Figure 1.2B that neurons characteristically are complex, branched structures with a central body or soma from which there arises the branching dendrites and the single axon. These form respectively what we may call the receptive pole and the transmitting pole of the neuron (Bodian 1967), but this terminology must not be taken too literally because the soma also participates in the receptivity of the neuron and usually in the transmissive function as well. In fact, the dendrites can be regarded simply as extensions of the soma and we shall see that they have a similar cytoplasmic composition. It is otherwise with the axon, right from the site of its origin from the soma.

Many illustrations of later chapters show special morphological features of neurons. In the mammalian brain the largest neurons may have a soma 70 μm across whereas the smallest would be less than 5 μm. Dendrites usually arise as multiple branches from the soma (cf. Figures 1.2B and 1.3) often with a dominant *apical dendrite* as in cells B, C, D, and E of Figure 1.2B. But the most elaborately branched dendrite, that of the Purkinje cell, arises unipolarly (cf. Figure 13.12, Section 13.4.3). With pyramidal cells of the cerebral cortex (Figures 1.2B; 1.3B; and 15.16, Section 15.7) the apical dendrite extends up to the surface, giving off several branches en route (Ramón y Cajal 1911; Szentágothai 1969). In most species of neurons the axon arises from a conical projection of the soma, the initial segment (Figure 1.3A). It may continue for millimeters to a meter as a transmission line from the soma, being then termed a *nerve fiber*. Usually it gives off several branches soon after its origin, *the axon collaterals*. The most numerous neurons are those with short axons. Usually, as in Figure 1.2B, cells *A*, *F*, and *J*, the axons of these neurons terminate in profuse branching soon after their origin (the so-called Golgi cells). In other cases the axon may extend for a few millimeters with very restricted branching. The most notable example is the granule cell of the cerebellum (cf. Figure 13.12, Section 13.4.3), where the axon dichotomizes to form a parallel fiber with a trajection of several millimeters (Ramón y Cajal 1911; Fox 1962; Eccles *et al.* 1967). These illustrative examples of neuronal morphology are offered with the warning that this morphology is very various and often very disarrayed by random branchings. There has been much effort in attempting to systematize neurons purely on the basis of gross morphology, but it is doubtful if much functional meaning can be attached to shapes *per se*. Rather, have we to consider the diverse structures as providing opportunities for communication.

An important generalization is that all neurons are completely ensheathed by a surface membrane which is about 7 nm thick and is composed of a bimolecular leaflet of phospholipid molecules with the hydrophilic polar groups pointing both outward and inward. The detailed structure with the associated proteins will be treated in a later section of this chapter (Figure 1.16).

1.2.2 Synapse

Since a neuron is a biological unit completely enclosed by the surface membrane, we are confronted by the problem of how neurons communicate to each other. This communication is the essence of all brain action, from the simplest to the unimaginably complex. It happens by means of the fine branches of axons of other neurons that make contact with its surface and end in little knobs scattered all over its soma and dendrites, as indicated in Figures

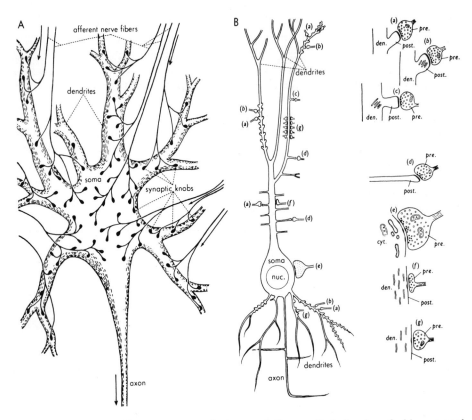

Figure 1.3 Synaptic endings on neurons. A. A general diagram. B. A drawing of a hippocampal pyramidal cell to illustrate the diversity of synaptic endings on the different zones of the apical and basal dendrites, and the inhibitory synaptic endings on the soma. The various types of synapses marked by the letters a to g are shown in higher magnification to the right. (From Hamlyn 1962.)

1.3A and 1.3B. It was Sherrington's concept (1906) that these contact areas are specialized sites of communication, which he labeled synapses from the Greek word *synapto*, which means to clasp tightly. (For a historical review, reference can be made to Chapter 1 of Eccles 1964b.)

Diagrammatically, the neuronal connectivity is indicated in the model of a neuron suitably enlarged in Figure 1.3A. From the soma or cell body, there are shown just the stumps of the dendrites which would be running a long way off in all directions, and projecting downward from this soma there is the single axon that eventually ends in many branches. The arrows indicate that messages or impulses come to the neuron by the many fine nerve fibers making contacts on the surface of the soma and dendrites. The sole output line is along the axon. Figure 1.3B shows the varieties of contacts made by nerve fibers on the dendrites and soma of a pyramidal cell of a special part of the cerebral cortex, which is similar to the pyramidal cells of Figure 1.2B. Many contacts are made on small spines that branch out from the dendrites.

Numerous small spines can be seen projecting from the dendrites of the cells in Figure 1.2B. The total number of synapses of all kinds on central cortical cells is estimated to be about 40,000 per neuron (Cragg 1975).

Electron micrographs show that central synapses resemble neuromuscular synapses (Figure 3.2, Section 3.2.2) in their essential features. The presynaptic nerve fiber expands to a terminal bulb, or it may traverse several synaptic zones, transiently expanding in each, forming the "boutons de passage." In the presynaptic fiber there are synaptic vesicles that are concentrated toward the synaptic cleft, often in discrete clusters (Figure 1.4B). The numerous mitochondria are more remotely-placed. The synaptic cleft has a remarkably uniform width of about 20 nm. The presynaptic and postsynaptic membranes bounding the cleft tend to be darkly stained. On the dendrites of some neurons, for example pyramidal cells of the cerebral cortex (Figure 1.3B), there are numerous spines, each of which is the postsynaptic element of a synapse (Gray 1959). These spine synapses can be extremely numerous—up to 20,000 on a large pyramidal cell of the cerebral cortex (Figure 15.16, Section 15.7) and up to 200,000 on a Purkinje cell of the human cerebellum (Figure 13.12, Section 13.4.3).

The synaptic knobs cover the surface far more densely than is shown in Figure 1.3A. For example, in the reconstruction of Figure 1.4A a synaptic scale provides an almost complete coverage of the surfaces of the soma, the dendritic stumps, and even the axonal origin. On the axon there is shown the beginning of the myelin sheath that will be discussed later in this chapter. When the synaptic knobs are highly magnified by electron microscopy (Figure 1.4B), their location at the ends of fine nerve fibers (Figure 1.3A) is not often shown because the fine nerve fiber is usually not in the section. The most important finding is that the synaptic knobs can be seen to be separated from the surface of the dendrite by an extremely narrow space. This synaptic cleft, as it is called, completely separates the nerve terminal from the underlying cell surface. The neuron theory of Ramón y Cajal is thus corroborated. The whole communication apparatus—the knobs, the space, and the subjacent neuronal surface—was named the synapse by Sherrington, and the derivative terms are synaptic knob and synaptic cleft. Synapses such as those of Figure 1.4B are the structural basis of the chemical transmission whereby the messages that come down the fine fibers to synaptic knobs on the recipient cell act on that cell by means of the secretion of minute amounts of specific chemical substances. The mode of operation of synapses will be described in Chapters 3 and 4.

In Figures 1.3B and 1.4A synapses are seen to vary widely in size, which is in agreement with the recent electron microscopic observations of Conradi (1969), who describes five types of synapses on motoneurons. The very large synapses (cf. Figure 1.4A) presumably are the endings from afferents of annulospiral endings giving the monosynaptic excitation that is diagrammed in Figure 4.1 (Section 4.1) and on them Conradi finds some small synaptic knobs, which form a distinctive type of synapse that will be referred to in Chapter 4 (Figure 4.18, Section 4.8). Figure 1.4A gives some feeling for the busy life that a nerve cell has with incessant bombardment by impulses activating its dense

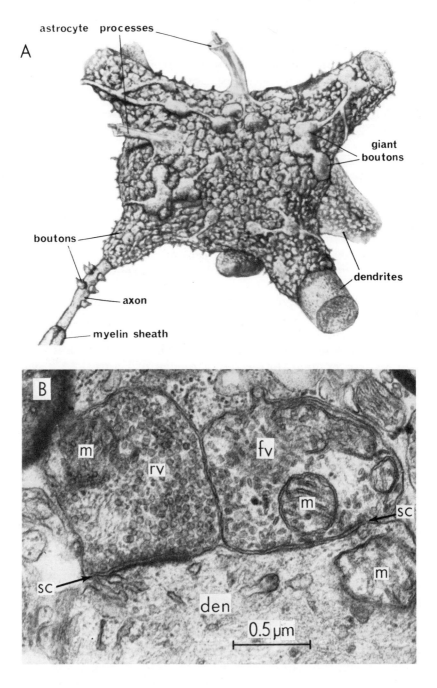

Figure 1.4 Synapses on surface of motoneurons. A. Reconstruction from serial electron micrographs of a motoneuron with its tightly packed surface scale of synaptic knobs (boutons), some being large. (Poritsky 1969.) B. Electron micrograph of two synaptic knobs on a motoneuron in a fish spinal cord; that to the left is excitatory with spherical vesicles (rv) and dense staining on each side of the synaptic cleft (sc), and that to the right is inhibitory with ellipsoid vesicles (fv) and a much lighter staining of the membranes on each side of the synaptic cleft. Mitochondria are labeled *m*. (From Gray 1970.)

coverage by synapses, which are even on the axon at its origin. We shall see, however, that the neuron has some relief because, as first recognized by Sherrington (1906), many of these synapses, the inhibitory synapses, are specialized to prevent the neuron from firing impulses in response to the excitatory synapses. The neuron is in the hands, as it were, of the two opposing operations of excitation and inhibition, which is the theme of much of Chapter 4.

In electron micrographs it is now usually possible to distinguish between excitatory and inhibitory synapses. Figure 1.4B illustrates a reliable criterion that was discovered by Uchizono (1967). When fixed in glutaraldehyde, the vesicles of some synapses (cf. the synapse on the spine in Figure 1.5B) have a fairly uniform spherical shape. In other synapses the vesicles are smaller and tend to be ellipsoid.

Uchizono has shown that the ellipsoid-type vesicles occur in synapses that have been identified as being inhibitory on physiological grounds, while the spherical vesicles characterize excitatory synapses. This proposed differentiation of synapses by vesicular shape has been subjected to intensive experimental tests and now has some general acceptance. For example Figure 1.4B shows two adjacent synapses on a spinal neuron, one with spherical (rv) the other with ellipsoid (fv) vesicles. Before this criterion was discovered, there was evidence that the two types of neuronal synapses described by Gray (Whittaker and Gray 1962) fitted the two functions of synapses, the so-called Types 1 and 2, being respectively excitatory and inhibitory. It now seems that this identification is usually correct as tested by vesicular shape, but there are exceptions. It must be recognized that the ellipsoid shape is an artifact of the fixation technique.

As shown diagrammatically in Figure 1.5A, Type 1 is distinguished from Type 2 by three criteria: (1) the postsynaptic membrane thickening is denser than the presynaptic, hence Type 1 synapses are referred to as asymmetric and Type 2 as symmetric; (2) the synaptic cleft of Type 1 is wider and includes a band of dense material; (3) with Type 1 the specialized contact area is large and continuous, whereas with Type 2 the specialized dense areas tend to be small and fragmented. With Types 1 and 2 there is an accumulation of synaptic vesicles in close relation to the dense areas and the presynaptic dense structures can be seen in favorable preparations.

A third specialized contact area between neurons resembles the desmosomal contacts between epithelial cells. These desmosomes are characterized by symmetrical membrane thickening as with Type 2 synapses, but there are no associated synaptic vesicles; hence, these structures are assumed to function as attachment sites and not as channels of communication.

Spine synapses such as those labeled a to d in Figure 1.3B are almost invariably Type 1, as shown in Figure 1.5B, and have spherical vesicles; hence, an excitatory identification is indicated. Synapses e and f of Figure 1.3B are of Type 2. The vesicles in Figure 1.3B are all shown spherical, as they would be because glutaraldehyde fixation was not employed. It is now recognized that the synapse illustrated by e is made by basket cells and is inhibitory with

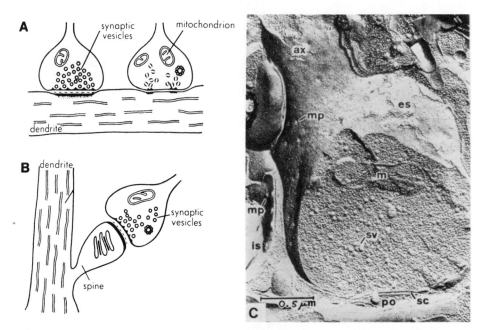

Figure 1.5 Synaptic structure. A. Diagrams of two synapses on a dendrite, the left and right being respectively Type 1 with spherical vesicles and Type 2 with ellipsoid vesicles and other characteristic features of inhibitory synapses. B. A synapse on a spine of a pyramidal cell (cf. Figure 1.3B) or of a cerebellar Purkinje cell with all the features of an excitatory synapse (Whittaker and Gray 1962). C. A photograph of a synapse as revealed by the freeze-etching technique. (From Akert *et al.* 1969.)

ellipsoid vesicles. Synapse f also may be inhibitory, but g is excitatory from the Schaffer collaterals (Figure 15.14, Section 15.6).

The small round vesicles that occur at peripheral cholinergic synapses, such as at the neuromuscular synapses, appear also to characterize central synapses mediated by acetylcholine (Figures 5.5, Section 5.2; and 5.6, Section 5.3.2). On the other hand, the vesicles of GABAnergic nerve endings in the deep cerebellar nuclei (Figure 6.1, Section 6.2) and the substantia nigra (Figures 6.4, Section 6.3; 13.23, Section 13.5.4) are flattened. It is possible that the multitudes of synapses operating by as yet unidentified excitatory transmitters also have small round vesicles (cf. Figures 1.3B, 1.4B, 1.5A, and 1.5B) and that glycine and other inhibitory transmitters have flattened ones. Pleomorphic morphology is also seen, for example in vesicles of dopaminergic (Figure 8.5C, Section 8.2.2) and serotonergic (Figure 9.4B, Section 9.2) nerve endings. In terminals of neurosecretory cells, in chromaffin cells, and in peripheral noradrenergic varicosities, vesicles of different sizes exist that often contain a dense core (Cooper *et al.* 1974). Other varieties of synaptic vesicles have been recognized, e.g., coated vesicles, complex vesicles (Gray 1964).

A remarkable new technique for displaying the ultrastructure of synapses is the freeze-etching technique illustrated in Figure 1.5C (Akert *et al.* 1969).

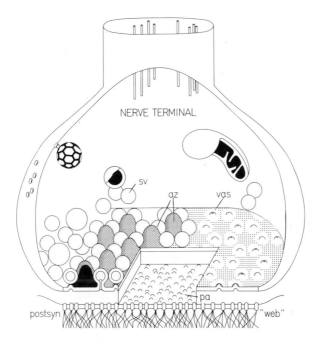

Figure 1.6 Schema of the mammalian central synapse. The active zone (az) is more complex and allows far more vesicle attachment sites (vas) per square unit of surface than the motor end plate. The postsynaptic aggregation of intramembranous particles is restricted to the area facing the active zone. The connection between the particles and the "web" of De Robertis is hypothetical. sv is synaptic vesicles, pa is particle aggregations on postsynaptic membrane (*postsyn*). (From Akert *et al.* 1975.)

Synaptic vesicles (sv) and a mitochondrion (m) are in the cut surface of the synaptic knob, that also has its external surface (es) and its axon (ax) displayed. Below is the narrow synaptic cleft (sc) separating the knob from the postsynaptic element (po).

By means of this freeze-etch technique Akert and associates (Akert 1973; Akert and Peper 1975) have examined the inner and outer aspects of the surface membrane surrounding a synaptic knob. The cutaway perspective drawing of Figure 1.6 shows that the synaptic vesicles (SV) tend to be in a hexagonal array separated by dense structures (az) (cf. Figures 1.5A, 1.5B) (Gray 1964). To the right, the etch technique has removed the vesicles and dense structures, disclosing the underlying small protuberances (vas) into the synaptic knob from the synaptic cleft. These correspond to the attachment sites of the vesicles through which they may discharge their contents into the synaptic cleft. Below is a cutaway area of the knob revealing the outer face of the postsynaptic membrane encrusted with fine particles (pa) which may be the postsynaptic receptor sites (cf. Figure 3.13, Section 3.2.10). This illustration gives some indication of the cellular machinery associated with the movement and discharge of synaptic vesicles and with the postsynaptic action of the transmitter, but the mode of operation of this machinery is as yet unknown.

It is now recognized that the plane of cleavage in Figures 1.5C and 1.6 is not the synaptic cleft, but is along the middle of the bimolecular leaflets that form the cell membrane (Figure 1.16) (Akert and Peper 1975). This reinterpretation of the cleavage planes, however, does not materially affect the general description given here. Similar fine structures occur at the neuromuscular junction, but there the attachment sites of the synaptic vesicles are arranged in longitudinal bands that are congruent with the folds on the postsynaptic membrane.

Only brief reference need be made to electrical synapses, where transmission is by electrical currents flowing from the presynaptic to the postsynaptic component. Structurally, such synapses are identified by the very narrow cleft, 20–40 nm, and consequently they are called gap junctions (Pappas 1975). Electrical synapses are important special sites in invertebrate and lower vertebrate nervous systems. Gap junctions rarely occur in the mammalian brain and appear to have little or no functional significance; hence, they will not be further considered in this book. Two other specialized synapses will be treated in later chapters, *viz.* axoaxonic synapses giving presynaptic inhibition, and the reciprocal synapses, excitatory in one direction and inhibitory in the other.

1.2.3 Nucleus[1]

Neurons have nuclei that are unusually large for the cell body or soma, but it must be remembered that the bulk of the neuron (soma plus dendrites and axon) is many times larger than the soma. Usually the nuclei are approximately spherical, and sections of typical nuclei can be seen in the electron micrographs of Figures 1.7, 1.8, and 1.9. The chromatin is dispersed in a fine granular state, and there is the fenestrated bounding membrane forming the nuclear envelope (Figures 1.9, 1.10). The nuclei of neurons are unique in that mitotic competency is lost at the time of the differentiation into the immature neurons that ultimately develop into neurons (Chapter 12).

The process of differentiation is complete in most mammalian species at birth. But at this stage one of the most miraculous processes in all of nature develops. The axons of billions of neurons begin to grow out, acting under a set of instructions that is not yet even vaguely understood (Chapter 12). They develop precise linkages with predetermined neurons that may lie many centimeters away. The complexity of the process could be likened to that of a spontaneous wiring together of the components of a giant computer that is enormously more vast than anything yet attempted electronically. By comparison, the process taking place in other organs such as the liver, where cells duplicate each other's function, would be equivalent to connecting a bank of lights at a football stadium.

[1]General reference: Palay and Chan-Palay 1972.

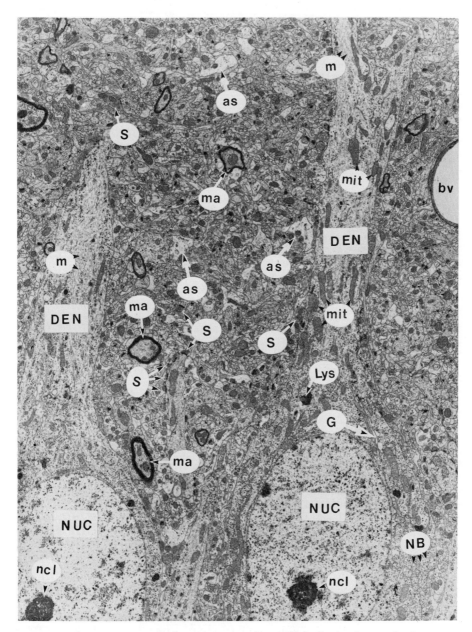

Figure 1.7 Electron micrograph of two hippocampal pyramidal cells, showing proximal segments of their apical dendrites. Processes are identifiable as dendrites because of absence of myelin sheaths and presence of synapses along the surfaces. Nuclei (NUC) are at the bottom of the photomicrograph, with both showing a prominent nucleolus (ncl). The Nissl bodies (NB) are represented by highly ordered rough endoplasmic reticulum with associated ribosomes attached to long and shallow cisterns. Lysosomes (Lys) are seen throughout the cytoplasm. One representative of a Golgi apparatus (G) can easily be identified. Mitochondria (mit) are spread throughout the cells. They tend to be elongated, with their long axes oriented in the direction of the dendrites. Microtubules (m) can be seen as faint structures along dendrites. Some appreciation of the incredible complexity of the neuropil can be gained from examining the multitudinous processes in between the cell bodies. Several myelinated axons (ma), astrocytic processes (as), and synapses (s) can be identified. (Courtesy of A. Rose, Simon Fraser University.)

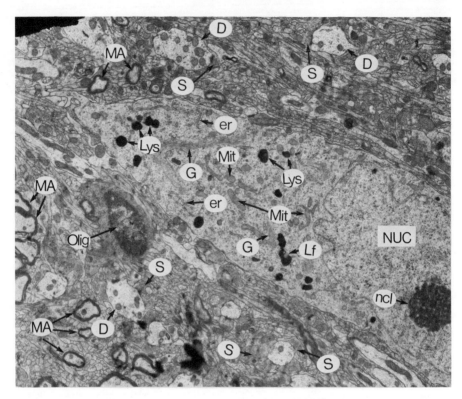

Figure 1.8 Perikaryon of presumed dopaminergic neuron of rat substantia nigra with surrounding neuropil. Nucleus (NUC) shows characteristic indentations in its envelope. Prominent nucleolus (ncl) is visible. Particularly prominent in the cytoplasm are many dark lysosomes (Lys) which have been stained for horseradish peroxidase. The peroxidase was transported to the cell body by the process of retrograde axoplasmic transport (Figure 1.14), and can also be seen by light microscopy (Figures 1.15A, 1.15B, 1.15C). Note also the mitochondria (Mit), Golgi apparatus (G), and rough endoplasmic reticulum (er) (Nissl substance). The dense neuropil shows many prominent dendrites (D) in cross section with easily detected synapses (S). Numerous myelinated axons (MA) are also visible, as well as an oligodendroglia (Olig). (Courtesy of T. Hattori, University of British Columbia.)

The flow of information responsible for this wiring commences in the nucleus. Despite the fact that neurons have lost their capability to divide, they appear to retain their full complement of DNA. Indeed, some large neurons of the mammalian brain, such as the cerebellar Purkinje cells, are tetraploid. But much of the DNA must remain permanently repressed, presumably by the action of histones, because only that necessary for neuronal functioning is utilized. The DNA exists in two strands, joined together like a ladder twisted in a helical fashion.

The concept of a double helix structure for DNA by Watson and Crick was one of the most fruitful single hypotheses in modern biology. As a consequence of this structure, they proposed that a molecular sequence of transcription, through translation to enzyme activity, takes place in all cells

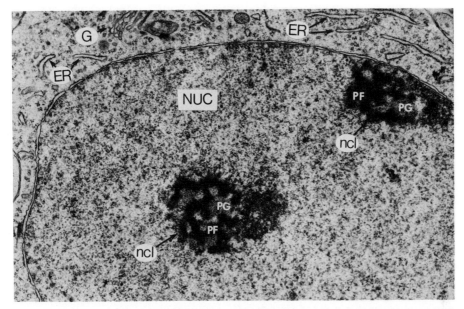

Figure 1.9 Nucleus (NUC) of immature neuron of neonatal rat cerebral cortex, showing primarily the double-layered nuclear membrane. Note the many gaps in the membrane and the apparent attachment of ribosomes to the external surface. The gaps represent holes in the spherical membrane. Ribosomes on the nuclear surface are thought to be newly formed. Two nucleoli (ncl) can be seen showing the pars granulosa (PG) and pars fibrosa (PF) segments. They participate in the formation of ribosomes. Many ribosomes are seen in the cytoplasm, including some in association with rough endoplasmic reticulum (ER). A Golgi apparatus (G) is also present. (Courtesy of T. Hattori, University of British Columbia.)

(Watson 1970). Sections of the double helix briefly part, and the genetic information which is to be transcribed is read off one of the strands. The structure of the initiator and terminator sites on DNA, and the mechanisms for selecting the strand to be read, remain unknown. For messenger RNA synthesis an enzyme, RNA polymerase, attaches to the initiator site on the single strand and catalyzes synthesis until the terminator site is reached. This corresponds to a precise sequence of amino acids in a protein later to be synthesized in the cell cytoplasm by ribosomes (Figure 1.11).

Two other forms of RNA, ribosomal and transfer, need to be produced before the translation process can take place. Ribosomal RNA is thought to be synthesized as a consequence of nucleolar activity. Transfer RNA is a relatively small molecule of about 70 nucleotides that comes from the nucleus, and its job is to bring amino acids in the cytoplasm to the ribosome.

The nucleolus is usually single and darkly stained with the conventional dyes used for light microscopy, as well as by the heavy metals used for electron microscopy. It is so conspicuous on the background of the poorly stained karyoplasm that it provides an identifying feature of neurons. There may be more than one nucleolus, as Figure 1.9 shows.

The ribosomes are complex bodies of RNA and protein which are probably assembled on the surface of the nuclear membrane where they are particularly abundant in immature neurons (Figure 1.9). The many holes in the nuclear membrane (Figures 1.9 and 1.10) make possible the easy exit of macromolecules.

Once the process of transcription has taken place in the nucleus, the process of translation can commence in the cytoplasmic organelles.

1.2.4 Organelles

The cytoplasm of a neuron is filled with structures called organelles that are not only the basis of translation, but also of all the biochemical processes that are associated with its extremely high metabolism (Figure 1.10). Neurons can be considered as cauldrons of metabolic activity. They have one of the highest metabolic rates of mammalian cells and have very specialized chemical enzyme systems. This intense metabolism is required for operating the ionic pumps across the surface membrane so as to maintain the correct ionic composition inside the cell (cf. Figure 1.17), and it is also required for the manufacture of the chemical transmitting substances (Chapters 5–11) and for the manufacture and transport of specific substances along the axon that we shall be considering in Chapter 12 (cf. Figures 1.12, 1.13, 1.14).

The ultrastructure of the common neuronal organelles and inclusions as they appear in electron micrographs are shown diagrammatically in Figure 1.10. The principal ones are the following:

Rough Endoplasmic Reticulum (Nissl Bodies, Nissl Substance, and the Associated Ribosomes. [2]　　The rough endoplasmic reticulum is present in large amounts in the perikaryon, as can be well seen in Figures 1.7, 1.8, and 1.9. It forms the characteristic Nissl bodies (NB in Figure 1.7) or rough endoplasmic reticulum (ER in Figures 1.8 and 1.9) that are so conspicuous in light microscopy of neurons. In Figure 1.7 the endoplasmic reticulum consists of flattened cisterns having parallel courses, but there are many variants in different neurons and even in the same neuron (cf. Figures 1.7, 1.8, 1.9). Many ribosomes are attached to the outer surface of these cisterns, but most occur between the cisterns with a characteristic polysomal array in rosettes of five or six granules. The basic combination of cisterns of endoplasmic reticulum with attached or free ribosomes in polysomal array occurs in all neurons.

The process of translation is concentrated in the rough endoplasmic reticulum. It commences when messenger RNA becomes attached to a ribosome. The messenger RNA is composed of groups of three nucleotides called codons. Each amino acid that is to be included in the protein is specified by one or more of these codons. During the process of translation, transfer RNA fetches the amino acids in sequence as required by the messenger RNA attached to the ribosome. Upon return of the transfer RNA/ acid complex, the new amino acid is connected to the previous one in the chain, which up to this

[2]General references: Palay and Chan-Palay 1972; Peters et al. 1976.

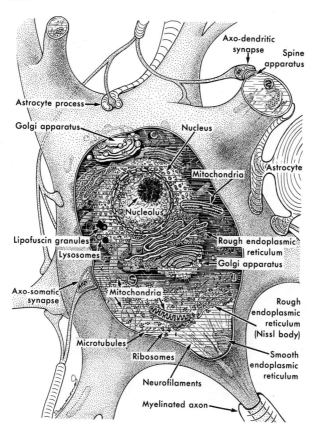

Figure 1.10 Schematic diagram showing many features of a neuron as seen in electron micrographs. Several dendritic processes are shown with the single myelinated axon appearing at the lower right of the diagram. Astrocytic processes are shown enveloping synapses on the surface of the cell body. The number of synapses and complexity of the neuropil is tremendously reduced over that seen in Figure 1.7 and presented diagramatically in Figure 1.4A. Microfilaments and microtubules are shown running down the dendrites and axons. Such subcellular organelles as Nissl bodies (rough endoplasmic reticulum), smooth endoplasmic reticulum, mitochondria, the Golgi apparatus, lysosomes, lipofuscin granules, and ribosomes are shown. The perforated nature of the nuclear membrane is depicted. One section is cut away so that the nucleolus can be seen.

point has been complexed with its own transfer RNA. On completion of the chemical coupling of the amino acids, the previous transfer RNA molecule is released. The ribosome then "moves" along the messenger RNA to "read" the next codon (Figure 1.11). The peptide chain increases in length by one amino acid at a time and the RNA message is read sequentially, three nucleotides at a time, until completion (Watson 1970).

The rough endoplasmic reticulum and the ribosomes extend out along the dendrites (Figures 1.7, 1.8), often having a characteristic longitudinal orientation. Structurally, the larger dendrites resemble the soma (Figures 1.7, 1.8), but there is a progressive diminution of the organelles with distance

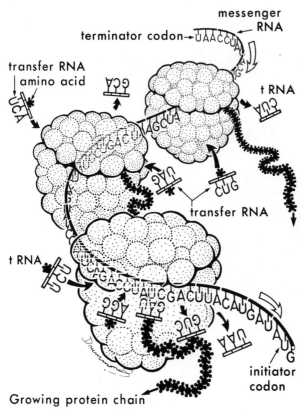

Figure 1.11 Schematic diagram of the synthesis of protein chains by ribosomes. Each ribosome consists of a 30 S ribosome subunit (smaller portion) and a 50 S ribosome subunit (larger portion) joined to make a complete (70 S) ribosome. Messenger RNA (mRNA), which consists of codons comprising triads of nucleic acid bases, threads between the ribosomal subunits. Transfer RNA (tRNA) with complementary base sequences exists in the cytoplasm capable of binding appropriate amino acids. Synthesis of the growing protein chain commences at a chain initiation codon where a 30 S ribosome subunit joins *N*-formyl methionyl-tRNA. Next a 50 S ribosome subunit attaches to complete the assembly of the ribosome. The chain grows when an aminoacyl-tRNA molecule corresponding to the next codon after initiation binds to the ribosome at the adjacent site on the messenger RNA. Peptidyltransferase attaches the amino acids together and releases the spent tRNA back to the cytoplasmic pool. Translocation then occurs as the ribosome moves one codon down the mRNA molecule. Repetition of the binding-bond formation-translocation sequence continues until a specific terminator codon is reached, releasing the synthesized protein. mRNA can join together clusters of ribosomes (in this case, three). Protein chains grow out from each of the ribosomes.

from the soma. In remote dendrites ribosomes are rare, occurring only as free polysomal arrays.

The origin of the axon, often from an axon hillock (cf. Figure 1.4A), is distinguished by the sparsity of the rough endoplasmic reticulum. Ribosomes are also greatly diminished. They still occur in the initial segment, but not along the whole length of the myelinated axon. Except in certain special neurons such as Purkinje cells, synapses are rare on the initial segment (cf. Figure 1.4A).

The axon constitutes a distinct metabolic compartment of the neuron with no protein-synthesizing capacity other than a very limited contribution by mitochondria. The dendrites, too, have a diminished capacity, particularly far out from the soma. It is not known how this ribosomal distribution is controlled. Is there some block on ribosomal migration from the perikaryal region, or is the ribosomal population depleted by a high level of destruction? Whatever the reason, the tips of the axons and dendrites must still be supplied with proteins synthesized by these ribosomes. The proteins are transported to these neuronal outposts by a fascinating but little understood process called axoplasmic flow (Figures 1.12, 1.13, 1.14, and Chapter 12).

Agranular or Smooth Endoplasmic Reticulum. The agranular endo-plasmic reticulum is so called because it is devoid of ribosomes. It is abundant in the neuronal perikaryon, where it constitutes the Golgi apparatus that encir-cles the nucleus at some distance, as indicated by the structures labeled G in Figures 1.7 and 1.8. Probably the proteins manufactured by the ribosomes are transferred to the Golgi apparatus for further transmission (cf. Figure 1.14), but much has yet to be learned about the metabolic machinery in the peri-karyon. The agranular reticulum also runs out into the dendrites and the axon, often close to the surface membrane (Figure 1.14, Axl) and may sub-serve protein transport along the dendrites and axon. It is unfortunate that as yet these organelles have not been isolated from homogenates as subcellular fractions for detailed study.

Mitochondria. The mitochondria are the power plants of the cell. They have dimensions of 1000 to 3000 nm, in comparison with the 20-nm diameter of a ribosome. A single mitochondrion may contain as many as 15,000 com-plexes of respiratory chain enzymes, each of which contains a dozen or more separate enzymes (Figure 1.10). The inner mitochondrial membrane is not merely an inner skin, but is a sheet of multienzyme systems. These are specialized for stripping carbon chains and extracting their energy. The energy is banked in the form of high-energy phosphate bonds, particularly ATP. Carbon dioxide and water are formed by this Krebs cycle oxidative process (Figure 7.2, Section 7.2.1) which is coupled to phosphorylation. The mitochondria contain other enzymes. Monoamine oxidase is one which will be considered in the several chapters dealing with neurotransmitter metabolism.

Neuronal mitochondria are generally similar to their counterparts in other cells. In Figures 1.7 and 1.8 mitochondria can be seen as numerous small ovoid organelles with the characteristic internal structure. They are often clustered together in the interstices between the endoplasmic reticulum and the Golgi apparatus. In Figure 1.7 the mitochondria are seen to be linearly arranged along the dendrite, and this orientation becomes progressively more evident further along the dendrites. It is also very evident along the axon. Mitochondria are concentrated in axonal terminals where there is the intense metabolic activity associated with the synthesis, storage, release, and reuptake of neurotransmitters (cf. Figures 1.4B, 1.5A, 1.5B, 1.5C).

Neurofilaments and Microtubules. These structures run from the perikaryon along the dendrites and the axon (Figures 1.14, 2.12A, Section 2.5) and probably are used for transport of macromolecules, although it has also been suggested they supply structural support. Neurofilaments are smaller than microtubules, but both are of indefinite length. Neurofilaments are not solid, but tubular. The outside diameter is about 10 nm, formed by a dense wall which leaves a clear center about 7 nm in diameter. The wall is composed of globular subunits from which short, spokelike processes radiate. Neurofilaments are particularly prominent in large axons. They run down to the synaptic knobs, often forming a loop. They are readily stained by silver, and probably account for the so-called neurofibrils of classical light microscopy. Neurofilaments also pass out from the soma along the dendrites (cf. Figure 1.10) (Peters *et al.* 1976).

Like neurofilaments, microtubules are also tubular structures, but have a larger diameter of 20–40 nm. In contrast to the neurofilaments, they are found most prominently in axons of small diameter. They are also found in all dendrites (cf. Figures 1.5A, 1.5B). Some experiments have suggested an interconnection between neurofilaments and microtubules. In immature animals nerve fibers contain many microtubules but few neurofilaments. With increasing age there is a reversal in some axons, but even then the neurofilaments still tend to be aggregated. After treatment with colchicine, a mitotic spindle inhibitor, neurons initially lose their neurotubules while neurofilaments proliferate (Peters *et al.* 1976). Sometime afterward the process reverses. Since the diameter of a neurofilament is greater than the component filaments of the microtubule wall, it is supposed that any transformation would have to involve some rearrangement of common subunits. Colchicine has the property of strongly binding to solubilized microtubular protein. Since microtubules with the same composition as neuronal microtubules function in the nucleus to separate the DNA components during mitosis, it is not surprising that colchicine should be such a powerful antimitotic compound. It also leaves neuronal microtubules in disarray, thus halting the process of axoplasmic flow.

Axoplasmic Flow.   The method by which materials produced in the perikaryal region reach axonal and dendrite tips is called axoplasmic flow (Ochs 1974). It was first described by Weiss and Hiscoe (1948).

The earlier experiments of Weiss (1970) demonstrated flow along nerve fibers by observing the damming up of material proximal to a ligature. In that way it was possible to derive approximate measures for flow of material along axons and to derive a rate of about 1 mm/day, which was supported by radiotracer labeling experiments. Much faster transportation rates have now been measured by improvements in technique, particularly in giving localized radiotracer injections close to the nerve cells and by keeping the nerve fibers under good physiological conditions. We will base our account largely on the attractive story that emerges from the investigations of Ochs on mammalian nerve.

Ochs injected tritiated leucine or tritiated lysine into a dorsal root ganglion of the spinal cord (arrow at G in Figure 1.12B) and showed that the ganglion cells picked up the amino acid and built it into proteins and polypeptides. These tritiated macromolecules were followed by radiotracer techniques and shown to move along the nerve fibers in both directions from the ganglion, out to the periphery and centrally up the spinal cord. A series of experiments in which there was termination at the indicated hours is plotted in Figure 1.12A. Counts per minute were measured for the successive 2-mm lengths into which the nerve was cut and are plotted on the logarithmic scale. The arrows indicate the sites of the wave fronts at the postinjection times of 2, 4, 6, 8, and 10 h. In this way the wave front of increased radioactivity was observed

A

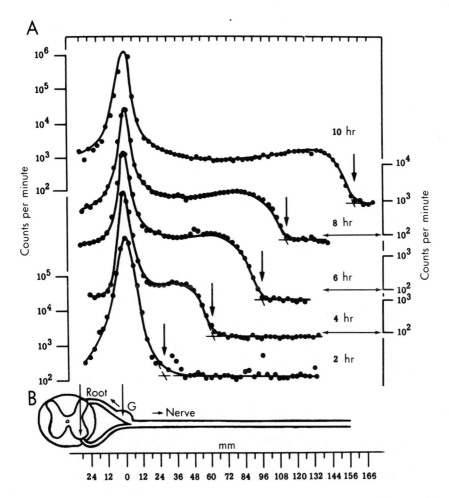

Figure 1.12 Fast axon transport down sensory fibers. A lumbar dorsal root ganglion (G) of the cat was injected with tritiated leucine and the radioactivity was measured in separate experiments at 2, 4, 6, 8, and 10 h later at the distances shown along the sciatic nerve. The radioactivity is plotted on a logarithmic scale in counts per minutes (CPM), with separate scales as shown for each of the curves. (From Ochs 1972b.)

to travel along the nerve at the surprisingly fast velocity of 400 mm/day, and this speed is the same for mammals with a size range from rat to goat. It seems to be a basic biological speed in these axons and of quite extraordinary efficiency. So proteins and other macromolecules manufactured in the perinuclear region of the soma (Figure 1.10) are transported down along the axon. We can assume that this transport is concerned with establishing and maintaining the chemical coding all the way along the nerve fiber. The transport is not down the spaces between the nerve fibers because the labeled macromolecules are shown by autoradiography to be inside the nerve fibers.

The remarkable feature of Figure 1.12 is that the wave front moves bodily forward hour by hour. If it were due to some diffusional process, the wave front would become progressively more spread out. Evidently there is some highly organized transport mechanism. It has long been known that nerve fibers are not just tubes filled with some protein jelly, but that they contain many fine structures, neurofibrils and neurotubules that run along their length (cf. Figure 2.12A, Section 2.5). It is postulated by Ochs that after manufacture in the cell body the macromolecules are loaded onto these tubules and fibrils and travel along by some sliding mechanism, hence the uniform rate. Thus the rapid transport of 400 mm/day is envisaged as being due to a "conveyor belt" operation.

It is interesting that this transport goes at about the same velocity central from the ganglion cells and up the tract fibers (cf. Figures 2.1, Section 2.1; 15.1, Section 15.2). There is a similar transport in motor nerve fibers. We can predict that it will be found to occur in all nerve fibers, both peripheral and central.

An important discovery was that this transport was dependent upon metabolism of the nerve. It was not due to some pressure exerted from the nerve cell, but, as would be expected from the sliding filament hypothesis, each segment of the nerve had to provide the energy for the transport. Figure 1.13A shows that local anoxia of the segment results in a failure of transport and a banking-up proximal to the block. The same enzyme poisons—cyanide, azide, and dinitrophenol—block the transport along the fiber and the ionic pumping across its membrane. In Figure 1.13B poisoning by cyanide along the whole length of the nerve fiber at 3 h after the injection completely blocked the transport. The distribution of radioactivity remained frozen at the 3-h position (cf. Figure 1.12A). Transport can also be blocked by administration of the antimitotic agents colchicine and vinblastine, which interfere with the microtubules and microfilaments. A thorough experimental study of this fast transport shows that in every respect the observations are in accord with the hypothesis that the fast transport is due to active transport machinery that is dependent on energy provided by adenosine triphosphate, and that it is distributed along the whole length of nerve fibers.

In addition to this fast transport along nerve fibers, many other transport mechanisms have been described, some being almost as fast, but others much slower—at velocities measured in a few millimeters per day. There is need of much more experimental investigation, but in general it appears that there are three main modes of transportation along nerve fibers. The best investigated is the fast, as illustrated in Figures 1.12 and 1.13. The slow tends to be about 100 times slower and moves soluble proteins and various structural proteins such as those depicted on the phospholipid membrane in Figure 1.16. Finally, there is the reverse transport from axon to nerve cell.

One possible overview of transport is shown in Figure 1.14. According to this scheme of Droz's (1975), there is transport from the rough endoplasmic reticulum (Erg) of synthesized proteins to the Golgi apparatus (Go) and so down the smooth endoplasmic reticulum (REL) to the axon and synaptic knob (MPS). An additional transport is via mitochondria, while the polysomal ribo-

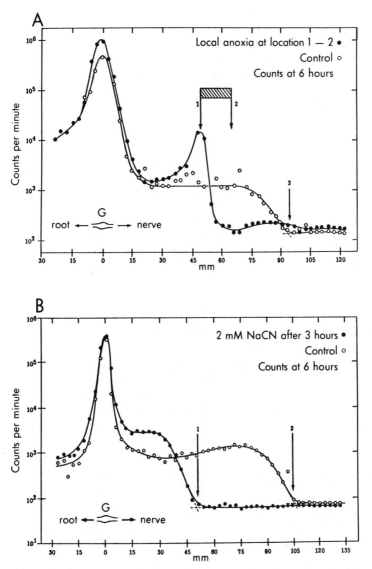

Figure 1.13 Metabolic blockade blocking fast axon transport. A. Experiment as in Figure 1.12, but 2 h after the injection, excised roots, ganglia, and nerves of both sides were placed in the chamber in moist 95% O_2, 5% CO_2 at 38°C. A length of 2 cm of one sciatic nerve was made anoxic by covering it with petrolatum jelly. At 6 h after the injection counts were done on both nerves. B. Experiment as in *A*, but nerves were excised at 3 h, one being in a 2-mM solution of NaCN, the other in the normal moist chamber. Both were counted 6 h after the injection. (From Ochs 1972a.)

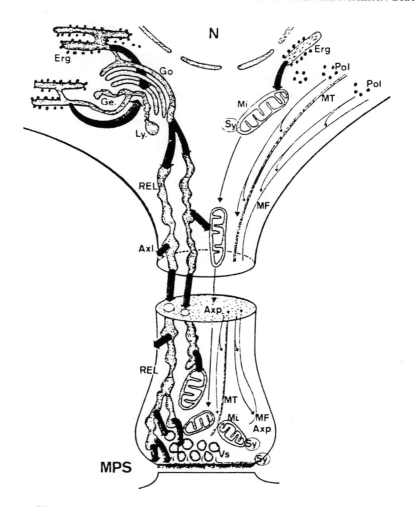

Figure 1.14 Diagrammatic representation of the contribution of the axonal migration of macromolecules to the maintenance of nerve cell processes and synapses. On the left side, the thick arrows indicate the part taken by the fast transport. Polypeptide chains synthesized in the ergastroplasm of the Nissl body (Erg) are transferred to the Golgi apparatus (Go) and give rise to protein and glycoprotein sequestered in the smooth endoplasmic reticulum (REL). Passing into the axon with the axonal endoplasmic reticulum, they are transported at a high speed and yield new membrane components to the axolemma (Axl) and mitochondria (Mi). They accumulate in the terminal part of the axon and ensure the renewal of constituents of synaptic vesicles (Vs) and presynaptic plasma membranes (MPS). Thus, the fast transport purveys new membrane components and the logistical support for the conduction and (mainly) transmission of nerve impulses. On the right side, the thin arrows point to the elements transported with the slow axonal flow. Polypeptide chains are released from free polyribosomes (Pol) into the neuronal cytoplasm and migrate slowly into the axoplasm (Axp). They may assemble as protein subunits giving rise to microfilaments (MF) and microtubules (MT). Mitochondria receive by transfer the great majority of their own proteins; these organelles are displaced along the axon and continue to exhibit a slight synthesis of hydrophobic polypeptides (Sy). The retrograde axonal transport has not been represented in this figure. (From Droz *et al.* 1974.)

somes (Pol) utilize microtubules (MT) and neurofilaments (MF) for transport. Many other interpretations are possible.

Droz (1975) gives an excellent summary of the dynamic design involved in the activity of the nerve ending:

> The neuron has to cope with a paradoxical situation. On one hand, the synaptic terminal synthesizes, stores, and releases the neurotransmitter; on the other hand, the metabolism of the neurotransmitter is controlled by specific proteins that are genetically programmed in the nucleus and synthesized by the perikaryal machinery. Thus, to replenish the axon terminal, the neuron has to dispatch new enzymes, membrane components, and mitochondria; simultaneously, disused synaptic constituents have to be cleared up.

The process of clearing up involves retrograde axoplasmic flow. Less is known about retrograde than anterograde flow because the techniques of study are considerably more difficult. Moving pictures of axoplasmic flow, taken by means of time-lapse cinematography, show peripheral nerves pulsating with activity, with particles darting in both anterograde and retrograde directions. Of the two, retrograde seems the more active by this technique. It may be that the retrograde flow process also serves as an intracellular signaling system by which the axons and dendrites may keep the nucleus and perikaryal cytoplasm informed of the state of biochemical affairs in the neuronal outposts.

Lysosomes. Part of the neuronal waste disposal system involves lysosomes. These are characteristically between 0.3 and 0.5 μm in diameter but they may be considerably larger. The name reflects the localization of hydrolytic enzymes, particularly acid phosphatase. This latter enzyme can be identified histochemically, providing a characteristic marker for the organelle. Many lysosomes appear to become converted to lipofuscin granules, which accumulate with age and are regarded as neuronal refuse (Figure 1.8).

Material from the axons and nerve endings is returned to the perikaryon by retrograde axoplasmic flow for disposal by the lysosome. A much-used technique in neuroanatomy takes advantage of this principle (La Vail 1975). Horseradish peroxidase is injected into the area of nerve endings (Figure 1.15A). The nerve endings take up the foreign protein and return it by retrograde axoplasmic flow to lysosomes in the cell body (Figures 1.8, 1.15B, 1.15C). Here it can be identified by its characteristic enzyme activity, providing positive identification of the cells of origin.

1.2.5 Surface Membrane of Neurons and Nerve Fibers

In the light of intensive investigation, the very thin membrane around the nerve fiber and nerve cell can be pictured as in Figure 1.16, much as was originally conjectured by Davson and Danielli (1943). The main components of the 7-nm membrane are arranged as a bimolecular leaflet of phospholipid molecules, with hydrophilic polar groups pointing both outward and inward and being about 4 nm apart. The fatty acid tails are both saturated and

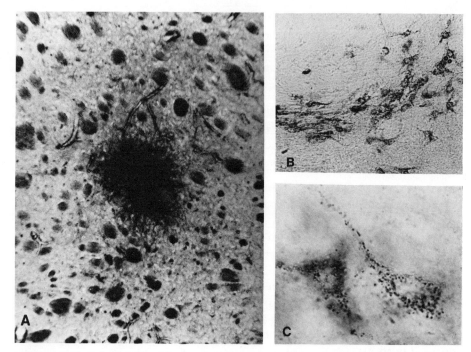

Figure 1.15 A. Injection site of horseradish peroxidase in the rat caudate–putamen. Material is picked up by nerve endings and transported by retrograde axoplasmic flow to the cell bodies. Staining for peroxidase is by a diaminobenzidine oxidation reaction (Figure 5.3, Section 5.2). B. Low-power photomicrograph of the zona compacta cells of origin for caudate–putamen nerve endings. This nigrostriatal dopaminergic pathway is discussed in detail in Chapters 8 and 13. C. High-power photomicrograph of two zona compacta cells. Note the distinctly granular appearance of the peroxidase stain caused by its concentration in lysosomes. Electron micrograph of comparable cell shown as Figure 1.8. (Courtesy T. Hattori, University of British Columbia.)

unsaturated. If all were saturated, the membrane would be rigid. The larger the proportion of unsaturated fatty acids, the more fluid is the membrane and the less orderly is the packing. Such membranes can be made artificially and in many of their properties the artificial membranes resemble natural cell membranes. In addition, protein molecules are associated with the membrane (Figure 1.16). Some on the inner and outer surfaces give the membrane stability. Some proteins make ensembles across the membrane that can be the basis of channels for ion penetration; one is shown to the right. In addition there are specific proteins on the outer side of the membrane that are concerned in the recognition of other cell surfaces or of chemical substances of which the synaptic transmitter substances are a special example (Chapters 3 and 12). There is great diversity in the protein composition of membranes, particularly those concerned in molecular recognition. It is conjectured that the proteins are incorporated into preformed phospholipid membranes (Fox 1972).

In the first place it will be realized that the bimolecular phospholipid leaflet is very highly resistant to electrical current and has the considerable capacity of about 1 μF/cm^2, a value that accords with a dielectric coefficient of

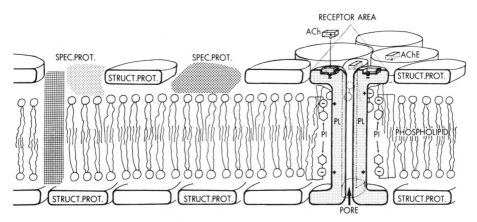

Figure 1.16 Macromolecular organization of surface membrane. The basic structure is a bimolecular leaflet of phospholipid molecules that is stabilized by structural proteins (Struct. Prot.) applied to both surfaces. In addition there are various specific proteins (Spec. Prot.) or other macromolecules applied to the outer surface or penetrating through the membrane. Finally, a transmembrane pore is shown with controls by receptor sites for acetylcholine (ACh). (Modified from De Robertis 1971.)

5 and a plate separation of 5 nm. The surface membrane of a neuron or nerve fiber may be likened to a wrapping around it of a leaky condenser, with a resistance of about 1000 Ω-cm^2 because ions pass through these many channels across the membranes that are formed by proteins.

The surface membranes of all nerve fibers and nerve cells have essentially the same structure and properties throughout the whole of the invertebrate and vertebrate kingdoms. The cell membrane was a basic innovation at a very early stage of evolution, and it was such a good discovery that it can be seen to be essentially the same over a wide range of invertebrates and with all vertebrates. Furthermore, the nerve impulse itself was also a very early innovation and in its essentials it has been preserved right up to the vertebrates and the mammals. No engineer could design or imagine anything so beautiful, efficient, and effective as a nerve impulse and the communication that it gives along nerve fibers, and so from one neuron to others.

The mammalian motoneuron will be used in Chapter 4 for the exposition of the electrical and ionic properties of the surface membrane because the theme of this book is the mammalian, and especially the human brain, and the motoneuron has been the most studied neuron. As shown in Figure 1.17 the surface membrane of a neuron separates two aqueous solutions that have very different ionic compositions. The external concentrations are approximately the same as for a protein-free filtrate of blood plasma. The internal concentrations are derived more indirectly from investigations of the equilibrium potentials for some physiological processes that are specifically produced by one or two ion species. Within the cell, sodium and chloride ions are at a lower concentration than on the outside, the ratios being about 10:1 and 14:1. On the contrary, with potassium there is an even greater disparity—almost 30-fold—but in the reverse direction. Under resting conditions potassium and

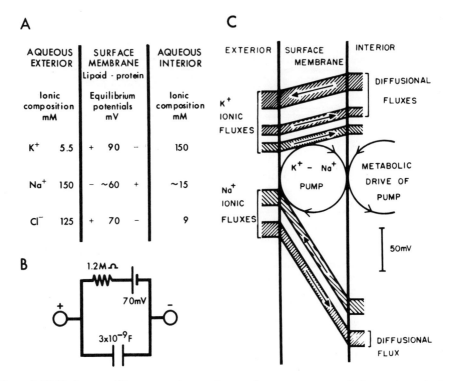

Figure 1.17 Various conditions across the membrane of a cat motoneuron. A. Approximate ionic compositions inside and outside, and the respective equilibrium potentials. B. A formal electrical model for an average motoneuron as measured by a microelectrode in the soma. C. Ionic fluxes across the membrane for K and Na ions under resting conditions. These fluxes are in part diffusional down the electrochemical gradients as indicated, and in part due to specific ion pumps driven by metabolism. The fluxes due to diffusion and the operation of the pump are distinguished by the cross-hatching, and the magnitudes are given by the respective widths of the channels. (From Eccles 1957.)

chloride ions move through the membrane much more readily than sodium. Of necessity the electrical potential across the membrane influences the rates of diffusion of charged particles between the inside and outside of the cell. The potential across the surface membrane is normally about −70 mV, the minus sign signifying inside negativity, as demonstrated in Figure 2.5, Section 2.2).

The equilibrium potentials in Figure 1.17A are derived from the concentrations by the Nernst equation. The equation simply gives the electrical potential that just balances a concentration difference of charged particles, such as ions, so that there is equality in their inward and outward diffusion rates across the membrane.

In Figure 1.17C across the membrane there is shown a large electrochemical potential difference for sodium ions (130 mV) which is derived from 70 mV for the membrane potential plus 60 mV from the concentration difference as calculated by the Nernst equation. This large electrochemical gradient causes the inward diffusion of sodium ions to be more than 100 times faster than outward. Fortunately, the resting membrane is much more impermeable to sodium ions than to potassium and chloride ions, else the cell would rapidly be swamped with sodium ions. But, of course, there must be some other factor concerned in balancing sodium transport across the membrane even at an unbalanced slow diffusion rate. This is accomplished by a kind of pump that uses metabolic energy to force sodium ions uphill (up the electrochemical gradient) and so outward through the cell membrane, as is diagrammatically shown in Figure 1.17C. This diagram further shows that there is an excess of diffusion outward of potassium down the electrochemical gradient of about 20 mV for potassium, and again the transport of potassium ions is balanced by an inward pump. In fact, as shown, the sodium and potassium pumps are loosely coupled together and driven by the same metabolic process.

With the squid axon all ionic concentrations are several times larger because they are isotonic with the sea water in which the animal lives; nevertheless, the ratios are similar, being for inside/outside, 20:1 for potassium, 1:9 for sodium, and 1:14 for chloride. Thus, the diagrammatic representations in Figure 1.17C may be assumed also for the squid axon, both for electrochemical gradients and for ionic fluxes.

As shown in the formal electrical diagram of Figure 1.17B, under resting conditions the surface membrane of the nerve cell and its axon resembles a leaky condenser charged at a potential of about -70 mV (inside to outside). If this charge is suddenly diminished by about 20 mV, i.e., to -50 mV, there is a sudden further change in the membrane potential—a nerve impulse has been generated, as will be described in Chapter 2.

1.2.6 Pathological Changes in the Neuron[3]

Since such essential metabolic processes as protein (and enzyme) synthesis are concentrated in the perikaryal region, it is to be expected that any parts of the neuron severed from this vital region will rapidly die, as indeed they do. This forms the basis of histological techniques (Nauta and Fink-Heimer) for discovering the distribution of the axonal branches of neurons (Guillery 1970). The degenerative changes in the synaptic knob will be treated in Chapter 12 (see Figure 12.15, Section 12.2.3). The region central to the section also shows changes, however. Presumably because such a large fraction of the neuron has been amputated with a consequent depletion of transport, there is disorganization and degeneration of the Nissl bodies. These changes have been termed chromatolysis or the axon reaction, and are the basis of the Gudden method for studying the axonal path from a neuron. There is a great variation in the susceptibility of neurons to axotomy, neuronal death being common in young animals. The usual outcome, however, is that the neuron recovers over some weeks, the stump meanwhile sprouting and regenerating connections if it is in a peripheral nerve. Besides the characteristic nuclear changes there are also changes in the surface membrane of the soma with glial

[3]General reference: Bodian 1967.

displacement of synaptic knobs from their postsynaptic attachment sites (Blinz-inger and Kreutzberg 1968).

When neurons are deprived of a considerable fraction of their synapses, atrophic changes set in which result in diminution in their size or in death, a process called anterograde transneuronal degeneration. The review of Cowan (1970) surveys the degenerations observed in a wide range of central struc-tures subjected to surgical deprivation of synapses. The atrophy is more se-vere in young animals, and the severity also depends on the degree of com-pleteness of the synaptic deafferentation. There is much variability in severity between different cerebral nuclei, however, and with different species of ani-mal. Of the various explanations suggested, two may be preferred: atrophy is due to disuse; or atrophy is due to removal of some transneuronal trophic substance that is supplied by the synapses. As we shall see in Chapter 12, there is now good experimental evidence for transsynaptic transport of mac-romolecules.

1.3 Neuroglial Cells[4]

Cell counts in brain suggest that glial cells may outnumber neurons. Yet very little is really known about the function of these glial cells (Kuffler and Nicholls 1976).

Two major classes of neuroglial cells, oligodendroglia and astroglia, have the same ectodermal derivation as neurons. The microglia are mesodermal in origin and can act as phagocytes.

The oligodendroglia have rounded nuclei and few short processes and are concerned with the formation and maintenance of the myelin sheaths of nerve fibers. In the adult it is difficult to demonstrate any connection between the oligodendroglial cytoplasm and the myelin lamellae in a manner compar-able with that observed for Schwann cells and peripheral axons (cf. Figure 2.12, Section 2.5). The role of oligodendroglia in myelination is clearly shown in electron micrographs of the developing brain, however (Bunge *et al.* 1961). As would be expected from their function, oligodendroglia are much more numerous in the white matter than in the gray matter of the brain, this population distribution matching the respective populations of the myeli-nated nerve fibers.

The astroglia have a much wider range of functions. The relatively small perikaryal region gives rise to many widely branching processes that make a meshwork in intimate relationship with the neurons. The nuclei are large with finely distributed chromatin. The cytoplasm is pale with very few organelles, although some cells are distinguished by the extensive bundles of fibrils in the cytoplasm. Doubtless the interlacing astroglial processes provide a structural support to the neurons as was originally surmised, but astroglia are now given much more significant roles that will be detailed below and in Chapter 12.

Of particular importance is the relationship of the astroglia to blood

[4]General references: Gray 1964; Palay and Chan-Palay 1972.

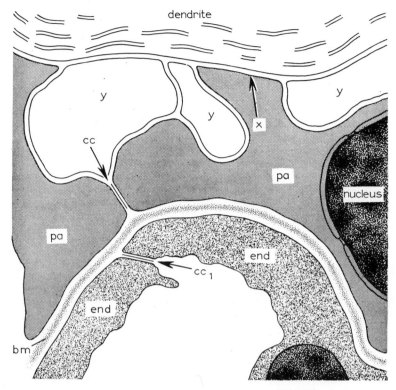

Figure 1.18 Diagram to show the pericapillary relationships of processes in the cerebral cortex. Key to labeling: y, neuronal and glial processes; x, extracellular zone; cc, closed contacts; cc_1, diffusion barrier between endothelial cells; pa, protoplasmic astroglia; end, endothelial cells; bm, basement membrane. (From Gray 1964.)

capillaries. As diagrammatically illustrated in Figure 1.18, capillaries are always separated from neurons by the astroglial sheath surrounding the capillary. The diagram illustrates the structures concerned in protecting the neurons from potentially harmful substances in the blood. The blood–brain barrier (BBB), as it is called, will be the subject of the next section.

Figure 1.18 illustrates a still more important astroglial function, namely, to act as an intermediary in providing neurons with nutrient substances from the capillaries.

As we shall see in Chapter 12, astroglia play an important role in guiding the growths of nerve fibers during development (Rakic 1971, 1972). Possibly they also function during the regeneration in the mammalian brain by triggering the sprouting of nerve fibers and guiding the sprouts to the vacant synaptic sites (Raisman and Field 1973; Eccles 1976a).

Glial cells can take up and metabolize some neurotransmitters (Chapters 6 and 8), and it has even been suggested that under special circumstances they may release transmitters. Glial metabolism of GABA and the monoamines will be discussed in later chapters. It is one of the methods by which transmitter action at synapses can be terminated.

In this connection the astroglia can also act as a water reservoir. The large bulk of their "watery" cytoplasm excellently fits them for this role. Disorders of their water reservoir function may result in cerebral edema, the brain swelling being specifically due to the excess uptake of water by astroglia.

Finally, astroglia function as phagocytes in removing degenerated fragments of neurons, and all metabolic waste production from neurons have to pass through astroglia in order to be carried away by the bloodstream.

1.4 The Blood–Brain Barrier

It has long been known that the relationship of the brain to the blood circulating through it is radically different from that of any other organ. For a wide range of substances there is a barrier to diffusion from blood capillary to neurons. Full accounts of investigations on the blood–brain barrier (BBB), as it is called, are available in several recent reviews (Davson 1976; Oldendorf 1975; Brightman and Reese 1975; Fenstermacher 1975). Quantitative studies have been made on a wide range of substances, and correlations have been made with predicted penetrations through cell membranes having a phospholipid structure (cf. Figure 1.16). Figure 1.18 illustrates the two possible sites for the barrier. First, the endothelial cells of capillaries differ from those elsewhere in that they are sealed together in overlapping fashion. The oblique tight junction (CC_1) probably blocks the passage out of the capillary of all substances that do not penetrate the capillary endothelial cells (Oldendorf 1975). Second, beyond the permeable basement membrane there is an enclosure of the capillary by processes of astroglia that are sealed tightly to one another, the intercellular gap (CC) being extremely narrow. The problem now is to determine the blocking properties of these two "tight" junctions, CC_1 and CC. There is experimental evidence (reviewed by Oldendorf 1975) that the tight sealing of the capillary endothelium is the principal factor concerned in the blood–brain barrier. On the other hand an important function can be given to the astroglia that insert themselves between the capillary endothelia and the neurons. In contrast to a negative blocking function they can operate positively in regulating the supply to the neurons of such essential metabolic substances as sugars, amino acids, and salts (Fenstermacher 1975; Davson 1976). Also, they can aid in the transport of excretory substances to the capillaries. There is evidence, however, that the capillary endothelium may be principally concerned in active transport requiring metabolic energy (Oldendorf 1975). It must not be assumed that the BBB has properties specifically protecting the brain from harmful organic materials. The penetration that is not carrier mediated is related to lipoid solubility and not to potential toxicity. It could be predicted that such substances as ethanol, caffeine, heroin, and nicotine score well in lipoid solubility so that they can penetrate rapidly through the capillary endothelial cells without any impediment by the BBB.

2

Signaling in the Nervous System

2.1 Signaling by Nerve Impulses

Chapter 1 was devoted to giving an account of what a neuron is in itself. Now we come to consider how neurons are concerned in receiving and in giving signals. In Figure 2.1A the diagram of the spinal cord is partly transverse and partly longitudinal, and below there is a muscle which has a stretch receptor or annulospiral ending (cf. A-S in Figure 4.1, Section 4.1). When you give a brief pull to a muscle, as when you make a knee jerk, impulses run up in the primary afferent fibers from the stretch receptors and excite nerve cells in the spinal cord (the so-called motoneurons) which fire impulses out to the muscles that are thus made to contract, the reflex circuit being indicated by arrows. That is the knee jerk. It is the very simplest central reflex pathway, and we will be dealing with it in Chapter 4.

In Figure 2.1B to the right there is shown diagrammatically a cutaneous receptor that is excited by touch or pressure or irritation on the skin (Figure 2.2A). The impulses run up the primary afferent fiber into the spinal cord, then via an interneuron to motoneurons and so to the flexor muscles that bend joints; three such muscles are shown. The resulting flexor reflex withdraws the limb from the irritating stimulus.

On the surface of the hairy skin of a cat there are very sensitive tactile receptors giving responses like those shown diagrammatically in Figure 2.1B. In Figure 2.2A steady indentations of the indicated amounts are put upon such a tactile receptor, causing impulses to be discharged along its nerve fiber, which are seen as brief potentials. With 88 μm of indentation the receptor fires fast at the start, then it slows up but keeps going during the indentation. This same fast-slow sequence can be seen with the larger indentations—154, 374, and 706 μm. The receptor fires more frequently the stronger the indentation. This is a typical signal system. All our touch sensations come like this from special cutaneous receptors firing along the primary afferent nerve fibers. In Figure 2.2A it can be seen that all of the impulses have the same size. With

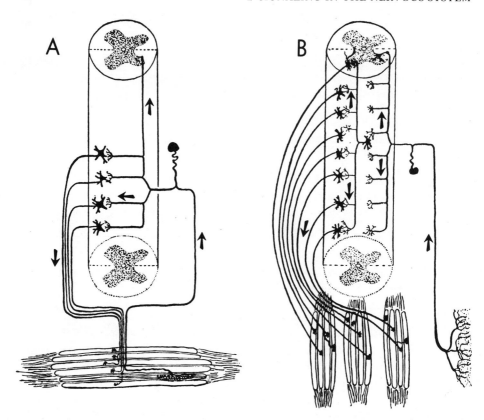

Figure 2.1 Reflex pathways for monosynaptic reflex arc. A. With afferent fiber from annulospiral ending. B. For polysynaptic flexor reflex with cutaneous afferent fiber. The three muscles in B represent flexors of hip, knee, and ankle. (Modified from Ramón y Cajal 1909.)

stronger indentation there is a faster train of impulses. It is a universal property of the nervous system that signaling is by coded information like this. Trains of similarly sized impulses signal intensity by frequency. It is like a Morse code with dots only.

In Figure 2.2B are impulses discharged from a quite different receptor, a light receptor in the eye of an invertebrate. When the intensities of light increase by factors of 10, that single unit fires more and more frequently. Again, intensity is coded by frequency and the impulses always have the same size.

In Figure 2.3A, a pull upon a muscle excites a stretch receptor (cf. AS in Figure 4.1A, Section 4.1) that discharges during the whole duration of the stretch. Some receptors, called phasic, only fire at the onset of the stimulus. Others, as in Figures 2.2 and 2.3A, discharge throughout the whole duration of an applied stimulus and are called tonic. There are all varieties in between.

The same general principle of signaling occurs on the output side of the nervous system to muscles. For example, in Figure 2.3B the firing of a motoneuron to an eye muscle is associated with a downward movement of the

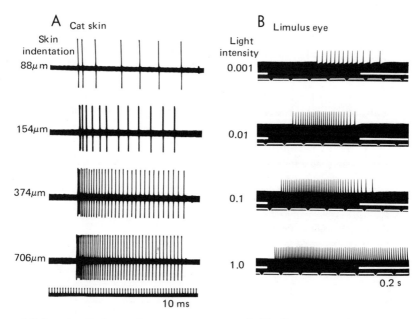

Figure 2.2 Impulse discharges from receptor organs. A. Tactile receptor of cat hairy skin discharging into a single afferent fiber in response to the indicated skin indentations. (From Mountcastle 1966). B. Photoreceptor of limulus eye discharging into a single afferent fiber in response to a l-s flash of light signaled by the gap in the white bar and of the indicated intensity. (Hartline 1934.)

eye, which is signaled by the lower trace. A wide range of frequencies of impulse discharge is related to the amount of downward eye movement. With strong upward movement the motoneuron is silent. Note in Figure 2.3B (b) that the discharge begins at about position 0 and progressively increases with further downward movement.

Throughout the whole central nervous system—with the complexities of organization suggested in Figures 1.2 (Section 1.2.1) and 1.3 (Section 1.2.2)—the signaling is by coded information of uniformly sized impulses. Intensity is signaled by frequency as in Figures 2.2 and 2.3. By studying the firing patterns of single nerve cells that have this coded language in all their responses to controlled sensory inputs we learn to understand the mode of operation of some parts of the brain. In fact, nerve impulses are the only language used in the brain for communication at a distance.

All peripheral receptor organs generate the firing of impulses along the nerve fiber that projects from them by reducing the electrical potential across the surface membrane of the terminals of that afferent nerve fiber—a process called depolarization. This has been established by most thorough investigations on many species of receptor organs, though the retina is an exception, but their consideration is beyond the scope of a book concentrating on the brain.

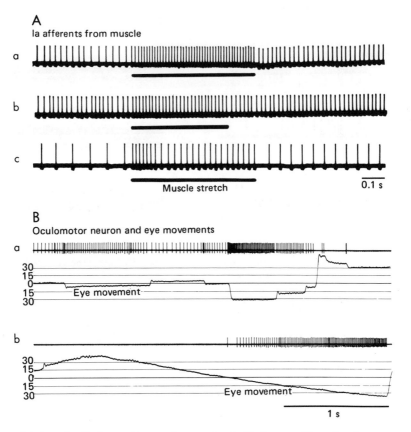

Figure 2.3 Impulse discharges from and to muscle. A. Discharges of impulses from an annulo-spiral ending of cat soleus muscle in response to a slowly augmenting stretch (2.9 mm/s) during the bars. The muscle spindle was under strong influence from a gamma motoneuron discharge (cf. Figure 13.8, Section 13.3.5) in *a,* less in *b,* and zero in *c.* (Jansen and Matthews 1962.) B. In the upper traces of *a* and *b* are discharges from an oculomotor neuron of a cat and in lower traces are the simultaneously recorded eye movements given in degrees of angle up or down. (Schiller 1970.)

2.2 The Nerve Impulse[1]

We now come to the question: What is this impulse that is the basis of all signaling? We have regarded it so far as an extremely brief message that runs along the nerve fibers. The frequency of firing may be higher or lower, but the impulse is always full sized. It is the universal currency of the nervous system. It is the only currency that the nervous system knows for any action at a distance. All signals from one nerve cell to another are conveyed by impulses. A nerve cell not firing impulses is mute. It is not communicating. (There are a few minor exceptions to this generalization with action at short

[1]General reference: Hodgkin 1964.

distances. They will be referred to in the fourth chapter; Figures 4.13 and
4.14, Section 4.6.)

In experimental efforts to study the nature of the nerve impulse the giant
axon of the squid has been of inestimable value. It has been known since the
1930s and has been utilized enormously, most fruitfully by Hodgkin (1964)
and Huxley (1959), for which they received the Nobel Prize in 1963. Figure
2.4A shows a squid with its tentacles. From the stellate ganglion several nerves
run out to the mantle musculature conveying impulses that bring about its
contraction. Contraction of the mantle powerfully ejects water, and the squid
jets in the reverse direction (to the left in Figure 2.4A)—a true jet propulsion!
When you look at one of those mantle nerves from the stellate ganglion in
section, the most prominent feature is the giant axon, which is almost 500 μm
across in Figure 2.4B. In the longitudinal section (Figure 2.4C) the giant axon
is seen to run straight and to have smooth walls, so they can easily be isolated
by dissection to give a beautifully clean axon. It turns out that this giant axon
has all the properties, essentially, of our own nerve fibers, and because of its
enormous size it can be investigated much more effectively.

In Figures 2.4B and 2.4C it appears that the fiber has a uniform core
surrounded by a thin membrane. When studied by electron microscopy, the
essential part of the membrane is only 7 nm thick, so it is very tenuous indeed
(cf. Figures 1.10, Section 1.2.4; 1.16, Section 1.2.5). The content of the axon
has the consistency of a jelly, and for most purposes you can substitute an
appropriate salt solution without deteriorating impulse conduction by the
fiber. For example, Baker and Shaw were able to squeeze out the contents of a

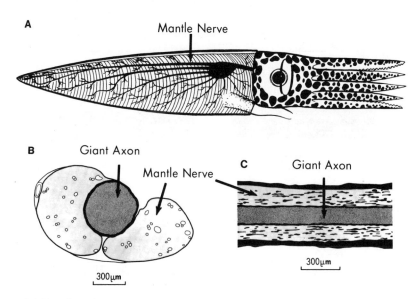

Figure 2.4 Drawings showing squid with the mantle nerves radiating from its stellate ganglion. In
the transverse and longitudinal sections the single giant axon is seen embedded in the mantle
nerve.

giant axon with an open end by a kind of microroller, leaving a collapsed, flattened axon that appeared destroyed. Yet when they reinflated it by an appropriate salt solution (a potassium salt) the fiber was restored and conducted well for hours. The refill should be isotonic and with the normal high potassium for its cationic content, but the anion is unimportant (Baker *et al.* 1962).

Figure 2.5 illustrates a simple yet fundamental experiment on a single giant axon immersed in a salt solution resembling the blood of the squid. The axon (black) is shown in two segments so as to indicate a long distance between the stimulating and recording electrodes. As mentioned earlier, stimulation by an electric current is effective when it takes charge off the membrane, depolarizing it and so setting up impulses. In Figure 2.5, a brief current was applied through the stimulating electrode just on the surface of the axon. As shown in the trace below the diagram, when the recording microelectrode was outside there was zero potential against the indifferent electrode; then, suddenly, as the recording electrode was advanced to penetrate the membrane, there appeared a membrane potential of −80 mV, which is the voltage across the surface membrane from inside to outside (cf. Figure 1.17, Section 1.2.5). This immediate change is quite dramatic. Since the microelectrode has a tip diameter of less than 1 μm, it can either be outside or inside because the membrane is only 7 nm across, which is only about 1% of the tip diameter.

In Figure 2.5, after intracellular recording had been established, three brief currents (S) of increasing intensity were applied in the direction to increase the membrane potential. Even quite a large current had no effect except for the brief downward artifacts that are seen to grade with the inten-

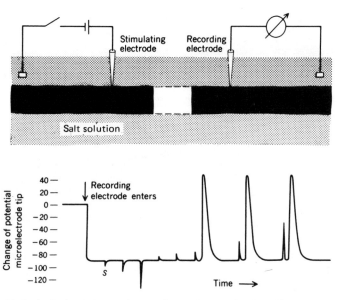

Figure 2.5 Method of stimulating and recording with a single giant fiber. (From Katz 1966.)

sity of the current. With the reverse direction the two weakest currents also were ineffective, but, at a certain intensity of the brief depolarizing current, a full-sized impulse was recorded—nothing less. With two further increases in the applied currents there was no increase in the response, only in the artifact. This is a good illustration of the all-or-nothing nature of the response. The explanation will be given later. Already we have seen examples of all-or-nothing impulses in Figures 2.2 and 2.3. In fact, all that can be propagated along an axon is a response that is full sized for the condition of that axon. The alternative is nothing. And this sharp antithesis holds whether the axon is fired by receptors (Figures 2.2 and 2.3A), by nerve cells (Figure 2.3B), or by stimulating electrodes (Figure 2.5).

A further important discovery is illustrated in Figure 2.5. During the impulse the membrane potential does not just go up to zero. It attains 30 mV or more in reverse. This discovery in 1939 by Cole and by Hodgkin and Huxley falsified all the theories up to that time about the nature of the impulse, which was thought to arise as a brief removal of the membrane charge. The observed reversal gave rise to most interesting investigations. The impulse became a more complicated phenomenon than had been thought. The elegant biological mechanism that was disclosed in the subsequent investigations can be understood in terms of the nature of the surface membrane of the nerve fiber and of the ionic concentration differences across it, which have already been treated in Chapter 1 (Figure 1.17, Section 1.2.5).

2.3 Ionic Mechanism of the Nerve Impulse

The essentials of the ionic mechanism of the impulse are shown in Figure 2.6 where the time scale is in thousandths of a second or milliseconds (ms). To the right is the membrane potential scale showing that the resting potential is about -65 mV. The time course of the membrane potential change during the impulse is given by the broken line (V) rising up to a reversal potential of almost $+30$ mV and then rapidly coming down again. The explanation of this sequence of potential changes is provided by the measured time courses of the changes in the sodium and potassium conductances (see scale to left) across the membrane (g_{Na} and g_K).

This explanation is based upon investigations by Hodgkin, Huxley, Katz, and Keynes on the squid giant axon using such refined techniques as voltage clamping of the membrane and measurements of ionic fluxes by radiotracers. These investigations have been carried out under conditions where the internal and external ionic concentrations have been widely varied. For a full description of these elegant investigations reference should be made to Hodgkin (1964).

In Figure 2.6 the sodium conductance (g_{Na}) is initially so dominant that the membrane potential (V) by an extremely steep rise approaches toward the sodium equilibrium potential (V_{Na}) at about $+50$ mV. Quite quickly, however, the potential ceases to rise and then falls toward the initial resting level. This

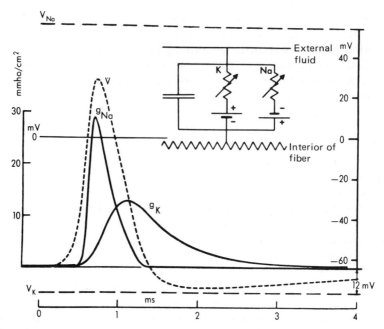

Figure 2.6 Theoretical action potential (V) and membrane conductance changes (g_{Na} and g_K) obtained by solving the equations derived by Hodgkin and Huxley for the giant axon of the squid at 18.5°C. V_{Na} and V_K are the equilibrium potentials for sodium and potassium across the membrane. The inset shows an element of the excitable membrane of a nerve fiber. Note the constant capacity, the channel for K^+, and the channel for Na^+. (Modified from Hodgkin and Huxley 1952.)

occurs for two reasons. First, even if the membrane potential is clamped at a depolarized level, the sodium conductance is quite transient, rapidly declining to the very low level of the resting state. More important is the increase in potassium conductance (g_K) that begins a little later than the increase in g_{Na} and runs a slower time course. The large membrane potential change during the impulse (V) moves the membrane far from the equilibrium potential for potassium ions (V_K at about -75 mV in Figure 2.6), establishing a high gradient for the outward flux of potassium ions across the membrane and counteracting the effect of sodium influx. At first it slows, then halts, the rise of the potential (V), and then causes it to fall to the initial level and even reverses it for some milliseconds. Meanwhile, this restoration of the resting membrane potential has accomplished the turning down of both g_{Na} and g_K to their resting levels. The depolarization of the impulse (V) occupies only about 1 ms in Figure 2.6, but for a mammalian nerve cell or fiber the changes are still faster, an impulse having a total duration of only 0.5 ms (Figure 2.10).

The inset diagram of Figure 2.6 shows an electrical model of the sodium and potassium conductances across the membrane with the capacity also shown as in Figure 1.17B (Section 1.2.5). The essential mechanism concerned in the rising phase of the impulse is a kind of autocatalyzing reaction. An

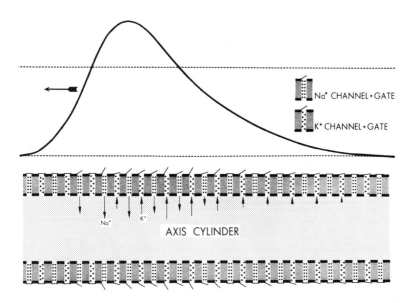

Figure 2.7 Opening of sodium and potassium gates at an instant during the propagation of a nerve impulse. In the upper part of the diagram there is a plot of the membrane potential along the fiber, showing reversal as in Figure 2.6. In the lower part, the opening of the two species of gates is symbolically represented by the angles of the respective gates. The thicknesses of the membranes on either side are greatly exaggerated with respect to the axis cylinder.

initial depolarization leads to an increase in sodium conductance, which in turn leads to entry of sodium and that to more depolarization, and so on up to the peak of the impulse. This explains why the impulse is all or nothing. It is the product of a self-regenerative or explosive reaction. Once it starts, it rises to its full effect, the energy being provided by the ions running down their electrochemical gradients.

Figure 2.7 is a diagram of an impulse showing the sodium and potassium channels through the membrane with their control by gates. The membrane is shown with separate channels for the sodium and potassium ions because these channels and their gates have quite distinctive properties (the key convention is given in the inset). To give an example, not only do the gates open at different times in response to a depolarization (Figure 2.6) but the sodium channels are selectively blocked by the poison TTX (tetrodotoxin) when applied on the outside, while the potassium channels are selectively blocked by TEA (tetraethylammonium) injected on the inside. There are various other distinctions between them, so they are depicted as separate channels with their own gates. Presumably each species of channel is formed by some specific protein and enzyme structure across the membrane resembling that shown in Figure 1.16 (Section 1.2.5) intercepting the bimolecular leaflet.

Figure 2.7 shows that at a certain stage of membrane depolarization the sodium gates open widely in the autocatalyzed reaction already described and then rapidly close. The potassium gates begin to open a little after the sodium

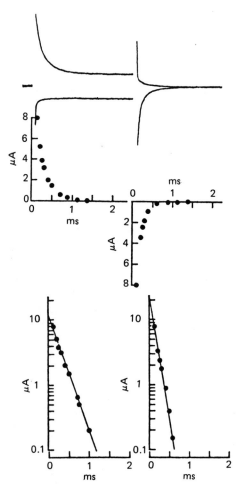

Figure 2.8 Asymmetry of the displacement current on application of ±120 mV voltage-clamp pulses. Axon perfused with 300 mM CsF + 600 mM sucrose and bathed in Na- and K-free saline containing 10 mM Ca and 1000 nM tetrodotoxin; holding potential, −100 mV; temperature 6°C. The results of graphical addition of the current records are plotted beneath the pair of tracings, both on the same scale linearly and on a logarithmic scale. (Keynes and Rojas 1974.)

gates and are fully open during the falling phase of the impulse. Their full closure is seen to be delayed until after the end of the impulse, as in Figure 2.5.

Recent work of Armstrong, Keynes, and their associates on the squid axon shows that the opening of the sodium gates, *per se,* is associated with an outward current across the membrane that is called the gating current (Bezanilla and Armstrong 1975; Rojas and Keynes 1975; Hille 1976). It could be uncovered only when the movement of ions across the membrane was blocked by the powerful tetrodotoxin (TTX) for Na^+ and tetraethylammonium for K^+. Furthermore, in Figure 2.8 the squid axon was bathed in Na^+- and K^+-free solution. In the upper traces of Figure 2.8 there are at the outset the recorded potentials when equal and opposite (±120 mV) voltage clamps are applied to the squid axon. The traces are asymmetrical, the upper being larger and with a slower time course. Symmetry would be expected if these two traces were simply due to the currents changing the charge on the capacity of the membrane (cf. Figure 1.17B, Section 1.2.5). Prior treatment of the axon prevents variations in both the Na^+ and K^+ resistances that are indicated in the inset of Figure 2.6. This asymmetry is plotted below, both linearly and exponentially, by the simple addition of the two traces and can be seen to have a rapid exponential decay. The removal of the voltage clamp induces the opposite asymmetry, as seen in the two upper traces to the right, while below this inward current is seen also to exhibit a rapid exponential decay.

The simplest explanation of these currents is that the voltage change across the membrane effects a movement of three charged particles or dipoles from a blocking position to an open position in the Na^+ channel (Hille 1976). Thus they constitute an integral part of the membrane that is concerned with the opening and closing of the Na^+ gates; hence the name—gating currents. In this theory the quantities of electricity involved in the on-and-off response should be identical, as indeed they are. Also, the quantity should reach a ceiling or saturation with sufficiently large pulses, as indeed it does. The observed temperature independence is also in accord with this identification. It is possible to estimate that the density of the sodium gates on the membrane is almost $500/\mu m^2$ for the squid axon. This figure is in very good agreement with the density determined for these gates by measuring the radioactivity when the gates are blocked by combination with tritiated TTX. There is as yet no evidence that there is a related process for potassium gates. Its existence is theoretically predictable, but it has not been demonstrated for two reasons. First, it is not possible to block the K^+ currents by tetraethylammonium as effectively as it is to block Na^+ currents by TTX. Second, from other evidence there is reason to believe that the density of the K^+ gates is smaller by an order of magnitude.

Comparable observations on the sodium gates have been made on the node of Ranvier (Figure 2.12) of frog myelinated nerve fibers (Nonner *et al.* 1975). The results suggest that the density of the sodium channels $(5000/\mu m^2)$ is 10 times higher than in the squid axon, but with a similar value of Na^+ conductance per channel. It is of interest that the local anesthetic procaine reversibly reduced both the changes in Na^+ conductance and the amplitude of the gating currents. It seems that the way may be open to an explanation of local anesthetic action in terms of the interaction between the gating particles and the phospholipids (cf. Figure 1.16, Section 1.2.5) of the membrane (Keynes 1976).

2.4 Conduction of the Nerve Impulse

After this detailed study of the events at one site on the nerve fiber, we are in the position to answer the question: How does the impulse travel along the fiber? So far we have considered in detail the happenings in one segment of the nerve fiber as the nerve impulse passes along it. The scale of Figure 2.6 is in milliseconds. Figures 2.7 and 2.9A are diagrams with a scale in length, not time. In Figure 2.9A the axon is shown in longitudinal section with the charge across the membrane (negativity inward) ahead of the impulse and the reversal of charge during the impulse brought about by the influx of sodium. Later, the efflux of potassium restores the charge.

When we come to consider the propagation of this impulse along the fiber in the direction of the arrow, we have to introduce the concept that, in addition to the ionic mechanisms of the surface membrane, the fiber as a whole is acting like a kind of cable. This cable property of the nerve, however, is terribly poor by any engineering standards. In electrical transmission along a submarine cable there is a conducting core and an insulating sheath. In the nerve fiber the specific conductance of the core is about 10^8 times worse than the copper core that the electrical engineer would use. Moreover, the sheath is about 10^6 times leakier than that of a good cable. So the cable-like performance of a nerve fiber is about $10^8 \times 10^6$ times poorer. Nevertheless, in evolutionary design this very discouraging performance of the biological cable was circumvented by a device also used in cable transmissions over long distances, where attenuation becomes serious. Boosters are inserted at intervals to lift the attenuated signals.

In the nerve fiber a booster mechanism is built in all the way along the surface. In Figure 2.9B the impulse is drawn as if frozen in its propagation along the fiber, from right to left. It is shown in Figure 2.9C that ahead of the impulse there is a passive cable-like spread with current leaving the outer surface of the fiber and circling back toward the impulse zone. Thus, charge is taken off the membrane by this quite limited cable-like spread ahead of the impulse, with the result that this zone becomes depolarized sufficiently to open the sodium gates (cf. Figure 2.6) and thus to turn on the self-regenerative process of sodium conductance that leads to the full-sized potential change of the impulse. All that the cable-like property of the nerve fiber has to do is to transmit the depolarization for a minute distance along the surface. Then at a critical level of depolarization the booster mechanism of the membrane, i.e., the self-regenerative sodium conductance, takes over and builds the full impulse. The process goes on seriatim, giving the indefinite propagation of a full-sized impulse at a uniform conduction velocity.

Figure 2.9 thus gives in diagrammatic form the essential features of the ionic mechanism of nerve transmission that occurs for invertebrates and vertebrates. It is the explanation of the universal currency of the nervous system, and undoubtedly it is the most efficient biological mechanism that could have been developed for the fast transmission of messages over the relatively long distances required in large animals.

We have now explained how the impulse has moved along the fiber, and how recovery occurs in its wake. The whole process has been accomplished with the loss of some potassium and the gain of some sodium. This ionic exchange has been measured by radiotracer techniques and is in good agreement with prediction. There is the further requirement that ionic recovery has to be brought about by ionic pumps such as those diagrammed in Figure 1.17C (Section 1.2.5). It has been found that the more sodium there is inside, the stronger the pumps work, so there is an automatic recovery mechanism

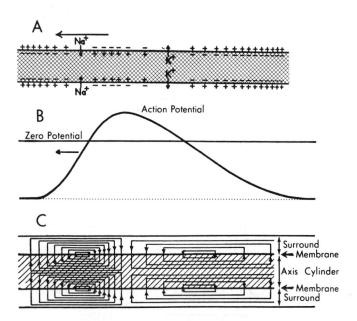

Figure 2.9 Ionic fluxes and current flow at an instant during the propagation of a nerve impulse. A. The ionic fluxes. B. The membrane potential, as in Figure 2.7 . C. The lines of current flow between the axis cylinder and the surround, illustrating the cable properties of the nerve fiber. (From Eccles 1953.)

governed by feedback control. But there is no urgency in this ionic restoration because nerve fibers contain enormously more ions than they need for one impulse. A squid giant axon only loses about one-millionth of its potassium per impulse, but the thinnest nerve fibers, with a much larger surface/volume ratio, may lose as much as one-thousandth. Nevertheless, there is an ionic reserve for hundreds of impulses, and meanwhile the ionic composition is continuously being restored by metabolically operated ionic pumps. As would be expected, conduction of nerve impulses can continue for hundreds of impulses in the anaerobic state.

Figure 2.9C illustrates a remarkable property of impulse transmission. The cable-like spread ahead of the impulse is accomplished by electrical currents that flow outward from the membrane and thence through the surrounding medium to enter the membrane at the zone of the impulse, and complete the circuit by returning up the conducting core of the nerve fiber. In accord with predictions, alterations in the conductivity of the "surround" change the effectiveness of the cable-like spread, the conduction velocity being faster with an increase in conductance and slower with a decrease. It might be thought that this use of the surround for return current flow is an undesirable feature of design because an impulse in one fiber might spread to adjacent fibers (cross talk) under the usual conditions of close packing. Reference to Figure 2.9C shows, however, that the privacy of the fibers is safeguarded by virtue of the directions of current flow into adjacent fibers of

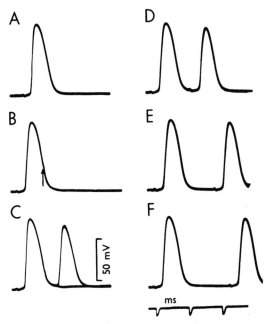

Figure 2.10 Refractoriness of nerve fiber: responses to double stimulation of a motor axon. Intracellular recording from a motor axon in the cat spinal cord that was stimulated about 20 mm away, there being double stimulation in B to F (note small stimulus artifacts). (Coombs *et al.* 1957a.)

the surround. There is an initial strong inward current into them before the outward current associated with the peak of action potential, and finally there is a terminal inward current. Thus, an impulse travels along a nerve fiber and adjacent fibers are subjected to the depressant action of anodal current before and after the cathodal current; hence, cross talk does not occur even with the closest packing of normal nerve fibers. Injured regions of nerve fibers, however, can be zones of cross talk and this can result in sensory disturbances, as in causalgia and neuralgia.

It has long been known that a nerve impulse is followed by a period during which it is impossible to set up a second impulse (the absolutely refractory period) and a further period during which a stronger stimulus is required to set up an impulse (the relatively refractory period), which in addition is smaller in size. The usual illustrations from the older literature give a confused picture because they are recorded from nerve trunks often containing thousands of fibers. In contrast, Figure 2.10 is an intracellular recording from a motor nerve fiber in a cat spinal cord of action potentials evoked by double stimulation of that fiber after it has left the spinal cord (cf. Figure 2.1). At the shortest stimulus interval, Figure 2.10B (0.7 ms), the second stimulus was ineffective, at Figure 2.10C (0.8 ms) it was sometimes effective, the second action potential being depressed. As the stimulus interval was progressively lengthened to 2.3 ms (Figures 2.10D to 2.10E to 2.10F), the action potential recovered almost to full size.

This rapid recovery is typical of the nerve fibers in the brain. As a consequence nerve fibers can carry high frequencies of discharge. In Figure 2.3B the highest frequencies of the motoneuronal discharge were over 200 per second. Neuronal discharge frequencies also exceeded 200 per second in Figures 2.19A and 2.19B. But more remarkable are the extremely high frequencies often attained by the discharges of many species of interneurons in the central nervous system. For example, in Figure 4.15B (Section 4.7.2) the frequency of the Renshaw cell discharge was initially at 1600 per second; yet its axon carried this high frequency of impulses, though with a diminished size, and its synapses operated effectively for each successive impulse (Figure 4.15C). The absolutely refractory period for mammalian nerve fibers is usually about 0.5 ms. Evidently nerve fibers have a functional design that gives a potentiality for frequency response adequate for the most extreme demands made by intensely discharging neurons or receptor organs.

2.5 Conduction Velocity and the Myelinated Fibers[2]

Now we come to a disadvantage that is inherent in the initial invertebrate design of the nerve fiber as a conducting device: the conduction velocity is slow. If the impulse is to glide along the fiber in the manner of successive invasion by the cable transmission and supplementary boosting as in Figure 2.9, the progress is slow. For example, with a large crab axon 30 μm in diameter the conduction velocity is only 5 m/s. The cable transmission is more expeditious if the fiber diameter is larger, but, if other factors are unchanged, it can be shown theoretically that the reward in speed is disappointing, for it is proportional to the square root of the diameter. Thus, in accordance with expectation, increasing the diameter by 16 times, as from crab axon to squid axon (30 μm to 500 μm), gives only a fourfold increase in velocity (i.e., from 5 to 20 m/s).

And so the invertebrates such as the squid were caught in a dilemma. As they developed in size and needed to retain their quickness of response, progressively more of their bulk had to be devoted to nerve fibers. For example, a doubling of bodily dimensions would entail an increase of 4-fold in diameter and 16-fold in cross-sectional area of nerve fibers to give the same conduction time for a doubling of conduction distance. Thus, for a linear doubling of dimensions, the mass of the animal would go up 8 times and the bulk of a nerve fiber by 32 times. Evidently, with the development of such giant nerve fibers the squid has gone about as far as is evolutionarily possible in order to effect a contraction of the mantle musculature, its escape mechanism, with a maximum of expedition. And all the squid can afford is one of these giant fibers in each mantle nerve (Figure 2.4B), with the consequence that it has an all-or-nothing movement with no gradation.

The situation was transformed by the brilliant innovation of coating the nerve fibers with a thick insulation that was interrupted at intervals. Actually this innovation was foreshadowed by an invertebrate, the prawn, that achieved the velocity of the squid giant axon with a fiber diameter of only 20

[2]General reference: Hodgkin 1964.

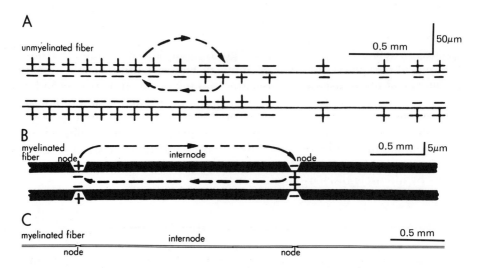

Figure 2.11 Propagation of impulse along a nerve fiber. A. There is propagation as in Figure 2.9C, but with only one line of current flow drawn in. B. The propagation is in a myelinated fiber, with the current flow restricted to the nodes. The dimensions in B are transversely exaggerated as shown by the scale, but are correctly shown in C. (Modified from Hodgkin 1964.)

μm. But the vertebrates perfected the design. Figure 2.11B illustrates the essential features of impulse conduction in a nerve fiber coated by the insulating myelin sheath, in contrast to the unmyelinated fiber in Figure 2.11A. In Figure 2.11A the impulse glides along the fiber by smooth progressive cable invasions as in Figure 2.9. In Figure 2.11B only the spaced nodal interruptions of the myelin sheath are active in the propagation, and the cable transmission is by currents flowing from the inactive node at the left into the active node with its reverse membrane potential. There is thus activation in succession of node to node without any active contribution from the long internodal zones. Actually, the myelinated fiber in Figure 2.11B is drawn distorted by the large transverse magnification shown by the scale. In the undistorted drawing of Figure 2.11C the relative lengths of nodes and internodes can be appreciated for an ordinary large vertebrate nerve fiber.

Extraordinary efficiency is achieved by having the impulse hop from node to node (the so-called saltatory transmission), with the long internodal section passive. It is often mistakenly thought that the myelin sheath achieves this remarkable result simply because it is an insulator preventing the flow of current across the nerve membrane of the internode. At least as important a contribution by the myelin sheath, however, is given by the enormous reduction (to less than 1%) in the electrical capacity between the axis cylinder and the surround. This reduction simply results from the increased plate separation of the cylindrical condenser enveloping the fiber in the internode. It is for this reason that the brief currents shown in Figure 2.11B spread so effectively from node to node, even with the long distances shown in Figure 2.11C. No more than half of the inward current at an active node is lost by leakage

through the resistance and capacity of the myelinated membrane of the internode.

Figure 2.12A shows diagrammatically the myelin wrapping around the
internode and the manner of its interruption at the node, where the bare
nerve membrane has channels for Na^+ and K^+ ions controlled by gates, just as
for the surface of an unmyelinated fiber (Figure 2.7), but at a considerably
greater density, about 10 times. Since the booster operations by ionic fluxes
during activity are restricted to less than 1% of the surface area of the nerve
fiber, the total ionic fluxes are greatly reduced; hence, there is an enormous
metabolic advantage in myelination.

Figures 2.12B, 2.12C, and 2.12D illustrate the remarkable manner in
which the spiral wrapping of myelin is applied to the nerve fiber. There is one
enveloping cell called a Schwann cell around each internode (Figure 2.12B).
Then it starts a spiral migration around the fiber, wrapping its surface membrane around in layer after layer. All the Schwann cell cytoplasm is eventually
squeezed out as the wrapping continues (Figures 2.12C and 2.12D) so that just
the myelin sheath is left, with layer after layer of two opposed Schwann cell
membranes, each of about 8.5 nm thickness. Around a large nerve fiber there
may be more than 100 of the double membranes. Thus, the myelin wrapping
of each internode is the creation of a single Schwann cell. It is remarkable that

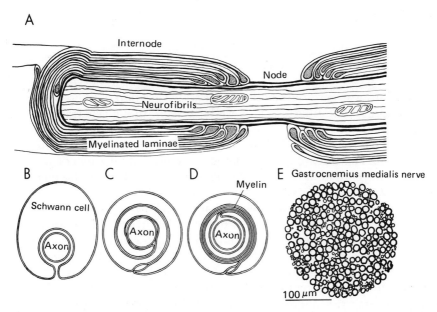

Figure 2.12 Myelinated nerve fibers. A. A node is shown diagrammatically in a longitudinal
section. The terminals of the laminated myelin wrapping are shown applied to the surface of the
nerve fiber adjacent to the node. Also shown are the neurofibrils and three mitochondria. B, C,
and D. Sequences in the formation of the myelin sheath by rotary migration of the Schwann cell.
(Robertson 1958). E. The tight packing of myelinated fibers of a wide range in size in a muscle
nerve of the cat. (Eccles and Sherrington 1930.)

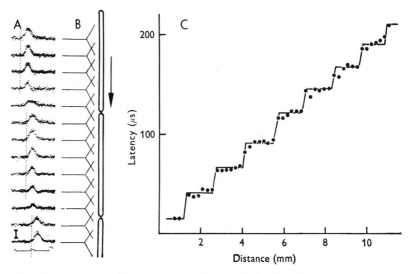

Figure 2.13 Saltatory conduction along a myelinated fiber. The fiber is shown in B with two nodes, the arrow giving the direction of propagation. The fiber is in a very fine filament of a spinal cord root and recordings of the impulse are made by two fine electrodes about 350 μm apart that are applied as shown at various positions along the nerve fiber. Note in B the two forks of the Ys whose stems point to the recorded potentials in A. In A the vertical dotted lines define three sets of records with simultaneous onsets and with steps of 25 μs between. In C are plotted measurements showing eight steps in another fiber. (Rasminsky and Sears 1971.)

with respect to the direction of the spiral wrapping there is no collusion between Schwann cells of adjacent internodes.

Figure 2.12E is a transverse section of a mammalian nerve showing the dark myelin sheath around the fibers, the wide variation in fiber size, and the close packing of the fibers. Yet, as we have seen, though its transmission is by electric currents, there is no risk of cross talk. The conduction velocity is approximately proportional to the fiber diameter, the ratio being about 6:1 for mammalian nerves. Thus, a large nerve fiber of 20 μm in diameter (including the myelin) would conduct impulses at about 120 m/s. It should be mentioned that all large nerve fibers in the brain also have a myelin sheath. Impulse conduction is comparable to that in peripheral nerve. The only difference is that the myelin is made by the oligodendroglia, which are a special variety of glial cells that were referred to in Chapter 1.

Figure 2.13 gives an elegant demonstration of the way in which a nerve impulse hops from node to node, as was first discovered by Tasaki (1939) and then fully investigated by Huxley and Stampfli (1949). Figure 2.13B shows a nerve fiber with two nodes in a rat dorsal rootlet. Recording is by two electrodes on the nerve fiber at the positions indicated by the two stems of the "Y," about 350 μm apart, and these are moved progressively along the fiber as shown. The impulse propagates downward (arrow). In the four upper records of Figure 2.13A there is no change in the time of the electrical response until the brief delay as the electrodes record from the next internode, then again a constant time until crossing the third internode. Plotting latency against distance (Figure 2.13C) shows this stepwise delay in the impulse from node to node. The average time between the successive nodal invasions is very brief (25 μs). Since the distance between nodes was 1.25 mm, the conduction velocity would be about 50 m/s.

This technique is very valuable in disclosing local defects in the myelin sheath. They can be produced experimentally by diphtheria toxin. With mild damage there is delay at each node so that the internodal time may increase from 25 μs to as long as 300 μs under conditions just critical for blockage. This diphtheria damage provides an experimental model for the state of the nerve fibers in the spinal cord pathways to and from the brain in the disease called multiple sclerosis. For an as yet unknown reason the myelin of the nerve fibers disintegrates, hence the failure of impulse transmission and the resultant severe incapacity. The nerve fibers are not killed, so there is hope that the disease could be cured if they can have their myelin wrapping restored. The prospects for success are much brighter than for diseases where the nerve cells and fibers have died.

A remarkable feature of mammalian nerves is the extremely wide range of fiber diameters (cf. Figure 12.12E). This wide variation also occurs in the brain and in the pathways to and from the brain. This finding has to be considered in relation to impulse-conduction velocity, the velocity in meters per second being approximately equal to six times the diameter in microns. Thus, the range in velocities for the normal variation of myelinated fiber diameter, 2–20 μm, would be 12–120 m/s. In addition, there are extremely numerous unmyelinated fibers from 0.2 to 1.0 μm in diameter that have much slower velocities—0.2 to 2 m/s. There is much evidence for the attractive postulate that the fiber velocity is related to the urgency of the information it is called upon to transmit. The fastest fibers are concerned in the control of movement, both the afferent input from the muscles (cf. Figure 2.3A) and up to the brain, and the descending pathways from the brain to muscles. Some of the pathways of cutaneous sense (cf. Figure 2.2A) are also fast. At the other extreme, pathways carrying visceral information are not urgent and are slowly conducting: they are even unmyelinated over much of their course. In general it can be postulated that, in the evolutionary design of the nervous system, nerve fiber size is related to the urgency of the information it carries. More simply put: Nerve fibers are no larger than they should be!

2.6 Metabolic Considerations

It is useful now to provide a general survey of the movements of Na$^+$ and K$^+$ across the nerve membrane that were already diagrammed in Figure 1.17C (Section 1.2.5). It is important to recognize that there is a very clear distinction between the passive (diffusional) and active (metabolically driven) transports across the membrane, because they travel by entirely distinct pathways. Table 2.1 summarizes these differences (Keynes 1976). It is to be noted

TABLE 2.1
**Evidence of the Independence of the Sodium Pump
from the Na and K Excitability Channels**[a]

	Na and K channels	Na pump
Direction of ion movements	Down the electrochemical gradient	Against the electrochemical gradient
Source of energy	Preexisting concentration gradient	ATP
Voltage dependence	Regenerative link between potential and Na conductance	Independent of membrane potential
Blocking agents	TTX blocks Na at 10^{-9} M; tetraethylammonium blocks K at 10^{-3} M Ouabain has no effect	TTX and tetraethylammonium have no effect Ouabain blocks at 10^{-7} M
External calcium	Increase in Ca raises threshold for excitation; decrease lowers threshold	No effect
Selectivity	Li^+ not distinguished from Na^+	Li^+ pumped much more slowly than Na^+
Number of channels or pump sites	Rabbit vagus has 27 TTX-binding sites per μm^2; squid axon has $500/\mu m^2$	Rabbit vagus has 750 ouabain-binding sites per μm^2; squid axon has 4000 per μm^2
Maximum rate of movement of Na	10^{-8} mol/cm^2-s during rising phase of spike	6×10^{-11} mol/cm^2-s at room temperature
Metabolic inhibitors	No effect: electrical activity is normal in axon perfused with pure salt solution	CN^- (1 mM) and DNP (0.2 mM) block as soon as ATP is exhausted

[a] From Keynes 1976.

that the passive properties are for both Na and K channels whereas the active are for Na only, though as indicated in Figure 1.17C (Section 1.2.5) the K pump is coupled with the Na pump and both depend on energy provided by ATP.

Evidently, the propagation of nerve impulses must be associated with the production of heat, initially due to the flow of the electricity (action currents) and probably to other events immediately associated with the ionic mechanisms during the impulse. Then, later, there must be heat production associated with the operation of the ionic (Na^+ and K^+) pumps restoring the original ionic composition of the nerve fiber. The detection of this heat had to await the very sensitive techniques developed by Downing, Gerard, and Hill in 1926. For several decades progressive improvement in techniques by Hill (1960) and his associates has revealed complications which still are not understood. At least the initial output of heat is recognized for all varieties of nerves, unmyelinated and myelinated, but with the slower processes occurring in the unmyelinated nerves, the initial output (positive) is followed by a prolonged negative heat. Approximate values for the initial heat are 14×10^{-6} cal/g/

impulse for unmyelinated and 0.8×10^{-6} cal/g/impulse for myelinated. Thus an impulse in a myelinated fiber results in a rise of temperature of no more than 10^{-6}°C. These differences reflect the increased efficiency of transmission in myelinated fibers (cf. Figures 2.11 and 2.12). No figures are available, however, for the heat produced by the slow recovery processes associated with the pumping of ions.

It may now be asked: How far does the heat production associated with impulses account for the heat production of the brain? Kety (1961) reports that the oxygen consumption of the normal human brain is 3.3 ml per 100 g per minute, or 46 ml/min for the whole brain. This gives an energy consumption of almost 20 W. The heat production would be 2.75×10^{-3} cal/g of brain per second. Actually this figure is too low for the cerebral cortex, which has about five times the metabolic rate of the white matter of the brain. Creutzfeldt (1975) utilities this value of Kety's in order to determine the relationship between the total energy production of the brain and that attributable to generation of impulses. By making probable assumptions for the number and sizes of nerve cells in the human cerebral cortex (2×10^6 per cm^2 of surface), Creutzfeldt calculates that the heat production would be 7×10^{-5} cal/g/s. The impulse discharge rate is at the reasonable average level of 10 per second. This value amounts to no more than 2%-3% of the total heat production, which is not surprising. It would be expected that by far the greater energy utilization would be in operating all the other activities of nerve cells that can be envisioned by reference to Figures 1.7 (Section 1.2.3), and 1.14 (Section 1.2.4). And in any case this impulse energy figure is only for the initial energy of the impulse and not for the recovery processes. Even the synaptic mechanisms are not taken into account. Nevertheless, these calculations are of interest in showing the orders of magnitude of oxygen consumption and energy production. It is further of interest that Kety's measurements showed no significant fall in oxygen consumption of the brain during sleep. This matches the finding that even neuronal discharges are not much depressed in sleep (Evarts 1961). Some are depressed and others aroused. On the other hand, in surgical anesthesia and in coma of various kinds there is a reduction of the oxygen consumption of the brain, even down to half.

Brief reference should be made to the remarkable studies of Ingvar (1975) on the patterns of brain activity revealed by measurements of regional cerebral blood flow. Radioactive ^{133}xenon is given into the carotid artery of human subjects and the pattern of radioactivity is measured by a battery of 32 detectors over the left cerebral hemisphere. Increases in blood flow of up to 30% indicated the areas which were activated during the carrying out of various tasks—talking, reading, calculating, hand movements, etc. In general, the activated areas were in accord with expectations from anatomical and chemical studies. It has even been found (Carmon *et al.* 1975) by the same techniques that verbal activity increased the flow more in the left than in the right hemisphere, whereas with musical stimulation the right exceeded the left. Regional brain metabolism is now being studied by an autoradiographic technique that has great potentiality.

2.7 Impulse Propagation in Neurons: Somata and Dendrites[3]

Hitherto attention has been restricted to nerve fibers, which are geometrically the simplest components of neurons, being long uniform cylinders in the cases investigated (Figures 2.7, 2.9, 2.11, 2.12). Figure 1.2B (Section 1.2.1), however, reveals that many neurons have short axons which branch profusely. It is assumed that impulses also propagate along these branches to their terminals, and this has been established in special cases (cf. Figure 3.7, Section 3.2.6) that give favorable opportunity for investigation. More complicated situations arise in the propagation of impulses over the neuronal somata and the dendrites branching therefrom (Eccles 1957, 1964a).

Some of the principal features of the spike potentials generated by a neuron are illustrated in Figure 2.14. The impulse is initiated by stimulation of the axon of this motoneuron in the ventral root and it propagates antidromically as indicated by the arrow in Figure 2.14B, where the three specifically responding zones are delineated as M, IS, and SD. Recording is from the intracellular microelectrode through which steady current could be passed to vary the membrane potential, as indicated to the left of the series of responses in Figure 2.14A. This motoneuron had a high membrane potential (-80 mV), and it was displaced over the range -87 mV to -60 mV, as indicated. The antidromic spike potential is observed to have three distinct components, each all or nothing in character. There is first the very small spike (about 5 mV) that is seen alone sometimes at -82 mV and always at -87 mV, and which is shown by threshold differentiation (lowest record of Figure 2.14A) to be generated by an impulse in the motor axon of the impaled motoneuron. Second, there is the larger spike of about 40 mV that is always set up at membrane potentials of -80 mV or less, and also sometimes at -82 mV. Finally, the full-sized spike is superimposed at membrane potentials of -77 mV or less, and rarely at -78 mV. Figure 2.14B shows the antidromic pathway together with the regions of the motoneuron (M, IS, and SD) in which the three components of the spike potentials are believed to be generated.

This identification has been firmly established by many investigations, particularly the elegant analysis by Terzulo and Araki (1961) using simultaneous recording from a motoneuron by intracellular and extracellular microelectrodes. Furthermore, they recorded simultaneously with one microelectrode inserted into a dendrite and the other into the soma of the same motoneuron (Figures 2.14C, 2.14D). The dendritic potential showed a delayed summit by about 0.3 ms in Figure 2.14C and 0.5 ms in Figure 2.14D, and it had a slower time course. It was full sized in Figure 2.14C, but in the more remote location of Figure 2.14D the size was reduced and the time course much lengthened. Analysis of extracellular fields indicates that impulses conduct for at least 200 μm along dendrites at the very slow velocity of 1 m/s. It seems that propagation does not continue along the fine dendritic branches. Thus, in Figure 2.14B the arrows indicate the zones of blockage:

[3]General reference: Eccles 1964a.

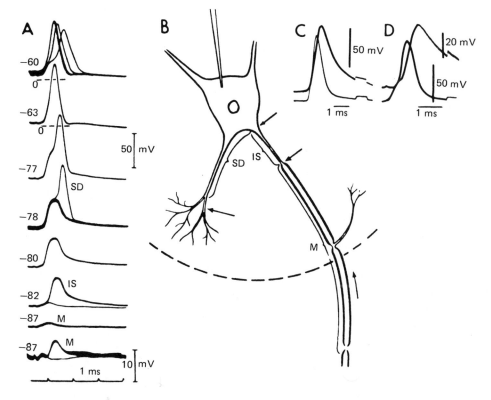

Figure 2.14 Intracellular responses evoked by an antidromic impulse, indicating stages of blockage of the antidromic spike in relation to the initial level of membrane potential. A. Initial membrane potential (indicated to the left of each record) was controlled by the application of extrinsic currents. Resting potential was at −80 mV. The lowest record was taken after the amplification had been increased 4.5 times and the stimulus had been decreased until it was just at threshold for exciting the axon of the motoneuron. (From Coombs *et al.* 1955a.) B. Schematic drawing of a motoneuron showing dendrites (only one drawn with terminal branches), the soma, the initial segment of axon (IS), and the medullated axon (M) with two nodes, at one of which there is an axon collateral. The three arrows indicate the regions where delay or blockage of an antidromic impulse is likely to occur. The regions producing the M, IS, and SD spikes are indicated approximately by the labeled brackets. (From Eccles 1957.) C. and D. Simultaneous intracellular traces of dendritic and soma spike potentials from the same motoneuron. (From Terzuolo and Araki 1961.)

between M and IS; between the IS and the SD; between the large dendrites and their terminal branches. The location of this last blocking site is believed to be much more variable than the others.

It can be objected that this antidromic invasion provides an artificial test situation that may not be relevant to the normally generated neuronal impulses. Figure 2.15 answers this objection for motoneurons and cortical pyramidal cells, at least for the IS and SD components. Three methods can be employed to elicit the motoneuronal spike potential, antidromic in Figures 2.15A and 2.15D, synaptic excitation in 2.15B and 2.15E, and direct stimulation by an applied current through the recording microelectrode (not illus-

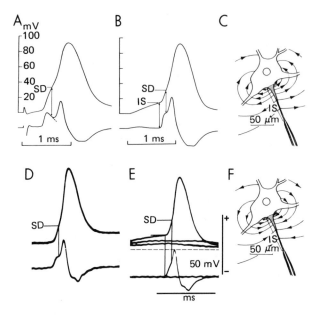

Figure 2.15 Impulse generation in the region of axonal origin from neurons. A–C. Tracings of intracellulary recorded spike potentials evoked by antidromic (A) and monosynaptic (B) stimulation of a motoneuron, respectively. The lower traces show the electrically differentiated records. Perpendicular lines are drawn from the origins of the IS and SD spikes, as indicated in the differentiated records, the respective threshold depolarizations being thus determined from the potential records, and indicated by horizontal lines labeled respectively IS and SD. C. and F. A diagram of the lines of current flow that occur when a synaptically induced depolarization of the soma-dendritic membrane electronically spreads to the initial segment to generate an IS impulse there (C), and when, secondarily, the spike potential of the initial segment depolarizes the soma-dendritic region to evoke the SD spike (F) (Coombs *et al.* 1957a). D and E. Intracellular records of a pyramidal tract cell of the motor cortex that is excited antidromically from the pyramidal tract in D, and orthodromically in E. Note that the EPSP in E excited a spike potential only in the trace where it was largest. The EPSP began before the illustrated response, the broken line giving the resting membrane potential. (Koike *et al.* 1968.)

trated). Discrimination between the initial IS and the later SD spike potentials is aided in the lower records of each frame which represents the electrically differentiated records. The IS–SD separation is more distinct with antidromic invasion (Figures 2.15A, 2.15D), but it is quite clear in the differentiated records for the spike potentials in Figures 2.15B and 2.15E. In the traces of Figures 2.15B and 2.15E the vertical lines mark times of origin of the IS and SD spikes, and show that in 2.15B the IS spike was generated at a membrane depolarization of 15 mV, whereas the threshold depolarization for the SD spike was 32 mV. With the motor pyramidal cell in Figure 2.15E the respective thresholds were 12 mV and 25 mV. Thus, the normal sequence of events for synaptic excitation of these species of neurons is, first, the synaptic depolarization of the soma and dendrites; then, by the current flow indicated in Figure 2.15C, the IS membrane is depolarized. When an impulse is generated there the current flow reverses, which is the third stage (Figure 2.15F). This current flow greatly adds to the synaptic depolarization of the SD membrane so that the much higher threshold is attained and the SD spike generated.

With cat motoneurons the average threshold depolarizations for IS and SD spikes are respectively 10 mV and 25 mV. An important functional consequence is that, with the generation of impulse discharge in the IS region, there is the best possible arrangement for integration of the whole synaptic excitatory and inhibitory action on that neuron. All these influences are electrotonically transmitted to the initial segment and there algebraically summed, as will be described in Chapter 4. Many species of neurons exhibit a low threshold zone in the initial segment comparable to that in the motoneurons, for example, the neurons in the spinal cord projecting up to the brain and also the motor pyramidal neurons of the cerebral cortex as illustrated in Figures 2.15D and 2.15E (Oshima 1969). However, there seems to be no such threshold discrimination with short axon neurons in the brain or spinal cord (Eccles 1966).

Definitive evidence of spike generation in dendrites has been presented by Andersen and associates in a systematic investigation on pyramidal cells of the hippocampus (cf. Figure 1.3B, Section 1.2.2). They took advantage of the sharply defined dendritic zone for the excitatory synapses from the Schaffer collaterals, as is indicated in Figure 2.16 (synapses labeled g in Figure 1.3B). These collaterals were selectively activated by stimulation of the entorhinal cortex by the pathway: perforating fibers → granule cells → mossy fibers →

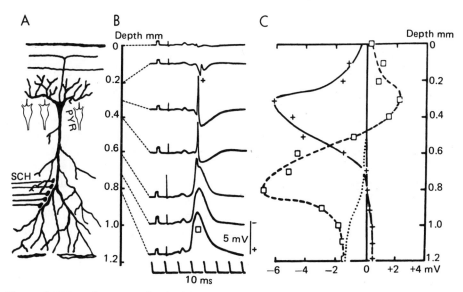

Figure 2.16 Impulse generation in hippocampal pyramidal cells. A. A typical hippocampal pyramidal cell, shown inversely from that in Figure 1.3B, with synapses of Schaffer collaterals (Sch). The somata of other pyramidal cells are also faintly indicated. Note depth scale to right. B. Extracellular field potentials generated by a volley in the Schaffer collaterals and recorded at the depths indicated. C. The depth plottings of the extracellular EPSPs (open squares in B and C) and the spike potentials (crosses in B and C). (Andersen *et al.* 1966a.)

CA3 pyramidal cells → Schaffer collaterals (Figure 4.16, Section 4.7.3). Figure 2.16B shows the intracellular records of field potentials generated by CA1 pyramidal cells in response to entorhinal stimulation. At depths of 0.8–0.6 mm there is a large late negative potential that has a spike superimposed at 0.6 mm depth. More superficially, at 0.4–0.3 mm there is a large extracellular spike potential. Measurements of these two potentials, as indicated by the symbols, are plotted in Figure 2.16C on the same depth profile as the drawing of the CA1 pyramidal cell in Figure 2.16A. The negative potential at 0.6–0.8 mm can clearly be interpreted as the extracellular counterpart of the synaptic excitatory action of Schaffer collaterals (Sch in Figure 2.16A). The depth profile corresponds, as may be seen by comparing A, B, and C in Figure 2.16; and the long latency is, of course, attributable to the two additional synaptic relays plus the long conduction pathway. Evidently the spike potential is produced by Schaffer collateral excitation. At depths of 0.3–0.5 mm, much faster recording than that illustrated in Figure 2.16B shows that this spike is conducted from the region of the Schaffer collateral synapses toward the soma with a velocity of 0.4 m/s. The relationship of the extracellular spike to the slow potential (the sequence from depths of 0.8–0.6 mm in Figure 2.16B) suggests that this spike is generated in the dendrites just on the somatic side of the Schaffer synapses. Intracellular recording corroborates this inference from depth profile recording as in Figure 2.16, namely, that full-sized impulses were generated in the dendritic tree close to the synapses of Schaffer collaterals and propagated down to the soma at a velocity of about 0.4 m/s. Sometimes small spikes were generated that did not propagate (Andersen and Lømo 1966).

Probably neocortical pyramidal cells resemble hippocampal pyramidal cells, but it has not yet been possible to perform sufficiently discriminative experiments because of the much greater structural diversity of the neocortex, and also because the pyramidal cells are located over a wide range of depths. The Purkinje cells of the cerebellum give excellent opportunity for the detection of dendritic spike potentials. Analysis by depth profile recording shows, however, that the antidromic invasion of the mammalian Purkinje cells only extends as far as the bases of the large dendritic trees, even when facilitated by strong synaptic excitation of the dendrites (Eccles *et al.* 1971a).

In contrast to normal mammalian motoneurons, an active role of dendrites is exhibited by motoneurons that are suffering from chromatolysis on account of severance of their axons some weeks previously (Eccles 1960) (Figure 2.17). Synaptically induced depolarizations (cf. Chapter 4) evoke partial spike responses of localized regions of the soma or dendrites (Figures 2.17B and 2.17C). The additional depolarization may then cause further partial spikes (Figures 2.17F to 2.17J) and eventually a full propagated spike may arise either at the initial segment of the axon or elsewhere, in the soma or up some dendrite (Figures 2.17F to 2.17H). In Figures 2.17A to 2.17D there is in addition recording from the motoneuron axon in the ventral root, the arrows indicating an impulse discharge just after the large intracellular spike (Figures 2.17A, 2.17D).

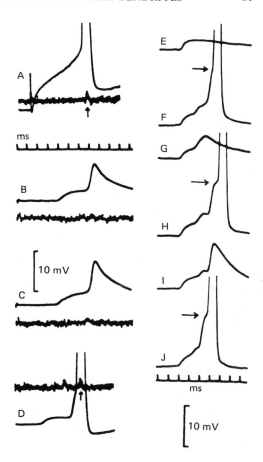

Figure 2.17 Partial responses and the discharge of impulses. Upper tracings in A–D are intracellular records from a flexor digitorum longus motoneuron (spike potential, 84 mV) whose motor axon had been severed 16 days previously in L7VR. Lower tracings are from the filament of L7VR that contained the motor axon. A. Response to a depolarizing current pulse that generated a spike potential. B–D. Responses at same amplification and sweep speed, but evoked by a group I maximum afferent volley in the nerve to flexor digitorum longus. E–J. Responses as in B–D, but in another motoneuron, which are selected to show the wide range of variability in the partial responses that are superimposed on the EPSP (E) and evoke full spikes when the depolarization attains the critical level (about 13 mV) shown by the arrows in F, H, and J. (Eccles *et al.* 1958b.)

These demonstrations of the existence of dendritic spike potentials, both small and full sized, with some neurons raises the expectation that this is a much wider phenomenon, particularly with neurons that have large, long dendrites. Remote excitatory synapses alone would suffer from the grave disability of a severe electronic decrement before these depolarizing potentials would reach the site of impulse generation in the initial segment. The brief Na$^+$ conductance of partial spikes would act as a local booster mechanism aiding in the spread of depolarization to the generating site, as is illustrated in Figure 2.17. There has been much speculation about the opportunities for local integrative actions of excitatory and inhibitory synapses so that a large neuron can be assumed to function as a conglomerate of many such interacting zones. It has to be recognized, however, that a neuron is a unity by virtue of the fact that it has only one axon along which it can fire. In any case, we have the overwhelming challenge of trying to understand the operation of the immensely complicated circuits of neuronal action, as we will see in Chapters 4, 11, and 14, that are based on the neuron operating as a unit. Attempts to theorize on a possible fragmentation of that unity are certainly out of place at this time.

2.8 Afterpotentials

The spike potential of the squid giant axon (Figure 2.6) was followed by a phase of hyperpolarization that is attributable to the continuance of a relatively high level of K^+ conductance across the membrane. Similarly, the spike potential of a motoneuron is followed by an afterhyperpolarization (AHP) usually reaching a maximum of 5–10 mV at about 10 ms and continuing for some 100 ms (Figures 2.18A, 2.18D). Its duration is longer—up to 150 ms—for motoneurons innervating slow muscle fibers than for those innervating fast fibers, usually 60–80 ms (Eccles *et al.* 1958a). There is evidence for two phases of K^+ conductance, the initial phase associated with the decline of the spike potential, analogous to that in Figure 2.6, and a later phase that is partly separated, which gives the AHP (Eccles 1957; Gustafsson 1974).

In Figure 2.18A, varying the membrane potential by a steady background current as in Figure 2.14 had a remarkable effect on the AHP following an antidromic spike potential. Because of blocked antidromic transmission it was not possible in this experiment to investigate membrane potentials larger than -87 mV, but interpolation of the linear curve relating size of AHP to membrane potential gave an equilibrium potential of -90 mV. When Na^+ was injected electrophoretically into a motoneuron, thus reducing its K^+ concentration, the AHP was temporarily inverted as in Figures 2.18C and 2.18D. In Figure 2.18C, recovery to the normal level occurred exponentially with a time constant of about 100 s. Reference to Figure 1.17B gives the explanation, namely, that the linked Na^+-K^+ pump is restoring the normal ionic concentration, and in particular building up the K^+ concentration to its normal level with an equilibrium potential of -90 mV. Thus, it can be postulated that the AHP of mammalian spinal motoneurons is due to an increased K^+ conductance of the neuronal membrane. This hypothesis of Coombs, Eccles, and Fatt (1955a; Eccles 1957) has been called into question, but a recent systematic study (Gustafsson 1974) has decisively corroborated the original hypothesis.

Immediately following the spike potential there is a slower decline of depolarization that after some milliseconds leads to the AHP. This phase is called the afterdepolarization and is probably due to a residuum of raised Na^+ permeability, particularly that persisting in the dendrites, as indicated by the slow dendritic spike potentials (cf. Figures 2.14C, 2.14D). There is much variability in this afterdepolarization, depending particularly on the amount of injury inflicted on the motoneuron by the impalement. When a motoneuron is in excellent condition, as in Figure 2.18E, there is a large afterdepolarization, the reversal to the AHP taking several milliseconds.

The electrophoretic injection of Na^+ into a motoneuron, with the raising of internal Na^+ and depletion of internal K^+, changes the spike potential (Figure 2.18B), the rise and decline being greatly slowed. This is exactly in

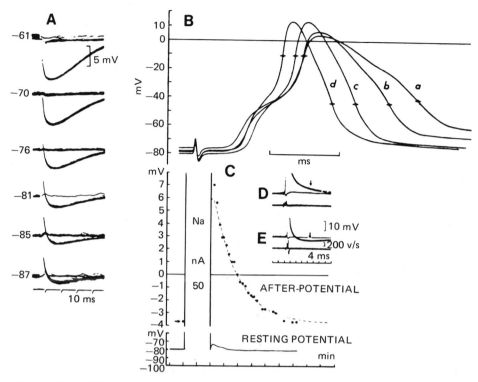

Figure 2.18 A. Afterhyperpolarizations of a motoneuron, occurring at various levels of membrane potential as controlled by extrinsic current. For each record, the stimulus applied to the ventral root was adjusted to the critical strength at which the axon of the particular motoneuron was sometimes excited and other times when it was not. The membrane potentials in millivolts at which the action potentials were evoked are given alongside each record. The resting potential varied from −76 mV to −79 mV. The spike component of the antidromic action potential does not appear in these records, the amplification being too high and the sweep too slow to display it. (From Coombs *et al.* 1955a.) B. Superposition of antidromic spike potentials recorded at 17 (*a*), 52 (*b*), 151 (*c*), and 348 s (*d*) after a sodium injection into an anterior biceps motoneuron by passing a current of 50 nA for 60 s. (Ito and Oshima 1964.) C–E. Changes in the afterhyperpolarization following the indicated sodium injection from a NaCl-filled single microelectrode inserted into gastrocnemius-soleus motoneuron. The measurements were made at the points indicated by the arrows in D and E. The interrupted line through the plotted points is an exponential curve with a time constant of 104 s. D and E. Specimen records of the membrane potential (upper traces) obtained at 36 (D) and 342 s (E) after a sodium injection. (Ito and Oshima 1964.)

accord with the hypothesis that the spike potential of neurons is produced by the linked Na^+–K^+ permeability increase as illustrated in Figures 2.6 and 2.7 for the squid axon. Recovery is complete after about 6 min, which corresponds well with the recovery of the AHP (Figure 2.18C). This correspondence is in accord with the hypothesis that both recoveries are due to a restoration of the normal ionic composition by the operation of the linked Na^+–K^+ pump.

Gustafsson (1974) has shown that the cells of origin of the dorsal spinocerebellar tract (DSCT) in the spinal cord differ from motoneurons in that the AHP is smaller and reaches an earlier maximum than occurs with motoneurons (cf. Figures 2.18A, 2.18E). Otherwise, the AHP had similar properties—for example, an associated conductance increase to K^+ ions and summation with repetitive stimulation. There is evidence that all neurons that have been sufficiently investigated tend to be grouped into two classes: on the one hand are the large alpha motoneurons as above, the large pyramidal tract cells of the motor cortex, the trochlear and hypoglossal motoneurons; and on the other hand are the DSCT neurons and other tract neurons, the rubrospinal and the vestibulospinal. To these two classes there should be added a third, a wide-ranging class comprising interneurons of various types. These are characterized by very high frequencies of firing—up to 1000/per second or even higher. It is evident that a large AHP is not conductive to such high frequencies. Further reference to these classes of neurons will be made when considering repetitive responses.

2.9 Repetitive Neuronal Firing

In the normal operation of brain synapses, there is rarely a sharp single action by a synchronized volley of impulses, as in Figure 2.15B, but instead prolonged repetitive inputs would give the responses (cf. Figure 2.3B). Hence the question now arises: How effective is a long, continued membrane depolarization in causing a prolonged impulse discharge? This question can be answered most directly by investigating the effects of steady depolarizing currents, instead of using synapses to fire a neuron. In Figure 2.19A a microelectrode was inserted into a motoneuron and steady currents were passed to take charge off the surface membrane, while at the same time the responses of the neuron were recorded. It is seen that a current of 7 nA just caused the neuron to fire, but the firing was slow and irregular. As the steady current was progressively increased, it produced more and more depolarization of the surface, and a faster and faster firing of the cell. Finally, with four times the threshold current, the neuron fired initially at the very high rate of at least 200 per second. In all cases the initial frequency slowed during the steady current application; nevertheless, this experiment on prolonged impulse discharge reveals the immense range in the responses of neurons. Continuous synaptic bombardment is active in the same way. There are many species of neurons in the central nervous system that can be driven by intense synaptic bombardment to fire at 500–1000 per second. This is evidence of real synaptic power, and these neurons correspondingly exert strong synaptic power at their synapses.

In Figure 2.19B a cell in the cerebral cortex is giving similar responses. The weakest current caused only one discharge. Again, frequency of response increased sharply with intensity, and with a current just over four times threshold the initial frequency was over 200 per second.

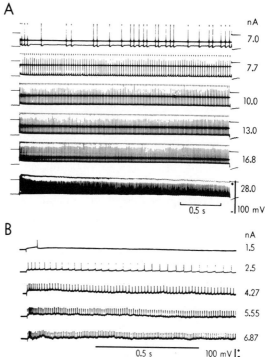

Figure 2.19 Generation of impulse discharges in neurons by prolonged depolarizing currents. A. Intracellular responses of a rat motoneuron to prolonged (2.5-s) rectangular depolarizing currents applied through the same electrode. Strengths of currents are indicated to the right as nanoamperes. (Granit *et al.* 1963.) B. Intracellular responses of a pyramidal tract cell of the motor cortex to currents indicated in nanoamperes for each trace. (Oshima 1969.)

It is immaterial whether depolarization is due to a steady synaptic bombardment or to an applied current. There is essentially the same response, depolarization generating impulse discharge, in a manner illustrating that intensity is coded by frequency, just as has been seen with receptor organs in Figures 2.2 and 2.3A. The synapses and the neurons of the central nervous system are well adapted metabolically to respond continuously at frequencies as high as 100 per second and with peak responses as high as 500 per second. This is a performance far superior to that of neuromuscular synapses. It used to be thought that one of the distinguishing properties of synapses was fatigue (Sherrington 1906), but this is belied by the performance of synaptic mechanisms in the normal functioning of the brain.

The AHP following the spike potential (Figure 2.18) exerts an important contributing influence on the frequency of neuronal firing in response to a steady depolarizing current. After each spike potential there is a phase of increased membrane potential, the AHP, the next discharge occurring when the depolarizing influence has overcome this hyperpolarization, the membrane depolarization again reaching threshold (cf. Eccles 1957; Schwindt and Calvin 1972; Calvin and Schwindt 1972; Gustafsson 1974). In the upper trace of Figure 2.19B the AHP can be seen after the single spike. The second trace also illustrates the origin of each successive spike response on the rising phase of depolarization following the AHP. With stronger depolarizing currents in Figure 2.19B the firing frequency is so high that this relationship is obscured, but it has been studied in detail by Schwindt and Calvin (1972). With a current intensity of 7.7 nA in Figure 2.19A the spikes can be seen to arise from a phase of increasing depolarization following each AHP. It has already been mentioned that the motoneurons supplying slow tonic muscles have a longer AHP than those supplying fast muscles. This is a good physiological arrangement, for it ensures that slow muscles are activated at a slow frequency. The motoneurons are thus tuned to the frequencies appropriate for the muscles, slow discharge rates for slow muscles and fast rates for fast muscles.

When considering the mechanism of impulse generation by a motoneuron it is important to recognize the two distinct zones of the neuron (Figure 2.14). There is, first, the soma-dendritic zone (SD) that is subjected to the influences of synapses, the excitatory depolarizing and the inhibitory hyperpolarizing, as described in Chapter 4. It also generates the AHP after each SD spike discharge. Second, there is the initial segment (IS) that is the trigger zone for firing impulses, but in which the firing of an impulse produces a negligible AHP (cf. Eccles 1957, Figure 2.7). The normal sequence is: (1) synaptic depolarization of the *SD* and *IS* membranes to the threshold level for initiating the IS spike; (2) generation of an SD spike by the IS spike (Figures 2.15B, 2.15E); (3) generation of the AHP by the SD zone. Thus, there is a division of function between the IS and the SD zone of the neuron. The IS zone is simply a firing zone. The SD zone is a controller of this firing by virtue of the potential changes it transmits electrotonically to the IS zone, namely, the algebraically summed, synaptically generated depolarizations and hyperpolarizations and the AHP autogenously initiated by the SD spike response. With repetitive discharge there is summation of the AHPs, sometimes with decrement. Baldissera and Gustafsson (1974a, 1974b) have developed computer models on the basis of these concepts and have shown that they account very satisfactorily for all the details of the effects produced by steady depolarizing currents in producing repetitive discharge as illustrated in Figure 2.19.

In Figure 2.19 it can be seen that the intracellular spike potentials were reduced by the depolarizing currents. The more the depolarization, the greater the reduction. This effect is more obvious in Figure 2.14A, where there was a large reduction of spike height when the membrane potential was depolarized to -60 mV. Similar reduction of spike potential by depolarization occurs with impulses in nerve fibers (cf. Figure 3.15, Section 3.3), and is attributable to some interaction of the gating mechanism. It is of general pharmacological interest because spike reduction is observed when various putative synaptic excitants are tested. When there is a powerful pharmacologically induced depolarization, there may even be suppression of the spike discharge, which can be a source of confusion.

3

Chemical Synaptic Transmission at Peripheral Synapses

3.1 The Discovery of Chemical Synaptic Transmission

In the preceding chapter it was shown how transmission along a nerve fiber occurred by brief all-or-nothing impulses and that this impulse propagation depended on two main factors: first, the cable-like properties of the nerve fiber, the conducting core being ensheathed by a relatively resistant membrane; and second, the large and momentary increase in sodium conductance of this membrane when its resting potential was suddenly diminished. As a consequence, the extremely poor cable-like transmission along the nerve fiber is amplified at each segment in a self-regenerative manner. The all-or-nothing character of the propagation derives from the amplification that each segment of the nerve gives to the attenuated signal transmitted in the manner of a cable from the adjacent active region.

In contrast to this impulse propagation along nerve fibers in a decrementless manner there has long been the idea that some special mechanism was responsible for transmission from one conducting element to the next, e.g., from nerve fiber to muscle fiber. The concept of the neuron was developed by the neuroanatomists, His, Forel, and Ramón y Cajal, in the latter part of the 19th century and led to the concept of the synapse by Sherrington, namely, specialized transmitting structures by which one neuron affected another (cf. Figures 1.3, 1.4, 1.5, and 1.6, Section 1.2.2). Much earlier there had been evidence by Claude Bernard that the junctional region between nerve and muscle was blocked by curare, and in 1877 Du Bois Reymond is reputed to have suggested that neuromuscular transmission was in part chemically mediated. Unfortunately, he based this on the mistaken notion that there was a protoplasmic continuum (Krnjevic 1974). At the turn of the century there were only the first beginnings of the hormonal story, where the so-called humoral transmission occurred by "chemical messengers, or hormones," which are specific chemical substances that are carried by the bloodstream and act slowly for prolonged periods.

In the first three decades of this century it was shown that transmission across the synapse at the neuromuscular junction is very fast, with a delay of at most 1 ms in transmission from nerve impulse to muscle impulse. This contrast between the slowness of the chemical transmission by hormones and the extremely fast and brief transmission at neuromuscular junctions and synapses made it appear that chemical transmission was too impossibly slow. Hence the concept arose that these latter transmissions were electrically mediated, i.e., that the electrical currents responsible for transmission along the nerve fiber were themselves responsible for exciting the nerve cell or muscle fiber across the synaptic junction. A great difficulty confronted this hypothesis because of the extraordinary mismatch in the electric properties between the nerve fiber and the muscle fibers. This was largely overlooked, however, because of the inadequacy of the evidence on the respective electrical parameters.

The first suggestion of chemical transmission was put forward in 1904 by T. R. Elliott, a Cambridge University student. He was struck by the similarity in effects of injection of adrenaline and stimulation of the sympathetic nervous system. It was an Oxford student, W. E. Dixon (1906), who followed up on the concept with regard to the parasympathetic nervous system. Dixon's experiments drew attention to the comparable action of injected muscarine and stimulation of parasympathetic fibers. The idea of chemical transmission was not well received and Dixon unfortunately became discouraged by his inability to extract an active principle.

Some years later Dale (1914) broke open the field of autonomic pharmacology. He recognized the attractiveness of the speculation for adrenaline mediation of sympathetic nerve impulses and for acetylcholine (ACh) mediation of parasympathetic impulses. The great potency of ACh demonstrated by Hunt and Taveau (1906) and its blockade by atropine greatly impressed Dale. In 1914, he identified a whole series of agents that mimicked electrical stimulation of autonomic fibers which he termed either parasympathomimetic or sympathomimetic. He was on the verge of substantiating the hypothesis of chemical transmission in the autonomic nervous system. Not only that, he was working with the transmitters themselves. Nevertheless, this crowning step eluded him, just as it had Elliott and Dixon.

In retrospect, Dale (1938) spoke in his characteristically vivid style of this preliminary stage of the chemical transmitter hypothesis:

> Such was the position in 1914. Two substances were known, with actions very suggestively reproducing those of the two main divisions of the autonomic system; both for different reasons were very unstable in the body, and their actions were as a consequence of a fleeting character; one of them was already known to occur as a natural hormone. These properties would fit them very precisely to act as mediators of the effects of autonomic impulses to effector cells, if there were any acceptable evidence of the liberation at nerve endings. The actors were named, the parts allotted, and almost forgotten; but only direct and unequivocal evidence could ring up the curtain and this was not to come until 1921.

Dale was referring to the experiment of Otto Loewi, establishing the first satisfactory proof of chemical transmission. Loewi's elegantly simple experiment with two frogs' hearts set up in series will go down as one of the classical experiments of all time. A dramatic flourish to the occasion has been given by Loewi's (1960) own recollection of the event almost 40 years later:

> The night before Easter Sunday of that year I awoke, turned on the light, and jotted down a few notes on a tiny slip of thin paper. Then I fell asleep again. It occurred to me at 6:00 in the morning that during the night I had written down something most important, but I was unable to decipher the scrawl. The next night, at 3:00 A.M., the idea returned. It was the design of an experiment to determine whether or not the hypothesis of chemical transmission that I had uttered seventeen years ago was correct. I got up immediately, went to the laboratory, and performed a simple experiment on a frog heart according to the nocturnal design.... the hearts of two frogs were isolated, the first with its nerves, the second without, both hearts were attached to Straub cannulas filled with a little Ringer solution. The vagus nerve of the first heart was stimulated for a few minutes. Then the Ringer solution that had been in the first heart during the stimulation of the vagus was transferred to the second heart.... if carefully considered in the daytime, I would undoubtedly have rejected the kind of experiment I performed.... it was good fortune that at the moment of the hunch I did not think but acted immediately.

Loewi described the active material released by stimulation of the vagus nerve as "vagusstoff." In 1926, he and his co-worker Navratil presented evidence that it was identical with acetylcholine. The year before Loewi had also reported on "acceleranzstoff," obtained from stimulation of accelerator fibers to the heart. This experiment pointed to adrenaline as the neurotransmitter for sympathetic nerves.

The choice by Loewi of the frog was providential. The cholinesterase content in this organ is small compared to that in mammalian hearts, the low temperature favors stability of acetylcholine, and the saline medium avoids the cholinesterase in erythrocytes and plasma. Thus, Loewi was successful, whereas Elliott, working on the same logical principle but under less favorable circumstances, had been unsuccessful. Loewi was also fortunate in choosing the frog to demonstrate his "acceleranzstoff" because in the spring, the season of Loewi's accelerator experiments, the adrenaline content is very high. Loewi and others could not know that the frog is an exception in that it liberates adrenaline from sympathetic nerves whereas most other species release noradrenaline.

The beautifully simple experiments of Loewi (1933) "rang up the curtain." For the first time it was shown experimentally that nerve impulses act by chemical transmission at synapses, albeit synapses of a very specialized character, from nerves onto heart muscle.

In the following decade a stream of outstanding experiments was performed by the physiologists of the time, all experts on the peripheral autonomic nervous system. Feldberg and Krayer (1933) confirmed Loewi's experiment by vagal stimulation of the dog's heart, using the recently developed leech muscle bioassay to demonstrate the active principle. Dale and Dudley

(1929) isolated acetylcholine (ACh) and identified it chemically from extracts of horse spleen. Feldberg and Gaddum (1934) established that peripheral ganglia used ACh as their transmitter and demonstrated that adding ACh to the perfusing fluid mimicked the effects of preganglionic stimulation. Dale proposed in 1933 that those fibers which liberate ACh be called cholinergic and those which liberate adrenaline or allied substances be called adrenergic.

The techniques developed by Dale, Feldberg, Gaddum, Brown, and Vogt were next applied with positive results to the neuromuscular junction (Dale *et al.* 1936). The chemical transmissions in the autonomic system were very slow, with times of action measured in tenths of seconds, hence for many years there was a controversy with respect to Dale's hypothesis that the very fast neuromuscular transmission was mediated by ACh. There is a recent account of this controversy (Eccles 1976b).

Originally the hypothesis was proposed by Dale and his colleagues (cf. Dale 1935, 1937, 1938) on the basis of experiments which showed that when motor nerve fibers were stimulated, minute quantities of ACh were liberated into fluid perfused through the muscle, and that intraarterial injection of very small amounts of ACh evoked impulses in muscle fibers and hence caused their contraction. In addition there was evidence that curare-like substances blocked neuromuscular transmission by depressing the response of muscle to injected ACh, and not by depressing the liberation of ACh. The action of inhibitors of cholinesterase provided further supporting evidence, for it was predicted that under such conditions the ACh effect would be prolonged and intensified, and this was observed. The excitatory effect of a single nerve volley was prolonged to give a brief waning tetanus of the muscle, and in curarized muscles there was restoration of neuromuscular transmission. This pioneer work demonstrated that impulses in nerve fibers excited muscle by the liberation of ACh. Stemming from this hypothesis are the amazingly elegant investigations during the last three decades by Katz, Kuffler, and their associates that will be described below. It has turned out that chemical transmission is a much more complicated biological process than Dale had supposed.

In the peripheral synapses the essential physiological problem that was solved in their functional design was the transmission of impulses across the synaptic gap with the minimum of delay and distortion, and also with a safety factor high enough to give reliability. These chemically transmitting synapses were designed to compensate for the electrical mismatch between the presynaptic and postsynaptic components of the synapse, e.g., the very small nerve terminal and the large area of the muscle fiber membrane with its high capacity. Though these same functions are fulfilled at the synapses of the central nervous system, a quite different functional design is called for in relation to the general problem of connectivity. Major differences occur because of the integrative function of the nervous system, as will be described in Chapter 4.

3.2 Neuromuscular Transmission

3.2.1 Introduction

Before embarking on the attempt to study neuronal mechanisms in the brain, it is important to study the transmission at peripheral synapses because the investigations on some of these provide the basis for our understanding of synapses in the brain. Essentially the same biological processes go on in the peripheral and central synapses, but those in the brain are more elusive when it comes to the precise investigations that will be discussed in the next chapter. Of the many peripheral synapses that have been studied, there are two of particular value in our attempts to study brain synapses: neuromuscular synapses, especially in the frog and snake; and the giant synapses in the stellate ganglion of the squid. These two exemplars of peripheral synapses form an adequate experimental base for the understanding of the essential features of central synapses.

Figure 3.1A shows diagrammatically a motoneuron in the spinal cord with the course of its axon down to the muscle fibers which it innervates. Limb muscles of the cat, for example, are innervated by some hundreds of motoneurons and the large motor fibers (cf. Figure 3.1B) stemming therefrom. Usually the motor axon branches to innervate some hundreds of muscle fibers in a particular muscle, not the five shown here, and usually a muscle fiber receives only one such innervation. Thus, a motoneuron makes its contribution to movement by discharging impulses down its axon and so exciting its own group of muscle fibers to contract. The whole executive ensemble is called a motor unit. It is a true unitary component of movement because usually the discharged impulse is effective in setting up impulses and the ensuing contraction in all the innervated muscle fibers. The total number of motoneurons with their dependent motor units is about 200,000 for the human spinal cord. That number is responsible for the contractions of all the muscles of the limbs, body, and neck, i.e., for our total muscular performance except for the head.

Figure 3.1C shows the isometric contraction tensions produced by tetanization of a motor unit over a wide range of frequencies. In the cat, the contraction tension produced by rapid repetitive activation of a single motor unit ranges from 50 to 100 g for the large extensor muscles, and from 10 to 20 g for smaller muscles (Devanandan *et al.* 1965). Within limits, the higher the frequency of the nerve cell discharge, the larger the resulting contraction. There are three rather separate twitches of the muscle when the frequency of discharge from the single motoneuron is 10 per second. At 20 per second the contractions tend to fuse and this tendency increases at 32 and at 50, until at 80 per second there is a smooth, strong contraction. It is perhaps somewhat surprising that a single cat motoneuron can command a contraction of 100 g in one of the leg muscles. In man the unit contraction would be higher, but it is not accurately known. Stronger actions of its muscles are secured by the

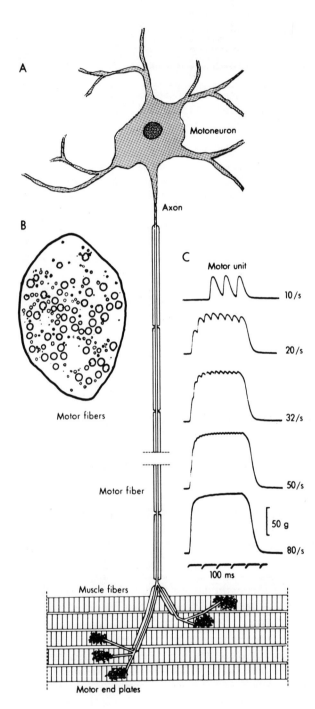

Figure 3.1 The motor unit. A. A motoneuron with its axon passing as a myelinated nerve fiber to innervate muscle fibers. B. Transverse section of motor fibers supplying a cat muscle, all afferent fibers having degenerated. (Eccles and Sherrington 1930.) C. Isometric mechanical responses of a single motor unit of the cat gastrocnemius muscle. The responses were evoked by repetitive stimulation of the motoneuron (cf. A) by pulses of current applied through an intracellular electrode at the indicated frequencies in cycles per second. (Devanandan *et al.* 1965.)

faster firing of its motoneurons. But it is more effective to bring into action many more motoneurons. Any large muscle has many hundreds of motoneurons innervating it. In Figure 3.1B there are about 60 large motor axons in the transverse section of a nerve branch to a cat muscle.

3.2.2 Structural Features of the Neuromuscular Synapse

In the mammal the nerve terminal is a tightly clustered structure on the surface of the muscle fiber, as indicated in Figure 3.1A and as drawn in microscopic section in Figure 3.2A, where the axon (ax) makes three small contacts on the surface of the muscle fiber (mf), but it does not fuse with it. There is the separation that is characteristic of all chemical synapses and that has been displayed in brain synapses in Figures 1.4, 1.5, and 1.6 (Section 1.2.2). The special structural enlargement of the sarcoplasm (sarc) of the muscle fiber at the junctional region is called a motor end plate. The problem is that the nerve impulse travels down to the nerve ending and after only a thousandth of a second of transmission time a new impulse starts up in the muscle fiber at the region of the end plate and runs along it in both directions, so setting up the contraction.

Light microscopy, however, good as it was, was not adequate to reveal the essential structures of the nerve–muscle synapse. It was only with the advent of electron microscopy that the structural bases of chemical transmission were revealed. The frog neuromuscular synapse is much less compact than that in the mammal. Fine nerve fibers run for hundreds of microns in longitudinal grooves on the surface of the muscle fiber, as is well illustrated in the drawings from a photomicrograph in Figure 3.4, where two such fibers are shown. In

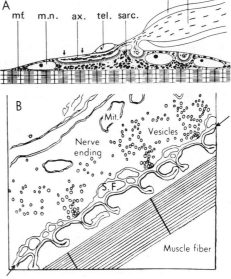

Figure 3.2 Microscopic structure of neuromuscular synapses. A. Schematic drawing of a motor end plate: ax., axoplasm with its mitochondria; my., myelin sheath; tel., teloglia (terminal Schwann cells); sarc., sarcoplasm of muscle fiber with its mitochondria; m.n., muscle nuclei; m.f., muscle fiber. The three (one separate) terminal nerve branches lie in "troughs." (Couteaux 1958.) B. Drawing of electron micrograph of part of neuromuscular junction from frog sartorius. Longitudinal section of the muscle. (Birks *et al.* 1960.)

Figure 3.2B there is a drawing showing an electron micrograph of a longitudinal section of a segment of one such long nerve fiber in the frog. The muscle can be recognized by the longitudinal striations with two transverse striations. The nerve terminal can be recognized by the dense aggregation of vesicles, with mitochondria (Mit.) lying further back from the muscle. The surface of the muscle fiber is shown by the dense line with frequent foldings. Opposite three of the foldings are clusters of synaptic vesicles, but in several places small projections (SF) from the enveloping Schwann cell intrude between the nerve and muscle surfaces. Elsewhere there is separation by a cleft of about 40 nm across, as marked by the arrow on each side. The numerous spherical vesicles are about 50 nm in diameter and are characteristic of all chemically transmitting synapses (cf. Figures 1.3B, 1.4B, 1.5, and 1.6, Section 1.2.2), hence they are appropriately called synaptic vesicles. The folds run transversely to the elongated nerve terminal (Figure 3.4) and vesicle attachment sites (cf. Figure 1.6, Section 1.2.2) are revealed by the freeze-etch technique to be congruent with the folds (Akert and Peper 1975).

There are two essential structural features of a chemical synapse—for example, as is diagrammatically illustrated in Figure 3.2B—first the synaptic vesicles and then the synaptic cleft, the space (note arrows) across which transmission has to occur. As illustrated in Figures 4.9A and 4.9B, this synaptic cleft provides the essential space for currents to flow that are generated by the action of the transmitter substance on the membrane across the synapse—the postsynaptic membrane—in this case, the membrane of the motor end plate.

The essential mechanism of the synapse is that the nerve impulse causes some of the synaptic vesicles to liberate their contents into the synaptic cleft. The vesicles at the neuromuscular junction contain prepackaged acetylcholine, about 5000 molecules of acetylcholine in each, according to latest estimates (see below), and they squirt their contents into the synaptic cleft. By diffusion across the cleft the acetylcholine acts on the membrane of the motor end plate, which is folded in order to increase its area. The transmitter opens the ionic gates and ions stream through, causing a change in membrane potential of the muscle fiber in the direction of a depolarization. When this reaches a critical level, a muscle impulse is generated that propagates along the muscle fiber and sets off the complex sequence of events responsible for a muscle contraction. We now turn to consider in some detail the experiments by which this synaptic mechanism has been revealed. Special reference should be made to two books by Katz (1966, 1969).

3.2.3 Physiological Features of the Neuromuscular Synapse

In Figure 3.3A there is a nerve fiber terminating on the motor end plate, with the fine nerve terminals running along the muscle fiber as in Figure 3.4A. We can stimulate the nerve fiber by a brief electrical pulse through the applied electrode and so send an impulse down to the nerve terminal as described in Chapter 2. Recording from the muscle fiber is, as shown in

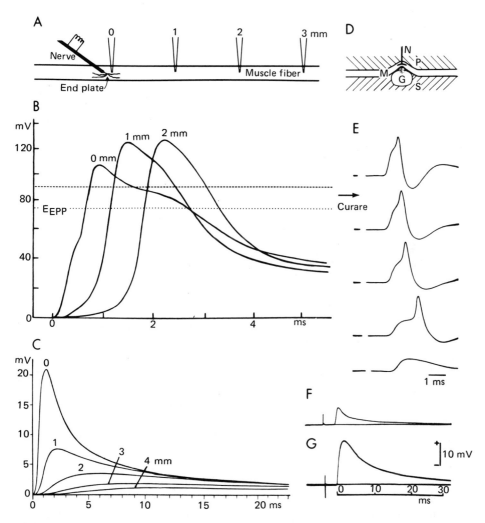

Figure 3.3 End-plate potentials and muscle action potentials. A. Nerve fiber with its terminal branches innervating a frog muscle fiber. B. Action potentials evoked by a single nerve impulse and recorded as in A at 0, 1, and 2 mm from the motor end plate. The broken line gives the level of zero membrane potential, the resting membrane potential being -90 mV. The dotted line is at the equilibrium potential for the end-plate potential (E_{EPP}). C. Action potentials evoked and recorded as in B, but after blockage of neuromuscular transmission by curare. (Fatt and Katz 1951.) D. Isolated single nerve (N)–muscle (M) preparation that was mounted on a recording electrode. E. Series of potentials recorded as in D showing progressive action of curare. (Kuffler 1942.) F and G. EPPs of curarized muscle recorded at end-plate region as in C, before and after addition of 10^{-6} M prostigmine. (Fatt and Katz 1951.)

Figure 3.3A, by an intracellular microelectrode either right at the motor end plate or 1, 2, or 3 mm away (Fatt and Katz 1951).

At all sites the potential across the muscle membrane is about 90 mV inside negative, which is the normal value for a frog muscle fiber. Just as with the giant nerve fiber in Figure 2.5 (Section 2.2), when the microelectrode crosses the membrane, a membrane potential of -90 mV is immediately registered. In Figure 3.3B with recording at the motor end plate (0 mm), after a brief latency a nerve impulse is evoked, a complex potential change with humps on the rising and falling phases that will be explained later. At 1 and 2 mm away the response looks simpler. It rises rapidly and smoothly, and starts to go down just like an ordinary nerve impulse (Figures 2.6 and 2.7, Section 2.3), but it is greatly slowed in its latter part. Two changes have occurred in moving away from the motor end plate. One is that the impulse is later. That is because the muscle impulse started at the end plate and moved along, just as with an impulse in a nerve fiber, except that it is much slower, only about 1 m/s. The other is that the humps have disappeared, and the summit is higher. How can this be explained? What is happening at the end-plate zone that reduces the muscle action potential and adds the humps before and after? The answer is that a special operation due to the chemical transmitter effects these changes at the end plate.

The initial hump is in fact due to depolarization by the transmitter, and at a critical level of this depolarization (about 50 mV) it can be seen to fire the impulse. But this impulse is much smaller than at 1 or 2 mm from the end plate. Evidently the transmitter action on the end-plate membrane is preventing the impulse from reaching as high a reversal point as elsewhere. The transmitter depolarization has an equilibrium potential (E_{EPP}) at about -15 mV (dotted line), so it effectively pulls the muscle impulse potential down toward this value. The ionic mechanism of the impulse would tend to give a membrane potential of at least $+40$ mV, as has been seen with the nerve impulse in Figures 2.6 (Section 2.3) and 2.9 (Section 2.4). The net result is the compromise at about $+15$ mV for the summit of the action potential at the end plate (0 mm). Finally, the depolarization by the residual transmitter action satisfactorily explains the hump on the declining phase of the impulse at the end plate. These end-plate effects are very slight at 1 mm and virtually absent at 2 mm (Figure 3.3B).

If the size of the end-plate potential could be reduced so that it failed to fire an impulse, it would be much simpler to investigate the events at the neuromuscular synapse. This can be done by poisoning with curare. In recent years, curare or related substances have become of great clinical value because in this way the anesthetist can greatly depress neuromuscular transmission so that the surgeon can operate with the advantage of having the patient with perfect muscular relaxation.

Figures 3.3D and 3.3E gives a good illustration of progressive curarization. In the early 1940s, about 10 years before intracellular recording, Kuffler (1942) dissected out a single nerve–muscle fiber, seen in Figure 3.3D (N.M), and held it between paraffin (P) and saline (S) while recording by a fine platinum wire supported by a curved glass rod shown in transverse section (G). It was the first example of an investigation on a single nerve–muscle preparation in isolation. When the nerve is stimulated and the recording is from the end-plate zone, there is a local potential followed by an action potential (E), just as in Figure 3.3B. After curare is added to the saline there is a progressive change, as is shown in the subsequent records of E. The initial potential was progressively reduced and fired the impulse progressively later, eventually failing. The muscle is now paralyzed. Until that happened the muscle action potential was full sized. It is a beautiful example of the all or nothing. The nerve impulse either fires a full-sized muscle action potential or it does not fire at all.

As will be described below, curare works simply by preventing the acetylcholine from effectively depolarizing the muscle at the end plate. When the muscle impulse fails, there remains the end-plate potential (EPP), as in the lowest trace of Figure 3.3E. In Figure 3.3C there are superimposed traces of such end-plate potentials recorded intracellularly as in Figure 3.3A at the end-plate zone (0) and at 1, 2, 3, and 4 mm distal thereto. The end-plate potential is decremented progressively with distance and becomes slower in time course. It is spreading simply by the cable properties of the muscle fiber, approximately halving every millimeter, so it becomes very much depleted from 0 to 1 to 2 to 3 to 4 mm. Cable transmission is even poorer in muscle fibers than in nerve fibers.

3.2.4 Pharmacological Properties of the Neuromuscular Synapse

We have already discussed how it was established that the neuromuscular transmitter is acetylcholine. Though the work of Dale and his colleagues seems pretty crude by present standards, it was a remarkable achievement at that time.

A great improvement came with the electrophoretic injection of acetylcholine out of a fine micropipette that could be manipulated into very close proximity to a motor end plate. Recording was by an intracellular electrode in a preparation that was curarized so that only an end plate potential was evoked, much as in Figure 3.3C at 0. With the best location, Krnjevic and Miledi (1958) found that electrophoretic injection of 1.5×10^{-14} g of ACh just outside the motor end plate gave a potential very similar to that produced by a nerve impulse that was estimated to liberate 1.5×10^{-15} g at the end plate. This discrepancy by a factor of 10 is as good as can be expected when it is remembered that the injection could only be at one side of the cleft, whereas the acetylcholine liberated by the nerve impulse would be directly onto the end-plate surface (cf. Figure 3.2) and would be acting over the whole of this surface.

A still higher level of technical excellence has been achieved recently by McMahan, Kuffler, and their associates (Peper and McMahan 1972). In Figure 3.4A the drawing is from a photomicrograph of a living muscle fiber identified by its cross striations, and on its surface can be seen two terminal

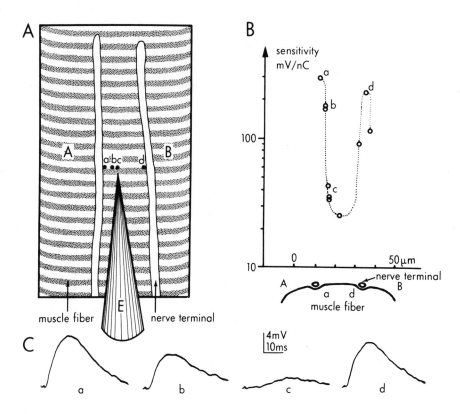

Figure 3.4 Distribution of ACh receptors in the vicinity of nerve terminals on skeletal muscle of the frog. (From Peper and McMahan 1972.)

nerve fibers running longitudinally. This beautiful display is due to a special technique called Nomarski interference microscopy, which allows the living tissue to be studied with very high magnification and with sharp optical sections. It was a particularly good arrangement for the experimenter that the two nerve fibers were so nicely parallel and terminating together. Below the graph (Figure 3.4B) you can see the muscle fiber in section with the two nerve terminals in little grooves (a and d). The muscle fiber has a recording intracellular electrode that is out of the picture in A, but at a distance that is quite small relative to the length constant of the muscle fiber so that it would record with little distortion the potential changes across the muscle membrane.

  Electrophoretic injections of ACh were made by the micropipette (E) that is shown accurately applied at a point (c) on the muscle membrane between the two nerve terminals. The injecting current was 32 nA for 1 ms, and the applied points are shown as black dots along the line between the two nerve terminals. When the point of application (a in Figure 3.4A) was close to the visualized nerve terminal, it produced in Figures 3.4B and 3.4C the large depolarization labeled a. The same injection only about 5 μm away (b in Figure 3.4A) gave the smaller depolarization b, while, when injection was between the two nerve terminals (c in Figure 3.4A), the ACh hardly affected the membrane at all (c). Finally, when the injection was near (d in Figure 3.4A) to the other nerve terminal, there was the large response (d) again. The results of several additional injections as well are plotted in the graph and show how subtle and delicate this performance is, the two nerve terminals being shown below Figure 3.4B on the same scale.

Now what does this show? That only the muscle membrane under the nerve fibers is sensitive to ACh. The remainder is less sensitive by about two orders of magnitude. The effects at other sites are largely due to diffusion of the injected ACh onto the sensitive sites at the nerve contacts. The same decrement of response, as from a to c, is observed if the tip of the electrode is moved vertically away from the nerve fiber area. It can be concluded that the sensitivity to ACh is exactly at the actual groove, just where the nerve contact is.

A still more exquisite experiment has recently been carried out by Kuffler and Yoshikami (1975a, 1975b) on the synapses on snake muscle fibers. Pretreatment by collagenase to remove fibrous tissue allows the nerve terminal with the whole presynaptic apparatus to be lifted gently off the postsynaptic sites, which remain as "craters" directly accessible to the electrophoretic application of ACh (cf. Figure 3.10D). High sensitivity to ACh is found to be restricted precisely to the "craters." Quantitative details follow after the next section.

3.2.5 Quantal Liberation of Acetylcholine

We now come to an investigation relating the liberation of ACh to the vesicles that are seen in the nerve terminal in Figure 3.2B. It is based largely on the classical experimental work of Katz, Fatt, Miledi, and their associates. In fact, at least two years before the vesicles were recognized by electron microscopy there was evidence that the ACh was liberated from nerve terminals in packages (Fatt and Katz 1952).

In Figure 3.5A the microelectrode was inserted into the muscle fiber quite close to the motor end plate. In 1952 when Fatt and Katz first detected the sequence of irregularly recurring small potentials, as shown in the upper traces of Figure 3.5A, they did not realize at first that this "biological noise" was a major discovery! The accurate location of the recording electrode at the end plate is shown by the action potential with an initial EPP set up by a nerve impulse in the lowest trace of Figure 3.5A, which resembles that in Figure 3.3B at 0 mm. When the recording electrode was 2 mm away as in Figure 3.5B, there was the simple action potential (cf. Figure 3.3B) and almost no "biological noise," just traces attributable to cable transmission.

How can it be shown that this "biological noise" in Figure 3.5A is due to ACh? First, there is the action of curare which depresses the end-plate potential and also the "biological noise" in the same way. Figure 3.5C shows another test. An electrophoretic injection of ACh at the end-plate region caused a brief depolarization, as in Figure 3.4. The lower traces of Figure 3.5C show

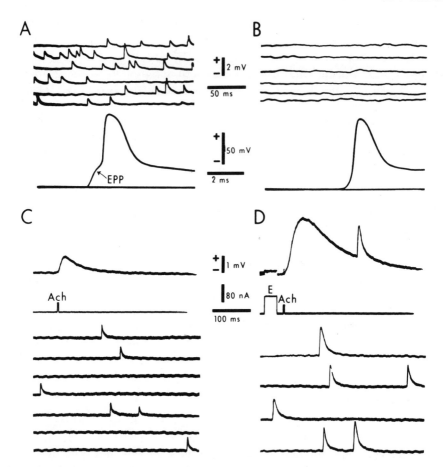

Figure 3.5 Spontaneous miniature end-plate potentials. A. Intracellular recording at an end plate. B. Recorded 2 mm away in the same muscle fiber. Upper portions were recorded at low speed and high amplification; they show the localized spontaneous activity at the end-plate region. Lower records show the electric response to a nerve impulse, taken at high speed and lower gain (cf. Figure 3.3B). (Fatt and Katz 1952.) C and D. Upper traces show the potentiating effect of edrophonium (physostigmine) (applied by pulse E) on end-plate depolarization produced by a brief pulse of ACh. In lower traces it is seen that the spontaneous min. EPPs in C are greatly increased and prolonged in D during steady electrophoretic application of edrophonium. (From Katz 1969.)

"biological noise" as in Figure 3.5A, but at a much slower frequency. In Figure 3.5D the electrophoretic injection of ACh is preceded by a substance, physostigmine, that very rapidly inactivates acetylcholinesterase, the enzyme that rapidly destroys ACh. Physostigmine works so rapidly that when injected electrophoretically at E just before the ACh, the depolarization produced by the ACh jet was increased and prolonged because the ACh was not being destroyed by the acetylcholinesterase. At the same time, in the lower traces of Figure 3.5D it can be seen that the units of the biological noise were increased and prolonged. This is good evidence that the "biological noise" is due to brief jets of ACh that are acting on the end-plate membrane of the muscle fiber. In every respect these miniature potentials have the properties of the end-plate potential (EPP), hence the justification of their designation as miniature end-plate potentials (min. EPP).

Synaptic vesicles (cf. Figure 3.2B) were recognized by electron microscopy of nerve terminals just after the min. EPPs were discovered. There was such an obvious and attractive correlation that suspicions became fashionable! After over two decades of investigation it still seems reasonable that the synaptic vesicles are preformed quantal packages of the transmitter substance, and that under resting conditions vesicles discharge their contents into the synaptic cleft at relatively infrequent intervals, thus producing the min. EPPs. A further remarkable property discovered by Katz and his colleagues (Katz 1966) is that the spontaneous release of the quanta is timed in a strictly random manner.

If the transmitter is packaged in the synaptic vesicles, it would be expected that the release of transmitter by a nerve impulse would also have a quantal composition. Normally, however, the EPP has a size that indicates composition of 100 to 300 quanta so a "quantal grain" could not be detected. By reducing the calcium in the bathing solution and adding some magnesium, a reduction of the EPP can be effected even to the point where it resembles the min. EPPs. For example, in the 18 traces of Figure 3.6A the nerve impulse evoked a very small EPP only four times, at the time of the second dotted line. The two spontaneous min. EPPs seen late in two traces resemble three of these EPPs, the fourth clearly being a double. Figure 3.6B shows another series with a less severe depression of neuromuscular transmission (Liley 1956). In two traces there was no EPP, in two it had the size of the min. EPPs, and in three it was double that size. Evidently these observations show that when the size of the EPP is sufficiently reduced, it exhibits a quantal grain corresponding to that of the min. EPPs.

The series of double traces of Figure 3.6C were recorded by an extracellular electrode in close proximity to the end-plate region, the quantal composition of the EPP being reduced by low calcium (Katz and Miledi 1965b). An extracellular recording is much more localized than an intracellular, yet it can be seen that there is a wide range of latencies of the quantal EPPs. In contrast, there is no latency variation in the initial diphasic spike produced by the nerve impulse. The latency variation of the EPP is shown in the histogram of Figure 3.6D for a large number of observations, a few of which are illustrated in Figure 3.6C. It can be concluded that the quantal liberation of transmitter by a nerve impulse has a considerable range in latency. An explanation of this variance will be given later.

Statistical analysis using Poisson's theorem (Katz 1958, 1962) gives a precise confirmation of this quantal composition of the EPPs, and shows that the quanta are identical to those randomly released to give the min. EPPs. Indeed, detailed studies by Katz and associates have shown that the quantal emission of ACh conforms with predictions from Poisson's theorem for the independent release of small numbers of quanta (Katz 1966, 1969).

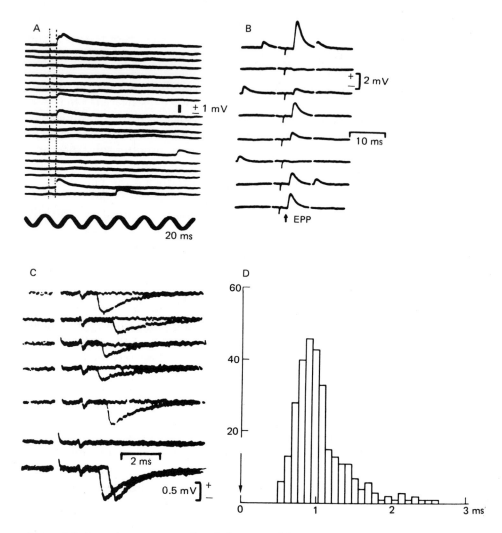

Figure 3.6 Quantal components of end-plate potential. A. Intracellular recording from frog neuromuscular synapse in calcium-deficient and magnesium-rich medium. Times of nerve stimulus and of EPP onset are indicated by the pair of dotted lines. (del Castillo and Katz 1954.) B. Intracellular records from a muscle fiber of the rat diaphragm with stimulation of the phrenic nerve (note artifact) also in a calcium-deficient and magnesium-rich medium. (Liley 1956.) C. Extracellularly recorded EPPs of frog muscle fiber with calcium level adjusted so that only about half the nerve impulses evoked EPPs. D. Histogram for latencies of EPPs for the experiment partly illustrated in C. (Katz and Miledi 1965b.)

It has now been established that a nerve impulse at the neuromuscular junction generates an EPP by causing the liberation of a number of quantal packages of ACh. Two quantitative questions now arise. First, how many quanta are released at an end plate in a normal calcium environment? The estimates range from 100 to 300 (Hubbard 1973). Second, how many ACh molecules are there in one quantum? There is a large range in the estimates (cf. Hubbard 1970, 1973), but, as described below, Kuffler and Yoshikami (1975b) have experimentally determined that the upper limit is 10,000 molecules, which would give a moderately hypertonic solution in a vesicle. It was originally thought that the quantal packages were very uniform in size, but it now seems that the largest may be double the smallest.

3.2.6 Factors Controlling Quantal Emission from the Nerve Terminal

In the further inquiry into the manner in which the nerve impulse causes the liberation of quanta from the nerve terminal, it is important to consider if the nerve impulse actually does invade the nerve terminal, or if, on the contrary, propagation ceases short of the terminal, there being merely an electrotonic spread of depolarization into the cable-like terminal. Investigations such as that illustrated in Figure 3.7 (Katz and Miledi 1965a) establish that the impulse normally travels along the fine terminal nerve fibers of the frog neuromuscular junction, and that this propagation is essential for its effectiveness. In Figure 3.7A the recording electrodes are seen to be on a terminal nerve fiber close to its origin (a) and on its extreme end (b). In Figure 3.7B (prox.) the nerve action potential has the typical triphasic configuration (positive–negative–positive) for an impulse propagating past a surface recording electrode, while in Figure 3.7B (dist.) it is only positive–negative, which would be expected for an impulse propagating to the end of a fiber. The faster recording of Figure 3.7C enables the conduction velocity to be calculated for this terminal fiber, a value of about 0.4 m/s being so obtained.

The inset diagram above Figure 3.7D shows the experimental conditions for a further testing experiment by Katz and Miledi (1968b). The nerve (N) is stimulated and the three electrodes are placed along a fine terminal nerve fiber. Accurate location of quantal release is ensured by employing a very unfavorable Ca–Mg ratio in order to abolish all quantal emission except in the immediate vicinity of the recording ($CaCl_2$) micropipettes (Ca_1 and Ca_2) that gave a controlled release of Ca^{2+}. Under such conditions the nerve impulse in Figure 3.7D evoked an EPP at both the proximal (1) and distal (2) recording sites. Then an emission of the nerve blocking agent tetrodotoxin (TTX) was effected from the middle pipette, resulting in Figure 3.7E in complete failure of the EPP at the distal electrode (2), the proximal being unchanged. In Figure 3.7F it is seen that after the TTX blockage had passed off, the distal EPP was restored. This experiment shows that blocking of the impulse less than 100 μm from the distal electrode caused a complete failure of the EPP. With present techniques it is only possible to establish that nerve impulse transmission to within 50 μm of the release sites is necessary for effective action.

It is important now to correct a possible misunderstanding. Though normally the propagation of the impulse into the vicinity of the transmitting site is required in order to cause the release of the transmitter from the presynaptic terminals, Figure 3.8 shows that the impulse, *per se*, has no specific

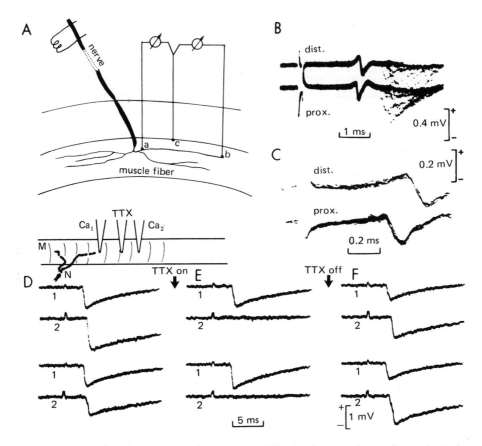

Figure 3.7 Nerve impulse and transmitter release. With stimulating and recording as in A, the nerve impulse is recorded in B and C both at proximal (a) and distal (b) sites on the nerve fiber lying in the groove on the muscle fiber. In B are also seen the extracellular EPPs at both distal and proximal recording sites (Katz and Miledi 1965a.) The inset diagram above *D* shows the experimental arrangements for *D* to *F*. N is nerve fiber supplying muscle fiber M, there being three extracellular micropipettes, Ca$_1$, TTX , and Ca$_2$, applied at intervals of 93 and 74 μm along a nerve fiber branch lying in a groove on the muscle fiber, but not drawn in as in A. The artifacts of the nerve stimulus are seen as a brief upward deflection in all traces of D to F, and 1 and 2 identify the traces recorded by Ca$_1$ and Ca$_2$ micropipettes, respectively. (Katz and Miledi 1968b.)

action (Katz and Miledi 1967a). Nerve impulse transmission was blocked by TTX, and the experimental arrangement shows that one of the current-passing electrodes was placed on the nerve as it enters the muscle and the other on it more remotely. When strong currents were passed (8.6 or 4.1 μA), depolarizing at the former site, intracellular recording reveals large depolarizations of the muscle membrane that resemble EPPs. Weaker currents (0.93 μA) were ineffective or less effective (1.85 μA). Evidently the applied depolarization had spread electrotonically in a cable-like manner, depolarizing the presynaptic terminals and so causing the release of transmitter, exactly as would occur for a nerve impulse. It can be concluded that the nerve impulse

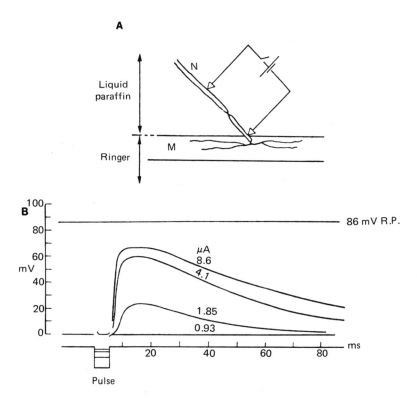

Figure 3.8 EPP production evoked by electrotonic depolarization of terminal nerve fibers. A. Electrode arrangement for applying a depolarizing pulse to the nerve fiber as it enters on terminal branching on the muscle fiber, nerve impulse transmission being blocked by TTX. The intracellular electrode is for recording the EPP. B. Traces of records so obtained, three intensities of applied current pulses evoking EPPs are shown with the current intensities indicated. The weakest current pulse (0.93 μA) was ineffective. (Katz and Miledi 1967b.)

triggers the transmitter release by virtue of its depolarization and not as a result of some other influence, for example, by the inward sodium flux during the rising phase of the nerve impulse. In fact, sodium ions are not at all concerned with the transmitter release, which can be very effectively produced in a sodium-free medium. On the other hand, there is a wealth of experimental evidence that extracellular calcium is essential for transmitter release at all types of chemical synapses. In investigations illustrated in Figures 3.6 and 3.7 advantage has been taken of the fact that transmitter release can be greatly depressed or even abolished by eliminating the extracellular calcium. Figure 3.9 illustrates a more analytical investigation on this key role of calcium.

3.2.7 The Essential Role of Calcium in Quantal Release

The experimental arrangements for Figure 3.9A are shown below the three series of superimposed traces evoked by stimulation of the nerve (Katz

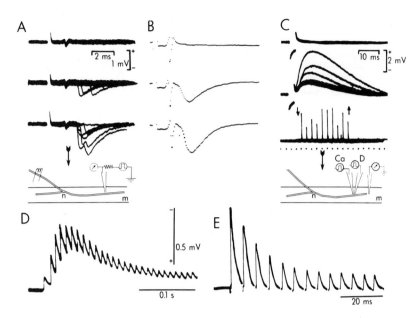

Figure 3.9 Extracellular calcium and neuromuscular transmission. The experimental arrange-ment for A and B is identified by the arrow below A. The extracellular recording was by a micropipette filled with 0.5 M CaCl$_2$. (Katz and Miledi 1965c.) The experimental arrangement for C is identified by the arrow below C. (Katz and Miledi 1967a.) D and E are described in the text.

and Miledi 1965c). The extracellularly recording micropipette was filled with 0.5 M CaCl$_2$, but diffusion outward of Ca^{2+} ions was prevented by a restrain-ing negative voltage in the upper trace of Figure 3.9A. The nerve impulse was completely ineffective because there were the same highly unfavorable condi-tions for release of transmitter, namely a Ca-free bathing solution containing 0.84 mM Mg^{2+}. In the second and third traces the negative bias on the mi-cropipette was reduced in two steps so that some outward diffusion of Ca^{2+} could occur: the Ca^{2+} level in the vicinity of the recording site was raised sufficiently to allow some quantal release by the nerve impulse. This effec-tiveness was lost as soon as the original restraining voltage was restored, at which point there was a return to the zero response illustrated in the first trace. In Figure 3.9B are the averages of 600 traces of the corresponding records in Figure 3.7A. It is interesting to note that the triphasic nerve im-pulse potential remains unaltered by the increase in calcium; it is only the release of transmitter that is changed.

The essential role of extracellular calcium for transmitter release can also be demonstrated in the absence of nerve impulses (Katz and Miledi 1967a). The experimental arrangements are indicated below Figure 3.9C. Elimination of nerve impulses has been secured by TTX, and two micropipettes are lo-cated close together on one of the nerve terminals on the muscle fiber that is being intracellularly recorded from as shown. The bathing solution is Ca-free with 1.7 mM Mg^{2+}. Brief applications of depolarizing currents by the D electrode caused no transmitter release, as shown in the upper trace of Figure 3.9C. Then the restraining negative voltage on the Ca^{2+} electrode was reduced, allowing Ca^{2+} to diffuse around the nerve terminal, with the consequence that the same depolarizing pulse now released transmitter, as shown by the large EPPs of the second trace. The third trace shows the same phenomenon at a much slower sweep speed. The depolarizing pulses are indicated by the dots at 0.5-s intervals below the trace and the successive EPPs appear as vertical lines. The bias on the Ca^{2+} pipette was released between the two upper arrows. It is seen that immediately there was a production of EPPs by each pulse, though there was a considerable variation in their quantal content. On restoration of the restraining voltage there was again a complete failure of EPP production.

So far we have concentrated on the action of single impulses at the neuromuscular synapse. Movements are brought about by repetitive discharge of impulses to muscles, however, as in Figure 2.3B (Section 2.1). This rapid, repetitive activation raises many new problems, of which two are illustrated in Figures 3.9D and 3.9E, where there was recording from the endplate zone of intact muscles. In Figure 3.9D a curarized frog nerve–muscle preparation was tetanized at 100 per second, and the successive EPPs increased at first and then declined. The reason for the increase appears to be that calcium moved in across the membrane of the nerve terminals (cf. Figure 3.13) with each impulse in the process of transmitter liberation by that impulse, and that some remained inside and so boosted the transmitter liberation of the next impulse, and so on for the several successive impulses at the start (Katz and Miledi 1968a). But then the EPPs declined, an effect which is attributed to exhaustion of transmitter, or at least of those synaptic vesicles readily available for ejecting their contents into the synaptic cleft. In Figure 3.9E, with a curarized mammalian preparation there was as usual an immediate decline with no initial phase of calcium potentiation. Under such conditions there was no significant depression of the end-plate depolarizations evoked by directly applied ACh. Evidently transmitter depletion is dominant, but it should be pointed out that the frequency of stimulation (200 per second) is higher than the highest frequencies of discharge of mammalian motoneurons to limb muscles. It has been found that at frequencies above a certain value (about 100 per second) the rate of ACh liberation is constant, the amount per impulse being inversely proportional to frequency.

3.2.8 Molecular Action of Acetylcholine (ACh)

Hitherto the action of ACh has been investigated by the membrane depolarization produced either by the electrophoretic application of quite large numbers of ACh molecules—about 20 million in Figure 3.4—or by the quantal emission of some thousands of molecules, as in Figures 3.5 and 3.6. Recently Katz and Miledi (1972, 1973) have accomplished a fantastic advance—in fact, to the ultimate level—in their identification of the size and time course of the depolarizations produced by single ACh molecules. This identification

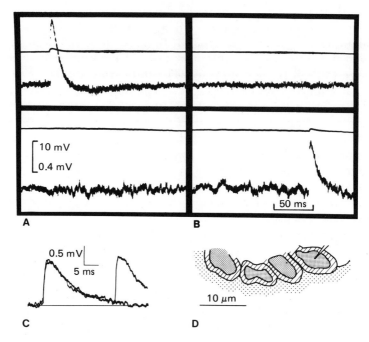

Figure 3.10 Intracellular recording from an end-plate region in frog sartorius. A and B. 21°C. In each block, the upper trace was recorded on a low-gain DC channel (scale, 10 mV); the lower was simultaneously recorded on a high-gain condenser-coupled channel (scale, 0.4 mV). The top row shows controls (no ACh); the bottom row shows membrane noise during ACh application, by diffusion from a micropipette. In the bottom row of records, the increased distance between the low- and high-gain traces is due to upward displacement of the DC trace because of ACh-induced depolarization. Two spontaneous min. EPPs are also seen (Katz and Miledi 1972.) C and D. Comparison of pipette-evoked ACh potential with min. EPPs in a snake muscle fiber. One-ms pulses were passed through an ACh pipette on the crater to the right. D. The four boutons were removed exposing the craters. Two iontophoretically produced responses (left) are superimposed; during one of them a spontaneous min. EPP occurred (right). Rise time (from 10%–90%) of the response to the pipette-applied ACh is about 1.1 ms, and of the min. EPP about 0.75 ms. The preparation had been lightly treated with collagenase. This treatment removes some but not all of the AChE. This accounts for the slightly slower than normal time course of both potentials, although no anticholinesterase was used. (Kuffler and Yoshikami 1975b.)

is accomplished by spectral frequency analyses of the electrical noise that the application of ACh evokes across the postsynaptic membrane (Figure 3.10A, 3.10B). The computer analysis of the noise reveals that it is made up of a random assemblage of extremely minute elements, each about 0.3 μV, and of a very brief duration, about 1 ms. It will be recognized that this noise is at least 1000 times less than the "biological noise" that was discovered by Fatt and Katz (1952), that turned out to be miniature EPPs. Exquisite testing procedures and rigorous controls establish that these minute potentials are produced by single ACh molecules, that momentarily (about 1 ms) attach to receptor sites on the postsynaptic membrane and open ionic gates, presumably by a brief conformational change in the protein structure of the pore, as will be described in detail in Chapter 5 (cf. Figure 1.16, Section 1.2.5). Since the average min. EPP of frog muscle is about 1500 times larger, it could be concluded that about 1500 ACh molecules participate in generating a min. EPP, and hence that there are about 1500 molecules of ACh in a quantum (or synaptic vesicle). This would be a minimum number, however, because an appreciable fraction—at least 30%—of the quantal population is hydrolyzed by cholinesterase before it can reach a receptor site, as indicated by the ACh–AChE interaction in Figure 3.13B.

This very interesting and important feature, quantal number, has been studied in a quite different manner by Kuffler and Yoshikami (1975a, 1975b). The bared "craters" of snake neuromuscular junctions were subjected to electrophoretically applied jets (1-ms duration) of ACh. With extremely close apposition the injection produced a depolarization with a remarkable similarity to the min. EPP, as illustrated in Figure 3.10C. They utilized a refined biological assay for the estimation of the numbers of ACh molecules actually injected electrophoretically, and arrived at a quantal number of 10,000. They regard this as an overestimate, however, because the injection site over the "crater" of necessity could not be in such a snug relationship as the nerve terminal. A possible value of 5000 molecules for the snake is not so far removed from the minimum of 1500 molecules derived from Katz and Miledi's results for the frog. It is interesting that 5000 ACh molecules in a spherical vesicle 45 nm in diameter give an approximately isomolar concentration (Hubbard 1973). On the basis of a quantal number of 5000 and the average size of a min. EPP (1.5 mV) it can be calculated that a single ACh molecule causes the opening of an ionic channel through which pass 6000 univalent cations in 1 ms. Katz and Miledi (1973) calculated a much higher value, 50,000, for the frog, but this is based on the very low estimate of 1000 for the quantal number.

We have given this recent work in some detail because it illustrates so well the advance from neurobiology to molecular neurobiology. It is only 40 years since Dale and his associates made the startling suggestion that neuromuscular transmission was accomplished by the secretion of ACh from the nerve terminals. In addition to the discoveries outlined above, this molecular level of investigation has led to a detailed study of the gate-opening times of analogs of ACh and of the effects of temperature, blocking agents, and anticholinesterases. It was not unexpected to find that curare and prostigmine, for example, had no effect on the unitary ACh action. Those receptors not occupied by curare behaved quite normally to ACh molecules.

3.2.9 Postsynaptic Events in Neuromuscular Transmission

It is now necessary to deal with the ionic conductances participating in the end-plate current, that in turn gives rise to the EPP. The only ion species in sufficient abundance to participate significantly in this conductance are sodium, potassium, and chloride. The voltage-clamping technique (Takeuchi and Takeuchi 1960) is employed to give the end-plate current (EPC) traces of

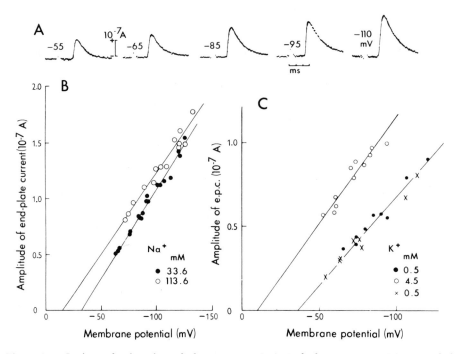

Figure 3.11 Ionic mechanisms in end-plate currents: I. A. End-plate currents (EPCs) recorded from a frog neuromuscular synapse under voltage-clamp conditions in response to a single nerve impulse. The membrane potentials are indicated for each record in mV. (Takeuchi and Takeuchi 1959.) B. Relationship between amplitude of EPC and membrane potential as in A, but obtained from an end plate in two different concentrations of external sodium as indicated. C. As in B but for an end plate in Ringer with low, then high, then low potassium, as indicated. (Takeuchi and Takeuchi 1960.)

Figure 3.11A. The membrane potential was changed by passing a DC pulse across it, the actual membrane potentials being given for each trace. As indicated by the plotted open circles of Figure 3.11B the EPC is linearly related to the membrane potential, and extrapolation shows that it is zero at −15 mV membrane potential. In Figure 3.11B reduction of the external sodium by about 70% displaced the curve to the right (solid circles), the equilibrium potential for the EPC being changed from −15 to −32 mV. Decrease of the equilibrium potential for sodium can also be effected by increasing the intracellular sodium, and this likewise caused the same displacement of the equilibrium potential for the EPC. These results indicate that Na^+ ions carry a considerable part of the EPC. In Figure 3.11C reduction of the extracellular K^+ concentration similarly displaced the equilibrium potential for the EPC about 28 mV in the depolarizing direction. This effect likewise indicates that K^+ ions carry a considerable part of the EPC.

In contrast to these experiments with cations, variation of the external chloride concentration, even its elimination in Figure 3.12A, caused no appreciable change in the equilibrium potential of the EPC, the two lines cross-

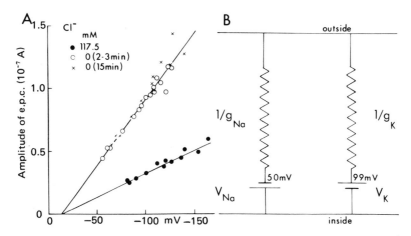

Figure 3.12 Ionic mechanisms in end-plate currents: II. A. Plot of EPCs for an end plate under voltage-clamp conditions as in Figure 3.11 and plotted as in Figures 3.11B and 3.11C, but for normal Ringer and then in Ringer with replacement of chloride by glutamate as indicated, and also with a reduction in the tubocurarine concentration. B. Schematic diagram of the assumed changes that occur in sodium (g_{Na}) and potassium (g_{K}) conductances during an EPC. (Takeuchi and Takeuchi 1960.)

ing the base line at the same point. The lines have different slopes because of the diminution in tubocurarine concentration after the first series. Figure 3.12B shows diagrammatically the channels for Na$^+$ and K$^+$ ions across the subsynaptic membrane with their respective equilibrium potentials of +50 and −99 mV. The action of the transmitter, ACh, is to increase greatly the ionic conductances in these two channels. There is evidence that in the end-plate membrane there are separate ionic gates for sodium and potassium, i.e., that the two channels of Figure 3.12B are distinct. They can be varied in relative conductance over quite a wide range, and procaine changes the time course of the sodium conductance channel independently from the potassium conductance (Maeno 1966; Gage and Armstrong 1968). This independence is still controversial, however (Hubbard 1973; Krnjevic 1974). It can be calculated that the ionic channels opened by the transmitter have very high conductances, the values being transiently as high as 10^{-10} mho for the conductance change induced by a single molecule of ACh, and 5×10^{-8} mho for a quantum giving a miniature EPP (Gage and McBurney 1972).

3.2.10 Diagrammatic Representation of Neuromuscular Synapses

Figure 3.13 illustrates diagrammatically the essential elements of the hypothesis of transmission at the neuromuscular synapse (cf. Hubbard 1973). In the nerve terminal (Figure 3.13A) there are seen to be synaptic vesicles with the membrane around them, and packed tightly inside are the ACh molecules, about 5000 per vesicle. One vesicle is liberating its transmitter into the synaptic cleft—a quantal emission. We do not know how this quantal

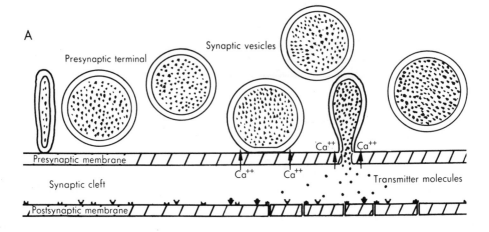

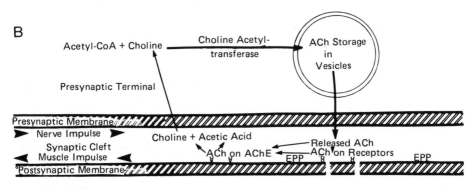

Figure 3.13 Essential elements in neuromuscular transmission. A. Portion of synaptic cleft with synaptic vesicles in close proximity to the presynaptic membrane, one actually discharging transmitter molecules (ACh) into the synaptic cleft. Some molecules are shown combined with receptor sites on the subsynaptic part of postsynaptic membrane with the consequent opening up of pores through that membrane. Arrows denoted Ca show movement of Ca^{2+} through the presynaptic membrane when it is depolarized. Also shown by different symbols are the molecules of the enzyme AChE attached to the cleft side of the postsynaptic membrane. B. Schematic representation of the sequence of events in neuromuscular transmission.

emission is brought about, only that the influx of calcium ions is essential. And, of course, the vesicle itself is not ejected into the synaptic cleft—only its contents—and then it is assumed to be refilled with ACh and so to be available for reuse. To the left is shown a vesicle that has been previously squeezed out and now is being recharged with ACh. An approximate estimate with neuromuscular synapses is that the reserve of synaptic vesicles is enough for

2000 to 5000 impulses, which is only a few minutes' supply with ordinary muscular activity. Evidently an efficient and fast replenishment is essential. The local new formation of synaptic vesicles or their transport from manufacturing sites in the motoneuronal soma would be much too slow.

Also shown on the postsynaptic membrane of Figure 3.13A are two types of receptor sites that have different molecular configurations. One type is related to transmembrane channels for ions, and these are shown opened up when an ACh molecule is attached. The other type is an acetylcholinesterase site that destroys the attached ACh. A more detailed and sophisticated diagram of these two types of receptor sites has been published by De Robertis (1971), and is illustrated in Figure 1.16 (Section 1.2.5).

Figure 3.13B illustrates other features of the neuromuscular synapse. There is, first, the vesicle with stored ACh, and arrows indicate its release and subsequent fate. Part goes to acetylcholinesterase (AChE) and is very rapidly hydrolyzed into choline and acetic acid, and part goes on to the receptor sites on the postsynaptic membrane, staying there momentarily to open the ionic gates and so produce the end-plate potential (EPP). Within 1 to 2 ms it leaves these sites and is rapidly destroyed by the AChE. The effectiveness of this destruction is illustrated in Figures 3.3F and 3.3G where inactivation of the AChE resulted in the large increase of the curarized EPP from Figure 3.3F to 3.3G. It is also shown in Figure 3.5D, where the fast-acting anticholinesterase, edrophonium, caused a large increase in the depolarization produced both by injected ACh and quantally liberated ACh. Figure 3.13B further indicates the operational sequence: ACh on receptor sites; the end-plate potential, EPP; the generation of a muscle impulse; the muscle contraction. The diagram, of course, shows only the surface membrane on one side of the muscle fiber.

A nerve impulse releases about 10^6 molecules of ACh into the synaptic cleft. On the postsynaptic membrane there are about 2×10^7 receptor sites for ACh in the rat or mouse junction and also 2×10^7 active centers of AChE. There is thus a large excess of both receptor and destroying sites relative to quantal liberation of ACh (Hubbard 1973).

Also shown in Figure 3.13B is the recycling of choline after it has been produced by hydrolysis of ACh. This will be discussed in more detail in Chapter 5 (Hubbard 1973).

The key factor in the emission of ACh by the vesicles is the entry of calcium into the presynaptic terminal (Figure 3.13A). Because the quantal release of ACh is approximately proportional to the fourth power of the external Ca^{2+} concentration, it has been postulated that four Ca^{2+} ions are necessary for one vesicle to burst (Dodge and Rahamimoff 1967). In order to have this concentration of Ca^{2+} ions in proximity to one vesicle it is necessary to have an immense excess of entering calcium—actually, about 10,000 times as much. Even with this large excess there is often a little delay in the assemblage of the necessary four Ca^{2+} ions, which presumably accounts for the variability in the timing of quantal emission in Figures 3.6C, 3.6D, and 3.9A (Katz and Miledi 1967c, 1969b).

3.3 Transmission across the Giant Synapse of the Squid Stellate Ganglion

Further insights into the physiological events occurring in chemical synaptic transmission have been given by investigations on the giant synapse in the squid stellate ganglion (cf. Figure 2.4, Section 2.2). The synapse is truly of gigantic dimensions, there being a contact up to 1 mm long between a presynaptic and a postsynaptic fiber, each of several hundred micrometers in diameter. There is not, however, a continuous synaptic contact over this whole area of approximation, but a multitude of small contacts, as indicated in Figure 3.14H. This synapse gives unique opportunities for investigation. For example, electrodes can be inserted into the presynaptic fiber for direct recording of its membrane potential and its action potential, or for passing currents to alter the membrane potential, a facility that is not possible with the neuromuscular synapse because the presynaptic nerve terminal is far too small.

That it is a chemically transmitting synapse is established by much evidence: there is the characteristic synaptic delay between the presynaptic and postsynaptic responses; no direct electrical transmission of an electrotonic potential can be detected in either direction (Hagiwara and Tasaki 1958); Ca^{2+} and Mg^{2+} ions play mutually antagonistic roles, just as with neuromuscular transmission (Takeuchi and Takeuchi 1962); and there are miniature synaptic potentials analogous to the min. EPPs, but much smaller (Miledi 1967).

The transmitter has not been identified with certainty. There was evidence that it was glutamate or a near relative, but a more recent investigation by Miledi (1969) has raised serious, but perhaps not insurmountable, difficulties. Fortunately, this uncertainty does not concern our present inquiry into the ionic mechanisms responsible for generating the postsynaptic potentials.

Figure 3.14 illustrates a remarkable experiment by Katz and Miledi, which shows the way in which the large dimensions of this giant synapse can be utilized to disclose special features of chemical synaptic transmission. Impulse transmission was blocked by TTX and presynaptic depolarization was effected by intracellular application of a brief current pulse (see a in inset of Figure 3.14G), as shown in the upper traces of the series of Figures 3.14A to 3.14F. Below are shown the presynaptic (Pre) and postsynaptic (Post) membrane potentials. In Figure 3.14A the presynaptic depolarization had no postsynaptic effect, there being only brief artifacts at "on" and "off." In Figure 3.14B there was a slight depolarizing action and in Figure 3.14C to 3.14F it was more and more. The plotted points in Figure 3.14G show that the more you depolarize the terminal (plotted as abscissas) the more transmitter is liberated, as indicated by the postsynaptic depolarization (ordinates). Before any transmitter was liberated, however, there had to be almost 40 mV presynaptic depolarization (note the more sensitive plotting in the inset of Figure 3.14G).

Figure 3.15 illustrates essentially the same results under conditions where presynaptic currents were employed to change the size of the presynaptic impulse in a preparation not poisoned by TTX. In the frame labeled 0 no current was passed and the large presynaptic action potential (upper trace)

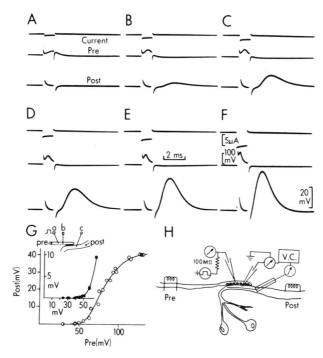

Figure 3.14 Synaptic transmission, stellate ganglion. Presynaptic depolarization and transmitter release with impulses eliminated by TTX. *H* is a drawing of the whole synaptic structure of the stellate ganglion of the squid (cf. Figure 2.4, Section 2.2), with experimental arrangements: Pre, presynaptic terminal; Post, postsynaptic giant axon. In the upper diagram of *G* are the special arrangements for this experiment. Length of synaptic contact, 0.8 mm, a, current-passing electrode; b, prerecording electrode; c, postrecording electrode. A–F. Sample recordings in six frames: upper trace, applied current pulse; middle and lower, presynaptic and postsynaptic potentials recorded through electrodes b and c, respectively. G. Input–output relation obtained with 1-ms current pulses. Abscissa, peak of presynaptic depolarization; ordinate, size of postsynaptic response. Inset, initial part of curve in greater detail. Temperature in these experiments was about 10°C; concentration of tetrodotoxin 2×10^{-7} g/ml. (Katz and Miledi 1966.)

evoked a later postsynaptic potential like an end-plate potential. It is called a synaptic potential or, in accordance with usual terminology, an excitatory postsynaptic potential, EPSP. When the presynaptic impulse was increased by making the membrane potential larger, the output of transmitter was increased, as indicated by the EPSPs in the frames labeled +3 to +8. If, on the other hand, the presynaptic impulse was decreased by reducing the membrane potential (frames −3 to −9), the EPSPs reveal that the liberation of transmitter was reduced, even to zero.

It is a remarkable finding in Figures 3.14 and 3.15 that a threshold depolarization of about 40 mV is necessary for transmitter liberation from the presynaptic terminal. This can be regarded as the threshold for opening the calcium gates on the presynaptic terminal so that Ca^{2+} can enter the presynaptic terminal (cf. the Ca_D gates in Figure 3.16). Just as with the neuromuscular synapses, there is direct experimental evidence that the influx of Ca^{2+} ions is essential for the quantal liberation of the transmitter, as has been diagrammed

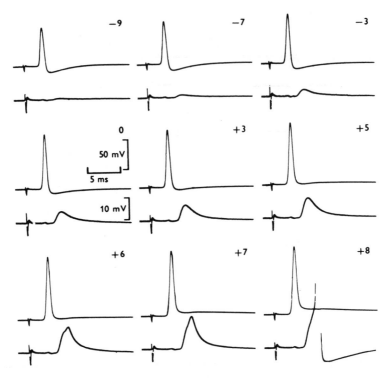

Figure 3.15 Presynaptic depolarization and transmitter release. The experimental arrangements are shown in Figure 3.14H. Prolonged currents of the intensities indicated for each frame in relative units were passed through the electrode inserted into the presynaptic element. This electrode was also used for recording the presynaptic action potentials (upper traces of each frame) evoked by a presynaptic stimulus (note stimulus artifact). The lower traces are the simultaneously recorded postsynaptic responses. (Miledi and Slater 1966.)

in Figures 3.13A and 3.16B. When the Ca^{2+} of the bathing fluid was greatly reduced synaptic transmission was blocked, just as with the neuromuscular synapse (Figures 3.9A, 3.9B, 3.9C). It is restored by injection of Ca^{2+} ions close to the external surface of the synapse even at a rate as low as 10^{-14}–10^{-15} mol of Ca^{2+} ions in 1 s. The role of Ca^{2+} ions in causing the emission of transmitter has recently been elegantly displayed by Miledi (1973), who has injected Ca^{2+} ions electrophoretically into the presynaptic terminal in a squid giant synapse pretreated by TTX. The Ca^{2+} injection resulted in an enormous output of transmitter, which was in the form of quantal release.

3.4 Ionic Channels across the Presynaptic and Postsynaptic Membranes of a Chemically Transmitting Synapse

Figure 3.16 gives a diagrammatic display of the various ionic channels across the postsynaptic membrane of the neuromuscular junction and of the

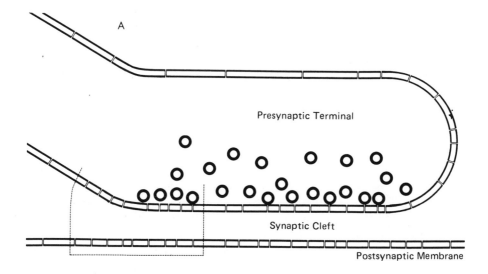

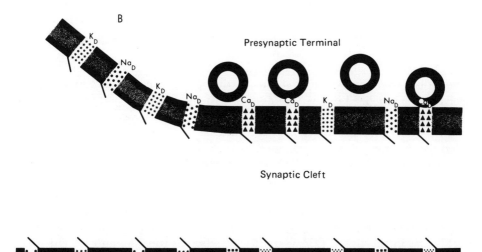

Figure 3.16 Chemically transmitting synapse. A. Presynaptic terminal separated by the synaptic cleft from the postsynaptic membrane, and containing many spherical synaptic vesicles, some close to the membrane fronting the synaptic cleft. Note the various channels through the membrane. B. Segment outlined by the broken line in A is shown enlarged to give fine detail. Note that all channels have a gate control, and that symbols in the channels indicate five different kinds of chemical constitution for the carrier molecules: Na_D, K_D, Ca_D, K_T, and Na_T. The subscripts D and T indicate opening by depolarization, and by the transmitter substance, respectively.

squid giant synapse, there being separate Na_T^+ and K_T^+ channels for the post-synaptic currents (cf. Figure 3.12B). Also shown in the presynaptic nerve fiber are the Na^+ and K^+ channels with gates that are opened by depolarization, and that are responsible for the rise and fall of the action potential. They are labeled Na_D and K_D to signify their relationship to depolarization. Though selectively permeable to sodium ions, the Na_D channel is also effective for Li^+ transport (Hodgkin 1964), and is even slightly effective for K^+, Rb^+, and Cs^+ (Chandler and Meeves 1965). Ammonium and some substituted ammonium ions can pass through the Na_T^+ channel (references in Eccles 1964b, p. 52). The Ca_D channel is also permeable to Sr^{2+} and Ba^{2+}, but less effectively (Dodge *et al.* 1969). It is important that TTX blocks the Na_D channel selectively, not affecting the K_D and Ca_D channels (Katz and Miledi 1969a), while intracellular TEA blocks only the K_D channel. The Ca_D channel is distinctive in that it is very effectively blocked by external Mn^{2+} and Mg^{2+} (Katz and Miledi 1969b). When the Na_D and K_D channels are eliminated by TTX and TEA respectively, the Ca_D channel becomes so effective that it even exhibits a re-generative depolarizing response (Katz and Miledi 1969b). By an extremely elegant fluorescence method for detecting calcium, Baker *et al.* (1970) have shown that membrane depolarization is associated with a two-phase influx of Ca^{2+}, one with about the same time course as Na^+, and the other slower in onset and continuing during the depolarization. The former is blocked by TTX and is plausibly attributed to a slight Ca^{2+} transport through the Na_D gates. The other is TTX resistant and is blocked by Mn^{2+} and Mg^{2+}, corresponding precisely to the Ca_D channel of Figure 3.16.

Figure 3.16 illustrates two features of the five types of ionic channels there displayed. There are, first, the relative specificities of the channels themselves, as symbolized by the various ions principally concerned under normal conditions. Presumably these specificities have chemical bases, and the discovery of these chemical mechanisms would be a great scientific achievement. Because of their different properties it is postulated that each type has a unique chemical composition. Second, there are the gates controlling these channels, which are opened by depolarization for the Na_D and K_D channels, and by a chemical transmitter for the Na_T and K_T channels—ACh at the neuromuscular synapse, and possibly glutamate at the squid synapse. A further difference is that the Na_D and K_D channels are blocked by extracellular TTX and intracellular TEA respectively, whereas the Na_T and K_T channels, opened by transmitter action at the neuromuscular and squid synapses, are unaffected. For this reason these channels in Figure 3.16 are shown filled by different chemical mechanisms, as indicated by the different symbolic shadings and by the labels Na_T and K_T.

Figure 3.16 further illustrates a special feature of the presynaptic membrane, namely, the presence of specific calcium channels operated by membrane depolarization, Ca_D. The influx of Ca^{2+} ions along these channels is essentially concerned in the discharge of transmitter into the synaptic cleft (Figures 3.7, 3.9). In accordance with the evidence of Katz and Miledi (1965a; Figures 3.7A to 3.7C) the Na_D and K_D channels are shown along the nerve terminal in Figures 3.16A and 3.16B, but these channels are not shown in the subsynaptic component of the postsynaptic membrane, where there are exclusively the Na_T and K_T channels. It is not known how far the Na_D and K_D channels are interspersed with the Na_T and K_T channels, but at least they are in close proximity. It is of interest that the Ca_D channels are opened at approximately the same levels of depolarization (Figures 3.14, 3.15) as are the Na_D and K_D channels associated with the action potential. The Ca_D channels shown in Figure 3.16 are restricted to the presynaptic membrane and are not blocked by TTX (Katz and Miledi 1969a), which is in contrast to the findings on Ca_D channels of some excitable membranes (Baker *et al.* 1970).

The neuromuscular synapse and the squid giant synapse exemplify very well the mechanism of chemical transmission, and are seen to be very similar in essentials, though differing in respect to the transmitter substances and the dimensional relationships. They also differ in the ionic conductance ratios for the Na_T and K_T channels of the postsynaptic membrane. A ratio of 1.3 was determined by Takeuchi for the neuromuscular synapse (cf. Figures 3.11, 3.12), but with the squid synapse the ratio is probably in excess of 5 (Gage and Moore 1969). In the diagram of Figure 3.16B this would be symbolized by the relative number of Na_T and K_T channels. In place of the ratio of about 1.3 for the neuromuscular synapse there would be, for example, a ratio of 5 for the squid synapse.

3.5 Concluding Statements

We have finished the story of the investigations on peripheral synapses. In the principal example a nerve impulse propagates to the nerve terminals of a synapse and there acts with extraordinary efficiency and speed to inject ACh into the exceedingly narrow cleft between the nerve fiber and the muscle fiber. The electrical activity of the nerve impulse has been transformed into a chemical injection. Then, in turn, the chemical transmitter momentarily opens ionic gates on the postsynaptic membrane and so reconstitutes an electrical process—an end-plate potential leading on to a muscle impulse—on the far side of the synapse. It may be asked: What is the point of this double transformation? Could not the electrical energy of the nerve impulse be utilized to excite the muscle fiber directly? The answer is easy and obvious. There is a large electrical mismatch between the very fine nerve fiber and the large muscle fiber, which means that the currents responsible for the propagation of the nerve impulse (cf. Figure 2.9, Section 2.4) would be far too small to excite the muscle fiber no matter how efficiently the nerve–muscle junction would be arranged. The chemical transmitting mechanism overcomes this extreme disability by introducing an amplification of an order larger than 100-fold. In fact, there is a safety factor of about five in transmission from nerve impulses to muscle fibers. This is a very important safeguard for our muscular performance, even under extreme conditions when, for example, the supply of transmitter may be depleted. Muscle remains a reliable servant in carrying out the commands of the nervous system, as is illustrated in Figure 3.1.

In the next chapter there will be an account of chemical synapses in the brain, where we shall see that they are essential for the operation of complex nervous systems. The alternative synaptic mechanism, by electrical transmission, apparently is used only at special sites in the brains of lower animals where accurate timing is essential, but not to any significant extent in the mammalian brain.

4

Synaptic Transmission in the Central Nervous System

4.1 Introduction

In the preceding chapter there was an account of the transmission across the neuromuscular synapse whereby a nerve impulse generates a muscle impulse and thus a muscular contraction. In this chapter on synaptic transmission in the central nervous system, it is convenient to start with some simple concepts of the mode of action of the neuron that operates the motor unit, i.e., the motoneuron, because this study links with the classical reflex work (cf. Creed *et al.* 1932) and because the motoneuron has been very extensively studied by modern analytical techniques.

At the beginning of this century Sherrington (1906) made integration the basic theme in his classical book *The Integrative Action of the Nervous System.* He developed the concept of central integration as a consequence of a study of reflexes selected in order to give insight into one or another aspect of integration. Specific sensory inputs were employed and observations were made on the muscle contractions produced thereby. It had already been established that the muscle was a reliable servant of the nervous system, the observed contractions faithfully reporting the output signals from the central nervous system, which of course are the impulse discharges from motoneurons. In this manner Sherrington was able to define the performance of simple arrangements of neurons (as, for example, in Figure 2.1, Section 2.1), the so-called reflex pathways, and to relate it specifically to the properties of the synapses between neurons. Since these synaptic contact areas were very small, Sherrington surmised that many had to be activated at about the same time in order to cause neurons to discharge on impulse; hence the concepts of convergence and summation. A complementary property derives from the profuse branching of nerve fibers with the consequent extensive dispersion of activity entering the nervous system by some afferent pathway. The

101

functional principle of divergence, as it was called, insures that any particular input has at least the potentiality for a widespread influence on reflex discharges. The most revolutionary concept introduced by Sherrington was that some of the synapses on a membrane had an inhibitory action, preventing the discharge that otherwise would have been evoked by the excitatory synaptic action.

In a later publication, *Reflex Activity in the Spinal Cord* by Creed, Denny Brown, Eccles, Liddell, and Sherrington (1932), there was an account of the later analytical study of reflexes by the Sherrington school of neurophysiology at Oxford. Already the concept of the motor unit had been developed and quantitatively studied (shown diagrammatically in Figure 3.1, Section 3.2.1). A motor unit is defined as the axon of a motoneuron and the muscle fibers innervated thereby (Sherrington 1929). By extensive axonal branching this number of muscle fibers is usually in excess of 100 and usually each is exclusively innervated by a single motoneuron. The 1932 book was essentially an attempt to describe a variety of reflexes of the spinal cord in terms of their motor-unit composition. Reflex contractions were quantitatively explained by the number and temporal arrangements of the individual motor units, and in fact this statement can be extended to all behavioral reactions involving the contractions of striated muscle.

After the Sherrington era there were great transformations in our detailed understanding of the central nervous system, but his work provided the foundation for all this new building. Great advances have been made possible by technical developments that sharpened the focus with which the brain could be studied in its basic structure and activity. Intracellular recording from neurons gave important insights on synaptic mechanisms. Examples are the mode of action of excitatory and inhibitory synapses and their interaction, and also the special synaptic mechanisms of presynaptic inhibition and of reciprocally acting synapses. Electron microscopy provided wonderful insights into the structural detail of synapses and neurons as illustrated in Chapter 1. Radiotracer techniques and improved histological methods have given new insights into the communication systems of the brain and spinal cord. Most recently, horseradish peroxidase has been of great value in these respects, as also have electrophysiological studies. And overarching all these achievements there is the immense range of neurochemical and neuropharmacological studies that will be the theme of many chapters of this book. In each section of the book there will be reference to the historical aspects of the studies.

Figure 4.1A shows the simplest excitatory and inhibitory pathways through the central nervous system from muscles that act antagonistically at the knee joint. Annulospiral stretch receptors (AS) of the knee extensor muscle (E) are connected to the spinal cord by large fibers (group Ia) that exert a powerful synaptic excitatory action directly on the motoneurons of that muscle (E) and related muscles. If there is a strong enough pull on the muscle, the annulospiral endings by their discharges along their fibers will produce an excitation of motoneurons powerful enough to cause the discharge of im-

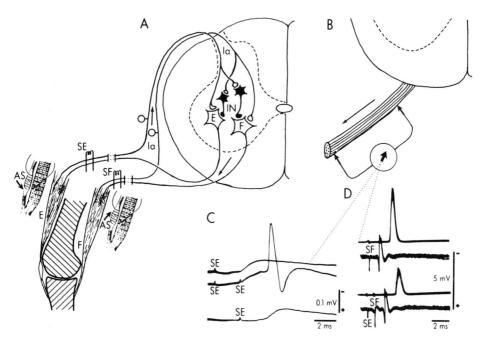

Figure 4.1 Simple reflex pathways and responses. A. A diagrammatic representation of the pathways from and to the extensor (E) and flexor (F) muscles of the knee joint. The small insets show the details of the origin of the Ia afferent fibers from the annulospiral endings (AS) of muscle spindles. Thence the afferents run via the dorsal roots to enter the spinal cord and form excitatory synapses on motoneurons (E, F) supplying the muscle of origin and on interneurons (IN) that inhibit synaptically the motoneurons of the antagonistic muscle. The pathway from motoneuron to muscle is as for the motor unit in Figure 3.1 (Section 3.2.1), the end plates being shown enlarged in the insets. B. A ventral root by which motor axons leave the spinal cord. On it are recording electrodes, one close to the spinal cord, the other on the cut end. C. Double stimulation of the Ia fibers (SE electrode in A) evokes an impulse discharge, whereas single stimulation failed and only produced a slow depolarizing potential that was electrotonically propagated from the motoneurons. D. This is the opposite action. A single stimulation of flexor Ia fibers (SF electrode in A) evoked the monosynaptic reflex discharge in the upper trace. A preceding stimulation of extensor Ia fibers (SE electrode in A) inhibits this discharge to about half. Lower traces show nerve volleys recorded from dorsal root.

pulses along their axons, so that the muscle contacts. This is the familiar tendon jerk or stretch reflex of neurophysiology and the classical exemplar of monosynaptic reflexes. The brief pull produced by a tap on the tendon gives a tendon reflex such as the knee jerk or ankle jerk. A prolonged steady pull evokes the prolonged contraction of the stretch reflex (Creed *et al.* 1932). Figure 4.1C gives an example of temporal facilitation in this monosynaptic pathway. Double stimulation of the Ia fibers by the SE electrode (Figure 4.1A) evokes the discharge of an impulse from many motoneurons (the spike response), whereas either stimulus alone was ineffective.

In Figure 4.1A there is another pathway to the E motoneuron, that from the annulospiral endings (AS) of the antagonistic knee flex (F) muscle. Each of these afferent fibers gives off collaterals in the spinal cord that then form excitatory synapses on small nerve cells (interneurons, IN) that in turn form inhibitory synapses on the extensor motoneurons, either preventing their reflex discharge or reducing the probability of that discharge. This reduction appears as a diminution in the monosynaptic reflex response (Figure 4.1D) recorded in Figure 4.1B as a "population spike" from the whole assemblage of motor axons supplying the muscle. The spike response to SF stimulation (Figure 4.1A) was reduced to less than half by a preceding SE stimulus. Figure 4.1A diagrams the simplest pathway for an inhibitory action from the periphery, and further shows the reciprocal relationship of the pathways from the antagonistic muscles, the extensors and flexors of the knee joint. The afferent fibers from the annulospiral endings are excitatory at all of their synapses, whether on motoneuron or other neuron. In order to have an inhibitory action there is interpolated a special class of interneurons (IN) that are exclusively inhibitory at their synapses. It was first found by Lloyd (1946) that the afferent fibers from a muscle have an inhibitory action on the motoneurons of antagonistic muscles, as indicated by a reduction of the population spike recorded from the motor axons in the ventral root that is shown in Figure 4.1D. Only much later was it established that there was interpolation of the inhibitory neuron, as diagrammed in Figure 4.1A (cf. Eccles 1957, 1969a).

It must be understood that the diagram of Figure 4.1A is greatly simplified both in the monosynaptic excitatory path to a single motoneuron and in the inhibitory path through a single interneuron. There are more than 100 annulospiral endings in a large muscle, which of course is an assemblage of hundreds of motor units (Creed *et al.* 1932). Each afferent fiber from an annulospiral ending branches extensively in the spinal cord (the principle of divergence mentioned above), so that it innervates a large number of motoneurons with their attendant motor units. Reciprocally, there is a large convergence of excitatory fibers (usually 10–50; Kuno 1964) on each motoneuron of the hundreds of motoneurons innervating the muscle. There would be this divergence and convergence at each stage of the inhibitory pathways in Figure 4.1A, which likewise have to be conceived as having many lines in parallel. Figure 4.1A indicates correctly, however, that despite this massive in-parallel arrangement, there is only a single interneuron (IN) in series in the inhibitory pathway; hence this inhibitory pathway is disynaptic.

4.2 Excitatory Synaptic Action[1]

The experimental investigation of synaptic transmission in the simplest pathways of the spinal cord (cf. Figure 4.1) was transformed in 1951 by the technique of recording electrically from the interior of nerve cells, using for this purpose the intracellular glass microelectrode and a cathode-follower amplifier, as had been used a year or so previously by Fatt and Katz in

---

[1]General reference: Eccles 1964b.

investigating neuromuscular transmission (Brock *et al.* 1952). With the motoneuron there was the added difficulty that it had to be located and penetrated without visual control, and of course the necessary rigid fixation was harder to secure; nevertheless, even the earliest experiments with primitive techniques were surprisingly successful, and soon we had secured new insight into the nature of excitatory and inhibitory synaptic action on motoneurons (cf. Eccles 1957). After a few minutes many become stabilized to give, even for hours, responses which appeared to be unaffected by the impalement. The most satisfactory electrodes are those with a very gradual taper and with a tip diameter of about 0.5 μm, which gives them a resistance of 10–15 MΩ when filled with 3 M KCl.

Figure 4.2 schematically shows the stimulating electrode on the many large nerve fibers from the annulospiral endings of muscle (cf. Figure 4.1A) that converge on and give excitatory synapses to a motoneuron of that muscle. In Figures 4.2B to 4.2J, a progressively stronger stimulation has been applied through the stimulating electrode, and the number of fibers so excited has been monitored by the recording electrode placed so as to record the impulses

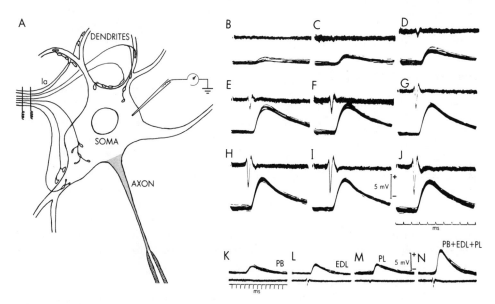

Figure 4.2 Monosynaptic excitation of motoneurons by the group Ia afferent pathway. A. A drawing of a motoneuron showing the central dendritic regions, the soma, the initial segment (IS) of the axonal origin (cf. Fig. 2.14C) and the beginning of the axonal medullation. On the dendrites and soma are shown the excitatory synaptic endings of seven group Ia afferent fibers that have an applied stimulating electrode (actually in the peripheral muscle nerve). The intracellular microelectrode recording is shown diagrammatically. B–J. The upper traces give the size of the afferent volley as it enters the spinal cord, and the lower the simultaneously recorded EPSPs. All records are formed by the superimposition of about 25 faint traces. K–M. The EPSPs recorded in another motoneuron (peroneus longus) in response to maximum group Ia volleys in the nerves to three muscles—peroneus brevis, extensor digitorum longus, and peroneus longus. N. All three muscles combined. (Eccles *et al.* 1957.)

as they enter the spinal cord (upper traces of Figures 4.2B to 4.2J). In the lower traces are the intracellularly recorded potentials produced by the synaptic excitatory action. They are seen to resemble the EPPs described in the preceding chapter (Figures 3.3, Section 3.2.3; 3.5, Section 3.2.5) in being in the depolarizing direction, there being a temporary diminution of the resting membrane potential that is about -70 mV. The size, but not the time course, is changed by an increase in the number of activated fibers, as indicated by the spike potentials in the upper traces. This is a good illustration of the way in which convergence of afferent fibers onto a neuron gives summation of the excitatory synaptic potentials which are recorded across the postsynaptic membrane, hence their name, excitatory postsynaptic potentials or EPSPs. After Figure 4.2G further increase in the stimulus did not result in any increase in the EPSP. In Figure 4.2G there was stimulation of all the group Ia fibers, the increase in afferent spike in Figures 4.2H to 4.2J being due to the stimulation of group Ib fibers that do not have a monosynaptic excitatory action (Eccles et al. 1957). Usually there is a linear relationship between the amount of monosynaptic excitation and the size of the EPSP. This is illustrated in Figures 4.2K to 4.2N, where EPSPs were produced in a motoneuron by the convergence of Ia afferent fibers in the nerves to three different muscles (Figures 4.2K, 4.2L, 4.2M) and when synchronized (Figure 4.2N) the EPSP was exactly the sum of the three component EPSPs. According to the formal electrical diagram of Figure 4.5B an appreciable deviation from the linearity would be expected only when the EPSPs were very large. Thus, it may be assumed that the observed EPSPs arise by a simple summation of the EPSPs produced by each synapse.

The EPSPs evoked in a motoneuron monosynaptically by a group Ia afferent volley (Figures 4.2B to 4.2N) represent the sum of several unitary EPSPs, each produced by the synaptic knobs stemming from a single Ia fiber (cf. Figure 4.2A). The unitary EPSP produced by single presynaptic impulses can easily be recognized when there is a random bombardment of the motoneuron that gives what we may term "synaptic noise" (cf. Brock et al. 1952). For example, in Figure 4.3A two distinct types of unitary EPSPs can be recognized in independent rhythmic series, one very brief, the other prolonged. Their respective enlargements are shown in Figures 4.3B and 4.3C on the same time scale for the EPSP produced by a Ia volley in Figure 4.3D. The unitary EPSP in Figure 4.3B is exceptionally brief and would be produced by a small synaptic excitation on or close to the motoneuronal soma, which is the presumed recording site, whereas the unitary EPSP in Figure 4.3C exhibits the slowing of time course consequent on electrotonic transmission from a rather distant synaptic site on a dendrite. The composite EPSP of Figure 4.3D is 10 times larger than the unitary EPSP of 4.3C, 9 mV as opposed to 0.9 mV, and it has a rather faster time course. Evidently it is formed by the summation of many unitary EPSPs that mostly are generated distantly on the dendrites. A whole spectrum of unitary EPSPs has been recognized (Burke 1967), the two of Figures 4.3A to 4.3C being extreme examples.

Over several years Rall has made a detailed quantitative study (reviewed

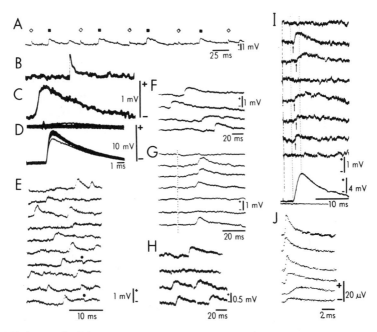

Figure 4.3 Unitary and miniature excitatory postsynaptic potentials. A. Spontaneously occurring unitary EPSPs of a plantaris motoneuron, with two types indicated by symbols. B and C. These types are shown enlarged on the same time voltage coordinates. D. An EPSP of that same motoneuron generated by an almost maximal Ia volley as in Figure 4.2, with the same time scale but with a much lower amplification. (Burke 1967.) E. Spontaneous unitary synaptic potentials of a motoneuron. (Blankenship and Kuno 1968.) F and G. Unitary EPSPs in a frog motoneuron, spontaneous in F and by a just threshold afferent stimulus at the vertical broken line in G. H. Spontaneous miniature EPSPs of a frog motoneuron after prolonged soaking in a potassium-rich solution. (Katz and Miledi 1963.) I. The responses of the eight upper traces were evoked by stimulation probably of a single Ia afferent fiber. The lowest record of I is the EPSP evoked by a large Ia afferent volley, recorded at the same speed but at a lower amplification. (Kuno 1964.) J. Averaged recordings of EPSPs produced by impulses in the same Ia fiber terminating on six different motoneurons. (Mendell and Henneman 1971.)

in 1970) of the modifications produced in intracellularly recorded EPSPs when the excitatory synapses are at dendritic sites progressively more remote from the soma (cf. Eccles 1957, p. 46). For a rigorous mathematical modeling of nerve cells, consult Chapter 7 of Jack *et al.* (1975).

Figure 4.3E represents a more complex series of spontaneous unitary EPSPs recorded from a motoneuron after all dorsal roots had been cut, so these EPSPs were produced by synaptic bombardment by interneurons. There is seen to be a very diverse assortment of unitary EPSPs. Of special interest are the two brief hyperpolarizations marked by dots that are identified as unitary inhibitory postsynaptic potentials, described later in this chapter.

Strictly speaking, on analogy with the end-plate potential the term miniature (min.) EPSP should be applied not to the unitary EPSPs of Figures 4.3A to 4.3C and 4.3E, but to the EPSPs produced by quantal emission of transmitter from single synaptic vesicles. Such min. EPSPs have been observed when all impulse transmission has been eliminated by soaking an isolated frog spinal cord in high potassium solution (Katz and Miledi 1963). For example, in Figure 4.3F there are the unitary spontaneous EPSPs as in Figure 4.3A, and also in Figure 4.3G unitary EPSPs evoked by a very small presynaptic volley, possibly of only one impulse, as in Figure 4.3I, and with several failures. Then the spinal cord was soaked in a Ringers solution with high potassium (60 mM). Figure 4.3H shows unitary EPSPs that are as large as those of Figures 4.3E and 4.3F and that must be true min. EPSPs. True min. EPSPs have been recognized in mammalian motoneurons after all impulse transmission had been eliminated by tetrodotoxin (Blankenship and Kuno 1968).

Figure 4.3I resembles Figure 4.3G, but is for a normal cat motoneuron upper traces and hence shows strictly unitary EPSPs (Kuno 1964). The failures in three traces and the variable latencies of the unitary EPSPs in the other five traces recall the miniature EPC series of Figure 3.6C (Section 3.2.5). After analysis of extensive series of EPSPs of this unitary type with their frequent failures, Kuno (1964) concluded that monosynaptic transmission onto motoneurons occurred in quantal steps, and that one impulse in a single afferent fiber released on the average only one quantum of transmitter. This contrasts with the 100–200 quanta at the normal neuromuscular junction, but is not unexpected in view of the small size of the synaptic knobs relative to the expanded motor nerve terminal on a muscle fiber. Evidently, therefore, the unitary EPSPs of Figures 4.3A, 4.3B, 4.3E, 4.3F, 4.3G, and 4.3I must often be min. EPSPs.

A still more refined study by Mendell and Henneman (1971) is illustrated in Figure 4.3J where the same group Ia fiber produced the EPSPs in six different motoneurons. Some of the extremely small unitary EPSPs of Figure 4.3J would certainly have been missed before the introduction of the averaging techniques. Amplitudes of EPSPs varied from 17 to 600 μV, with an average of 102 μV. The wide variation in time course is attributable to the location of the synapses on the soma or dendrites, as already discussed. With this very sensitive technique it was found that a single Ia fiber gave excitatory synapses to nearly all of the 300 motoneurons supplying the muscle (cat gastrocnemius medialis) from which it arises. A large Ia fiber may give as many as 4000 synapses to motoneurons, and it also gives synapses to other neurons in the spinal cord. This widely distributed synaptic potency is certainly impressive.

It can be regarded as established that, like the EPP, the EPSP is generated by a chemical transmitter, whose identification is as yet unsure but is probably glutamate or aspartate (Curtis and Johnston 1974). There are, for example, several distinguishing characteristics: the synaptic delay seen in Figure 4.2 between the presynaptic impulse and the EPSP; the min. EPSPs of quantal character illustrated in Figure 4.3H; and the effects of varying the membrane potential on the size and direction of the EPSP (Figures 4.5A, 4.5C). As with the EPP (Figure 3.11A, Section 3.2.9) voltage clamp experiments show that

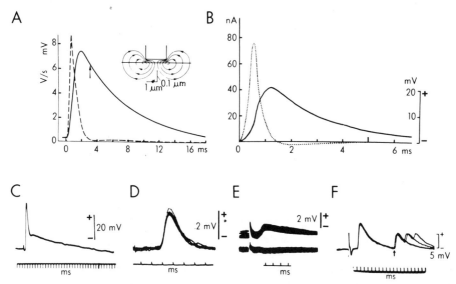

Figure 4.4 Excitatory postsynaptic potentials of a variety of neurons. A. The mean curve of the EPSP of a motoneuron recorded as in Figure 4.2 is plotted as a continuous line and analyzed on the basis of the membrane time constant in order to determine the time course (the broken line) of the postsynaptic currents that generate this potential change. The flow of this current relative to an activated synapse is shown diagrammatically in the inset for a synapse in which the vertical scale is exaggerated tenfold relative to the horizontal. (Curtis and Eccles 1959.) B. A similar figure for the EPSP of a neuron of Clarke's column of the spinal cord, which is a cell of origin of the dorsal spinocerebellar tract. (Kuno 1969.) C. An EPSP (not spike) recorded intracellularly from a Renshaw cell (Figure 4.15) and evoked by a ventral root volley (Eccles 1961b.) D. An EPSP from a cell of origin of the ventral spinocerebellar tract. (Eccles *et al.* 1961a.) E. An EPSP evoked by stimulation of the lateral geniculate body and recorded intracellularly in a neuron of the visual cortex. (Toyama, unpublished observations.) F. An EPSP evoked by a climbing fiber impulse in a cerebellar Purkinje cell. (Eccles 1966.)

the EPSP is produced by a relatively brief depolarizing current (the excitatory postsynaptic current, EPSC), which depolarizes the postsynaptic membrane as shown in the inset of Figure 4.4A. The slow exponential recovery phase is due to the rebuilding of the membrane potential by the operation of the membrane battery (F_M) shown in Figure 4.5B in series with the membrane resistance (R_M) and in parallel with the membrane capacity (C_M). The membrane time constant has been directly measured for motoneurons (cf. Eccles 1961) and in Figure 4.4A it has been used for calculating the EPSC (broken line) that gives the EPSP of Figure 4.4A. At the most there is only a minute residuum of transmitter action after 2–3 ms.

In Figure 4.4 there is an assemblage of EPSPs produced monosynaptically in a wide variety of neurons of the central nervous system. There is always the characteristic brief synaptic delay, then the graded depolarization with a rapid rising phase and a slower decay that often has the temporal characteristics of the neuronal membrane as in Figure 4.4A and 4.4B. With some neurons, however, there is a prolonged transmitter action that is seen in Figure 4.4C to give a very prolonged decline. In this case the transmitter is known to be ACh, and apparently it is removed only very slowly by AChE. Figures 4.4B and 4.4D show the EPSPs for the cells of origin of the dorsal and ventral spinocerebellar tracts. These EPSPs, as well as of the large pyramidal cells in the motor cortex (Figure 4.4E), have a characteristic sharp rise and slower decay, so that by 10 ms there is little or no residuum of depolarization. Figure 4.4F shows that even with the massive unitary potentials produced in a cerebellar Purkinje cell by a single impulse in a climbing fiber, the time course has the usual features of a sharp rise and a slow, approximately exponential, decay. At the arrow there are the EPSPs produced by a later burst of one, two, or three impulses in the same climbing fiber.

Figure 4.5C gives an example of a very interesting synaptic arrangement of red nucleus neurons. There are two quite distinct locations of the excitatory synapses, those on the soma from the interpositus neurons (IP) and those on the rather remote dendrites from the cortical pyramidal cells (CP). This differential location is shown by the following features. The former has a sharp rise and slow decay, and is greatly increased by hyperpolarizing the neuron, as shown with currents of 15, 34, and 46 nA. The CP EPSPs exhibit a very slow rise, attributable to electrotonic distortion, and are much less affected by the neuronal hyperpolarization, as is well demonstrated by the plotted points of Figure 4.4D. The composite EPSPs of Figure 4.5C thus resemble the unitary EPSPs of Figures 4.3A and 4.3B, the IP and CP synapses corresponding to types illustrated in Figures 4.3A and 4.3B respectively (Tusukahara et al. 1975b). It was possible to model the responses of the red nucleus neuron in Figures 4.5C and 4.5D with two sets of synaptic depolarizations, one close to the recording site and the other remotely on equivalent "dendrites."

By investigating the effect of variation of the membrane potential on the EPC (end-plate current) it was shown in Figures 3.11 and 3.12 (Section 3.2.9) that it reversed at a membrane potential of about -15 mV and this equilibrium potential was explained as a compromise between the equilibrium potentials of sodium and potassium ions, whose conductances were both greatly increased by the action of the synaptic transmitter on the receptor sites of the postsynaptic membrane. Under favorable circumstances it is possible to change the membrane potential of a motoneuron sufficiently to reverse the EPSP (Figure 4.5A). The equilibrium potential is at about 0 mV membrane potential, which suggests that sodium conductance is larger relative to potassium conductance than at the motor end plate. The sodium ion conductance, however, is less dominant than has been observed for the giant synapse of the squid.

By electrophoretic injection through an intracellular electrode large changes can be made in the ionic composition of motoneurons. These injections of anions or cations may cause a considerable decrease in the membrane potential, and the EPSP is then diminished correspondingly, but is not otherwise changed (cf. Eccles 1964b). In the light of the recent evidence that anions are not appreciably concerned in generation of the end-plate potential

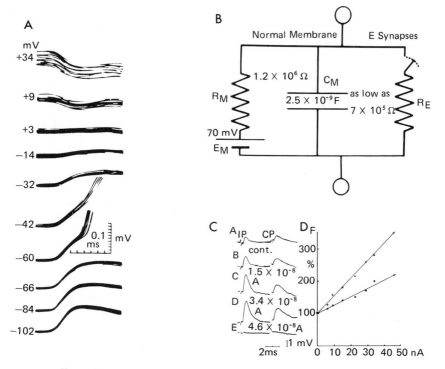

Figure 4.5 Effect of membrane polarization on EPSPs. A. EPSPs set up in a cat biceps-semitendinosus motoneuron at various levels of membrane potential as indicated. Each record is formed by the superimposition of about 20 faint traces. The membrane potential was shifted to the indicated values from its resting value of −66 mV by steady currents through the other barrel of the double microelectrode. (Coombs *et al.* 1955c.) B. Formal electric diagram of a postsynaptic membrane with areas of excitatory synapses as shown on the right side: R_M 1.2 MΩ; C_M, 2.5 nF; E_M, −70 mV. C. EPSPs generated in a red nucleus neuron by stimulation of the cerebellar pathw: (IP) and of the cerebral pathway (CP), both of which monosynaptically excite the neuron. Note that hyperpolarization of the neuron by steady currents of 15 nA, 34 nA, and 46 nA greatly increased the former, but much less the latter. D. The EPSP amplitudes are plotted against the respective current intensities, crosses for IP-EPSPs and filled circles for CP-EPSPs. (Tsukahara *et al.* 1975b.)

(EPP), further investigation is desirable, particularly with isolated prepa-rations where it should be possible to change the extracellular ions in the way that was done in investigations by the Takeuchis on the EPP (Figures 3.11 and 3.12, Section 3.2.9).

Figure 4.5B gives the formal electrical diagram for the operation of the excitatory synapses on a mammalian (cat) motoneuron. All the values, except the minimum given for R_E, represent the means of investigations on many cells. Since the equilibrium potential for the excitatory synapses is about 0 mV, the conductance channel could be shown as a simple shunt with a variable resistance under control of the synaptic transmitter. From this diagram it can be shown that if there were a series of small shunting conductances in parallel, representing individual synaptic actions, there would be an approximate summation of these conductances as long as the total conductance change is not above 20% of the background conductance across the membrane. The observations of Figures 4.2K to 4.2N are thus accounted for.

4.3 Impulse Generation by Synaptic Action

As shown in Figures 4.6A to 4.6D if the EPSP attains a critical threshold level (Figure 4.6B), it causes the neuron to discharge an impulse, the latency being briefer the larger the EPSP (Figures 4.6B to 4.6D). In Figures 4.6B to

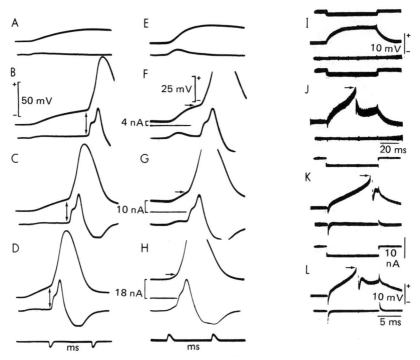

Figure 4.6 Impulse generation by membrane depolarization. A–D. Intracellularly recorded potentials of a cat gastrocnemius motoneuron (resting membrane potential, −70 mV) evoked by monosynaptic activation that was progressively increased from A to D. The lower traces are the electrically differentiated records, the double-headed arrows indicating the onsets of the IS spikes in B–D. E–H. Intracellular records evoked by monosynaptic activation that was applied at 12.0 ms after the onset of a depolarizing pulse whose strength is indicated on each record in nanoamperes. A pulse of 20 nA was just below threshold for generating a spike. E shows control EPSP in the absence of a depolarizing pulse. Lower traces give electrically differentiated records. Note that spikes are truncated. (Coombs *et al.* 1957b.) I–L. A pyramidal tract cell of the motor cortex was subjected to a depolarizing current of an intensity and duration signaled by the uppermost traces. The middle traces are the intracellular potential changes and the lowermost the extracellular control records. Peaks of spike potentials are off the record, but the voltages at the spike origins are indicated by horizontal arrows. Note that K and L were taken with a much faster sweep. (Koike *et al.* 1968.)

4.6D this was brought about by increasing the size of the presynaptic volley, i.e., by increasing the number of activated synapses on the motoneurons (cf. Figure 4.2A). As would be expected, a subliminal EPSP can also be made to generate an impulse by procedures that change the membrane potential toward the critical threshold level of depolarization. For example, in Figures 4.6F to 4.6H the same EPSP as in Figure 4.6E was made effective by the operation of a background depolarizing current which was commenced 12 ms before, with an intensity shown in each record in nanoamperes. The shift in membrane potential is indicated to the left. The impulse arose at the arrows when the total level of depolarization was about 18 mV, which was made up in varying proportions by a conditioning depolarization and the superimposed EPSP. The threshold level of depolarization may also be attained by superimposing the EPSP on the depolarization produced by a preceding EPSP, which provides a sufficient explanation of the reflex phenomenon known as temporal facilitation—a single afferent volley fails to evoke a reflex discharge, but two such volleys in a short interval are effective, as in Figure 4.1C. Comparable observations have been made with impulse generation by the EPSPs of a wide variety of nerve cells.

In Figures 4.6I to 4.6L the critical level of depolarization for generation of an impulse is displayed by a cortical pyramidal cell that was subjected to depolarizing pulses of increasing size, as shown in the upper traces. In Figure 4.6I the depolarization did not reach threshold. In Figures 4.6J to 4.6K the impulse discharge was generated progressively earlier, but, as shown by the arrows, at approximately the same level of depolarization of about 13 mV.

There is now general agreement that the synaptic excitatory action is effective in generating the discharge of an impulse solely by producing an EPSP that attains the threshold level of depolarization.

4.4 Inhibitory Synaptic Action[2]

There is a second class of synapses that opposes excitation and tends to prevent the generation of impulses by excitatory synapses; hence, these are called inhibitory synapses. There is general agreement that these two basic modes of synaptic action govern the generation of impulses by nerve cells. Figure 4.1A shows that the simplest inhibitory pathway in the spinal cord goes via a type of neuron that is specialized to exert an inhibitory synaptic action and nothing else. In 1946 Lloyd discovered that the Ia afferent fibers of a muscle inhibited motoneurons of the antagonist muscle. But the mechanism of this action was not known, nor was the inhibitory pathway diagrammed as in Figure 4.1A. There had been developed an electrical theory of inhibitory synaptic action, and eventually in 1951 the theory was tested by inserting a microelectrode inside the motoneuron (Figure 4.2A) in order to see what inhibition did to the membrane potential. The electrical theory was proved

[2]General references: Eccles 1964b, 1964d.

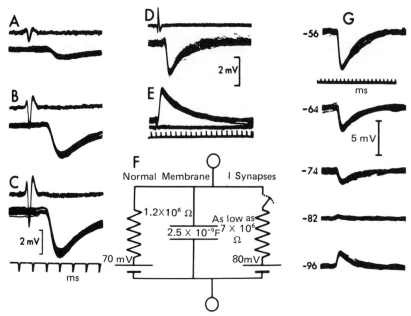

Figure 4.7 Inhibitory postsynaptic potentials of a motoneuron. A–C. Graded responses as described in the text. D and E. Comparison of EPSP and IPSP (time below E in milliseconds). (Curtis and Eccles 1959.) F. To the left there is the formal electrical diagram of the membrane of a motoneuron (cf. Figure 1.1B, Section 1.1), while to the right the mode of operation of inhibitory synapses is symbolized by the closing of the switch. G. Resembles Figure 4.5A in showing effect of varying membrane potential on the IPSP. (Coombs *et al.* 1955b.)

wrong by the discovery that the inhibitory synapses produced a generalized increase of charge on the neuronal membrane, an effect called hyperpolarization (Brock *et al.* 1952).

In Figures 4.7A to 4.7C there was graded stimulation of the group Ia fibers of the nerve to the antagonistic muscle (cf. Figure 4.1A); the upper traces show the action potential of these fibers just as in Figures 4.2B to 4.2J. There was an associated gradation of the inhibitory hyperpolarization of the membrane in the lower traces. This hyperpolarization we termed the inhibitory postsynaptic potential, IPSP. Its size is dependent upon the number of converging inhibitory synapses, just as with the excitatory synapses and the EPSPs in Figures 4.2B to 4.2G. Figures 4.7D and 4.7E show that the EPSP and IPSP of the same motoneuron are nearly mirror images. In both cases there is a brief flow of synaptic current for about 1 ms that effects the membrane potential change. Subsequently, the membrane potential is restored to the initial level by ionic fluxes across the motoneuronal membrane in general (cf. Figure 1.17B, Section 1.2.5). This explains the similarity in time courses of the EPSP and IPSP. It is also a regular feature that the IPSP has a central latency about 1 ms longer than the EPSP, which is the time for traversing the inhibitory interneuron that is interpolated on the pathway (cf. Figure 4.1A).

The inhibitory action of Ia impulses from a muscle on the motoneuron of an antagonist muscle was the first example of an inhibitory mechanism studied by intracellular recording in the central nervous system. It has turned out that the same general principles obtain for almost all inhibitory actions in the brain, with the exception of presynaptic inhibition (to be discussed later; cf. Figure 4.18). The techniques used in the spinal cord have been applied at all levels of the central nervous system. In many laboratories there is now a great ongoing enterprise of identifying the excitatory and the inhibitory neurons in all the regions of the brain, and of determining the manner of their interaction in complex circuits. We have to discover at all these levels, not only the neuronal pathways, but also the mechanisms at all synaptic relays, in particular the synaptic transmitter substances at these relays and the ionic gates that cause the different postsynaptic actions, and so on. It is remarkable that we are more advanced in answering these questions with respect to inhibition than to excitation.

Figure 4.8 gives a kind of overall view of excitatory and inhibitory synaptic action as we now understand it. In Figure 4.8C the neuron is shown with an intracellular microelectrode. It has been discovered that the motoneuronal excitatory synapses (E) tend to be out on the dendrites as drawn here, while the inhibitory synapses (I) tend to be concentrated on the soma. Moreover, there are structural differences, as already illustrated in the two synapses of Figures 1.4B and 1.5A (Section 1.2.2), where, with glutaraldehyde fixation, the synaptic vesicles of the inhibitory synapse to the right have an ellipsoid configuration while in the excitatory to the left they are spherical. In addition, the membranes on either side of the synaptic cleft are less dense with inhibitory than with excitatory synapses. In essential structural features, however, the two kinds of synapses are similar.

In Figure 4.8A there are records of pure IPSPs and EPSPs. In the former, charge is put on the membrane, increasing the membrane potential from −70 to about −73 mV. The EPSP has an antagonistic action, decreasing the charge across the membrane. If you superimpose the IPSP on the EPSP, as in the fourth frame, the IPSP is seen to be very powerful in counteracting the EPSP, which is also shown alone in the same frame to aid comparison. In the third frame of Figure 4.8B the IPSP very effectively suppresses the impulse discharge generated by the EPSP alone, which is illustrated in the last frame.

Figure 4.8D is a good illustration of the EPSP generating an impulse discharge (cf. Figures 4.6B to 4.6D) that travels down the axon to be recorded as the brief spike in the lower trace. In Figure 4.8E an initial inhibitory action reduces the EPSP so that it no longer generates an impulse, and the spike in the axon is also suppressed. Evidently, inhibitory synapses achieve their effectiveness by generating the IPSP that directly counteracts the depolarizing action of the EPSPs. So we have the question—What are inhibitory synapses doing?—solved at that level. There are, however, many more questions to answer.

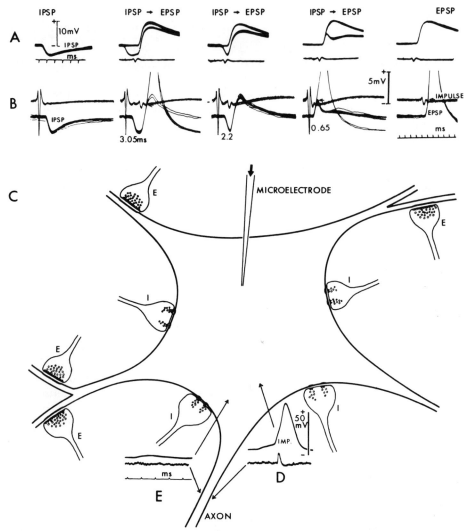

Figure 4.8 Summarizing diagram of excitatory and inhibitory synaptic action. In C the excitatory (E) and inhibitory (I) synapses are shown with characteristic histological features (cf. Figure 1.4B, Section 1.2.2).

Figure 4.9B shows diagrammatically the flow of current under an activated inhibitory synapse—outward through the subsynaptic membrane, along the synaptic cleft, and circling back to hyperpolarize the postsynaptic membrane by inward flow over its whole surface, which is the reverse of that for an excitatory synapse in Figure 4.9A. The outwardly directed current across the inhibitory subsynaptic membrane could be due to the outward movement of a cation such as potassium or the inward movement of an anion like chloride, or to such a combination of anionic and cationic movements that there is a net

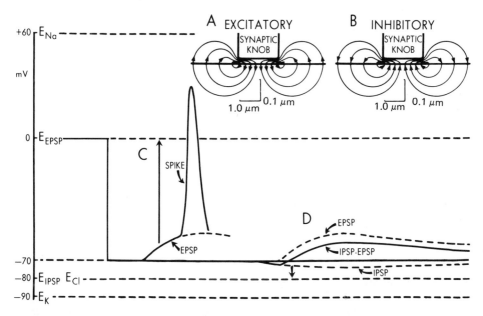

Figure 4.9 Excitatory and inhibitory synaptic action. A. Diagram showing an activated excitatory synaptic knob. As indicated below, the synaptic cleft is shown at 10 times the scale for width as against length. The current is seen to pass inward along the cleft and in across the activated subsynaptic membrane. Elsewhere, as shown, it passes outward across the membrane, so generating the depolarization of the EPSP. B. Similar diagram showing the reverse direction of current flow for an activated inhibitory synaptic knob. C. Diagram for cat motoneuron showing the equilibrium potentials for sodium (E_{Na}), potassium (E_K) and chloride (E_{Cl}) ions (cf. Figure 1.14, Section 1.2.4.5) together with the equilibrium potential for postsynaptic inhibition (E_{IPSP}). The equilibrium potential for the EPSP (E_{EPSP}) is shown at zero. To the left an EPSP is seen generating a spike potential at a depolarization of about 18 mV (see Figures 4.6B to 4.6D). To the right of the diagram an IPSP and an EPSP are shown alone (broken lines) and then interacting (continuous line). As a consequence of the depressant influence of the IPSP, the EPSP that alone generated a spike (left diagram) no longer is able to attain the threshold level of depolarization, i.e., the inhibition has been effective.

outward flow of current driven by a battery of about −80 mV in series with a fairly low resistance (Figure 4.7F).

Figure 4.9 serves to illustrate the simplest findings on the EPSP and the IPSP and their interaction. The approximate equilibrium potentials for sodium, chloride, and potassium ions are shown by the horizontal lines, the equilibrium potential for chloride ions being assumed to be identical with the E_{IPSP} (cf. Figure 4.12E). In Figure 4.9C the EPSP is seen to be large enough to generate a spike potential, the course of the EPSP in the absence of a spike being shown by the broken line. In Figure 4.9D there is an initial IPSP (continuous line) which is seen to diminish the depolarization produced by the same synaptic excitation so that it no longer is adequate to generate a spike. We can now ask: What is the ionic mechanism responsible for the outward current across the subsynaptic membrane of the inhibitory synapse? A first step is to determine the equilibrium potential in the manner illustrated in Figure 4.7G. The IPSP observed at the resting membrane potential (-74 mV) was increased by depolarizing the membrane (-64 mV and -56 mV) and was inverted by hyperpolarizing the membrane to -82 mV and still more at -96 mV. So in this way one can determine the point at which there is zero ionic flux during the IPSP. This is the equilibrium potential, E_{IPSP}, which is about -80 mV (Figures 4.9C and 4.9D). Under these conditions, when the ionic gates are open the ionic fluxes carry a zero net charge.

Experimental investigations on the ionic mechanisms of inhibitory synaptic action involve altering the concentration gradient across the postsynaptic membrane for one or another species of ion normally present, and in addition employing a wide variety of other ions in order to test the ionic permeability of the subsynaptic membrane. With the inhibitory synapses on invertebrate nerve and muscle cells the investigations are usually performed on isolated preparations.

Changes in relative ionic concentration across the postsynaptic membrane are readily effected by altering the ionic composition of the external medium, as illustrated in Figures 3.11 and 3.12 (Section 3.2.9). This method is not suitable for mammalian neurons because they have to be observed under conditions of blood circulation, and the cardiac contraction would be greatly disturbed or suppressed by the necessary large changes in the ionic composition of the blood. Instead, the procedure of electrophoretic injection of ions out of the impaling microelectrode has been employed to alter the ionic composition within the postsynaptic cell. For example, the species of anions that can pass through the inhibitory membrane have been recognized by injecting one or another species into a nerve cell and seeing if the increase in intracellular concentration effects a change in the inhibitory postsynaptic potential. These injections are accomplished by filling microelectrodes with salts containing the anions under investigation. When the microelectrode is inserted into a nerve cell, a given amount of the anion can be injected electrophoretically into the cell by passing an appropriate current through the microelectrode.

In Figure 4.10 the IPSP in 4.10A was changed to a depolarizing potential, 4.10B, by the addition of about 5 pica equivalents of chloride ions to the cell, which would more than triple the concentration, whereas after more than twice this injection of sulfate ions into another cell the IPSP was unchanged (Figures 4.10E and 4.10F). This simple test establishes that, under the action of the inhibitory transmitter, the subsynaptic membrane momentarily becomes permeable to chloride ions, but not to sulfate. In Figures 4.10I and 4.10J it is seen that with two types of inhibitory synaptic action in the mammalian spinal cord the inhibitory membrane was permeable to nitrite ions, and recovery from the effect of the ionic injection was complete in 2–3 min.

It is essential to recognize that Figures 4.10I and 4.10J exemplify two quite distinct processes of ionic exchange. First, the specialized subsynaptic areas under the influence of the inhibitory transmitter momentarily develop a specific ionic permeability of a high order, thus giving the greatly increased

Figure 4.10 Ionic injection into a motoneuron. A and C are IPSPs and EPSPs generated in a biceps-semitendinosus motoneuron by afferent volleys as in Figures 4.7D and 4.7E, respectively. A and C were first recorded, then a hyperpolarizing current of 20 nA was passed through the microelectrode, which had been filled with 3 M KCl. Note that the injection of chloride ions converted the IPSP from a hyperpolarizing (A) to a depolarizing response (B), while the EPSP was not appreciably changed (C and D). Passing a much stronger hyperpolarizing current (40 nA for 90 s) through a microelectrode filled with 0.6 M K_2SO_4 caused no significant changes (E to F) in either the IPSP or the later EPSP. G and H represent the assumed fluxes of chloride ions across the membrane before (G) and after (H) the injection of chloride ions, which is shown greatly increasing the efflux of chloride. I to J show the effects of electrophoretic injection of NO_2^- ions into motoneurons. In I IPSP in a motoneuron is evoked by quadriceps Ia volley. J shows a Renshaw IPSP in a motoneuron (cf. Figure 4.15C), the innervation of which was not identified, induced by a maximal L_7 ventral root stimulation. Records in the top row show control IPSPs evoked before the injection. Records from the second to the ninth rows illustrate IPSPs at the indicated time (identical in I and J) after the injection of NO_2^- ions by the passage of a current of 50 nA for 60 s. The bottom records are IPSPs at the end of recovery. Note different time scales of I and J. All records were formed by the superimposition of about twenty faint traces. (Araki *et al.* 1961.)

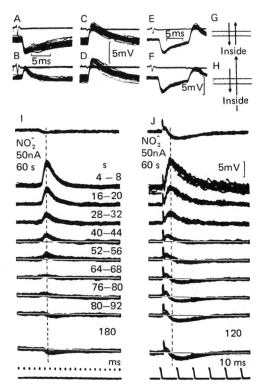

conductivity of the resistance shown to the right of Figure 4.7F. This increased ionic conductance for 1–2 ms is responsible for the ionic fluxes that carry the inhibitory subsynaptic currents shown diagrammatically in Figure 4.9B. Second, the ionic permeability of the *whole* postsynaptic membrane controls the intracellular ionic composition of the neuron and is responsible for its restoration to normal after it has been disturbed by the ionic injection. This loss by outward diffusion of the injected ion occupies rather more than 2 min in Figures 4.10I and 4.10J.

Experimental tests such as those illustrated in Figure 4.10 have disclosed that 12 anions can pass through the gates opened by the inhibitory transmitter. In Figure 4.11, these permeable anions (solid bars) are seen to be distinguished by having small diameters in the hydrated state, whereas the impermeable anions are larger. The formate ion is the only exception to this generalization; otherwise, the activated inhibitory membrane is permeable to all anions that in the hydrated state have a diameter not more than 1.14 times

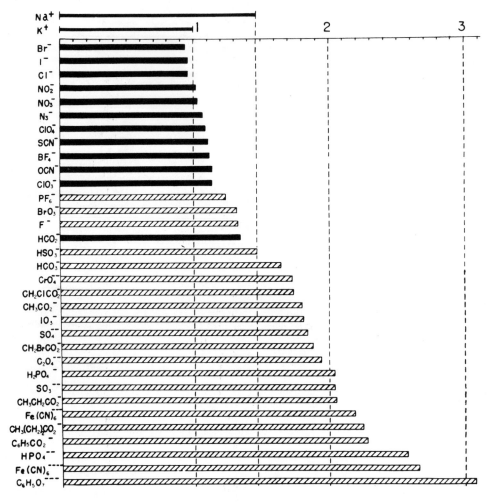

Figure 4.11 Correlation between the ion sizes in aqueous solution and the effects of their injection upon the IPSP. Length of bands indicates ion size in aqueous solutions as calculated from the limiting conductance in water. The black bands are for anions effective in converting the IPSP into the depolarizing direction, as in Figures 4.10I and 4.10J, and the hatched bands for anions not effective, as in Figures 4.10E and 4.10F. Hydrated sizes of K^+ and Na^+ are shown above the length scale, the former being taken as the unit for representing the sizes of other ions. (Ito *et al.* 1962.)

that of the potassium ion, i.e., not more than 0.29 nm, which is the size of the hydrated chlorate ion. The ion diameters in Figure 4.11 are derived from limiting ion conductances by Stokes's Law, on the assumption that the hydrated ions are spherical. Possibly the hydrated formate ion may have an ellipsoid shape and hence be able to negotiate membrane pores that block smaller spherical ions. Similar series of permeable and impermeable ions, even to anomalous formate permeability, have been observed in comparable investigations on inhibitory synapses in fish, toads, and snails. It will certainly be remarkable if the ionic mechanism of central inhibition is almost the same throughout the whole animal kingdom (cf. Eccles 1964b, 1964d). Recently Eccles *et al.* (1977) found the very large IPSPs of hippocampal pyramidal cells (Figure 4.16) to display a discrimination between anions comparable to that of Figure 4.11.

Since potassium is normally at such a high level in nerve cells, the ion injection procedures cannot produce large changes in its concentration, for the hypertonicity produced by its increase immediately results in the influx of water across the cell membrane. Hence, injection procedures have been indecisive in providing evidence for or against potassium ion permeability as a contributory factor in production of the IPSP. Nevertheless, there is indirect evidence for a relatively large contribution from potassium ion permeability: the equilibrium potential for potassium is about 20 mV more hyperpolarized than the resting membrane potential (Figures 1.17, Section 1.2.5; 2.18, Section 2.8); the equilibrium potential for inhibition is similarly in the hyperpolarizing direction, but less so, probably about 6–10 mV (cf. Figure 4.9C). These considerations suggested that the activated inhibitory membrane was permeable for potassium ions as well as for chloride ions. The simplest assumption is equality, that permeability is determined solely by hydrated ion size, but for cations as well as anions. This size criterion is, of course, sufficient to exclude the large hydrated sodium ions, as indicated by the upper line of Figure 4.11. The equilibrium potential of -80 mV for the IPSP could be attained only if the pores in the inhibitory membrane were small enough to effect a virtually complete exclusion of sodium ions; otherwise the sodium ionic flux resulting from the synaptic action would produce depolarization and excitation. In fact, the fundamental difference between excitatory and inhibitory synapses is that sodium permeability is high with the former and negligible with the latter.

Figure 4.12 gives models of inhibitory synaptic action. Figures 4.12A and 4.12B show two ionic channels through the bimolecular leaflet of the surface membrane as illustrated in Figure 1.16 (Section 1.2.5). One gate (Figure 4.12A) is closed and one (4.12B) opened by the steric actions of the transmitter molecule that has become attached to the receptor site. With inhibitory action in the mammalian central nervous system, it has been established that the inhibitory transmitters are amino acids, at some synapses glycine and at others γ-aminobutyric acid or GABA (Chapter 6). It is further believed that these transmitters are packaged in the synaptic vesicles in inhibitory synaptic knobs (cf. Figure 1.4B, Section 1.2.2) and liberated into the synaptic cleft in a quantal manner. By diffusion, a transmitter molecule would find a steric site as in Figure 4.12B and so be able to effect a conformational change, which effectively opens the gate for about 1 ms. According to the original theory (Coombs *et al.* 1955b) K^+ and Cl^- ions move by diffusion through the ionic channel so opened, and that produces the subsynaptic current that generates the inhibitory postsynaptic potential.

The ionic channels for inhibitory action seem to be lacking in chemical

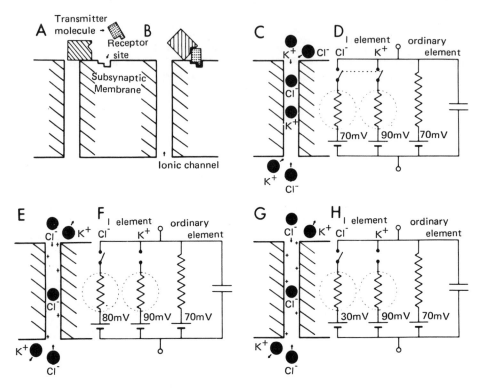

Figure 4.12 Diagrams summarizing the hypotheses relating to the ionic mechanisms employed by a variety of inhibitory synapses in producing postsynaptic membrane potentials.

specificity for anions. They merely have a size discrimination. For example, it has been shown that all 11 species of anions smaller than a critical size in the hydrated state (0.29 nm) will traverse the channels regardless of their chemical properties, while all those larger than a critical size are excluded. Thus, the activated inhibitory postsynaptic membrane would act merely as a sieve with a critical pore size, as is diagrammed in Figure 4.12C. In Figure 4.12D is shown an electrical model incorporating the equilibrium potentials for the Cl^- and K^+ ions as given in the normal membrane model of Figure 1.17H (Section 1.2.5).

Lux, Loracher, and Neher (1970) and Lux (1971) have criticized the model shown in Figures 4.12C and 4.12D on the basis of evidence that there is an outward Cl^- pump across the neuronal membrane. The principal evidence was that during intravenous infusion of ammonium acetate the inhibitory synapses were affected, the equilibrium potential of the IPSP approaching the membrane potential. At the same time there was a slowing in the rate of decline of internal Cl^- after a test injection. They postulated that normally there is an outward chloride pump across the membrane, that the equilibrium potential for Cl^- is as a consequence about -80 mV, and that this pump is reversibly inactivated by ammonium ions. Lux thus was able to explain the hyperpolarization of the IPSP as due solely to the opening of Cl⁻ gates, K⁺ ions playing no significant role. It has recently been found, however (Allen *et al.* 1977), that Lux's experimental evidence with motoneurons cannot be obtained even to the slightest extent with the very large IPSPs of the hippocampus (cf. Figure 4.16), which are unaffected by ammonium. Also, with the hippocampal pyramidal cells there is no evidence for an outward Cl⁻ pump that is dependent on the internal concentration of Cl⁻ as postulated by Lux; hence, the linked chloride–potassium ionic fluxes of the Figure 4.12D probably still obtain for the very important inhibitory synapses at higher levels of the brain.

The ionic mechanism for generation of motoneuronal IPSPs according to the Lux hypothesis is illustrated in Figures 4.12E and 4.12F, where the battery on the Cl⁻ channel in Figure 4.12F is shown at -80 mV in accord with the value postulated for E_{Cl} as a consequence of the outward Cl⁻ pump. The ionic channel is shown as being impermeable to K⁺. Figures 4.12G and 4.12H show in the same convention the mode of action of the presynaptic inhibitory transmitter (also GABA) in producing a depolarization (cf. Figure 4.18). A further description will be given later. Not shown in Figure 4.12 are many invertebrate inhibitory synapses where the opening of the ionic gates results in no change in the normal resting membrane potential. In the absence of a Cl⁻ pumping mechanism there is Cl⁻ equilibrium, but the inhibition is still effected because the depolarization of excitatory synapses is counteracted by the ionic mechanism that acts to hold the membrane potential at the resting level. It has been suggested that if there are fixed positive charges on the channel, the anions will go through and the cations will be repelled, as is shown in Figures 4.12E and 4.12G.

The models of central inhibitory action illustrated in Figures 4.12D and 4.12F can be regarded merely as provisional. Much more quantitative work is needed to test the various components of the Lux hypothesis. There are many anomalous features, particularly the strange effectiveness of intracellular acetate in diminishing and even reversing the IPSPs of hippocampal pyramidal cells (Allen *et al.* 1977). Suggestions of further levels of complexity come from the admirable studies by Kehoe (1972) on the pleural ganglion cells of *Aplysia*. The IPSPs generated either by synaptic action or by application of the transmitter ACh display a double time course. The fast component is due to an increased Cl⁻ conductance, the slow to an increased K⁺ conductance, the respective equilibrium potentials being -60 mV and -80 mV. In this elegant investigation there was complete separation between these two actions of the same transmitter, and each ionic conductance was in good accord with predictions from the ionic composition on the Nernst equation (Figure 1.17A, Section 1.2.5). In a related investigation Gerschenfeld and Paupardin-Tritsch (1974) have shown that, on other mollusk neurons in *Helix* and *Aplysia*, the inhibitory transmitter is serotonin (cf. Chapter 9), and it likewise generates fast and slow IPSPs due respectively to increases in conductance for Cl⁻ and K⁺.

4.5 General Features of Transmission by Postsynaptic Action in the Brain[3]

Let us now return to the general diagram of chemical transmitting synapses in Figure 3.16A (Section 3.4). The wonderful biological process of evolution was not so uncompromisingly innovative. When good solutions were developed, they were retained by natural selection. They had survival value, and certainly chemical synaptic transmission had good survival value. It was invented long ago in relatively primitive invertebrates. And even the transmitter substances at mammalian synapses mostly have an invertebrate lineage. The detailed features of Figure 3.16A apply to both excitatory and inhibitory synapses of the central nervous system (cf. Figures 1.3B, 1.4B, and 1.5A, Section 1.2.2). There are the vesicles in the presynaptic terminals, a synaptic cleft of similar dimension, and a postsynaptic membrane with sensitivity to the transmitter substance that opens ionic gates, as indicated in the enlargement in Figure 3.16B for excitatory synapses.

Essentially the same events occur with the excitatory synapses in the central nervous system as at the neuromuscular junction. The all-or-nothing impulse propagates to the terminal with the sodium Na_D and potassium K_D gates opening at a critical level of depolarization (Figure 2.6, Section 2.3). At the terminal there is, in addition, opening of Ca_D gates so that calcium ions enter and participate in the process whereby the vesicles discharge their quantal content of transmitter into the synaptic cleft. This transmitter then diffuses to the subsynaptic membrane as in Figure 3.13A (Section 3.2.10) and momentarily lodges on steric receptor sites (cf. Figure 1.16, Section 1.2.5) that allow it to open the sodium (Na_T) and potassium (K_T) gates that are shown in Figure 3.16B (Section 3.4). Unfortunately, the excitatory transmitters at brain synapses are still largely unknown. It is ACh in a few cases (Chapter 5), but it could be glutamate or aspartate (Chapter 7) in many, and there probably are other substances.

The ionic fluxes through the Na_T and K_T channels depolarize the postsynaptic membrane enough to open the sodium and potassium gates concerned in impulse transmission, i.e., the Na_D and K_D are gates shown open in Figure 3.16B. The Na_D gates for impulse transmission in the brain are blocked in the same way by tetrodotoxin (TTX), and it is presumed that the potassium gates (K_D) are similarly blocked by intracellular tetraethylammonium (TEA). All evidence indicates that impulse transmission in the central nervous system is essentially the same as in peripheral nerves. Likewise, it is assumed that the central depressant actions by manganese and magnesium are due to their effects in preventing Ca^{2+} ions from moving in through the Ca_D channels.

With central inhibitory synapses the only difference is that the transmitters are the amino acids, glycine or γ-aminobutyric acid, GABA (Chapter 6), and that these transmitters open up channels across the subsynaptic membrane for chloride or for chloride and potassium, and strictly not for sodium.

[3]General reference: Eccles 1964b.

Nevertheless, the essence of the story is the same. The presynaptic impulse opens the calcium gates at a certain level of depolarization, and the inward movement of Ca^{2+} ions liberates the transmitter quantally.

The transmitter diffuses across the synaptic cleft and opens ionic gates with the consequent ionic fluxes that increase the charge across the subsynaptic membrane, causing the flow of current (Figure 4.9B) that hyperpolarizes the postsynaptic membrane.

4.6 Inhibition by Reciprocal Synapses[4]

In the schematized diagrams of Figure 4.17 it is assumed that synaptic excitation of the inhibitory neurons generates impulse discharges (cf. Figure 4.15B) and that the inhibitory action is consequent on the passage of these impulses to the inhibitory synapses and the liberation of transmitter therefrom, in the manner that has been described in detail in Chapter 3. It was shown there, however, that a presynaptic impulse was not essential for the liberation of transmitter from a presynaptic terminal (Figures 3.8, Section 3.2.6; 3.14, Section 3.3). It was sufficient merely to produce an equivalent depolarization of the terminal by an applied current so that the Ca_D gates would be opened and the quanta of transmitter released. It has now been shown that in certain inhibitory neurons the liberation of transmitter similarly is produced by an electrotonically spreading depolarization, and not by the depolarization of an impulse.

Rall and his colleagues (1966) and Nicoll (1969) have provided convincing evidence, both physiological and histological, that in the olfactory bulb (Figure 4.13A) the granule cells (g) inhibit mitral cells (m) by quite a unique synaptic arrangement. By electron microscopy it has been shown that the secondary mitral cell dendrites (SD) have synapses on the gemmules that bud off from the branches of the granule cells in the external plexiform layer (EPL); and these same gemmules in turn give synapses to the mitral cell dendrites, as is illustrated in the graphical reconstruction of Figure 4.13B. Where the mitral cell dendrite (m) contains vesicles, it is presynaptic to the gemmule; the consequent synaptic transmission is in the direction of the upper arrow. The lower synapse is in the reciprocal direction. Analysis of field potentials recorded in depth through the laminated structure of the olfactory bulb (cf. the technique illustrated in Figures 2.16B, 2.16C, Section 2.7; and 4.16D, 4.16E) reveals that the former synapses are excitatory; i.e., the mitral cell dendrites directly excite the gemmules of the granule cells. Analysis of field potentials and also intracellular recording from the mitral cells (Figures 4.13C and 4.13D; Nicoll 1969) show that, after a delay of about 3 ms, mitral cell excitation results in an IPSP of mitral cells, even of those not initially excited. This inhibition apparently occurs in the absence of impulse discharge in granule cells.

[4]General reference: Shepherd 1974.

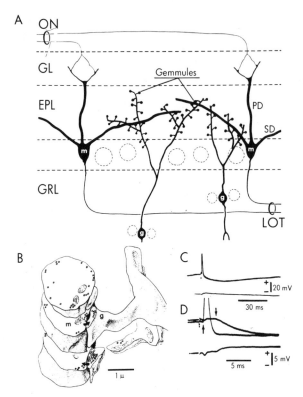

Figure 4.13 A. Layers and connections in the olfactory bulb (adapted from Ramón y Cajal 1911). GL, glomerular layer; EPL, external plexiform layer; GRL, granular layer; ON, olfactory nerve; LOT, lateral olfactory tract; g, granule cell; m, mitral cell; PD, primary mitral dendrite; SD, secondary mitral dendrite. B. Graphical reconstruction of a granule synaptic ending (g) on a mitral secondary dendrite (m). The granule ending is shaped like a gemmule and arises from a granule dendrite lying approximately perpendicular to the mitral dendrite. Within a single ending are two synaptic contacts with opposite polarities (indicated by arrows). The reconstruction was made directly from a series of tracings of 23 consecutive electron micrographs; no sections are omitted in showing cut surfaces that separate a reciprocal synaptic region into three slices. Microtubules and endoplasmic reticulum are not shown. (Rall *et al.* 1966.) C and D. Antidromic responses of mitral cells on stimulation of the LOT. C and D are at low and high amplification and slow and fast speeds, as indicated. In D the stimulus was just at threshold for the axon of the mitral cell which was excited on one trace and not on the other. Lower traces in C and D give control extracellular potentials. (Nicoll 1969.)

Thus, the proposed inhibitory pathway from mitral cell to mitral cell is by a very short linkage through the inhibitory granule cells, even being within a single gemmule. It is pointed out that this reciprocal synaptic action may occur in the absence of impulses. First, impulses in the mitral cell somata depolarize the secondary dendrites by an electrotonic spread of depolarization, there being as a consequence a liberation of excitatory transmitter with depolarization of the gemmules (upper arrow of Figure 4.13B). Second, this depolarization of the gemmule-studded branches of the granule cell causes the liberation of inhibitory transmitter at their inhibitory synapses (lower arrow of Figure 4.13B) with the consequent hyperpolarization of the secondary mitral dendrites. Moreover, it is assumed that the electrotonic spread of depolarization in the granule cell dendrite is adequate for the depolarization of one gemmule to spread to others that in turn inhibit secondary dendrites, often from many mitral cells, as is indicated in Figure 4.13A. The physiological significance of this reciprocal synaptic mechanism can be understood when it is recognized that a dense meshwork is made by the gemmule-studded branches of the granule cells and by the secondary dendrites of mitral cells. It is thus possible for mitral cells very effectively to inhibit each other via the reciprocal synapses. GABA is the inhibitory transmitter of the granule cells (Ribak *et al.* 1977a).

This remarkable new development gives an additional manner in which inhibitory interneurons can be effective in producing postsynaptic inhibition. Furthermore, similar reciprocal synapses are observed with horizontal and amacrine cells in the retina (Dowling and Boycott 1966), which likewise are inhibitory.

These fundamental discoveries in the synaptic mechanisms of the olfactory bulb and the retina have opened the way to an understanding of many perplexing observations on the synaptic structures in relay nuclei on afferent pathways to the cerebral cortex. There are in these nuclei short-axon interneurons (Golgi type II neurons) that in general resemble the granule cells of Figure 4.13 in that they act as a local inhibitory mechanism, their dendrites forming inhibitory synapses on the dendrites of the relay neurons, as indicated in Figure 4.14 for the lateral geniculate nucleus (Hámori *et al.* 1974). There are three differences in detail: first, the Golgi cells (GO) are synaptically excited in the usual manner by impulses in the afferent fibers (Ret); second, there is often a triadic synaptic arrangement as shown in the electron micrograph of Figure 4.14 and encircled in the diagram below; third, the axon of the Golgi cell gives inhibitory synapses to the dendrites of the relay cell (Rel). The principal mode of operation is that an afferent impulse in Ret excites both the relay cell dendrite and the protrusion from the Golgi cell dendrite, and this latter in turn inhibits the relay cell dendrite. This arrangement is presumably to give a very effective depolarization of the Golgi cell protrusion with, as a consequence, a very effective inhibitory action by it. But the Golgi cell dendrite is excited at other sites besides the triadic. It can be assumed that, as with the mitral cells of the olfactory bulb, electrotonic transmission along the Golgi cell dendrite is adequate for pooling the depolarizations in order to inhibit many relay cells. And there is also a third inhibitory mechanism, as in the conventional manner, by the axodendritic synapses shown from the GO cell to the soma and dendrites of the Rel cell.

Besides the lateral geniculate nucleus of the monkey (Hámori *et al.* 1974) essentially similar inhibitory mechanisms have been identified in the medial geniculate nucleus of the cat (Morest 1971), in both the medial and lateral geniculate nuclei of the cat (Jones and Powell 1969a), in the ventrobasal thalamic nucleus of the cat (Ralston and Herman 1969), and in the superior colliculus (Lund 1969). In addition, Sloper (1971) has reported dendrodendritic synapses as occurring, but rarely, in the deeper laminae of the primate motor cortex. Unfortunately, there have been no complementary electrophysiological studies that give a quantitative evaluation of the inhibitory effectiveness of dendrodendritic synapses relative to the standard synapses made by axons of the inhibitory neurons (cf. Figure 4.14). In the olfactory bulb there is no such complication, and in fact the primary discovery was

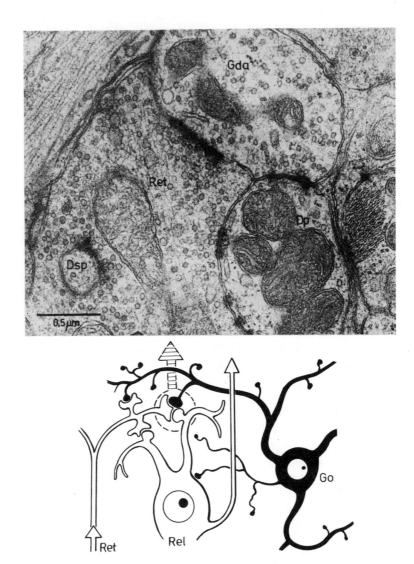

Figure 4.14 Triadic coupling of the subcortical sensory relay nuclei. Diagram at the bottom shows neuronal arrangement in the LGB, where a retinal (Ret) afferent terminal contacts both a dendritic protrusion of a geniculate cortical relay (Rel) neuron and the large dendritic appendage of a Golgi (Go) type II neuron. Electron micrograph at the top shows ultrastructural arrangement of the three neuronal constituents. Retinal afferent terminal is presynaptic (by contacts with round vesicles and asymmetric membrane attachments) to filamentous dendritic protrusions (Dp) of the relay cells and to the dendritic appendages (Gda) of Golgi type II neurons. The latter are again presynaptic (by contacts with flattened vesicles and asymmetric membranes) to the same relay cell dendritic protrusion. Dsp is an invaginated dendritic spine. Scale: 0.5 μm. (Hámori *et al.* 1974.)

electrophysiological. It has been authoritatively reviewed by Shepherd (1974). He also gives a comprehensive review of the retinal and thalamic dendrodendritic synapses.

4.7 Simple Neuronal Pathways in the Brain

We now come to the neuronal pathways in the brain. There are many questions to be answered. How are the properties of individual synaptic actions effective in the linking of neuronal assemblages into some meaningful performance? What principles of neuronal organization can we define? How is information from receptor organs transmitted up to higher centers of the brain? In Chapter 11 we will treat the complementary problem of how the higher centers of the brain act on lower centers eventually to bring about motoneuronal discharge and thus movement.

4.7.1 Pathways for Ia Impulses

Figure 4.1A is a very simple pathway, where a Ia afferent fiber from a muscle enters the spinal cord and branches so that the discharges of this annulospinal receptor organ can exert both an excitatory and an inhibitory action. The inhibitory pathway has to be relayed through an interneuron because the afferent neuron itself can only act in an excitatory manner, presumably because it can only make the excitatory transmitter substance. So the transformation comes via an interneuron that has a different biochemical competence to make a different transmitter that has an inhibitory synaptic action. One can assume that this inhibitory interneuron has a genome specified for this purpose, and that this dates back to its origin in a differentiating mitosis, as will be discussed in Chapter 12. So far as we know, all afferent fibers entering the spinal cord are excitatory in their action. For central inhibitory action to occur there must be the transformation via interneurons specialized for inhibition, as shown for the Ia input in Figure 4.1A. This requirement explains many of the design features in the brain. Figure 4.1A is an example of the simplest pathway producing excitation and reciprocal inhibition via a feed-forward inhibitory pathway.

4.7.2 Renshaw Cell Pathway

In Figure 4.15 there is a simple example of a feedback inhibitory pathway. In Figure 4.15A a motoneuron is shown with its axon going all the way out to the muscle, as has already been illustrated in Figure 3.1 (Section 3.2.1), and it acts there by the transmitter ACh. The broken lines of the axon signify that it goes many centimeters after it leaves the spinal cord before it reaches the muscle. But in the spinal cord this motoneuronal axon sends off collateral branches that make synapses on special neurons called Renshaw cells, after Birdsey Renshaw, a distinguished American neurobiologist who died from

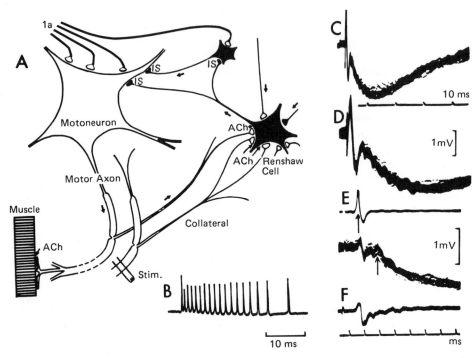

Figure 4.15 Feedback inhibition via Renshaw cell pathways. IS signifies inhibitory synapses and ACh excitatory synapses operating with ACh as transmitter. (Eccles *et al.* 1954.)

polio when he was very young. He discovered these cells and recorded their unique responses, but unfortunately did not work out their mode of operation, which was done much later (Eccles *et al.* 1954). Both the identification and the location of these neurons have been placed beyond all doubt by the beautiful technique of injecting the dye procion yellow into neurons recognized as Renshaw cells by their characteristic intracellular responses (Jankowska and Lindstrom 1971). There is a complete agreement between these anatomical results and the original findings with electrophysiological techniques.

When by electrical stimulation impulses are set up in the axons of motoneurons, the Renshaw cells fire with a long burst, often at extremely high frequency, as in Figure 4.15B. All kinds of pharmacological tests have established that the excitatory transmitter is ACh (cf. Chapter 5), which is the same as at the neuromuscular synapses that are made by these same motor axons. This is in precise accord with Dale's principle, which is defined by stating that at all the synapses formed by a neuron there is the same transmitter. For example, all synapses made by a motoneuron have ACh as the transmitter despite great differences in the location (central or peripheral), in the mode of termination, and in the target cells. The Renshaw cell sends its axon to form inhibitory synapses on motoneurons. Figures 4.15C and 4.15D show the large

IPSP (at two different sweep speeds) produced in the motoneuron by the rapid Renshaw cell discharge (Figure 4.15B). The latent period for the IPSP is shown by comparison of the extracellular record of the motoneuron (Figure 4.15F) with the intracellular (lower trace of Figure 4.15E), the IPSP thus being shown to begin at the arrow, which is only 1.1 ms after the volley in the motor axons which is at the arrow in the upper trace of Figure 4.15E (note the faster time scale for Figures 4.15E and 4.15F). This latency shows that there is just time for the pathway via Renshaw cells to the motoneuron that is drawn in Figure 4.15A.

Furthermore, the Renshaw cell is also seen in Figure 4.15A to give inhibitory synapses to an inhibitory interneuron. It so happens that in many cases this is the same inhibitory interneuron that is on the Ia inhibitory pathway to motoneurons. The action of the Renshaw cell on this inhibitory interneuron is to effect an inhibition of its inhibitory action, i.e., to diminish its inhibitory action, which is the equivalent of excitation. Removal of inhibition thus causes excitation by a process that is called disinhibition (Wilson and Burgess 1962).

There is a lot of nice tricky physiology in this circuitry, but we shall concentrate on the feedback operation. What is the point of it? The more intensely this motoneuron in Figure 4.15A fires impulses, the more it turns itself off by negative feedback via the Renshaw cells. You might well say: "This is an example of poor design and is wasteful of neuronal activity." But it is not to be understood simply in the oversimplified diagram of Figure 4.15A. Motoneurons activate many Renshaw cells via their axon collaterals, and each of these Renshaw cells inhibits many motoneurons. It is a very good example of the principles of divergence and convergence mentioned earlier in this chapter. Any one Renshaw cell will be inhibiting an array of motoneurons largely regardless of function. When one group of motoneurons is firing strongly, it will exert a strong feedback inhibition via Renshaw cells on the whole ensemble of motoneurons in their neighborhood no matter what they are doing. As a consequence there is suppression of all weak discharges. Only the strongly excited neurons survive this inhibitory barrage. In this way the actual motor performance is made much more selective by the negative feedback eliminating the stray weakly responding motoneurons that would frequently cause disorder in the movement.

4.7.3 Hippocampal Basket Cell Pathway

Figure 4.16 is another example of a feedback pathway to a pyramidal cell in the hippocampus, which is a primitive part of the cerebral cortex. The neuron labeled p in Figure 4.16F is the same kind of neuron as that in Figure 1.3B (Section 1.2.2), but it is shown inversely, with its apical dendrite pointing downward and its axon upward. From the axon there is a collateral just like that of the motoneuron in Figure 4.15A and it makes an excitatory synapse on a neuron (b) that we can liken to a Renshaw cell because its axon (ba) branches profusely to give inhibitory synapses on the somata of this and adjacent

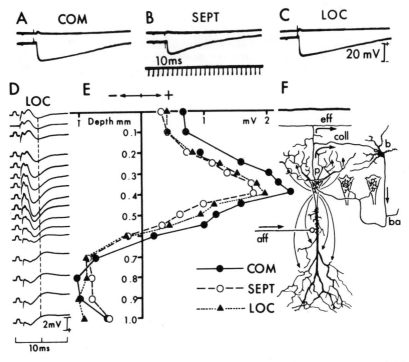

Figure 4.16 Feedback inhibition via basket cells in the hippocampus. (Andersen *et al.* 1964.)

pyramidal cells. This is another example of feedback inhibition. The more this pyramidal cell fires, the more it activates the feedback inhibition.

These inhibitory neurons (b) were called basket cells by Ramón y Cajal (1911) because of the basket-like embracement of the somata of the pyramidal cells. He thought they were excitatory but, following investigations by Kandel and his co-workers (Kandel *et al.* 1961), it was established that they were inhibitory by a series of tests that are partly illustrated in Figure 4.16 (Anderson *et al.* 1964).

The very large IPSPs produced in a hippocampal pyramidal cell by the basket cell synapses can be seen in the lower traces of Figures 4.16A, 4.16B, and 4.16C, the upper traces being the controls just outside the cell. Recordings COM, SEPT, and LOC signify three different locations of stimulation for exciting the basket cells. Extracellular recording at the indicated depths gives the field potential profile in Figures 4.16D and 4.16E, which indicates that the IPSP is generated at the level of the somata. The plottings in Figure 4.16E show the field potentials recorded (4.16D) at the depths corresponding to the drawing of the pyramidal cell in Figure 4.16F and measured at the time of the vertical broken line in Figure 4.16D. The maximum positivities occur at the depth of the somata in Figure 4.16F, which indicates that the inhibitory synapses are located there. Reference to Figure 4.9B reveals that the inhibitory synaptic site is the source of current flow to the rest of the cell, as shown by the arrows in Figure 4.16F.

In this way it was established that the basket cells of Ramón y Cajal were inhibitory, and hence the location of inhibitory synapses was defined for the first time (Figure 4.16F). Their clustered position on the soma of the hippocampal pyramidal cell is strategically an excellent place because their function is to stop the cell firing. This location on the soma provided the opportunity to identify inhibitory synapses in electron micrographs and to define their morphology, as has already been shown in Figure 1.3B(e) (Section 1.2.2).

It has also been shown by Anderson, Holmqvist, and Voorhoeve (1966b) that the numerous spine synapses on the dendrites, types a, b, c, and d in Figure 1.3B, are excitatory. These syanpses are very powerful and can generate the discharge of impulses that propagate down the apical dendrite, over the soma, and down the axon (Figure 2.16, Section 2.7). The inhibitory synapses are also very powerful, however, as may be seen in the large IPSPs of Figures 4.16A to 4.16C. Their location on the soma at the axonal origin gives them the final say as to whether the impulse is to be allowed to propagate down the axon. That is good strategical design to have the inhibitory synapses located at the optimal site. If, for example, they were located on the dendrites, impulses generated nearer to the soma would be able to fire down the axon without effective inhibitory constraint. With rare exceptions inhibitory synapses in all species of neurons are located on or close to the somata.

4.7.4 Operative Features of Inhibitory Pathways

In Figure 4.17A there is a general diagram of the feedback pathways whose operations were illustrated in Figures 4.15 and 4.16. In Figure 4.17B there is the feed-forward pathway that has been illustrated for the Ia afferent fibers from muscle receptors (Figure 4.1A), where collateral branches of the excitatory afferent fibers excite inhibitory interneurons that inhibit neurons in the forward direction. Both of these inhibitory pathways are now known to have more than a general value in keeping down the level of excitation and thus suppressing discharges from all weakly excited neurons. In addition, they participate very effectively in neuronal integration, molding and modifying the patterns of neuronal responses. We can think that inhibition is a sculpturing process. The inhibition, as it were, chisels away at the diffuse and rather amorphous mass of excitatory action and gives a more specific form to the neuronal performance at every stage of synaptic relay. This suppressing action of inhibition can be recognized very clearly at higher levels of the brain—particularly in the cerebellum, as we shall see in Chapter 13.

The other point that we wish to emphasize is that inhibitory cells are uniquely specified neurons, as has already been stated. They act via unique transmitter substances, which in the spinal cord is almost always glycine, while at the supraspinal levels, in the higher levels of the brain, it is almost always

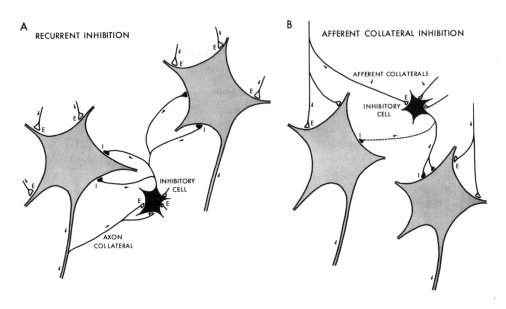

Figure 4.17 The two types of inhibitory pathways. A. Feedback. B. Feed-forward.

γ-aminobutyric acid (GABA) (Chapter 6). These two transmitters are very similar in their action, opening the same ionic gates, i.e., the gates for chloride ions or for potassium and chloride ions, as in Figures 4.12E and 4.12F, or 4.12C and 4.12D. It would be expected that substances which interfere with inhibitory synaptic action would cause unfettered excitatory action of neuron onto neuron and so lead to convulsions. This is indeed the case. In fact, as shown by Curtis and his associates, there are two classes of convulsants: those like strychnine that selectively block glycine transmission, and those like bicuculline and picrotoxin that selectively block GABA transmission (Curtis and Johnston 1974).

The unfolding of the central inhibitory story of the brain has been a great success, and it is now better understood than the central excitatory story. Over 10 supraspinal inhibitory pathways are now known to work by GABA and in almost as many spinal pathways the transmitter is shown to be glycine. As opposed to this impressive list of 24 purely inhibitory pathways in the mammalian central nervous system (Eccles 1969a), there is no example of a nerve cell in the mammalian brain that is ambivalent, exercising excitatory action by some of its axonal branches and inhibitory action by others. Originally it was envisaged that the inhibitory interneurons were interpolated merely to effect a transformation in the synaptic transmitter substance, being just a commutator-like device, to use Granit's phrase, and that they were very localized in their action with short axons restricted to the gray matter, as indeed they were in the examples then known in the spinal cord. But now very extensive investigations, particularly at higher levels of the nervous system and, par excellence, the cerebellum (Eccles *et al.* 1967), have disclosed examples of complex integrating functions of inhibitory neurons (cf. Chapter 11). Some inhibitory neurons have axons that are measured in centimeters in the descending direction (Wilson *et al.* 1970; Ito *et al.* 1970) or in the ascending direction (Ekerot and Oscarsson 1975). A further remarkable discovery was that inhibitory neurons could inhibit inhibitory neurons, so effecting by disinhibition an apparent excitatory action. This was first established by Wilson and Burgess (1962) for the Renshaw cell system where activation of some Renshaw cells can cause a depolarization of motoneurons by inhibiting the background discharge of other inhibitory neurons (cf. Figure 4.15A). Several other disinhibitory pathways have since been recognized, particularly in the cerebellum where basket cells and stellate cells inhibit Purkinje cells that in turn are inhibitory (Eccles *et al.* 1967).

In attempts to unravel the neuronal pathways in the brain it is a great help to recognize that the first big step is to identify the excitatory and inhibitory neurons. This criterion of a sharp classification is of essential importance in trying to discover the mode of operation of cell assemblages in the nuclear and cortical structures of the brain. In the first instance one need not bother about asking the more confusing question: Are some species of neurons both inhibitory and excitatory? The possibility of exceptional neuronal operation, however, always has to be kept in mind. We now come to the nearest approach to such an exception. It occurs in the special synaptic mechanisms giving what is called presynaptic inhibition.

4.8 Presynaptic Inhibition

The other method of inhibitory control in the central nervous system was first recognized by Frank and Fuortes (1957), but it took several years before there was an understanding of the mechanism of presynaptic inhibition, and an appreciation of its preeminent role in negative feedback control of the sensory pathways (Eccles 1964c; Schmidt 1971). The inventiveness of the evolutionary process is well illustrated by this quite different neuronal mechanism for depressing synaptic excitatory action.

The mode of operation of presynaptic inhibition is illustrated above the neuron in Figure 4.18. The presynaptic inhibitory fiber is drawn making a synapse on the excitatory fiber, and many such synapses have been seen in electron micrographs (Gray 1962; Conradi 1969; Kojima *et al.* 1975). Presynaptic inhibition is explained by the following sequence of events. By a chemical transmitter action the excitatory synaptic knob is depolarized (Figure 4.18C); consequently, a spike potential in this knob is diminished and the output of excitatory transmitter substance is depressed, as has been shown for peripheral synapses (cf. Figure 3.15, Section 3.3). It is thus postulated that inhibitory action is exerted on excitatory presynaptic terminals and not at all on the postsynaptic membrane, so it is called presynaptic inhibition. Presynaptic inhibition is exerted on the central synaptic terminals of all varieties of large afferent fibers that have been examined so far. At the first synaptic relay

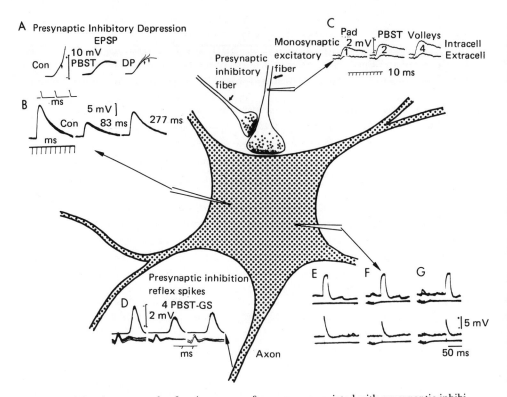

Figure 4.18 Specimen records of various types of responses associated with presynaptic inhibition. An excitatory synapse is shown on the neuron, with a presynaptic inhibitory ending on it, and by intracellular recording (note electrode) the excitatory fiber is shown to be depolarized—primary afferent depolarization (PAD). Note the difference between the intracellular and extracellular traces in C for one, two, and four volleys in the presynaptic inhibitory pathway. In A and B are the effects of presynaptic depolarization in reducing the excitatory synaptic action (EPSP), recorded intracellularly. Note large diminution of the EPSPs in B with respect to the control (CON) when tested at 83 and 277 ms after a brief tetanic stimulation of the presynaptic inhibitory pathway. In A presynaptic inhibition from two different muscle nerves depresses the EPSP so that it often fails to generate a spike, as it regularly does in the control (CON) at the arrow. In D the resultant diminution of reflex spike discharge is recorded by the population spike (cf. Figure 4.1B) in the ventral root, the first being the control response. The lower trace in E gives the control monosynaptic EPSP, and in F and G its diminution by the presynaptic inhibitory action of a conditioning train of volleys (6 at 250 per second) beginning 45 and 95 ms earlier. In the upper traces are the potentials produced by an intracellular current pulse of 15-ms duration under conditions identical with the EPSPs of the corresponding lower traces. (Eide *et al.* 1968.)

presynaptic inhibition is usually of greater potency than postsynaptic inhibition. All large afferent fibers exert a presynaptic inhibitory action on the central terminals of afferent fibers, but to a considerable extent the distribution is dependent on the respective modalities of the afferent fibers (Eccles 1964c; Schmidt 1971).

Presynaptic inhibition can be displayed by various experimental procedures. A direct test is to record from the excitatory fiber by an intracellular electrode and observe the depolarization directly, as in the records of Figure 4.18C, where the upper traces are the observed intrafiber potentials produced by one, two, or four stimuli to the conditioning afferent nerve. Another method is to record intracellularly from a neuron that is being synaptically excited and show that the excitatory postsynaptic potential is reduced because the excitatory transmitter output is depressed, as may be seen in Figures 4.18A and 4.18B. In Figure 4.18A the EPSP depression results in the delay or prevention of the generation of the impulse discharge that occurred at the arrow in the control, CON. In Figure 4.18B, after a conditioning tetanus the EPSP is seen to be greatly reduced relative to the control, but not altered in its time course. A very convenient way of evaluating presynaptic inhibition is provided by the diminution in the discharge of impulses evoked by a test excitatory input, as is shown in Figure 4.18D, the first response being the control population spike. This test is similar to that for postsynaptic inhibition in Figure 4.16D, and is not discriminative. All of these tests have been applied systematically. Another way of demonstrating presynaptic inhibition (technically the simplest) is to test the excitability of the excitatory fibers subjected to the presynaptic inhibition, that is, of the primary afferent fibers such as the excitatory presynaptic fiber in Figure 4.18. When depolarized by the presynaptic inhibitory action as in Figure 4.18C, these become more excitable. This method alone, however, is unreliable because the excitability may be enhanced by other depolarizing mechanisms, for example by an increase in extracellular potassium.

In the initial report on presynaptic inhibition Frank and Fuortes (1957) stated that the postsynaptic neuron exhibited no change in its electrical properties such as membrane potential or excitability. A very important confirmation by Eide, Jurna, and Lundberg (1968) is illustrated in Figures 4.18E to 4.18G. The lower traces show the control monosynaptic EPSP (Figure 4.18E) and its diminution in Figures 4.18F and 4.18G by presynaptic inhibition at two test intervals. The upper traces show that there was no accompanying postsynaptic change in conductance when tested by the response produced by a brief pulse applied intracellularly. Furthermore, no change was detectable in the time course of the EPSP during presynaptic inhibition (cf. Figure 4.18B; Eccles 1964c; Eide *et al.* 1968); hence, all precise experimental testing is in agreement with the hypothesis that presynaptic inhibition is indeed exclusively presynaptic (Eccles 1964b, 1964c; Eide *et al.* 1968; Schmidt 1971). There is no need to consider further the alternative suggestion that it is due to postsynaptic inhibitory synapses acting so remotely on the dendrites of the postsynaptic neuron that no effect can be detected with intracellular recording from the soma. Another confusion arises from attempts to account for the presynaptic inhibitory effect by the extracellular accumulation of potassium, which of course does occur with intense neuronal activity. Nicoll (1975), however, has shown that presynaptic inhibition has a quite different time course from the potassium accumulation, which in any case exerts a very much smaller depression.

There is now good evidence that the presynaptic inhibitory transmitter is

GABA. GABA resembles presynaptic inhibitory synapses in producing a depolarization of the presynaptic terminals of primary afferent fibers. Both actions (either of GABA or the presynaptic synapses) are selectively depressed by bicuculline and picrotoxin. GABA effects the depolarization of the presynaptic terminals or dorsal root ganglion cells largely by opening Cl^- gates (Nicoll 1975; Nishi *et al.* 1974; Gallagher *et al.* 1975) as indicated in Figures 4.12G and 4.12H. The equilibrium potential for this ionic mechanism is about -30 mV, which indicates that there is an inward Cl^- pump in the presynaptic fibers that causes the E_{Cl} to be about 40 mV in the depolarizing direction from the resting membrane potential. The intracellular Cl^- concentration must be as high as 50 mM. In the frog this E_{Cl} value obtains even as far peripherally as the dorsal root ganglion cells. This is consistent with the fact that GABA opens Cl^- gates in its action both in postsynaptic and in presynaptic inhibition. Finally, presynaptic terminals have been shown by immunohistochemistry to contain the GABA-synthesizing enzyme glutamic acid decarboxylase (GAD) (Figure 7.5, Section 7.2.4).

One generalization about presynaptic inhibition is that it has no patterned topography. It is widely dispersed over the afferents of a limb with but little tendency to focal application. For example, presynaptic inhibitory action by the afferent fibers of a muscle is effective on the afferents from all muscles of that limb regardless of function. It is not selective on one class of muscle, or on the muscles acting at any one joint of a limb. This widespread, nonspecific character is exactly what would be expected for the general suppressor influence of negative feedback. Nevertheless, there is organization or pattern in the distribution of presynaptic inhibition; this pattern depends on the class or modality of the afferent fiber on which the presynaptic inhibition falls, the cutaneous afferent fibers being the most strongly affected (cf. Eccles 1964c; Schmidt 1971).

Another generalization is that presynaptic inhibition is much more effective at the primary afferent level than at the higher levels of the brain. It has been shown to exercise an important inhibitory influence on pathways through the thalamus and lateral geniculate body, however, and it is utilized by descending pathways from the cerebrum for inhibitory action on synapses made by primary afferent fibers either in the spinal cord or in the dorsal column nuclei (Schmidt 1971). So far there has been no evidence for presynaptic inhibition at the highest levels of the mammalian brain, i.e., the cerebellar cortex and the cerebral cortices—both the neocortex and the hippocampus. An interesting evolutionary problem is raised by this rejection of presynaptic inhibition at higher levels of the brain in favor of postsynaptic inhibition.

4.9 Principles of Neuronal Operation

What can we now say as a result of this study of individual nerve cells in the brain and of some of the simplest organizations of them (there will be further examples in the chapters of Part III)? Can we define principles of operation? Our understanding of the brain has now advanced so far that we can enunciate a number of general principles in a quite dogmatic manner. We now proceed to do this seriatim because it will clear the way for advancing in our further attempts to understand the brain in the remaining chapters.

First, in the brain all transmission at a distance is by the propagation of nerve impulses. These are the all-or-nothing messages that travel along fibers and which were the main theme of Chapter 2. There is a minor reservation with respect to synaptic actions at close range of the type illustrated in Figures 4.13 and 4.14 where it can be cable-like transmission. It is to be understood that this exclusion principle does not apply to the chemical transport of macromolecules along nerve fibers that was considered in Chapter 1.

Second, we have the principle of divergence, as already defined. There are numerous branchings of all axons with a correspondingly great opportunity for wide dispersal because the impulses discharged by a neuron travel along all its axon branches to activate the synapses thereon. The divergence may be as low as ten, but values may be in the hundreds, as will be illustrated in Chapter 13.

Third, we have the complementary principle of convergence. All neurons receive synapses from many neurons, usually of several different species, as illustrated for the hippocampal pyramidal cell in Figure 1.3B (Section 1.2.2); also, they receive both excitatory and inhibitory synapses. It is doubtful if any neuron in the brain receives only excitation, and certainly there is no example of a purely inhibitory reception. The numbers of synapses on individual neurons are usually measured in hundreds or thousands, the highest recorded being around 100,000 (Chapter 13).

Fourth, there is the successive transmutation from electrical to chemical and back to electrical in each synaptic transfer. From this process there arises the integrational properties of the nervous system. There is the necessity for convergence of many synaptic excitations before a neuronal discharge is evoked, and there is the further opportunity for synaptic inhibitory action to prevent this discharge. In Chapter 3 we concentrated on the principles of chemical synaptic transmission: the presynaptic impulse, the chemical transmitter substances, their quantal liberation by impulses, their action on the postsynaptic membrane opening ionic gates with the consequent transmutation to an electrical event. This is the basic principle of communication in the brain. We are stressing it because it is fashionable for neuroscientists these days to propose that there is another kind of communication by what they call "the slow potential microstructure." They endow these potential fields with integrative or synthetic properties in their own right, whereas we can with assurance ascribe their production to the flow of currents generated by synaptic action in the ordinary process of synaptic transmission, as diagrammed in Figures 4.9A and 4.9B and revealed in the depth profile in the hippocampal cortex (Figures 4.16D and 4.16E). There is no neurobiological evidence that these currents are anything more than a spinoff from chemical synaptic transmission. Doubtless, if large enough, electrical fields would slightly modify the production of impulse discharge by synaptic action. But this would be diffuse and carry no significant information. Neurophysiological investigations strongly emphasize the almost exclusive role of impulses in coding information transmission both in the peripheral nervous system and in the brain, intensity being coded by frequency of firing. Figures 4.13 and 4.14 illustrate the rare exceptions. That principle can be extended indefinitely for sequence after sequence of synaptic transmission.

Fifth, in the brain there is almost always background firing of neurons. If there is coding by frequency of firing, then it has to be remembered that this coding is superimposed on a background of incessant, irregular discharge. Even when asleep the neurons of one's cerebral cortex are firing impulses. Actually some fire even faster when one is asleep than when one is awake. The problem is to extract a reliable performance out of the nervous system, considering that it has so much background noise. This is done by having many convergent lines in parallel, all carrying much the same signals. The columnar arrangement of input areas in the cerebrum (Chapter 15) does just this, and there are other examples now being discovered where neurons of similar connectivities are arranged in clusters, for example, in the cerebellar cortex and nuclei (Chapter 11). The same kinds of neurons are organized together, receiving the same kinds of message on the whole and transmitting the same kind of coded output to another cluster of neurons. Because of the incessant background noise the responses of one neuron are lost. The neurons have to "shout together," as it were, to get the message across and so make a reliable signal above all the background noise. This will be one of our problems in Chapters 12, 13, and 15, namely, to see how signals are lifted out of noisy backgrounds by collusion of many cells lying together in clusters and working together in parallel.

Sixth, there is the whole question of inhibition coming in and sharpening signals and controlling neuronal discharges by the feed-forward, the feedback, and the presynaptic inhibitory circuits that have been fully described in this chapter.

Seventh, we would stress that the activity in the brain has a basis in neuronal events and can be measured in terms of signals which by synaptic operation fire neurons, and so on in most complex organizational patterns. Chapter 13 will be an introduction to some attempts to illuminate some simpler levels that are concerned with the control of movement. We are attempting to understand some simpler levels—some of the simpler jigsaw components of the immense ensemble of complexity—and so gradually to understand more and more complex levels of brain action. Finally, in Chapter 16 we will be engaged on the most important and challenging problem of all—the events in the cerebral cortex, the highest level of the brain—which again have to be understood as the weaving by impulses of the complex spatiotemporal patterns that have been likened by Sherrington to the weaving of an enchanted loom. But again all we will postulate at the neuronal level is impulse or electrotonic transmission and the excitatory and inhibitory synaptic mechanisms responsible for controlling the impulse discharges.

II

Specific Neuronal Participants and Their Physiological Actions

As we learned in the first series of chapters, each neuron manufactures and stores one chemical transmitter. This becomes its biochemical fingerprint and its principal method of communicating with other neurons. Unfortunately, the transmitter is not known for the vast majority of neurons in the CNS. This is at once the greatest shortcoming in our understanding of the brain, and at the same time the greatest challenge. Far more is understood of the anatomical connections and physiological functions of neurons than is understood of their individual neurotransmitters. And yet when this information has become available, impressive horizons have frequently opened up. New understandings of the integrated action of the brain, new insights into disease states, and new approaches to drug development have resulted.

In this next section we describe the known neurotransmitters, the types of cells that use them, and the physiological actions that can be observed. We also describe another method of neurotransmission, not covered in the first section, which involves a second messenger in the postsynaptic cell. This is covered in detail in Chapter 8. The second messenger, a cyclic nucleotide possessing an unusually powerful high-energy phosphate bond, triggers off a series of chemical reactions with a variety of results. They may be short term, altering the membrane potential, or longer term, bringing about trophic and plastic changes.

Neurotransmitters, therefore, fall into two categories. The first we refer to as *ionotropic* because the principal function of the transmitter is to open ionic paths across the postsynaptic membrane. The action is rapid and produces large conductance changes. Acetylcholine at the neuromuscular junction and at Renshaw cells (Chapter 4); glycine in the spinal cord (Chapters 4 and 6); GABA at higher levels in the CNS (Chapters 4 and 6); and glutamate as a putative excitatory transmitter (Chapter 7) are some examples.

The second we refer to as *metabotropic* because these neurotransmitters stimulate cyclic nucleotide production and consequent metabolic changes in the postsynaptic cell (Figures 8.9 and 8.11, Section 8.2.3; and 8.13, Section 8.2.4). They are slower acting and do not produce large conductance changes in the neuronal membrane. Dopamine in SIF cells and in the striatum (Chapter 8); noradrenaline (Chapter 8) and serotonin (Chapter 9) at most CNS sites; and acetylcholine (Chapter 5) at some muscarinic sites are examples. Information on the physiology of metabotropic neurotransmitters is scanty compared with the ionotropic ones, as we shall see in succeeding chapters. It is probable that some transmitters function in both capacities depending upon the nature of the receptors, i.e., acetylcholine in the superior cervical ganglion (Figures 8.9 and 8.11, Section 8.2.3; and 15.9, Section 15.4).

Each transmitter has an extraordinarily sophisticated and delicate task to perform. It is the final end product of all the efforts of a neuron. The complex anatomical connections faithfully produced during embryological development and the intricate firing patterns precisely executed by a neuron can be utterly defeated by the failure of the transmitter to finish the job. On the other hand, fine tuning of transmitter action through the use of drugs can sometimes compensate for neuronal malfunctioning in disease states.

Transmitter action can be modified in many ways. Its synthesis, storage, or release may be impaired. Its receptor sites may be blocked, or fooled by a competitor. Its synthesis may be stimulated or its destroying enzymes impaired. The reuptake mechanism which sweeps it away from receptor sites may be interfered with or the receptor sites themselves may be occupied by other, more effective agents. Knowledge of the chemistry and pharmacology of neurotransmitters can thus be a powerful tool in the hands of the neuroscientist or physician.

The first step is to know whether a compound is a transmitter at all. In the past, attempts have been made to set down definite criteria for neurotransmitter status. These may work well in some situations, however, but be of little value in others. All that really can be said is that many tests may be applied to determine whether a material is *the* specific molecule stored in vesicles and released from the terminals of a given neuron. When a certain number of tests are satisfied, then the molecule can be regarded as a bona fide transmitter. If there is reasonable doubt, it is described only as a putative transmitter. If the information is speculative, it is usually referred to as a transmitter candidate.

The following are the more important tests:

Anatomical

1. The substance must be present in the nervous system in reasonable concentration. This is an obvious but necessary condition. There are many compounds that can be isolated from the nervous systems of submammalian species, or from other tissues of mammalian species, which have biological activity. But they are of no interest if they are absent from brain or occur only in miniscule concentrations.

2. The substance should occur in the nerve ending fraction of subcellular homogenates (Figure 5.8, Section 5.3.3). Since a transmitter must be stored in nerve terminals for there to be an adequate physiological supply, failure to find a concentration in this compartment is regarded as important negative evidence.

3. The substance may be distributed unevenly in brain. This is important because it suggests an association with particular neurons. Such uneven distribution was a key point in the identification of dopamine, glycine, and several other transmitters.

4. There may be a drop in concentration of a substance following lesions to known or suspected brain pathways. Since degeneration always takes place distal to a lesion, and also proximally whenever there are no sustaining axon collaterals (chromatolysis), such a concentration change is strong presumptive evidence of association of a substance with a pathway. The technique is almost invariably used in establishing the biochemical nature of long-axoned pathways, but is of no value for short-axoned neurons.

5. A histochemical or immunohistochemical method for the substance may establish its localization to specific neurons and pathways of the brain. This is extremely strong evidence. Histochemistry was the crucial technique for the establishment of the catecholamines and serotonin as transmitters (Chapters 8 and 9). Unfortunately, histochemical methods are extraordinarily difficult to develop and they exist for very few compounds in the CNS. Immunohistochemistry is possible if antibodies can be developed for the material as, for example, with substance P (Chapter 10).

Chemical

1. The presence of enzymes for synthesis and destruction should be identifiable. A method must exist for supplying large quantities of the transmitter to the nerve ending, and then rapidly disposing of it. The synthetic enzymes obviously need to be more specific than the destroying enzymes, and need to be localized to the terminal fraction. The destroying enzymes may be in dendrites or glial cells. The synthetic enzymes should also behave as does the transmitter by dropping in concentration following lesions. Tyrosine hydroxylase, the key enzyme for catecholamine synthesis, for example, is found in the nerve ending fraction, and decreases in concentration following lesions to known pathways, while catechol-O-methyltransferse (COMT), one of the destroying enzymes, is found in the soluble fraction and is not affected by the lesions.

2. Immunohistochemical localization of the synthetic enzymes to specific neurons may be possible. Immunohistochemistry is a powerful new technique which works well with enzymes because of their high molecular weight and the relative ease of developing antibodies to them. It can be effective at the electron microscopic level. As with the transmitters themselves, this cellular and subcellular localization can constitute the most persuasive evidence of all. Immunohistochemical localization of glutamic acid decarboxylase (GAD), and

choline acetyltransferase (CAT), was critical for the establishment of GABA and ACh pathways in brain.

3. The material, or its precursor, should be taken up into nerve endings by a pump, or high-affinity uptake system (Figure 5.7, Section 5.3.3). This permits it to be accumulated against a concentration gradient, and is necessary if the nerve ending is to obtain an adequate supply for physiological purposes. Pumping is a selective, but not absolutely specific, process, so that nontransmitters may be pumped.

4. The material should be released from suitable tissue preparations by a K^+ stimulation in a Ca^{2+}-dependent process. These conditions stimulate *in vivo* nerve stimulation. K^+ is a nonspecific neuronal depolarizer, while Ca^{2+} mobilization in the nerve ending is necessary for transmitter release.

5. The material should be bound to intrasynaptosomal vesicles. Since the binding is reversible, being used only for temporary storage purposes, this is sometimes hard to demonstrate.

6. The molecule, as well as its synthetic enzymes, may be transported by the process of axoplasmic flow. Techniques now exist for demonstrating this in central as well as peripheral pathways, but this is one of the less valuable methods.

Physiological

1. A transmitter must have action at receptor sites. Therefore the material should show definite physiological action when administered by various techniques. Ideally, it should exactly duplicate the effects of stimulation of the nerve or pathway in which it is believed to be the transmitter. Classical tests of neurotransmitter status using peripheral systems required such duplication (Chapter 4), but it is almost impossible to achieve centrally. Iontophoretic application is the most widely used technique, and, while quantitative comparisons are seldom possible, excellent correlation can often be found in postsynaptic neurons between stimulation of a pathway and the iontophoretic application of the presumed transmitter. The iontophoretic technique has been instrumental in turning up GABA, glycine, glutamic and aspartic acids, and many transmitter candidates.

2. A transmitter must be released upon nerve stimulation. Again, such release is relatively easy to achieve peripherally by perfusing isolated organs or ganglia and collecting the released agent. Centrally, it is much more difficult. Cups applied to the surface of the brain, however, or push–pull cannulae embedded in tissue, have been successfully applied to collecting increased quantities of transmitters such as ACh or dopamine following appropriate stimulation.

Pharmacological

1. It should be possible to find agents which interfere with the transmitter at any of the stages of synthesis, storage, release, or action at receptor sites.

Specificity is critical, and yet extremely hard to prove. Selective blockers were extremely valuable in confirming GABA and glycine as transmitters and would be extremely valuable if they could be found for putative excitatory transmitters such as glutamic acid.

2. It should also be possible to find agents which mimic the proposed transmitter at receptor sites, or indirectly enhance its actions in other ways. This could be by stimulating its synthesis or release, or by inhibiting its reuptake or destruction.

In each chapter of Part II the manner in which transmitters came to the attention of neuroscientists is described, followed by the anatomical, chemical, physiological, and pharmacological evidence upon which their proposed roles are based. There are many chapters in such a section which cannot yet be written because it is only possible to present an account of those few transmitters and few pathways which have been biochemically fingerprinted. We have tried to do this with some perspective, realizing that present information is poorly balanced. For example, there are probably fewer than 20,000 catecholamine neurons in the rat brain, and at most a million in the human brain. Yet the data regarding them are immense. On the other hand, the transmitters for most of the 10 billion cells of the human neocortex are unknown. Because there is so much more to learn, stress has been placed on how the known transmitters have been discovered and what techniques have been used to verify their roles.

Although we have grouped neurons according to the transmitter they supposedly manufacture and use, this is a generality which must be pursued with caution. The cells themselves may be very different in nature. For example, there is little in common from a functional point of view between the large, long-axoned neurons of the substantia nigra and the small, intensely fluorescent (SIF) interneurons of the superior cervical ganglion. Yet, because they both use dopamine as their neurotransmitter, they may find common ground in the action of drugs which affect dopamine. Here detailed knowledge can be extremely helpful in understanding the totality of action of a drug.

The section starts with Chapter 5 on acetylcholine, the compound which "rang up the curtain" on chemical transmission in the nervous system. It is the classical ionotropic excitatory neurotransmitter, although, as we will describe, it is by no means certain that it always acts in this fashion in the brain. ACh receptors vary considerably, and there is good evidence for metabotropic action as well.

Chapter 6 is a brief chapter on the ionotropic excitatory amino acids glutamate and aspartate. The evidence that these compounds are neurotransmitters is slim, but they may be the "workhorses" that operate the great majority of the excitatory synapses.

Chapter 7 deals with the classical inhibitory amino acids GABA and glycine. Their ionotropic mechanism of action is reasonably well understood.

Chapter 8, on the catecholamines, focuses on the completely different aspects of neuronal activity suggested for metabotropic neurotransmitters. They do not seem to obey the microphysiological rules followed by the previously described ionotropic transmitters, but on the other hand they stimulate second messenger production in the postsynaptic cell. Their neurons have tremendous divergence (each dopaminergic neuron of the substantia nigra has roughly 500,000 axonal varicosities for releasing transmitter) so that interference with the transmitter produces widespread effects.

Chapter 9 on serotonin neurons presents even more puzzling data on neurotransmitter action, again emphasizing a possible metabotropic role.

Chapter 10 describes the promising peptides. For modern neuroscientists this is a "hot" field of investigation because the diversity of agents being discovered lends interest to the speculation that many of the neurons not yet fingerprinted may turn out to be peptidergic.

Chapter 11 deals with putative transmitters and mentions a large number of transmitter candidates. The evidence that these compounds are neurotransmitters ranges from very strong in the case of histamine to highly doubtful in many others. Nevertheless, it is from such data that future transmitters will emerge.

5

Cholinergic Neurons

5.1. Introduction

In Chapters 3 and 4 we described how acetylcholine (ACh) was the first neurotransmitter to be identified, thus ushering in the concept of the transposition of electrical transmission within neurons to chemical transmission between neurons. We discussed how this was achieved, with the quantal release of ACh at nerve endings of the neuromuscular junction (Figure 3.13, Section 3.2.10) serving as the model. We detailed the effect ACh has in opening up ionic channels across the postsynaptic membrane (Figure 3.16), and the way it induces an EPSP (Figure 3.12, Section 3.2.9). These we have defined as ionotropic neurotransmitter functions. We say in Chapter 4 that the classic ionotropic actions of ACh were observed only with nicotinic receptors. The picture with muscarinic receptors was much less clear, opening the possibility of metabotropic action.

We must now explore some further concepts of the cholinergic neuron, including the varying responses of the cholinoceptive cell. Approximately 99% of all cholinergic synapses in brain are of the muscarinic type, indicating that this mode of ACh transmission is clearly of major importance in the CNS.

The chemistry of the cholinergic neuron has not yet been considered but it seems to be identical no matter what the nature of the cholinoceptive cell. Quastel and associates (1936) were the first to demonstrate the capability of brain tissue to synthesize ACh. They obtained a material they referred to as "choline ester" from brain slices treated with eserine and incubated with glucose and oxygen. It was indistinguishable from ACh by bioassay. Following this, Stedman and Stedman (1939) described the isolation of ACh from beef brain and, in subsequent papers, Quastel and associates attributed their activity to ACh. These early papers established the principal requirements for ACh synthesis and for almost two decades little new information was added.

It was recognized from early times that ACh existed in tissue in a free and bound form but the nature of the binding was obscure. At first it was supposed that the binding was to mitochondria because early differential centrifugation studies showed that bound ACh appeared in a crude mitochondrial fraction. But De Robertis and Bennett's discovery (1955) of vesicles in nerve endings led to the proper explanation of tissue binding of ACh and triggered a series of highly productive investigations in a number of laboratories. They speculated that these synaptic vesicles could be the site of storage and synthesis of transmitter substances.

Shortly thereafter, del Castillo and Katz (1956) developed the concept of a quantized release of ACh at the neuromuscular junction. The notion that the two might be related led to attempts to correlate electron microscopic and biochemical experiments. Working independently, De Robertis and his colleagues in Argentina and Whittaker and his team in England achieved fundamental breakthroughs using subcellular fractionation techniques.

Homogenized brain tissue was subjected to centrifugation at high speeds (up to $100,000 \times g$) through solutions of different density in order to separate some of the subcellular organelles discussed in Chapter 1 (Figure 1.10, Section 1.2.4). One fraction, which initially sedimented with the mitochondria but could be separated off by further centrifugation on sucrose gradients, was of intense interest. This contained nerve endings (Figure 4.8C, Section 4.4), also called synaptosomes. The homogenization procedure, while breaking down the cell bodies and snapping off axons and dendrites, preserved the nerve endings and even seemed to seal off their axonal stumps. The synaptosome frequently contained a portion of the postsynaptic membrane as if the presynaptic terminal had been tightly glued to the postsynaptic dendritic spine (De Robertis et al. 1962, 1963) (Figure 5.7A).

Within the nerve endings there could be seen synaptic vesicles. The next step was to concentrate the vesicles. This was done by placing the synaptosomes in hypotonic medium in order to explode them. The vesicles were not disrupted by the process and could be centrifuged down by spinning the solution at ultrahigh speeds. Electron micrographs of pellets from such centrifugations showed a high concentration of intact synaptic vesicles (Figure 5.7B).

It was found that both choline acetyltransferase (CAT), the enzyme which synthesizes ACh, and bound ACh were concentrated in the synaptosomal fraction while acetylcholinesterase (AChE), the enzyme which metabolizes ACh, was associated with membrane fragments. When the synaptosomes were disrupted, bound ACh was primarily associated with the sedimented vesicles.

This pioneering work not only verified previous theories regarding the association of ACh with synaptic vesicles, it led to the understanding of the synaptosome, or pinched-off nerve ending. It paved the way for similar subcellular work on every material suspected of being a neurotransmitter. Such subcellular work is now a standard test for neurotransmitter candidacy and will be referred to in many subsequent chapters.

Despite the fact that ACh is the premier neurotransmitter for model physiological studies, the explosion of chemical and pharmacological information that has developed in recent years around such other transmitters as the catecholamines and serotonin has not taken place with ACh. This is undoubtedly due to the fact that good methods for chemical assay of ACh have not been available. Much of the classical work on ACh was done using delicate bioassay techniques. These included the leech dorsal muscle, the frog rectus abdominus muscle, the guinea pig ileum, and cat blood pressure. While these methods are reliable and sensitive, they are also painstaking and foreign to the biochemist's background. Simple chemical methods for ACh assay have only recently become available. They include gas chromatographic, isotope derivative, and fluorescence techniques.

This lack of simple chemical techniques has undoubtedly also delayed fundamental studies on the chemical systems for synthesizing ACh. For example, it was not until very recently that complete purification of CAT was reported (Chao and Wolfgram 1973; Singh and McGeer 1974a). It can be anticipated that these technical advances will permit substantial progress to be made in the not too distant future on the precise anatomy of central cholinergic neurons, as well as on their physiology and pharmacology.

5.2 Anatomy of Cholinergic Neurons

It is now part of classic neuroanatomy that in the spinal cord all anterior horn cells supplying nerves to voluntary muscles and all lateral horn cells supplying preganglionic nerves to autonomic ganglia are cholinergic. It is also part of classic neuroanatomy that ganglion cells giving rise to postganglionic parasympathetic fibers are nearly all cholinergic. In addition, those cranial nerves containing voluntary motor fibers (III, IV, V, VI, IX, XI, and XII) and those containing preganglionic parasympathetic fibers (III, VII, IX, and X) also have cholinergic cell bodies. The reader is referred to standard textbooks of neuroanatomy for the details of the tracts and to standard pharmacology texts for the classical evidence upon which the cholinergic assignment has been made. The sympathetic postganglionic fibers to sweat glands are also cholinergic, and there is suggestive evidence, which needs to be pursued, of an even more widespread distribution of such postganglionic cholinergic fibers in the sympathetic system.

The main challenge at the present time is to understand the distribution of cholinergic pathways in the central nervous system. Here work has scarcely begun.

The distribution of CAT in many areas of human, baboon, and monkey brain is shown in Table 5.1. The values are remarkably similar considering the different species and somewhat varying techniques used by the three laboratories. The table illustrates the widespread distribution of CAT in brain tissue and indicates that the few pathways which have been defined (Table 5.4) must include only a small fraction of all the cholinergic cells in brain. Even areas having relatively low values of CAT nevertheless possess a synthesizing capacity for ACh comparable to the highest synthesizing capacities

TABLE 5.1
Choline Acetyltransferase Levels in Brains of Primates[a]

Area	CAT levels		
	Human[b] (Av. age 55)	Baboon[c]	Monkey[d]
Cortex			
Frontal	0.71	0.56	0.44
Motor	0.64	0.74	0.65
Occipital	0.34	0.64	0.33
Pyriform	—	2.22	1.86
Limbic nuclei			
Hippocampus	0.87	2.28	1.32
Amygdala (medial)		4.49	2.62
Amygdala (basal)	2.14	5.22	2.62
Amygdala (central)		9.15	2.62
Septal area	1.39	6.20	—
Nucleus accumbens	8.35	—	—
Basal ganglia			
Caudate head	10.70	9.50	7.28
Caudate body	—	16.50	9.00
Putamen	11.45	16.70	11.10
External globus pallidus	1.50	1.74	9.30
Internal globus pallidus	0.74	1.22	9.30
Diencephalon			9.30
Preoptic area	1.60	0.91	1.39
Medial hypothalamus	0.78	0.57	1.39
Mammillary bodies	0.49	1.51	1.39
Anterior thalamus	0.78	1.56	1.49
Medial thalamus	0.83	3.10	2.39
Ventrolateral thalamus	0.56	1.06	1.27
Habenula	—	4.43	—
Lateral geniculate	1.02	2.64	—
Midbrain			
Substantia nigra (compacta)	0.45	0.71	—
Substantia nigra (reticulata)	0.63	0.34	—
Red nucleus	1.01	2.86	—
Interpeduncular nucleus	—	26.60	—
Raphe (linearis)	0.59	6.70	3.60
Superior colliculus	1.28	5.62	3.52
Inferior colliculus	0.59	0.63	1:05
Pons–Medulla			
Locus coeruleus	1.50	3.29	—
Periaqueductal gray	—	3.26	—
Gracilis nucleus	—	3.01	—
Dorsal nucleus vagus nerve	—	3.05	—
Cerebellum			
Cortex	0.36	—	0.08
Vermis	0.72	0.25	—
Dentate nucleus	0.36	1.41	—
White Matter			
Corpus callosum	0.29	—	0.14
Internal capsule	0.96	—	—

[a] Choline acetyltransferase values expressed as micromoles of ACh synthesized per 100 mg protein per hour, except for the baboon where it was per gram wet weight.
[b] E. G. McGeer and McGeer 1976.
[c] Kataoka *et al.* 1973.
[d] Yamamura *et al.* 1974.

for the catecholamines (Table 8.4, Section 8.3) or serotonin (Table 9.1, Section 9.2) which exist in brain.

Distribution data on ACh are more limited and of lower reliability. ACh is a labile material which can be rapidly destroyed unless it is sequestered from AChE. As a result, values for ACh often vary considerably from laboratory to laboratory depending on the methods of preparing tissues and extracts for analysis. The most reliable values for brain are now thought to be those obtained when animals are sacrificed by microwave irradiation. This technique permits very rapid heating of brain tissue, so that within a second the enzymes become inactivated. Since ACh is stable at the temperatures generated, it is unaffected by the treatment. Unfortunately, the instrument is relatively expensive so that not too many laboratories have yet obtained data using this technique. Some idea of the variability in ACh and choline values that can be obtained, however, has been given by Jenden (1975) (Table 5.2). For whole rat brain a value of 24.8 nmol/g was obtained by microwave irradiation, while only 14.9 was obtained with standard cervical dislocation followed by the normal dissection, which takes about 5 min. Choline, on the other hand, increased from 26.3 to 148.4 nmol/g during this period.

Some values for ACh in rat brain using the microwave irradiation technique are shown in Table 5.3. These values are higher than those usually quoted using techniques that would only measure bound ACh. Nevertheless, they follow the same general pattern and parallel to a considerable degree the relative CAT values.

Evidence for central ACh pathways has largely been obtained through lesion and immunohistochemical methodology. In the immunohistochemical technique antibodies are made to a protein marker characteristic of a cell type. Since the enzymes which are responsible for synthesizing the neurotransmitter are unique to that transmitter cell type, they are the trademark of a cell and therefore are ideal compounds for immunochemistry. In this case CAT was purified from human neostriatum and injected into rabbits so that anti-human CAT antibodies were developed in the rabbit serum. For the technique to be successful, the enzyme must be absolutely pure and the antibodies produced to it monospecific. Figure 5.1 illustrates this for CAT where an

TABLE 5.2
Choline and ACh Concentrations in Whole Rat Brain Following Varying Methods of Sacrifice[a]

Method of sacrifice	Concentration (nmol/g of brain)	
	Choline	ACh
Microwave: immediate homogenization	26.3	24.8
Microwave: homogenization after 5 min	25.4	26.1
Decapitation: homogenization after 5 min	156.7	17.2
Cervical dislocation: microwave after 5 min	148.4	14.9

[a] From Stavinoha and Weintraub 1974.

TABLE 5.3
ACh Content of Rat Brain[a]

Area	ACh content (nmol/100 mg protein)
White matter	
Internal capsule	19
Corpus callosum	7
Limbic nuclei	
Olfactory tubercle	20
Nucleus accumbens	60
Septal nucleus (triangularis)	27
Amygdala (anterior)	37
Amygdala (basalis)	31
Amygdala (corticalis)	49
Basal ganglia	
Caudate	60
Globus pallidus	16
Diencephalon	
Hypothalamus (anterior)	20
Thalamus (anterior)	29
Thalamus (lateral)	18
Thalamus (medial)	40
Habenula (lateral)	43
Mammillary bodies (lateral)	26
Midbrain	
Substantia nigra	11
Red nucleus	43
Raphe nucleus (dorsal)	72
Raphe nucleus (medial)	13
Reticular formation	27
Interpeduncular nucleus	127
Pons–Medulla	
Locus coeruleus	34
Dorsal nucleus of the vagus nerve	36
Cerebellum	
Dentate nucleus	17

[a] From Cheney *et al.* 1976.

Ouchterlony-type double-diffusion plate shows the results of a neostriatal extract containing many proteins being run against the rabbit serum. When the rabbit serum diffusing from one well in the plate meets the brain proteins diffusing from the center well, a precipitate is formed when the serum antibody meets the antigen. The fact that only one precipitin line forms is a strong indication that the antibodies are monospecific and can therefore be used for tissue localization of CAT.

In essence, the immunohistochemical technique is a reproduction of the double-diffusion plate at the tissue level. Additional reactions are added to provide a marker that can easily be detected through the microscope. One

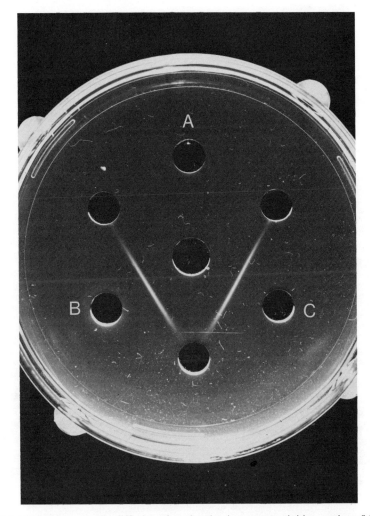

Figure 5.1 Ouchterlony double-diffusion plate showing immunoprecipitin reaction of CAT with rabbit serum antibodies. Inner well contained CAT protein. Outer wells at lower left and lower right contained serum from two rabbits immunized against CAT. Outer well at top contained serum from a normal rabbit. Note single precipitin line with each immune serum and absence of reaction with normal serum. (P. L. McGeer *et al.* 1974c.)

type of marker is the intensely fluorescent molecule fluorescein, which can be chemically coupled to serum proteins. With this technique tissue CAT is reacted with rabbit anti-CAT to form the first-stage tissue complex. The tissue complex is then treated with fluorescein-linked goat anti-rabbit serum (Figure 5.2A) which attaches to the complex, forming a sandwich which is visible under a fluorescence microscope. Figure 5.2B shows anterior horn cells of beef spinal cord fluorescing after this treatment, establishing that they contain CAT.

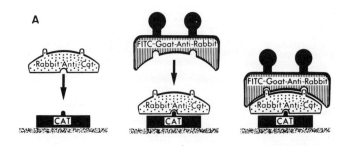

Figure 5.2 A. Sandwich technique for immune fluorescent tagging of tissue proteins. Tissue CAT acts as the antigen which couples with immune rabbit serum (left). This is reacted with fluorescein-tagged goat anti-rabbit serum (center). The complete sandwich is shown at right. B. Anterior horn cells of beef spinal cord fluorescing following the treatment described in 2A. (From P. L. McGeer *et al.* 1974b.)

The most sensitive immunohistochemical technique is Sternberger's ingenious peroxidase–antiperoxidase marker (Figure 5.3) (Sternberger *et al.* 1970). With this somewhat complicated technique, horseradish peroxidase is injected into an animal such as a rabbit and an antiserum made to the enzyme. This serum is then reacted in a test tube with peroxidase to produce an enzymatically active peroxidase–antiperoxidase complex. Brain tissue to be tested for CAT (or any other enzyme) is first reacted with the enzyme antiserum. This tissue complex is treated with goat anti-rabbit serum as shown in Figure 5.3, and then with the PAP complex. The tissue CAT at the bottom of

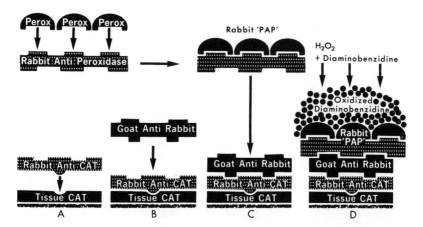

Figure 5.3 Schematic diagram of the peroxidase–antiperoxidase multiple sandwich technique of Sternberger *et al.* 1970). A. A similar reaction of tissue CAT with rabbit anti-CAT serum as in Figure 5.2A. B. The complex is reacted with goat anti-rabbit serum to serve as a binding agent for the rabbit peroxidase–antiperoxidase (PAP) complex. This is formed by reacting horseradish peroxidase with a rabbit serum made immune to the enzyme protein. C. The multiple sandwich is completed when the rabbit PAP is reacted with the sandwich containing goat anti-rabbit serum. D. The reaction of horseradish peroxidase with hydrogen peroxide and diaminobenzidine is shown.

the multiple sandwich is revealed by using peroxidase at the top of the sandwich as the marker. It catalyzes the splitting of hydrogen peroxide, which then oxidizes diaminobenzidine. A brown reaction product which can be seen by both light and electron microscopy is produced. The cleverness of the technique lies in its amplification. For each tissue molecule of CAT, many thousand diaminobenzidine molecules are oxidized.

The results using this technique for striatal tissue are shown in Figures 5.4 and 5.5. Figure 5.4A shows striatal interneurons staining positively for CAT. Some cell bodies and dendrites are prominently stained by the test procedure, indicating that they contain CAT. Others are clear, indicating that at least one population of noncholinergic striatal neurons exists. Many of these may be GABA containing (Chapter 6) or substance P containing (Chapter 10). The cells are 7–15 μm in diameter, typical of striatal interneurons. A considerable amount of background staining can also be seen. This presumably represents structures of the neuropil which are more clearly defined in the electron micrographs of Figure 5.5. Figure 5.5A shows a positively staining cell by electron microscopy. The nucleus is lightly positive while the cytoplasm is more intensely and diffusely stained. This is particularly true of the endoplasmic reticulum, shown as an enlarged insert. Positively staining processes can be identified in the surrounding neuropil. Dendrites and axons are also stained (Figures 5.5B and 5.5C), particularly over neurotubules. A curious shortcoming of the procedure is that it stains nerve endings rather poorly, yet they should be the most heavily stained structures. Most of those which do stain have been clearly damaged by the work-up procedure. Others (Figure 5.5B) are more intact, and in such cases the CAT seems to be associated with

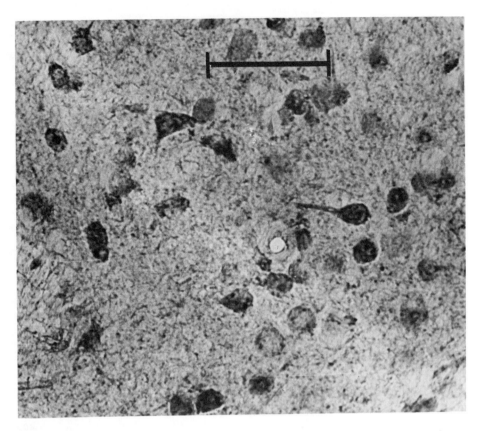

Figure 5.4 Light microscopic immunohistochemical micrograph of guinea pig neostriatum stained by the PAP technique. Numerous positively staining cells of intermediate size are seen. Note reticular network in the background which probably represents staining of cellular processes. These are seen by electron microscopy in Figure 5.5 (Hattori *et al.* 1976b). Calibration bar, 50 μm.

the vesicles. These are round and of similar morphology to those found at the myoneural junction. The reason for the poor staining of nerve endings is not known, but it could be related to the large molecular size of the PAP complex, which would decrease its penetration into small structures.

The presence of CAT in dendrites (Figure 5.5B), and even in dendritic spines, raises a fundamental question: Is ACh synthesized in these structures, and if so, why? As will be discussed in Chapter 8 on catecholamines, specific histochemical staining for these amines reveals their presence in dendrites as well. This seems to answer the question as to whether a neurotransmitter can exist in a dendritic structure. Its purpose in such a location is another matter because no current theory suggests any general dendritic role, although dendrodendritic synapses exist (Figures 4.13, 4.14, Section 4.6). It could be that neurotransmitters generally promote dendritic conductivity, or it could be that by retrograde release into the synaptic cleft, they "talk back" to the synaptic bouton and influence normal axodendritic synaptic activity.

Table 5.4 lists those pathways for which satisfactory evidence of cholinergic tracts have so far been obtained. Within the brain, neostriatal interneurons, the habenulo-interpeduncular tract, and the septal-hippocampal pathway are the best established.

The neostriatum contains some of the highest levels of ACh, CAT, and AChE of any major structure in the nervous system (Tables 5.1, 5.3). The majority of neurons are small Golgi type II interneurons. But large neurons (Kemp and Powell 1971) as well as smaller neurons comparable in size to the Golgi type II neurons send efferents to caudal structures (Bunney and Aghajanian 1976). The neostriatum receives a massive input from the cerebral cortex (Chapters 7 and 13), as well as a substantial input from the midline thalamic group of nuclei (Chapter 13). It receives dopaminergic input from the substantia nigra (Chapter 8), and serotonergic input from the rostral raphe system (Chapter 9). Lesioning of all these known afferents fails to bring about any major reduction in ACh, CAT, or AChE of the neostriatum, indicating that these inputs could not be mainly responsible for the cholinergic activity of the nucleus. Furthermore, lesioning the known efferents to the globus pallidus and substantia nigra also fails to bring about a reduction in cholinergic levels in either the neostriatum or the target areas, indicating that the large as well as the smaller efferent neurons are also noncholinergic (P. L. McGeer et al. 1971). Thus, the small neurons staining for CAT in Figures 5.4 and 5.5 must be interneurons (Hattori et al. 1976b).

A second cholinergic pathway in brain which has also been studied by immunocytochemistry is that from the medial habenula to the interpeduncular nucleus. Also termed the fasciculus retroflexus, this is a major tract in the brainstem. The medial habenula has a very high CAT activity which is substantially greater than the lateral habenula. The interpeduncular nucleus has the highest CAT activity of any structure in brain (Table 5.1), and it drops dramatically if the habenula is lesioned (Kataoka et al. 1973). The cell bodies of the medial habenula stain strongly for CAT. Again, the nerve endings in the interpeduncular nucleus stain very poorly, but some axonal staining can be seen (Hattori et al. 1977).

The septal-hippocampal pathway is the third reasonably well-established cholinergic tract in brain. The pathway is via the dorsal fornix, the alveus, and the fimbria. Lesions of the fimbria or septal area cause a large decrease of AChE and CAT in the hippocampus on the operated side (Lewis et al. 1967; E. G. McGeer et al. 1969; Fonnum 1970). It has also been shown that ACh is released from the hippocampus after septal stimulation (Smith 1972).

Cerebral cortical cells have also been shown to stain for CAT (P. L. McGeer et al. 1974b), but details of the connections of these and most other cholinergic cells must still be worked out.

The great majority of cholinergic pathways have not yet been identified. Clues as to their nature can be obtained by examining the distribution of

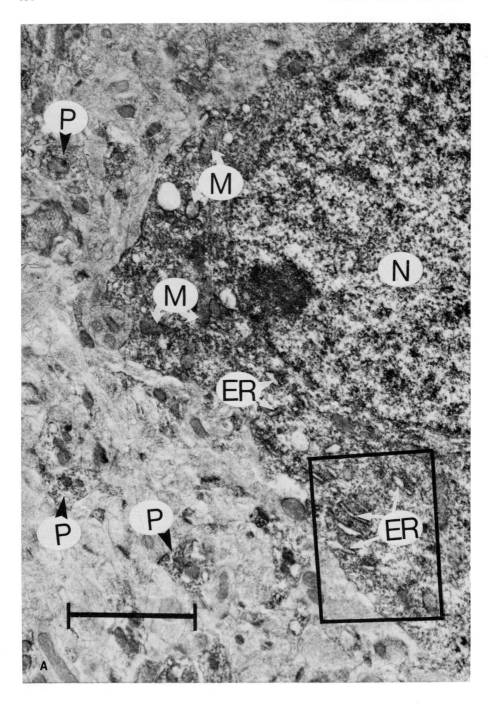

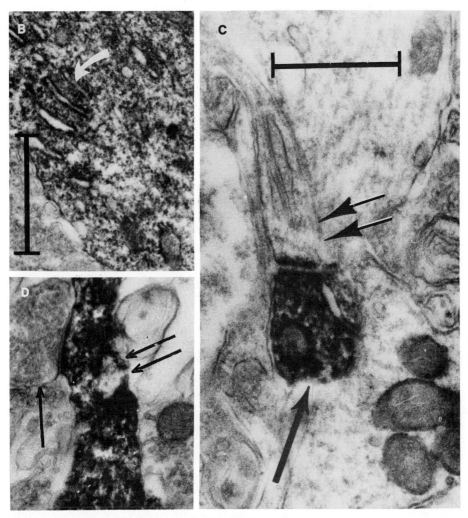

Figure 5.5 Striatal interneurons staining for CAT by electron microscopy. A. A low-power electron micrograph of CAT-containing neostriatal neuron. The nucleus (N) is lightly positive while the cytoplasm is more intensely and diffusely stained. Mitochondria (M) are unstained. Note intense positive stain on the outer surface of the endoplasmic reticulum (ER). Positively staining processes (P) can be identified in the surrounding neuropil. An enlargement of the inset (B) is shown with higher magnification of the rough endoplasmic reticulum showing particularly intense staining. Calibration bar, 1.5 μm (inset, 0.75 μm). C. Intact nerve ending (↑) staining positively for CAT. Note the intensity of stain around the vesicles and the asymmetrical axospi-nous contact with nonstaining dendritic spine (↑↑). Calibration bar, 0.5 μm. D. Dendrite staining positively for CAT (↑↑) receiving asymmetrical synapse (↑) from nonstaining nerve ending which contains round, mildly pleomorphic vesicles. (From Hattori *et al.* 1976b.)

TABLE 5.4
Presumed ACh Pathways

Anterior horn cells to all voluntary muscles and to Renshaw cells
Lateral horn cells to all autonomic ganglia
Nuclei of cranial nerves III, IV, V, VI, VII, IX, X, XI, XII
Postganglionic parasympathetic fibers
Occasional postganglionic sympathetic fibers
Septal-hippocampal tract
Habenulo-interpeduncular tract
Striatal interneurons
Cortical neurons

so-called cholinergic markers. These include CAT (Table 5.1), ACh (Table 5.3), high-affinity choline uptake, binding to ACh receptors by specific agents such as quinuclidinyl bromide (QNB), and AChE. A comparison of CAT, choline uptake, QNB binding, and AChE in various areas of monkey brain from one study is shown in Table 5.5.

5.3 Chemistry of Acetylcholine

5.3.1 Synthesis of ACh

The synthesis of ACh involves the single step shown in equation (1), where the reaction of acetyl-coenzyme A and choline is catalyzed by CAT (EC 2.3.1.6):

$$CH_3C\overset{O}{\underset{SCoA}{\diagup}} + HOCH_2CH_2\overset{+}{N}(CH_3)_3 \xrightarrow{CAT} CH_3C\overset{O}{\underset{OCH_2CH_2\overset{+}{N}(CH_3)_3}{\diagup}}$$

$$+ CoASH \qquad (1)$$

As described in the previous section, it is the enzyme CAT that characterizes cholinergic neurons. It is synthesized in the cell bodies and transported by axoplasmic flow to nerve endings (Hebb and Waites 1956). It also appears in dendrites (Figures 5.4 and 5.5), but the major synthetic action appears to be in the nerve endings where ACh storage vesicles are located. The intrasynaptosomal localization of CAT is not certain. CAT is easily solubilized by high salt concentrations. This would suggest that it is associated with the synaptosomal sap. *In vivo*, however, it could have some loose association with the surface of storage vesicles as seems to have been implied by the electron microscopic localization (Figure 5.5C).

The substrates for ACh synthesis are widely available and are not characteristic of the cell. Choline is produced in the liver and is used for many purposes such as the synthesis of glycerophosphorylcholine, phosphoryl-

TABLE 5.5
Regional Distribution of Specific [³H]-QNB Binding, [³H]Choline Uptake,
CAT and AChE Activities in Rhesus Monkey Brain Regions[a]

Region	[³H]QNB binding	[³H]Choline uptake	CAT	AChE
Cerebral hemispheres				
Frontal pole	439	1.72	4.4	15.7
Occipital pole	578	1.16	3.28	14.8
Precentral gyrus	483	1.73	6.47	18.6
Postcentral gyrus	516	2.46	7.95	30.5
Cingulate gyrus	546	2.48	6.35	18.5
Pyriform cortex	474	5.97	18.6	29.3
White matter areas				
Corpus callosum	107	0.94	1.4	11.7
Corona radiata	87	0.55	3.49	18.4
Optic chiasm	34	—	3.60	32.9
Cerebral peduncles	140	1.2	3.2	44.2
Limbic cortex				
Amygdala	496	—	26.2	122.0
Hippocampus	502	6.25	13.2	45.2
Hypothalamus	241	9.67	13.9	57.4
Thalamus				
Anterior	285	7.29	14.6	59.1
Medial	369	7.20	23.9	73.1
Lateral	360	3.96	12.7	59.8
Extrapyramidal areas				
Caudate nucleus (head)	976	28.1	72.8	281.0
Caudate nucleus (body)	1061	56.1	90.0	172.0
Putamen	1126	35.0	111.0	354.0
Globus pallidus	168	2.4	9.3	72.0
Midbrain				
Superior colliculi	381	10.5	35.2	121.0
Inferior colliculi	279	4.58	10.5	46.1
Raphe area	149	4.54	36.0	61.1
Cerebellum–lower brainstem				
Pons	212	2.70	9.2	28.5
Cerebellar cortex	125	1.45	0.8	54.3
Floor of fourth ventricles	96.5	5.05	21.4	45.3
Medulla oblongata	114	4.29	10.2	36.5
Inferior olivary nucleus	47.6	3.85	2.4	68.8
Pontine tegmentum	146	—	12.6	63.8
Spinal cord				
Cervical cord	47.6	3.19	8.1	22.2

[a] Specific [³H]quinuclidinyl benzilate [³H]QNB binding values are given in pmol/g protein. Velocities of choline uptake are given in nmol/g protein/4 min. Choline acetyltransferase (CAT) activities are given in nmol ACh synthesized/mg protein/h. Cholinesterase (AChE) activities are given in nmol ACh hydrolyzed/mg protein/min (Yamamura *et al.* 1974).

choline, and phosphotidylcholine. It is concentrated in cholinergic nerve endings by a pumping or high-affinity uptake system. Acetyl-CoA is a general substrate used in many metabolic reactions and thus is common to a variety of cells. Cholinergic cells do generate their own acetyl-CoA, and the process is somewhat complicated. It is important to be aware of the fact that acetate is not a precursor, as is so often stated in textbooks. The remote source is glucose, as was shown from the earliest experiments on ACh synthesis (Quastel *et al.* 1936). The more proximate source is its breakdown product, pyruvate. A complex pyruvate dehydrogenase system exists in mitochondria which ultimately produces acetyl-CoA. It has been extensively studied in *Escherichia coli* bacteria, pigeon breast, beef kidney, and other tissues as well as brain. Details of the system can be obtained by reference to standard textbooks or reviews (Reed and Cox 1966; Stryer 1975). The pyruvate dehydrogenase system requires thiamine pyrophosphate. Gibson and Blass (1976) have proposed that deficiencies in glucose, or in the pyruvate dehydrogenase system, affect cholinergic neurons because of their high acetyl-CoA requirements.

CAT has been completely purified from a variety of sources. These have been the human neostriatum (Singh and McGeer 1974a; Roskoski *et al.* 1975), the human placenta (Roskoski *et al.* 1975), and the bovine caudate (Chao and Wolfgram 1975). The human enzyme has a molecular weight of about 67,000 while the bovine enzyme, which is made up of 808 amino acid residues, has a slightly higher molecular weight of 89,000 (Chao 1976). Roskoski *et al.* (1975) found the placental enzyme to be identical with brain enzyme. The occurrence of CAT in the placenta is an oddity which has never been explained. Otherwise, it occurs only in cholinergic neurons and sperm (Harbison *et al.* 1974).

The antibodies produced against CAT purified from human neostriatum cross-react with CAT from other mammalian species but not submammalian species such as birds and fishes (Singh and McGeer 1974b). This might indicate some evolutionary specialization in the development of the cholinergic neuron. The cholinergic neuron is phylogenetically ancient, but does not occur in the most primitive species. It first appears in species such as the flatworm which show some degree of cephalization.

In any event, the availability of purified mammalian enzyme should greatly speed the development of knowledge regarding this important synthetic reaction. A complex can be isolated by column chromatography after the enzyme comes in contact with acetyl-CoA, but no such complex can be detected after contact with choline. This suggests that the enzyme first forms a complex with acetyl-CoA and then choline is transferred to the acetyl-CoA after occupying an adjacent site on the enzyme. To be active, CAT seems to require free thiol and imidazole groups, because agents such as cupric ions and photooxidized methylene blue which interact with these prosthetic groups inhibit the reaction.

5.3.2 Destruction of ACh

ACh is disposed of in a single fashion. It is destroyed by the action of cholinesterases which hydrolyze the ester bond, as shown in equation (2). The products are choline and acetate.

$$CH_3C\overset{O}{\underset{OCH_2CH_2N(CH_3)_3}{\diagup}} + H_2O \xrightarrow{\text{AChE}}$$

$$CH_3COOH + HOCH_2CH_2\overset{+}{N}(CH_3)_3 \quad (2)$$

AChE (EC 3.1.1.7) is contained in all cholinergic neurons and, evidently, all cholinoceptive neurons. In addition, there is a wider series of enzymes known as the "pseudo" cholinesterases which can be found in most organs of the body and which also carry out the reaction. There must be a considerable biological advantage to destroying ACh because the AChE and the pseudo-cholinesterases are among the most active enzymes in the body. Their blockade by so-called "nerve gases" produces rapid paralysis and death. Purified AChE has been found capable of hydrolyzing the amazing total of 150 g of ACh per milligram of protein per hour. It is also incredibly stable, retaining catalytic activity after the harsh treatment of formaldehyde fixation. There is little wonder that ACh is absent from the urine and bloodstream and only trace amounts are found in the CSF.

AChE has been isolated but the structure is still rather uncertain. Several active forms have been isolated from pig brain with molecular weights in multiples of approximately 60,000 ($\times$ 1, 2, 3, 4, 6, and 7) (McIntosh and Plummer 1973).

Like other proteins, AChE is transported by axoplasmic flow from the cell bodies, where it is synthesized, to the nerve endings. It has the distinction of being the first enzyme to be studied by axoplasmic flow (Hebb and Waites 1956). Ligature experiments on the sciatic nerve have established that it can travel at the very fast rate of 100–400 mm/day (Chapter 1).

Subcellular fractionation studies show that AChE is not only contained within synaptosomes but is also concentrated in the membrane fragments. These two facts suggest that the enzyme is distributed to the remote regions of cells where it becomes attached to surface membranes.

AChE can be localized in brain tissue at the light microscopic level. The method is to incubate partially fixed tissue with the ACh analog acetyl-thiocholine. AChE hydrolyzes the analog to produce thiocholine, which is then precipitated by reaction with cupric ions. The enzyme is present in so many cell bodies, dendrites, and axons of neurons that a confusing picture normally results. But an ingenious approach by Lynch et al. (1972), whereby the irreversible AChE inhibitor diisopropyl fluorophosphate (DFP) is administered in vivo to the animal before sacrifice, has permitted individual neurons and their processes to be visualized. Existing AChE is poisoned and must be replaced by newly synthesized AChE. This appears rapidly, and by

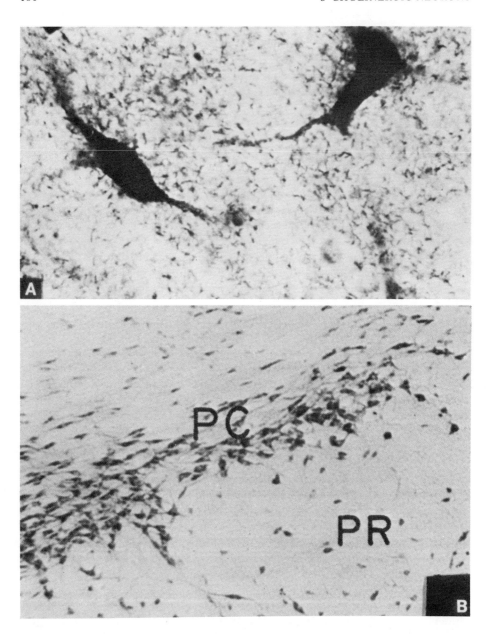

Figure 5.6 Striatal and nigral neurons staining for AChE. A. High-power photomicrograph of AChE-staining cells in the caudate nucleus 48 h after the systemic administration of DFP. (From Lynch *et al.* 1972.) B. Low-power photomicrograph of AChE-staining cells in the substantia nigra 5 h after the intramuscular injection of 1.5 mg/kg of DFP. Note staining of cell bodies and dendrites in the pars compacta (PC) with no staining in the pars reticulata (PR). (From Butcher and Bilezikjian 1975.)

waiting for a few hours an appropriate amount of newly synthesized material can be stained. Using this technique, Lynch *et al.* were able to observe individual AChE-containing cells in the caudate (Figure 5.6A) which are presumably cholinergic. But Butcher and Bilezikjian (1975), using the same technique, were able to stain dopaminergic cells of the zona compacta of the substantia nigra (Figure 5.6B). These are the same cells demonstrated by retrograde axoplasmic transport of horseradish peroxidase in Figures 1.8 (Section 1.2.3) and 1.15 (Section 1.2.4) and by histofluorescence staining for dopamine in Figure 8.4 (Section 8.2.1). They presumably contain AChE because they are cholinoceptive, although a cholinergic input to them has not yet been substantiated. The important point is that AChE is not a reliable marker for cholinergic cells even though it parallels ACh and CAT concentration in many areas of brain and provides a satisfactory index of cholinergic activity in some areas such as the hippocampus (Storm-Mathisen 1970).

Shute and Lewis (1963) described a series of cholinesterase-containing pathways in rat brain, including ascending dorsal and ventral tegmental pathways. The relationship of these to cholinergic pathways, however, remains to be established in view of the results cited above.

The pseudocholinesterases complicate the picture even more. They occur in brain and many other tissues and can only be distinguished from AChE by their preference for substrates such as butyrylcholine and benzoylcholine and by their differing sensitivity to various inhibitors.

5.3.3 Storage, Release, and Turnover of ACh

The principal area for cholinergic action is at the nerve endings of cholinergic cells. These can be concentrated by using the subcellular fractionation techniques of Whittaker *et al.* (1964) or De Robertis *et al.* (1962, 1963). The scheme of Whittaker is shown in Figure 5.7.

Since their original isolation from homogenates of cerebral cortex, cholinergic synaptosomes from a wide variety of nervous tissues and from many different species have been investigated. The cerebral cortex, neostriatum, and cervical ganglion from mammals, the optic lobes of the octopus, the electric organ of the torpedo, and the ganglion of the squid have been particularly utilized.

The characteristic features of cholinergic synaptosomes are their high content of ACh and CAT as well as their system for high-affinity uptake of choline. Synaptosomes from the cerebral cortex, for example, contain 50%–70% of the total CAT and bound ACh in the cortex, but only 35% of the total AChE. Of the ACh within the synaptosomes, about 60% is contained in the storage vesicles. The remainder, referred to as extravesicular, can only be measured on subcellular fractionation if AChE is first inhibited.

The information presented so far is represented by the schematic diagram shown in Figure 5.8 of a cholinergic synapse. Such tiny bodies must be capable of functioning in many metabolic ways independently of the nucleus,

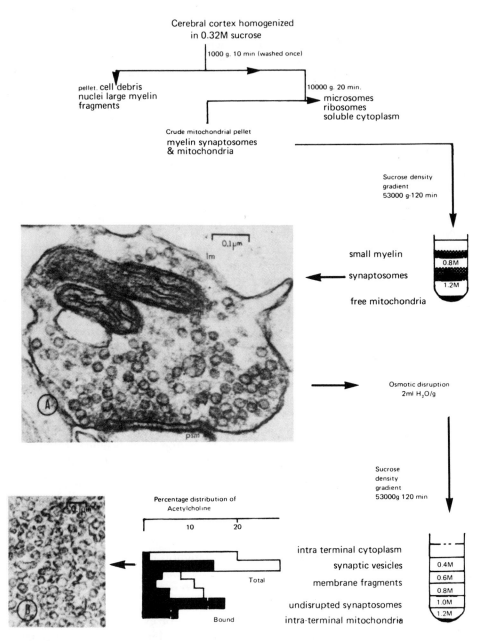

Cerebral cortex homogenized
in 0.32M sucrose

1000 g. 10 min (washed once)

pellet. cell debris
nuclei large myelin
fragments

10000 g. 20 min.
microsomes
ribosomes
soluble cytoplasm

Crude mitochondrial pellet
myelin synaptosomes
& mitochondria

Sucrose density
gradient
53000 g-120 min

small myelin

0.8M

synaptosomes

1.2M

free mitochondria

Osmotic disruption
2ml H₂O/g

Sucrose
density
gradient
53000g 120 min

Percentage distribution of
Acetylcholine

10   20

intra terminal cytoplasm

0.4M

synaptic vesicles

0.6M

Total

membrane fragments

0.8M

1.0M

undisrupted synaptosomes

1.2M

Bound

intra-terminal mitochondria

Figure 5.7 Schematic diagram for subcellular fractionation of cholinergic nerve endings following the procedure of Whittaker *et al.* (1964). Insert A shows an isolated nerve terminal; lm, limiting membrane apparently fully sealed; m, mitochondria; sv, synaptic vesicles; psm, an adhering piece of postsynaptic membrane. Insert B shows the appearance of the synaptic vesicles after isolation. Total ACh in the density gradient is that found in the presence of an acetylcholinesterase inhibitor, whereas bound is that which does not require esterase inhibition for its conservation. Free ACh is regarded as the difference between total and bound ACh. (Diagram from Marshbanks 1975.)

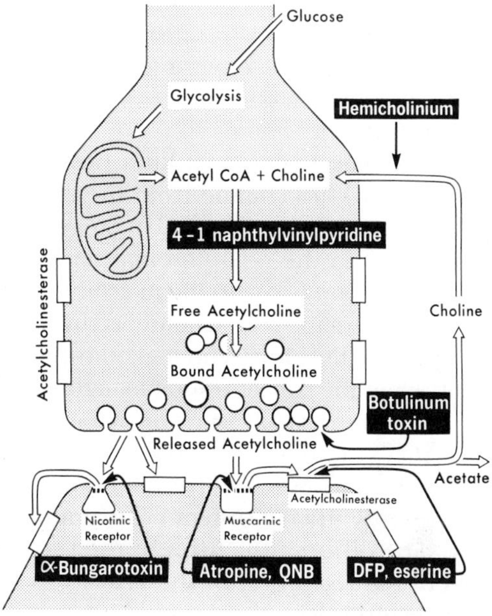

Figure 5.8 Schematic diagram of hypothesized cholinergic synapse. Light rectangles and light arrows show the sites of action of endogenous cholinergic elements. Black rectangles and black arrows show the sites of action of some common drugs which interact with the cholinergic system (Table 5.6). On the postsynaptic side, both nicotinic and muscarinic receptors are shown, although this would not be expected to occur on the same dendritic spine.

almost as miniscule cells. They utilize glucose and oxygen, and are capable of a net synthesis of ATP and phosphocreatine. Their major job, however, is to synthesize, store, and release the neurotransmitter. The nerve ending depicted in the diagram is shown converting glucose to acetyl-CoA and picking up choline from the extracellular space for ACh synthesis. ACh is synthesized by CAT and then sequestered from AChE by vesicular binding. It is released in quantal fashion from vesicles by the process of exocytosis (Figures 1.6, Section 1.2.2; 3.13, Section 3.2.10), where it crosses the synaptic cleft. It can be destroyed directly by AChE or reach a receptor site. Depending on the cholinoceptive cell the receptor site may be nicotinic, muscarinic, or both (e.g., nonadrenergic cells of the superior cervical ganglion; Figures 8.9, Section 8.2.3; 15.9, Section 15.4). The sites of action of various drugs which will be described later are also shown.

We must now consider precisely how this machinery controls the supply of ACh to the receptor sites. Such information is vital if we are to have methods of modifying the action of cholinergic pathways. As with all neurotransmitters there must be a system of regulating the supply to equal the demand. Since this can obviously vary widely according to the rate of neuronal stimulation, the capacity to synthesize must considerably exceed the normal requirements. Otherwise we should never be able to set Olympic records or escape from a burning building! Thus, a generous oversupply of enzyme must be present, along with a regulating mechanism, like the throttle on a car, to limit the supply of transmitter.

One distinct possibility for this regulation is the supply of choline. The amount of free choline measured in brain is only 25–45 nmol/g (25–45 μM), depending on the species and methods of measurement. This is far less than is apparently required to satisfy CAT. The concentration to half-saturate the enzyme in test tube experiments (K_m) was only 5 μM for acetyl-CoA but 250 μM for choline in one set of experiments using the human enzyme (Singh and McGeer 1977).

Cholinergic nerve endings possess a pumping mechanism which allows them to pick up choline from the extracellular fluid even when the concentrations are in the micromolar range and transport it against a concentration gradient into the synaptosomal sap. As a consequence, if labeled choline is added to brain tissue in the micromolar concentration range, almost all of the label appears in ACh. Choline is diverted to other uses, such as the synthesis of glycerophosphorylcholine, phosphorylcholine, and phosphotidylcholine, only when the concentration is high enough so that the priority requirements of ACh synthesis have first been looked after. This uptake into nerve endings for ACh synthesis is referred to as the high-affinity uptake system, while that for other purposes in different nervous system structures is referred to as the low-affinity uptake system (Yamamura and Snyder 1973).

The different features of the two affinities are illustrated in Figure 5.9, where the reciprocal of the velocity of [³H]choline uptake is plotted against the reciprocal of the choline concentration. Such double-reciprocal (Lineweaver-Burke) kinetic plots are routinely used in enzyme and drug studies. With simple kinetics, a single straight line is obtained, which can be used to calculate K_m, the concentration of substrate required to half-saturate the active site. In Figure 5.9, which is a Lineweaver–Burke plot using a synaptosomal fraction of rat corpus striatum, two straight lines are obtained with sharply differing slopes. These correspond to the K_ms of the high- and low-affinity uptake systems and have values of 1.43 and 93.4 μM respectively.

Other areas of brain known to contain cholinergic nerve endings have a similar high-affinity uptake system, but the cerebellum, which is apparently devoid of significant cholinergic input, has only a low-affinity uptake system.

The high-affinity uptake system requires sodium, is sensitive to various metabolic inhibitors, and is associated with considerable [³H]ACh formation. The low-affinity system, which presumably is associated with other metabolic

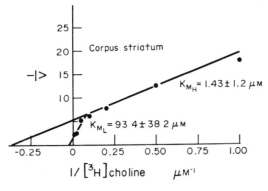

Figure 5.9 Double reciprocal plot of the uptake of [³H]choline into homogenates of rat corpus striatum. The data can be resolved into two linear components representing high-affinity uptake (K_m 1.43 ± 1.2 μM) and low-affinity uptake (K_m 93.4 ± 38.2 μM). Velocity (V) is expressed as μM of [³H]choline accumulated per gram of protein in 4 min. (Yamamura and Snyder 1973.)

pathways involving choline, is not associated with rapid [³H]ACh synthesis, is not sodium dependent, and is not affected by the same inhibitors.

The high-affinity uptake system can thus be a cholinergic marker. For example, lesioning of the known cholinergic pathway from the septum to the hippocampus causes a sharp loss in high-affinity uptake but no change in low-affinity uptake (Kuhar et al. 1973). This would be expected if cholinergic nerve terminals were degenerated by the lesion.

As described in the introduction to Part II, a high-affinity uptake system with similar kinetic characteristics exists for most neurotransmitters. An important question with respect to cholinergic nerves is whether or not the supply of choline provided by the uptake system actually governs the synthesis of ACh. An argument in favor of this hypothesis is the fact that hemicholinium-3 (HC-3), which competes with choline for high-affinity uptake, is an extremely effective inhibitor of ACh synthesis. Guyenet et al. (1975) found that the maximal velocity of high-affinity choline uptake in striatal synaptosomes (130 nmol/g of protein/min) approaches the maximal rate of synthesis of ACh in that system (150 nmol), strongly supporting the concept. Figure 5.8, therefore, shows choline, formed by the action of AChE on released ACh, being recycled to the nerve ending to replenish the supply. Many other mechanisms for control of ACh synthesis could be suggested. Among these are the hypotheses that the ratio of CoA to acetyl-CoA is critical for synthetic control, that there is feedback inhibition by ACh, and that intrasynaptosomal ACh inhibits the entry of choline into the nerve terminal. While the choline hypothesis is presently preferred, the question is by no means settled.

Synthesis of ACh is, of course, intimately associated with its release and consumption. But they are not directly linked because synthesized ACh is first stored in vesicles prior to its release. This sequence of synthesis → storage → release is depicted in Figure 5.8, but a further complication is that not all ACh is handled equivalently by the nerve ending. There are different pools or compartments.

The original concept of compartmentalization was suggested by the now classical studies of Birks and McIntosh (1961). Using the isolated perfused superior cervical ganglion, they identified a readily releasable pool of ACh which comprised some 20% of the total releasable amount. A larger component, termed the reserve pool, was then released upon further stimulation. They identified a third pool that they termed "surplus" ACh which may have been artifactual due to the use of anticholinesterases in their perfusate.

Many investigators have undertaken experiments since then whereby ACh is labeled by means of adding labeled choline or acetyl-CoA to a preparation and then measuring the newly synthesized material along with the unlabeled pool already present in the tissues. All have drawn the same conclusion: newly synthesized ACh is released in preference to old ACh.

Marchbanks (1975) has even suggested on the basis of such data that the quantal release theory does not apply and that the tightness of binding, and thus the ease of release of a particular ACh molecule, is related to how close to the center of the vesicle it is bound. This theory does not agree, however, with the elegant work of Katz and colleagues (Figure 3.7, Section 3.2.6) on the myoneural junction. A more reasonable explanation is that some vesicles are much more active than others by virtue of their proximity to the attachment cleft on the presynaptic surface (Figure 1.6, Section 1.2.2).

The difficulty of multiple pools of ACh may account for the wide variation in data on turnover rates in the literature (for a review, see Hanin and Costa 1976). The turnover rate is the amount of substance resynthesized per unit of time to replace that metabolized. To be valid, a calculated turnover rate presumes steady-state conditions so that the appearance of new material exactly balances the disappearance of old. In most studies, radioactive precursors such as [^{14}C]choline, phosphoryl-Me-[^{14}C]choline, or [^{14}C]glucose have been used, and the ratio of labeled to unlabeled ACh measured at various time intervals after the administration of the precursor.

Saelens *et al.* (1973) obtained values for ACh turnover of 20.9 nmol/g/min for rat cortex and 5.0 for brainstem. These compare with values of 3.3 and 1.5 by Racagni *et al.* (1974) for these same areas, and 21.7 for the striatum. Domino and Wilson (1972), however, using the nonisotopic method of inhibiting ACh synthesis by intraventricular administration of HC-3, obtained values of only 0.15–1.2 nmol/g/min for whole rat brain. Szerb *et al.* (1970), measuring the ACh released from exposed cat cortex, calculated a similarly low value of 0.15 nmol/g/min. It seems probable that the higher values are more reliable in that they involve more sophisticated techniques. Unfortunately, they are unable to distinguish between ACh synthesized and released for physiological purposes, and that which might be synthesized and destroyed by AChE presynaptically without ever being bound. Comparison of the turnover rates with steady-state values for ACh (Table 5.2) suggests that the brain possesses only a few minutes' supply of ACh.

5.4 Physiological Action of ACh and the Cholinergic Receptor

5.4.1 Types of Cholinergic Receptors

Two types of cholinergic receptors have been distinguished for years. The classical descriptive terms for these two types, "muscarinic" and "nicotinic," derive from the fact that the physiological effects of ACh at certain sites can be mimicked by nicotine, an alkaloid from *Nicotiana tobacum,* while others can be stimulated by muscarine, a drug obtained from the fungus *Amanita muscaria.* We will present the hypothesis here that nicotinic receptors mediate ionotropic actions of ACh, while muscarinic receptors mediate metabotropic actions.

The nicotinic receptor occurs in ganglia of both the sympathetic and parasympathetic systems in cells innervated by the preganglionic cholinergic fibers. α-Bungarotoxin is the specific blocking agent. The neuromuscular junction, innervated by cholinergic fibers from anterior horn cells, also has a nicotinic-type receptor. The muscarinic receptor occurs in smooth muscle innervated by postganglionic fibers of the parasympathetic system. This also includes cardiac muscle and certain exocrine glands. Atropine is the classic blocking agent.

Central synapses have muscarinic- and nicotinic-like receptors, which again respond somewhat differently to various agents. The thalamus and cerebellar cortex are generally regarded as having central nicotinic receptors, while most other areas of brain are thought to have muscarinic receptors. It is estimated that brain has a 100-fold predominance of muscarinic receptors. It is known that some cholinoceptive cells, such as the noradrenergic cells of the superior cervical ganglion, possess both nicotinic and muscarinic receptors (Figures 8.9 and 8.11, Section 8.2.3; and 15.9, Section 15.4) so that a mixed action is possible on the cholinoceptive cell. It is not known how commonly this may occur.

The basic question which needs to be asked is why there should be two different responses to ACh, particularly on the same cell. What different physiological purposes are being achieved? For answers to such questions we have to look beyond the simple ionotropic mechanisms discussed for ACh in Chapters 1–4 and consider the metabotropic mechanisms mentioned in the introduction to Part II.

5.4.2 Metabotropic Actions of ACh

The nicotinic actions of ACh have been thoroughly discussed in Chapters 3 and 4. What about the muscarinic? Since the pioneering work of Sutherland great attention has been focused on the role of cyclic nucleotides as the intracellular second messengers for the action of many hormones on their target cells (for reviews, see Rall 1972; Drummond and Ma 1973). This concept has been extended by Greengard and his colleagues to include their role as intracellular second messengers for certain neurotransmitters. For example, cyclic adenosine-3',5'-monophosphate (cAMP) has been suggested as the second messenger for dopamine (Chapter 8) while cyclic guanosine-3',5'-monophosphate (cGMP) has been suggested as the second messenger for the muscarinic actions of ACh (Greengard 1976). The formulae for the cyclic nucleotides are shown in Figure 5.10.

The hypothesis is that the neurotransmitter (in this case, ACh) does not directly open ionic channels. Instead, it induces a conformational change on the external surface of the membrane (Figure 8.8, Section 8.2.3) which activates a generator of the second messenger (in this case, guanyl cyclase) on the inside of the membrane. The second messenger, cGMP, then initiates a series of chemical reactions. These commence with the phosphorylation of protein kinases that are dependent on the cyclic nucleotide. It is this family of protein kinases that is actually responsible for the variety of actions that might take place in the postsynaptic cell. The specificity of action would then depend upon the kinases and their substrate proteins.

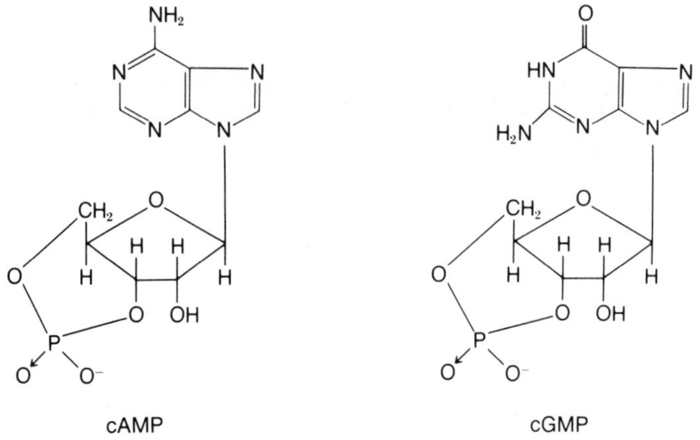

Figure 5.10 Structures of cyclic 3′,5′-adenosine monophosphate (cAMP) and cyclic 3′,5′-guanosine monophosphate (cGMP).

Among these substrate proteins could be ones controlling membrane permeability and altering either the ion conductance or the electrogenic pump. Thus, a metabotropic neurotransmitter could indirectly produce ionotropic effects of either an excitatory or an inhibitory nature, although the effects would have to be slower in onset because of the chemical method of initiation. Termination of the action of cGMP is brought about by phosphodiesterase hydrolysis and that of protein kinases by phosphoprotein phosphatases.

Most of the chemical components of the proposed system have been identified, but it still remains a speculation as to whether ACh or any other transmitter actually operates through such a metabotropic scheme.

Lee *et al.* (1972), however, using slices of either mammalian cerebral cortex, heart ventricle, or ileum, found that the concentration of cGMP in the tissues was increased two- to three-fold in a 3-min incubation by the addition of 1 μM ACh to the medium. These are all tissues with muscarinic- as opposed to nicotinic-type ACh receptors. Cholinomimetic agents with a predominantly muscarinic action, such as methacholine, pilocarpine, or bethanechol, also caused an increase in cGMP, while the nicotinic cholinomimetic tetramethylammonium had no effect. The increase was blocked by the muscarinic antagonist atropine but not by the nicotinic antagonist hexamethonium. These procedures had little or no effect on the content of cAMP, which is believed to be associated with catecholamine receptors (Chapters 8, 14, and 15). These results are all compatible with the hypothesis that nicotinic actions of ACh are ionotropic while the muscarinic ones are metabotropic, with cGMP as the postsynaptic mediator. While such data are by no means conclusive, they do help to interpret what would otherwise be extremely puzzling results on the iontophoretic effects of ACh on brain cells.

5.4.3 Iontophoretic Effects of ACh on Central Neurons

The iontophoretic action of ACh on brain cells varies widely. It may cause an excitation with rapid onset (nicotinic), an excitation with slow onset (muscarinic), or an inhibition with slow onset (muscarinic).

The classical example of nicotinic action of ACh on CNS cells is that of motor axon collaterals on Renshaw cells. The Renshaw cells respond with an extremely short latency to stimulation by ventral roots. This can be duplicated by iontophoretic application of ACh or nicotine. The activity is blocked by the nicotinic antagonists dihydro-β-erythroidine or hexamethonium. Other areas where somewhat similar effects can be observed are the medulla, areas of the thalamus and hypothalamus, and the cerebellum. Occasional cortical cells also respond similarly (see reviews by Krnjevic 1974 and Phillis 1970).

A much more commonly observed response to iontophoretically applied ACh on CNS cells is a slow and prolonged excitatory action which is blocked by atropine. With extracellular application of ACh, and intracellular recording of the response, it can be quite clearly demonstrated that this depolarizing effect is not associated with the decrease in membrane resistance that would accompany the opening of typical sodium and other ionic channels. Instead, the membrane resistance *increases*, with the depolarizing action having a reversal level close to -100 mV. This result can best be explained by a decrease in the conductance of either Cl^- or K^+, which have equilibrium potentials in this vicinity. Since intracellular injections of Cl^- cause large positive shifts in the Cl^- equilibrium potential but do not change the character of response to ACh, it must be concluded that a reduction in K^+ conductance (G_K) is responsible for the effect. Such slow, excitatory effects, blocked by atropine, are seen with cells in the caudate, hippocampus, pyriform cortex, lateral and medial geniculate bodies, ventrobasal thalamus, and other areas. Even Renshaw cells have some of these muscarinic receptors (Krnjevic 1974).

ACh can also depress the action of a wide variety of CNS cells in the cortex, hypothalamus, pons, medulla, and other areas (Phillis 1970; Krnjevic 1974). This inhibitory action is also clearly muscarinic because it is slow in onset and can easily be blocked by atropine. In this case, however, the membrane changes are thought to be due to an *increase* in potassium conductance rather than a decrease, as is observed with vagal inhibition of the heart (Trautwein and Dudel 1958). A direct correlation between these slowly developing membrane changes and cGMP formation has never been made; however, the two have been separately observed by techniques suitable for their measurement and it is not unreasonable to hypothesize that they should be causally related.

5.4.4 Separation of the ACh Receptor

Efforts have been made to define the nature of the macromolecules which constitute cholinergic receptors with some initial success. The most advanced work has come from using the nicotinic receptor of the electric organ of the torpedo or electrophorus. Certain snake venoms, particularly α-bungarotoxin, bind very strongly and specifically to this receptor. [^{125}I]-labeled toxin has been mixed with electric organ membranes and separation achieved by following the label through various purification procedures. The receptor appears to be a protein with a molecular weight of 35,000 to 45,000 (Schmidt and Raftery, 1973; De Robertis 1975).

The injection of purified cholinergic receptor from the fish electrophorus into rabbits has resulted in production of an antiserum which will precipitate the receptor protein on double-diffusion plates and which will inhibit the response of the electric organ to the cholinergic agonist carbamylcholine. Furthermore, the antibodies seem to attach to the neuromuscular junction in the injected rabbits. After a few weeks they develop a flaccid paralysis (Patrick and Lindstrom 1973; Heilbronn and Mattson 1974). This can be temporarily corrected by the use of AChE inhibitors such as neostigmine. The response of the rabbits is similar to that observed in human cases of myasthenia gravis, where it is believed a deficient receptor mechanism exists, possibly because of an autoimmune phenomenon.

It is known that the cholinergic receptor proteins are distributed across the postsynaptic membrane in a lattice-like fashion (Figure 1.16, Section 1.3). The receptor is a hydrophilic protein of elongated shape, ideal for traversing the thickness of the membrane. It is made up of subunits (most likely four), each of which has a molecular weight of about 40,000 and binds one α-bungarotoxin molecule.

In the past, it has often been suggested that AChE corresponds to the ACh receptor. This is clearly incorrect. AChE can be solubilized from synaptic membranes by high salt concentrations while the receptor is tightly bound. The binding agent α-bungarotoxin interferes with cholinergic transmission, but not with AChE *in vivo*. Its action is confined to binding with and blocking the nicotinic cholinergic receptor, which it does in both mammalian and sub-mammalian species. Furthermore, it has been shown by autoradiography that labeled α-bungarotoxin is attached to the surface of the folded myoneural junction, whereas AChE seems to be distributed throughout the postsynaptic membrane, including the depths of the folds (Fertuck and Salpeter 1974).

Work on isolation of the muscarinic cholinergic receptor is less advanced, although atropine (Heilbronn 1975), quinuclidinylbenzilate bromide (QNB) (Yamamura and Snyder 1974), and propylbenzilylcholine mustard (Burgen *et al.* 1974) have all been found to be specific binding agents.

5.4.5 Overall Central Nervous System Effects of ACh

A number of functions have been proposed for central cholinergic neurons which extend considerably beyond our knowledge of definite cholinergic pathways and their specific role. They are based upon general observations of animals (in most cases rats) following either systemic administration of cholinergic drugs, or local injection into the brain or cerebral ventricles. At this stage they can only be regarded as rough clues upon which more precise data might be built in the future.

Cholinergic mechanisms are involved in extrapyramidal function (Figures 13.21, 13.22, and 13.24, Section 13.5.4; and 13.25, Section 13.5.5). Anticholinergic agents help to counteract the rigidity of both idiopathic and drug-induced Parkinsonism. The cholinomimetic tremorine produces tremors, as the name implies.

Cholinergic mechanisms may play a role in the alerting process. Cholinergic drugs promote desynchronization of the EEG. Eserine will induce the effect and atropine will block it. ACh secretion from the cortex increases during paradoxical sleep (Celesia and Jasper 1966). Desynchronization can be induced by the local injection of ACh or carbachol into the area of the reticular activating system in the brainstem (Jouvet 1975).

Much investigation has taken place regarding the possible role of cholinergic neurons in the learning and memory process. The data are somewhat contradictory, but there seems to be a consensus that cholinergic stimulation facilitates learning, although higher doses of anticholinesterases seem to produce the reverse effect (Karczmar 1975). It has been speculated that this latter phenomenon is due to interference with the retrieval system rather than any difficulty with the acquisition, consolidation, or retention process.

Many other central roles, particularly ones involving behavior, have been proposed for cholinergic neurons, but until more is known about their precise anatomy, or more delicate pharmacological agents are available, these conclusions will remain highly speculative.

5.5 Pharmacology of ACh

The armamentarium of pharmacological agents available to interact with ACh is rather uneven. For example, an incredible number of AChE inhibitors exists, but there are no really effective CAT inhibitors. No specific agents have been found for inhibiting ACh binding in vesicles or promoting its release. Because of the widespread peripheral and central distribution of ACh, agents which dramatically disturb its functioning tend to have highly toxic effects. Some of the better known agents are listed in Table 5.6, with their site of action shown in Figure 5.8.

5.5.1 Nicotinic Antagonists

Most such agents are neuromuscular blockers and have a devastating physiological effect. D-Tubocurarine, the chemical warfare agent discovered by primitive South American Indian tribes, is the classic agent. It stabilizes the receptor by combining with cholinoceptive sites at the postjunctional membrane, thereby blocking the transmitter action of ACh. The muscles will no longer respond to applied ACh or to nerve stimulation. They will respond,

TABLE 5.6
Representative Drugs Interacting with Cholinergic Systems

Drug	Mechanism of action
D-Tubocurarine	Neuromuscular blocker
Gallamine	Neuromuscular blocker
Decamethonium	Neuromuscular depolarizer
Succinylcholine	Neuromuscular depolarizer
Tetraethylammonium (TEA)	Ganglionic blocker
Hexamethonium	Ganglionic depolarizer
Nicotine	Ganglionic and neuromuscular agonist
Dimethylphenylpiperazinium (DMPP)	Ganglionic agonist
Phenyltrimethylammonium (PTMA)	Neuromuscular agonist
Atropine	Parasympathetic blocker
Scopolamine	Parasympathetic blocker
Muscarine	Parasympathetic agonist
Methacholine	General cholinergic agonist
Arecoline	General cholinergic agonist
Carbachol	General cholinergic agonist
Oxotremorine	General cholinergic agonist
α-Bungarotoxin	Nicotonic binding agent
Quinuclidinyl benzilate	Muscarinic binding agent
Propylbenzilylcholine mustard	Muscarinic binding agent
Physostigmine (eserine)	Competitive cholinesterase inhibitor
Neostigmine	Competitive cholinesterase inhibitor
Diisopropylphosphofluoridate (DFP)	Irreversible cholinesterase inhibitor
Pyridine-2-aldoxime methiodide (2-PAM)	Regenerator of cholinesterase
4-Naphthylvinylpyridine	Choline acetyltransferase inhibitor
Hemicholinium-3 (HC-3)	Choline uptake inhibitor
Botulinus toxin	Inhibitor of acetylcholine release
Black widow spider venom	Promoter of acetylcholine release
Dihydro-β-erythroidine	Cholinergic antagonist at Renshaw cells

however, to the depolarizing action of K^+ ions or to direct electrical stimulation, indicating that the muscle cell itself is unaffected by the drug.

A decisive experiment was performed by Smith and associates in 1947 to establish that D-tubocurarine exerts almost no central effects. Smith permitted a dose 2.5 times that necessary to produce total muscular paralysis to be administered to him intravenously. Consciousness, mentation, memory, the EEG, pain threshold, general sensorium, and other central functions were completely normal, while the typical peripheral paralysis, including total loss of muscular control, inability to swallow or speak, and inability to breath without artificial respiratory support, were experienced. Such complete helplessness in the presence of total consciousness contributes to the feeling of terror reported by almost all individuals who have experienced the effects of large doses of neuromuscular blocking agents. D-Tubocurarine crosses the blood–brain barrier poorly, and in any event there are relatively few nicotinic receptors in brain.

Decamethonium and succinylcholine have similar effects to D-tubocurarine, but have a slightly different mechanism of action. They combine with the receptor in such a way as to leave it depolarized for a prolonged period. The action can be partially antagonized by D-tubocurarine but, unlike D-tubocurarine, it is not antagonized by anticholinesterase agents. Furthermore, K^+ ions cannot antagonize the block because the muscle fiber is already depolarized.

Hexamethonium represents yet another class of nicotinic blockers which act preferentially on ganglionic receptors. The effects can be predicted on the basis of knowledge as to which division of the autonomic system, sympathetic or parasympathetic, has dominant control of the smooth muscles. The most prominent result is a reduction in blood pressure, which has an important practical significance in the therapy of hypertension. Cardiac rate and cardiac output are also decreased. The tone and motility of the gastrointestinal tract are reduced and the bladder becomes atonic. Sweating is reduced, there is mydriasis, and a loss of accommodation. Although most ganglionic blocking agents penetrate the blood–brain barrier poorly, those that do also have little central effect. This again emphasizes the lack of dominance in the central nervous system of so-called nicotinic receptors. Dihydro-β-erythroidine is a nonquaternary blocking agent with a spectrum of action similar to D-tubocurarine.

5.5.2 Muscarinic Antagonists

In contrast to nicotinic antagonists, muscarinic blockers have central as well as peripheral actions. Atropine and scopolamine are the classic agents. They cause dilation of the pupil (mydriasis), decreased salivation, decreased secretions of the pharynx and respiratory tract, increased heart rate, decreased secretory activity, decreased motility of the gut, and decreased bladder tone. They also cause drowsiness, euphoria, amnesia, fatigue, and dreamless sleep. Higher doses result in excitement, restlessness, hallucinations, and delirium. They have a mild anti-Parkinsonian effect (Chapter 13), and reverse the tremor caused by the central cholinergic agonist tremorine. Methylscopolamine does not cross the blood–brain barrier and thus gives only the peripheral signs of muscarinic blockade.

5.5.3 Nicotinic Agonists

Nicotine is, of course, the principal agent. While it has no therapeutic application, its presence in tobacco testifies to its medical importance. Its action in the periphery is biphasic. Stimulation rapidly turns to blockade and paralysis of the neuromuscular junction as the dose increases. It causes tachycardia and vasoconstriction. Centrally, it produces tremors followed by convulsions.

Dimethylphenylpiperazinium (DMPP) and phenyltrimethylammonium

(PTMA) are two synthetic agents with a somewhat narrower spectrum of action.

5.5.4 Muscarinic Agonists

Muscarine, the active agent in the *Amanita* genus of mushrooms, has well-known physiological effects. They include salivation, sweating, miosis, dyspnea, abdominal pains, watery diarrhea, vertigo, confusion, weakness, coma, and, at sufficient doses, death. Pilocarpine and arecoline are two cholinomimetics with primarily muscarinic action.

Methacholine and carbachol are two ACh-like agents that are more slowly hydrolyzed by AChE than ACh. Their actions are also more heavily weighted on the muscarinic as opposed to the nicotinic side.

5.5.5 Synthesis Inhibitors

Cholinergic function can be impaired by either synthesis or release inhibitors. HC-3 limits synthesis by interfering with choline uptake. Most searches for specific inhibitors of CAT have been disappointing. Pyridine derivatives seem to be promising, with 4-(1-naphthylvinyl)pyridine being the most specific (Goldberg *et al.* 1971). They are much more effective *in vitro* than *in vivo*, suggesting that ample enzyme exists in tissues and that choline uptake is a more sensitive locus for chemical attack. Haloacetyl derivatives of ACh and acryloylcholine, an unsaturated ester that occurs naturally in various snails, are among a variety of other reported inhibitors (Haubrich 1976).

5.5.6 Release Inhibitors and Promoters

Botulinum toxin, the causative agent in certain types of food poisoning, inhibits the release of ACh from all types of nerve fibers. The mechanism is unknown. Death is by respiratory paralysis.

Black widow spider venom promotes the release of ACh. It is not a specific agent, however, and it induces morphological changes in synaptic vesicles.

5.5.7 Anticholinesterases

AChE inhibitors have a broad spectrum of action. The inhibitors are divided into two classes: reversible and irreversible. The outstanding representative of the reversible group is physostigmine (eserine). The best known of the irreversible AChE inhibitors is diisopropylphosphofluoridate (DFP). This is one of a class of organophosphorus compounds which were discovered shortly before World War II. They were used initially as agricultural insecticides but later were developed as chemical warfare agents. Organophosphorus compounds react with AChE, phosphorylating the enzyme apparently on a serine OH group and releasing HF. The enzyme is thereby rendered inactive, unless the catalytic site is regenerated. This can be achieved through the use of pyridine-2-aldoxime methiodide (2-PAM).

The effects of AChE inhibitors are manifold. They cause marked miosis, watery nasal discharge, wheezing respiration due to bronchiolar secretions, nausea and vomiting, cramps and diarrhea, involuntary urination, excessive sweating, involuntary twitchings, and weakness. Centrally, the effects are slurred speech and confusion, followed by loss of reflexes, convulsions, coma, and death. The effects can be countered by huge doses of atropine and, in the case of DFP poisoning, by 2-PAM. Therapeutically, anti-AChE compounds have only three uses: myasthenia gravis, bladder atony, and glaucoma (for reviews, see Main 1976; Koelle 1970), although they have been tried in Huntington's disease.

5.6 Summary

ACh was the first compound proved to be a neurotransmitter and thus was responsible for establishing the concept of chemical transmission in the nervous system. Through elegant physiological and pharmacological experiments, now part of classical neurobiology, it was established that cholinergic neurons supplied all preganglionic fibers of the autonomic nervous system, all postganglionic fibers of the parasympathetic system, all fibers of the musculoskeletal system, and a few fibers of the postganglionic sympathetic system.

All components of the cholinergic system are widely distributed within the brain but little information so far exists regarding precise pathways. The septal-hippocampal and habenulo-interpeduncular tracts are cholinergic, and there are striatal and cortical cholinergic neurons, but many more pathways remain to be identified.

Two broad classes of cholinergic receptors exist: nicotinic and muscarinic. Nicotinic receptors are excitatory and operate by opening ionic channels. They are therefore ionotropic. Muscarinic receptors appear to be linked to the formation of cGMP and thus may be metabotropic. Membrane alterations are interpreted in relation to K^+ conductance changes. They are slow in onset and may be excitatory or inhibitory.

Nicotinic receptors include all ganglionic receptors, all receptors of the musculoskeletal system, and a few central receptors. Muscarinic ones include all smooth muscle and exocrine gland receptors of postganglionic parasympathetic fibers, and the overwhelming majority of brain receptors.

Synthesis of ACh is a relatively simple process involving AcCoA, produced in brain mitochondria, and choline, produced in liver. The synthetic enzyme CAT has been completely purified, antibodies to it produced in rabbits, and an immunohistochemical method developed for its localization in brain.

Regulation of ACh production is not well understood, but choline uptake into cholinergic nerve endings may be the limiting factor. The choline uptake inhibitor HC-3 is an effective blocker of ACh synthesis *in vivo* and *in vitro*. There is an excess of enzyme for normal rates of *in vivo* synthesis.

ACh is destroyed by the extraordinarily active enzyme AChE. This enzyme is found not only in cholinergic neurons, but also in cholinoceptive neurons such as the dopaminergic neurons of the substantia nigra. Thus, AChE is an unreliable marker of cholinergic neurons. Pseudocholinesterase, which exists in many organs of the body, including brain, can also destroy ACh.

Subcellular fractionation techniques have been developed for brain tissue using the cholinergic system as the model. Most of the ACh and CAT is found to be localized to nerve endings or synaptosomes. AChE, on the other hand, is associated primarily with pre- and postsynaptic membranes. Within the synaptosome, CAT is associated with the soluble material, while ACh is primarily located in storage vesicles.

Chemically, ACh is found in both bound and free states. These pools probably correspond with the cytoplasmic and vesicular fractions identified in synaptosomes. Newly synthesized ACh is preferentially released on stimulation.

Some progress has been made toward isolating cholinergic receptors by means of binding agents. The greatest success to date has been with the nicotinic receptor of the electroplax organ of the electric eel or electrophorus, to which the snake venom α-bungarotoxin binds tightly. The protein has been purified and injected into rabbits, and specific antibodies produced. These antibodies seem to attack the rabbit's own neuromuscular receptors. A paralysis develops, reminiscent of myasthenia gravis.

There are agents such as D-tubocurarine which specifically block the nicotinic receptor of the neuromuscular junction by stabilizing the membrane. Others, such as decamethonium, block the receptors by means of sustained depolarization.

Different antagonists, such as tetraethylammonium and hexamethonium, exist for the nicotinic receptor of ganglion cells.

Still others, such as atropine and scopolamine, block muscarinic receptors. These are the blockers which produce the greatest central effects.

All nondepolarizing blocking agents can be at least partially antagonized by AChE inhibitors. These are reversible and competitive, or, in the case of organophosphorus compounds such as DFP, irreversible. The AChE inhibitors themselves, after initial excitatory effects due to enhanced ACh action, can produce a blockade of receptors as a result of prolonged ACh depolarization.

A number of ACh mimetic agents exist, such as methacholine, arecoline, tremorine, and carbachol. They have varying degrees of central effects depending on their access to brain tissue.

Black widow spider venom promotes ACh release while botulinum toxin prevents it.

A number of physiological roles for central cholinergic neurons have been proposed which remain somewhat speculative. They include a role in promoting learning, stimulating thirst, increasing body temperature, promoting aggressiveness, desynchronizing the EEG in paradoxical sleep and in the alert state, and acting to balance dopaminergic tone in extrapyramidal function (Chapter 13).

The cholinergic neuron has been the most important one of all for establishing key neurobiological principles, but our understanding of it remains in a primitive state.

6

Putative Excitatory Neurons: Glutamate and Aspartate

6.1 Introduction

It is clear from Chapter 5 that cholinergic neurons could be responsible for only a small fraction of the ionotropic excitatory neurons of the brain. The transmitter is unknown for the vast majority of such cells. Glutamate and aspartate, however, are extremely promising candidates for such a role. They are present in high concentration in brain tissue, and possess the requisite physiological properties.

The excitatory effects of glutamate and aspartate on cerebral cortical cells were first demonstrated after direct application of intracarotid injections by Japanese workers studying epileptic phenomena (Okamoto 1951; Hayashi 1952).

Hayashi actually anticipated the iontophoretic method with his technique of injecting small volumes of glutamate and other substances directly into the cortex through a small metal tube, but he did not anticipate that glutamate (or GABA, which he was also investigating) would later be considered neurotransmitters. Subsequent iontophoretic work confirmed the excitatory potency of glutamate, but it was still dismissed as a serious neurotransmitter candidate because it did not have all the properties that were anticipated for such a material. It was not until the systematic studies of Krnjevic and Phillis (1963) on cortical cells that it was realized that most of the appropriate microphysiological criteria for neurotransmitter action were being met by glutamate and aspartate.

Chemical evidence for a neurotransmitter role for these excitatory amino acids was largely missing until very recent years because of the lack of any means of differentiating the "transmitter" amino acid pool from the larger proportion used in energy metabolism. A major breakthrough on the chemical side came when Wofsey et al. (1971) demonstrated high-affinity uptake of glutamic and aspartic acids into an unique population of synaptosomes from rat brain and spinal tissue. Even more convincing evidence has recently appeared relating this high-affinity uptake to specific neurons. Its decrease after certain lesions suggests the existence of major "glutamatergic" or "aspartatergic" tracts in mammalian brain. It seems probable that the next few years will see the rapid accumulation of evidence supporting the importance of these amino acids in many or perhaps most of the classical excitatory tracts in the brain.

6.2 Anatomy and Distribution

The anatomical criteria that need to be met for neurotransmitter status (Introduction to Part II) are: presence in the nervous system in reasonable concentration, uneven distribution, presence of high-affinity uptake, drop in concentration following lesions, and localization to specific neurons through histochemistry or immunohistochemistry. Glutamate and aspartate would appear to meet all but the last of these anatomical criteria.

While glutamate and aspartate are present in abundance (Table 6.1) they serve many roles in the central nervous system. Glutamic acid, in particular, is incorporated into proteins and peptides, is involved in fatty acid synthesis, contributes (along with glutamine) to the regulation of ammonia levels and the control of osmotic or anionic balance, serves as precursor for GABA and for various Krebs cycle intermediates, and is a constituent of at least two important co-factors (glutathione and folic acid). In view of these many roles, it is not surprising that L-glutamic acid should be the most plentiful amino acid in the adult CNS (Table 6.1). Brain has three to four times the concentration of glutamate as taurine, glutamine, or aspartate, the three amino acids that are next most abundant.

The distribution of glutamate and aspartate in various brain areas is shown in Table 6.2 for human postmortem tissue (Perry *et al.* 1971a). In the areas measured, there is little more than a twofold variation in either sub-

TABLE 6.1
Content of Some Amino Acids
in Whole Rat Brain
(in μmol/g wet wt.)

Serine	1.4 ± 0.1
Glutamine	4.4 ± 0.2
Aspartate	3.7 ± 0.2
Glutamate	13.6 ± 0.4
Glycine	1.7 ± 0.1
Alanine	1.1 ± 0.1
GABA	2.3 ± 0.1
Lysine	0.4 ± 0.0
Taurine	4.8 ± 0.3

TABLE 6.2
Glutamine, Glutamate, and Aspartate Concentrations
in Human Brain (in μmol/g wet wt.)[a]

Area	Glutamate	Glutamine	Aspartate
Frontal cortex	8.93	3.84	1.92
Temporal cortex	10.78	4.41	1.89
Occipital cortex	8.64	3.75	2.56
Cerebellum	9.64	5.17	2.29
Corpus callosum	5.54	2.56	1.16
Thalamus	8.37	3.70	2.48
Putamen/globus pallidus	10.54	3.72	1.84
Caudate	10.48	3.18	1.30
Amygdala	9.36	4.65	1.73
Substantia nigra	5.08	2.65	2.59
Red nucleus	4.95	2.38	1.96
Hypothalamus	5.77	3.23	2.26

[a] From Perry *et al.* 1971a.

stance. This is weak evidence for specific pathways, but would nevertheless be anticipated if the amino acids were serving generally as excitatory transmitters as well as filling other metabolic roles.

An exception is the case of the spinal cord where L-glutamate is more concentrated in the dorsal than in the ventral roots (Table 6.3). It is also more concentrated in the dorsal than in the ventral medulla. These observations led to a proposal by Graham *et al.* (1967) that glutamate might be the excitatory transmitter released by terminals of some primary somatic afferent fibers (see also Rizzoli 1968; Johnston 1976; Johnson 1977). Aspartic acid is particularly concentrated in the ventral gray (Table 6.3; Graham *et al.* 1967), and the suggestion has been made that aspartate is the excitatory transmitter of certain spinal interneurons (Davidoff *et al.* 1967). Aortic occlusion causes a loss of these spinal interneurons but not primary afferents. Following such treatment

TABLE 6.3
Concentration of Aspartate, Glutamate, and
Glutamine in Cat Spinal Cord (μmol/g)

Area	Aspartate	Glutamate	Glutamine
Sensory nerve	0.7	1.8	1.8
Dorsal root ganglion	2.3–4.4	3.2–4.5	5.1
Dorsal root	0.7–1.5	2.8–4.6	0.9–2.5
Dorsal white	1.1–1.9	3.7–4.8	3.6–4.0
Dorsal gray	2.1–3.9	5.9–6.5	5.3–5.5
Ventral white	1.3–2.5	3.3–3.9	3.7–3.8
Ventral gray	3.1–4.8	4.4–5.3	4.8–5.4
Ventral root	0.7–1.3	1.7–3.1	0.7–1.9
Motor nerve	0.9	2.0	2.2

there is a loss of aspartate and glycine (Chapter 7) but not GABA or glutamate in gray matter of the cord.

The demonstration of a specific, high-affinity, sodium-dependent uptake system for glutamate and aspartate in nerve endings (Wofsey *et al.* 1971) represented a major advance. It implied that nerve endings could accumulate these amino acids selectively.

The high-affinity uptake system is distinct from those accumulating GABA or glycine, both on the basis of competitive uptake and subcellular fractionation studies. Accumulation appears to be into a population of synaptosomes in the cerebral cortex which are somewhat heavier than those accumulating GABA (Wofsey *et al.* 1971; Wenk *et al.* 1976). The existence of a high-affinity uptake into glial cells, which so far seems indistinguishable from that into synaptosomes (Balcar *et al.* 1977; Stewart *et al.* 1976), however, means that high-affinity uptake cannot be taken as a definite marker of excitatory amino acid neurons in the absence of other data. These other data have been supplied in certain cases, providing the strongest evidence so far of definite glutamatergic and aspartatergic pathways. Lesioning of specific brain pathways has resulted in a drop in this high-affinity uptake in the suspected terminal areas.

TABLE 6.4
Some Proposed Glutamate and/or Aspartate Pathways

Pathway	Proposed transmitter	References
Corticostriate	Glutamate	Divac *et al.* 1977; P. L. McGeer *et al.* 1978
Entorhinal-hippocampal	Glutamate	Nadler *et al.* 1976; Storm-Mathisen 1977
Retinotectal	Glutamate	Henke 1976a
Cerebellar granule cells	Glutamate	Young *et al.* 1974a
Primary sensory afferents	Glutamate	
Spinal interneurons	Aspartate	
Hippocampal commisural	Aspartate	Nadler *et al.* 1976
Hippocampal ipsilateral	Aspartate and/or glutamate	Storm-Mathisen 1977; Taxt *et al.* 1977
Dentatohippocampal (mossy fibers)	Aspartate and/or glutamate	Taxt *et al.* 1977
Hippocampo/subiculoseptal/ mammillary	Aspartate and/or glutamate	Storm-Mathisen 1978
Corticocortical	Aspartate and/or glutamate	Reiffenstein and Neal 1974
Olfactory bulb–olfactory cortex	Aspartate and/or glutamate	Harvey *et al.* 1975; Bradford and Richards 1976
Visual corticotectal	Aspartate and/or glutamate	Lund-Karlsen and Fonnum 1978
Visual corticogeniculate	Aspartate and/or glutamate	
Corticothalamic	Aspartate and/or glutamate	Fonnum 1978

Some pathways proposed to date are shown in Table 6.4. They are highly tentative of course, but they serve as a base upon which future data may be built. For glutamate they include the corticostriate, entorhinal cortex-hippocampal, retinotectal, and primary sensory afferent pathways as well as cerebellar granule cells. For aspartate they include hippocampal commissural fibers and spinal interneurons. Other possible aspartate and/or glutamate pathways are ipsilateral hippocampal, dentatohippocampal, hippocampo-subiculomammillary, corticocortical, corticothalamic, visual corticotectal, and olfactory bulb/olfactory cortex.

P. L. McGeer et al. (1978) and Divac et al. (1977) have both lesioned the massive, excitatory corticostriate pathway (Figure 13.20, Section 13.5.2). These groups either ablated or undercut the cortex on one side and found a drop of 40%–50% in high-affinity glutamate uptake into the synaptosomal fraction of the ipsilateral striatum; GABA and dopamine uptake were not affected. Both laboratories suggested that the input was probably glutamatergic rather than aspartatergic on the basis of the relative concentration of the amino acids (Table 6.2), the reported receptor binding (Table 6.5), and physiological studies in the striatum.

Since the glutamate and aspartate high-affinity uptake systems are highly similar, high-affinity uptake by itself, or after lesioning, cannot be used as a method for distinguishing between glutamate and aspartate pathways. Other methods must be found. In the case of the corticostriate pathway, the concentration of glutamate is 5.8 times as high as aspartate in the striatum, (Table 6.2) as opposed to only 3.7 for whole brain (Table 6.1). The rigid analog of glutamate, kainic acid (Figure 6.5), which is thought to bind relatively specifically to glutamate receptors, has higher binding in the striatum than any other area (Table 6.5). Diethyl glutamate antagonizes both the excitation of striatal neurons induced by iontophoresis of glutamate and that induced by cortical stimulation (Spencer 1976). As discussed later, however, the specificity of diethyl glutamate as a physiological antagonist of glutamate has been questioned (Clarke and Straughan 1977).

Figure 6.1 shows rat neostriatal terminals labeled by axoplasmic transport (Figure 1.14, Section 1.2.4) following the administration of [3H]proline to the neocortex. The terminals are presumably glutamatergic and show the common round vesicles and asymmetric contacts typical of type 1 excitatory synapses (Figures 1.4B and 1.5, Section 1.2.2).

Hippocampal slices also demonstrate high-affinity uptake of glutamate and aspartate. In this region terminal fields of afferents have distinct localizations. The autoradiographic distribution of glutamate and aspartate uptake are alike and highly suggestive of a localization in the axon terminals of pyramidal cells on the same (ipsilateral) and contralateral (commissural) sides, as well as in terminals of granular cells (mossy fibers) and afferents from the entorhinal cortex (perforant path) (Figures 6.2; 15.14, Section 15.6). Lesion of the ipsilateral plus commissural pyramidal cell axons resulted in 85% loss of glutamate and aspartate uptake in the target hippocampal area, most of which was due to the ipsilateral fibers, and to a 40% loss of endogenous

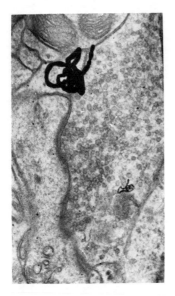

Figure 6.1 Presumed glutamatergic nerve ending in the rat caudate–putamen. [³H]Proline was injected into the cerebral cortex and transported by axoplasmic flow to the caudate–putamen. The nerve ending contains two silver grains indicative of transport of radiolabel. Note the small round vesicles and asymmetrical synaptic contact characteristic of excitatory synapses. (Electron micrograph courtesy of T. Hattori, University of British Columbia.)

glutamate and aspartate. Perforant path lesions led to a 52% loss of glutamate uptake from the target zone in the area dentata (Storm-Mathisen 1977). Both the perforant path and the pyramidal axons account for a very large proportion of the terminals in their respective target zones. Since the normal level of uptake was about three times higher in the latter than in the former target zone, the avidity of acidic amino acid uptake must be much greater in the pyramidal than in the perforant path axon terminals. If the two pathways had shared terminal fields the uptake in the perforant path would therefore have been difficult to detect. Electron microscopic autoradiography demonstrated glutamate uptake in mossy fiber terminals as well as in small terminals in the pyramidal and perforant path terminal fields (Taxt *et al.* 1977; Storm-Mathisen 1978) (Figure 6.2).

Calcium-dependent release of endogenous aspartate and glutamate was measured on slices from lesioned and control animals. With perforant pathway lesions, glutamate but not aspartate release was affected, but the opposite was the case with commissural lesions. Thus, the perforant pathway could be glutamatergic while the commissural pathway could be aspartatergic.

A third region demonstrating high-affinity uptake of glutamate and aspartate is the optic tectum (Henke *et al.* 1976a). Here synaptosomal localization has also been shown by autoradiography (Beart 1976), with approximately 11%–15% of the synaptosomes appearing to be labeled. The uptake is reported to be reduced after retinal ablation in both the chick (Bondy and Purdy 1977) and the pigeon (Henke *et al.* 1976a). On this basis it has been suggested that the retinotectal pathway is glutamatergic, although the data have not been confirmed (Beart 1976). Further evidence that has been cited is that the tectal terminals labeled by fast axoplasmic transport after intraocular injection of [³H]proline (Chapter 1) have sedimentation characteristics identical to those of the particles accumulating glutamic acid in uptake exper-

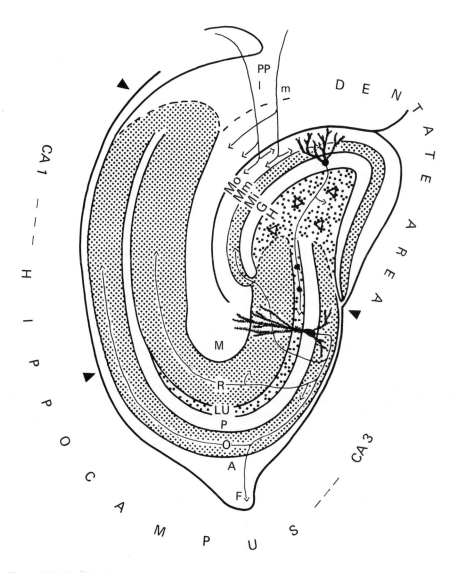

Figure 6.2 Possible glutamatergic or aspartatergic neurons in the hippocampal formation. Fine stippling (R, O, Mi) indicates zones with highest glutamate- and aspartate-uptake activity, which are the same as those receiving axon terminals from pyramidal cells (layer P) in hippocampus CA1 and CA3. Coarse stippling (LU, H) indicates labeled mossy fiber boutons, i.e., axon terminals of granular cells (layer G) of area dentata. Perforant path (PP) fibers from the lateral (l) and medial (m) parts of the entorhinal area terminate in the superficial molecular layers of area dentata (Mo, Mm) and hippocampus (M), which have a moderate uptake activity for glutamate and aspartate. (From Storm-Mathisen 1978.)

iments (Henke *et al.* 1976b). Another possible cortical projection pathway that is glutamatergic is to the thalamus (Fonnum 1978).

Harvey *et al.* (1975) reported a reduction in glutamate and aspartate uptake in the olfactory cortex after removal of the olfactory bulbs. Bradford and Richards (1976) found enhanced release of glutamate on stimulation of bulbocortical fibers *in vitro*. These data also argue in favor of glutamate and/or aspartate fibers serving the pathway from the olfactory bulb to the olfactory cortex.

An association of glutamate with cerebellar granule cells has been suggested by mutant and viral studies which selectively deplete these cerebellar interneurons and reduce high-affinity glutamate uptake. "Staggerer" and "weaver" mutant mice that have granule cell dysgenesis show lowered glutamate levels (McBride *et al.* 1976; Hudson *et al.* 1976). Viral infections in hamsters that destroy these cells also bring about reduced endogenous glutamate as well as reduced high-affinity glutamate uptake (Young *et al.* 1974a).

The data with respect to spinal cord are less definite. The accumulation of glutamate by the high-affinity uptake system is some twofold greater in the dorsal roots than in the ventral roots of spinal cord (Roberts and Keen 1973), which is in accord with the difference in endogenous levels reported in Table 6.3. The failure of Roberts and Keen (1974), however, to find any correlation between the loss of primary afferent fibers after dorsal root section and the uptake of glutamate into a synaptosomal preparation from the spinal cord argues against the suggestion that some of these afferent fibers may have a glutamatergic component.

The original physiological studies on the effects of glutamate were on cortical neurons, and it would be presumed on the basis of these studies that some cortical cells receive glutamate fibers.

From electron microscopic studies of the uptake of glutamate by synaptosomal preparations of rat cerebral cortex, it has been estimated that some 14%–15% of the terminals take up this amino acid by a high-affinity uptake process (Beart 1976a). In comparison, the proportions of terminals in such cerebral homogenates labeled by exogenous noradrenaline or GABA have been estimated as <1% and 30%, respectively. P. L. McGeer *et al.* (1978) found that glutamate uptake in the cortex was reduced after undercutting, but this is not definitive evidence of afferents to the cortex.

While the anatomical data at this stage must still be regarded as highly tentative, it can be said that glutamate and aspartate meet all the generally accepted anatomical criteria for neurotransmitter status except histochemical localization to specific neuronal pathways.

6.3 Chemistry

Both glutamic and aspartic acids are nonessential amino acids which are synthesized from glucose and other precursors by means of the tricarboxylic acid cycle (Figure 6.3).

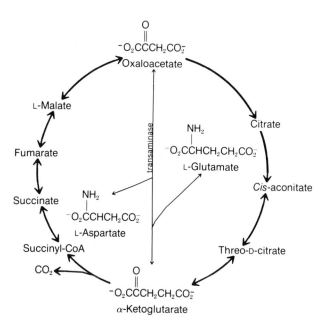

Figure 6.3 Intermediates of the Krebs cycle showing the formation of L-glutamate and L-aspartate by transamination reactions.

Synthesis of the corresponding molecules, α-ketoglutarate and oxaloacetate, takes place in mitochondria in all tissues. Brain is no exception in this regard. The amino acids are formed by transamination as shown by equations (1) and (2):

$$\underset{\text{glutamate}}{\text{HOOCCH}_2\text{CH}_2\overset{\overset{\text{NH}_2}{|}}{\text{C}}\text{HCOOH}} + \text{RCOCOOH} \xrightarrow[\text{B}_6]{\text{transaminase}}$$

$$\underset{\alpha\text{-ketoglutarate}}{\text{HOOCCH}_2\text{CH}_2\text{COCOOH}} + \text{R}\overset{\overset{\text{NH}_2}{|}}{\text{C}}\text{HCOOH} \qquad (1)$$

$$\underset{\text{aspartate}}{\text{HOOCCH}_2\overset{\overset{\text{NH}_2}{|}}{\text{C}}\text{HCOOH}} + \text{RCOCOOH} \xrightarrow[\text{B}_6]{\text{transaminase}}$$

$$\underset{\alpha\text{-ketosuccinate}}{\text{HOOCCH}_2\text{COCOOH}} + \text{R}\overset{\overset{\text{NH}_2}{|}}{\text{C}}\text{HCOOH} \qquad (2)$$

As will be discussed in Chapter 7, glial cells have a very active synthetic system for converting glutamate into glutamine (equation 3). The key enzyme is glutamine synthetase. The glutamine formed by this process can be transferred to neurons, and appears to be a reservoir for glutamate in this location.

$$\underset{\text{glutamate}}{\text{HOOCCH}_2\text{CH}_2\overset{\text{NH}_2}{\overset{|}{\text{CH}}}\text{COOH}} + \text{NH}_3 + \text{ATP} \xrightarrow[\text{synthetase}]{\text{glutamine}}$$

$$\underset{\text{glutamine}}{\text{H}_2\text{NOCCH}_2\text{CH}_2\overset{\text{NH}_2}{\overset{|}{\text{CH}}}\text{COOH}} + \text{ADP} + \text{P}_i \qquad (3)$$

Some neurons have an extremely active system for reconverting glutamine to glutamate by means of the enzyme glutaminase (equation 4):

$$\underset{\text{glutamine}}{\text{H}_2\text{NOCCH}_2\text{CH}_2\overset{\text{NH}_2}{\overset{|}{\text{CH}}}\text{COOH}} + \text{H}_2\text{O} \xrightarrow{\text{glutaminase}}$$

$$\underset{\text{glutamate}}{\text{HOOCCH}_2\text{CH}_2\overset{\text{NH}_2}{\overset{|}{\text{CH}}}\text{COOH}} + \text{NH}_3 \qquad (4)$$

Whether or not this occurs in glutamatergic neurons remains to be established.

Glutamate, of course, could also be formed from α-ketoglutarate by transamination (equation 1) or by nitrogen fixation (equation 5) using the enzyme glutamic acid dehydrogenase.

$$\underset{\alpha\text{-ketoglutarate}}{\text{HOOCCH}_2\text{CH}_2\text{COCOOH}} + \text{NH}_4^+ + \text{DPNH} \xrightarrow[\text{dehydrogenase}]{\text{glutamic acid}}$$

$$\underset{\text{glutamate}}{\text{HOOCCH}_2\text{CH}_2\overset{\text{NH}_2}{\overset{|}{\text{CH}}}\text{COOH}} + \text{DPN}^+ + \text{H}_2\text{O} \qquad (5)$$

The scheme shown in Figure 6.4 is the one currently believed to apply to glutamate nerve endings. In the figure, free glutamate can be formed from either glutamine or α-ketoglutarate. It is then bound, and released upon stimulation. Glutamate may be recaptured by the nerve ending or taken up by glial cells. In glial cells, it is recycled into glutamine by glutamine synthetase, and then transferred back to the nerve ending.

Aspartate is apparently synthesized by the transamination route, with glutamate being an indirect precursor.

Beyond synthesis, all that would be required for a glutamatergic or aspartatergic neuron to function would be the existence of appropriate vesicular storage and uptake mechanisms. The latter, at least, appears to exist in all areas of brain tested.

Several chemical compartments or pools have been identified in brain for both glutamate and aspartate through metabolic studies (Van den Berg et al. 1975). Unfortunately, it has not been possible so far to associate these pools with distinct functioning of glutamatergic or aspartatergic nerve endings. Nevertheless, these amino acids meet the chemical criteria for neurotransmit-

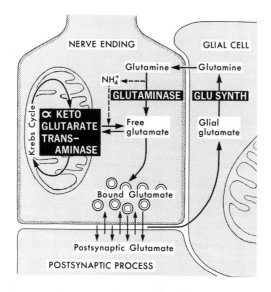

Figure 6.4 Schematic diagram of glutamate nerve ending.*

ters, with some exceptions. No immunohistochemical localization of enzymes has been reported, and neither vesicular binding nor axoplasmic flow has been demonstrated.

6.4 Physiology of Glutamate and Aspartate

The most convincing evidence that L-glutamate and L-aspartate should be neurotransmitters comes from their iontophoretic actions. Both these dicarboxylic amino acids powerfully excite virtually all neurons with which they come in contact. The characteristics of glutamate excitation are its extraordinary sensitivity (under optimal conditions cells can be excited with as little as 10^{-15} mol), its instantaneous onset, and the rapid termination of its action. This is accompanied by a marked fall in membrane resistance and an increase in sodium and other ion permeability comparable to that observed when ACh is applied near the muscle end plate (Figure 3.3, Section 3.2.3) (Krnjevic 1974). The action is not blocked by tetrodotoxin so that it is a specific excitation of receptor sites (Figure 3.16, Section 3.4) and not a stimulation of ionic conductance as in Figure 2.7 (Section 2.3). Therefore, it cannot be mediated by the sodium channels responsible for the propagated action potential (Zieglgansberger and Puil 1973). In most respects L-aspartate has an iontophoretic action comparable to glutamate, although it is generally less potent.

As mentioned in Chapter 4, significant changes in the internal concentration of cations cannot be produced by electrophoretic injection. Thus, it has

*Recent work does not support the presence of appreciable glutaminase in glutamate nerve endings [McGeer and McGeer, *Brain Res. 166* 431–435 (1979)].

not been possible to test for the ionic mechanisms of excitatory synaptic transmission of putative synaptic excitants in the way that was done by the Takeuchis for neuromuscular transmission (Figures 3.11 and 3.12, Section 3.2.9). The testing of the ionic mechanisms operated by putative synaptic excitants must await the development of tissue slice techniques with mammalian brain.

Nevertheless, L-glutamate and L-aspartate meet the basic physiological requirement of ionotropic transmitter action in that they duplicate the effects of stimulation of most excitatory CNS pathways. There may be many excitatory transmitters discovered in the future, but our present understanding of the mechanism would not require that there be a large variety: glutamate and aspartate could suffice. Thus, they could be the "workhorses" of excitatory ionotropic action, just as GABA and glycine (Chapter 7) appear to be the "workhorses" of inhibitory ionotropic action.

The other physiological criterion these substances need to meet to be considered neurotransmitters is release upon nerve stimulation. This is extremely hard to demonstrate in the CNS. Release has been shown *in vivo* from the surface of cat cerebral cortex, however, and is increased by brainstem stimulation (Jasper and Koyama 1969). Furthermore, many groups have demonstrated glutamate and aspartate release from slices or synaptosomes by electrical or chemical stimulation (Curtis and Johnston 1974).

It might be anticipated that neurons would show differing sensitivity to glutamate and aspartate if these amino acids were acting as transmitters of separate pathways. Some evidence exists, but it is still fragmentary. It is now recognized that glutamate is the probable transmitter in certain invertebrate synapses (Kravitz *et al.* 1970). *Aplysia* neurons have been reported to have specific aspartate receptors, specific glutamate receptors, and receptors that

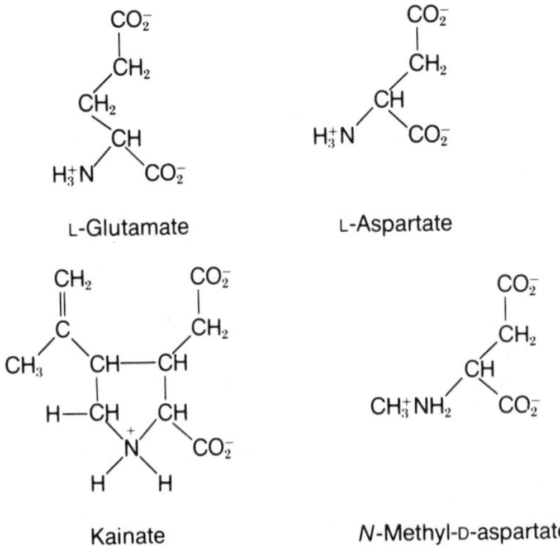

Figure 6.5 Structural formulae for L-glutamate, L-aspartate, kainate, and N-methyl-D-aspartate.

TABLE 6.5
Regional Distribution of Specific
[³H]Kainic Acid Binding in Rat Brain[a]

Area	Cpm/mg protein
Striatum	273 ± 9
Hippocampus	151 ± 8
Cerebral cortex	141 ± 13
Cerebellum	106 ± 11
Thalamus	74 ± 8
Midbrain	52 ± 8
Pons/medulla	48 ± 7

[a] From Simon *et al.* 1976.

respond to both amino acids. In this primitive mollusk, aspartate excites some cells and hyperpolarizes others (Yarowsky and Carpenter 1976). In the CNS, differential sensitivity of certain selected neurons to glutamate and aspartate has also been reported (Morgan *et al.* 1972; Krnjevic 1974), but the data are not substantial.

A promising approach is through receptor binding. It is possible that this type of binding can be used to distinguish between aspartate and glutamate "receptors" (De Robertis and de Plazas 1976) as well as to estimate the relative amounts of such receptors in various brain regions (Simon *et al.* 1976). This binding differs from synaptosomal or glial uptake since it is not sodium dependent and is inhibited by drugs which do not inhibit the sodium-dependent high-affinity uptake. It can also be demonstrated in tissue after freezing and thawing while high-affinity uptake requires fresh preparations with intact synaptosomes and glia. Since glutamate and aspartate are flexible molecules both of which can apparently fold into configurations suitable for either receptor, binding can best be done by using more rigid structural analogs of the two natural amino acids. De Robertis and de Plazas (1976) found that kainate (a more rigid analog of glutamate, Figure 6.5) inhibits binding of L[¹⁴C]glutamate. N-Methyl-D-aspartate on the other hand, preferentially inhibits L-[¹⁴C]aspartate binding.

A partial regional distribution of kainate binding indicates the widespread nature of supposed glutamate receptors but also indicates regional variation. For example, the striatum shows more than any other brain region studied (Table 6.5).

6.5 Pharmacology of Glutamate and Aspartate

Table 6.6 lists some of the pharmacological agents which interact with glutamate and aspartate, along with their presumed mechanism of action. They include agonists, antagonists, and uptake inhibitors. As can be seen from the table, the agents are relatively nonspecific as far as glutamate and

TABLE 6.6
Some Agents Affecting Glutamate and Aspartate Action

Agent	Effect
Folate	Uptake inhibitor
B-N-Oxalyl-L-α,β-diaminopropionate	Uptake inhibitor and agonist for glutamate
DL-Homocysteic acid	Agonist and uptake inhibitor
N-Methyl-D-aspartate	Agonist, possibly preferential for aspartate
Kainic acid	Agonist, possibly preferential for glutamate
N-Methyl-DL-glutamate	Agonist, possibly preferential for glutamate
Quisqualic acid	Agonist
Ibotenic acid	Agonist
Diethyl glutamate	Antagonist, possibly preferential for glutamate
DL-α-Methylglutamate	Antagonist, possibly preferential for glutamate
1-Hydroxy-3-aminopyrrolidone	Antagonist

aspartate are concerned. This is particularly true of antagonists. Highly specific blockers for glutamate and aspartate would be of inestimable value in judging whether these agents were true transmitters, as opposed to being neuronal excitants. Diethyl glutamate is a fairly general blocker, and while it may show some preference for glutamate as opposed to aspartate, its specificity has been questioned (Clarke and Straughan 1977).

Systemically administered L-glutamate (Mushahawa and Koeppe 1971) or L-aspartate (Johnston 1973) produces degeneration of cells in the retina and the hypothalamus in immature animals. Seizures may also occur in immature animals with somewhat higher doses of both glutamate (Johnston 1973; Stewart *et al.* 1972) and aspartate, and the same result is produced in mature animals with intraventricular administration (Crawford 1963). It seems likely that the depolarizing action of these excitatory amino acids on central neurons is the basis of their toxic effects (Olney 1976).

Microinjections of glutamic acid or other excitatory amino acids into various brain areas produce an acute reaction which selectively destroys certain neurons in the area. The order of potency of four excitatory amino acids—kainic acid > N-methyl-DL-aspartic acid > L-glutamic acid > DL-homocysteic acid—parallels their excitatory potency. Electron microscopic examination shows that there is a primary effect on dendrites and postsynaptic constituents, progressing rapidly to neuronal necrosis. There is little or no primary effect on presynaptic elements, and there is a total sparing of selected neurons within the lesioned area. These observations have both theoretical and practical implications as they suggest that central neurons can be "excited to death" when overstimulated (Olney *et al.* 1975). The materials themselves may serve as useful chemical lesioning agents. For example, kainic acid injections into the striatum cause a loss of neurons with presumed glutamate receptors and reproduce many of the biochemical findings of Huntington's chorea (E. G. McGeer and McGeer 1976a; Coyle and Schwarcz 1976) (Chapter 13).

Very little pharmacological data are available on glutamate or aspartate uptake inhibitors although they might be expected to have potent pharmacological effects. It has recently been shown, however, that β-N-oxalyl-L-α, β-diaminopropionic acid is a competitive inhibitor of the high-affinity uptake of glutamate into synaptosomes (Lakshmanan and Padmanaban 1974). This is a potent neurotoxin which can be isolated from the seeds of *Lathyrus satidus* and is implicated in the etiology of human neurolathyrism caused by eating the seeds. This compound produces convulsions in young rats when injected intraperitoneally and precipitates hindleg paralysis in monkeys when introduced into the CSF through the lumbar route.

It might be anticipated from these data that high doses of glutamate and aspartate might be toxic in humans, but there are no firm data to support this supposition. Caution should be exercised with infants, however, where the blood–brain barrier might not be fully developed. At one time, glutamate was described as an adjuvant in the control of petit mal and psychomotor epilepsy (Price *et al.* 1943) and, as with aspartic acid (Evans 1968), was supposed to increase performance efficiency and a sense of well-being in normal individuals. For a while, glutamic acid was used in the treatment of mental retardation (Zimmerman *et al.* 1946), and there was at least one report that it significantly increased maze learning in rats (Zimmerman and Ross 1944). After a number of years, however, resistance to the dosage of 25–60 g/day and the advent of control studies put an end to the popularity of glutamate as a treatment for mental deficiency (Lombard *et al.* 1955). No reports of neural toxicity from this glutamate therapy appeared, but Kwok (1968) described the strange sensations he incurred after eating in Chinese restaurants. This "Chinese Restaurant Syndrome" involves the production of burning sensations, facial pressure, and chest pains in individuals particularly sensitive to glutamate ingestion. The effects are probably at least partially peripheral.

Glutamate and aspartate both stimulate the formation of cAMP in guinea pig cerebral cortex (Shimizu *et al.* 1974, 1975) or cerebellar (Schmidt *et al.* 1977) slices. Glutamate also stimulates an increase in cAMP (Kinscherf *et al.* 1976) in all tissues studied except the cerebellum. Whether these increases are receptor linked is hard to estimate in the absence of specific blockers. Shimizu *et al.* (1975) report that α,ω-diamino-carboxylic acids inhibit the glutamate elicited accumulation of cAMP, but specificity has not been demonstrated.

6.6 Summary

Glutamate and aspartate are amino acids which meet many of the anatomical, chemical, physiological, and pharmacological criteria for ionotropic neurotransmitter status. They open sodium and potassium ionic channels and bring about a rapid, powerful excitatory response. Glutamate is the more abundant and powerful of the two dicarboxylic acids. They are both present in all brain cells because of their involvement in many metabolic roles. A high-affinity uptake system exists for them in nerve endings. On the basis of changes in this high-affinity uptake following lesions, a number of specific glutamate and aspartate pathways have been proposed. These include, for glutamate: cerebellar granule cells and the corticostriate, entorhinal-hippocampal, and retinotectal pathways; and for aspartate: hippocampal commissural fibers and spinal cord interneurons.

Few pharmacological agents have been discovered which specifically affect glutamate and aspartate action. This is a handicap in resolving some of the still undecided questions regarding their role. Powerful agonists such as kainic acid are neurotoxic, suggesting the interesting possibility that overexcitation of neurons can have deleterious effects.

7

Inhibitory Amino Acid Neurons: GABA and Glycine

7.1 Introduction

In Chapters 5 and 6 we discussed excitatory ionotropic transmitters. While ACh is well established in this regard, the cholinergic neurons described in Chapter 5 could account for only a small percentage of the total excitatory neurons in brain. It is possible, as suggested in Chapter 6, that many of the remainder are glutamatergic or aspartatergic. We must now consider what transmitters could represent the inhibitory ionotropic neurons, because these are probably as plentiful as the excitatory ones. GABA and glycine are proven inhibitory transmitters that might account for the majority of such cells. Both were ignored as potential transmitters even after their presence in nervous tissue had long been established. It is therefore instructive to recall what indications prompted the critical experiments which led to our appreciation of them as neurotransmitters.

Eugene Roberts and Jorge Awapara independently discovered the presence of GABA in brain tissue in 1950. Roberts, working with the then new technique of paper chromatography, came across a metabolite present in high concentration in brain which had not been found in other tissues. His summer project at Bar Harbor had been to study the amino acid content of tumors, including neuroblastomas, and his examination of normal brain was undertaken merely because a co-worker suggested he use this tissue as a control for the neuroblastomas. The newly discovered material turned out to be GABA, a compound then so obscure that only one source for it could be found in the United States.

Roberts completed his preliminary identification in St. Louis and journeyed to the Federation Meetings in Atlantic City in 1950 to report his findings. By chance, his roommate at that meeting turned out to be another young neurochemist, Jorge Awapara of Houston. Awapara had also been working on a cancer project and had been using the same paper chromatographic techniques. He too had examined brain tissue and had independently discovered the presence of GABA, although in his abstract he had not yet named the new substance.

The two young investigators realized they had an important new discovery in GABA but they had also each discovered high quantities of another amino acid in brain, taurine. It was decided at the meeting that Roberts should pursue work on GABA while Awapara should follow up on taurine. Their independent discoveries were reported in greater detail in the same issue of the *Journal of Biological Chemistry* (Roberts and Frankel, 1950; Awapara *et al.* 1950). As discussed in Chapter 11, only recently has taurine become of interest as a potential inhibitory transmitter.

In an entirely separate line of investigation, Florey (1953) discovered that extracts of mammalian brain had an inhibitory action on the slowly adapting neuron of the abdominal stretch receptor organ of the crayfish. He and his co-worker McLennan termed their substance Factor I. Bazemore, Elliott, and Florey, working together at McGill University, set out to identify the material. Matching chemical fractionation and bioassay techniques, they concluded that GABA was the material mainly responsible for the Factor I activity (Bazemore *et al.* 1957). Meanwhile, Kuffler and Edwards (1958), working at Johns Hopkins University, produced evidence that GABA duplicated the action of the physiological transmitter in one of the crayfish cord synapses, providing the first sound evidence that GABA might be an inhibitory neurotransmitter.

At the same time, studies were underway in a number of centers on the metabolism of GABA. Roberts and others discovered that the routes for the formation and destruction of GABA fitted in with a shunt of the well-known energy-producing Krebs cycle. The synthetic enzyme, glutamic acid decarboxylase (GAD), required vitamin B_6 (pyridoxine) as a co-factor. The concentration of GABA in brain, being in micromoles per gram of wet tissue, was far higher than for other known or suspected neurotransmitters, and was of the same order of magnitude as many energy-producing metabolic intermediates. But a special shunt for brain seemed out of place because it was less efficient than the Krebs cycle. This confusing state of affairs led neurochemist Heinrich Waelsch to remark at one stage that GABA was probably a metabolic waste basket.

Considerable scientific interest in GABA metabolism was aroused, however, by a rather tragic error in the preparation of a commercial babyfood. It was deficient in vitamin B_6, and several hundred children raised on the food developed convulsions. The seizures were promptly abolished by the administration of B_6 (pyridoxine). Although it was suspected that deficient production of GABA might have been responsible for the seizures, evidence for a relationship between deficient GABA and seizures was not to come until Killam and Bain (1957) reported that convulsant hydrazides, which have an anti-B_6 action, reduced GABA levels in brain.

The groundwork was thus laid for the first international conference on

GABA held at the City of Hope Medical Center in Duarte, California, in 1959 (Roberts 1960). Evidence of the inhibitory activity of GABA had been established, its metabolism worked out, and its relationship to epilepsy suggested. Interpretation of the findings presented at that historic conference were, however, cautious. There was doubt that GABA exactly duplicated the properties of the Factor I of Florey and McLennan. In mammalian spinal cord, GABA appeared to be a nonspecific neuronal depressant rather than an inhibitory neurotransmitter, and its role in epilepsy seemed doubtful. One of those attending remarked ruefully that GABA had gone into the conference as a rich neurotransmitter and had emerged as a poor metabolite.

Further critical experiments were required to establish more definitely the true role of GABA. Kravitz and colleagues (1963) showed that lobster inhibitory neurons contained far greater concentrations of GABA than excitatory ones; and that it was released in response to stimulation (Otsuka et al. 1966). Krnjevic and Schwartz (1966), using advanced microiontophoretic techniques, made a significant breakthrough by presenting evidence that GABA mimicked the activity of inhibitory neurons of the mammalian cerebral cortex. It increased the conductance to chloride and hyperpolarized the membrane potential. Meanwhile, Ito and Yoshida (1964), working on the neuronal connections of the cerebellum, showed that Purkinje cells displayed classical inhibitory action on cerebellar nuclear cells. On this basis, Obata et al. (1967) demonstrated first that GABA applied iontophoretically mimicked the inhibitory action of Purkinje cells, and that it was released into the fourth ventricle when Purkinje cells were stimulated (Obata and Takeda 1969). This constituted the first really convincing proof that GABA was an inhibitory neurotransmitter in mammalian brain. Confirming evidence of a different kind soon followed. Fonnum et al. (1970) demonstrated that cutting Purkinje axons resulted in a disappearance of glutamic acid decarboxylase, the synthesizing enzyme, from the terminals. Elegant final proof of GABA's role in Purkinje cells was developed by the Roberts team; they purified glutamic acid decarboxylase, developed antibodies against it, and showed by the immunohistochemical method that it was localized to Purkinje cell terminals (Saito et al. 1974a; McLaughlin et al. 1974). Finally, the movement of GABA by axoplasmic flow from Purkinje cell bodies to the terminals was shown by McGeer and colleagues (P. L. McGeer et al. 1975b).

By the time the second International Conference on GABA was held at Santa Ynez, California, in 1975 (Roberts et al. 1976), the role of GABA as an inhibitory transmitter in the mammalian central nervous system had been firmly established. Some information was available on specific GABA pathways in brain, and definite evidence of its deficiency in one disease state, Huntington's chorea, was presented. As might be expected, however, more questions were raised than answered. The metabolic role of GABA is still in doubt. Its relationship to convulsive states has not been clarified, and much remains to be learned about its physiological action and association with particular neuronal systems.

The discovery of glycine followed a different route—one that illustrates the advantages to be gained from close collaboration between neurochemists and neurophysiologists. Glycine is the simplest of the amino acids in structure, and is involved in a multitude of metabolic pathways. As a consequence, it was not seriously considered as a neurotransmitter candidate at the time its inhibitory neurophysiological properties were first noted. Purpura *et al.* (1959) had shown that it inhibited cortical electrical activity in cats when applied topically. Curtis *et al.* (1961), in their comprehensive survey of agents affecting spinal cord neurons, had also recognized the inhibitory properties of glycine but considered it only a weakly active agent.

From 1950 to 1965, leading workers in the field strongly stressed that a transmitter role was unlikely for dietary amino acids such as glycine. But Aprison, a neurochemist, and Werman, a neurophysiologist, seeking ways of collaboration at the Institute of Psychiatric Research at Indiana University, decided that a method for identifying potential new transmitters in the central nervous system was to determine the relative distribution of various amino acids in different areas. The first published study on spinal cord showed that glycine had the predicted distribution of the postsynaptic inhibitory transmitter (Aprison and Werman 1965). Davidoff, a postdoctoral fellow, and two graduate students, Graham and Shank, joined Aprison and Werman on their research program, and the group discovered the uneven distribution of GABA, aspartate, and glutamate within the spinal cord of the cat. This latter study was quickly followed by another which showed that after temporary aortic occlusion there was a significant loss of glycine and aspartate but not of GABA or glutamate in the gray matter of the spinal cord; moreover, the loss of glycine and aspartate correlated with the amount of interneuronal disruption seen in histological sections (Davidoff *et al.* 1967).

A rigorous neurophysiological comparison showed that glycine iontophoresed onto motoneurons duplicated the action of the inhibitory transmitter released on stimulation of these spinal interneurons. This was not the case for GABA, although it has since been shown that GAD-positive terminals exist on these motoneurons. It had been shown, however, that the convulsions caused by the well-known agent strychnine were due in part to blockade of the inhibitory actions of spinal interneurons on anterior horn cells. Strychnine was shown to have no effect on the action of GABA, but blocked the inhibition of glycine on the motor neurons (Werman *et al.* 1968; Curtis *et al.* 1967).

Confirmatory chemical evidence came when it was shown that there was a high-affinity uptake system for glycine in slices and homogenates of cat or rat spinal cord (Logan and Snyder 1971) and that glycine was preferentially localized to synaptosomes on subcellular fractionation studies (Arregni *et al.* 1972).

7.2 Anatomical Distribution of GABA Pathways in Brain

The brain content of GABA is 200- to 1000-fold greater than that of other neurotransmitters such as dopamine, noradrenaline, ACh, and serotonin. It is widely distributed, as might be anticipated for a transmitter suspected of serving the inhibitory interneurons found in almost all areas of brain. Uptake studies with labeled GABA suggest that 25%–45% of nerve endings, depending on the brain area, may contain this neurotransmitter. These facts suggest that a significant proportion of all CNS neurons may use GABA as their neurotransmitter, although it is quite possible that GABA is taken up in other than GABAnergic neurons. Where are these GABA neurons located?

Table 7.1 is a list of those neuronal systems currently considered to be GABA-containing, along with the techniques that have been utilized to establish their identity. It is based upon only partially adequate methodology and is obviously far from complete. It can be anticipated that there will be many additions and possibly some deletions in future years.

The cerebellar Purkinje cell (Figure 13.12, Section 13.4.3) is the classical example of a cell that has been proved to exert an inhibitory action by release of GABA. The evidence is the following:

1. Stimulation of the Purkinje cells results in hyperpolarization of postsynaptic cells in the deep cerebellar nuclei and Deiters nuclei. This hyperpolarization is reversed when the resting membrane potential of the postsynaptic cell is increased beyond the equilibrium potential for the IPSP.
2. The postsynaptic membrane is permeable to chloride and other anions of comparably small size during transmitter action.
3. Picrotoxin and bicuculline, which are classical GABA antagonists, block the effects of stimulation while strychnine, which is not a GABA blocker, is ineffective.
4. Iontophoretic application of GABA on single Deiters neurons mimics Purkinje cell stimulation. The effect is blocked by picrotoxin and bicuculline but not by strychnine.
5. There is a large difference in GABA concentration in different neuronal groups in Deiters nucleus. Those in the dorsal part, that receive nerve endings from Purkinje cell neurons, have much higher GABA concentrations than cells in the ventral part of the nucleus.
6. When the cerebellar cortex is removed by suction, or the axons undercut, there is a decrease in GABA and GAD content in the dorsal, but not the ventral part of Deiters nucleus.
7. Stimulation of Purkinje cells results in the release of detectable amounts of GABA into the perfusion fluid of the fourth ventricle or into the output fluid of a push–pull cannula inserted into the area of the deep cerebellar nuclei.
8. GAD has been detected by immunohistochemistry in Purkinje cell terminals in the deep cerebellar nuclei. This is the most convincing evidence of all.

TABLE 7.1
Proposed Neuronal Pathways Involving GABA

Anatomical System	Evidence
Purkinje cells	Inhibitory action,[a] GABA release,[b] pharmacological correlation,[a] lesion,[c] axoplasmic flow,[d] immunohistochemistry[e]
Cerebellar Golgi cells	Uptake,[f] inhibitory action,[a] immunohistochemistry[e]
Cerebellar basket cells	Uptake,[f] inhibitory action,[a] immunohistochemistry[e]
Cerebellar stellate cells	Uptake,[f] inhibitory action[a]
Hippocampal basket cells	Localization,[g] uptake,[h] inhibitory action,[i] lesion,[g] immunohistochemistry[q]
Neostriatal interneurons	Uptake,[j] lesion[k]
Striatonigral neurons	Lesion[l]
Pallidonigral neurons	Uptake,[j] lesion,[j] axoplasmic flow[m]
Retinal interneurons	Localization,[r] uptake[h]
Spinal cord interneurons	Uptake,[o] inhibitory action,[p] localization,[p] immunohistochemistry,[q] physiological action[r]
Cortical interneurons	Uptake,[s] inhibitory action[t]
Olfactory bulb interneurons	Immunohistochemistry,[u] physiological action[v]

[a] Ito and Yoshida 1964.
[b] Obata and Takeda 1969.
[c] Fonnum et al. 1970.
[d] E. G. McGeer et al. 1975.
[e] McLaughlin et al. 1974.
[f] Hokfelt and Ljungdahl 1972.
[g] Storm-Mathisen and Fonnum 1971.
[h] Iversen and Schon 1973.
[i] Anderson et al. 1964.
[j] Hattori et al. 1973.
[k] P. L. McGeer and McGeer 1975.

[l] Okada et al. 1971.
[m] P. L. McGeer et al. 1975a.
[n] Kuriyama et al. 1968.
[o] Iverson and Bloom 1972.
[p] Miyata and Otsuka 1972.
[q] Wood et al. 1976.
[r] Eccles et al. 1963.
[s] Bloom and Iversen 1971.
[t] Krnjevic and Schwartz 1966.
[u] Ribak et al. 1977a.
[v] Nicoll 1969.

Eugene Roberts and his colleagues purified GAD and, by injecting the purified protein into rabbits, developed monospecific antibodies against the protein. Lightly fixed slices of tissue were then treated with the anti-GAD rabbit serum, and the reaction product further treated with goat anti-rabbit serum which had been tagged with peroxidase. The anti-GAD rabbit serum became the "meat" in a sandwich and the upper layer, containing peroxidase, became a marker when it was reacted with hydrogen peroxide and diaminobenzidine (cf. Figure 5.3, Section 5.2). The brown reaction product could be visualized not only at the light, but also at the electron microscopic level (Saito et al. 1974a; McLaughlin et al. 1974). Figure 7.1 shows the localization of GAD in Purkinje cell terminals in the deep cerebellar nuclei which provides final proof that it is a GABA system.

When tritiated GABA was applied to the exposed cerebellar surface of rats in vivo and the animals killed one hour later, radioactive GABA was found

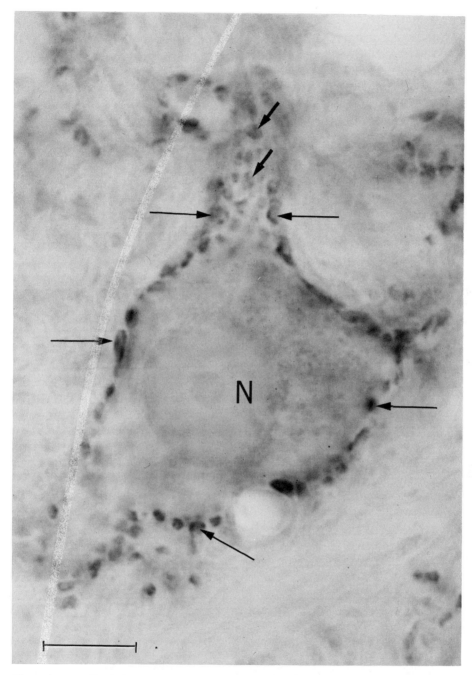

Figure 7.1 Localization of glutamic acid decarboxylase (GAD) in Purkinje cell terminals. A neuron (N) from a deep cerebellar nucleus treated with anti-GAD serum is surrounded by profiles of dense, punctate, GAD-positive Purkinje cell axon terminals (long arrows). Numerous GAD-positive terminals are demonstrated on the surface of one of the grazed neuronal processes (short arrows). Bar represents 10 μm. (From Barber *et al.* 1978.)

to be concentrated in the deep cerebellar nuclei. Electron microscopic autoradiograms of the deep nuclei showed the radioactivity to be primarily in nerve endings, indicating that the GABA was transported in the Purkinje cells by the process of axoplasmic flow (Figure 7.3).

7.2.1 Cerebellar Interneurons

Other cells in the cerebellum also appear to use GABA as the transmitter substance. The most obvious of these are the basket cells. Basket cells lie close to the Purkinje cell layer (Figure 13.12, Section 13.4.3). They are excited by granule cell parallel fibers just as are Purkinje cells. But the basket cells send their axons at right angles to the Purkinje cells, and act to inhibit rows of Purkinje cells which lie parallel to the row being primarily excited (Figure 13.14, Section 13.4.4). The basket cells accomplish this action by clustering their terminals in a pouchlike fashion around the axon hillocks of the Purkinje cells.

Figure 7.2 is a photomicrograph of a cerebellar folium using the definitive immunohistochemical method for GAD. It shows prominent staining in areas expected to have a concentration of nerve terminals of basket cells and other inhibitory Golgi type 2 interneurons. Radioactive GABA can also be used to mark GABAnergic neurons. When brain slices are incubated with [^{3}H]-GABA, the labeled material is pumped into cellular elements. The tissue can then be fixed, sliced, and prepared for light or electron microscopy. It is coated with a photographic film and left for several weeks. The radioactive molecules expose the film, creating an autoradiograph.

GABA is a particularly good material to work with in this regard because it reacts with the fixative to make an insoluble derivative which apparently remains at the original site. After brief incubations, the unchanged amino acid is accumulated without any significant amounts of labeled metabolites being found. Most of the activity under these circumstances is found in nerve terminals. The labeling is of an "all-or-none" type, i.e., some nerve terminals show evidence of several radioactive GABA molecules being contained within them while others are devoid of activity.

Figure 7.2 Localization of GAD in basket cell terminals and stellate cells. A. A cerebellar folium treated with anti-GAD serum showing a glomerulus (g) surrounded by punctate, GAD-positive terminals of Golgi II axons (small arrows). The Purkinje cell somata (P) are surrounded by GAD-positive basket cell axon terminals (white arrows). In the molecular layer there are numerous GAD-positive terminals (long arrows) associated with Purkinje cell dendrites (d) having the expected distribution of stellate cell axon terminals. Some stellate cell somata (s) appear to be outlined by GAD-positive product (arrow heads). Bar represents 10 μm. B. A cerebellar folium treated with nonimmune serum. No specific staining is observed. Purkinje cell layer (P). (From Barber *et al.* 1978.)

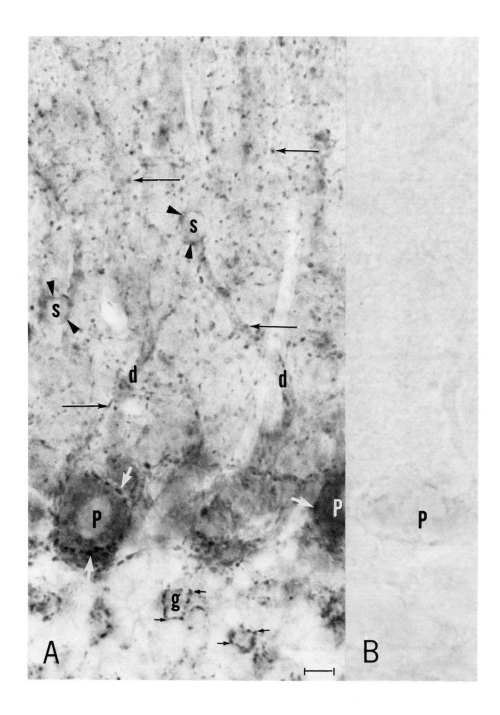

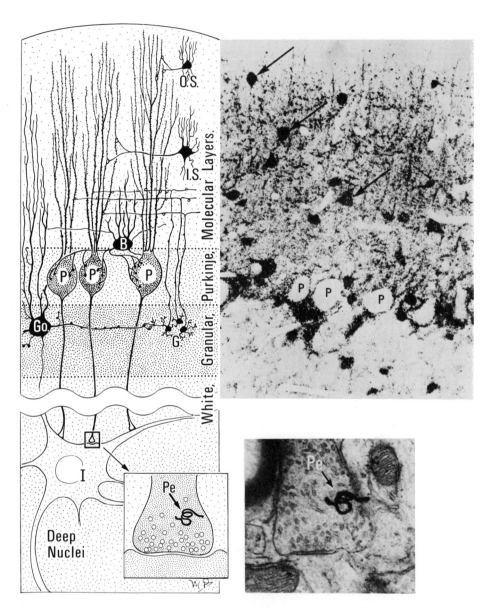

Figure 7.3 Uptake of [³H]-GABA into basket, stellate, and Golgi cells, and axonal transport of [³H]-GABA to Purkinje cell terminals in interpositus nucleus. *Left upper side:* Diagrammatic representation of cellular organization of cerebellar cortex. Layers delineated by dotted lines: molecular, Purkinje, granular layers, and white matter. Cells: O.S., outer stellate; I.S., inner stellate; B, basket; P, Purkinje; Go, Golgi; G, granule cells. Granule cells and their processes do not accumulate GABA. *Right upper side* (from Hokfelt and Ljungdahl 1972): Light microscopic autoradiograph of area comparable to that shown in the diagram following the injection of [³H]-GABA. Some cell bodies with high activity denoted by arrows. Note correspondence of cell bodies accumulating radioactivity with areas occupied by inner and outer stellate cells, basket, and Golgi cells. Note also dot and fiber-like accumulations along tracks of

Uptake studies in the cerebellum using radioactive GABA clearly show heavy labeling in the basket, stellate, and Golgi cell regions (Figure 7.3). The uptake in basket cell terminals seems to be so concentrated that the Purkinje cells themselves, which would normally be expected to concentrate GABA, appear relatively clear. Thus, in the cerebellum, only the granule cells are non-GABAnergic, and these appear to be glutamatergic (Table 6.4, Section 6.2).

7.2.2 Hippocampus

The hippocampus is a layered structure which in some respects parallels the cerebellum. The most prominent feature is a layer of large pyramidal cells lying in a sheet approximately 0.5 mm deep to the surface (Figure 15.14B, Section 15.6). Stimulation of the known inputs to the hippocampus, i.e., from the hippocampal commissure and the septum, as well as local stimulation, produce identical inhibitory effects (Figures 4.16A to 4.16C, Section 4.7.3). Such a physiological action could take place only if special inhibitory interneurons were excited by all three inputs and with these interneurons ramifying broadly to the pyramidal cells. Basket cells, lying just superficial to the pyramidal cells (i.e., in the region of the basal dendrites) fulfill these criteria. They are activated by all of the inputs to the hippocampus and discharge just prior to the onset of the inhibition of pyramidal cells. The basket cells are also driven by axon collaterals of pyramidal cells, which in turn inhibit surrounding pyramidal cells. The wide distribution of the ramifying axons of basket cells suggests that each one may inhibit hundreds of pyramidal cells.

Storm-Mathisen and Fonnum (1971, 1972) studied the GAD activity in the various layers of the hippocampus and found the activity to be highest in the area of the pyramidal cells. This GAD was not altered by lesioning any of the known hippocampal afferents and therefore had to be associated with interneurons. The high concentration in the molecular and pyramidal cell layers would be consistent with GAD being contained in basket cell terminals. Okada and Shimada (1975) confirmed this result by finding a high concentration of GABA in the pyramidal cell layer. In physiological studies GABA has been found to inhibit pyramidal cells in the hippocampus, an effect which is blocked by bicuculline. Iversen and Schon (1973) found 42% of the terminals were labeled when slices of hippocampus were incubated with [³H]-GABA, the highest percentage of terminals so labeled in the six areas of the brain that they studied. Finally, immunohistochemistry for GAD has

---

their processes. Purkinje cell bodies have low activity even though they use GABA as their neurotransmitter. Magnification × 500. *Left lower side:* Continuation of the diagram of Figure 7.3 to include neuron of the nucleus interpositus. *Center:* A 100-fold enlargement of the insert area. *Right lower side:* An actual electron microscopic autoradiograph of a Purkinje cell nerve ending (Pe) in the nucleus interpositus following injection of [³H]-GABA into the cerebellar cortex. Transport of GABA to the nerve ending was by axoplasmic flow *in vivo.* Arrow points to silver grain of exposed film. (From P. L. McGeer *et al.* 1975a.) Magnification × 54,000.

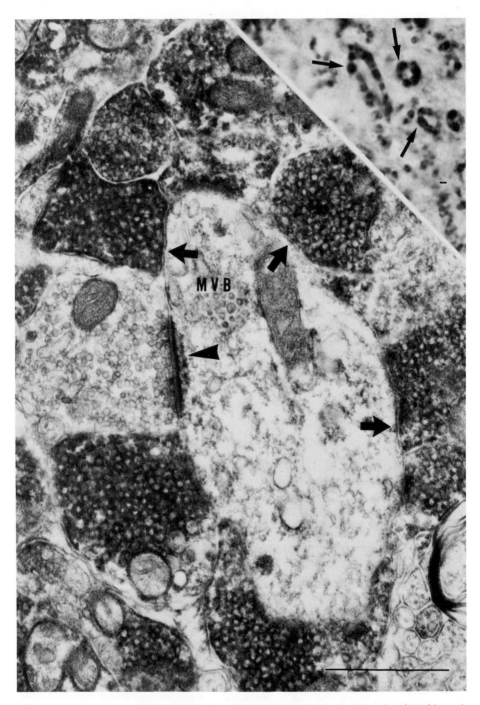

Figure 7.4 GAD-containing terminals in the substantia nigra from specimens incubated in anti-GAD serum. The inset shows a semithin (1 μm) section of the pars reticulata with obliquely and transversely sectioned dendrites which are encircled by punctate structures containing GAD-

shown localization to nerve terminals around the somata of pyramidal cells in keeping with the localization of basket cell terminals. All of these results argue strongly that the inhibitory basket cells of the hippocampus use GABA as their neurotransmitter agent.

7.2.3 Basal Ganglia

Some of the highest levels of GAD and GABA are found in the basal ganglia, particularly in the globus pallidus and substantia nigra. It might be expected, therefore, that GABA-containing neuronal pathways are associated with extrapyramidal function. Electrolytic lesions of the globus pallidus, hemitransections of the brain at the level of the subthalamus, or destruction of the striatum by suction produced significant decreases in GABA and GAD in the substantia nigra. This suggests that long GAD-containing neurons extend from the basal ganglia to the substantia nigra. The supposition that the substantia nigra has such GAD-containing nerve endings is greatly strengthened by experiments showing that GABA is avidly taken up by slices of substantia nigra and is highly localized to nerve endings. Many of these nerve endings have been shown by immunohistochemistry to contain GAD (Ribak *et al.* 1977b) (Figure 7.4).

Double labeling experiments have suggested that at least some of the descending GABA fibers originate in the globus pallidus and terminate on dendrites of dopaminergic neurons in the substantia nigra. The dopamine cells and their dendritic processes were first labeled by the intraventricular injection of 6-hydroxydopamine. [³H]Leucine was then injected into the globus pallidus and, by the process of axoplasmic flow, labeled protein was transported to substantia nigra nerve endings. Electron microscope autoradiography revealed that these nerve endings were making contact with the degenerating dendrites of the dopaminergic neurons (Figure 13.23, Section 13.5.4) (Hattori *et al.* 1975). [³H]-GABA was also shown to pass along this pathway (P. L. McGeer *et al.* 1974d). There is thus compelling evidence in favor of a descending pallidonigral GABA-containing pathway.

Several groups of investigators have shown in addition that there is a neuronal tract extending from the neostriatum to the pars reticulata of the substantia nigra. Lesions of the striatum have also been reported to cause decreases in the GAD and GABA content of the substantia nigra, although not nearly as large as with pallidal lesions or hemitransections. Thus, there is

---

positive reaction product (arrows). The accompanying electron micrograph shows an obliquely sectioned dendrite in the substantia nigra which is surrounded by many axon terminals filled with GAD-positive reaction product that are equivalent to the puncta observed in the inset. Some of these terminals form symmetric synapses (arrows), whereas the unstained terminal contains round synaptic vesicles and forms an asymmetric synapse (arrowhead) with this dendritic shaft. On the basis of previous studies, it is likely that many of the GAD-positive axon terminals in the substantia nigra arise from neurons located in the neostriatum and globus pallidus. MVB, multivesicular body. Each marker indicates 1 μm. (From Ribak *et al.* 1977b.)

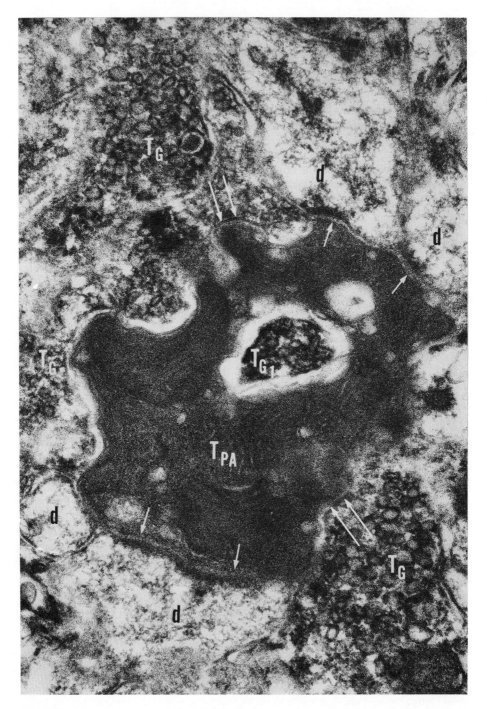

Figure 7.5 Immunocytochemical localization of the GABA-synthesizing enzyme glutamic acid decarboxylase in laminae II–III of rat lumbar spinal cord. Ipsilateral dorsal root lesions were performed 24 h prior to tissue preparation for immunocytochemistry. A dense, degenerating

evidence that this well-known striatonigral pathway may be at least partially GABA containing (Okada *et al.* 1971; Fonnum *et al.* 1974). It also has substance P fibers (Chapter 10).

7.2.4 Spinal Cord

Presynaptic inhibition in the spinal cord and its role in negative feedback control of sensory pathways has already been discussed (Chapter 4, Figure 4.18, Section 4.8). The spinal cord interneurons responsible for this effect are GABAnergic. Figure 7.5 shows GAD-containing nerve endings surrounding a primary afferent terminal. The dorsal root has been sectioned, permitting the afferent terminal to be identified because it is degenerating. The degenerating terminal is postsynaptic to GABAnergic terminals and presynaptic to surrounding dendrites.

7.2.5 Olfactory Bulb

The reciprocal dendritic inhibitory synapses of granule cell interneurons in the olfactory bulb have already been described (Chapter 4, Figure 4.13, Section 4.6). GAD has been identified in the granule cells and in the gemmules of their dendrites, establishing that these unusual inhibitory interneurons are GABAnergic (Ribak *et al.* 1977a).

7.2.6 Retina

GABA inhibits the firing of retinal ganglion cells (Noell 1959; Straschell and Perwein 1969). GABA and GAD have been detected in the retina but not in the optic nerve, iris, or ciliary body of the eye. The retina is a layered structure, and the content of the GABA system varies widely according to which layer is investigated. By far the highest levels are found in the areas near ganglion cells (Graham 1972). Since the optic nerve is deficient in GABA, it seems unlikely that the origin of the high activity could be the ganglion cells themselves, or their projections to the optic nerve. It is also unlikely that any excitatory interneurons would be GABA-containing since in the majority of systems GABA is inhibitory. Amacrine cells are appropriately located, however, and have the correct physiological action. They can form

---

primary afferent terminal (T_{PA}) is surrounded by GAD-positive axon terminals (T_G) and by dendritic profiles (d). One of the GAD-positive axon terminals (T_{G_1}) invaginates into the primary afferent terminal from a tissue level not included in this sectioning plane. Two of the GAD-positive terminals appear to be presynaptic (double arrows) to the degenerating primary afferent terminal. These axoaxonal synaptic relationships are consistent with the evidence that implicates GABA as one of the neurotransmitters which mediates presynaptic inhibition in the dorsal spinal cord. In addition to being postsynaptic to GAD-positive terminals, the primary afferent terminal is also presynaptic (single arrows) to some of the surrounding dendritic profiles. Magnification × 79,000 (Electron micrograph courtesy of James E. Vaughn, City of Hope National Medical Center, Duarte, California). (Barber *et al.* 1978.)

multiple synapses with ganglion and bipolar cells. They are morphologically similar to basket cells in the cerebellum. Uptake (Iversen and Schon 1973) and immunohistochemical (Wood *et al.* 1976) studies suggest one population of amacrine cells is probably GABA-containing; a second population may use glycine as the transmitter. Further work is needed to establish precisely the role played by GABA in the retina.

7.2.7 Other Systems

Further ideas regarding the localization of GABA-containing neurons can be gained from studies of the general distribution of GABA and GAD in brain as shown in Table 7.2. The table illustrates that these materials are present in relatively high concentrations in nearly all areas of gray matter.

An indication of how these quantities might be related to numbers of

TABLE 7.2
Distribution of GABA and GAD in Monkey and Human Brain

Area	GABA, μmol/g tissue	GAD, μmol/g/hr	
	Monkey[a]	Monkey[b]	Human, average age 53[c]
Cerebral cortex			
Frontal cortex	2.10	9.1	5.7
Motor cortex	2.09	11.2	6.2
Visual cortex	2.68	14.6	5.7
Limbic System			
Amygdala	—	9.1	3.5
Hippocampus	—	4.9	3.0
Septum	—	8.6	5.4
Extrapyramidal System			
Caudate	3.20	13.2	4.6
Putamen	3.62	13.0	4.8
Globus pallidus	9.54	47.9	8.3
Substantia nigra	9.70	27.2	9.3
Pars compacta	—	—	14.0
Diencephalon			
Hypothalamus	6.19	—	6.4
Anterior thalamus	2.50	—	4.6
Lateral thalamus	2.68	—	3.5
Midbrain			
Superior colliculus	4.19	—	4.5
Inferior colliculus	4.70	—	4.2
Cerebellum			
Cortex	2.03	8.1	6.0
Deep nuclei	4.30	8.7	6.6
White matter	0.31	0.43	0.1

[a]Fahn and Cote 1968b.
[b]Enna *et al.* 1975.
[c]P. L. McGeer and McGeer 1976.

TABLE 7.3
Localization of [³H]-GABA Uptake
in Nerve Terminals of Rat Brain[a]

Area	% total terminals labeled
In vitro	
Cerebral cortex	27
Cerebellum	
Molecular layer	14
Granular layer	46
Substantia nigra	51
In vivo	
Locus coeruleus	44
Hypothalamus	39
Caudate	27

[a]From Iversen and Schon 1973.

nerve endings is given in Table 7.3, which shows the percentage of nerve endings accumulating GABA following exposure to the labeled material. It must be remembered, however, that GABA uptake may not be completely specific. Therefore, the data of Table 7.3 should be regarded with caution.

More detailed distribution studies of GABA have been done on a micro basis in a few special areas of brain. The spiral ganglion cell layer of the inner ear contains high levels of GABA (Tachibana and Kuriyama 1974). This is in sharp contrast to the outer hair cell and inner hair cell layers, as well as the spiral ligament and vascular stria. Since spiral ganglion cells are excitatory in nature, there is speculation that the GABA may be due to interneurons in the area (Fisher and Davies 1976). Other areas containing particularly high levels of GABA in the brain are the dorsal part of the cochlear nucleus, the lateral hypothalamus, and the superficial gray layer of the superior colliculus. In each case it has been speculated that there are a large number of GABA neurons in the area.

7.3 Metabolism of GABA

The first experiments that were undertaken after the discovery of GABA in 1950 were to establish how it was formed and metabolized. Newly developed radiolabeling techniques greatly facilitated these studies which soon revealed that glutamic acid was the immediate precursor of GABA, and that glutamic acid could be formed from either glutamine or α-ketoglutarate. By incubating slices of brain with various other radiolabeled starting materials a whole complex of metabolic interrelationships involving GABA was revealed, the complete significance of which is not understood to this day.

Glucose, the main energy source of the brain, is a particularly efficient precursor of GABA and is probably the principal *in vivo* source. Pyruvate and several other amino acids can serve by entering the Krebs cycle. Eventually,

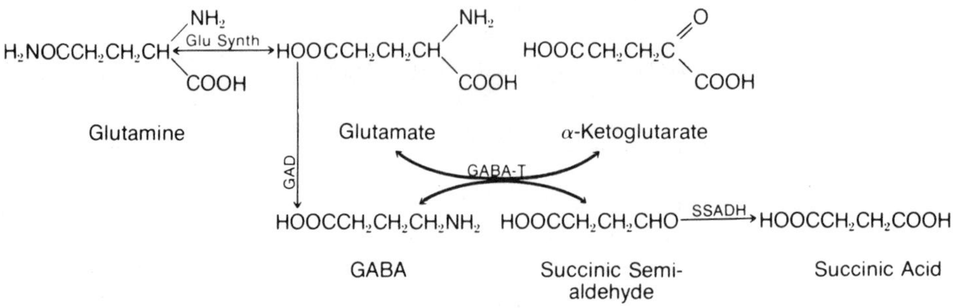

Figure 7.6 Schematic diagram showing the reactions of the GABA shunt.

the more remote precursors must be converted via α-ketoglutaric acid to glutamate or glutamine, which together constitute 60% of the free α-amino nitrogen in brain tissue. It is thought that one of their principal functions is to serve as a reservoir system for producing GABA.

Central to the system is the so-called GABA shunt (Figure 7.6). The GABA shunt is actually a closed loop which acts in such a way as to conserve the supply of GABA. The first step in the shunt is the transamination of α-ketoglutarate, an intermediate in the Krebs cycle, to glutamic acid. Glutamic acid is then decarboxylated to form GABA. The next step is the key. GABA is transaminated to form succinic semialdehyde, but the transamination can only take place if α-ketoglutarate is the acceptor of the amine group. This transforms α-ketoglutarate into the GABA precursor, glutamate, thereby guaranteeing a continuity of supply. Thus, a molecule of GABA can only be destroyed metabolically if a molecule of precursor is formed to take its place. The succinic semialdehyde formed from GABA is rapidly oxidized to succinic acid, which is a normal constituent of the Krebs cycle.

Three enzymes are involved in the shunt: glutamic acid decarboxylase (EC 4.1.1.15) (GAD) which converts glutamic acid to GABA, GABA-α-oxoglutarate transaminase (EC 26.1.19) (GABA-T) which converts GABA to succinic semialdehyde, and succinic semialdehyde dehydrogenase (EC 1.2.1.24) (SSADH) which returns the metabolic remnant to the Krebs cycle. From an energy point of view, the GABA shunt yields three ATP molecules. It is therefore less efficient than that portion of the Krebs cycle which converts α-ketoglutarate through the usual pathway to succinic acid. This direct route, used principally by brain as well as exclusively by all other tissues, yields three molecules of ATP and one of GTP.

Figure 7.7 is a schematic diagram showing intracellular localization of the components of the GABA shunt. GABA-T and SSADH are attached to mitochondria. Glutaminase and GAD are free in the intraneuronal cytoplasm. Glutamine synthetase (Glu Synth) occurs in glial cytoplasm. An internal segment of the shunt operates entirely within the nerve ending. A molecule of α-ketoglutarate leaves the Krebs cycle (which is confined to the mitrochon-

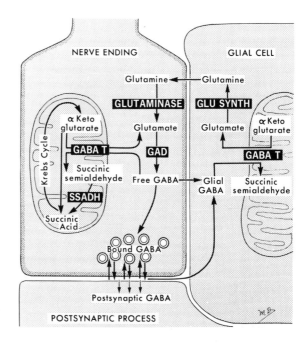

Figure 7.7 Schematic diagram showing relationship of GABA nerve ending, a postsynaptic process, and a glial cell. Enzymes are shown in black rectangles and endogenous intermediates in white.

drion) to form a molecule of glutamate. The reaction is catalyzed by GABA-T and the amine group required is obtained from GABA. Replacement GABA is formed by the decarboxylation of glutamate, a reaction catalyzed by GAD occurring exclusively in GABA nerve endings. Released GABA can be picked up by glial cells, where transamination by GABA-T can take place in the external or glial segment. The glutamate formed cannot, however, be converted to GABA in the glia since they lack GAD, and it is apparently transformed by glutamine synthetase into glutamine before being returned to the nerve ending. In the nerve ending, the enzyme glutaminase converts the glutamine back to glutamate.

Many studies have been performed attempting to learn whether the shunt is primarily for the purpose of producing GABA or whether it also serves as an alternative source for generating the brain's energy. It is still not certain to what extent glucose, the principal fuel for the brain, is diverted through the shunt rather than being stripped through the standard Krebs cycle. Some experiments have indicated that as much as 44% of the glucose might be funneled through the shunt (McKhann *et al.* 1960), but figures based on *in vivo* turnover rates of GABA as compared with Krebs cycle intermediates suggest as little as 10% (Baxter 1976). Since it is not known how much GABA is actually utilized for neurotransmitter function, it is impossible to determine whether any of that which is produced is designed primarily as a contribution to the energy pool. It may be that GABA is being continuously formed in nerve endings at a rate well in excess of that required for neurotransmission, with the surplus being metabolized presynaptically and winding up as a supplemental energy source.

The key enzyme in the formation of GABA, L-glutamate decarboxylase, has recently been purified from mouse brain (Wu *et al.* 1973). It has a molecular weight of 85,000 and requires pyridoxal phosphate as a co-factor. The K_m, or concentration of substrate required to half-saturate the enzyme, is 0.7 mM for glutamate and 0.05 mM for pyridoxal phosphate. Tissue levels *in vivo* of these materials are so high that the enzyme should normally be fully saturated with its substrate and co-factor. The enzyme has a slight capacity (2%) to decarboxylate aspartic acid, but accepts no other amino acid as a substrate. It is highly localized to the synaptosomal fraction of subcellular homogenates of brain. This indicates that production of GABA is concentrated in the nerve endings rather than in the cell bodies or dendrites of neurons, or in the glial cells. Within the synaptosomes it is in a soluble form. Thus, it is not attached to the mitochondria of nerve endings as are the Krebs cycle enzymes (Wu 1976). Antibodies to the purified enzyme have been produced in rabbits, and, as already shown (Figures 7.1, 7.4, 7.5), localization to nerve endings has been confirmed submicroscopically by the immunocytochemical technique.

GABA is destroyed by the enzyme GABA-T. The reaction involves a reversible transamination of GABA with α-ketoglutarate. No transamination of GABA takes place with any other α-keto acid, and only β-alanine can substitute for GABA as a substrate. As with GAD, GABA-T requires pyridoxal phosphate as a coenzyme. The enzyme from mouse brain has been purified to homogeneity and antibodies to it prepared (Saito *et al.* 1974b). It has a molecular weight of 109,000. The concentration of GABA required to half-saturate the enzyme is 1.1 mM (Schousboe *et al.* 1973). It is clear that the availability of α-ketoglutarate plays an important role in the destruction of GABA. For example, when respiration ceases at death, the level of α-ketoglutarate rapidly declines. GABA cannot then be destroyed, although it can still be formed from glutamate. There is, therefore, a rapid increase in brain GABA levels postmortem accompanied by a rapid decline in glutamate.

GABA-T has a different subcellular localization from GAD. It is attached to free mitochondria instead of being localized to the synaptosomal sap. The mitochondria containing GABA-T appear to be in three locations. They are in GABA-containing synaptosomes, in glial cells, and in postsynaptic neuronal processes. In the first location they participate in maintaining appropriate presynaptic concentrations of GABA, while in the latter two they participate in its destruction after release.

The final enzyme in the GABA shunt, SSADH, is closely coupled to GABA-T and is similarly distributed in brain. The high tissue content required to saturate the enzyme relative to that for GABA-T probably explains why succinic semialdehyde has never been detected as an endogenous metabolite in neural tissue. That which is formed from GABA is very rapidly converted to succinic acid, which then enters the Krebs cycle.

Most of the metabolic studies on the formation and destruction of GABA have involved the exogenous addition of various radioactively labeled precursors. Depending on the precursor used, and the conditions of the experiment, the amounts of labeled glutamate and GABA produced relative to the endogenous tissue levels vary widely. This has given rise to the concept of "pools" or "compartments" of metabolites. It is important to understand that this is a chemical definition and not a morphological one. A compound, such as glutamate or GABA, has a distinct metabolic rate in one chemical compartment independent of other compartments or pools. Although such kinetic data are obtained without any physical concept of where such compartments might be, it is obvious that there must be some cellular or subcellular basis to chemical compartmentation.

In the case of GABA, evidence for compartmentation comes, in part, from attempts to estimate the half-life. Labeled GABA has been injected directly into the cerebral ventricles and its rate of disappearance measured. The data suggest one rate of disappearance of GABA with a half-life between 0.5 and 1 h, and another with a half-life exceeding 1 h. Similar conclusions have been reached by inhibiting GABA-T with aminooxyacetic acid and measuring the rate of increase of GABA. A much shorter half-life for GABA, however, has been estimated by giving the GAD inhibitor, 3-mercaptopropionic acid, and measuring the decline of GABA. This compound, when administered *in vivo* to rats, will cause convulsions within about 7 min. The decrease in GABA concentration immediately before the onset of convulsions is about 35%. These data suggest a half-life of GABA of only about 11 min (Karlsson *et al.* 1974).

There are several physical possibilities for the GABA compartments suggested by these varying chemical kinetics. Much GABA is obviously held in nerve endings, where it can exist either free or bound to synaptic vesicles. But it is also taken up by glial cells, and in between must have a finite time in the extracellular space. Some probably also finds its way into postsynaptic neuronal processes.

Each of these could represent a chemical pool and there might easily be many more. For example, other routes have even been described for the formation of GABA in brain. It has recently been shown that GABA can be formed from putrescine in a series of steps. Isotopic studies have also shown that γ-hydroxybutyrate can be converted to GABA *in vivo*. Moreover, GABA may be a precursor of other significant brain metabolites. It is known that homocarnosine, the dipeptide derivative of GABA and histidine, exists in brain. Similarly, the conversion of GABA to γ-hydroxybutyrate has been shown. Other possible conversions are to γ-aminobutyrylcholine, γ-butyrylbetaine, γ-guanidinobutyric acid, and several others. None of these alternate routes has however, been established as being of major importance either metabolically or physiologically (Baxter 1976).

GABA is metabolized pre- and postsynaptically in neurons, and in glial cells. In each location, GABA-T attached to a mitochondrion is required to capture a molecule of GABA, and after transamination to release a molecule of glutamate into solution. But the glutamate must find its way back, either unchanged or as glutamine, to the presynaptically located enzyme GAD in order to be reformed into GABA.

One suggested mechanism is that when GABA is released into the synaptic cleft, it is primarily taken up by glial cells. GABA is metabolized there by GABA-T, but the glutamate formed is not returned as such to the nerve endings. Instead, it is converted to glutamine and the glutamine transported to nerve endings. In support of this concept is the fact that glutamine synthetase, the enzyme which forms glutamine, is highly concentrated in glial cells but not in nerve endings. The mechanism would imply the existence of glutamine and glutamate pools of different sizes in glial cells and nerve endings.

It is still not known how the synthesis of GABA is governed, and what controls the physiologically active pool. Turnover studies cannot establish which molecules of GABA are metabolized presynaptically, which reach glial cells, and which are consumed by postsynaptic neurons. In each case GABA-T produces the same metabolite. Presumably, GABA is continuously being formed presynaptically at a rate much higher than is ordinarily required for neurotransmission. In that case the excess would merely be cycled through the shunt within the nerve ending. Stimulation of the nerve ending would release bound GABA into the extracellular space, where it would be metabolized by glial cells or postsynaptic neurons, or pumped back into the nerve ending. The deficit in bound GABA would be made up from the excess continuously being synthesized.

More detailed studies on the formation and destruction of GABA may prove that models such as the ones outlined are naive, but they form working hypotheses for future experiments.

In summary, metabolic studies have established the probable route for the formation and destruction of GABA. Large quantities of GABA are synthesized from glutamate in nerve endings, probably in vesicles, until it is released, presumably by nerve stimulation. The sites for destruction of GABA are localized in mitochondria, apparently in glial cells and postsynaptic dendritic elements as well as nerve endings. The destruction of GABA causes a new molecule of glutamate to form from α-ketoglutarate. The newly synthesized glutamate, which is not necessarily at a site where it can be utilized by GAD, may be converted to glutamine before it reenters the nerve ending. These conversions must take place almost entirely within the nervous system because GABA levels are below ordinary detection limits in the CSF and in the periphery of the body.

7.4 Pharmacology of GABA

Pharmacological agents exist which are capable of interacting with GABA in all of the classical areas for manipulation of neurotransmitters. These are

TABLE 7.4
Some Drugs Which Affect GABA Action

Drug	Presumed mechanism of action	Physiological effect
Synthesis		
Allylglycine	GAD inhibitor	Convulsant
3-Mercaptopropionic acid	GAD inhibitor	Convulsant
High-pressure oxygen	GAD inhibitor	Convulsant
Isonicotinyl hydrazide	B_6 antagonist	Convulsant at high doses
Thiosemicarbazide	B_6 antagonist	Convulsant at high doses
Release		
Tetanus toxin	Inhibitor of GABA and glycine release	Convulsant
Pump		
2,4-Diaminobutyric acid	GABA neuronal pump inhibitor	Convulsant at high doses
Nipecotic acid	GABA neuronal pump inhibitor	
β-Alanine	GABA glial pump inhibitor	
Haloperidol	Monoamine blocker and weak GABA neuronal pump inhibitor	Antipsychotic
Chlorpromazine	Monoamine blocker and weak GABA neuronal pump inhibitor	Antipsychotic
Imipramine	Monoamine and GABA pump inhibitor	Antidepressant
Destruction		
Aminoxyacetic acid	B_6 antagonist for GABA-T	Sedative
n-Dipropylacetate	GABA-T inhibitor	Anticonvulsant
Hydrazinopropionic acid	GABA-T inhibitor	Sedative
Gabaculine	GABA-T inhibitor	Anticonvulsant
γ-Acetylenic GABA	GABA-T inhibitor	
Antagonist		
Bicuculline	GABA antagonist	Convulsant
Picrotoxin	GABA antagonist	Convulsant
Agonist		
Muscimol	GABA agonist	Psychotomimetic
Imidazoleacetic acid	Weak GABA agonist	Weak sedative
Homotaurine	GABA agonist	
Lioresal	Possible GABA agonist	Muscle relaxant
γ-Hydroxybutyrate	Possible GABA agonist	Weak sedative
Diazepam	Glycine agonist and stimulator of GABA cell firing	Minor tranquilizer, anticonvulsant, muscle relaxant

sites of synthesis, storage, extraneuronal release, presynaptic reuptake, postsynaptic destruction and postsynaptic activation.

Some of the better known agents are listed in Table 7.4. So far few of them have found their way into clinical medicine. Many of them interact prominently with other systems, so that their specificity is limited. Moreover, inconsistencies appear in a number of cases between the presumed mechanism of action and the overall physiological effects. In general, however, drugs which diminish GABA activity cause convulsions, while those that enhance it cause sedation.

7.4.1 Administration of Precursors

So far no way is known for increasing GABA levels by administering either the material itself or substrates for its synthetic enzyme. Since neither glutamate nor GABA crosses the blood–brain barrier with ease, administering these materials is without effect beyond the neonatal period. The administration of either glutamine or vitamin B_6 is also ineffective, even though they do cross the blood–brain barrier. Apparently, precursor availability is not critical to GAD activity, and normal tissue levels of both precursor and co-factor are sufficient to saturate the enzyme.

7.4.2 Synthesis Inhibition

GABA levels can, on the other hand, be lowered by a variety of agents which inhibit GAD synthesis. The so-called carbonyl trapping agents, which act against the co-factor pyridoxal phosphate, form the largest family. Classical members are such hydrazides as isoniazid, semicarbazide, and thiosemicarbazide which will inhibit GAD in the concentration range of 10^{-4} M to 10^{-3} M *in vitro* and *in vivo*. Carbonyl trapping agents affect both GAD and GABA-T, however, since both enzymes depend upon pyridoxal phosphate. Thus, if GABA-T is more severely inhibited than GAD by a pyridoxal phosphate antagonist, GABA levels may increase rather than decrease. Some agents, such as isonicotinyl hydrazide, will both raise and lower GABA levels depending upon the time after administration of the drug and the dose. Those carbonyl trapping compounds which appear to inhibit GAD more specifically than GABA-T have the greatest tendency to produce seizures in animals.

There is a question as to whether the convulsant action of the hydrazides is really due to inhibition of GAD. At the time when convulsions first commence, it has been shown that GABA levels are not sharply reduced and GAD is not completely inhibited.

There is much evidence, nevertheless, that lowered GABA levels correlate with convulsions. 3-Mercaptopropionic acid inhibits GAD almost totally at a concentration of 10^{-3} M. Following administration of this agent to rats, convulsions commence in about 7 min. At that time GABA levels are reduced by about one-third (Karlsson *et al.* 1974). Allylglycine has a similar effect (Alberici *et al.* 1969). Furthermore, high-pressure oxygen, which also inhibits GAD activity and reduces GABA levels, produces the same effect (Wood 1975).

7.4.3 Storage and Release Mechanisms

While it is presumed that GABA is stored in presynaptic vesicles in the same fashion as the catecholamines and serotonin, direct evidence has been hard to accumulate. Subcellular fractionation studies have failed to show GABA in synaptic vesicles, although the reason for this might be the failure of GABA to remain bound during the lengthy centrifugation procedures. Drugs which will specifically affect the binding of GABA have not yet been found.

Tetanus toxin, a convulsant protein, acts on the CNS apparently by reducing the synaptic release of the inhibitory transmitters GABA and glycine (Curtis *et al.* 1973). When injected directly into the cerebellum, it suppresses basket cell inhibition of cerebellar Purkinje cells. It does not influence the postsynaptic inhibitory effect of GABA when administered microelectrophoretically, however, indicating that it has no effect on receptor sites and must therefore be affecting the presynaptic release of GABA.

7.4.4 Pump Inhibition

After GABA is released into the synaptic cleft, it is taken up again into cells by a so-called high-affinity uptake system. This simply means that GABA is pumped back into cells against a concentration gradient even when the extracellular concentration is very low (i.e., 10^{-5} to 10^{-8} M). This process is dependent on sodium ions and actively consumes energy. One of the confusing aspects of the GABA pump system is that there seems to be one not only for synaptosomes but for glial cells as well. The two processes are not identical because they are affected by different inhibitors. β-Alanine, for example, is 200 times more potent as an inhibitor of glial as opposed to neuronal uptake. 2,4-Diaminobutyric acid, on the other hand, is 10-fold more active at inhibiting neuronal as opposed to glial uptake. Unfortunately, diaminobutyric acid also affects glutamic acid uptake and may also interact with pyridoxal phosphate, so it is not completely specific (Iversen and Kelly 1975). Nipecotic acid is a relatively powerful inhibitor of GABA uptake, but it is also a weak inhibitor of L-proline uptake and an antagonist of glycine-induced depression (Johnston 1976). It is not yet known whether it is specific enough to be of pharmacological interest.

Other GABA pump inhibitors include GABA analogs such as 2-hydroxy-, 2-chloro-, and 4-methyl-GABA as well as agents with major effects on aromatic amine systems such as chlorpromazine, imipramine, and haloperidol (Chapter 14).

It is probable that the pump system serves, in a fashion similar to that in aromatic amine systems, as an important means for terminating the physiological action of released GABA.

The iontophoretic effect of pump inhibitors is to enhance and prolong the inhibitory action of GABA. The physiological effects of these compounds vary according to their action on other systems. It is not yet known what effects a pure GABA pump inhibitor would produce.

7.4.5 Inhibition of Metabolism

Levels of GABA in brain can be increased by inhibiting the metabolizing enzyme system involving GABA-T. The inhibitors are relatively nonspecific, and in some cases highly toxic. Thus, they are of limited clinical usefulness. In animals they tend to produce sedation. This is in contrast to GAD inhibitors which lower GABA levels and produce convulsions.

As mentioned previously, GABA-T requires pyridoxal phosphate. Thus, carbonyl trapping agents and all other anti-B_6 compounds affect this enzyme system as well as GAD, which also requires B_6. The effect is often a differential one. Agents such as aminooxyacetic acid and hydroxylamine, which have a greater effect on GABA-T than on GAD, succeed in raising GABA levels in animals, but due to convulsions produced at higher doses are not used in humans. Isonicotinic acid hydrazide, which may produce a rise in GABA, or at high convulsant doses a slow decline followed by a rise, will only bring about these changes in rats at levels two to three times higher than the maximal therapeutic doses used in man. Other MAO inhibitors also preferentially reduce GABA-T activity and raise GABA levels in animals, but always at doses beyond those used clinically. Therefore, it is doubtful whether the antidepressant action of these agents in man is in any way related to changing GABA levels.

Hydrazinopropionic acid, a structural analog of GABA, also inhibits the GABA-T system. It raises GABA levels in animals and produces somnolence (Van Gelder 1969). It also inhibits GAD to a lesser extent and, at high doses, produces convulsions.

The most useful GABA-T inhibitor so far developed is sodium n-dipropylacetate. This agent produces a dose-dependent rise in GABA in animals. Clinically, it has proved to be effective in treating grand mal, psychomotor, and certain types of petit mal epilepsy. It also causes somnolence, and in some instances nausea and vomiting (de Boilley and Sorel 1969). More recent additions are γ-acetylenic GABA (Jung *et al.* 1977) and gabaculine (Matsui and Deguchi 1977), both of which produce long-lasting GABA-T inhibition *in vivo.*

7.4.6 Receptor Interaction

Antagonists.   There is considerable neurophysiological evidence that picrotoxin and bicuculline are potent blockers of GABA whether the GABA is applied iontophoretically or released by suitable stimulation from synaptic terminals. Accordingly, these two agents are regarded as the classical GABA antagonists. Both produce convulsions. Bicuculline has a mild anticholinesterase action but is nevertheless regarded as being highly specific (Curtis *et al.* 1970).

Strychnine, another convulsant, blocks glycine inhibitory receptors but not GABA receptors. Thus, strychnine and bicuculline are an important combination for the study of inhibitory mechanisms. Strychnine-sensitive, bicuculline-insensitive inhibitory synapses are usually glycinergic, whereas strychnine-insensitive, bicuculline-sensitive ones are GABAnergic. Strychnine also antagonizes the action of taurine and β-alanine throughout the CNS.

Agonists.   Muscimol, a psychotomimetic isoxazole isolated from the mushroom *Amanita muscaria,* is a potent agonist of the bicuculline-sensitive, strychnine-insensitive receptors of spinal cord characteristic of GABA inhibitory sites. The reason for its seemingly unrelated pharmacological action is unknown. Several other compounds, including two aminocyclopentanecarboxylic acids, have been shown through screening procedures to have actions similar to GABA at receptor sites, but their pharmacological profiles are unknown (Johnston 1976b). Homotaurine also has GABAmimetic properties.

Imidazoleacetic acid, a naturally occurring metabolite of histamine, satisfies some of the criteria. It acts like GABA on the stretch receptor neuron of the crayfish (E.G. McGeer *et al.* 1961) and on cortical neurons when directly applied. It is absorbed from the gut and, unlike GABA, readily penetrates the blood–brain barrier. Pharmacologically, it resembles a minor tranquilizer with sedative, muscle-relaxing, and hypothermic properties.

β-(*p*-Chlorophenyl)-GABA (Baclofen) is a GABA derivative which penetrates the blood–brain barrier. It is used clinically as a muscle relaxant, but the extent to which it is a GABA agonist is undetermined. It does not activate GABA receptors on cortical neurons and does not affect GABA transport. It does depress the firing rate of spinal interneurons and Purkinje cells, however, and may block substance P.

γ-Hydroxybutyric acid (γ-OH) is another GABA-like agent in clinical use as a basal anesthetic, or, at lower doses, a mild sedative. It readily crosses the blood–brain barrier. It inhibits the release of dopamine from nerve endings in the neostriatum, apparently by inhibiting the firing of dopaminergic cell bodies in the substantia nigra (Roth and Suhr 1970). These cell bodies are thought to receive inhibitory GABA fibers (Figure 13.23, Section 13.5.4). γ-OH is apparently formed in minute quantities from GABA in brain tissue but relatively massive amounts are required to bring about detectable pharmacological changes.

It has been suggested (Costa *et al.* 1975) that benzodiazepines, a class of minor tranquilizers and muscle relaxants which has achieved amazing popularity, may exert their effects on the CNS by facilitating the action of GABA by a process not yet understood.

7.5 Anatomical Distribution of Glycine

The other known ionotropic inhibitory neurotransmitter is glycine. As a dietary amino acid, it is found in all tissues. Its presence *per se* is therefore not indicative of a neurotransmitter role. Nevertheless, distribution studies have offered valuable clues because the glycine concentrations in spinal cord, medulla, and pons are much higher than in other areas of the central nervous system (Table 7.5). Apart from its uneven distribution the existence of a high-affinity uptake system with a K_m of approximately 10^{-5} M is regarded as strong evidence of a neurotransmitter role. It is confined to the spinal cord, pons, and medulla, and is characteristically dependent on sodium (Neal 1971; Johnston and Iversen 1971; Logan and Snyder 1971). It is easily distinguished from the low-affinity uptake system (K_m approximately 10^{-4} M) which is shared by other small amino acids, is present in all CNS areas, and is presumably concerned with other aspects of glycine metabolism.

The distribution within the spinal cord is illustrated in Figure 7.8 for the cat and is, again, in accord with uptake studies using tritiated glycine. These have shown preferential accumulation in synaptosomes of the ventral horn gray matter, particularly around the perikarya of motor neurons where nerve

TABLE 7.5
Glycine Concentration in Various Brain Regions
(in μmol/g)[a]

Region	Monkey	Cat	Rat	Frog	Snake
Lumbar cord	—	4.40	4.42	3.96	3.02
Thoracic cord	—	1.82	3.42	3.50	3.06
Cervical cord	—	3.68	3.96	3.70	3.26
Medulla	—	3.46	3.75	3.28	4.19
Midbrain	1.54	1.98	1.58	1.38	1.38
Diencephalon	3.29	1.60	0.98	0.85	0.84
Cortex	2.68	1.28	0.87	0.79	0.65
Cerebellum	2.90	0.79	0.82	1.17	1.32

[a]From Aprison *et al.* 1970.

endings for spinal cord interneurons are concentrated (Figure 7.8) (Ljungdahl and Hokfelt 1973).

The synaptosomes accumulating glycine contain flattened synaptic vesicles by glutaraldehyde fixation (Ljungdahl and Hokfelt 1973), which agrees with current concepts that inhibitory synapses contain vesicles of such morphology (Figure 1.5, Section 1.2.2). There is no uptake into synaptosomes containing round vesicles. When spinal cord homogenates are incubated simultaneously with tritiated glycine and tritiated GABA, about half the synaptosomes are labeled. With tritiated glycine alone only about 25% of the synaptosomes are labeled. This suggests that the nerve endings accumulating glycine are different from those accumulating GABA, with each constituting

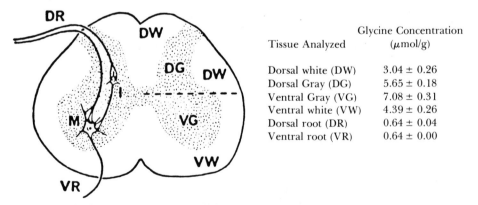

Tissue Analyzed	Glycine Concentration (μmol/g)
Dorsal white (DW)	3.04 ± 0.26
Dorsal Gray (DG)	5.65 ± 0.18
Ventral Gray (VG)	7.08 ± 0.31
Ventral white (VW)	4.39 ± 0.26
Dorsal root (DR)	0.64 ± 0.04
Ventral root (VR)	0.64 ± 0.00

Figure 7.8 Glycine distribution in cross section of cat lumbar spinal cord. The synapses between dorsal root fibers and an interneuron ("direct" or disynaptic inhibitory pathway) and a motoneuron (monosynaptic excitatory pathway) are illustrated. Glycine concentrations in roots and various cord areas are given. (From Aprison *et al.* 1970.)

about 25% of all terminals in the spinal cord (Iversen and Bloom 1972). This is considerably higher than the 10%–15% estimate for dopaminergic terminals in the corpus striatum, which is the area of highest concentration of this transmitter.

Glycine occurs in higher concentrations in the cervical and lumbar enlargements than in the thoracic portion of the cord. The enlargements of the spinal cord are in the areas of major motor outflow to the extremities and greatest complexity of neuronal organization. The marked localization of glycine in these areas correlates with the large number of synaptic junctions and interneurons. In species such as the fish and snake which do not have limbs, no differences are found in glycine concentration in various levels of the spinal cord.

The exact location of glycine in the medulla and pons is unknown but it has been suggested that glycine may be a transmitter in some of the descending fibers of supraspinal origin. These fibers inhibit the firing of Renshaw cells and their inhibitory action can be blocked by strychnine (Curtis et al. 1961). Based on neurophysiological studies using iontophoresed glycine and/or the blocking action of strychnine against inhibitory neurotransmission, it has been suggested that glycine may be the inhibitory transmitter for inhibition of hypoglossal motoneurons by the glossopharyngeal nerve, for reticulospinal neurons in the medullary reticular formation (Trebecis 1973), for the commissural inhibition of vestibular neurons (Precht et al. 1973), in Deiters nucleus (Obata et al. 1970), in the cuneate nucleus (Galindo et al. 1967), and in the nucleus isthmi of the pigeon projecting to the tectum (Cuenod et al. 1976). It may also be the inhibitory transmitter of some cells of the retina. Accumulation of radioactive glycine into a population of amacrine cells in the rat retina has been demonstrated (Marshall and Voaden 1974) and some of the inhibitory cells in the rabbit retina appear to be blocked by strychnine (Wyatt and Daw 1976). Furthermore, light-evoked release of glycine from the retina of cats can be demonstrated both *in vivo* and *in vitro*.

7.6 Chemistry of Glycine

In mammals glycine is a nonessential amino acid. Glycine makes up 1%–5% of typical dietary proteins, but it can also be synthesized from glucose and other substrates in nervous tissue. It crosses the blood–brain barrier with ease and thus may be transported to brain and spinal cord from blood. It is incorporated into peptides, proteins, nucleotides, and nucleic acids, and its fragments participate in other metabolic sequences. Thus, it is one of the more versatile compounds in brain.

The immediate precursor of glycine is serine. Studies using radioactive precursors suggest that most of the glycine in brain is derived from *de novo* synthesis from glucose via serine (Figure 7.9) and not from transport.

The enzyme serine hydroxymethyltransferase (SHMT; E.C.2.1.2.1) is responsible for converting serine to glycine. It requires tetrahydrofolic acid,

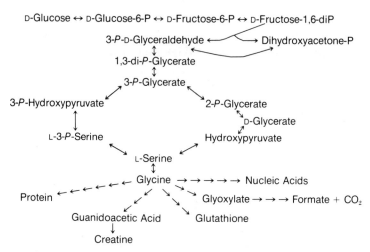

Figure 7.9 Probable synthetic pathway and possible metabolic routes for glycine in nervous tissue.

pyridoxal phosphate, and manganese ions (Figure 7.9), and is strongly inhibited by antipyridoxal compounds such as aminooxyacetic acid (McBride *et al.* 1973).

Unfortunately, the activity of SHMT cannot be used as an index of the presence of glycinergic neurons. The rate of formation of glycine from serine appears constant from area to area and thus there is no distinct correlation with the presumed presence of glycinergic cell bodies or nerve endings.

The metabolic disposal of glycine is unclear. It has been shown in test tube experiments that glycine can be converted to glutathione, guanidinoacetic acid, glyoxylate, or even back to serine via a cleavage mechanism (Figure 7.9).

Again, however, there is no correspondence between the activity of this cleavage system or other disposal mechanisms and the probable occurrence of glycine neurons.

Obviously, much still remains to be discovered about the chemical reactions important for glycine synthesis and disposal, as well as the control of glycine levels and the chemistry of the glycinergic receptor.

7.7 Pharmacology and Physiology of Glycine

Glycine fulfills the physiological criteria expected of the transmitter released by inhibitory interneurons activated by Ia muscle afferents (Chapter 4). Its iontophoretic action duplicates the effects of synaptic activation. It hyperpolarizes the membrane, brings about a large fall in membrane resistance, and increases Cl^- permeability. In these respects its action cannot be distinguished from that of GABA. As mentioned previously, however, glycine is blocked by strychnine, and GABA is blocked by bicuculline. The action of spinal inhibitory interneurons is prevented by the former, but not the latter agent.

Furthermore, in central and cerebellar cortical neurons where high-affinity uptake of glycine is absent and where it is not a transmitter, it is only 1/20 as potent as GABA at increasing membrane conductance to Cl⁻.

Relatively few specific pharmacological agents have been found to interfere with the glycine neuronal systems. This is probably due to a lack of information regarding the factors controlling the synthesis and metabolism of glycine in these neurons. The only types of specific agents which have been definitely identified are receptor blockers and uptake inhibitors.

Strychnine is the best known of the receptor blockers. The regional localization of strychnine binding correlates closely with endogenous glycine concentrations (Young and Snyder 1973), suggesting that the binding is specific for glycine receptors. Additional evidence is the fact that glycine will displace the bound strychnine.

Although strychnine is highly specific, there are several other substances which block the action of glycine but not GABA on spinal neurons. Examples are brucine, thebaine, and 4-phenyl-4-formyl-N-methylpiperidine (Curtis *et al.* 1967).

It has been suggested that the benzodiazepines may be weak glycine agonists. They are among the most widely prescribed of all drugs. The most important members are diazepam (Valium) and chlordiazepoxide (Librium) (Chapter 14, Figure 14.4, Section 14.5). They are used for the relief of anxiety and muscular spasm and to some extent for the amelioration of convulsive states. In one study, the rank order of potency of 21 benzodiazepines in displacing specific [³H]strychnine binding correlated with a variety of pharmacological tests that supposedly predict clinical efficacy (Young *et al.* 1974b). An alternative suggestion for the action of benzodiazepines is that they enhance presynaptic inhibition in the spinal cord and thus mimic the action of GABA.

Some compounds have been identified which inhibit glycine uptake, and, if analogies can be drawn with other neurotransmitters, might therefore enhance glycine action. Parachloromercuriphenylsulfonate is the most potent inhibitor so far found, but imipramine, chlorpromazine, and hydrazinoacetic acid all have some inhibitory action (Aprison *et al.* 1970).

As far back as 1932 glycine was used in the treatment of muscular dystrophy. Apparently the treatment was ineffective, although in some cases up to 60 g of glycine were reported to be ingested in a day without apparent neurotoxicity. It must therefore be considered that glycine is not a particularly potent pharmacological compound. At low doses it protects dogs from the effects of several emetic drugs, and on this basis a specific action on the medullary or vomiting center has been proposed.

There are at least two disorders of metabolism in which glycine accumulates in appreciable amounts in the blood. These neonatal conditions are characterized by developmental retardation, lethargy, hypothermia, and convulsions. Since administration of glycine does not intensify the symptoms, it seems unlikely that these diseases are due simply to an excess of glycine.

Glycine concentrations in brain appear to be unaffected by hypoxia, insulin-induced hyperglycemia, deprivation of protein, or infusion of ammonia.

It is clearly evident that our understanding of glycine neurons is extremely limited, and that almost nothing is known about how to modify their functioning with pharmacological agents.

7.8 Summary

GABA neurons have been identified as one of the two classical ionotropic inhibitory types in the mammalian CNS. The other is glycine. GABA, released from nerve endings, hyperpolarizes the postsynaptic neuron by opening membranous pores or gates so as to permit Cl^- ions to flow from the extracellular fluid into the postsynaptic cell (Figure 4.12, Section 4.4). GABA and its synthesizing enzyme GAD are widely distributed throughout the CNS in high concentrations, suggesting that a sizeable proportion of all neurons use GABA as their transmitter. This view is supported by uptake studies which indicate that 25%–40% of all nerve endings accumulate GABA. The techniques of biochemical localization, lesion, release, physiological and pharmacological action, uptake, and immunohistochemistry have all been used to identify GABA neuronal systems. A list, far from complete, has been accumulated which includes Purkinje cells, cerebellar and hippocampal interneurons, descending neurons from the basal ganglia to the substantia nigra, and retinal, spinal cord, olfactory bulb, and cerebral cortical interneurons.

GABA is synthesized in nerve endings by GAD and destroyed pre- and postsynaptically by GABA-T. Despite much detailed knowledge of the metabolism of GABA, little is known about the control of synthesis and release of physiologically active molecules. A number of pharmacological agents have been discovered which interact with GABA at most of the known sites of neurotransmitter regulation. So far few have found their way into clinical medicine, possibly because of low specificity, either for the GABA system as a whole, or for subtypes of neurons within the GABA system. Those agents which impair GABA functioning to a major degree generally seem to produce convulsions, while those that enhance its action tend to decrease excitability of nerve cells.

The action of glycine cannot be distinguished from that of GABA except by the use of specific blocking agents. GABA is blocked by bicuculline and glycine is blocked by strychnine. As might be anticipated, both blockers are convulsant agents.

Glycine is established as the transmitter for inhibitory spinal interneurons activated by Ia muscle afferents. It is thought to be the transmitter for inhibitory fibers of the glossopharyngeal nerve to hypoglossal motoneurons, for reticulospinal neurons in the medullary reticular formation, for commissural inhibitory fibers of vestibular neurons in Deiters nucleus, in the cuneate nucleus, and in the retina. It is not considered to play a transmitter role in the cerebral or cerebellar cortices. The concentration of glycine is highest in the spinal cord, particularly in the ventral gray area, and the medulla. In these areas, a high-affinity uptake system is present for glycine, and glycine has been shown to accumulate by autoradiography in nerve endings in these areas.

The metabolic routes by which glycine is formed and destroyed in nerve endings are uncertain. Apart from specific blockers, which have no therapeutic application, appropriate pharmacological agents to interact specifically with glycine have not been found.

8

Catecholamine Neurons

8.1 Introduction

Our discussion now shifts to a group of transmitters whose action cannot properly be understood by applying the principles of ionotropic neurotransmitter action discussed in earlier chapters. It is necessary to take into more detailed account some of the principles of metabotropic action referred to briefly in the Introduction to Part II and in Chapter 5. It must be appreciated that only the barest outlines of what may be involved in metabotropic action are yet available. Nevertheless, the principles help to account for the very different anatomical arrangements and overall physiological action of transmitters such as those of the catecholamine family.

An enormous literature exists on the catecholamines. The reader might infer from this that catecholamine neurons are quantitatively important in the brain. On the contrary, they constitute only a minute fraction of the total, numbering in the human brain probably no more than a million cells as opposed to the more than 10 billion in the neocortex alone.

The importance of catecholamine neurons, then, cannot reside in numbers, or in simple excitatory or inhibitory synaptic activity. It must be found elsewhere. One probable source is in divergence. Some catecholamine cells have the largest divergence numbers in the nervous system. For example, each dopaminergic neuron in the rat substantia nigra is estimated to have about 500,000 boutons in the neostriatum, and in the human it may be as high as 5 million (Chapter 13). The other may be in the nature of their postsynaptic effects. The time sequence of overall catecholamine action is much slower than that of many other cells, and their receptor action seems linked to cyclic nucleotides. Thus, they may modulate the reaction of cells to more specific, ionotropic transmitters, or initiate other types of postsynaptic activity.

The complete story of the catecholamines cannot be told because the most important discoveries, i.e., their detailed actions on postsynaptic cells, have not yet been made. As a starting point in providing perspective on this important group of compounds, we shall give an account of the evolution of our present state of knowledge.

The term *catecholamine* is derived from a structure which couples an amine side chain with a dihydroxyphenyl (catechol) ring (Figure 8.1). The catecholamine family is comprised of dopamine, noradrenaline, and adrenaline. They are formed from the dietary amino acid tyrosine by the following sequence of reactions:

L-tyrosine → L-dihydroxyphenylalanine → dopamine
→ L-noradrenaline → L-adrenaline

The only transient intermediate in the sequence is L-dihydroxyphenylalanine (Dopa). As will be described later, however, it can be employed as a pharmacological agent and thus has a special importance in its own right.

The final product in the chain, adrenaline, was the first body hormone to be isolated in crystalline form. Oliver and Schafer (1895) used the cardiovascular actions of extracts of adrenal medulla as the assay system. This led to its isolation and finally to its synthesis by Stolz (1904). As described in Chapter 3, a Cambridge University student, T. R. Elliott, initiated the idea of chemical transmission by proposing that adrenaline was the effector agent released from sympathetic nerve endings upon stimulation. As with the proposal made shortly after by Dixon for ACh, this was not taken seriously until the experiments of Dale a few years later. He and Barger lent credence to the hypothesis with their classic examination of a series of synthetic amines which they termed "sympathomimetic." Their observation that some of these sympathomimetic derivatives were superior to adrenaline in mimicking the action of sympathetic stimulation was the first clue that noradrenaline and not adrenaline might be the true transmitter.

Otto Loewi, whose most important contribution has already been described in Chapter 4 on ACh, carried the work to the next stage by obtaining

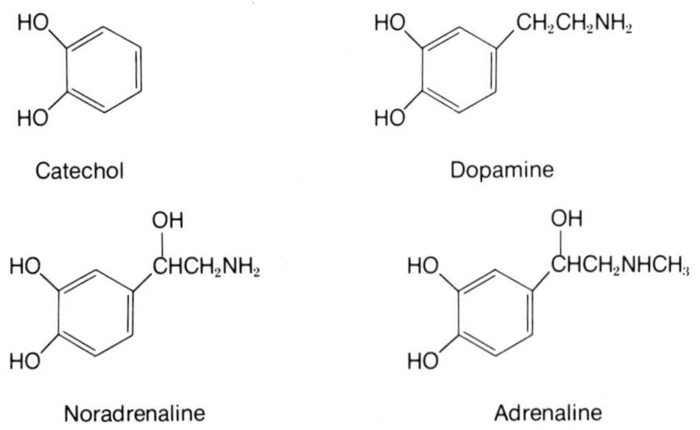

Figure 8.1 Formulae for catechols.

"acceleranzstoff" from the frog's heart (the frog is unique in releasing adrenaline rather than noradrenaline as the transmitter agent), showing that stimulation of sympathetic nerves also led to the release of neurotransmitters.

The next chapter in the unfolding story of the catecholamines was written by Cannon (1921). He had found evidence of a "special and unknown factor," different from adrenaline, which was released following hepatic nerve stimulation (Cannon and Uridil 1921). More thorough experimentation a dozen years later led him to propose that his "sympathin" obtained by stimulation of autonomic nerves was really composed of sympathin E and sympathin I. Bacq (1934), working in Cannon's laboratory, suggested that the inhibitory sympathin I was identical with adrenaline, while the excitatory sympathin E was identical with noradrenaline. The insight of Bacq was not to be proved until a dozen years later when von Euler showed that extracts of sympathetic nerves did indeed contain noradrenaline.

For years there had been speculation on the biosynthetic origin of adrenaline. Casimir Funk (1911) synthesized its precursor Dopa, believing that it might reveal the route of synthesis. But the critical experiment linking Dopa to the catecholamines was not to be performed until 1939. It was then that Holtz (1939) incubated an extract from guinea pig kidney which contained Dopa decarboxylase, with its substrate. When the incubate was injected intravenously, a strong rise in blood pressure was obtained. It was dopamine that they isolated and presumed was the active principal. Thus, the third active member of the catecholamine family was biologically identified. In subsequent experiments, Holtz and his co-workers (1947) proved their hypothesis by injecting themselves with L-Dopa intravenously and isolating dopamine from their urine.

Blaschko (1939), noting that Dopa but not N-methyldopa was decarboxylated by mammalian tissue, reasoned that dopamine and noradrenaline must be the precursors of adrenaline. He correctly predicted the complete biosynthetic sequence starting from tyrosine. It was not until 1951, however, that Goodall (1951) finally demonstrated that dopamine actually existed in mammalian tissues.

Thus, the three catecholamines had all been identified in peripheral tissue. But their presence in brain was still not suspected. Furthermore, there was much confusion about whether dopamine and noradrenaline were physiologically active in their own right or merely precursors of adrenaline.

A major breakthrough occurred in 1954 when Marthe Vogt measured the "sympathin" content of brain (Vogt 1954). She discovered that both adrenaline and noradrenaline were present in amounts and distribution that could not be explained on the basis of brain vascularity. Regions containing very high concentrations were the midbrain and hypothalamus.

A series of highly significant discoveries then followed in rapid succession. The Rauwolfia alkaloid reserpine had been introduced into clinical medicine a few years previously. It was an ancient Indian remedy with no physiological explanation at that time as to why it possessed antihypertensive and tranquilizing properties. Moreover, there was no anticipation of an unexpected side effect, that of a Parkinsonian-like rigidity, which was noted by several clinical groups using the drug.

Pletscher, Shore, and Brodie (1955) gave investigators the scent by discovering that reserpine depleted the recently discovered amine serotonin from the gut. This was rapidly followed in 1956 by several reports showing that reserpine reduced both noradrenaline and serotonin concentrations in all tissues, including the brain, and that there was a failure in sympathetic transmission.

Carlsson and his group then developed a chemical assay for dopamine and showed that the distribution did not parallel that of the other catecholamines. Extraordinarily high concentrations of dopamine were found in the corpus striatum, and, like the other catecholamines and serotonin, it was depleted following the administration of reserpine. This correlation between high levels of dopamine in the corpus striatum, its depletion by reserpine, and the accompanying Parkinsonian-like side effects led Carlsson to propose that dopamine was involved in extrapyramidal function (Carlsson 1959).

The 1958 Catecholamine Symposium in Bethesda, Maryland, where Carlsson presented his brilliant speculation, was the same meeting at which Bernard Brodie made his proposal that noradrenaline was involved in the ergotrophic functions of the brain, while serotonin was involved in tropotrophic functions. This concept helped to explain the tranquilizing action of reserpine, which depleted the amines, as well as the phenothiazines which blocked their action (Brodie et al. 1959).

Almost simultaneously another clinical breakthrough was made. This was the discovery by Kline and co-workers that iproniazid, a monoamine oxidase inhibitor, was a mood elevator. It took little time to make the connection between enhanced levels of catecholamines and serotonin, and the psychological improvement (Loomer et al. 1957).

Birkmayer and Hornykiewicz (1961) followed up on the hypothesis of Carlsson regarding dopamine by measuring dopamine levels in the autopsy brains of a series of Parkinsonian patients. They found that there were sharply decreased levels. Since Carlsson had shown some years previously that the administration of L-Dopa would overcome the effects of dopamine depletion, it followed naturally that people should administer Dopa to Parkinsonian patients.

The first striking clinical results were reported by Cotzias and co-workers. They titrated their patients with very large doses of L-Dopa over periods of several weeks. This, of course, has since become the method of choice for treating Parkinsonian patients (Cotzias et al. 1969).

By the early 1960s the importance of dopamine and noradrenaline in brain had been firmly established. It was still not clear, however, whether these materials were contained in neurons or glia, or precisely what their function might be. Moreover, while the biosynthetic sequence for producing them seemed certain, there was one vital gap. That was the biological source of Dopa. This was indicated by the demonstration of E. G. McGeer et al. that labeled tyrosine injected directly into the neostriatum was converted to labeled catecholamines (E. G. McGeer *et al.* 1963). Thus, brain was capable of converting the dietary amino acid tyrosine to Dopa and catecholamines. Udenfriend and colleagues isolated and purified tyrosine hydroxylase and showed it was the rate-limiting step in the biosynthetic sequence (Udenfriend 1966).

The climactic event in the catecholamine story came with the extension by Falck *et al.* (1962) of the histofluorescence technique of Eranko (1955) into a method for detecting catecholamines and serotonin in nerve tissue. The Falck technique was rapidly exploited by Dahlstrom, Fuxe, and many others to lay out the pathways of dopamine and noradrenaline neurons in brain (Dahlstrom and Fuxe 1964a, 1964b).

The position of adrenaline in brain was less certain. Vogt and others had demonstrated by bioassay the presence of adrenaline, but chemical techniques made the finding seem less certain. P. L. McGeer and McGeer (1964) demonstrated the conversion of labeled noradrenaline to adrenaline by the hypothalamus, but again proof that there were distinct adrenaline-containing cells was lacking. The demonstration was to wait another decade until the elegant immunohistochemical procedures of Hokfelt and his colleagues were developed. Goldstein purified phenolethanolamine-N-methyltransferase (PNMT) from adrenal tissue and prepared antibodies to it. These were then used to establish by immunohistochemistry the presence of distinct adrenaline-containing neurons in the brainstem (Hokfelt *et al.* 1974).

8.2 The Dopamine Neuron

8.2.1 Anatomy

There is now an extensive literature on the anatomy of monoamine pathways in brain involving the use of a variety of techniques. The key, however, was the development of the histochemical fluorescence technique of Falck and Torp. It was an adaption of the discovery by the Finnish investigator Eranko (1955), who observed that adrenal medullary tissue fixed in formaldehyde revealed a bright yellow color under a fluorescence microscope. It was several years before Falck and his co-workers discovered how to apply this finding to the detection of catecholamine-containing neurons in nervous tissue. To prevent diffusion and intensify the reaction, they prepared the tissue by freeze-drying before exposing it to dry formaldehyde vapor. Catecholamines produced an intense apple green color and serotonin an evanescent yellow color when exposed to an appropriate ultraviolet light source by fluorescence microscopy. The reactions are as shown in Figure 8.2. Initially, the method was used to describe sympathetic structures in the periphery (Falck *et al.* 1962), but this was soon followed by the first rough map of catecholamine- and serotonin-containing neurons in the brain itself (Dahlstrom and Fuxe 1964a, 1964b).

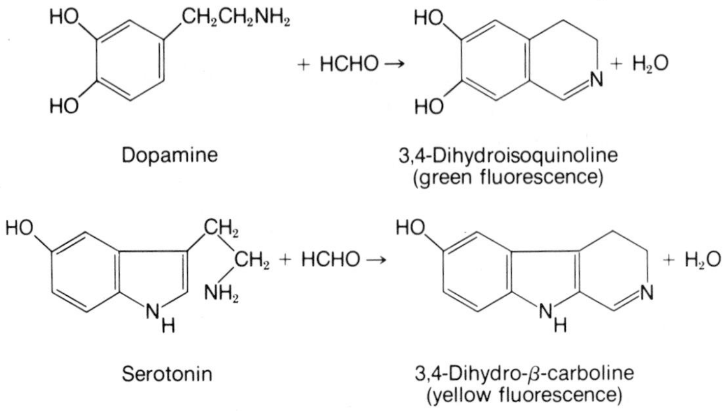

Figure 8.2 Reactions of dopamine and serotonin with formaldehyde gas to produce fluorescent derivatives suitable for histochemical detection.

The main disadvantage of this histochemical technique is that the reaction products of the various catecholamines cannot easily be distinguished. Thus, many additional techniques have been used to establish which catecholamine is associated with individual pathways. These include pharmacological manipulations, lesions, immunohistochemistry, enzyme histochemistry, autoradiography, electron microscopy, and axoplasmic flow.

The catecholamine neuronal systems emanate from a series of cell bodies in the brainstem that were designated A1 to A12 by Dahlstrom and Fuxe. Of these cell groups, A9, A10, and A12 are dopaminergic. They give rise to the nigrostriatal, mesolimbic-mesocortical, and tuberoinfundibular pathways respectively. These pathways are shown in horizontal and sagittal projections in Figures 8.3A and 8.3B. There are, additionally, dopaminergic interneurons in the brainstem, superior cervical ganglion (small intensely fluorescent, or "SIF" cells), retina, olfactory bulb, and carotid body (Table 8.1). It is entirely possible that other dopaminergic structures will be identified in the future.

The nigrostriatal dopamine system is now one of the most widely studied pathways in the brain. The A9 cell group, which gives rise to it, is located in the zona compacta of the substantia nigra. This is a cap of large multipolar cells which sits on the dorsal aspect of the substantia nigra (Figure 8.4A). By electron microscopy, these large cells have multiple indentations in the nuclei and a very prominent Golgi apparatus (Figure 1.8, Section 1.2.3). The dendrites extend into the zona reticulata where they are surrounded by nerve endings, many of which are GABAergic (Figure 7.4, Section 7.2.3). The axons from these cells are only lightly myelinated. They proceed rostrally in a prominent pathway that ascends in the lateral hypothalamus just dorsolateral to the medial forebrain bundle (Figure 9.3, Section 9.2). They enter the crus cerebri at the midhypothalamic level, intermingle with the myelinated fibers in the internal capsule, and then fan out through the globus pallidus to enter the caudate and putamen. Some dopamine nerve terminals which enter the

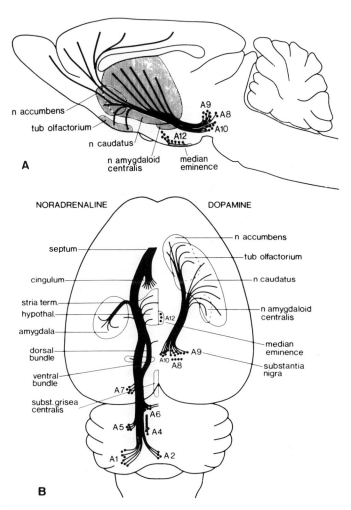

Figure 8.3 A. Sagittal projection of dopamine pathways in the rat. Striped areas indicate dense terminal fields. (Modified Ungerstedt 1971.) B. Horizontal projections of the ascending dopamine and noradrenaline pathways. Terminal fields in the cortex are not shown. (From Ungerstedt 1971.) See Figure 9.3A (Section 9.2) for midbrain transverse section.

amygdaloid nucleus seem to be an extension of the terminals in the caudate and putamen (Ungerstedt 1971).

The fine structure of the pathway has been studied by axoplasmic flow. [³H]Dopamine or [³H]noradrenaline injected into the substantia nigra are equally well transported to nerve endings in the caudate–putamen (E. G. McGeer *et al.* 1974). When autoradiography is attempted, however, dopamine is dislocated during the fixation process while noradrenaline is not. Therefore, [³H]noradrenaline has been used for autoradiographic localization. It

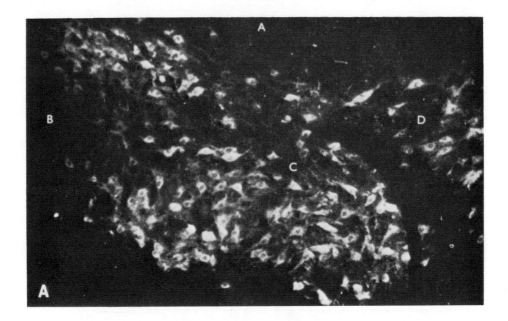

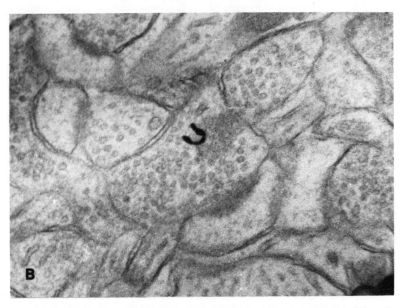

Figure 8.4 A. Transverse section of normal rat brain showing dopaminergic cells of the substantia nigra and adjoining ventral tegmental region. These cells showed a green fluorescence by the histochemical method of Falck *et al.* (1962) (Figure 9.2). Area marked by A corresponds to the medial lemniscus. C and D represent the medial regions of the ventral tegmentum. In this photomicrograph it is not possible to distinguish between cells of the substantia nigra zona compacta (A9) and those of the ventral tegmentum (A10). (From Dahlstrom and Fuxe 1964a.) B. Nerve ending in the corpus striatum containing radiolabel following injection of [³H] norepinephrine into the SN. Note the asymmetrical axospinous contact and mildly pleomorphic vesicles. (From E.G. McGeer *et al.* 1975.)

has made possible unequivocal identification of the morphology of dopaminergic nerve endings. Figure 8.4B shows a radioactive grain located in a typical nerve ending of the nigrostriatal tract. The nerve ending contains mildly pleomorphic vesicles and makes an asymmetrical contact with dendritic spines of neostriatal neurons.

The next most prominent tract is the mesolimbic-mesocortical pathway. These A10 cell bodies extend medially from the A9 cells, forming a cap over the interpeduncular nucleus. The axons ascend together with the axons of the nigrostriatal dopamine system following a slightly more medial course. They do not enter the crus cerebri but continue in a rostral direction just dorsal to the medial forebrain bundle to innervate limbic structures such as the nucleus accumbens, olfactory tubercle, and possibly the amygdala.

Cortical innervation is an extension of the complex. Thierry *et al.* (1973) originally presented evidence of a biochemical nature that a dopaminergic innervation of the neocortex must exist. More careful analysis of histochemical preparations established that the fibers sweeping along the medial aspect of the nucleus accumbens separate into a number of branches, the most abundant of which runs dorsally and rostrally into the deep layers of the frontal cortex. A second branch turns dorsally above the corpus callosum and moves caudally to innervate the anterior limbic cortex. Another branch innervates the septum, and the remaining portion gives terminals to the olfactory tubercle (Lindvall and Bjorklund 1974) (Figure 8.3).

The tuberoinfundibular dopamine system has cell bodies (A12) located within the arcuate nucleus of the hypothalamus. These cells innervate the external layer of the median eminence. There are also some interstitial neurons in the hypothalamus and it has been suggested that another dense group, A13, which lies close to the medial dorsal nucleus of the hypothalamus, might give rise to ascending axons into the thalamus (Ungerstedt 1971). Presumed dopaminergic axons from the A11 or A13 cell bodies have also been observed in the zona incerta (Lindvall and Bjorklund 1974).

A quantitative study of the nigrostriatal system in the rat (Anden *et al.* 1966) gives some idea of the amazing divergence number of these dopaminergic neurons. There are only about 3500 dopamine cell bodies in the zona compacta region. Each cell body gives rise to a single axon which branches so profusely that it contains about 500,000 varicosities (nerve endings or boutons *en passant*). The concentration of dopamine in one of these varicosities is estimated to be as high as 8000 μg/g (2.5×10^{-4} pg per varicosity) as compared with a cell body concentration of 60–200 μg/g. Since there are roughly 4 million neurons (not containing dopamine) in the rat neostriatum, an equal distribution of synapses would result in about 440 dopamine contacts per neostriatal neuron. Thus the dopaminergic input to the neostriatum is massive indeed, and of a field, rather than a discrete, nature.

Much more detailed information on dopaminergic innervation of the brain can be expected in the future as techniques of measurement improve.

TABLE 8.1
Presumed Dopamine-Containing Cells and Pathways

Nigrostriatal pathway (A9 cells)[a]
Mesolimbic pathway (A10 cells)[b]
Mesocortical pathway (A10 cells)[c]
Tuberoinfundibular pathway (A12 cells)[a]
Caudal hypothalamic (A11 cells)[c]
Dorsal hypothalamic (A13 cells)[c]
Anterior hypothalamic (A14 cells)[c]
Cylinder cells of the chemoreceptor trigger zone[d]
Interplexiform retinal interneurons[e]
Periglomerular olfactory bulb interneurons[f]
Small intensely fluorescing (SIF) interneurons of sympathetic ganglia[g]
Glomus cells of the carotid body[h]
Mast cells of some ruminant species[i]

[a]Dahlstrom and Fuxe 1964a, 1964b.   [f]Halasz et al. 1977.
[b]Ungerstedt 1971.   [g]Bjorklund et al. 1970.
[c]Lindvall and Bjorklund 1974.   [h]Bolme et al. 1977.
[d]Murphree and Domino 1968.   [i]Bertler et al. 1959.
[e]Ehinger 1977.

The glyoxalic acid method (Lindvall and Bjorklund 1974) seems to be an advance over the original Falck technique.

Another advance is the immunohistochemical method which has now been developed for tyrosine hydroxylase. It works both at the light and electron microscopic level (Pickel et al. 1975). Unfortunately, the method cannot distinguish between various types of catecholaminergic neurons because tyrosine hydroxylase is contained in all of them.

It is interesting that this immunohistochemical method reveals considerable amounts of tyrosine hydroxylase in dendrites. Histochemical fluorescence reveals that the dendrites also contain catecholamines. This parallels what has been found for cholinergic neurons.

Additional information on the localization of dopamine in brain can be found in Tables 8.1 and 8.4. In many locations it is simply the precursor of noradrenaline and adrenaline, but it exists in interneurons in such locations as the olfactory bulb, retina, superior cervical ganglion, and carotid body (Table 8.1). In addition, large quantities of dopamine are found in the liver, duodenum, heart, and lungs of certain ruminant animals where it seems to be correlated with a special type of mast cell (Bertler et al. 1959).

8.2.2 Chemistry of Dopaminergic Neurons

We have now discussed the anatomical pathways involving dopamine. The next step is to consider how it is manufactured, stored, and released. Although the enzymes for synthesizing dopamine exist in all parts of the neuron, the principal synthesis takes place in the nerve ending. It is in the nerve ending that vesicles for binding and storing dopamine exist, permitting high concentrations of the amine to be accumulated.

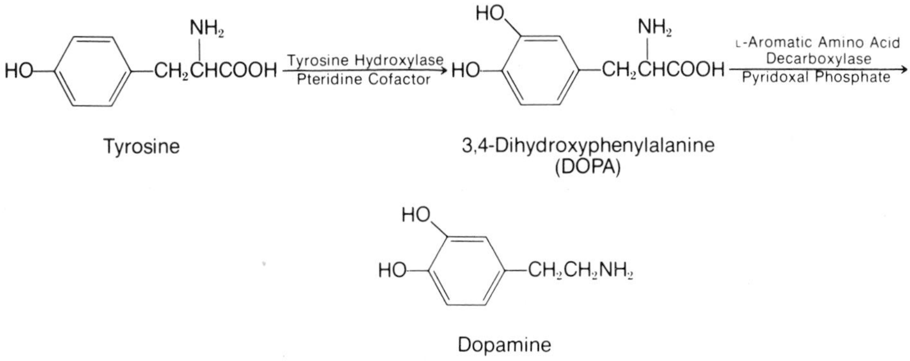

Figure 8.5 Synthesis of dopamine.

The synthesis itself takes place in the cytoplasm of the nerve ending. The synthesized dopamine is either stored in vesicles, destroyed presynaptically by intrasynaptosomal monoamine oxidase (MAO), or released into the synaptic cleft. After release, it can be recaptured by the nerve ending or destroyed by either of two enzymes: catechol-O-methyl transferase (COMT) or MAO. COMT is apparently located on postsynaptic membranes while MAO is on the surface of mitochondria, in glial cells and dendrites (Figure 8.12). The metabolic sequence is shown in Figure 8.5.

Tyrosine Hydroxylase (EC 1.14.16.2). There are two enzymes involved in the synthesis of dopamine: tyrosine hydroxylase and Dopa decarboxylase. This combination is not exclusive to dopaminergic neurons, for, as will be described later, the two also occur as precursors of noradrenaline and adrenaline in noradrenergic and adrenergic neurons, respectively.

There are three principal reasons why tyrosine hydroxylation is considered to be the slow step in the production of catecholamines. First, free tyrosine is detected in the tissues, but not L-Dopa. This implies that all tyrosine which is converted to Dopa is rapidly carried through the remaining steps. Second, catecholamine levels are not influenced by changing dietary levels of tyrosine or by parenteral administration of large doses of the amino acid. On the other hand, catecholamine levels are dramatically changed by oral or parenteral administration of L-Dopa. Finally, catecholamine levels in tissue are sharply reduced by the administration of tyrosine hydroxylase inhibitors but only slightly by Dopa decarboxylase inhibitors.

Tyrosine hydroxylase protein seems to be present in tissue in much lower concentrations than Dopa decarboxylase and dopamine-β-hydroxylase, which probably accounts for the fact that this initial step was the last in the biosynthetic sequence to be demonstrated experimentally. Indeed, convincing data showing the conversion of tyrosine to Dopa by catecholamine-containing tissues was not obtained until radiolabeled tyrosine of high specific activity became available.

Tyrosine hydroxylase can be studied in material from dopaminergic-, noradrenergic-, or adrenergic-cell-containing tissues. The two most commonly employed sources are the corpus striatum, which is rich in dopaminergic nerve endings, and the adrenal medulla, which is largely composed of adrenaline- and noradrenaline-secreting cells.

Tyrosine hydroxylase (TH) has been shown to be an enzyme readily solubilized from adrenal medulla. It requires molecular oxygen, ferrous ions, and a tetrahydropteridine co-factor (Nagatsu *et al.* 1964). The reaction sequence for the conversion of tyrosine to Dopa is presumed to be as follows:

$$TH\text{-}H_2 + \text{tyrosine} \rightarrow TH\text{--}H_2 \cdot \text{tyrosine} \tag{i}$$
$$TH\text{-}H_2 \cdot \text{tyrosine} + \tfrac{1}{2} O_2 \rightarrow TH + \text{Dopa} + H_2O \tag{ii}$$
$$\text{Pteridine-}H_2 + TPNH + H^+ \rightarrow \text{pteridine-}H_4 + TPN^+ \tag{iii}$$
$$\text{Pteridine-}H_4 + TH \rightarrow TH\text{--}H_2 + \text{pteridine-}H_2 \tag{iv}$$

where TPNH is reduced triphosphopyridine nucleotide and $TH \cdot H_2$ is reduced tyrosine hydroxylase.

It is thought that TH is active only in the reduced state. During the hydroxylation of tyrosine to Dopa, the $TH\text{-}H_2$ protein becomes oxidized to the inactive TH form (step ii). It is reactivated by a tetrahydropteridine molecule which donates its hydrogen atoms (step iv). The resulting dihydropteridine molecule must be regenerated by another enzyme, pteridine reductase, which donates hydrogens to convert the dihydropteridine back to tetrahydropteridine (step iii). Although it has been demonstrated that pteridine reductase occurs in both adrenal glands and brain, little is known about the TH regeneration system. While many synthetic pteridines act as co-factors for tyrosine hydroxylase *in vitro*, the co-factor *in vivo* is presumed to be the reduced form of biopterin (Brenneman and Kaufmann 1964). Proof will require the isolation of this material from catecholamine-containing brain tissues, but synthesis has been shown to occur in brain (Gal and Sherman 1976). This complicated regeneration system may be crucial to governing the production of catecholamines, but as yet it has received scant investigational attention.

The dependence of tyrosine hydroxylation on Fe^{2+} ions has also been challenged by authors who found that catalase or peroxidase could be substituted. Again, this is a significant detail of the hydroxylation step which requires further study.

Although tyrosine hydroxylation is clearly the rate-limiting step in catecholamine biosynthesis, the amount of enzyme protein in nerve endings must be far in excess of normal requirements for catecholamine production. Spano and Neff (1972) estimated the rate of formation of dopamine from tyrosine in the guinea pig caudate nucleus *in vivo* to be roughly 32 nmol/g/h. By contrast, the maximum velocity (V_{max}) levels of tyrosine hydroxylase in caudate have been variously reported as 100–200 nmol/g tissue/h (E. G. McGeer *et al.* 1967; Fahn and Cote 1968a), 400–500 (Coyle and Axelrod 1972), and as high as 1200 (Cicero *et al.* 1972), depending on the *in vitro* assay conditions used. Moreover, stimulation of peripheral sympathetic nerves leads to rapid increases in the conversion of tyrosine to catecholamines without a short-term change in the amount of enzyme protein (Weiner 1970).

Obviously, some other mechanism than the amount of TH protein must govern the minute-to-minute conversion. Udenfriend and colleagues (1965) showed that a variety of catechols, including naturally occurring catecholamines, were inhibitors of tyrosine hydroxylation *in vitro*. This led to the proposal that control of catecholamine synthesis *in vivo* is achieved by feedback inhibition on the pteridine co-factor. In other words, the limitation on catecholamine synthesis is not set by the amount of tyrosine, or by the total tyrosine hydroxylase protein, but by the amount of reduced tyrosine hydroxylase (TH-H$_2$) produced by the pteridine-H$_4$ co-factor (step iv). Pteridine-H$_4$, in turn, is limited by excess dopamine, which has its own capacity to oxidize the co-factor to pteridine-H$_2$. On the other hand, longer term stimulation of nerve or a deprivation of stimulation may change the tyrosine hydroxylase protein. The time course is measured in days rather than minutes. Drugs which block catecholaminergic receptor sites, such as chlorpromazine, or which deplete the nerve endings, such as reserpine, induce increased tyrosine hydroxylase protein (Nyback 1972; Segal *et al.* 1971). Increased nerve stimulation or prolonged administration of high doses of L-Dopa lead to decreased levels of tyrosine hydroxylase protein (Dairman *et al.* 1972).

In summary, short-term control of catecholamine synthesis is achieved without alteration in enzyme protein because there is a reserve of this protein which is temporarily suspended in an oxidized form. A reduced pteridine co-factor activates the protein as catecholamine stores deplete through excitation of the nerve ending. The amount of tyrosine hydroxylase protein nevertheless adapts over a period of days if there is a shift in overall demand for dopamine. Physiological or psychological stress, drugs which deplete the amine from nerve endings, or drugs which block the receptor sites will cause increases. Conversely, drugs which potentiate the action on the receptor cause decreases. The induction can be blocked by isolating the nerve ending from its presynaptic input or by blocking the protein and nucleic acid synthesis which takes place in the cell body. Clearly, the precise control afforded by these mechanisms is essential for the normal function of catecholamine-producing cells. This is a fruitful area for investigation in a variety of disease states where deranged catecholamine metabolism is thought to exist.

Tyrosine hydroxylase itself has been completely purified from adrenal medulla by several laboratories, and antibodies to it prepared. In common with many enzymes, it appears to exist in a number of molecular weight forms. Tyrosine hydroxylase in noradrenaline neurons seems to have a molecular weight of approximately 200,000, but a form has been found in dopaminergic neurons with a molecular weight of only 65,000. The lowest molecular weight form is approximately 34,000; it can be produced by treating the enzyme *in vitro* with trypsin. Tyrosine hydroxylase in sympathetic ganglia has a molecular weight of 130,000, which may be a tetramer of the active form. The higher molecular weight forms may be associated with noncatalytic moieties, such as RNA (Reis *et al.* 1975). Within the neuron, tyrosine hydroxylase appears to be largely in the soluble state. If tissues such as the caudate are selected for study, the tyrosine hydroxylase is found almost entirely within synaptosomes (P. L. McGeer *et al.* 1965). It can be readily solubilized, however, by the use of detergent agents (Kuczenski and Mandel 1972). This suggests that within the synaptosome the enzyme is either soluble or in a very loosely bound form.

Dopa Decarboxylase (EC 4.1.1.28). The second enzyme in the dopamine synthetic sequence is L-aromatic amino acid decarboxylase, more commonly called Dopa decarboxylase. It is a pyridoxal-dependent enzyme which catalyzes the conversion of Dopa to dopamine. It apparently also converts 5-HTP to serotonin. There have been some reports suggesting that these two decarboxylations are carried out by different enzymes but, as will be discussed in the chapter on serotonin, the existence of two enzymes has been strongly disputed. The decarboxylase is not confined to nervous tissue. It is also found in peripheral tissues such as the intestinal mucosa, liver, and finally the kidney. It was the kidney decarboxylase that was initially studied.

Dopa decarboxylase has been purified from hog kidney to complete homogeneity. The enzyme has at least one free sulfhydryl group, which is necessary for activity, and depends upon pyridoxal-5-phosphate as a cofactor. It readily decarboxylates 5-hydroxytryptophan to serotonin, but not histidine to histamine. The molecular weight is 112,000 daltons. Active fragments of molecular weight 21,000, 40,000, and 57,000 have also been obtained (Christenson *et al.* 1972). An antiserum to the purified hog kidney preparation has been obtained. It cross-reacts with Dopa decarboxylase from adrenal and brain of several mammalian species. The antiserum precipitates both Dopa decarboxylase and 5-HTP decarboxylase equally. This is further evidence of the identity of Dopa decarboxylase and 5-HTP decarboxylase. The same enzyme has also been purified from bovine adrenal glands (M. Goldstein *et al.* 1972), and antibodies have been prepared and used for immunofluorescent histochemical localization (Hokfelt *et al.* 1973). Positively staining cells were found in the known dopaminergic, noradrenergic, and serotonergic systems of the CNS. The nonspecificity of this enzyme makes it an unsuitable marker for any specific monoaminergic pathway.

We have described the synthetic sequence by which dopamine is formed in dopaminergic neurons. We must now consider the methods by which it is metabolized.

As described earlier, dopamine is metabolized by both MAO and COMT. Only MAO acts presynaptically. It produces dihydroxyphenylacetic acid, which diffuses outside the nerve ending where it may be converted to homovanillic acid. If COMT acts upon released dopamine, then 3-methoxytyramine is produced extracellularly. This may be acted upon by MAO in dendrites or glia also to produce homovanillic acid. The biochemical sequence is shown in Figure 8.6.

Monoamine Oxidase (EC 1.4.3.4). Since its discovery half a century ago, an enormous amount of work has been done on MAO (see review by

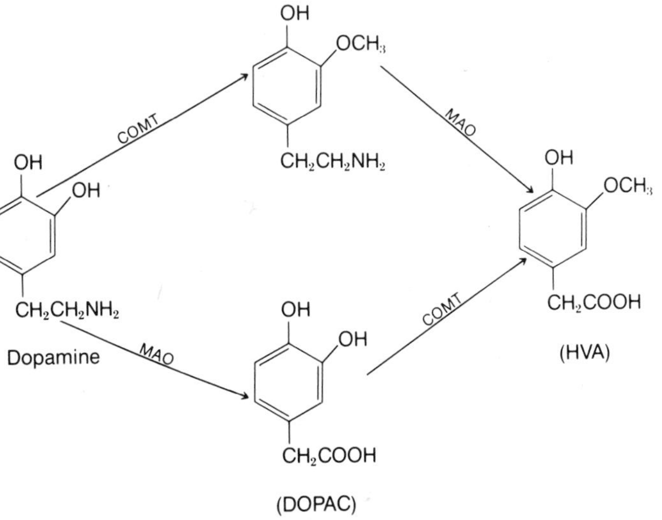

Figure 8.6 Metabolism of dopamine. DOPAC, dihydroxyphenylacetic acid; HVA, homovanillic acid.

Costa and Sandler 1972). The enzyme was first described by Hare (1928) and was called "amine oxidase" by Richter (1937). The term monoamine oxidase (MAO) was adopted later in order to distinguish the enzyme from others which attack diamines, such as histamine and spermine. Mitochondrial MAO is exclusively located on the surface of mitochondria. It is particularly abundant in the liver, kidney, and intestine, in addition to the central nervous system. It oxidizes a wide variety of aromatic amines, including the catecholamines, serotonin, tyramine, phenylethylamine, and others.

Mitochondrial MAO should not be confused with plasma amine oxidase which catalyzes the deamination of benzylamine, kynuramine, spermine, and spermidine. This latter type of MAO has also been found in connective tissues, such as the skin, bone, aorta, and dental pulp, and is supposed to participate in the cross-linking reactions of collagen.

Despite extensive study, mitochondrial MAO continues to be an enigmatic enzyme. Although it has been extensively purified from beef liver (Yasunobu *et al.* 1968) and from pig brain (Tipton 1968) there is little understanding of its various forms. No fewer than five isoenzymes have been reported from liver, and four from brain. The enzyme has a high molecular weight, and it has been suggested that liver MAO represents polymers of an active enzyme having a molecular weight of about 425,000.

Hartman and Udenfriend (1972) separated three forms of liver MAO but found that antibodies prepared to one of the fractions cross-reacted with the others. They concluded that the separable forms of liver MAO were antigenically identical. They could not precipitate immunochemically a portion of the brain MAO, however, which suggested that there might be an entirely different form making up part of the brain activity. This conclusion is in accord with the original report of Johnston (1968) that by using the selective MAO inhibitor clorgyline two forms of MAO could be distinguished in homogenates of rat brain. He designated them as A and B with enzyme A being inhibited by clorgyline. Inhibitors of enzyme B were subsequently found, the most effective of which is deprenyl.

Figure 8.7 indicates the manner in which MAO is inhibited by clorgyline using serotonin (type A substrate) and phenylethylamine (type B substrate) as indicators of enzyme activity. Two different curves are seen in the figure, indicating the presence of two different forms of the enzyme.

Dopamine is a substrate for both forms of MAO. It is not yet certain which type is found in dopaminergic nerve endings, in glial cells, and in postsynaptic neurons receiving dopaminergic boutons. It is quite conceivable that different forms of MAO might be located in these different sites.

Studies on sympathetic nerve endings indicate that type A enzyme is usually dominant in this location. It is also thought to make up approximately 90% of the MAO in the superior cervical ganglion. On the other hand, type B enzyme is dominant in the pineal gland (Neff and Goridis 1972). Glover *et al.* (1977) found human neostriatum to be preferentially inhibited by deprenyl compared with clorgyline when dopamine was the substrate. This implies a dominance of type B MAO, and suggests the possibility that MAO type B is in dopaminergic nerve endings. Undoubtedly, however, the situation is much more complex than the simple A/B classification so far suggested.

A wide variety of compounds has been described as MAO inhibitors. Most of these are broad spectrum. Since metabolism of the monoamines is critical to their physiological action, further investigation of the detailed properties of MAO and its inhibitors will be of great fundamental and clinical importance for the future.

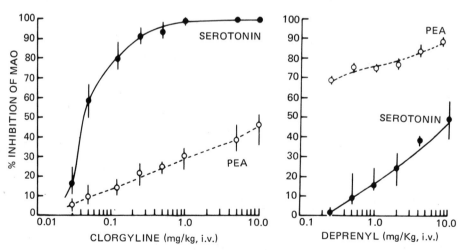

Figure 8.7 Blockade of rat brain MAO activity by increasing doses of clorgyline or deprenyl. Animals were killed 2 h after injection of the drugs and enzyme activity was assayed with serotonin or β-phenylethylamine (PEA) as substrate. Serotonin is a substrate for type A MAO and is preferentially blocked by clorgyline, while PEA is a substrate for type B MAO and is preferentially blocked by deprenyl. (From Neff *et al.* 1974.)

MAO does not convert its various substrate amines all the way to the corresponding acids, although this appears to be the usual end result of its action on dopamine. The oxidation only goes as far as the corresponding aldehyde. The aldehyde can then be oxidized further to the acid by aldehyde oxidase, but it can also be reduced to the alcohol by alcohol dehydrogenase.

Catechol-*O*-Methyltransferase (EC 2.1.1.6). The enzyme COMT was discovered and described by Axelrod and Tomchek (1958) following the finding of Armstrong, McMillan, and Shaw (1957) of vanillylmandelic acid in urine. It is the enzyme which methylates noradrenaline and adrenaline in addition to dopamine. It also methylates their corresponding alcohols and acids. The enzyme catalyzes the transfer of methyl groups from *S*-adenosyl methionine to the metahydroxy group of catechols. It is distributed widely in the cell sap of animal tissues, being particularly abundant in liver, kidney, and brain. COMT activity is present in homogenates of gliomas and cultured astrocytomal cells. This presence in nonneuronal brain tissue suggests that the enzyme is not primarily associated with presynaptic elements. It is generally believed that the COMT near synapses is either in glial cells or in or on the surface of dendritic elements of the postsynaptic neurons.

The enzyme has been completely purified (Creveling *et al.* 1973) and has a molecular weight of 23,000. Other forms of the enzyme have molecular weights of 11,250 and 37,000. Antiserum to COMT has been produced in rabbits.

O-Methylation is the major metabolic pathway for all extracellular catecholamines. It is assumed, therefore, that much released dopamine is converted first to 3-methoxytyramine before being attacked by MAO to form homovanillic acid (Figure 8.6). Homovanillic acid, 3-methoxytyramine, and 3,4-dihydroxyphenylacetic acid all appear in brain tissue and in CSF. The *O*-methyl derivatives of dopamine and other catecholamines have only a tiny fraction of the physiological activity of the catechols themselves. Evidently, the 3-methoxy group greatly reduces the capability of binding to receptor sites, which accounts for the deactivating capability of the enzyme.

It is not possible to determine from measuring the levels of homovanillic acid, the dominant metabolite of dopamine in tissue and urine, whether MAO or COMT acted first on the transmitter. It is of some importance to determine this, because dopamine metabolized presynaptically by MAO would have no physiological action.

Pyrogallol acts as a competitive inhibitor of COMT *in vitro* and *in vivo*, as do some tropolone and papaverine derivatives, but these do not have the physiological effects of MAO inhibitors.

8.2.3 Physiological Action of Dopamine and the Second Messenger Concept

The physiological action of dopamine cannot be considered without discussion of metabotropic action and the material which is thought to be its coupled messenger on the postsynaptic side, adenosine 3',5'-monophosphate or cyclic AMP (cAMP). The discovery of this interesting compound and the

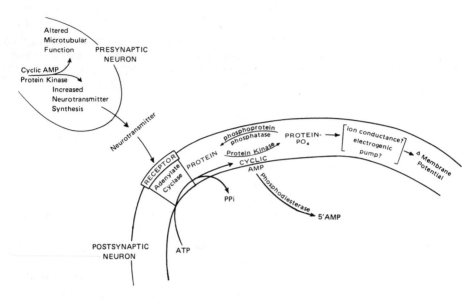

Figure 8.8 Proposed roles for cAMP and consequent protein phosphorylation in neuronal function. The most important proposed role is that of a second messenger for dopamine. Through a protein kinase system this second messenger affects the membrane potential of the cell. Other proposed roles include altered microtubular function and increased neurotransmitter synthesis. These are depicted as taking place in the presynaptic neuron. (From Greengard 1976.)

pioneering experiments on its peripheral actions were carried out by Rall and Sutherland. Their works resulted in an explosion of knowledge (for reviews, see the series *Advances in Nucleotide Research*). The essential feature of cAMP, and its companion compound guanosine $3',5'$-monophosphate (cGMP), is a labile, high-energy phosphate bond which can be utilized to initiate a variety of intracellular physiological events (Chapter 5, Figure 5.10, Section 5.4.2).

It appears that cAMP is involved in at least three distinct processes in the nervous system: the regulation of biosynthesis of certain neurotransmitters; the functioning of microtubules; and as a "second messenger" for certain neurotransmitters. The second messenger concept of these cyclic nucleotides is that they translate extracellular messages into an intracellular response (Chapter 5). We shall concentrate here on the role of cAMP as the second messenger for dopamine and thus the principal focus for its metabotropic action.

The mechanism by which it is hypothesized that dopamine (or in other situations noradrenaline, serotonin, or histamine) acts upon its receptor to stimulate cAMP formation is shown in Figure 8.8. As shown in the diagram, the receptor site on the postsynaptic side is attached to an adenylate cyclase which is activated when dopamine or another agonist attaches to the site. cAMP is formed on the postsynaptic cell membrane from ATP. cAMP then phosphorylates protein kinase, thus providing an activated enzyme. A variety of such cAMP-dependent protein kinases has been discovered in the periphery. They were first identified in muscle and then liver, before being identified in brain. The specificity of cAMP in its various sites of action can be accounted for in terms of the specificity of these protein kinases.

The hypothesis, as explained in Chapter 5, is that an activated protein kinase catalyzes the phosphorylation of a protein also present in the postsynaptic membrane, converting it from the nonphosphorylated to the phosphorylated state. Phosphorylation of this substrate protein leads, through either a change in ionic conductance or in the rate of an electrogenic pump, to a change in the membrane potential or postsynaptic potential. The termination of the sequence of events is through a phosphodiesterase which hydrolyzes the cAMP to 5'-AMP and a phosphoprotein phosphatase that converts the substrate protein back to its inactive, nonphosphorylated form. This leads to the termination of the postsynaptic potential.

The classic preparation for demonstration of the metabotropic effects of dopamine and ACh is the superior cervical sympathetic ganglion. A diagram of the principal synaptic connections of the ganglion with the appropriate neurotransmitters is indicated in Figure 8.9, along with the postulated site of involvement of cAMP and cGMP in the production of postsynaptic potentials.

ACh, released by preganglionic cholinergic fibers, activates nicotinic cholinergic receptors on the surface of postganglionic noradrenergic neurons, leading to the generation of EPSPs. When several EPSPs summate on a given postsynaptic neuron, an action potential is generated and is propagated down the nerve fiber (Chapters 2 and 3). Although there is no evidence that the generation of this EPSP involves a cyclic nucleotide, there are two further slow postsynaptic potentials which evidently modulate the excitability of the postganglionic neurons. These are designated s-IPSP and s-EPSP in

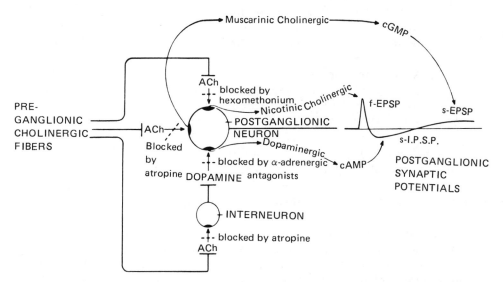

Figure 8.9 A schematic diagram of the principal synaptic connections in the mammalian superior cervical ganglion, and the postulated role of cyclic nucleotides in the genesis of the postganglionic synaptic potentials. (From Greengard 1976.) See also Figure 15.9, Section 15.4.

Figure 8.9, and appear to involve cyclic nucleotides. In one of these modulatory pathways, ACh released from preganglionic fibers activates a dopaminergic interneuron, one of the small intensely fluorescing (SIF) cells mentioned earlier in this chapter. The dopamine released from this inter-neuron activates a receptor on the postganglionic neuron leading, through cAMP, to the generation of the s-IPSP. In the other modulatory pathway, ACh released from preganglionic terminals also activates muscarinic cholinergic receptors on the surface of the postganglionic neuron leading, through cGMP, to the generation of the s-EPSP (Greengard 1976).

The evidence includes the following:

1. Electrical stimulation of the preganglionic cholinergic fibers causes an increase in the cAMP content of the ganglion. Similar results are ob-tained upon application of exogenous dopamine to ganglionic slices. Histochemical studies show that this cAMP occurs primarily or exclu-sively in the postganglionic neuron (Kebabian *et al.* 1975).
2. Dopamine-sensitive adenylate cyclase is present in homogenates of the ganglion.
3. Exogenous cAMP, applied to the ganglion, causes hyperpolarization of the postganglionic neurons, mimicking the hyperpolarization seen on either preganglionic stimulation or application of dopamine.
4. Phosphodiesterase inhibitors, applied to the ganglion, greatly increase the elevation of cAMP seen on preganglionic stimulation or upon application of dopamine. This also greatly increases the size and dura-tion of the dopamine-induced type of polarization.
5. Adrenergic and cholinergic agents affect the s-IPSP as well as the increase in cAMP in a fashion predicted by the scheme in Figure 8.9.

These studies indicate that cyclic AMP mediates dopaminergic transmis-sion and that cyclic GMP mediates the muscarinic component of cholinergic transmission in the superior cervical ganglion by s-IPSPs and s-EPSPs, respec-tively. They also suggest that the dopamine receptor is linked to an adenylate cyclase system, which dopamine stimulates to produce cAMP. Together, cAMP and cGMP modulate nicotinic cholinergic transmission through the ganglion.

Libet *et al.* (1975) found an additional property of dopamine in the superior cervical ganglion. It produces a specific and enduring enhancement of postsynaptic responses to ACh. They have argued that this system is an adequate model for the memory trace, a concept which will be discussed in considerably more detail in Chapter 15.

These studies on the superior cervical ganglion led to investigation of the possible existence of an adenylate cyclase in brain which was also stimulated by dopamine. The first region to be successfully investigated was the caudate nucleus (Kebabian *et al.* 1972), where it was found that cAMP production was stimulated by concentrations of dopamine in the 1- to 10-μm range. Low concentrations of the dopamine agonist apomorphine, but not the β-adrenergic agonist isoproterenol, also caused stimulation. On the other hand many known dopaminergic blockers, including the phenothiazines and butyrophenones, were found to block dopamine activation of the enzyme.

Figure 8.10 illustrates these effects. The adenylate cyclase activity of a subcellular fraction of rat caudate nucleus, rich in synaptic membranes, was measured in the presence and absence of dopamine as well as fluphenazine. Fluphenazine is a potent antipsychotic drug and a dopamine receptor antagonist. Dopamine caused a maximal increase in enzyme activity at about 40 μM, which was competitively inhibited by fluphenazine. With the concentration of fluphenazine at 10 μM, as in the figure, a concentration of dopamine considerably greater was required to produce a comparable stimulation of cAMP formation (Greengard 1976).

These studies support the conclusion that the dopamine receptor, in at least some regions of the brain, is similar to the superior cervical ganglion in possessing a dopamine-sensitive adenylate cyclase. Dopamine-sensitive adenylate cyclase has also been found in other structures of the central nervous system known to have high concentrations of dopaminergic terminals. These areas include the olfactory tubercle, the nucleus accumbens, the amygdala, and the cerebral cortex. The retina, which has dopaminergic activity, possibly

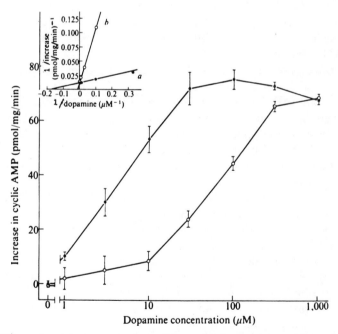

Figure 8.10 Effect of various concentrations of dopamine in the absence ($\bullet$) or the presence ($\bigcirc$) of 1×10^{-7} M fluphenazine on adenylate cyclase activity in a particulate fraction of rat caudate nucleus rich in synaptic membranes. Dopamine caused a sharp increase in cAMP formation which was substantially blocked by the low concentration of fluphenazine. The inset shows a double reciprocal plot of the cAMP increase as a function of dopamine concentration from 3 to 300 μm. a, control; b, 1×10^{-7} M fluphenazine. (From Clement-Cormier et al. 1975.)

from interneurons, is also active. In contrast, the cerebellum, which does not appear to receive significant dopaminergic innervation, has no detectable dopamine-sensitive adenylate cyclase activity (Horn *et al.* 1974).

It is known that adenylate cyclase, phosphodiesterase, protein kinase, protein kinase substrate, and phosphoprotein-phosphatase are present in very high concentrations in brain tissue fractions rich in synaptic membranes.

Proof that dopamine-sensitive adenylate cyclase is attached to the postsynaptic neuron comes from lesion experiments. When the dopamine-containing nerve terminals of the basal ganglia are destroyed by surgical means, or chemically by 6-hydroxydopamine treatment, the dopamine-sensitive adenylate cyclase remains in homogenates of the denervated striatal tissue, showing that the enzyme is not localized in dopaminergic nerve terminals. When the GABA cells and ACh cells of the rat basal ganglia are destroyed by means of local injections of kainic acid, then the adenylate cyclase drops to 15%–20% of normal and that part responding to dopamine stimulation drops to zero. Since glial cells are preserved by this kainic acid treatment, it means that the dopamine-sensitive adenylate cyclase must be attached to the types of neurons destroyed by kainic acid (E. G. McGeer *et al.* 1976).

As far as the other postulated actions of cAMP in nervous tissue are concerned, i.e., altered microtubular function and increased neurotransmitter synthesis, these can be accounted for by a comparable process whereby cAMP phosphorylates protein kinases associated with these activities (Figure 8.8). For example, highly purified microtubules from brain contain an intrinsic cAMP-dependent protein kinase which catalyzes the phosphorylation of a substrate protein which is an integral component of the microtubule system.

Similarly, cAMP has been shown to activate tyrosine hydroxylase. This is almost certainly mediated through a protein kinase, since the activation can be mimicked by a purified preparation of protein kinase and can be abolished by specific inhibitors of brain protein kinase.

There is overwhelming evidence that cAMP acts as a second messenger not only in postsynaptic cells in the central nervous system but also in various target organs in the periphery of the body. It is the key molecule in a common mechanism which has been adopted to serve many biological purposes. That common mechanism is the phosphorylation of a group of enzymes known as the protein kinases, converting them from an inactive to an active state. cAMP provides a phosphate bond of unusually high energy and thus is capable of activating systems that would be unaffected by such other donors of high-energy phosphate bonds as ATP and creatine phosphate. The significance can best be appreciated by considering the spectrum of actions of cAMP so far identified in the periphery of the body. It mediates the effect of gonadotrophins on ovum maturation, luteinization, and spermatogenesis. It is responsible for increased steroidogenesis by mediating the effects of ACTH on adrenocortical cells. When thyroid-stimulating hormone is administered to thyroid cells in tissue culture, cAMP is the second messenger which stimulates new RNA and protein synthesis. The cultured cells then organize themselves into follicles.

Thus, cAMP indirectly stimulates the translation process by which new RNA and protein is synthesized. It enhances the production of trophic compounds, such as the steroids, which themselves have profound effects on cellular processes.

Electrophysiology of Dopamine Receptors. With few exceptions the iontophoretic effects of dopamine have been inhibitory in practically every part of the CNS, but it is uncertain whether this technique is capable of precisely stimulating the same receptor sites that dopamine itself stimulates upon physiological release. It does not cause a progressive fall in spike amplitude such as when neuron firing is suppressed by local anesthetics. Instead, there is a more rapid time course of depression comparable to, but not as rapid as, that seen with hyperpolarizing agents like GABA. In the neocortex, retina, and amygdala, dopamine is generally more potent than other catecholamines and has a substantial depressant effect, but the overall potency is considerably less than that of the inhibitory amino acids GABA and glycine (Krnjevic 1974). There have been some reports of excitatory iontophoretic effects in the striatum (York 1970) and elsewhere, but these have been outweighed by a greater number of reports that the predominant action is one of inhibition.

In the mollusk, dopamine is the only catecholamine detected in appreciable amounts in the nervous system. There is, therefore, a strong reason for believing that it plays a neurotransmitter role in this species. But again there has been difficulty deciding whether that role is excitatory or inhibitory based on iontophoretic results. The vagaries of the technique are best illustrated by the report of Ascher (1973) that, in the mollusk, excitatory or inhibitory effects can be observed in the same neuron, depending on the positioning of the pipette.

The key to understanding the action of dopamine may not lie, however, in studying its short-term iontophoretic effects. Its longer term effects may be much more significant. For example, a brief dopaminergic stimulation of the superior cervical ganglion results in a prolonged enhancement of the response to cholinergic stimulation (Libet *et al.* 1975). Similarly, a brief stimulation of the substantia nigra produces a sharply increased response of caudate neurons to limb stimulation (York and Lentz 1976) which persists for many minutes.

Overall Physiological Effects of Dopamine. The overall physiological effects of dopamine can best be appreciated by considering its major anatomical pathways and its metabotropic actions which appear capable of altering responses to iontotropic transmitters. Dopamine seems able to exert a profound influence over a wide variety of physiological functions without being critical to any one of them.

The three major dopaminergic tracts are the nigrostriatal, mesolimbic, and tuberoinfundibular. The nigrostriatal tract is concerned with initiation and execution of movement. Loss of function of this tract produces the well-known akinesia and rigidity of Parkinson's disease. The greatest difficulty Parkinsonian patients encounter is that of initiating a movement. Once initiated, movements may proceed in a grudging fashion, but it is difficult to make modifications. The Parkinsonian launched in a festinating gait toward a wall may collide with it before it is possible either to arrest the movement or change direction. Animals do not appear to suffer from this disease although comparable motor deficits can be produced by lesions of the tract, destruction with the neurotoxin 6-OHDA, or pharmacological blockade of the tract. The role of the nigrostriatal tract in movement will be discussed in much more detail in Chapter 13.

The functions of the mesolimbic-mesocortical complex are not well understood in physiological terms. Presumably there is an emotional function as well as one of thought organization. Dopamine has been implicated in a number of behaviors including stereotypy, brain self-stimulation, and consummatory behaviors such as eating and drinking which may involve this complex. This system is also the one implicated in dopamine theories of schizophrenia. These behavioral aspects will be discussed in Chapter 14.

The tuberoinfundibular system plays a role in the coupling phenomenon between the hypothalamus and the pituitary. Specifically, dopaminergic stimulation seems to inhibit prolactin secretion, to stimulate growth hormone secretion, to inhibit ovulation, and possibly to play a role in the release of other pituitary hormones (for a review, see Costa and Gessa 1977). These neuroendocrine effects will be discussed in Chapter 10.

In addition to these major tracts, dopaminergic interneurons exist in several locations within the nervous system. They are found (Table 8.1) in the chemoreceptor trigger zone (Murphree and Domino 1968), the diencephalon (Lindvall and Bjorklund 1974), the olfactory bulb (Halasz et al. 1977), the retina (Ehinger 1977), sympathetic ganglia (Bjorklund et al. 1970), and the carotid body (Bolme et al. 1977). In these locations they may play a modulatory role, altering the response to input from other transmitters. This is in keeping with the proposed metabotropic function which is discussed in this chapter. The manner in which dopaminergic stimulation can radically alter input from other sources is described for movement in Chapter 13, and for the superior cervical ganglion in Chapter 15, where it is proposed as a model for memory and learning.

Turnover of Dopamine.   Several methods have been used to determine the rate at which dopamine is turned over within a given tissue. Turnover rates, of course, are considerably below the maximum synthetic rate of a tissue because excess capacity is always present. Turnover rates also do not necessarily reflect the speed at which some of the molecules are metabolized. It is a well-established principle of the nervous system that such rates can vary tremendously and that the most newly synthesized transmitter is usually the most rapidly metabolized. Different methods of measurement of turnover rates do not always yield the same answers, perhaps reflecting the fact that different molecules of the same transmitter are metabolized at different rates. Nevertheless, turnover rates do give some indication of the average length of survival of a neurotransmitter molecule and also establish the effects of various drugs on neurotransmitter utilization. The following methods have been used:

1. Administration of a labeled pulse of dopamine intraventricularly or directly into tissues followed by measurement of its rate of decline. This method suggests a half-life of dopamine in rat striatum of approximately 1.5 h. The weakness of this method is that the dopamine cannot be entirely limited to dopaminergic neurons since some of it will be taken up by noradrenergic neurons. Furthermore, the labeled dopamine may not enter into an "average" pool but may preferentially find its way into a storage pool different from the "newly synthesized" pool.

2. Rate of appearance or disappearance of labeled dopamine after administration of a pulse of labeled tyrosine or Dopa. If the rate of appearance following tyrosine is to be used, a decarboxylase inhibitor is often administered and only Dopa formation measured. This, of course, does not distinguish between dopamine and noradrenaline except by selection of dopaminergic- or noradrenergic-rich areas of brain. Both synthesis and disappearance methods suggest a half-life of striatal dopamine of 2.5–4 h.

3. Measurement of the disappearance of endogenous dopamine following the administration of α-methyl-p-tyrosine or other potent inhibitors of tyrosine hydroxylase. As with method (2), there may not be an adequate distinction between dopamine and noradrenaline neurons and the drug used may influence the metabolism of either amine. Nevertheless, the figures obtained by this method are generally comparable to those of other methods, being approximately 4 h for striatal dopamine.

4. Measurement of the tissue or fluid levels of dopamine metabolites. The appearance of labeled homovanillic acid (HVA) and/or labeled dihydroxyphenylacetic acid (DOPAC), following a pulse of labeled dopamine should be a reflection of the amount of dopmaine consumed. This method generally indicates turnover times of about 2 h in the striatum. Diffusion of the metabolites is a factor to consider.

In all methods utilizing radioactive material the cold pool sizes should be measured. This is sometimes neglected, which may account for some of the variability in estimates of dopamine half-life obtained by various laboratories. Nevertheless, all the estimates are of the same order of magnitude (1.5–4 h) and there is general agreement of the probability that dopamine turnover is somewhat faster than is noradrenaline turnover (Costa 1970; Iversen and Glowinski 1966; Persson 1970).

Axoplasmic Flow of Materials in Dopaminergic Neurons. The phenomenon of axoplasmic flow was mentioned in Chapter 1. The nigrostriatal dopaminergic tract has turned out to be a convenient model for studying some of the general properties of axoplasmic flow in a central neuron. Injection of [³H]leucine into the region of the substantia nigra will result in rapid uptake of the amino acid at the site of injection with synthesis at that site into protein. The protein is then transported along the nigrostriatal tract into terminals in the striatum.

Figure 8.11 illustrates the biphasic rate of arrival of labeled protein in the striatum following injections of radioactive leucine into the substantia nigra. The free labeled amino acid disappeared from the injection site within a few hours and was not detected in significant quantities in the target area or elsewhere in the brain. This suggests that some of the leucine was being synthesized into protein at the site of injection within a matter of hours, with the remainder being very widely circulated in the body for general metabolism (Fibiger *et al.* 1972).

The transported protein was retained in the striatum for at least three weeks (Fibiger *et al.* 1972). Autoradiography showed that the bulk of the label was in nerve endings, the predominant type of which is presumed to be dopaminergic. Increasing amounts of protein appeared in dendrites of the striatum as time passed, however, suggesting the possibility of transsynaptic movement of protein.

The nigrostriatal tract has also been used as a model for studying the flow of low-molecular-weight materials. Prominent among these is the neurotransmitter itself. The nigrostriatal tract transports both dopamine and nor-

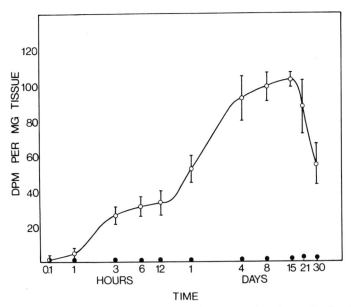

Figure 8.11 [¹⁴C]protein in the corpus striatum of rats at various intervals after injection of [¹⁴C]leucine into the left substantia nigra. ○ label in left corpus striatum; ● label in the right striatum. (From Fibiger *et al.* 1972.)

adrenaline (Figure 8.4B) but not the precursors tyrosine and Dopa. It also fails to transport serotonin, GABA, histamine, taurine, and glycine (E. G. McGeer *et al.* 1974). The process is thus selective, probably indicating a specificity in uptake and in the binding capacity of proteins carrying the transmitters. The amounts of transmitter carried by axonal transport are trivial by comparison with the amounts synthesized at the nerve ending. Therefore, this process is probably an accidental side effect of the transport of storage vesicles or the proteins that are contained in them. Further evidence in support of this conclusion comes from observing the effects of reserpine on transport of dopamine. Immediately following reserpine administration, transport of dopamine falls off considerably. After 24 h, however, it is greatly enhanced over control levels. This would be consistent with the well-known inactivation of vesicular storage processes by reserpine, followed by restoration commencing in the cell body and spreading to the nerve endings (Fibiger and McGeer 1973).

8.2.4 Pharmacology of Dopamine

The general importance of the three major dopaminergic pathways in brain, i.e., the nigrostriatal, the mesocortical-mesolimbic, and tuberoinfundibular, has already been discussed. They relate to extrapyramidal, behavioral, and neuroendocrine physiology, respectively. It is to be expected, therefore, that drugs which affect dopamine metabolism would influence these neuropsychiatric functions. Indeed, some of the drugs are among the most important in all of clinical medicine.

The various classes include toxins, receptor antagonists and agonists, vesicle storage depletors, false transmitters, pump inhibitors, synthesis inhibitors, dopamine precursors, and inhibitors of the metabolizing enzymes. Representatives are listed in Table 8.2, and their sites of action in Figure 8.12.

Toxic Agents.   A tool of inestimable value in elucidating the physiological effects of catecholamines is 6-OHDA (Johnsson *et al.* 1975). This widely used compound owes its selective neurotoxic action to specific uptake into catecholamine neurons. By destroying catecholamine pathways, it can reveal their function by the deficits produced.

The critical intraneuronal concentration which must be reached for 6-OHDA to be toxic has been calculated to be approximately 50 mM. Selectivity of the effect, therefore, depends upon the availability of 6-OHDA, as well as a strong pumping capacity to raise its intraneuronal concentration. Two hypotheses exist as to how destruction takes place at the molecular level. In one, either hydrogen peroxide or a superoxide radical, or both, generated by the *in vivo* oxidation of 6-OHDA is the causative agent. In the other, rapid nucleophilic reactions of the *p*-quinone of 6-OHDA or other quinone intermediates are responsible. Selective degeneration of dopaminergic versus noradrenergic neurons depends upon the conditions and site of administration. Intraventricularly administered 6-OHDA will attack most noradenergic and dopaminergic neuronal systems (and perhaps some noncatecholaminer-

TABLE 8.2
Some Drugs Which Affect Dopamine Metabolism

Drug		Presumed mechanism of action	Most prominent physiological effects
Antagonists			
Phenothiazines	—chlorpromazine[a] —fluphenazine —thioridazine	Receptor blockade	Tranquilization, antipsychotic and antinauseant effects
Butyrophenones	—haloperidol —spiroperidol		
Thioxanthenes	—flupenthixol —chlorprothixene		
Agonists			
Apomorphine		Receptor stimulation	Anti-Parkinsonian and emetic effects
Piribedil			
Bromocriptine			
Releasers			
Amphetamine		Release	Stimulant, appetite suppressant
Vesicular storage inhibitors			
Reserpine[a]		Depletion	Antihypertensive, tranquilization, and antipsychotic effects
Tetrabenazine[a]			

Pump inhibitors		
Benztropine	Reuptake inhibition	Anti-Parkinsonian effects
Amphetamine	Reuptake inhibition	Stimulant, appetite suppressant
Cocaine	Reuptake inhibition	Stimulant, euphoriant
Synthesis inhibitors		
α-Methyl-*p*-tyrosine[a]	Tyrosine hydroxylase inhibition	Depressant, akinesia
α-Methyl-Dopa	Dopa decarboxylase inhibition	Antihypertensive
Carbidopa	Dopa decarboxylase inhibition	Adjuvant for central Dopa
MAO inhibitors		
Iproniazid[a]	Broad-spectrum MAO inhibition	Antidepressant
Tranylcypromine[a]	Broad-spectrum MAO inhibition	Antidepressant
Clorgyline[a]	A type MAO inhibition	Antidepressant
Deprenyl	B type MAO inhibition	Antidepressant
COMT inhibitors		
Tropolone, pyrogallol[a]	COMT inhibition	Minimal effects
False transmitters		
α-Methyl-*m*-tyramine[a]	Replacement of transmitter	Mild tranquilization
α-Methyldopamine[a]		Antihypertensive
Toxin		
6-Hydroxydopamine[a]	Destruction of cells	Experimental
Precursor		
Dopa	Stimulates transmitter production	Anti-Parkinsonian and mild stimulant

[a] Also has prominent noradrenaline and/or adrenaline action.

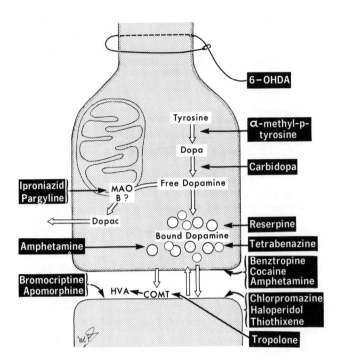

Figure 8.12 Site of action of dopaminergic drugs; schematic diagram of dopaminergic nerve ending. White rectangles and white arrows show endogenous metabolites and their location. Black rectangles and black arrows show the site of action of various drugs which interact with dopamine. See text and Table 8.2 for further details.

gic systems as well) because of its wide distribution. Direct injections into specific sites such as the substantia nigra will cause selective degeneration.

The nigrostriatal dopaminergic tract can thus be destroyed selectively by direct injection of 6-OHDA into the nucleus. The result of such bilateral injections is to produce animals with a profound motor deficit. In the rat, decreases in self-stimulation, eating, drinking, and other behaviors are noted. Animals so treated will eat some soft food, however, so that the desire for this or other consummatory behavior may be present without there being a motor capability of executing it. Attempts at selective destruction of other dopaminergic pathways in the brain have been made by combining 6-OHDA with a reuptake inhibitor for noradrenaline, such as desipramine, to protect noradrenergic pathways. The results have not been absolutely clear, so this technique, along with other drug studies, has failed to provide a clear identification of behaviors (other than motor function) with dopamine pathways as opposed to noradrenaline tracts.

Antagonists. Dopamine receptor blockers are the most important single group of dopaminergic drugs. They are used primarily in psychiatric medicine although they also have a place in general medicine because of their tranquilizing and antinauseant properties.

There is now overwhelming evidence that agents used to treat schizophrenia act as dopamine antagonists in the brain (Snyder *et al.* 1974a; Iversen 1975). Phenothiazines, butyrophenones, and thioxanthines are truly antipsychotic and have a specific effect in ameliorating the fundamental symptoms of psychosis. Their clinical effects will be discussed briefly in Chapter 14. Here we shall be concerned with their mechanism of action, remembering that other receptor sites than those for dopamine are affected by these agents.

The phenothiazines, butyrophenones, and thioxanthenes have comparable clinical and pharmacological effects and thus can be considered together. Within each of these families there are many representatives and a good number are in clinical use. There are definite variations in potency and properties, but these are less noticeable than their similarities.

Evidence that these agents are acting as dopamine blockers includes the following:

1. They antagonize the effects of iontophoretically applied dopamine.
2. They produce Parkinsonian-like extrapyramidal reactions.
3. They block dopamine-sensitive adenylate cyclase.
4. They produce an acceleration in the firing rate of dopaminergic cells due to compensatory transneuronal feedback stimulation.
5. They antagonize the effects of Dopa, of presumed dopamine receptor site stimulators, and of amphetamines.

The response of dopaminergic cells to systemically administered phenothiazines or butyrophenones is probably indirect. The firing rate dramatically increases, which is thought to be a transsynaptic effect in response to the blockade of endogenously released dopamine. Such an effect had been anticipated after the original report of Carlsson and Lindqvist (1963) of increased formation of the dopamine metabolite 3-methoxytyramine as well as normetanephrine following chlorpromazine or haloperidol. It was also shown that dihydroxyphenylacetic acid, another dopamine metabolite, was markedly elevated in the brains of rabbits treated with chlorpromazine or haloperidol. Next, it was shown directly with the use of [^{14}C]-labeled tyrosine that the synthesis of catecholamines as well as the turnover of dopamine in the corpus striatum was stimulated severalfold following treatment with neuroleptic drugs. This compensatory acceleration of synthesis of dopamine or noradrenaline by a centrally acting drug is now considered a requirement of receptor blockade. Phenothiazines, thioxanthines, butyrophenones, and dibenzoazepines all stimulate [^{14}C]dopamine and noradrenaline synthesis, but tricyclic antidepressants, sedatives, central stimulants, hallucinogens, and anticholinergics do not. Furthermore, phenothiazines lacking antipsychotic properties, such as promethazine, fail to stimulate dopamine synthesis.

Another method of observing the same effect is through the use of a cup implanted directly in the brain (Glowinski 1975). The release of [^{3}H]dopamine continuously synthesized from [^{3}H]tyrosine is greatly enhanced in superfusates reaching the cup following the systemic administration of neuroleptics. The effect is counteracted by transection of the nigro-striatal pathway indicating that it is a product of enhanced firing of dopaminergic cells and not some local action at the nerve ending.

Since loss of the nigrostriatal dopaminergic tract produces Parkinson's disease, it follows that dopaminergic blockers should produce extrapyramidal reactions. This is the case for all clinically effective butryophenones and thioxanthines and most phenothiazines. Some phenothiazines such as promethazine, which have strong antihistaminic action, lack this property and have little or no antipsychotic action. But others, particularly those with a piperazine moiety in the side chain, are strongly antipsychotic and are potent at producing extrapyramidal reactions.

The low incidence of extrapyramidal side effects in still other compounds such as thioridizine, and to a lesser extent chlorpromazine, may be associated with another pharmacological property which counteracts their dopamine-blocking action. That is a cholinergic-blocking action, which in itself has an anti-Parkinsonian effect. As previously described (Chapter 5), the striatum is rich in cholinergic receptor sites which are muscarinic in nature. These apparently play a balancing role to dopamine.

There is a direct dopaminergic–cholinergic linkup in the striatum (Chapter 13). Since the action of dopamine appears to be inhibitory, while that of ACh appears to be excitatory, the receptor sites for these two transmitters act antagonistically in that nucleus. Drugs which block both receptor sites should produce a different effect on movement than those which block only one.

The blocking potency of cholinergic as compared with dopaminergic receptors has been measured for a series of drugs (Iversen 1975). Benztropine, which can reverse drug-induced extrapyramidal reactions, has a ratio of cholinergic to dopaminergic-blocking potency of approximately 7.5 $\times$ 10^4. By contrast, agents with antipsychotic action, but only a weak tendency to produce extrapyramidal side effects, such as chlorpromazine and thioridazine, have a ratio near unity. Drugs with a powerful tendency to produce such side effects, like flupenthixol, may have a ratio as low as 5×10^{-5}.

As described earlier in this chapter, several studies have now shown that dopamine analogs stimulate adenylate cyclase activity in homogenates of striatum and other brain areas containing rich concentrations of dopaminergic nerve endings. Antagonists of dopamine block such stimulation. The results of four independent investigations are summarized in Table 8.3 (Iversen 1975) showing the relative potency of various agents to antagonize dopamine-sensitive adenylate cyclase. Striatal or retinal homogenates were stimulated by supramaximal concentrations of dopamine in the test system with or without various concentrations of the inhibitory (antipsychotic) agent. A variety of antipsychotic drugs inhibited the dopamine stimulation of adenylate cyclase. The classical drug chlorpromazine is less powerful (K_i 5-10 $\times 10^{-8}$ M) than such agents as flupenthixol and fluphenazine whose K_i values are in the range 10^{-9} M. Within families, such as the phenothiazines, the

TABLE 8.3
Inhibition of Dopamine-Sensitive
Adenylate Cyclase
by Antipsychotic Drugs

Drug	Inhibition constant K_i (nM)[a]		
α-Flupenthixol	1.0		
α,β-Flupenthixol	3.5		
Fluphenazine	4.3	8.0	7.1
(+)-Butaclamol	8.8		
α-Clopenthixol	16.0		
Trifluoperazine	19.0		
α-Chlorprothixene	37.0		
Trifluoperazine	44.0		
Chlorpromazine	48.0	66.0	48.0
Spiroperidol	95.0		
Prochlorperazine	100.0	55.0	
Thioridazine	130.0	55.0	
Pimozide	140.0	122.0	
(±)-Bulbocapnine	160.0		
Clozapine	170.0		
Haloperidol	220.0	38.0	
Chlorimipramine	420.0		
Ergotamine	430.0		
β-Chlorprothixene	950.0		

[a]The K_i values represent dissociation constants for binding of drugs to the dopamine sites of the dopamine-sensitive adenylate cyclase; the values were calculated from the drug concentrations needed to cause 50% inhibition of the dopamine response. Data assembled from several laboratories by Iversen (1975). Drugs with $K_i > 1000$ nM are not listed. None of them possesses antipsychotic potency. Results were largely obtained from rat corpus striatum, although calf retina was also used.

difference in potency in various members as inhibitors of the dopamine-stimulated cAMP formation is comparable to their clinical potency.

The thioxanthines offer a critical test of the hypothesis that antipsychotic action is related to the ability to block dopamine-sensitive adenylate cyclase in that they exist in isomeric form by virtue of the carbon–carbon double bond linking the side chain to the heterocyclic ring system. α-Flupenthixol has a side chain oriented *toward* the halogen substitution on the heterocyclic ring and has potent antipsychotic activity. The trans-isomer (β-flupenthixol) has its side chain oriented in the opposite direction and is virtually devoid of antipsychotic activity. α-Flupenthixol is an extremely potent inhibitor of dopamine-sensitive adenylate cyclase while β-flupenthixol is ineffective even at relatively high concentrations.

A weakness in the theory involves the relative potency of the butyrophenones. Drugs in this class, such as a haloperidol, spiroperidol, and pimozide, are among the most potent antipsychotic drugs known. Yet they are less potent than relatively weak phenothiazines such as chlorpromazine in inhibiting dopamine-sensitive adenylate cyclase. The reason for this quantitative discrepancy is not known, although qualitatively the rule still applies.

Most inhibitors of dopamine-stimulated adenylate cyclase have similar activities in various brain regions but sufficient variation apparently exists (Mishra *et al.* 1976) among the cortex, limbic system, and striatum to suggest that dopaminergic receptors in these locations are similar but not identical.

The adenylate cyclase related to dopamine receptors has pharmacological properties which clearly distinguish it from receptors related to other neurotransmitters such as noradrenaline. For example, the dopamine system is not activated by the β-adrenergic agonist isoproterenol, nor blocked by the β-adrenergic antagonist propranolol. It survives homogenization in hypotonic media which the adrenergic types does not. Nevertheless, as will be described later in this chapter, phenothiazine and other dopamine-blocking agents also affect other transmitters, particularly noradrenaline. Agents such as 6-OHDA, or any of the dopamine antagonists, produce supersensitivity in the dopamine receptor. Supersensitivity refers to a phenomenon in which a given amount of biologically active material produces a greater than normal biological response. It can apparently be produced by any treatment which results in chronic deprivation of normal stimulation; it is, for example, produced by reserpine or α-methyl-p-tyrosine (AMPT), but not by general depressants such as phenobarbital, diazepam, or promethazine.

Evidence of dopaminergic receptor supersensitivity in the striatum has been obtained from both behavioral and biochemical tests (Chapter 13). Increased sensitivity of striatal adenylate cyclase to dopamine in homogenates from rats subjected to electrolytic lesions of the nigrostriatal tract occurs, as well as a decreased threshold to the inhibitory effects of iontophoresed dopamine or apomorphine.

Agonists.   Apomorphine, piribedil, and more recently bromocriptine are the three most widely studied dopaminergic agonists.

Evidence for dopamine agonism is precisely the opposite of that for dopamine antagonism. Dopamine agonists stimulate dopamine-sensitive adenylate cyclase and antagonize phenothiazine blockade. They cause a decrease in dopamine turnover and reduce the firing rate of dopaminergic neurons (Figure 8.13). They counteract the effects of idiopathic and drug-induced Parkinsonism. They induce stereotypy in normal rats, which consists of sniffing and repetitive head and limb movements at low doses, and of continuous gnawing, biting, and licking at larger doses.

Their most important clinical action is to overcome the effects of Parkinson's disease by stimulating dopaminergic receptor sites in the striatum. The search for long-acting, selective dopaminergic agonists is an active one because it is presumed that the neostriatal receptor sites in Parkinson's disease

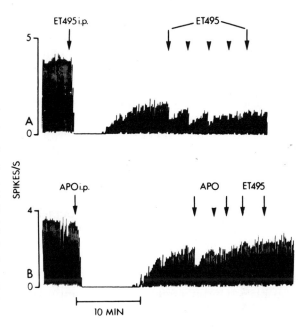

Figure 8.13 Typical effects of various dopamine agonists on the firing rate of dopaminergic cells located in the pars compacta of the substantia nigra. APO (apomorphine); ET 495 (piribedil). Drugs administered i.v. except where i.p. indicated. The firing of cells was completely inhibited by a single i.p. injection of either dopamine agonist with recovery to about 40% of the original activity after 10–15 min. Futher i.v. injections of drugs had only a transient further depressing effect. (From Walters *et al.* 1975.)

not only survive the disease process, but develop supersensitivity due to dopamine deprivation.

Dopamine agonists also stimulate receptor sites in the chemoreceptor trigger zone, inducing vomiting. This is hardly a clinical objective compatible with treating Parkinson's disease, although in the case of apomorphine, the classical dopaminergic agonist, this property has been capitalized upon in its clinical employment as an emetic agent.

Dopamine agonists all have the property of inducing hypothermia in mice, which may be related to actions on the mesolimbic system. They also block ovulation in immature rats and have other endocrine effects referable to the tuberoinfundibular system.

The effect of apomorphine on dopamine synthesis in rats that have had a transection of the nigrostriatal tract has helped to inspire the concept of autoreceptors on dopaminergic neurons (Kehr *et al.* 1975). When the tract is lesioned, dopamine synthesis in striatal nerve endings is increased. An explanation based on transneuronal feedback stimulation due to reduced activation of postsynaptic receptors is untenable inasmuch as the dopaminergic cell bodies are now disassociated from their nerve endings. An explanation based on simple leakage of dopamine might apply if it were not for the fact that apomorphine antagonizes this increased synthesis. Since apomorphine cannot be working on postsynaptic receptors, the conclusion is that it must be working on presynaptic or autoreceptors. Other evidence for autoreceptors on dopaminergic nigrostriatal neurons comes from the work of Bunney *et al.* (1973). They found that the firing of dopaminergic neurons in the substantia nigra was retarded by the iontophoretic application of dopamine or apomorphine. While this is evidence for autoreceptors on the cell body rather than the nerve ending, the result is in the expected direction and it is not unreasonable to suppose that autoreceptors should be distributed throughout the surface of the neuron.

Vesicle Storage Inhibitors.   Reserpine, which has already been mentioned several times in this chapter, is the classical intraneuronal storage inhibitor for dopamine, other catecholamines, and serotonin. Despite the fact that reserpine is one of the most widely used drugs in basic neurological science, it has still not been determined how it affects the vesicular storage process. The amounts required to displace amines are too low and too long lasting for the inhibition to be competitive, and yet preloading of a nerve terminal with an amine precursor will protect against depletion. Studies using platelets have shown that reserpine sticks to the platelet membrane, suggesting that it exerts its effects at the membrane level of the intraneuronal vesicle. Reserpine has many physiological effects, but the only ones clearly attributable to dopamine are those of Parkinsonian-like extrapyramidal reactions. Dopamine probably also plays a prominent part in the antipsychotic and tranquilizing effects. Reserpine will be further discussed in Chapter 14.

Tetrabenazine, the most useful of the benzoquinolizines, has an action comparable to that of reserpine, but of much shorter duration.

Pump Inhibitors.   Since reuptake may be the primary method of terminating the action of dopamine and other neurotransmitters, inhibitors of this process can be expected to have prominent physiological effects in the direction of enhancing transmitter function. Those inhibitors which appear to be preferential for dopamine, as opposed to noradrenaline or serotonin, have widely varying properties because of their other actions. The tricyclic antidepressants, which inhibit dopamine uptake to some extent, are stronger noradrenaline and serotonin pump inhibitors. They will be discussed in later sections. Benztropine, amphetamine, and cocaine are three well-known dopamine pump inhibitors. Each has prominent but differing alternative actions. Benztropine is a powerful muscarinic antagonist. Amphetamine is a releaser of dopamine, and to a lesser extent of noradrenaline, as well as being a competitive inhibitor for MAO. Cocaine is a noradrenaline pump inhibitor as well, and is a local anesthetic.

Several other agents that have an anti-Parkinsonian action are dopamine reuptake inhibitors (Coyle and Snyder 1969). Benztropine is the most effective, particularly against drug-induced extrapyramidal reactions, and is the most powerful pump inhibitor. Amphetamine and cocaine, on the other hand, have stimulant and euphoriant properties, and, at high doses, may induce psychotic reactions.

Precursors.   The precursors of dopamine are tyrosine and L-Dopa. As mentioned under metabolism, tyrosine loads do not influence dopamine levels even in the presence of an MAO inhibitor, but L-Dopa has a pronounced effect. Many books and articles now document the effectiveness of L-Dopa in clinical medicine and describe the nuances of its use (Calne 1973;

McDowell and Barbeau 1974). L-Dopa in doses of 1–8 g/day is now the treatment of choice in Parkinson's disease. Amounts of this order (50–100 mg/kg) given in acute parenteral doses to rats will more than double brain levels of dopamine. It is not known, of course, what tissue levels of dopamine are achieved in humans, but homovanillic acid levels in the CSF increase several fold after therapeutic doses of L-Dopa.

L-Dopa crosses the blood–brain barrier with ease, but the endothelial cells lining brain capillaries contain considerable quantities of Dopa decarboxylase. Thus, a portion of every parenteral dose is decarboxylated before it can reach catecholamine cells. It is presumed that in Parkinson's disease sufficient residual Dopa decarboxylase exists to convert Dopa to dopamine, but other hypotheses exist, including decarboxylation by brain capillaries. Some Dopa may be converted to dopamine by noradrenergic and serotonergic cells, and there is even the suggestion that Dopa itself may be active at receptor sites without conversion.

L-Dopa in high doses produces stereotypy and choreiform movements, particularly of the head and neck. Nausea is almost universal at high doses. These two effects are usually the limiting factor in achieving therapeutic results in Parkinson's disease and are referable to dopaminergic stimulation of receptors in the striatum and chemoreceptor trigger zone respectively.

In about 15% of cases L-Dopa will produce agitation, excitement, and even psychosis, presumably from stimulation of receptors in the cortex and limbic system. It will also stimulate the release of growth hormone and inhibit lactation, as expected from the effects of stimulating the tuberoinfundibular system.

Synthesis Inhibitors

Tyrosine Hydroxylase Inhibitors. Since tyrosine hydroxylase is the rate-limiting enzyme in catecholamine synthesis, it would be anticipated that inhibitors of the enzyme would greatly decrease catecholamine levels and produce physiological changes of practical benefit in clinical medicine. So far no agent has emerged that meets the tests of safety and effectiveness. The most commonly used inhibitor in experimental situations is α-methyl-p-tyrosine (AMPT), but it is toxic to liver and kidney.

Many compounds have been screened as potential inhibitors (for a review, see E. G. McGeer and McGeer 1973) and they generally fall into the categories of analogs of tyrosine, inhibitors of the pteridine co-factor, catechols, and chelating agents. In addition to AMPT, its methyl ester and 3-iodotyrosine have proved to be potent inhibitors.

In general, inhibitors which are successful in reducing catecholamine levels *in vivo*, such as AMPT, produce effects similar to reserpine and the phenothiazines even though serotonin action is not affected. AMPT, for example, causes sedation, reduces motor activity, decreases blood pressure, and decreases intracranial self-stimulation.

Dopa Decarboxylase Inhibitors. A number of effective Dopa decarboxylase inhibitors exist which are analogs of Dopa. Hydrazine derivatives such as carbidopa and MK 485 have found great clinical usefulness as Dopa adjuvants in the treatment of Parkinson's disease. They do not cross the blood–brain barrier and thus inhibit the decarboxylation of Dopa only in the periphery. The result is that the brain receives a much higher proportion of parenterally administered Dopa than if the decarboxylase inhibitor had not been administered. The same therapeutic result can be achieved with one-third to one-quarter the dose of Dopa, and the peripheral side effects can be substantially reduced. Dopa decarboxylase inhibitors are not as toxic as tyrosine hydroxylase inhibitors, though they are broader spectrum in that they affect 5-HTP decarboxylation as well. Some hydrazine-type Dopa decarboxylase inhibitors cross the blood–brain barrier and have central as well as peripheral effects. At very high doses they can completely inhibit catecholamine formation, and, as described earlier, even be used to estimate rates of catecholamine synthesis. Generally speaking, however, these agents are not capable of performing the equivalent of a chemical sympathectomy because the amounts of Dopa decarboxylase are so much higher than tyrosine hydroxylase.

An unusual compound is α-methylDopa (Aldomet), which is a competitive inhibitor of Dopa decarboxylase. It was originally thought that its sedative and antihypertensive properties were due to inhibition of this enzyme, but later it was realized that the decarboxylation products, α-methyldopamine and α-methylnoradrenaline, could act as false transmitters. Thus, its action overlapped with a separate class of catecholaminergic drugs.

False Transmitters. False transmitters are agents which indirectly affect transmitter action, substituting for the real transmitter in mechanisms of storage and/or release. Although there is an implication that such molecules are themselves less effective at receptor sites, and that failure of synaptic transmission is inevitably associated with their accumulation, this is not necessarily required by the definition.

α-MethylDopa and α-methyl-m-tyrosine are converted into methylated catecholamines and methylated m-tyramines, respectively. A number of similar phenylethylamines can also be taken up into catecholamine storage granules for release upon stimulation, including p-hydroxynorephedrine which can be made *in vivo* from amphetamine. Amphetamine and tyramine probably replace transmitters at the cytoplasmic level and are not stored to the same extent by cytoplasmic granules. Depending on the dose, time after administration, and strength of the false transmitter, the net effect may be excitatory or inhibitory to catecholaminergic action. Most of the properties of these agents are depressing and referable to the peripheral noradrenergic system. None is antipsychotic, and none is particularly useful in the treatment of Parkinson's disease. Amphetamine has a mild anti-Parkinsonian effect and is strongly stimulatory. It is also a pump and MAO inhibitor, and does not fall strictly within the definition of a false transmitter because it is not stored in granules *per se.*

MAO Inhibitors. As one of the neurotransmitters metabolized in substantial part by MAO, dopamine is markedly affected by MAO inhibitors. MAO, of course, is located in dopaminergic nerve endings, as well as glial cells and nondopaminergic postsynaptic dendrites. At least two forms of mitochondrial MAO have been identified, A and B. Dopamine appears to be a substrate for both. The types existing in various structures having to do with dopamine metabolism have not been determined. As a consequence, no information exists as to whether MAO inhibitors selective for dopamine, as opposed to the other catecholamines and serotonin, might be found. Broad-spectrum inhibitors such as ipronazid (irreversible type) or tranylcypromine (reversible type) produce a variety of actions (Chapter 14), some of which are undoubtedly attributable to enhanced dopamine activity. Too little information is available on such narrow-spectrum inhibitors as clorgyline and depre-nyl to know if they produce significantly different effects.

COMT Inhibitors.   Pyrogallol and tropolone are the two best known inhibitors of COMT. These agents have little physiological action, indicating the predominant role of MAO in metabolizing dopamine.

8.3 The Noradrenaline Neuron

The noradrenaline neuron differs from the dopamine neuron by possessing the enzyme dopamine-β-hydroxylase (DBH). DBH is associated with the noradenergic storage granule, which may explain why dopamine, even though it is available in generous quantity as a precursor, is not stored by noradrenergic terminals.

Noradrenergic pathways have been extensively mapped in brain by a variety of methods. Originally, the histochemical method of Falck and Torp was used, with differentiation from dopaminergic neurons achieved through the use of pharmacological agents. More recently, the glyoxylic acid method of Lindvall and Bjorklund (1974) and the immunohistochemical method for DBH of Hartman (Swanson and Hartman 1975) have been used. The newer techniques have resulted in some revisions in the original noradrenaline distribution described by Dahlstrom and Fuxe (1964a, 1964b).

Figure 8.14 shows a sagittal, and Figure 8.3B a horizontal projection of the ascending noradrenaline pathways according to Ungersted (1971) who employed the histofluorescence method. A comparable, but slightly different distribution is shown in Figure 8.15 using the immunohistochemical method of Swanson and Hartman (1975). Ungerstedt describes two main ascending noradrenergic pathways, a dorsal and a ventral bundle. It is the ventral noradrenergic pathway which has been questioned. It gathers fibers from the A1, A2, A5, and A7 cell groups of Dahlstrom and Fuxe in the medulla oblongata and pons. The axons ascend in the midreticular formation and are said to turn ventromedially along the lemniscus medialis and continue rostrally mainly within the medial forebrain bundle. The system gives rise to noradrenaline nerve terminals in the lower brainstem, mesencephalon, and diencephalon.

The route followed by the dorsal ascending noradrenergic pathway has been confirmed by immunohistochemistry. Cell bodies in the locus coeruleus contribute prominently to this system. Axons ascend in the dorsal tegmentum

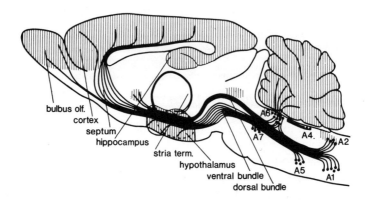

Figure 8.14 Noradrenaline projections by histochemistry; sagittal projection of the ascending NA pathways traced by the fluorescent histochemical method. The descending pathways are not included. Striped areas indicate major nerve terminal fields. (From Ungerstedt 1971). Horizontal projection of pathways is shown in Figure 8.3B. For transverse section at midbrain level, see Figure 9.3A, Section 9.2.

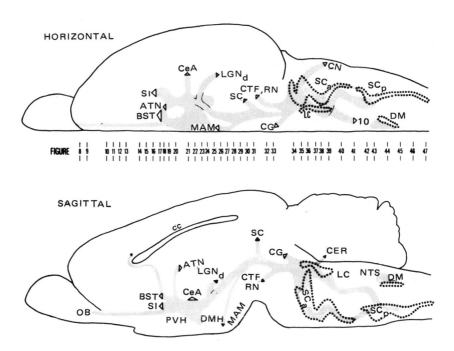

Figure 8.15 Noradrenaline pathways in horizontal and sagittal section traced by the immunohistochemical method for dopamine-β-hydroxylase. LC, locus coeruleus; SC, subcoeruleus complex; DM, dorsal medullary group. For other abbreviations, see the original paper of Swanson and Hartman (1975).

and at the diencephalic level turn ventrally to join the ascending dopamine axons in the medial forebrain bundle. Terminals are found in the diencephalon, limbic system, and cortex. The locus coeruleus also gives rise to noradrenergic terminals in the cerebellum, and, through a descending pathway, the lower brainstem nuclei. Thus, the locus coeruleus gives rise to noradrenaline terminals in practically all areas, and is in the position of influencing the brain in an unique way. The pathways are entirely uncrossed.

The A1 and A2 cell groups also give rise to descending pathways. They partly follow the anterior funiculus and ventral aspect of the lateral funiculus to terminate in the ventral horn, and partly follow the dorsal part of the lateral funiculus to terminate in the sympathetic lateral column and the dorsal horn. Some of these fibers appear to cross the midline.

The distribution described by Swanson and Hartman using DBH immunohistochemistry differs in a number of significant details. They define the noradrenergic system in the rat as consisting of approximately 10,000 neurons confined to three cell groups in the pons and medulla. The major group is in the locus coeruleus corresponding to the A4 and A6 groups of Dahlstrom and Fuxe. The second group, defined as the subcoeruleus complex, includes the A1, A3, A5, and A7 cell groups. The third, termed the dorsal medullary group and containing only about one-quarter as many cells as the locus coeruleus, corresponds to the A2 nucleus.

They find only one major ascending fiber system from these cell complexes which more or less follows the dorsal noradrenaline pathway of Ungerstedt. Ascending terminations are many. Proceeding rostrally, noradrenergic fibers sparsely innervate almost all structures in the medulla–pons area with a characteristic innervation of all nuclei of the raphe system. Innervation of the cerebellum is to all layers of the cortex and deep nuclei, but is sparse in density.

A major noradrenergic terminal field occurs in the localized zone of the mesencephalic central gray. Both inferior and superior colliculi are innervated, as are the lateral and medial geniculate nuclei of the thalamus. Most other nuclei of the thalamus are innervated with particularly rich terminal fields in the anteroventral and periventricular nuclei. The hypothalamus also receives rich innervation. Many fibers enter the paraventricular and supraoptic nuclei.

The septal area receives innervation from the fiber tracts that pass through it on their way to the neocortex via the genu of the corpus callosum. The hippocampus receives its input from caudally directed fibers passing through the retrosplenial cortex. The central anterior areas are the only parts of the amygdala to receive a significant noradrenergic input. This interpretation differs from that of Ungerstedt, who reported a dopaminergic input to the central nucleus and noradenergic input to the remainder. The neocortex is diffusely innervated with the largest number of noradrenergic fibers residing in layer 1 and with the cingulate cortex containing the densest innervation.

Areas receiving little noradrenergic innervation are the caudate-putamen, nucleus accumbens, globus pallidus, olfactory tubercle, and substantia nigra. All of these areas are innervated by dopaminergic neurons.

The most significant difference between the results of Swanson and Hartman using the immunological technique and those of Ungerstedt and others using the histofluorescence technique is the existence of a major noradrenergic component of the medial forebrain bundle at the level of the hypothalamus. Since this area is known to be richly innervated by ascending dopaminergic fibers, it is quite possible that the histofluorescence technique is incapable of distinguishing between a dense dopaminergic innervation and a noradrenergic one.

The peripheral sympathetic nervous system is, of course, primarily noradrenergic. The noradrenergic fibers are the postganglionic gray rami emanating from the paravertebral ganglia, which are fed by preganglionic fibers from the 12 thoracic and upper 2 lumbar segments. The noradrenergic cell bodies are concentrated in ganglia which lie in a paravertebral column. They are assembled in a somewhat irregular fashion so that certain ganglia, such as the superior sympathetic, are considerably larger than others. Nerve fibers feed the blood vessels, sweat glands, pilo-erector muscles of hair follicles, the eye, heart, intestinal tract, spleen, bladder, lungs, and sex organs. Fibers to the salivary glands and adrenal medulla, while anatomically arranged as part of the sympathetic system, are cholinergic rather than noradrenergic.

The relative content of noradrenaline in cell bodies versus axons and nerve endings can be estimated by comparing the content in sympathetic ganglia, postganglionic sympathetic fibers, and any of the various organs which are innervated by sympathetic nerve endings. Norberg and Hamberger (1964) have estimated the noradrenaline content of a sympathetic ganglion cell to be of the order of 10–100 μg/g. This compares with an estimated concentration in axons of 500 μg/g and in terminals 10,000 μg/g. The number of terminal varicosities per cell body is about 26,000. These 26,000 terminals contain in total approximately 300 times as much granule-bound noradrenaline as the cell body.

Keeping these rough approximations in mind, it is possible to gain some insight into the richness of noradrenergic innervation of a given tissue from the overall noradrenaline content. The highest concentration is found in the vas deferens (10 μg/g in the guinea pig), ciliary body and iris (5 μg/g in the rabbit), heart (2 μg/g in the rabbit), and salivary glands (1 μg/g in the rat). Most tissues also contain small amounts of adrenaline (from 2%–15% of the total catecholamine). In brain, the highest reported concentrations of noradrenaline are in the hypothalamus (1.3 μg/g in the human; 3.9 μg/g in the cat) and locus coeruleus (1.1 μg/g in the human).

Table 8.4 shows comparative tyrosine hydroxylase, Dopa decarboxylase, dopamine, noradrenaline, and homovanillic acid values for various areas of human brain.

TABLE 8.4
Values for Catecholamine Components in Human Brain

Brain area	TH[a,f]	DDC[b,f]	DA[c,g]	NA[d,h]	HVA[e,g]
Cortex	—	—	.05	—	—
Limbic System					
Nucleus accumbens	29.9	139	—	1.27	—
Septal area	11.4	—	0.30	0.24	—
Anterior perforating					
substance	28.6	107	—	0.33	—
Amygdala	3.7	61	0.60	0.06–1.37	0.67
Hippocampus	7.7	—	—	—	—
Basal ganglia					
Caudate	33.6	153	4.06	0.09	2.92
Putamen	35.3	141	5.06	0.12	4.29
Globus pallidus	17.5	101	0.42	0.10	1.93
Diencephalon					
Hypothalamus	19.8	125	0.80	1.25–1.92	—
Mammillary bodies	8.1	—	—	0.32	—
Midbrain-Pons					
Substantia nigra	52.1	177	0.38	0.23	—
Pars compacta	70.5	—	—	—	—
Pars reticulata	27.9	—	—	—	—
Locus coeruleus	—	—	1.22	1.10	—

[a]TH   — Tyrosine hydroxylase nm/h/100 mg protein.
[b]DDC — Dopa decarboxylase nm/h/100 mg protein.
[c]DA — Dopamine in μg/g wet tissue.
[d]NA — Noradrenaline in μg/g wet tissue.
[e]HVA — Homovanillic acid in μg/g wet tissue.
[f]P. L. McGeer and McGeer 1976.
[g]Lloyd 1975.
[h]Farley and Hornykiewicz 1977.

8.3.1 Biochemistry of Noradrenergic Neurons

Noradrenergic neurons are characterized by the presence of the copper-containing enzyme dopamine-β-hydroxylase (DBH, EC 1.14.17.1), which hydroxylates the ethylamine side chain of dopamine. Valence changes in copper evidently play an essential part in the activity of the enzyme because copper chelating agents inhibit its activity. One of the problems in measuring DBH activity in brain and other tissues is the presence of endogenous inhibitors which have this property. Thus, in order to assay the enzyme, it is necessary either to purify it partially or to add sulfhydryl agents which neutralize the endogenous inhibitors. The reaction which is catalyzed by DBH is shown in Figure 8.16.

During the oxidation, ascorbate reduces the cupric copper of the enzyme which is then oxidized when the reduced enzyme reacts with dopamine and oxygen. Thus, the enzyme-bound copper undergoes cyclic reduction and oxidation during the reaction. The K_m for dopamine is 6 mM.

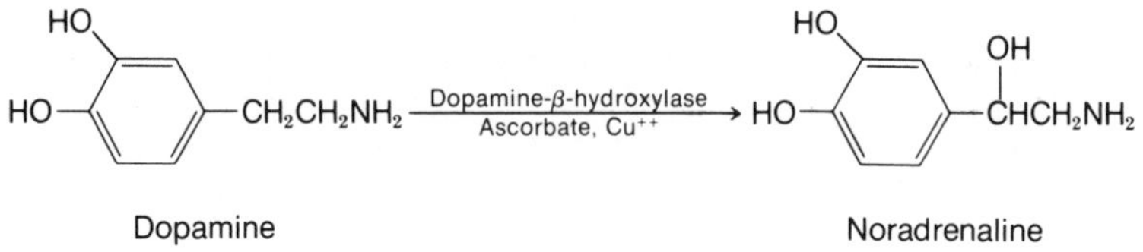

Figure 8.16 Synthesis of noradrenaline from dopamine.

The enzyme also catalyzes the side chain hydroxylation of several amines structurally related to dopamine, such as N-methyl-p-tyramine, α-methyl-p-tyramine, α-methyl-m-tyramine, p-hydroxy-N-methylamphetamine, α-methyldopamine, and mescaline.

DBH has been purified to homogeneity by several methods. Friedman and Kaufman (1965) estimated the molecular weight of their purified beef adrenal DBH to be 290,000. It has been more recently determined that it is a tetrameric glycoprotein of about 75,000 daltons (Wallace et al. 1973).

One of the striking features of DBH is its association with storage vesicles. Indeed, in subcellular fractionation studies, DBH has proved to be a reliable vesicle marker.

Noradrenaline is metabolized by the same enzymes that metabolize dopamine. These are MAO and COMT. But there is one significant difference in the manner in which MAO handles noradrenaline within the central nervous system. The aldehyde formed by the action of MAO is not converted to the acid but tends to be reduced to the alcohol. The sequence of reactions is as shown in Figure 8.17.

Thus, while the principal products of peripheral metabolism of noradrenaline are normetanephrine, 3,4-dihydroxymandelic acid, and vanillylmandelic acid (MHMA or VMA), the principal products of central metabolism are 3,4-dihydroxyphenylglycol and 3-methoxy-4-hydroxyphenylglycol (MHPG). Several studies in different species suggest that from 25%–60% of the urinary MHPG has its origins in the metabolism of brain noradrenaline, while virtually all of the urinary VMA results from the metabolism of peripheral noradrenaline. Studies of the fate of injected [³H]noradrenaline show that MAO plays only a secondary role to COMT in the metabolism of such exogenously administered catecholamines. Studies of the physiological effects of inhibitors of MAO and COMT, however, indicate that it is the former, not the latter enzyme, that is the most influential in modulating the action of the endogenously synthesized amine.

8.3.2 Physiology of Noradrenaline

Storage. The discovery that noradrenaline and other catecholamines were stored within highly specialized granular particles in sympathetic nerve endings and chromaffin cells represented a great conceptual advance in the 1950s. These granules were separated by subcellular fractionation studies and their makeup studied by a variety of techniques.

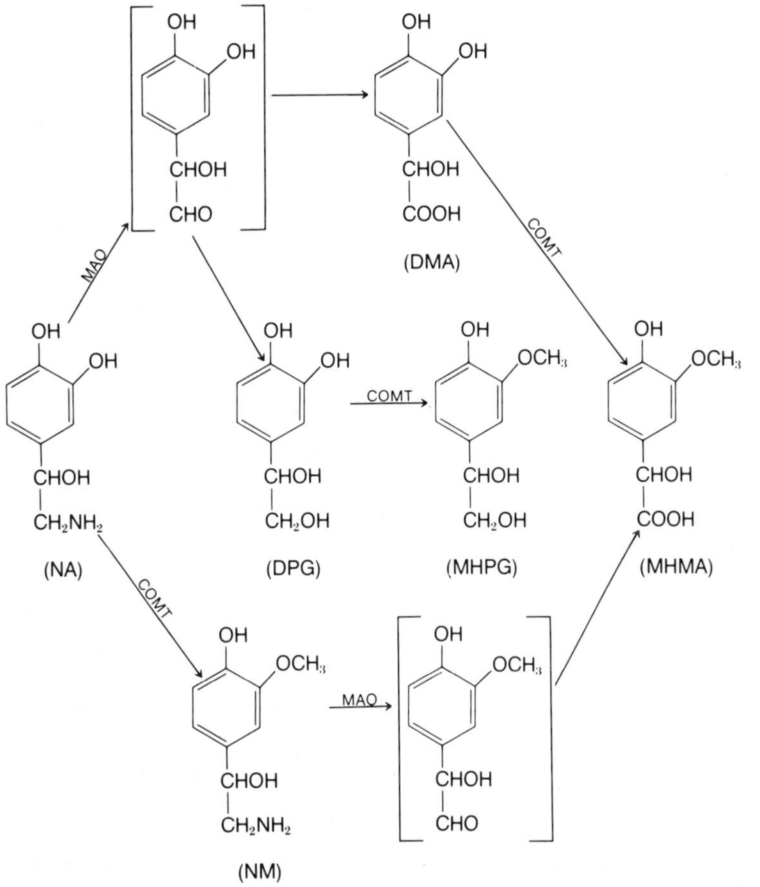

Figure 8.17 Metabolism of noradrenaline. DMA, dihydroxymandelic acid; DPH, dihydroxyphenylglycol; NM, normetanephrine; MHMA-VMA, vanillylmandelic acid; MHPG, 3-methoxy-4-hydroxyphenylglycol.

The concentration of catecholamines in the granules has been estimated to be 0.6 M—more than twice the osmolarity of mammalian body fluid and near the solubility limits. It is therefore of some considerable interest as to how this storage can be achieved. Granules have been found to contain adenine nucleotides, chiefly as ATP, in an approximate ratio of four molecules of catecholamine to one of the adenine nucleotide. The composition is roughly: water, 68.5%; catecholamine, 6.7%; adenine nucleotides, 4.5%; protein, 11.5%; lipids and other molecules, 7.8%. The principal protein in granules from adrenal chromaffin cells and splenic nerve is chromogranin, also believed to be involved in the storage process. The granules possess an outer limiting membrane which must play a role in taking up and storing noradrenaline. Obviously, noradrenaline cannot be in free solution inside the vesicle or else the vesicle would swell and burst due to hyperosmotic conditions. Therefore, some kind of a salt linkage between the anionic phosphate groups of ATP and the amine group of noradrenaline must exist. The granules contain an Mg^{2+}- and Ca^{2+}-dependent ATP-ase.

It is generally believed that the vesicles or components of them are formed in the neuronal cell body and are subsequently transported to the nerve terminal region by the process of axoplasmic flow (Figure 1.14, Section 1.2.4). It is conceivable that vesicles are formed from the outer membrane of the nerve terminal by the process of pinocytosis, but electron microscopic studies on the sciatic nerve just proximal to the ligated area reveals a substantial increase of granular elements, which favors the concept of transport. Additional evidence is that following reserpine administration noradrenaline begins to reappear in cell bodies, but not nerve endings, after a few hours. This implies a restoration of binding capacity through the synthesis of new proteins associated with vesicles. This is also consistent with the observation mentioned earlier of decreased transport of dopamine in the nigrostriatal tract immediately following reserpine, but greatly enhanced transport 24 h later when new binding material has been synthesized.

Noradrenergic storage vesicles differ from dopaminergic ones in possessing DBH. Most of the enzyme appears to be associated with the limiting membrane because, when the vesicles are subjected to membrane-disruptive procedures such as osmotic shock and then recentrifuged, only 10%–20% of the DBH is solubilized. This soluble component can evidently also be released upon stimulation. DBH is normally present in small quantities in the bloodstream, and the amount of DBH has been found to increase with nerve stimulation (Axelrod 1972). This is regarded not only as evidence of the association of DBH with noradrenaline storage vesicles, but also for the concept of exocytotic release of transmitters. If noradrenaline were released from the synaptosomal cytoplasm, with storage vesicles merely acting to maintain an appropriate intrasynaptic concentration of noradrenaline, then DBH would not be released upon nerve stimulation unless it too were coming from the cytoplasm. But if this were the case, then tyrosine hydroxylase, Dopa decarboxylase, and other soluble enzymes should also appear in the perfusates. This has never been observed.

Release.   The release mechanisms for noradrenaline appear to be comparable to those for dopamine, but they have been more carefully studied because the peripheral sympathetic system provides such an excellent model for this type of investigation. Isolated organs such as the spleen and intestine have been extensively investigated by preloading of the nerve terminals by prior intravenous administration of labeled noradrenaline. Following stimulation or treatment with drugs, labeled material is released into the perfusate, allowing quantitative estimation of the effects. The most striking aspect of sympathetic nerve stimulation is the relatively slow rate that is required to produce a significant increase in transmitter release. Most tissues respond to changes in the frequency of stimulation in the range of 1–3 per second and maximum responses can usually be obtained with stimulus frequencies of less than 20 per second. Using cat spleen, the highest amounts of noradrenaline release per stimulus are found at frequencies of the order of 10 per second and decline as the stimulation frequency increases to 30 per second. With higher frequencies of stimulation there is an overflow of noradrenaline into the venous circulation, but with low frequencies as much as 90% is inactivated by local processes. Nevertheless, the quantities of VMA and other metabolites of noradrenaline excreted in the urine suggest that the equivalent of 5–10 mg of noradrenaline is consumed each day in humans. The majority is derived from sympathetic nerve endings since the excretion of metabolites does not change appreciably after adrenalectomy (Iversen 1967).

Little is known about the details of how nerve impulses promote the release of noradrenaline from sympathetic nerves. It is clear that calcium plays a highly significant role because the amount of noradrenaline released upon sympathetic nerve stimulation is considerably reduced if the calcium content of the perfusing medium is lowered. In this respect, it seems analogous to the release of ACh from cholinergic nerve endings where it has been suggested (Katz 1962) that depolarization of the presynaptic terminal is accompanied by the attachment of vesicles to the inner surface of the presynaptic fibers. The transmitter packets attached to these sites discharge their contents to the exterior of the synaptic cleft, with calcium operating as the essential co-factor (Figure 1.6, Section 1.2.2; 3.7, Section 3.2.6). The impulse arriving at the nerve ending produces the necessary alteration in permeability to calcium, and the influx is the main stimulus responsible for catecholamine mobilization and secretion by the process of exocytosis. ATP, chromogranin, and a little dopamine-β-hydroxylase are also extruded into the synaptic cleft. Inactivation is primarily by reuptake of the noradrenaline into the nerve endings. Whether or not the protein which is also released by this process is recaptured again by the nerve ending is undetermined. If it is not recaptured, then the process of axoplasmic flow must be extremely vigorous to replace the lost material.

While it can be presumed that similar processes take place in the central nervous system, it must be remembered that duplicating the kinds of stimulation and perfusion experiments that can be easily performed in the periphery is virtually impossible in brain tissue. It has been possible to perform some crude experiments using perfusates, cortical cups, brain slices, and push–pull cannulae, and some interesting data have been obtained of a comparable nature.

Uptake. Methods by which the body disposes of released noradrenaline have already been described. They include reuptake into nerve endings, metabolism by MAO and/or COMT, overflow uptake into other cells, and overflow diffusion into the general circulation. By far the most important for terminating the physiological action of noradrenaline is uptake into nerve endings.

The system has been extensively studied, and the general characteristics are thought to apply to other transmitters, particularly other catecholamines. It is a highly sensitive pump which moves the transmitter back into the terminal against a huge concentration gradient. It requires membrane Na$^+$/K$^+$ ATPase, has a K_m in the range of 0.2–1.0 μM in most rat tissues, and is easily saturated. Data indicate that as much as 80% of the noradrenaline released by stimulation may be inactivated by the process. For example, inhibition of uptake with cocaine leads to approximately fourfold increases in the outflow of noradrenaline and its metabolites from isolated tissues following sympathetic nerve stimulation (Iversen 1973).

In the rat, this uptake is stereochemically selective for L-noradrenaline, although D-noradrenaline is also a weak substrate. Many other β-phenylethylamine derivatives can also act as substrates, including adrenaline, metaraminol, the α-methyl analogs of adrenaline and noradrenaline, dopamine, octopamine, etc. This uptake is potently inhibited by a number of drugs, particularly desipramine and other tricyclics. It can easily be distinguished from that for dopamine because desipramine is 1000 times less potent at inhibiting dopamine uptake. On the other hand dopamine, but not noradrenaline, uptake is potently inhibited by benztropine. Thus, the selective inhibition of noradrenaline and dopamine uptake sites in brain by desipramine and benztropine, respectively, can provide an indication of the presence of these two types of catecholaminergic neurons in the CNS.

The catecholamines are also taken up by a different transport system in various extraneuronal peripheral tissues such as vascular smooth muscle, cardiac, and certain granular tissues. At high perfusion concentrations a more rapid accumulation of catecholamines occurs in tissue than can be accounted for by uptake into nerve endings. This has sometimes been termed "Uptake 2." It is saturable, and has a much lower affinity for noradrenaline but a much higher capacity. It is not usually observed at low concentrations because uptake into nerve endings (uptake 1) is preferential. It shows no stereochemical specificity for D- or L-noradrenaline. Histochemical studies show that the sites are predominantly localized in muscle cells. Thus, it is extraneuronal in origin. This would be an uptake followed by a metabolic process rather than an uptake and retention process (Iversen 1973).

Noradrenaline Turnover. The same techniques used for measuring the turnover of dopamine have also been used for noradrenaline. The data are more extensive for noradrenaline and are generally considered to be more precise. Using these techniques, the mean half-life of noradrenaline in selected regions of brain has been reported by Iversen (1967) as 3.7 h for the hypothalamus, 3.6 h for the medulla, 3.0 h for the hippocampus, 2.7 h for the cortex, and 2.2 h for the cerebellum. This compares with turnover times reported by Costa (1970) of 14.3 h for noradrenaline in heart, 8.3 h for brain, 5.3 h for hypothalamus, and 2.1 h for the superior cervical ganglion.

Actions of Noradrenaline at Receptor Sites. As with the muscarinic actions of ACh at receptor sites (Chapter 5), and those of dopamine discussed earlier in this chapter, the actions of noradrenaline at receptor sites are far from clear. At least two types of receptor, α and β, have been defined in the periphery, but a comparable division cannot be made in the CNS. The metabotropic actions of noradrenaline are indicated by the involvement of cyclic nucleotides and at some receptor sites prostaglandins may act as adjuvants as well.

The concept of cAMP as a second messenger has already been discussed with respect to dopamine action. There is also an adenylate cyclase system in brain which is stimulated by noradrenaline and which can be distinguished from the dopamine system both by physical properties and by the effects of drugs. For example, the noradrenaline system operates best with slices of tissue while the dopamine system works well in homogenates. Dopamine stimulation is blocked by antipsychotic agents such as haloperidol and is mimicked by apomorphine, while the noradrenaline system is blocked by β-adrenergic antagonists such as propranolol and is mimicked by isoproterenol. The noradrenaline system is found in areas containing noradrenergic terminals such as the hypothalamus and the cerebellum, while the dopamine system is found in the striatum.

The cerebellum has been investigated as a model system (Hoffer *et al.* 1973) for noradrenaline. It slows the mean rate of discharge of Purkinje cells by interaction with β-receptors. Noradrenaline hyperpolarizes the membrane of these cells and this hyperpolarization is generally accompanied by increased membrane resistance, but never by increased membrane conductance. Noradrenaline action is blocked by iontophoretic application of prostaglandins of the E series and is potentiated by phosphodiesterase inhibitors that would prolong the action of cAMP. Thus, it has been proposed that the synaptic action of noradrenaline is mediated by interaction with adenylate cyclase much as is the case for dopamine. The effects of noradrenaline on Purkinje cells could be duplicated by stimulation of the locus coeruleus, indicating that endogenously released noradrenaline could duplicate the iontophoretic application.

A variety of responses can be obtained to iontophoretically applied noradrenaline in the central nervous system. The firing rate of some neurons is depressed, while others are excited and some show no effects of the iontophoretically applied material. The inhibitory response can be of two types, one of which is short lasting with rapid onset, and the other long lasting and often delayed in onset. Sometimes there is a biphasic response consisting of excitation preceded by inhibition. Many of the classical noradrenaline antagonists of the α and β types are ineffective against these iontophoretic effects. However, chlorpromazine and α-methylnoradrenaline consistently antagonize excitation caused by noradrenaline. Therefore, it is unlikely that all of these iontophoretic effects are artefactual although they certainly may not reflect the true postsynaptic actions of endogenously released noradrenaline (Bradley 1973).

The original concept of α- and β-receptors in the periphery was that they mediated excitatory and inhibitory noradrenergic actions, but this is clearly not a general rule. The division is properly based upon the action of specific blocking agents. The following are some of the better known tissues having α receptors: the radial muscle of the iris, most blood vessels, sphincter muscles of the stomach and intestine, the sphincter muscle of the urinary bladder, pilo-erector muscles and sweat glands of the skin, the salivary glands, and the capsule of the spleen. Included among the β-receptors are the ciliary muscle of the eye, the muscles of the heart, some coronary and pulmonary vessels, the bronchial muscle of the lung, muscles having to do with motility in the stomach and intestine, and the detrusor muscle of the bladder.

The mechanism of excitatory action has not been elucidated. There appear to be two types of inhibitory action. The first, blocked by α-antagonists, is associated with increased conductance, probably to K^+. The second, blocked by β-antagonists, is not associated with a conductance increase, and may involve cAMP. Alternative explanations offered have been binding of Ca^{2+} to the membrane and a reduction in Na^+ conductance (Krnjevic 1974).

8.3.3 Pharmacology of Noradrenaline

The same classes of pharmacological agents that interact with dopamine also interact with noradrenaline. The receptor sites and uptake systems differ, however, and there is an additional enzyme, DBH, in the biosynthetic sequence. These differences in detail permit some distinction between dopamine and noradrenaline pharmacology, although there is considerable overlap. The main types are listed in Table 8.5. Their sites of action coincide in many cases with those shown for dopamine in Figure 8.12.

Antagonists.   There are at least two types of peripheral noradrenaline receptors, α and β. The common adrenergic blocking agents dibenzyline, dibenamine, and phenothiazine act almost exclusively on α-receptors. β-Receptor blockers include dichloroisoproterenol and propranolol. Noradrenaline receptor blockers, such as phenoxybenzamine or chlorpromazine, will lead to large increases in the amount of noradrenaline released by nerve stimulation. This is comparable to the increased output of dopamine following blockade of the nigrostriatal tract. The blood levels of noradrenaline, for example, in patients on treatment with high doses of chlorpromazine can be more than 8–10 μg/liter under physical exercise, while untreated controls show only a level of about 2 μg/liter. Phenoxybenzamine will also cause an increase in the overflow secretion of noradrenaline into the blood following nerve stimulation which amounts to 4- to 10-fold, and which is not found after MAO and COMT inhibition. Since there is no decrease in the amount of noradrenaline taken up following intravenous administration after phenoxybenzamine, the uptake mechanism must be unaffected by the blockade. Therefore, it must be concluded that phenoxybenzamine and other blockers increase the amount of noradrenaline released per stimulation. A transneuronal feedback mechanism is possible for this effect, but it would have to work at some considerable distance in the periphery, i.e., via muscle feedback. A more acceptable interpretation is that autosomal noradrenaline receptors exist on nerve terminals, much as has been proposed for dopamine (Haggendal 1973).

Both α- and β-receptor blockers have a place in general medicine in the treatment of cardiovascular disorders, but neither has a place in neuropsychiatric medicine. The antipsychotic action of agents such as the phenothiazines, which have an adrenergic blocking action peripherally, may be because they also have a dopaminergic-blocking action, or because they block central noradrenergic receptors of a slightly different type than are affected by the peripheral blockers.

The directly acting sympathomimetic amines, or true noradrenaline agonists, can be divided into two groups: those which stimulate α-receptors and those which stimulate β-receptors. α-Receptor agonists include phenylephrine and clonidine. The classic β-receptor agonist is isoproterenol. Noradrenaline itself has rather feeble β-activity which is increased by the addition of the methyl group of adrenaline. But the isopropyl group of isoproterenol confers even greater activity.

Clonidine has a biphasic action on blood pressure. The initial pressor response is due to direct stimulation of peripheral α-adrenergic receptors. The more prolonged depressor response is due to an action in the CNS which is not yet understood.

Storage Inhibitors. The binding granules for noradrenaline are depleted of their transmitter by reserpine and tetrabenazine in the same way as other catecholamine and serotonin granules (Figure 8.12). Guanethidine is also a noradrenaline depletor, but its action is not clearly understood. It affects only peripheral stores, apparently because it does not cross the blood–brain barrier. Given to newborn rats, it depletes noradrenaline from cells of the superior cervical ganglion but not dopamine from SIF cells. Since guanethidine can itself be stored and released upon stimulation, storage granules may be affected by the drug. It is used in the treatment of hypertension, partly because it is also a ganglionic blocking agent with local anesthetic action.

Synthesis Inhibitors. Tyrosine hydroxylase and Dopa decarboxylase inhibitors influence noradrenaline synthesis as well as dopamine. Thus, the effects of such inhibitors are mixed as far as the catecholamines are concerned. Selective noradrenaline effects can only come about through the use of DBH inhibitors. DBH is inhibited by all copper chelators, especially FLA-63 (Anden and Fuxe 1971) and disulfuram, which is also an inhibitor of alcohol dehydrogenase. Fusaric acid is an effective inhibitor *in vitro* and *in vivo*, and also has antihypertensive action (Nagatsu *et al.* 1970).

Precursors. L-Dopa can act as a precursor for noradrenaline as well as dopamine. Most of the effects of L-Dopa are not believed to be due to synthesis of noradrenaline, although it is conceivable that production of excess dopamine in noradrenaline cells could account for some of the effects. Dihydroxyphenylserine (DOPS) can be decarboxylated directly to noradrenaline. Thus, it is a direct precursor, although again it cannot be stated with certainty that only noradrenaline in noradrenergic cells is formed after parenteral administration of DOPS. It is believed to cross the blood–brain barrier poorly and its effects have not been extensively studied.

Noradrenaline Releasers. Two potent releasers of noradrenaline are amphetamine and tyramine. These agents are taken up into catecholaminergic endings by active uptake processes. They promote release of noradrenaline by a replacement process which is not fully understood. The release cannot be primarily from granules into the intrasynaptosomal space as is the

TABLE 8.5
Some Drugs Which Affect Noradrenaline Metabolism

Drug	Presumed mechanism of action	Most prominent physiological effects
Antagonists		
Phenoxybenzamine	α-Receptor antagonist	Antihypertensive
Chlorpromazine[a]	α-Receptor antagonist	Tranquilizer
Propranolol	β-Receptor antagonist	
Agonists		
Phenylephrine	α-Receptor agonist	Hypertensive
Clonidine	α-Receptor agonist	Biphasic pressor response
Isoproterenol	β-Receptor agonist	Bronchiodilator, hypertensive
Releasers		
Amphetamine[a]	Releaser	Stimulant, euphoriant
Tyramine[a]	Releaser	Experimental
Storage inhibitors		
Reserpine[a]	Depletion	Antihypertensive, tranquilizer
Tetrabenazine[a]	Depletion	Antihypertensive, tranquilizer
Guanethidine	Peripheral depletion and ganglionic blockade	Antihypertensive
Pump inhibitors		
Desipramine	Reuptake inhibition	Antidepressant
Imipramine	Reuptake inhibition	Antidepressant
Amytriptylene	Reuptake inhibition	Antidepressant
Cocaine[a]	Reuptake inhibition, local anesthetic	Euphoriant, stimulant

Synthesis inhibitors		
α-Methyl-p-tyrosine[a]	Tyrosine hydroxylase inhibition	Depressant, akinesia
α-MethylDopa[a]	Dopa decarboxylase inhibition	Antihypertensive
Fusaric acid	DBH inhibitor	Antihypertensive
Monoamine oxidase inhibitors		
Iproniazid[a]	Broad-spectrum MAO inhibition	Antidepressant
Tranylcypromine[a]	Broad-spectrum MAO inhibition	Antidepressant
Clorgyline	Type A MAO inhibition	Experimental
COMT inhibitors		
Tropolone, pyrogallol[a]	COMT inhibition	Experimental
False transmitters		
α-Methyldopamine[a]	Replacement of transmitter	Antihypertensive
α-Methyl-m-tyramine[a]	Replacement of transmitter	Antihypertensive
Toxin		
6-Hydroxydopamine[a]	Destruction of cells	Experimental
Precursors		
L-Dopa[a]	Stimulates transmitter production	Counters Parkinsonism, mild stimulant
Dihydroxyphenylserine (DOPS)	Stimulates transmitter production	Experimental

[a]Also has prominent dopamine action.

case with reserpine because noradrenaline, rather than DOPAC, reaches the extracellular fluid. Therefore, amphetamine and tyramine must be displacing noradrenaline from storage granules into the extracellular space. They therefore act indirectly, but their stimulant effects wear off as supplies of the endogenous transmitter become depleted.

Pump Inhibition.   Noradrenaline pump inhibitors are clinical antidepressants and are discussed in Chapter 14. The best known of these is desipramine, the demethylated derivative of imipramine, which is the most potent uptake inhibitor so far described. In the isolated rat heart a concentration of 13 nM is sufficient to produce a 50% inhibition of noradrenaline uptake. Other well-known noradrenaline pump inhibitors are nortriptyline and amitriptyline. These tricyclic compounds differ only modestly in structure from the phenothiazines which block the postsynaptic site rather than the presynaptic uptake site. The middle ring is seven membered rather than six, which results in a molecule being bent out of its planar shape.

Cocaine also inhibits the uptake of noradrenaline. It produces a 50% inhibition at a concentration of 380 nM (Iversen 1967). Cocaine inhibits the accumulation of exogenously administered noradrenaline in sympathetic nerve endings, and following the administration of cocaine there is a substantial increase in the amount of noradrenaline overflowing into the venous circulation following sympathetic nerve stimulation.

Monoamine Oxidase Inhibitors.   MAO inhibitors are also clinical antidepressants. Sympathetic nerves contain predominantly type A MAO, and noradrenaline is a substrate for this type. Most MAO inhibitors in current use, such as phenylzine and tranylcypromine, are broad spectrum, in that they affect type A as well as type B. The half-life of MAO in brain is about 11 days and in peripheral tissue about 4 days. Thus, the length of time required to restore MAO activity to normal following irreversible inhibition with a hydrazine-type MAO inhibitor is several days.

The clinical effects of MAO inhibitors will be discussed in Chapter 14. Some patients appear clinically quite resistant to the antidepressant action of MAO inhibitors yet respond dramatically to tricyclic antidepressants. The reverse is also true.

False Transmitters.   The administration of α-methylDopa leads to long-lasting depletion of noradrenaline from peripheral sympathetic nerves caused by the accumulation of α-methylnoradrenaline in these tissues. Following stimulation of sympathetic nerves, α-methylnoradrenaline is released. Since this is less active at receptor sites than the true transmitter, hypotension results. A similar depletion is obtained following the administration of α-methyl-m-tyrosine which is accompanied by an accumulation of β-hydroxy-α-methyl-m-tyramine or metaraminol. This produces a similar physiological response to that following administration of α-methylDopa, i.e., decreased sympathetic action. They are employed clinically as antihypertensive agents (Iversen 1967).

COMT Inhibition.   Tropolones are the best known COMT inhibitors and, while they prolong the action of noradrenaline at receptor sites, they are of little clinical importance and are not particularly interesting agents for physiological studies.

Toxins. 6-Hydroxydopamine remains the classic toxic agent for catecholaminergic nerves. It is taken up well by noradrenergic as well as dopaminergic cells. When sufficient intracellular concentrations are reached there is destruction of the tissue. Details have been discussed under dopamine metabolism.

8.4 The Adrenergic Neuron

The adrenergic cell is characterized by the presence of phenylethanol-amine-*N*-methyl transferase (PNMT, EC 2.1.1.28), which converts nor-adrenaline to adrenaline. Peripherally, adrenaline is contained in chromaffin cells, but not all chromaffin cells are adrenergic in nature; some are nor-adrenergic. Nearly all chromaffin cells are contained in the adrenal medulla, although some are scattered in other tissues. Chromaffin cells are not neurons, and peripherally released adrenaline acts on remote target cells.

The possibility that adrenaline might be released from central adrenergic sites was suggested by the findings of Gunne (1962), Vogt (1954), and others who reported the presence of adrenaline in brain. The concept was further advanced when P. L. McGeer and McGeer (1964) showed that noradrenaline could be converted into adrenaline *in vivo* and *in vitro* by cat hypothalamus. Later, it was confirmed that this conversion could take place *in vitro*.

Evidence of adrenergic neurons in mammalian brain has not been obtained by means of the histofluorescence technique even though adrenaline gives a characteristic fluorescence color that can be detected in neurons of frog brain. Thus, it was not until the immunohistochemical method for PNMT was employed by Hokfelt *et al.* (1974) that strong evidence for central adrenergic neurons was obtained. These authors identified two neuronal groups and several terminal fields staining for this enzyme. The more caudal cell group, designated as C1, is the larger of the two and is found in the rostral medulla oblongata lateral to the olivary complex. The distribution corresponds closely to that designated as group A1 by Dahlstrom and Fuxe. The other complex, designated C2, is located close to the midline in the dorsal part of the reticular formation and corresponds roughly to the A2 group.

At least one axon bundle arises from these two cell groups which ascends in the region of the reticular formation and more rostrally in the ventral tegmental area (Figure 8.18). It appears to be identical with part of the as-cending ventral noradrenergic pathway described by Ungerstedt (1971). Terminal fields of this ascending pathway appear to be in certain regions of the thalamus, hypothalamus, and possibly other, more rostral structures. Terminal fields exist in the pons-medullary region and there also appears to be a descending pathway to the spinal cord. The fact that PNMT-containing neurons and pathways correspond closely with those showing histofluores-cence for catecholamines argues strongly that these cells, pathways, and ter-

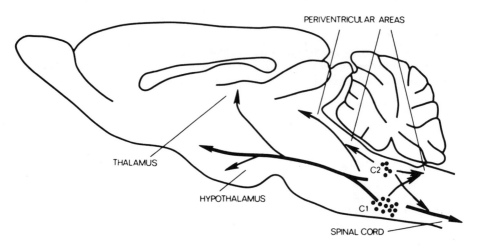

Figure 8.18 Sagittal projection of ascending and descending adrenaline pathways in rat brain. (From Hokfelt *et al.* 1974.)

minals genuinely represent adrenaline pathways. The other possibility is that PNMT cross-reacts with some other methylating enzymes whose purpose is other than to synthesize adrenaline. The distribution demonstrates that the presumed adrenaline cell bodies are reticular neurons innervating specific nuclei in the brainstem, most of which have periventricular localization. The sensitivity of the method is not high, so most of the data obtained to date must be regarded as preliminary. The low sensitivity, however, is in keeping with the low biochemical level of adrenaline found in brain by Gunne (1962), which is 5%–10% of the total catecholamines.

The distribution of PNMT by biochemical analysis is shown in Table 8.6. In general there is correlation with PNMT intensity as shown by immunohistochemistry. PNMT activity was high in the reticular formation and certain brainstem nuclei. It was also high in the anterior and posterior hypothalamus and locus coeruleus. Intermediate values were found in the substantia nigra, basal ganglia, amygdala, septum, and habenula (Lew *et al.* 1977).

Thus, biochemical distribution also supports the immunohistochemical result, lending further credence to the existence of adrenergic neurons in brain.

8.4.1 Biochemistry of Adrenergic Neurons

The enzyme which synthesizes adrenaline from its precursor noradrenaline, PNMT, is a soluble enzyme which requires S-adenosylmethionine to carry out the reaction:

Noradrenaline + S-adenosylmethionine → adrenaline + H^+
$$+ S\text{-adenosylhomocysteine}$$

The enzyme has been purified to homogeneity (Connett and Kirshner 1970; Joh and Goldstein 1973). It has a molecular weight of 40,000 but also

TABLE 8.6
Distribution of PNMT Activity in Monkey Brain[a]

Region	Apparent PNMT activity (units)[b]
N. reticularis lat. (C$_1$)[c]	10.0
N. tractus solitarius, N. vagi, N. commissuralis (C$_2$)[c]	3.0
Hypothalamus	
median eminence	9.0
paraventricular nucleus	9.6
ventromedial nucleus	8.2
dorsomedial nucleus	6.4
lateral nuclei	8.1
suprachiasmatic nucleus	6.6
mammillary nucleus	2.6
Preoptic nucleus	3.0
Locus coeruleus	3.1
Substantia nigra	3.3
Globus pallidus	4.4
Putamen	3.0
Caudate nucleus	2.1
Amygdala	3.2
Nucleus accumbens	1.0
Habenula	2.3
Olfactory tubercle	0.9
Frontal cortex	0.4
Cerebellum	0.2

[a] From Lew *et al.* 1977.
[b] One unit of enzyme activity is equal to 1 pmol of [^{3}H]-*N*-methylphenyl-ethanolamine formed per 10 mg of tissue and per hour of incubation. The results are the means from three animals ± 5%–15% S.E.M.
[c] Plus surrounding regions.

has two higher molecular weight forms of approximately 80,000 and 160,000. It will only methylate phenylethylamines with a hydroxyl group. L-Nor-adrenaline is the most active substrate with the lowest K$_m$, indicating it is much preferred over most other substrates such as normetanephrine, phenyl-ethanolamine, and octopamine. Inhibition by phenethylamines and amines related to tranylcypromine has been demonstrated.

Adrenaline is metabolized by COMT and MAO by the same sequence of reactions as noradrenaline. The only product difference is that metanephrine rather than normetanephrine is formed by the action of COMT (Figure 8.19).

8.4.2 Physiology of Adrenergic Neurons

Adrenaline is secreted from the adrenal medulla in times of stress following sympathetic stimulation. It increases blood pressure, cardiac output, blood flow to muscles, and breakdown of liver glycogen. But some chromaffin cells of the adrenal medulla secrete only noradrenaline and the basis of the difference has been somewhat of a mystery. It has also been a mystery as to why

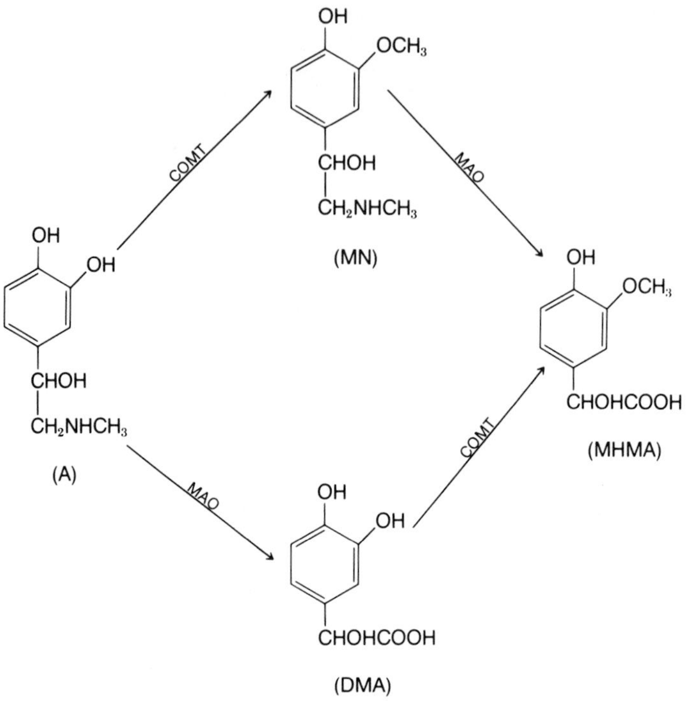

Figure 8.19 Metabolism of adrenaline. MN, metanephrine; MHMA, 3-methoxy-4-hydroxyman-delic acid; DMA, dihydroxymandelic acid.

in mammals the medulla should become surrounded during embryonic life by a cortex derived from coelomic epithelium which secretes steroids. Recent work by Coupland, Axelrod, and Wurtman has pointed out at least one major reason why these two organs should be so closely related: Glucocorticoid hormones, secreted from the adrenal cortex and delivered preferentially to the medulla via the adrenal portal vascular system, induce the synthesis of PNMT in chromaffin cells. The suggestion is that the mammalian adrenal medulla contains only a single clone of chromaffin cells whose members do or do not produce PNMT, and therefore adrenaline, depending upon the extent to which they are stimulated by glucocorticoids (Wurtman *et al.* 1972).

Coupland, the anatomist, first noticed that in species such as the rabbit where the adrenal cortex does not form a complete envelope around the medulla, little or no adrenaline is present in the portion of the medulla not adjacent to cortical cells. A similar situation exists in the dogfish. Wurtman, the physiologist, observed that patients suffering from pituitary disease had an unusual and prolonged sensitivity to insulin. These patients responded poorly to ACTH, the pituitary hormone stimulating adrenal cortical cells. Thus, a fast-acting factor, presumably adrenaline, was missing. Working with the biochemist Axelrod, Wurtman removed the pituitary in rats and noted that within a week PNMT activity fell to only 15%–25% of normal. It could be restored by treating rats with ACTH. Activity could also be restored with very large doses of dexamethasone, but not by normal replacement doses of cortical steroids. Thus, it was evident that only the extremely high concentrations of cortical steroids found in the corticomedullary portal circulation, which are at least 100-fold greater than that present in systemic arterial blood, were adequate to induce PNMT. The effect can be blocked by concurrent administration of actinomycin D or puromycin, indicating that protein synthesis is required for the effect to be observed.

In the frog, both adrenaline and PNMT activity exist in substantial quantities in the brain, heart, and spleen, as well as the adrenals. After hypophysectomy in the frog, PNMT activity remains unchanged in all frog organs, indicating that the mechanism of PNMT induction is different in that species. In addition, mammalian pheochromocytomas are capable of synthesizing adrenaline in the absence of glucocorticoid stimulation, indicating that more than one mechanism may exist in mammals (Wurtman *et al.* 1972).

As with dopamine and noradrenaline, adrenaline seems to work, at least in part, through cyclic AMP as a second messenger. In fact, cAMP was discovered in the course of investigation on the glycogenolytic action of adrenaline in liver. The proposition was then advanced that cAMP was an intracellular mediator of this glycogenolytic action (Rall 1972).

The actions of adrenaline peripherally are discussed in more detail in Chapter 14. Very little can be said about possible central actions. The areas innervated by PNMT-containing terminals offer some speculative clues, however. The dense innervation of the nucleus periventricularis suggests possible effects on oxytocin. Innervation of the perifornical area and dorsomedial hypothalamus indicates a possible influence on food and water intake. Effects on body temperature and gonadotropin secretion are suggested by hypothalamic innervation, and effects on blood pressure and respiration by innervation of the dorsal motor nucleus of the vagus nerve, the nucleus of the tractus solitarius, and the sympathetic lateral columns of the spinal cord. The dense innervation of the locus coeruleus suggests that adrenaline might influence noradrenergic pathways coming from the locus coeruleus and thus have some role in controlling sleep and wakefulness.

Drugs which interact exclusively with adrenaline but not noradrenaline and dopamine are yet to be reported. Thus, no separate discussion of adrenaline pharmacology will be attempted. It might be noted, however, that adrenaline is a much more powerful β-receptor agonist than noradrenaline. Therefore, agents which powerfully interact with β-receptors may be modifying the action of adrenaline more than noradrenaline. At this stage such an interpretation is only a speculation, and a promising field for future catecholamine research will be to distinguish more clearly between noradrenaline and adrenaline physiology.

8.5 Summary

The catecholamines constitute an interesting family in that the biosynthetic sequence proceeds from dopamine through noradrenaline to adrenaline. Each catecholamine is represented by neurons centrally and peripherally. But dopamine and noradrenaline also exist in some neurons as precursors. This greatly complicates investigation of the physiology and pharmacology. Nevertheless, the existence of discrete neurons of these separate catecholamines implies a distinct and definite function even if current techniques are unable precisely to define them.

In the central nervous system there are three principal dopaminergic pathways: the nigrostriatal, the mesolimbic-mesocortical and the tuberoinfundibular. Scattered interneurons also exist in the central nervous system, the retina, and sympathetic ganglia. In ruminant species dopamine also exists in organs as a component of mast cells.

The actions of dopaminergic neurons relate to extrapyramidal, behavioral, and endocrine function. The roles of interneurons are less certain but seem to be related to the chemoreceptor trigger zone, the visual system, and peripheral sympathetic activity.

The dopamine receptor appears to involve the action of an adenylate cyclase system which produces cAMP as a second messenger. The iontophoretic effects of dopamine are predominantly inhibitory, although under some conditions have been reported to be excitatory, or to produce biphasic excitatory-inhibitory sequences. The overall pharmacological effect seems to be one of inhibiting the release of transmitters, particularly ACh, from postsynaptic neurons.

Dopamine agonists such as apomorphine or bromocriptine are able to counteract Parkinsonian-like extrapyramidal reactions, but they also produce nausea. Administration of the precursor L-Dopa produces a similar, but better therapeutic effect. Antagonists of dopaminergic receptors are antipsychotic agents but also produce Parkinsonian-like extrapyramidal reactions. Phenothiazines, thioxanthines, and butyrophenones are all capable of such activity.

Amphetamine is a dopamine releaser, a stimulant, and, in high doses, an agent which produces paranoid psychosis. Pump inhibitors of dopamine counteract drug-induced extrapyramidal reactions but are not antidepressants.

Noradrenaline neurons differ from dopamine neurons in possessing the enzyme dopamine-β-hydroxylase (DBH). This enzyme is associated with the noradrenergic storage granule. Upon stimulation of a noradrenergic nerve, both noradrenaline and DBH are released into the extracellular space. This is regarded as one confirmation of the theory of exocytosis for transmitter release.

Several groups of noradrenergic cells are found in the brainstem. Dahlstrom and Fuxe described seven cell groups, but Hartman consolidated them into three. The principal cell group is in the locus coeruleus. One ascending pathway, the dorsal noradrenergic bundle, has been found by both the histofluorescence and immunohistochemical methods. A second, the ventral noradrenergic bundle, is controversial because it has not been confirmed by immunohistochemistry. Terminal fields of the ascending noradrenergic pathways heavily innervate the hypothalamus, with lesser innervation of the thalamus and limbic system. There is light innervation of the cerebral cortex. A descending noradrenergic pathway connects the more caudal cell groups with the dorsal, ventral, and lateral horns of the spinal cord.

The peripheral sympathetic nervous system is, of course, primarily noradrenergic. Exceptions are fibers to the salivary glands and adrenal medulla, but nerve fibers to the blood vessels, sweat glands, pilo-erector muscles of hair follicles, the eye, heart, intestinal tract, spleen, bladder, lungs, and sex organs are noradrenergic.

Peripheral noradrenergic receptors are divided into α and β types which are roughly, but not exactly, divided into excitatory and inhibitory receptors, respectively. Classic α- and β-receptor blockers such as phenoxybenzamine and propranolol are not antipsychotic. Chlorpromazine, however, which primarily blocks α-adrenergic receptors is antipsychotic. It may be that this blocking action is unconnected with its antipsychotic activity.

Iontophoretically applied noradrenaline shows both excitatory and inhibitory characteristics. The action of noradrenaline at receptor sites probably involves the participation of cAMP as a second messenger and possibly, at some sites, the prostaglandins as adjuvants. Noradrenaline stimulates a brain adenylate cyclase which is concentrated in different areas of brain than that for dopamine-mediated adenylate cyclase. This cyclase is blocked by β-receptor antagonists.

Noradrenaline pump inhibitors act as antidepressants, which distinguises them from dopamine pump inhibitors. Monoamine oxidase inhibitors, which enhance the action of both dopamine and noradrenaline, are also antidepressants. COMT inhibitors, on the other hand, have little physiological action, suggesting that this enzyme plays only a minor role in physiological inactivation of noradrenaline. Evidence points to uptake as being the predominant mechanism.

The adrenergic neuron is characterized by the presence of PNMT, which converts noradrenaline to adrenaline. Peripherally, adrenaline is contained in chromaffin cells. PNMT can be induced in adrenal medullary chromaffin cells by the high concentrations of glucocorticoids released in the adrenal portal circulation. Thus, the production of adrenaline is mediated in the periphery by ACTH from the pituitary.

There is no evidence that central adrenergic neurons are under such control. These neurons have been identified by immunohistochemistry using antibodies to PNMT. Two groups of cells in the brainstem have been located, with ascending and descending pathways. The descending pathways innervate areas of the spinal cord, while the ascending pathways terminate primarily in the hypothalamus but possibly also in more rostrally located structures. There seems to be little pharmacology which is unique to adrenergic cells, although adrenaline is a more powerful activator of β-receptors than is noradrenaline.

9

The Serotonin Neuron

9.1 Introduction

Despite an enormous amount of research, serotonin remains a compound whose basic functions are not understood at either the cellular or the gross level. The fact that it is a neurotransmitter is not in dispute. It is synthesized, stored, and released by central neurons and its synapses have conventional morphology (Figure 9.4), yet little can be said with regard to its precise action on postsynaptic neurons. The same state of affairs exists with respect to its overall actions.

While hints can be obtained from the action of drugs, particularly ones affecting mood and behavior (Chapter 14), many of these same agents also interact with the catecholamines, making it difficult to separate out a serotonin component.

Perhaps it is fitting that serotonin should first have been isolated from blood, where it is present in high concentration in platelets but has no clearly identified role. For years, Irvine Page of the Cleveland Clinic had been investigating materials in blood that caused vessel constriction. Rapport, an organic chemist, came to Page directly from graduate school and Page assigned him the task of isolating the pharmacologically active material that had been first detected in 1868. After two years Rapport got his first crystals. But they turned out to be the creatinine sulfate complex of serotonin, which gave a formula analysis that was too complicated to decipher (Rapport *et al.* 1948). It was not until 1949 that Rapport, by then working at Columbia University in New York, separated the complex into the picrates of serotonin and creatinine. The structure was then readily deduced, and rapidly confirmed by chemical synthesis (Rapport 1949; Hamlin and Fisher 1951.)

Shortly afterward, another young investigator, Betty Twarog, arrived at the Cleveland Clinic. She was assigned the job of studying the distribution of the new amine. Using the clam heart as a bioassay system, she reported in 1953 an unusually high concentration of the active substance in brain

(Twarog and Page 1953). Despite the fact that brain has only a minor proportion of total body serotonin (over 95% is in the gastrointestinal tract and blood platelets), this unexpected finding triggered a startling series of pharmacological investigations into mood and behavior.

Amin, Crawford, and Gaddum (1954) followed up on Twarog's discovery by studying exactly where in the nervous system serotonin was located. The highest concentrations were found to be in the hypothalamus and limbic system. This tied in with Woolley and Shaw's (1954) observation that antimetabolites of serotonin, particularly LSD, seemed to disrupt behavioral patterns. They were so impressed by the consequence of tampering with this material that they proposed that either an excess or a deficiency of serotonin was responsible for mental illness. Woolley's hypothesis remains to this day, with no clear-cut evidence emerging to refute it, and some interesting observations that tend to support it.

A few years later, Brodie, of the National Institute of Health, inspired by the finding that reserpine caused depletion of serotonin from body stores, speculated that serotonin was concerned with trophotropic functions in brain, and noradrenaline with ergotropic functions (Brodie *et al.* 1959).

Subsequent pharmacological investigations continued to link serotonin, as well as the catecholamines, with mental function. Phenothiazines were found to block the action of serotonin. Monoamine oxidase (MAO) inhibitors, the first psychic energizers, were found to raise serotonin levels in brain, and tricyclic antidepressants were shown to be inhibitors of serotonin neuronal reuptake. Thus, the major groups of psychopharmacological agents involved serotonin in their spectrum of action along with the catecholamines. Only recently have more specific agents become available to help with the difficult task of sorting out the discrete functions of these substances.

The role of serotonin as a neurotransmitter in brain was clearly established when it was discovered that the Falck technique for catecholamines also revealed serotonin. Cell bodies possessing a characteristic evanescent fluorescence were found to extend throughout the length of the raphe system, and terminals were detected in those diencephalic and telencephalic areas previously shown to have high concentrations of serotonin (Dahlstrom and Fuxe 1964a).

Although the discovery of serotonin in brain shifted the major focus of attention to the CNS, interest also continued to develop around its peripheral actions on smooth muscle. The Italian investigator Erspamer had detected "enteramine," a hormone secreted by argentaffin cells of the gut, and isolated it from the salivary glands of the octopus. It was only after the structure of serotonin had been worked out by Rapport and the synthetic material had become available that Erspamer realized the two substances were identical (Erspamer and Asero 1952). Lembeck (1953a) next demonstrated the presence of large amounts of serotonin in carcinoid tumors, which are derived from argentaffin cells. These tumors were well known to be associated with diarrhea and hot flushes, just as had been detected immediately following the administration of large doses of reserpine. Some aspects of the "carcinoid syndrome" were soon correctly attributed to release of large amounts of serotonin.

In studies completely independent of the work on serotonin, Lerner was studying extracts of pineal gland which caused aggregation of melanin granules in frog skin, thus causing it to lighten. It took him some years to purify his material because he did not suspect its instability to light and kept losing activity. He named his hormone melatonin and finally established its chemical identity as 5-methoxy-N-acetylserotonin (Lerner *et al.* 1959).

Lerner's report immediately closed another gap. It explained the high concentration of serotonin in the pineal gland and led Axelrod to conclude that the pineal must have an enzyme capable of methylating serotonin in the 5 position. He initiated a brilliant series of experiments on the functions of the pineal which contributed in part to his award of a Nobel Prize in 1969. Working with Wurtman, he established that in the rat there was a circadian rhythm in melatonin production, that it inhibited gonadal function, and that its production was inhibited by light (Axelrod and Wurtman 1968). The functions of the pineal gland in man are still not known, although destructive tumors apparently produce precocious puberty, and functional ones delay puberty, in males.

Falck, one of the developers of the histochemical method which has provided such sensational advances in our understanding of biochemical neuroanatomy, has reported the presence of neurons in the brainstem with a fluorescence characteristic of indoleamines but different from serotonin (Bjorklund *et al.* 1970, 1971a, 1971b). It seems possible, therefore, that there is a great deal yet to be learned not just about the distribution and function of serotonin neurons in brain, but also about the existence and function of other indoleamine systems.

9.2 Anatomy of Serotonin Neurons

Although a great deal of effort has been spent in the past decade attempting to map serotonin neuronal systems precisely, present knowledge is still incomplete and fragmentary. The evanescent nature of the serotonin fluorescence by the Falck technique has been the major limitation. The effectiveness of the method has been extended somewhat by the use of drugs, but now that tryptophan hydroxylase has been purified and antibodies to it prepared, an alternative immunohistochemical method may produce more sophisticated data (Pickel *et al.* 1976).

So far it can be said that most cell bodies of serotonin neurons are located in the raphe and reticular systems of the brainstem where they appear to be present in even fewer numbers than catecholamine cells. The more caudal ones give rise to descending pathways to the spinal cord, whereas the more rostral ones give rise to ascending pathways to the diencephalon and telencephalon. Dahlstrom and Fuxe (1964a) described nine major groups of serotonin cells in the rat, designated B1 to B9. These are shown in sagittal

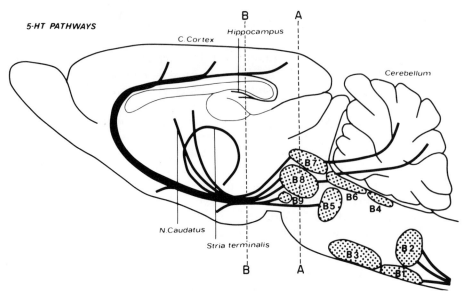

Figure 9.1 Schematic diagram of the central serotonergic cell groups and projections in sagittal section from rat brain. (From Fuxe and Johnsson 1974.) A and B are the cross-sectional planes shown in Figure 9.3.

section in Figure 9.1 (Fuxe and Johnsson 1974). Two of the cell groups shown in the figure, B3 and B9, are in the reticular formation, while the remainder are in the raphe system.

The axons of groups B1 and B2 descend to the spinal cord to terminate mainly in the lateral horn, but some also terminate in the anterior horn and a few terminate in the dorsal horn. The cerebellum is mainly innervated by group B7, but is probably also innervated by groups B5 and B6.

The axons of groups B5–B9 send their axons rostrally in three major pathways. These, as well as the pathways into the cerebellum, are shown in horizontal section in Figure 9.2.

The most medial ascending pathway into the forebrain innervates the hypothalamic, preoptic, and septal areas. It follows the medial forebrain bundle. Some fibers of this ascending system come from cell groups B5 and B6, but most come from groups B7 and B8.

The next pathway, slightly more lateral, innervates the cerebral cortex. It also runs through the medial forebrain bundle, then sweeps dorsally along the cingulate gyrus and curves laterally into the hippocampus. Along the way, branches are given off to all cortical areas. Again, the cells bodies are mainly in the midbrain raphe of groups B7 and B8, but some may also come from groups B5 and B6.

The third system primarily innervates the corpus striatum. This pathway is in the region slightly lateral to the medial forebrain bundle. It originates from group B9 with some innervation possibly coming from groups B7 and B8.

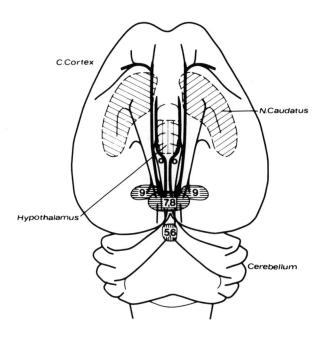

Figure 9.2   Horizontal projection of ascending serotonin pathways. (From Fuxe and Johnsson 1974.)

The arrangement of cell bodies and axon pathways in the brainstem is shown in Figure 9.3 for two cross-sectional levels. The most caudal of these, cross section A, shows the arrangement of cell groups B7–B9 in the midbrain. It also shows how the axonal pathways develop slightly lateral to cell groups B7 and B8 as well as around the periphery of the crus cerebri. The more rostral cross section B, at the level of the diencephalon, shows the arrangement of the axons as they proceed into the forebrain structures.

As with catecholamine neurons, serotonin neurons are relatively few in number and the cells are largely confined to the discrete areas in the midbrain, pons, and medulla shown in Figure 9.1. But the fiber tracts ramify broadly, suggesting that their purpose is to exert a tonic influence on remote regions of the brain.

As yet even less is known about serotonin terminals than about the cell bodies and axons. The best information about their detailed morphological nature comes from uptake studies of [³H]serotonin in various brain areas. An electron microscopic autoradiogram of the neostriatum following the administration of [³H]serotonin intraventricularly in rats is shown in Figure 9.4 (Hattori *et al.* 1976a). The labeled terminal is presumably the serotonin type, although absolute proof is lacking since terminals with other morphology are also occasionally labeled in this type of experiment. The typical terminal has a symmetrical contact, is axospinous in nature, and contains sparse and some-

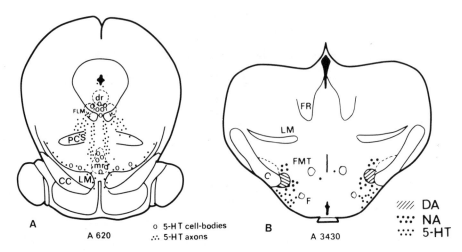

Figure 9.3 Serotonin pathways in transverse section. See Figure 9.1 for approximate location of sectional planes. A 620 (A) and A 3430 (B) refer to the distance in microns anterior to the inter-aural line. A. Schematic illustration of serotonin cell bodies and pathways in the midbrain (O, cell bodies; ●, axon pathways). B. Schematic illustration of the ascending monoamine bundles in the lateral hypothalamic area. The serotonin (5-HT) fibers are found mainly in the ventrolateral part and the noradrenaline (NE) fibers in the dorsolateral part of the lateral hypothalamic area. Dopamine (DA) fibers are just medial to the crus cerebri. Some noradrenaline fibers are just dorsal to the dopamine tract. Abbreviations: CC, crus cerebri; LM, medial lemniscus; FMT, mammilothalamic tract; FR, fasciculus retroflexus; dr, dorsal raphe; mr, median raphe; PCS, superior cerebellar peduncle; F, fornix. (From Fuxe and Johnsson 1974.)

what flattened vesicles. It differs from the typical catecholamine synapse which has characteristically an asymmetrical contact.

Table 9.1 shows the distribution of serotonin and tryptophan hydroxylase in some of the major regions of brain. The distribution probably contains further clues as to the detailed anatomy and functional purposes of serotonin in brain. But these are still only vaguely appreciated and important discoveries remain for future research.

It has been known for a long time, of course, that serotonin is contained in argentaffin cells. The function of serotonin there seems to be related to increasing gastrointestinal motility. The reason that serotonin is concentrated in platelets, where it is stored but not synthesized, is uncertain.

9.3 Metabolism of Serotonin

The sequence of reactions by which serotonin is created and destroyed is the following: The dietary amino acid tryptophan is converted to 5-hydroxytryptophan (5-HTP) by the enzyme tryptophan hydroxylase. 5-HTP decarboxylase then converts this intermediate amino acid to serotonin. Serotonin is metabolized both pre- and postsynaptically by the enzyme

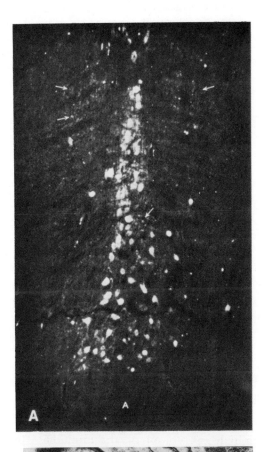

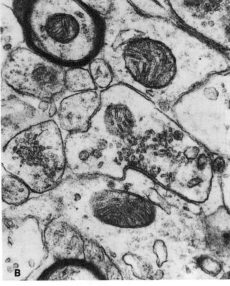

Figure 9.4 Demonstration of serotonin structures by histochemistry and electron microscopy. A. Serotonin cells fluorescing in the medial raphe nucleus just dorsal to the interpeduncular nucleus (A). A number of perpendicularly cut fluorescent axons are indicated by the arrows. Rat treated with reserpine and nialamide. Fluorescence reaction shown in Figure 8.2 (Section 8.2.1). (From Dahlstrom and Fuxe 1964a.) B. Presumed serotonin nerve ending in the rat neostriatum. Note the sparse content of pleomorphic vesicles and symmetrical synaptic contact. This type of nerve ending occupies 3%–5% of the total boutons in the neostriatum. It is preferentially attacked by the serotonergic toxin p-chloroamphetamine, and is readily labeled by the intraventricular administration of [³H]serotonin. (From Hattori *et al.* 1976a.)

TABLE 9.1
Concentrations of Serotonin and Tryptophan
Hydroxylase in Various Brain Areas[a]

Area	Serotonin μg/g tissue	Tryptophan hydroxylase nmol/h/g tissue
Amygdala	2.10	12.1
Hypothalamus	1.70	14.6
Septal area	1.50	16.0
Striatum	0.72	22.8
Hippocampus	0.65	4.3
Medulla	0.62	4.1
Thalamus	0.57	8.4
Pons	0.28	4.1
Cerebral cortex	0.17	0.9
Cerebellum	0.09	0.7
Midbrain raphe	—	17.3

[a] Serotonin values are for the dog (Bogdanski *et al.* 1957) and tryptophan hydroxylase levels for the cat (Peters *et al.* 1968); pineal tryptophan hydroxylase is reportedly at least 30-fold higher than that in any brain area shown.

monoamine oxidase (MAO) which produces the inactive metabolite 5-hydroxyindoleacetic acid (5-HIAA) (Figure 9.5).

9.3.1 Tryptophan-5-Hydroxylase (EC 1.16.4)

Tryptophan hydroxylase catalyzes the conversion of tryptophan to 5-hydroxytryptophan, the immediate precursor of serotonin. It has a very

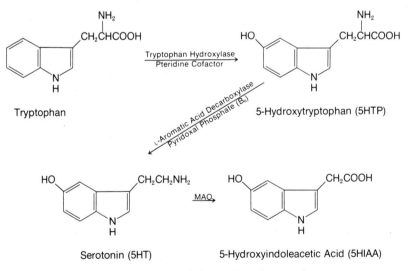

Figure 9.5 Synthesis and destruction of serotonin.

limited distribution, being found only in those cells specialized for synthesizing serotonin. It is also the rate-limiting enzyme, which means that the control of serotonin synthesis is exerted at this stage. Exactly how this is achieved is not known. It does seem certain that there is a considerable excess of enzyme capacity over normal synthetic needs. Presumably this is a reserve to meet peak demand. When the best possible conditions are chosen in test tube experiments, including sufficient oxygen, tryptophan, and pteridine co-factor, the maximum rate of tryptophan hydroxylation by whole rat brain homogenates is more than 12 nmol/g/h. It has been shown by a variety of techniques that the amount converted *in vivo* is only 0.5–1.9 nmol/g/h, however (Carlsson 1974; Gal 1974). It would appear, therefore, that only about 10% of the available capacity is normally being used for synthesis at any given time.

Tryptophan hydroxylase has absolute requirements for molecular oxygen and tryptophan as substrates and biopterin as a co-factor (Tong and Kaufman 1975). It is further stimulated by reduced pyridine nucleotide and dihydropteridine reductase. These facts suggest that the reaction might proceed by a comparable sequence to tyrosine hydroxylase (Chapter 8).

Several scientific groups have now purified tryptophan hydroxylase (Joh *et al.* 1975; Tong and Kaufman 1975; Youdim *et al.* 1974), but, despite this, there is still disagreement about many aspects of its nature as well as its mechanism of action. One team considers the molecular weight to be approximately 55,000 (Youdim *et al.* 1974) while another considers it to be 230,000 (Tong and Kaufman 1975). The difference of opinion centers around whether the enzyme is made up of four subunits of a molecular weight of approximately 55,000–60,000 or whether the higher molecular weight material is simply a protein aggregate formed artificially during purification.

A point of practical significance is whether there is adequate tryptophan in living brain to occupy the active centers on tryptophan hydroxylase. The concentration of tryptophan required to half-saturate the enzyme in test tube experiments has been estimated at 32 μM, while the overall tryptophan concentration in living rat brain has been estimated at 40 μM (Peters *et al.* 1968). These data would suggest that the enzyme is not fully saturated *in vivo,* and help to explain why serotonin production is sensitive to dietary tryptophan levels (Fernstrom and Wurtman 1974). Other data in the literature imply the enzyme is completely saturated *in vivo,* however, which would necessitate other explanations for the dietary effect.

Reduced serotonin synthesis has also been observed under conditions of severe oxygen deprivation (Davis *et al.* 1974), but this is a highly artificial circumstance, and it can be assumed that even under conditions of high altitude the enzyme is fully saturated with oxygen.

There are also considerations of substrate specificity. Strong similarities exist between phenylalanine, tryptophan, and tyrosine hydroxylases. This has led to speculation that these three enzymes may have a common genetic root. Although each is localized to separate structures, and is immunologically distinct, there is some capacity to accept the other aromatic amino acids as substrates. This is particularly true for phenylalanine and tryptophan hydroxylases. Liver phenylalanine hydroxylase has a significant capacity for hydroxylating tryptophan, while brain tryptophan hydroxylase has a significant capacity for hydroxylating phenylalanine. Using the natural co-factor tetrahydrobiopterin, Kaufmann found the maximum rate of hydroxylation of phenylalanine by tryptophan hydroxylase to be 86% of that observed for L-tryptophan itself. More than seven times as much L-phenylalanine was required to saturate the enzyme as tryptophan, however, which means that under normal physiological conditions the enzyme would be hydroxylating tryptophan almost exclusively. This may not, however, be the case in a disease such as phenylketonuria where the phenylalanine level is extremely high. There is deficient production of serotonin in phenylketonuria, with reduced urinary excretion of 5-HIAA. The explanation may simply be that the enzyme tryptophan hydroxylase has been sidetracked into hydroxylating phenylalanine. It is quite possible that other situations might be uncovered where an imbalance of amino acid substrates could lead to a reduction in the synthesis rate of the proper transmitter, or even the production of a false transmitter (Tong and Kaufman 1975).

An amino acid closely related to phenylalanine, *p*-chlorophenylalanine (PCPA), is a classic inhibitor of tryptophan hydroxylase. It has a complicated mechanism of action *in vivo* because it produces a depletion of serotonin which persists long after the drug has been metabolized. A short-term effect on the enzyme has been documented in the test tube (*in vitro*), where PCPA is a competitive inhibitor of tryptophan. But another explanation is needed to account for its long-term effect *in vivo*. One possibility is that PCPA is incorporated into the enzyme protein itself in such a way as to render the catalytic site inactive. Antibodies have been prepared to purified tryptophan hydroxylase by Gal. Following the administration of PCPA, Gal isolated a protein which was enzymically inactive but which nevertheless had cross-immunological reactions with normal tryptophan hydroxylase (Gal *et al.* 1975). This is a good demonstration of the principle that the catalytic site and immunological site of a protein may be completely different.

9.3.2 5-Hydroxytryptophan Decarboxylase (L-Aromatic Amino Acid Decarboxylase, Dopa Decarboxylase) (EC 4.1.1.28)

This pyridoxal-dependent enzyme catalyzes the conversion of 5-HTP to serotonin. Whether the identical enzyme is involved in the decarboxylation of Dopa is a question which remains unsolved. As the enzyme classification suggests, the present consensus is that it is a single enzyme. Sims *et al.* (1973), however, report different regional and subcellular distributions of 5-HTP and Dopa decarboxylase activities as well as different pH optima for the two substrates. Unfortunately, efforts to purify two distinct enzymes have been unsuccessful. As described in Chapter 8 on catecholamines, antibodies have been raised to a purified L-aromatic amino acid decarboxylase and immunohistochemical studies done. These studies are consistent with the idea that a single enzyme catalyzes the decarboxylation of both substrates since the enzyme is found in catecholaminergic as well as serotonergic cells. This does not constitute absolute proof, however. It is possible that two highly similar enzymes exist with cross-catalytic and cross-immunological properties. It is also possible that the two aromatic amino acids have separate affinity sites on the same enzyme. The problem may be comparable to that found for MAO where for many years it was thought that a single enzyme catalyzed the destruction of all aromatic monamines, but where it is now believed that at least two forms of the enzyme exist. Like MAO, the decarboxylase is not restricted to neurons synthesizing neurotransmitters. It is found, for example, in large quantities in the kidney, and it is this organ that was the source for most early studies on the enzyme.

The activity of 5-HTP decarboxylase is far in excess of tryptophan hydroxylase both *in vivo* and *in vitro,* with most values in the literature being 50- to 100-fold greater. How much of this activity is attributable to serotonergic nerve endings and how much to other sources is still in doubt. Aprison has reported very small quantities of 5-HTP in brain tissue (0.2 nmol/g vs. roughly 7 nmol/g for serotonin) (Aprison *et al.* 1974). This means that virtually all 5-HTP formed by the hydroxylation of tryptophan *in vivo* is immediately converted to serotonin by virtue of the excess 5-HTP decarboxylase present in serotonergic nerve endings.

Inhibition of tryptophan hydroxylase brings about a reduction in brain serotonin levels while moderate inhibition of 5-HTP decarboxylase is without effect. This is further evidence of the fact that tryptophan hydroxylase is the rate-limiting step in serotonin synthesis.

Although it is difficult to reduce serotonin levels by inhibiting the enzyme, the excess of 5-HTP decarboxylase has been used to advantage in increasing serotonin levels. 5-HTP crosses the blood–brain barrier with ease, and due to high levels of the decarboxylase, parenteral administration of this amino acid leads to sharp increases in brain serotonin. This is accompanied by decreases in catecholamine levels and a rather strange behavioral syndrome in animals. For example, when 30 mg/kg of 5-HTP was administered to cats most areas of brain showed more than a 30% increase in serotonin while dopamine dropped to about 25% of its initial level. Behavioral performance on conditioned escape or avoidance tests markedly deteriorated as the animals appeared to be detached from their surroundings (P. L. McGeer *et al.* 1963b). In view of the decline in catecholamines under these conditions it is difficult to be certain that all the serotonin formed was in serotonin nerve endings. Some might have been produced in catecholamine nerve endings, acting as a false transmitter. Alternatively, 5-HTP might have interfered with uptake of catecholamine precursors, producing the catecholamine deficit. In any event it cannot be assumed that the behavioral changes observed following administration of large doses of 5-HTP represent pure stimulation of serotonin receptors.

Brain L-aromatic amino acid decarboxylase activity is not completely specific for Dopa and 5-HTP; both tryptophan and tyrosine are decarboxylated to some extent, and the products, tryptamine and tyramine, have both been proposed as neurotransmitter candidates (Chapter 11).

9.3.3 Monoamine Oxidase (EC 1.4.3.4)

MAO catalyzes the conversion of serotonin to 5-HIAA. As with other amine oxidase reactions, it is a two-step process as shown in Figure 9.6. The first step is the oxidation of 5-HT to 5-hydroxyindoleacetaldehyde. The second is the metabolism of the aldehyde. It is normally oxidized to 5-HIAA, but under some circumstances it can be reduced to 5-hydroxytryptophol (5-HTOH). Following the ingestion of alcohol the metabolism of most aldehydes is retarded due to the formation of large quantities of acetaldehyde. As a result, the amount of 5-HTOH in the liver and urine is increased and 5-HIAA decreased (Feldstein *et al.* 1967). There is no indication that a similar shift takes place in brain, however. The tentative conclusion is that in brain the metabolism of serotonin always proceeds through the normal two-step conversion to 5-HIAA.

The properties of MAO have already been discussed in the chapter on catecholamines. The reader is also referred to excellent reviews and books on this enzyme complex (Costa and Sandler 1972; Ho 1972; Pletscher *et al.* 1966). Serotonin is evidently deaminated *in vivo* primarily by type A MAO. This form is inhibited by clorgyline while type B is inhibited by deprenyl. Since clorgyline at a dose of 1 mg/kg i.v. will increase brain serotonin and decrease 5-HIAA while deprenyl has no effect, it must be concluded that type A is

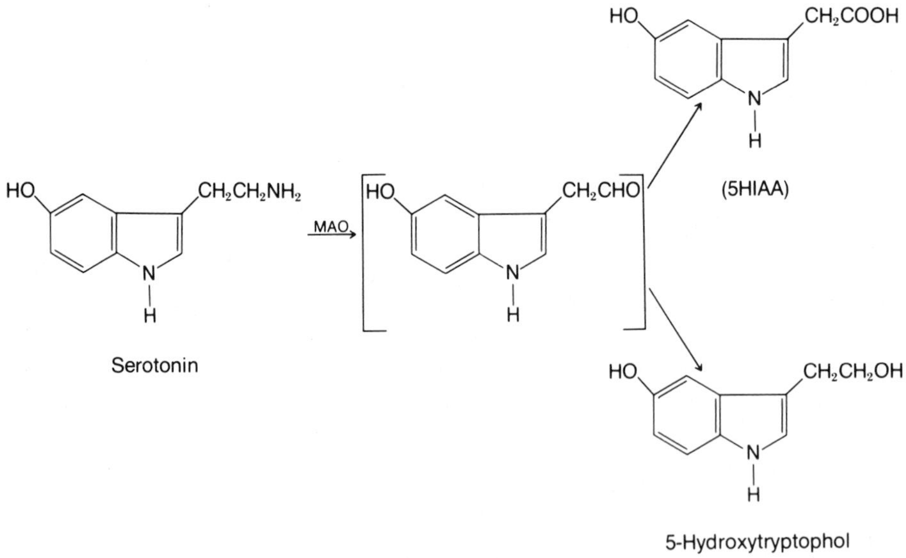

Figure 9.6 Two-step action of MAO on serotonin.

preferentially localized to presynaptic serotonergic neurons (Squires and Lassen 1975). Green and Youdim (1975) measured the effects of inhibition of type A (clorgyline), type B (deprenyl), or combined (tranylcypromine) MAO on hyperactivity and serotonin levels after a tryptophan load. Combined inhibition was required to achieve a maximal effect. They concluded that serotonin is metabolized by type A MAO preferentially, but when this is inhibited type B takes over. Before the behavioral syndrome can be observed, both types need to be almost totally inhibited. Thus, it may be that while type A is preferentially localized to serotonergic terminals, type B participates in postsynaptic destruction or even in glial disposal of serotonin.

MAO in general is widely distributed in brain and is found in the mitochondrial fraction, but so far little is known about the detailed regional and subcellular distributions of types A and B. Neff and Goridis (1972) found 90% type A and 10% type B in the superior cervical ganglion (which has a high concentration of noradrenergic cell bodies) but only 15% type A and 85% type B in the pineal (which has noradrenergic nerve endings and functional pinealocytes for producing melatonin). Thus, information is developing to indicate that the type of MAO may be related to the physiological requirements of a given region.

A minor route of metabolism of brain serotonin has been reported to be the formation of 5-HT-O-sulphate. Small amounts of this material have been isolated in brain following the administration of an MAO inhibitor plus probenecid to prevent elimination of the metabolite into the peripheral circulation (Gal 1972). It remains to be proved whether this route has any physiological significance.

An enzyme has also been reported in brain capable of N-acetylating serotonin. This is an important route in the pineal gland where N-acetylserotonin is the immediate precursor of melatonin. Melatonin is not produced in the brain so the functional significance of an N-acetylating enzyme is obscure. Nevertheless, N-acetylserotonin has been reported to be present in the cerebellum so that future research may attach some significance to this route (Chapter 11) (Bubenik et al. 1974, 1976a).

9.4 Control of Serotonin Synthesis

While the synthetic and destructive pathways for serotonin are well understood, the mechanisms by which the production of serotonin is controlled are still obscure.

Several methods have been developed for determining the amount of serotonin synthesized in vivo by brain. These include: (a) the administration of labeled L-tryptophan followed by the measurement of the specific activity of brain 5-HT; (b) the turnover in small regions of brain following the injection of labeled tryptophan intraventricularly or via an implanted cannula; (c) the immediate increase in brain 5-HT after MAO inhibition; (d) the accumulation of 5-HTP after 5-HTP decarboxylase inhibition; and (e) the accumulation of

5-HIAA following the administration of probenecid (Carlsson 1974; Neckers and Meek 1976).

These methods all give reasonably comparable values for the rate of synthesis of serotonin *in vivo,* and in the rat they are in the range of 1.6–2.2 nmol/h/g of brain tissue.

It has already been mentioned that the capacity of brain to synthesize serotonin is far in excess of this amount and that control must be exerted at the stage of tryptophan hydroxylation. So far there is no strong evidence that feedback inhibition by serotonin is a factor. Serotonin, even at levels of 10^{-3} M, does not inhibit tryptophan hydroxylation *in vitro,* nor do increased concentrations of brain 5-HT inhibit synthesis *in vivo.*

The availability of tryptophan is, however, important. Diets with little or no tryptophan lower brain serotonin while diets high in tryptophan increase brain serotonin. Following a detailed study of the factors controlling brain and plasma tryptophan levels, Fernstrom and Wurtman (1972) concluded that the main determinant of brain tryptophan was the ratio in plasma of this amino acid to other neutral amino acids such as tyrosine, phenylalanine, etc., which compete with it for uptake into the brain. The matter is complicated somewhat by the fact that most of the tryptophan in plasma is unavailable for transport into the brain because it is tightly bound to plasma albumin at tryptophan-specific sites. Once in the brain, tryptophan is taken up into synaptosomes by both a high-affinity and a low-affinity uptake system. Several amino acids are known to be competitive with these systems and it remains for future research to determine whether the uptake system for tryptophan by synaptosomes plays a critical role in the synthesis rate by limiting the supply of serotonin precursor.

Nerve stimulation is also a factor. Excitation of serotonin cell bodies in the raphe nucleus increases turnover in the septal area where terminal fields are located. LSD, which has a powerful inhibitory effect on serotonin neurons when applied directly, decreases the turnover (Aghajanian and Haigler 1973).

But serotonin production is not related entirely to nerve stimulation. Following acute spinal transection, which disconnects the nerve terminals in the spinal cord from their cell bodies located in the brainstem, the synthesis of serotonin is estimated to be decreased by only 50% in the first few hours following the lesion (Carlsson *et al.* 1973). The 5-HT, which continues to be formed despite the absence of nerve stimulation, is presumably being metabolized intraneuronally. Thus, serotonin can be synthesized and destroyed by presynaptic MAO without any reference to release and metabolism by postsynaptic MAO. This makes it difficult to judge what proportion of the 5-HT synthesized *in vivo* is actually consumed in nerve transmission and what proportion is merely wasted in presynaptic metabolism (Green and Grahame-Smith 1975).

Even though the amount of serotonin produced does not appear to be precisely geared to that consumed in neurotransmission, the synthetic system does respond to increased demand when the neuron is stimulated. Several authors have proposed that compartmentation of serotonin within the nerve ending is involved in this control. Most of the intrasynaptosomal serotonin is bound within vesicles, but some exists in a free state within the cytoplasm. Turnover studies suggest that the more recently synthesized molecules turn over more rapidly while others somehow find their way into a reserve capacity, perhaps in vesicles located at sites relatively remote from the areas of release. Grahame-Smith (1973) has proposed that an unbound prefunctional pool of serotonin is crucial to control, but there is still no explanation as to how tryptophan hydroxylation is held in check and allowed to vary in approximate proportion to the activity of the neuron.

9.5 Physiological Actions of Serotonin

The actions of serotonin on single neurons are variable and hard to interpret. The most general effect of iontophoretic application is a reduction in excitability. This is the case for cortical (Jordan *et al.* 1971), striatal (Herz and Zieglgansberger 1968), hypothalamic (Bloom *et al.* 1963), and cerebellar (Bloom *et al.* 1972) neurons. But serotonin also excites some neurons, for example, in certain areas of the thalamus (Phillis and Tebecis 1967) and bulbar reticular formation (Hosli *et al.* 1971). In other areas such as the lateral geniculate the effect is mixed (Satinsky 1967). In some situations, excitatory and inhibitory effects can even be observed in the same cell (Roberts and Straughan 1967), illustrating as with dopamine (Chapter 8) the vagaries of experimental conditions.

In view of the mixed iontophoretic actions, the large divergence of nerve endings from a relatively few cells, and the vague effects of drugs, it might be anticipated that serotonin would have prominent metabotropic actions. Serotonin does increase cAMP formation in brain slices (Kakiuchi and Rall 1968) but the effect is not prominent. It also stimulates cAMP formation in *Aplysia* (Cedar and Schwartz 1972), where it is believed to be a transmitter. There are many systems, however, where serotonin does not appear to be active which again raises speculation as to its mode of action.

Although the physiology of serotonin is by no means understood, evidence is accumulating suggesting it may play a role in sleep, sex, and mood. The data support the original proposal of Brodie (Brodie *et al.* 1959) that serotonin is involved in trophotropic function and plays an opposite role to that of the catecholamines.

9.5.1 Serotonin and Sleep

The cat normally spends about two-thirds of its time sleeping. The sleep is uneven, with 20- to 30-min periods of light sleep alternating with 5- to 10-min periods of deeper sleep. Light sleep is characterized by slow, synchronized cortical EEG activity, and muscle tone not greatly reduced from the waking state. It is thus referred to as slow-wave sleep. Deep sleep is characterized by a profound relaxation of peripheral muscle accompanied by rapid eye movements (REM) and a paradoxical activation of the cortical EEG. It is thus referred to as REM or paradoxical sleep. In man it has been established that this is when dreaming occurs. Jouvet and his colleagues reasoned that sleep was not just a passive process. Using the cat as an experimental model, they commenced searching for an area of brain that actively promotes sleep in the same way that the reticular activating system promotes wakefulness. They discovered that electrolytic lesions of the raphe system turned cats into insomniacs. There was a reduction in the time spent sleeping from 85% to 20% (Jouvet 1973).

Because of the known relationship of the raphe to serotonin systems, Jouvet next studied the effects of chemically reducing the activity of serotonin neurons. It had been known from the earlier experiments of Quay (1965) that the concentration of serotonin in brain showed a diurnal rhythm, being high during periods normally associated with rest and low during periods associated with activity. Jouvet found that one to two days following peritoneal administration of p-chlorophenylalanine (PCPA) the cats became even greater insomniacs than following electrolytic lesions. Administration of an appropriate dose of 5-HTP promptly reversed the effect. But in the case of electrolytic lesions 5-HTP did not reverse the effect. This would be expected if those pathways serving sleep were totally destroyed by the electrolytic lesion, so that synthesis of serotonin could not take place in appropriate nerve endings (Jouvet 1973).

Jouvet also investigated the role of catecholamines. Administration of L-Dopa produced a long-lasting arousal in the cat. Inhibition of catecholamine synthesis by the administration of α-methyl-p-tyrosine decreased waking in normal cats and totally suppressed the behavioral and EEG arousal which normally follows the injection of amphetamine.

Some efforts were made to distinguish which of these global catecholamine effects were attributable to dopamine and which to noradrenaline. Although destruction of the substantia nigra failed to interfere significantly with cortical arousal, it did strongly reduce the behavioral aspects of waking. Destruction of the A8 group of noradrenergic cells in the mesencephalon enhanced cortical synchronization. These results suggest that while movement is associated with dopamine, cortical arousal is associated with noradrenaline. Such a concept is in keeping, of course, with the known role of dopamine in extrapyramidal function, and the presumed role of noradrenaline in cortical arousal.

Jouvet proposed that catecholamine neurons were involved in the "executory mechanisms" of paradoxical sleep. He observed that when reserpine-pretreated cats were given L-Dopa they exhibited paradoxical sleep much earlier than animals not receiving L-Dopa. Destruction of the noradrenergic cells of the locus coeruleus suppressed all the central and peripheral components of paradoxical sleep (Jouvet 1973).

Unfortunately, somewhat different results were obtained in man. PCPA decreased REM sleep without affecting slow-wave sleep. 5-HTP not only reversed the PCPA effect, it actually elevated the time spent in REM sleep in normals. L-Dopa, on the other hand, reduced the duration of REM sleep, and its discontinuation was associated with a marked rebound effect. Thus, studies in man indicate that the catecholamines inhibit while serotonin enhances REM sleep (Wyatt *et al.* 1970).

Despite the species differences, there is general agreement that serotonin systems are somehow involved in the promotion of sleep while catecholamine systems are involved in the promotion of wakefulness.

9.5.2 Serotonin and Sex

Great interest surrounded the report that PCPA increases sex drive as evidenced by increased mounting activity among male rats (Shillito 1970). Much conflicting data then appeared, possibly because of the methods of defining enhanced sexual behavior. Experiments in monkeys showed no effect of PCPA. Furthermore, in humans being treated with PCPA for the carcinoid syndrome, or in volunteers, there was no obvious change in libido as a result of PCPA administration. Some have suggested that sexual behavior involves a balance between catecholamines and serotonin. PCPA, by reducing serotonin levels, may upset this balance and thus produce a mild influence on sexual behavior. There is positive evidence of increased libido in some patients treated with L-Dopa.

9.5.3 Serotonin, Mood, and Mental Illness

The most fascinating aspect of serotonin physiology is its possible involvement in mood and behavior. Ever since the original hypothesis of Woolley and Shaw (1954) that serotonin metabolism might be deranged in mental illness, there has been speculation that various psychopharmacologically active agents owe their efficacy to a serotonin interaction. This applies to hallucinogens as well as to antipsychotic and antidepressant drugs. Schizophrenia and depression are the two mental diseases around which specific serotonin theories have been developed. These aspects will be discussed in more detail in Chapter 14.

Serotonin has also been implicated in the elevation of body temperature (Myers 1973) and suppression of pain sensitivity (Harvey and Yunger 1973).

9.5.4 Melatonin and Other Indoleamine Pathways

Melatonin is the only indoleamine derivative other than serotonin known to have a physiological function. It is produced principally, if not exclusively, by the pineal gland.

The pineal gland is a legendary structure situated between the cerebral hemispheres just above the habenula. It has fascinated physiologists down through the centuries perhaps because it is the only unpaired structure within the cranium. The French philosopher Descartes made the most spectacular suggestion, proposing that it was the seat of the soul. But it was not until Lerner's identification of melatonin in 1958, and Axelrod's recognition of its route of synthesis in 1961, that any real understanding of its role began to emerge.

The pineal gland, although attached to the roof of the third ventricle by a vestigial and nonfunctional stalk, is actually outside the nervous system, being innervated by noradrenergic fibers of the superior cervical ganglion. It is made up of pinealocytes which contain the enzymes for producing serotonin as well as for converting it to melatonin. The enzymes necessary for the final steps are N-acetyltransferase (EC 2.3.1.5) and hydroxyindole-O-methyltransferase (EC 2.1.1.4) (HIOMT). The sequence of reactions is as shown in Figure 9.7.

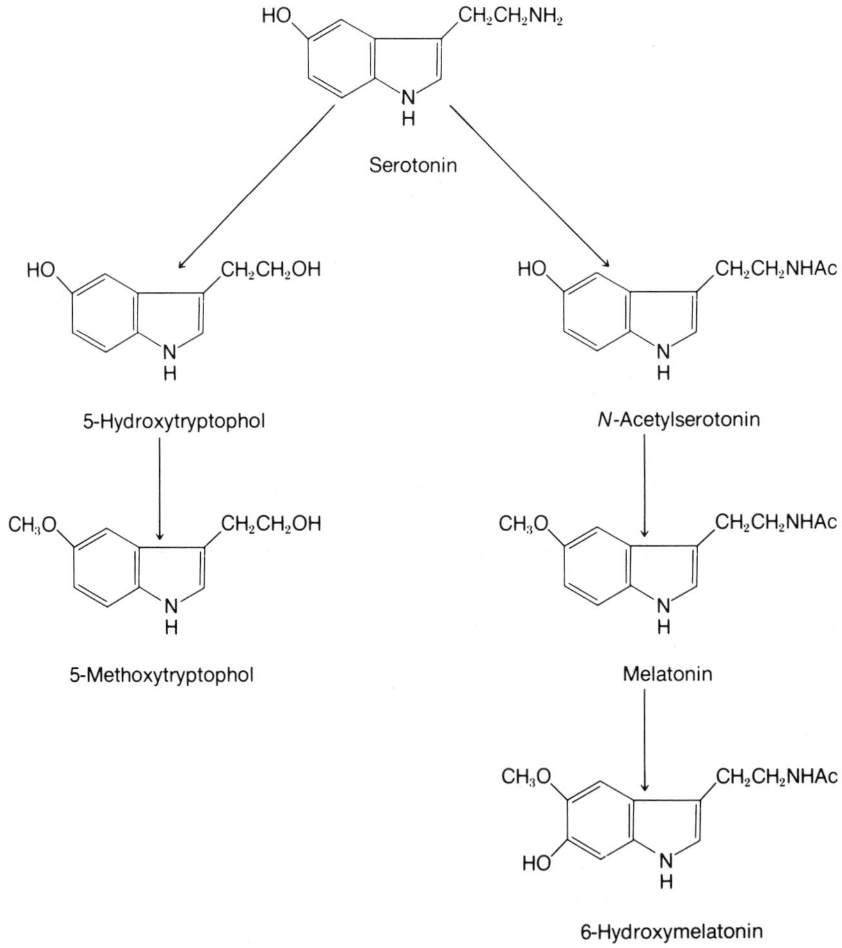

Figure 9.7 Metabolism of serotonin in the pineal, and formation of melatonin.

So far, the full range of physiological actions of melatonin is not known. In frogs it causes dispersion of melanocytes and a lightening of the skin, a property which was critical to tracing its chemical identification but which apparently has no relevance to mammals. Its major mammalian effect seems to be that of inhibiting gonadal function. One of the functions of the pineal, at least in small mammals, is to act as a neurochemical transducer, converting information about environmental lighting into chemical activity. Rats exposed to constant light have persistent estrus and larger gonads. Rats exposed to constant darkness are anestrus and have smaller gonads.

Axelrod and colleagues showed that the transducer pathway went from the retina through the inferior accessory optic tract, the medial forebrain bundle, and the lateral horn of the spinal cord to the sympathetic nervous system and the pineal. As a result of sympathetic nervous system activity, there is inhibition of the biosynthesis of melatonin (Axelrod and Wurtman 1968; Wurtman *et al.* 1968).

Under conditions of continuous light the activity of both HIOMT and *N*-acetyltransferase are markedly reduced compared with continuous darkness. The difference is about 2.5-fold in the case of HIOMT but up to 70-fold for *N*-acetyltransferase (Klein and Yuwiler 1973). Thus, melatonin production is greatly reduced during periods of light. The gonads are released from its inhibitory effects.

Serotonin synthesized as a precursor of melatonin by pinealocytes is also found stored in the noradrenergic nerve endings of the pineal. When noradrenaline synthesis is blocked, serotonin content in the nerve endings goes up. This illustrates what may be a broader principle, namely, that serotonin can be taken up and stored in noradrenergic nerve endings under favorable circumstances.

HIOMT activity has now been reported in the retina and Harderian gland, which is the only evidence so far that melatonin production may not be restricted entirely to the pineal (Bubenik *et al.* 1976b).

9.6 Pharmacology of Serotonin

A wide variety of pharmacological agents exists which specifically interact with serotonin. The spectrum is broadened where there is a high degree of similarity between serotonin and catecholamine mechanisms. This is the case for the decarboxylase and MAO enzymes, the vesicular storage mechanism, and to some extent the reuptake pump system. Table 9.2 is a list of the more prominent compounds. Figure 9.8 is a semidiagrammatic representation of how they may act at the synaptic level.

9.6.1 Agents Toxic to Serotonin Neurons

Three agents have so far been discovered which selectively damage serotonin neurons. They are 5,6-dihydroxytryptamine (5,6-DHT), 5,7-

TABLE 9.2
Some Drugs Which Affect Serotonin Action

Drug	Presumed mechanism of action	Physiological effects
5,6-Dihydroxytryptamine (5,6-DHT)	Toxin for serotonergic neurons	Not precisely determined
p-Chloroamphetamine (PCA)	Toxin for serotonergic neurons	Not precisely determined
p-Chlorphenylalanine (PCPA)	Tryptophan hydroxylase inhibitor	Antisleep, mild aphrodisiac?
NSD-1034	5-HTP decarboxylase inhibitor	None by itself
Clorgyline	Type A MAO inhibitor	Not precisely determined
Iproniazid	Broad-spectrum, irreversible MAO inhibitor	Antidepressant
Tranylcypromine	Broad-spectrum, reversible MAO inhibitor	Antidepressant
Chlorimipramine	Broad-spectrum amine pump inhibitor	Antidepressant
Nortryptilene	Serotonin pump inhibitor	Antidepressant
Clomipramine	Serotonin pump inhibitor	Antidepressant
Reserpine	Irreversible broad-spectrum storage inhibitor	Tranquilizer
Tetrabenazine	Reversible broad-spectrum storage inhibitor	Tranquilizer
Chlorpromazine	Broad-spectrum receptor blocker	Tranquilizer
LSD	Serotonin-receptor-site stimulator/blocker	Hallucinogen
Bufotenine	Serotonin analog (false transmitter?)	Hallucinogen
5-Methoxy-N,N-dimethyltryptamine	Serotonin-receptor-site stimulator	Hallucinogen

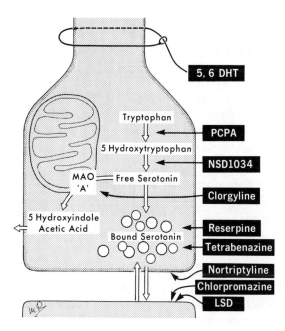

Figure 9.8 Mechanism of drug action at serotonergic synapse; schematic diagram of serotonin nerve ending. Light rectangles and light arrows show the site of action of endogenous serotonergic elements. Black rectangles and black arrows indicate various drugs which interact with serotonin and their proposed site of action. (See Table 9.2.)

dihydroxytryptamine (5,7-DHT) (Baumgarten and Lachenmayer 1972), and p-chloroamphetamine (PCA). These neurotoxic agents are accumulated within serotonin neurons by action of the membrane pump. The destructive effect of 5,6- and 5,7-DHT is probably due to the ease with which they are oxidized to materials that poison the respiratory enzymes of the cell. They share this characteristic with 6-OHDA which is pumped into catecholaminergic cells and is oxidized to a toxic derivative.

Uptake studies using labeled 5,6-DHT have shown that it competes with serotonin for uptake and is accumulated by brain slices preferentially *in vitro*. It does not compete with noradrenaline and dopamine uptake. 5,7-DHT is six times less potent than 5,6-DHT in inhibiting serotonin uptake, which may explain why it is a less potent toxin (Horn *et al.* 1973).

After a single intraventricular injection of 50–75 μg of 5,6-DHT in the rat, damage to unmyelinated axons can be observed as early as 2 h after the injection. About half the serotonin in the forebrain disappears after 10 days. Due to its general toxicity, it has so far not been possible to administer high enough doses of 5,6-DHT intraventricularly to bring about total depletion of serotonin in the brain, but well-placed local injections into ascending serotonin tracts can produce a 90%–95% reduction in the terminal areas (Bjorklund *et al.* 1974).

With sublethal doses of 5,6-DHT some regeneration of serotonin neurons seems to take place over a period of weeks. Apparently, the terminals go back into appropriate receptor areas because the supersensitivity to serotonin disappears as reinnervation takes place. This discovery of sprouting of serotonin and catecholamine neurons is causing a reevaluation of the classical concept that CNS neurons are incapable of regeneration (Fuxe and Johnsson 1974) (Chapter 12, Figure 12.15, Section 12.2.3).

PCA is also toxic to serotonergic neurons. 5-HT levels of brain are reduced to less than half the normal values a few days following the injection of a single i.p. dose of approximately 0.1 mmol/kg. It is a selective inhibitor of MAO. It does not inhibit 5-HT synthesis *in vitro*. The manner in which the selectively accumulated PCA damages 5-HT neurons is a mystery because it is not readily oxidized to potentially toxic metabolites as are 6-OHDA and 5,6-DHT (Sanders-Bush *et al.* 1974).

9.6.2 Inhibitors of Tryptophan Hydroxylase

The classic inhibitor of tryptophan hydroxylase is PCPA. Actually, Koe and Weissman (1966) discovered the effects of PCPA by examining its ability to inhibit liver phenylalanine hydroxylase so that its specificity is not absolute. When added to brain tryptophan hydroxylase preparations *in vitro* it inhibits the enzyme competitively with a K_i (the concentration required to produce 50% inhibition) of about 300 μM. But the inhibition *in vivo* is a persistent one which obviously does not depend solely upon competition with tryptophan for the active site on the enzyme. Evidently, PCPA is incorporated into tryptophan hydroxylase in such a way as to render the enzyme inactive. The inhibition is not reversed *in vivo* until new enzyme molecules are synthesized, a process which takes several days. The effect is nevertheless reversible because PCPA is not toxic to the neurons themselves. Reversible inhibition is produced by 6-halotryptophan derivatives, but due to the transient action they have never been intensively studied (E. G. McGeer *et al.* 1968).

9.6.3 Inhibitors of 5-HTP Decarboxylase

A variety of agents exists which inhibit 5-HTP decarboxylase but these are identical with ones that inhibit Dopa decarboxylase. Thus, they are not only nonselective, but are also ineffective at lowering brain serotonin because the decarboxylase is present in so much higher concentration than tryptophan hydroxylase. The better known inhibitors are NSD-1034, RO 4-4602, and MK 485. The latter two do not cross the blood–brain barrier and have achieved practical usage in the treatment of Parkinson's disease. They have also been used in conjunction with 5-HTP administration to reduce the amount of peripheral decarboxylation and thus increase the amount of 5-HTP reaching the brain.

9.6.4 Inhibitors of Monoamine Oxidase

The properties of MAO and the MAO inhibitors have already been extensively discussed in the chapter on catecholamines and elsewhere. Most of the MAO inhibitors affect both type A and type B MAO, although it is only type A which is believed to exist in presynaptic serotonin nerve endings.

These broad-spectrum MAO inhibitors are mostly the irreversible hydrazine type, but some, such as tranylcypromine, are reversible. Only recently have narrow-spectrum MAO inhibitors become available. Of these, the most specific one known for type A is clorgyline. Other type A inhibitors are harmaline, harmine, and α-ethyltryptamine, all of which are short-term competitive-type inhibitors. So far, it is not known whether such selective type A MAO inhibitors will produce more specific clinical effects than the non-selective MAO inhibitors which have been generally used.

9.6.5 Serotonin Pump Inhibitors

The so-called tricyclic antidepressants are believed to exert their therapeutic action by inhibiting neuronal reuptake pumps. The pump process is considered to be the major route for inactivation of many released neurotransmitters, and inhibition prolongs neurotransmitter action at the receptor site. As with the MAOI antidepressants, the tricyclics are broad spectrum, affecting both the catecholamine and serotonin systems. There is some variation, however. For example, chlorimipramine and imipramine are potent inhibitors of serotonin uptake while desipramine is not (Snyder *et al.* 1973). As further evidence of a preferential action, the former two compounds decrease serotonin synthesis while the latter does not. This would be expected if the prolonged action at receptor sites due to pump inhibition acted as a negative feedback system on serotonin synthesis. Nevertheless, desipramine is a powerful and effective antidepressant compound.

Nortryptilene and clomipramine are also effective serotonin pump inhibitors. The effective antidepressant, maprotiline, is only a weak serotonin pump inhibitor, however, but a relatively potent noradrenaline pump inhibitor. Maprotiline is not a tricyclic compound but, like desipramine, it illustrates the point that antidepressant activity is not directly related to serotonin pump inhibition. It may be more closely related to catecholamine (particularly noradrenaline) pump inhibition (Maitre *et al.* 1974).

9.6.6 Inhibitors of Storage

The rauwolfia alkaloids and tetrabenazine are classical depletors not just of serotonin but also the catecholamines. The mechanism which binds these amines to granules is so similar that selective agents acting at this particular level are unknown. For many years experiments have been undertaken to try to sort out which of the clinical effects of reserpine are due to serotonin and which to catecholamine depletion. Despite thousands of investigations, the question is not settled even today. The time frame for replenishing serotonin stores following a single dose of reserpine closely parallels that for the catecholamines. It corresponds to the time taken to synthesize new vesicle-binding proteins in the cell body and transport them to the nerve endings. The minimum time is 18–24 h, but it can take up to several days for complete replenishment.

Reserpine sedation can be temporarily overcome by an adequate dose of L-Dopa but not L-5-HTP. This clearly shows that the major tranquilizing effects are due to depletion of the catecholamines with an as yet uncertain role for serotonin. On the other hand, the diarrhea and hot flush, which are found shortly after reserpine administration, are most probably due to peripheral serotonin release since they closely parallel the symptoms of the carcinoid syndrome.

Tetrabenazine, unlike reserpine, does not bind irreversibly to the protein of storage vesicles. Thus, while its mechanism of action is similar, its effects are transient, persisting for only 24–36 h in man.

9.6.7 Receptor Site Agonists and Antagonists

Serotonin receptor sites are blocked by many of the same broad-spectrum compounds which block catecholamine receptor sites. These include the phenothiazines, thioxanthines, and butyrophenones. Again, the extent to which this serotonin-blocking action plays a role in the clinical effects is unknown. As previously described, there is strong evidence that blockade of dopamine is primarily responsible for the antipsychotic action, with evidence for noradrenaline coming next.

The dilemma created by our limited knowledge of serotonin receptor sites and the activities of agonist and antagonist compounds is best illustrated by the effects of LSD. Doses as low as 20–25 μg are capable of producing effects in susceptible individuals. The amounts reaching the brain are probably in the picomole range, indicating that the material would most likely be working directly at receptor sites. This agent has a definite serotonin-blocking action at some peripheral receptor sites. A similar blocking action is obtained with 2-bromo-LSD, methysergide (another LSD derivative), and such indole derivatives as gramine, harmine, and tryptamine. Serotonin antagonists have been classified as "musculotropic" or "neurotropic" depending on their relative ability to block the effects of serotonin on smooth muscle or to block peripheral neural responses. But LSD mimics rather than antagonizes serotonin at some classical receptors such as the mollusk heart. Clearly, there are many types of serotonin receptors (Gyermek 1966).

In the mammalian central nervous system LSD causes a slight rise in brain 5-HT and a fall in 5-HIAA. This would be expected if LSD excited rather than blocked receptor sites. Such a negative feedback on synthesis is also obtained with other indole alkylamine-type hallucinogens like psilocybin, dimethyltryptamine, and dimethoxydimethyltryptamine. Aghajanian et al. (1973) observed that neurons in the raphe system of the midbrain ceased firing when LSD was added parenterally. He also showed that LSD can mimic the effects of 5-HT when administered iontophoretically. Fuxe et al. (1972) have concluded that the common property of serotonin-like hallucinogens is that they stimulate 5-HT receptors and cause a decrease in 5-HT synthesis. It cannot, however, really be proved whether LSD produces its psychotomimetic effects by exciting serotonin receptor sites, blocking serotonin receptor sites, doing both, or affecting other sites.

There is still no clear picture of what really happens when serotonin receptor sites in the brain are stimulated. Administration of tryptophan plus an MAO inhibitor substantially increases intraneuronal serotonin stores, presumably making available excess serotonin for stimulating receptor sites. The effects of such a combination are comparable to those observed when large amounts of 5-HTP are administered, with or without an MAO inhibitor. Tremors, hyperthermia, and abnormal hyperactivity are observed in animals, and behavioral and mental abnormalities are documented in man. But the syndrome is not observed in animals if α-methyl-*p*-tyrosine, which decreases production of brain dopamine and noradrenaline, is given first (Youdim *et al.* 1975). Administration of L-Dopa under such circumstances restores the syndrome. Thus, the catecholamines are involved in what had first been believed was a pure serotonin effect.

(−)-Propranolol also blocks the excess serotonin behavioral response (Green and Grahame-Smith 1976). This is not due to an interference with serotonin synthesis since the same levels of serotonin are achieved with or without the drug following a tryptophan load in MAO-treated animals. (−)-Propranolol also blocks the syndrome when it is produced by 5-methoxy-*N,N*-dimethyltryptamine, a serotonin receptor site stimulator.

On the other hand (+)-propranolol has no effect on this syndrome and neither isomer blocks the motor response to L-Dopa plus tranylcypromine. This is regarded with some interest since (±)-propranolol is effective in the treatment of schizophrenia (Chapter 14) (Atsmon *et al.* 1972). If the behavioral syndrome brought about by excess serotonin can be blocked by agents which prevent dopamine synthesis or which block its receptor sites, then perhaps the antipsychotic action is due in part to compensation for excess serotonin. Propranolol would achieve the same result by direct blockade of the receptor sites.

9.7 Summary

Serotonin cell bodies in the CNS are confined to a remarkably limited anatomical area. Seven of nine groups which have so far been identified lie within the raphe system of the brainstem, while two groups (B3 and B9) are in the reticular formation. The axons of the more caudal neurons descend to the spinal cord to terminate mainly in the lateral horn, but terminals are also found in other spinal regions. The axons of the more rostral groups innervate the diencephalon, limbic system, and to a lesser extent the striatum and cortex. The divergence of terminals suggests that the role of serotonin neurons is to exert a generalized tonic effect rather than to have a highly selective action on a limited group of neurons.

The enzyme tryptophan hydroxylase, which converts tryptophan to 5-HTP, is contained exclusively in serotonin neurons. It controls the synthesis of serotonin by mechanisms not yet understood. 5-HTP decarboxylase, which converts 5-HTP to serotonin, is present in excess and is either identical or highly similar to Dopa decarboxylase. Serotonin is converted both pre- and postsynaptically to the inactive metabolite 5-HIAA by MAO. This identical pre- and postsynaptic enzymatic action has made it impossible to determine how much of the serotonin which is metabolized is wasted intraneuronally and how much is released extraneuronally to act on receptor sites.

The physiological actions of serotonin are not really known, although they seem to follow the pattern of emphasizing trophotropic functions. Serotonin promotes sleep, hyperthermia, and, at least peripherally, gut motility. It may have a mild depressing influence on sexual behavior and pain sensitivity, and it may also have a mild antidepressant effect. The possibility that serotonin may be involved in schizophrenia remains a speculation, based largely on the psychotomimetic action of LSD and other indolealkylamine derivatives of close structural similarity to serotonin (Chapter 14).

While serotonin and the catecholamines seem to be antagonistic in their physiological effects, it has nevertheless been extremely difficult to separate their actions through drug studies. Highly similar processes are involved in decarboxylation of the precursors 5-HTP and Dopa, in oxidative destruction of the amines, in their intraneuronal storage, in reuptake systems, and even in receptor activity. Thus, such important psychopharmacological agents as the phenothiazines, the rauwolfia alkaloids, the tricyclic antidepressants, and the monoamine oxidase inhibitors, all act on serotonin as well as catecholamine mechanisms (Chapter 14).

Melatonin is the only other indoleamine known to have a physiological function. It is produced by the pineal, shows a circadian rhythm, and has an inhibitory effect on the gonads.

10

Promising Peptides

10.1 Introduction

It is clear from a description of the anatomy of neurons served by the better known neurotransmitters that most neuronal systems in brain are still not biochemically fingerprinted. A promising area for exploration is in the peptides.

Powerful new techniques have recently become available for characterizing and assaying these materials. Many can be purchased commercially. As a result, an explosion of activity has taken place in the field. It has suggested to many that whereas the last 25 years has been the age of the monoamines in neuroscience, the next 25 years may be the age of the peptides. Because of the rapid accumulation of information our account must be a tentative one, subject to major revisions as new developments take place.

A number of peptides are widely distributed throughout the nervous system and may therefore be true neurotransmitters. Others appear to be largely confined to the hypothalamic–pituitary axis and may only act as neurohormones on pituitary cells or peripheral organs. These latter compounds are often referred to as hypophysiotropic peptides. It is beyond the scope of this book to deal with the complicated field of neuroendocrinology, but interest in the hypophysiotropic peptides is bound to grow, and the distribution and action of some of them may turn out to include other parts of the CNS.

The principal peptides having wide distribution in the CNS are the undecapeptide, substance P, and two pentapeptides, methionine enkephalin and leucine enkephalin. β-Endorphin, neurotensin, somatostatin, thyrotropin-releasing hormone (TRH) and angiotension II have also been reported to be concentrated in widely separated areas of the brain. In the hypothalamus two peptides, oxytocin and vasopressin, are produced in the supraoptic and paraventricular nuclei and transported along the supraopticohypophysial tract to the posterior pituitary. At least eight other stimulating and releasing peptides are formed in hypothalamic neurons which have nerve endings in the median eminence (Table 10.3). From this location, they can be released into the hypophyseal-portal circulation to influence cells of the anterior and intermediate pituitary lobes.

10.2 Substance P[1]

10.2.1 Anatomy of Substance P

The classic peptide from brain is substance P. It was given its oddly appropriate name by its discoverers, Von Euler and Gaddum (1931), long before its peptide structures became known. Von Euler and Gaddum obtained substance P by adding sulfuric acid to alcoholic extracts of horse intestine and brain. The precipitate they obtained had an ACh-like activity on smooth muscle which differed in not being blocked by atropine (Von Euler and Gaddum 1931). Early bioassay techniques (Pernow 1953; Amin et al. 1954) showed that substance P occurred in the central nervous system largely in gray matter and was particularly concentrated in the hypothalamus, basal ganglia, thalamus, and dorsal roots of the spinal cord. This latter finding plus its property of causing peripheral vasodilation inspired Lembeck (1953b) to propose that substance P was the primary sensory transmitter.

It had been known since 1876 that vasodilation was produced peripherally when the distal end of a sectioned posterior root was stimulated. This phenomenon came to be regarded as a special application of Dale's law, where it was supposed that the sensory transmitter released from axonal endings in the spinal cord would also be released by the peripheral dendrite upon antidromic stimulation. Thus, the tendency to produce vasodilation upon local injection became a screening test for sensory transmitters. Substance P was one of the most powerful agents at producing this effect, and there were no known blockers.

The substance P field really lay fallow until it was opened by a serendipitous discovery in 1970. Leeman and her co-workers at Brandeis University were involved in an attempt to isolate a corticotropin-releasing factor from bovine hypothalamic extracts. When the various fractions from one preparative column were screened for biological activity, Leeman noticed that one fraction stimulated salivary secretion when injected into anesthetized rats. This "sialogogic effect" was not inhibited by atropine. It proved to be a simple quantitative assay from which isolation and purification of the peptide could proceed. The Leeman group then realized that their purified "sialogen" had identical pharmacological actions and chemical properties (solubility in various solvents, molecular weight, isoelectric point, electrophoretic mobility, sensitivity to chymotrypsin and pepsin, and insensitivity to carboxypeptidases) to those previously reported for partially purified substance P. They concluded that sialogen and substance P were identical, having a undecapeptide struc-

---

[1]General Reference: von Euler and Pernow 1977.

H-Arg-Pro-Lys-Pro-Gln-Gln⤸Phe⤸Phe-Gly⤸Leu-Met-NH₂

Figure 10.1 Structure of substance P. Arrows indicate points of cleavage by endogenous enzyme activity.

ture (Chang and Leeman 1970; Leeman and Mroz 1974). The structure of substance P is shown in Figure 10.1.

Substance P has been synthesized (Yajima and Kitagawa 1973; Fisher *et al.* 1974) and made available commercially. Chemical, physical, and biological identity has been demonstrated (Studer *et al.* 1973) between the synthetic material and mammalian substance P isolated by classical methods from hypothalamus or intestine.

The chemical identification and synthesis of substance P permitted the development of a sensitive and specific radioimmunoassay as well as an immunohistochemical technique for its localization. Since it is a peptide of too low a molecular weight to be antigenic in itself, it was necessary to make it antigenic by coupling it to bovine gamma globulin. Antisera raised in rabbits and guinea pigs produced an assay sensitive to < 1 pmol of substance P (Powell *et al.* 1973).

The levels in whole rat brain of substance P are approximately 50 pmol/g wet weight of tissue; this means that the overall concentration of substance P in the brain is much less than those of other neurotransmitters such as dopamine, noradrenaline, or serotonin.

Most tissues show only trace amounts of substance P. Exceptions are the intestine, spinal cord, pineal, and parts of the brain. Particularly high levels occur in the dorsal horn, trigeminal nerve nuclei, substantia nigra, medial hypothalamus, and interpeduncular nucleus (Table 10.1).

The significance of such uneven distribution is, of course, a possible association with specific neurons. Such a possibility is further strengthened by the discovery that substance P is in the nerve ending fraction (Ryall 1962; Duffy *et al.* 1975) and is released by the same calcium-dependent process as are other neurotransmitters (Iversen *et al.* 1976; Schenker *et al.* 1976).

10.2.2 Association of Substance P with Central Pathways

Evidence of association of substance P with specific pathways comes from lesioning and immunohistochemical studies. Following lesions to the striatonigral or habenulointerpeduncular tracts, sharp drops in the substance P levels are found in the substantia nigra (Kanazawa *et al.* 1977a; Mroz *et al.* 1977) and the interpeduncular nucleus (Hong *et al.* 1976). Additional support for substance P association with a striatonigral tract comes from Kanazawa *et al.* (1977b), who found decreased nigral levels in Huntington's chorea. In this disease (Chapter 12) there is a loss of striatal neurons which should cause degeneration of striatonigral pathways. Similar losses of nigral substance P are found after injections of kainic acid into the caudate, particularly the anterior portion (Hong *et al.* 1977a; see Chapter 6).

Immunohistochemical studies (Nilsson *et al.* 1974; Hokfelt *et al.* 1975)

TABLE 10.1
Regional Distribution of Substance P (in pmol/g of tissue)
in the Rat Nervous System[a]

Somatosensory system		Hypothalamohypophyseal system	
Dorsal root ganglia	68	Medial hypothalamus	626
Dorsal horn	1070	Middle hypothalamus	514
Trigeminal nerve nucleus	1387	Lateral hypothalamus	488
Dorsal column	129	Posterior pituitary	71
Dorsal column nucleus	168		
Thalamic nucleus (tvd)	22		
Sensory cortex	19	Pyramidal system	
		Internal capsule	137
Visual system		Pyramis	81
Retina	66	Ventral horn	134
Optic nerve	64		
Lateral geniculate	80		
Visual cortex	23	Basal ganglia	
		Caudate	247
Limbic system		Globus pallidus	333
Olfactory bulb	62	Subthalamic nucleus	234
Olfactory tubercle	300	Substantia nigra	1730
Olfactory cortex	43		
Amygdala	382		
Hippocampus	37		
Habenula	376		
Interpeduncular nucleus	598	Cerebellar system	
Septum	405	Cerebellum	9
Nucleus accumbens	269	Thalamic nucleus (tv)	69
Mammillary body	207		
Thalamic nucleus (ant)	215	Frontal cortex	25

[a]Taken from Kanazawa and Jessell (1976). Similar but less detailed distribution data with high levels in the substantia nigra (370–390 pmol/g) hypothalamus (219 pmol/g) and pineal (277 pmol/g) have been reported by Brownstein *et al.* (1976), Duffy *et al.* (1975a), and Powell *et al.* (1973). Early distribution studies using bioassay techniques gave generally similar patterns (Pernow 1953; Amin *et al.* 1954).

have added further evidence regarding the association of substance P with specific neuronal systems. Substance-P-positive cell bodies occur in the medial habenula, and heavy staining exists in parts of the interpeduncular nucleus, periaqueductal gray, substantia nigra, and regions of the hypothalamus, amygdala, and thalamus. Nerve endings staining for substance P are seen in the human cortex (Hokfelt *et al.* 1976b).

The strongest evidence for a substance P neurotransmitter role is with respect to primary sensory afferents. Immunohistochemical, lesion, and pharmacological studies all support such a function although a definite association with a particular sensory modality is not yet certain.

Figure 10.2 illustrates the immunohistochemical evidence (Hokfelt *et al.* 1975). Figure 10.2A shows cell bodies in the rat spinal ganglion staining for substance P. Only about 20% of the cells contain this peptide, and they are of the smaller variety. Figure 10.2C shows immunofluorescent staining of the

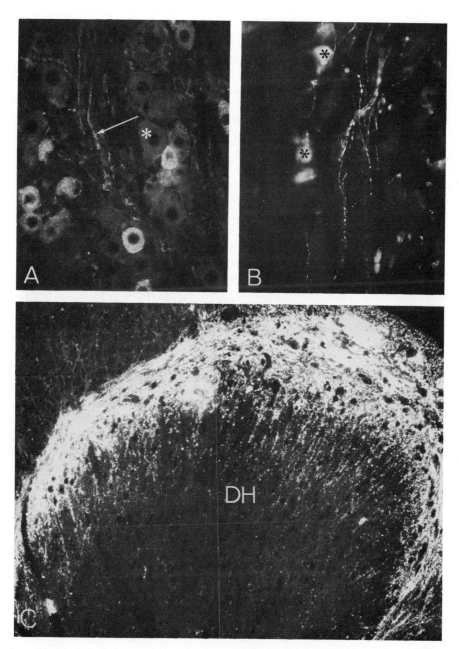

Figure 10.2 Localization of substance P by immunofluorescence in the rat. A. Dorsal root ganglion after incubation with substance P antiserum. Several small fluorescent cell bodies, as well as fibers (arrows) representing cell processes, can be seen. Most of the ganglion cells are nonfluorescent, as shown by the cell marked with an asterisk. B. Nasal mucosa showing substance P fibers. Cells marked with an asterisk in this photomicrograph are autofluorescent and not substance P-containing. C. Substantia gelatinosa of the dorsal horn (DH) of the spinal cord showing dense plexus of substance P positive fibers. (From Hokfelt *et al.* 1975.)

substantia gelatinosa of the dorsal horn of the cat spinal cord. Presumably this is the main terminal field of the substance-P-containing cells of the spinal ganglion (Figure 10.2A). Less dense terminal fields are seen in the ventral horns and around the spinal canal. By electron microscopic immunofluorescence, substance-P-containing nerve endings in the spinal cord have large dense core vesicles of 60–80 nm diameter (Pickel *et al.* 1977).

In the area of remote dendrites of these primary afferents, a dense plexus of positive structures is seen in such areas as the nasal mucosa and in the connective tissue of the skin, just beneath the epithelium. This positive staining of dendrites, cell bodies, and terminal areas of peripheral sensory neurons lends strong morphological support to Lembeck's hypothesis that substance P may play a role in sensory transmission.

Immunohistochemical evidence (Hokfelt *et al.* 1976a) suggests that two subpopulations of primary sensory neurons may exist: one containing substance P and the other containing somatostatin.

Otsuka and his colleagues provided physiological and additional biochemical evidence in favor of a neurotransmitter role for substance P in the spinal cord. They first detected the presence of a peptide capable of depolarizing motoneurons in extracts of bovine dorsal roots (Otsuka *et al.* 1972a). They next showed this peptide was pharmacologically (Otsuka *et al.* 1972b), chemically, and antigenically identical with substance P (Takahashi *et al.* 1974), and that synthetic substance P could depolarize motoneurons in their assay system (Konishi and Otsuka 1974). They measured substance P levels in various areas of spinal cord before and after lesioning of the dorsal root (Figure 10.3). In one area of the posterior horn, in the region of the substantia gelatinosa, the level dropped from 9.3 to 1.0×10^{-10} mol/g of tissue following the lesion. Ventral and lateral areas of the cord were unaffected. The content of substance P in the posterior root decreased between the cut section and the cord (distal to the lesion) but increased between the cut and the ganglion (proximal to the lesion) (Takahashi and Otsuka 1975).

These results are consistent with a synthesis of substance P in the spinal ganglia and then its transport in the dorsal root toward the intraspinal axon terminals, as might be expected for a neurotransmitter.

Similarly, substance P in rabbit auricular and sciatic nerves has been shown to rise proximal to a dendritic section i.e., between the dorsal root ganglion and the cut, and to fall in the distal parts (Holton 1960), suggesting transport in dendrites. Although it is not possible to associate substance P with certainty with any specific sensory modality, Henry (1976) has suggested that it is associated with input for pain. Its distribution in the spinal cord is appropriate and it overlaps with the distribution of opiate receptors and the enkephalins.

Both gross distribution studies and immunohistochemical work indicate that substance P is not confined to primary sensory neurons. Fibers surrounding the sweat glands are substance P positive, an indication that some substance-P-positive fibers may be associated with motor nerves (Hokfelt *et al.* 1975).

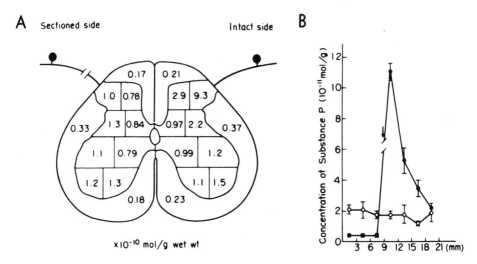

Figure 10.3 Distribution of substance P in the spinal cord and dorsal root after dorsal root ligation and sectioning. A. The distribution of substance P in the spinal cord of a cat 11 days after the unilateral section of the dorsal root. Numbers in the figure represent the concentration of substance P (10^{-10}mol/g wet weight). B. A comparison of the distribution of substance P in an intact dorsal root (○) and 11–12 days after dorsal root ligation (●). × axis is the distance from entry of the dorsal root into the cord. Distance from cord to ligation indicated by the arrow. (Takahashi and Otsuka 1975.)

10.2.3 Chemistry and Physiology

How substance P is formed in tissue is at present unknown. If this peptide is indeed a neurotransmitter, there should be a mechanism for synthesizing it within the nerve ending and there should also be a method for its rapid inactivation at the site of action. High-affinity uptake has not been demonstrated, and what uptake there is has been said to be by passive diffusion (Segawa *et al.* 1976). Benuck and Marks (1975) have reported the presence of an enzyme which inactivates substance P by cleavage of two internal peptide bonds (see Figure 10.1) with release of intermediate peptide products. It is a neutral endopeptidase similar to the enzyme which apparently inactivates LHRH and bradykinin (Marks and Stern 1974).

Henry *et al.* (1975) showed that the microiontophoretic application of substance P produced a strong, but slow and prolonged excitatory action on nearly half of the neurons tested in the cat lumbar spinal cord. The excitatory effects of iontophoresed substance P have also been demonstrated in the cuneate nucleus of cats (Krnjevic and Morris 1974) and in cerebral cortical Betz cells of rats (Phillis and Limbacher 1974). The depolarizations found, however, are slow in onset and long lasting, in contrast to the brief but sharp effect expected of ionotropic synaptic transmitters. This suggests the possibility of metabotropic action.

Since substance P was originally isolated from the intestine, it must be presumed to have some function in that location. It is present in only trace amounts in the aganglionic portion of the colon in Hirschsprung's disease (Ehrenpreis and Pernow 1952) but in large quantities in the proximal segments. Normal peristalsis of the gut does not occur in this disease because of the absence of Auerbach's plexus, a parasympathetic innervation of the gut. As a result, the bowel develops a constriction in the aganglionic portion, with enormous dilatation in the normally innervated proximal part. An electron microscopic study (Tafuri et al. 1974) confirmed the findings of Ehrenpreis and Pernow with regard to substance P's association with neurons of Auerbach's plexus and drew attention to the large (60–100 nm) granulated vesicles in the postganglionic unmyelinated fibers. Such large granulated vesicles were also associated with substance P in the spinal cord by Pickel et al. (1977).

Other peptides exist with the same or highly similar pharmacological properties to substance P, but none have been obtained from mammalian tissue. Physalaemin, isolated from the skin of a South American amphibian (Erspamer et al. 1964), and eledoisin, isolated from the salivary glands of a cephalopod (Erspamer and Anastasi 1962), are undecapeptides with actions highly similar to substance P, but only a portion of the amino acid sequence is the same.

A synthetic octapeptide, which lacks the H-Arg-Pro-Lys residue of substance P, is reported to be just as active as the undecapeptide. This suggests that the primary active site for substance P pharmacological actions may be at the Gly-Leu-Met-NH$_2$ end.

10.3 Enkephalins

Enkephalins are endogenous pentapeptides that bind to the same sites in brain as morphine and other opiates. Their discovery ranks as one of the most exciting chapters in contemporary neuroscience, and is unique in that their receptors were identified first.

Since antiquity, strong efforts have been made to discover the mechanisms by which opiates exert their extraordinary phsyiological effects. They are euphoriants, sedatives, hypnotics, antitussive and antidiarrheal agents. They are without equal in relieving pain. They are strongly addictive, so that their abuse is one of the major social problems of our times.

What types of physiological receptors could be responsible for these varied and powerful effects? And what materials might normally interact with such receptors? Until very recently answers simply could not be given to such questions.

Ingoglia and Dole (1970) opened a crack in the wall of ignorance when they injected labeled D- and L-isomers of the synthetic morphine substitute methadone intraventricularly. They found a 10% greater retention of the active L-isomer as compared with the inactive D-isomer. This suggested to them specific binding to an active opiate receptor.

Goldstein et al. (1971) then recommended a strategy for eliminating nonspecific binding to membranes so that these receptors could be identified.

Many technical developments contributed by Simon *et al.* (1973) and Pert and Snyder (1973a, 1973b) permitted true opiate binding to be identified and the distribution of the presumed receptor in brain to be determined. It is localized almost exclusively to neuronal elements. It has a heterogeneous distribution with the cerebellum and cerebral cortex containing little or none while the greatest concentrations are in the amygdala, the periaqueductal gray, hypothalamus, and caudate nucleus (Kuhar *et al.* 1973).

Autoradiographic studies using labeled diprenorphine, a potent morphine antagonist (Pert *et al.* 1975), showed concentrated grain densities in the dorsal portion of the interpeduncular nucleus, the medial habenula, the amygdala, the periaqueductal gray, patches of the caudate, and two catecholamine cell groups, the zona compacta of the substantia nigra and the locus coeruleus. Autoradiographic labeling in the spinal cord was restricted exclusively to the dorsal horn, particularly in the narrow band which includes the substantia gelatinosa.

The question of what endogenous materials might be intended for the receptors remained, for the moment, unanswered. The idea had first been advanced by Collier (1972), and later by many investigators, that it was unlikely on evolutionary grounds that an opiate receptor which mediated a physiologically important effect should exist in brain unless there were an endogenous ligand for the receptor. The search for the ligand commenced without there being the slightest idea of the type of compound that might be involved. Work by Terenius and Wahlstrom (1974, 1975a,b) using CSF, and Hughes *et al.* (1975a) using brain extracts, suggested that low-molecular-weight peptides were involved.

It was only a few months later that Hughes *et al.* (1975b) announced the structures of two pentapeptides responsible for the activity, methionine enkephalin (met-enk), and leucine enkephalin (leu-enk). It would be hard to overestimate the excitement generated among pharmacologists by this discovery. The structures were so simple that enkephalins were immediately synthesized and studies commenced in an amazing number of laboratories.

10.3.1 Anatomy of the Enkephalins

Met-enk has the structure *H*-tyrosine-glycine-glycine-phenylalanine-methionine-*OH*, while leu-enk has the structure *H*-tyrosine-glycine-glycine-phenylalanine-leucine-*OH*.

Hughes *et al.* (1975b) reported a 4:1 ratio of met-enk to leu-enk in porcine brain. Simantov and Snyder (1976) confirmed the presence of these two peptides in calf brain but found the ratio reversed. The significance of having two different peptides in brain with highly similar receptor binding but widely varying ratio in different species is not yet understood.

Antisera to these two peptides have been prepared by linking them with larger molecules such as polylysine or keyhole limpet hemocyanin. The antisera have been used for radioimmune assays and immunohistochemical localization. Although the antisera are highly specific to the enkephalins, most workers have observed some cross reactivity between the enkephalins them-selves. Thus, antiserum to met-enk reacts to some extent with leu-enk and vice versa. As a result, highly accurate data on the relative distribution of the two enkephalins in different brain areas of various species cannot be given at the time of this writing. A detailed distribution of met-enk in rat brain obtained by

TABLE 10.2
Regional Distribution of Met-Enkephalin in Rat Brain[a]

Region	Met-enkephalin (ng/mg protein)
White matter	
Corpus callosum	0.79 ± 0.15
Cerebral cortex	0.98 ± 0.12
Limbic cortex	
Amygdala	5.8 ± 0.8
Hippocampus	0.64 ± 0.12
Nucleus accumbens	10 ± 1.6
Olfactory bulb	1.3 ± 0.12
Olfactory tubercle	2.0 ± 0.28
Basal ganglia	
Caudate nucleus	10 ± 1.2
Globus pallidus	76 ± 16
Septum and Preoptic area	
Septum	2.7 ± 0.27
Interstitial nucleus of the stria terminalis	6.5 ± 1.6
Medial preoptic nucleus	6.5 ± 1.6
Lateral preoptic nucleus	4.2 ± 1.2
Hypothalamus	
Anterior hypothalamic nucleus	6.6 ± 1.3
Lateral hypothalamic nucleus	5.0 ± 1.5
Ventral medial nucleus	5.4 ± 1.1
Dorsal medial nucleus	3.7 ± 0.8
Posterior hypothalamic nucleus	3.5 ± 0.48
Mammillary body	2.2 ± 0.36
Thalamus	
Medial thalamic nucleus	1.6 ± 0.19
Lateral thalamic nucleus	0.69 ± 0.09
Ventral thalamic nucleus	0.75 ± 0.12
Habenula	1.4 ± 0.28
Midbrain	
Dorsal raphe nucleus	2.9 ± 0.44
Medial raphe nucleus	2.0 ± 0.24
Red nucleus	1.0 ± 0.12
Interpeduncular nucleus	5.8 ± 1.1
Substantia nigra	0.66 ± 0.07
Superior colliculi	0.5 ± 0.05
Inferior colliculi	0.8 ± 0.12
Periaqueductal gray	3.9 ± 0.06
Lateral pontine nucleus	0.32 ± 0.06
Cerebellum	0.47 ± 0.05
Medulla oblongata	2.5 ± 0.5

[a] From Hong *et al.* 1977b.

Hong *et al.* (Table 10-2), however, illustrates the wide variation in levels that exist. There is a high concentration in the periaqueductal gray, the hypothalamus, and most structures of the limbic system. By far the highest concentration, however, is found in the globus pallidus. This structure contains more than 10 times the concentration of the hypothalamus and 100 times that of the cerebellum.

At least five groups have utilized immunohistochemistry for visualization of enkephalins at the cellular level (Elde *et al.* 1976; Simantov *et al.* 1977; Watson *et al.* 1977a; Bloom *et al.* 1977; Sar *et al.* 1977). All have observed exclusive localization to neuronal processes and a highly similar distribution with respect to intensity of staining. The distribution is generally consistent with that previously determined for the opiate receptor (Simantov *et al.* 1977). It is also consistent with what would be anticipated from the known physiological actions of the opiates. Regions of intense staining include laminae I and II of the spinal cord where endings of primary pain fibers are localized. This does not imply that the pain fibers themselves are enkephalin containing. Evidence favors substance P for that role. Enkephalin-containing fibers may modify synaptic events in that region, however. The periaqueductal gray and certain nuclei of the thalamus involved in integrating pain perception also contain high concentrations of enkephalin-containing nerve fibers.

Intense staining has been observed in such limbic system structures as the lateral septum, central nucleus of the amygdala, hippocampus, and nucleus accumbens. These may relate to the euphoriant properties of opiates, as may the staining in the hypothalamus.

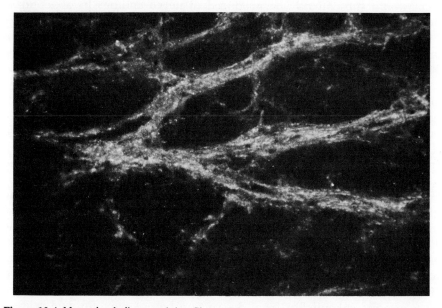

Figure 10.4 Met-enkephalin-containing fibers of the rat globus pallidus by immunohistochemistry. Cell bodies are presumably located in the caudate–putamen. Magnification × 180. (Photomicrograph courtesy of S. J. Watson. For details of methodology, see S. J. Watson *et al.* 1977a.)

The most puzzling area is the globus pallidus. Current evidence favors its primary role as a motor nucleus (Chapter 13). Considering the much higher levels of enkephalin in this structure as compared with any other area of brain, it might be anticipated that the opiates would have extremely prominent effects on motor performance. This is clearly not the case, and the role of the enkephalins in the globus pallidus and other structures of the basal ganglia remains a mystery. Most of the immunohistochemical staining in the globus pallidus appears to be in nerve endings from cell bodies located in the caudate–putamen (Figure 10.4).

Enkephalins are also located in cell bodies in the reticular formation and a number of brainstem nuclei. Obviously, there are many and varied "enkephalinergic" tracts in brain. In future years these will undoubtedly be traced in great detail.

10.3.2 Chemistry, Physiology, and Pharmacology of the Enkephalins

While the structure of the enkephalins is simple, the mechanism by which they are produced is still unknown. It may be that they are directly synthesized in nerve endings in the usual fashion for neurotransmitters. It is known that peptide synthesis can take place independently of ribosomes. On the other hand, it is possible that many peptides, including the enkephalins, are produced from larger, precursor peptides, that could be synthesized in Nissl bodies through the same ribosomal processes as proteins (Figure 1.10, section 1.2.4), or in nerve endings through other mechanisms. The active component would subsequently be produced in nerve endings by a peptidase which would split the molecule in the appropriate fashion.

Specific synthesizing enzymes for the enkephalins have not yet been found. By contrast, it has been noticed that the amino acid sequence of met-enkephalin corresponds to residues 61–65 of a larger peptide β-lipotropin.

β-Lipotropin is a 91-amino-acid molecule originally isolated from the pituitary by Birk and Li (1964). Although a role for β-lipotropin has never been determined, keen interest has developed in the possibility that it is a precursor for a number of active peptides. The structural sequence is shown in Figure 10.5. However, β-lipotropin has a different distribution in brain from enkephalins (S. J. Watson *et al.* 1977b), and it would seem as though the source of enkephalins is still to be determined.

The primary mode of degradation of enkephalins is enzymatic cleavage of the tyrosine–glycine amide bond. The high activity of this amino peptidase system accounts for the short biological half-life of the enkephalins. The enkephalin hydrolyzing activity is found in all subcellular fractions of brain homogenates. The highest concentrations are in the soluble and synaptosomal fractions. The synaptosomal localization indicates that some hydrolyzing activity is presynaptic, suggesting that a system, probably vesicular binding, must exist to protect enkephalins from destruction before synaptosomal release. Leucyl-β-naphthylamide inhibits the enzyme (Lane *et al.* 1977).

A feature of binding to opiate receptors, whether by enkephalins or other

Units 61–91 Fragment of β-Lipotropin

H-Tyr-Gly-Gly-Phe-Met₋Thr-Ser-Glu-Lys-Ser-Gln-Thr-Pro-Leu-Val-Thr₋
    61   65   70   75

Leu-Phe-Lys₋Asn-Ala-Ile-Ile-Lys-Asn-Ala-His₋Lys-Lys-Gly-Gln-OH
          80   85   90 91

Units 61–65 = Met-enkephalin   Units 61–76 = α-Endorphin

Units 61–79 = γ-Endorphin   Units 61–87 = C′ fragment

H-Tyr-Gly-Gly-Phe-Leu-OH   Leu-enkephalin

H-Tyr-Gly-Gly-Gly-Lys-Met-Gly-OH   A synthetic endorphin

Figure 10.5 Structures of some endorphins and possible relationship
to C fragment of β-lipotropin.

drugs, is the participation of sodium ions. Sodium selectively enhances antagonist binding and diminishes agonist binding. The reasons are not known,
but the hypothesis has been put forward that the opiate receptor can exist in
two conformations and that sodium is required to stabilize the antagonist
form. It is possible that the two-state model applies to the binding of other
neurotransmitters where sodium enhances antagonist, but diminishes agonist,
binding (Simantov and Snyder 1976).

The physiological action of enkephalins at receptor sites is uncertain.
They were originally isolated by Hughes *et al.* (1975b) on the basis of certain
peripheral actions. They produce a dose-related inhibition of electrically
evoked contractions of mouse vas deferens and guinea pig ileum. Met-enk is
three times as potent and leu-enk is one-half as potent as morphine on this
bioassay. The effect is completely blocked by the opiate antagonist naloxone.

Enkephalins reduce the accumulation of cyclic AMP by neuroblastoma
cells (Klee 1977). This, too, is blocked by naloxone.

Iontophoretically applied met-enk depresses the firing rate of neurons in
the cortex, neostriatum, and periaqueductal gray. The effect is mimicked by
morphine and blocked by naloxone (Frederickson and Norris 1976). Thus,
the overall action of enkephalins seems to be one of inhibition or depression.

The actions of enkephalins are extremely short lived. Even large doses
administered parenterally have minimal physiological effects. A transient
analgesic action is obtained with intraventricular or localized midbrain injections, however. Since this effect can be antagonized by naloxone, it is also
assumed to be receptor mediated.

Analogs to enkephalins are now being synthesized with the hope that they
will be degraded more slowly and therefore have more prolonged analgesic
and other effects. One such compound is N-allyl-(D-ala)²-met-enkephalin
(Pert *et al.* 1977).

While a definite role for the enkephalins has not yet been established,
their presence in specific neuronal tracts, their binding to opiate receptors,
and their analgesic and other physiological effects all argue for a neurotransmitter role. The possibility of developing synthetic analogs having

physiological actions similar to the opiates, but without their addicting properties, is a strong incentive for future pharmaceutical research.

10.4 Endorphins

The term "endorphin" was originally proposed by Simon, and later by Goldstein (1976) to apply to all endogenous morphine-like compounds. It commonly refers only to those endogenous peptides with chain length greater than the enkephalins, however. Three peptides with action at opiate receptors have been isolated from extracts of hypothalamus–pituitary (Guillemin et al. 1976). They are known as α-, β-, and γ-endorphin, and are reproduced as segments of β-lipotropin as shown in Figure 10.5. α-Endorphin is identical with residues 61–76 of β-lipotropin; β-endorphin, or "C-fragment," is identical with residues 61–91; and γ-endorphin is identical with residues 61–79. The requirement for activity is tyrosine as the initial amino acid in the chain.

β-Endorphin, the most potent of the three, is from 2 to 40 times more active than met-enk in opiate displacement and analgesic assays. When injected intraventricularly in rats, it produces a prolonged muscular rigidity and immobility similar to schizophrenic catatonia (Bloom et al. 1976). The effect is reversible by naloxone. It has been suggested that the more prolonged action of β-endorphin as compared with the enkephalins could be due to the longer chain length sequestering the active groups from peptidase degradation.

Antibodies to β-endorphin have been produced and used for preliminary distribution studies in rat brain by radioimmune and immunohistochemical techniques (Bloom et al. 1977). The distribution varies considerably from that of the enkephalins. For example, the pituitary has a much higher concentration and the striatum a much lower concentration. As with the enkephalins, the significance of these findings is uncertain. β-Endorphin is primarily a constituent of the pituitary, so it would be logical to presume that it would be primarily concerned with peripheral action. Nevertheless, its apparent occurrence in the brain itself, with unique distribution, certainly argues for a central role. Further work is required to be certain that it is not an artifact of the isolation procedure, and that the antiserum upon which the analytical studies are based is not cross-reacting with some other peptide. β-Lipotropin has been suggested as the precursor protein. There is a greater likelihood that this is the case than with met-enk because β-lipotropin and β-endorphin have been found in the same neuronal system in brain (Watson et al. 1977b).

Goldstein and his colleagues (Goldstein 1976) have purified a group of peptides from the pituitary, with opiate action, that do not seem to be fragments of a β-lipotropin. Thus, further variations on the opiate theme seem possible.

Only future research will reveal the true roles of the enkephalins and endorphins. Preliminary results to date suggest these will go far beyond blunting the sensation of pain. They could play some central role in a whole range of appetitive drives such as food, water, and sex, as well as affective states associated with the limbic system.

10.5 Neurotensin

Neurotensin is a tridecapeptide, isolated from bovine hypothalamus, which has been characterized by Carraway and Leeman (1975) as having the following structure: Glu-Leu-Tyr-Glu-Asn-Lys-Pro-Arg-Arg-Pro-Tyr-Ile-Leu-*OH*. It has many biological activities affecting glucoregulatory, hemodynamic, smooth muscle, neuroendocrine, thermoregulatory, and gastric secretory systems.

It is found in brain where it is concentrated in synaptosomal subcellular fractions. [^{125}I]-Labeled neurotensin binds to neuronal membranes in a fashion consistent with phsyiological receptor sites. Radioimmune assays show an unequal distribution throughout the nervous system (Kobayashi *et al.* 1977), and immunohistochemical localization shows neurotensin to be concentrated in specific neuronal tracts. These are found in highest concentration in lamina II, and to a lesser extent lamina I, of the spinal cord, the substantia gelatinosa of the spinal trigeminal nucleus, the midbrain tegmentum just dorsal to the interpeduncular nucleus, the preoptic area and a ventral band in the hypothalamus, the periventricular nuclei of the thalamus, the central nucleus of the amygdala, the ventrolateral portion and interstitial nucleus of the stria terminalis, the nucleus of the diagonal band of Broca, and the lateral part of the nucleus accumbens. Many other areas display a lesser intensity of staining. In the pituitary it is found in high concentration in cells of the anterior lobe and the posterior pituitary median eminence (Uhl *et al.* 1977).

Neurotensin fulfills the anatomical criteria for a neurotransmitter. It has potent physiological activity in the periphery, especially the gastrointestinal tract, and thus may soon be proved to be a central neurotransmitter.

10.6 Carnosine

Carnosine, the dipeptide of L-histidine and β-alanine, may be the neurotransmitter for primary olfactory neurons. The cell bodies of these bipolar neurons are embedded in the nasal mucosa. Axons penetrate the cribriform plate and terminate in the olfactory bulb (see Figure 4.13, Section 4.6). The concentration of carnosine in this terminal region of the bulb is in the range of 3–4 nmol/g as compared with 0.07–0.12 in other major regions of rat brain. Irrigation of the nasal mucosa with $ZnSO_4$ solutions causes a destruction of the primary chemoreceptor neurons and a loss of up to 98% of the carnosine content of the olfactory bulb. Destruction of the bulb, which causes retrograde degeneration of the bipolar neurons, leads to ten-fold losses of carnosine in the nasal mucosa (Margolis 1974; Ferriero and Margolis 1975; Margolis *et al.* 1974).

Additional evidence that carnosine is associated with primary olfactory neurons is that carnosine synthetase, the enzyme responsible for its synthesis, is also localized to the tract and suffers a similar fall after degeneration of the tract (Harding and Margolis 1976). A crucial question that remains to be answered is whether carnosine can duplicate depolarization of the mitral cells of the olfactory bulb in the same fashion as does stimulation of primary olfactory neurons.

The thalamus may be another region of interest in regard to a possible transmitter role for carnosine in that it has a concentration of 0.4 nmol/g in the cat as compared with only 0.03 nmol/g in the mesencephalon and less than 0.01 nmol/g in the cortex (Battistin et al. 1969).

A related compound, homocarnosine (γ-aminobutyryl histidine), also occurs in brain in concentrations ranging (in the human) from about 0.4 μmol/g in the corpus callosum and temporal cortex to slightly over 1 μmol/g in the thalamus, putamen–globus pallidus, and red nucleus (Perry et al. 1971a). Homocarnosine has been reported to have little effect on brainstem neurons in iontophoretic studies, which argues against a neurotransmitter role (Turnbull et al. 1972). On the other hand, injections induce hypothermia (Turnbull and Slater 1970), the concentrations in dorsal root ganglia are altered by electrical stimulation (Osborne et al. 1974), and evidence has been presented supporting a role for homocarnosine as the transmitter of some sensory neurons in submammalian species (Umrath and Fuerst 1976).

10.7 Angiotensin II

Angiotensin II (AII) is an octapeptide (Asp-Arg-Val-Tyr-Ile-His-Pro-Phe) derived from the decapeptide angiotensin I (AI) by the action of angiotensin-converting enzyme (ACE),* which hydrolyzes a terminal histidine–leucine residue. Angiotensin I is produced from a larger protein by the action of the proteolytic enzyme renin. The angiotensins were first discovered in the periphery, where AII serves as a potent vasoconstricter and stimulant for aldosterone secretion. All three components of the angiotensin system exist in brain (AI, AII, and ACE). The material is mostly in the form of inactive AI with the availability of ACE determining the amount converted to the active AII (Fischer-Ferraro et al. 1971; Ganten et al. 1971). AII is unevenly distributed in brain with the highest concentration found in the hypothalamus (Fischer-Ferraro et al. 1971). It is degraded by angiotensinase, a specific peptidase (D. J. Goldstein et al. 1972b). ACE activity is reduced to 20%–30% of control values in the caudate, putamen, and globus pallidus of patients dying from Huntington's chorea (Bennett et al. 1976). Striatal neurons are destroyed but glial cells multiply in this disease (Figure 13.24, Section 13.5.4).

---

*Recent evidence suggests that much of the ACE activity measured in brain homogenates may be associated with substance P or enkephalin neurons [McGeer and Singh, Neurosci. Letters, in press; Benuck and Marks, Biochem. Biophys. Res. Commun. 88 215–221 (1979)].

Synaptosomal uptake (Minnich *et al.* 1972) and possible receptor binding (Bennett *et al.* 1976) of AII have been demonstrated. These data are all supportive of a possible neurotransmitter role for AII, but obviously much more work is required before any conclusions can be drawn.

10.8 Hypophysiotropic Peptides and the Neuroendocrine System

The central nervous system exerts control over peripheral muscles, organs, and glands through three separate systems which differ enormously in their degree of complexity. The musculoskeletal system is the most highly developed, being voluntary and neocortical (Chapter 13). It has rapidly conducting fibers to striate muscles. The autonomic system is less sophisticated. It is integrated at the subcortical level and involves slowly conducting fibers to smooth muscle (Chapter 14). It is only partially under voluntary control. The neuroendocrine system is not under voluntary control at all. The hypothalamic-pituitary axis regulates this slowly responding system that does not utilize peripheral nerve fibers. Instead, circulating hormones reach target cells in peripheral endocrine glands. In turn, hormones released by these peripheral glands may return to brain via the bloodstream and may affect the activity of hypothalamic neurons controlling the pituitary. Thus, feedback loops exist (Figure 14.1, Section 14.2).

We shall not be concerned here with peripheral endocrine hormones. We shall only discuss the so-called releasing-factor peptides elaborated by hypothalamic neurons. It is these which control pituitary function. These peptides are also referred to as hormones on occasion, even though they are clearly synthesized in, and released by, nerve cells. It may be that they cannot be considered true CNS transmitters in this situation because most are released into the hypophyseal portal circulation with the target cells being in the pituitary. Nevertheless, the same peptides may exist in other brain locations where a classical neurotransmitter function is possible.

The pituitary gland is divided into three parts: the anterior, the intermediate, and the posterior lobes. Table 10.3 lists the pituitary hormones elaborated in each of these lobes, the target peripheral organ for each hormone, and, where applicable, the hypothalamic peptide or peptides which control release (McCann *et al.* 1974; Kastin *et al.* 1976).

The anatomical mechanism by which the hypothalamus controls hormonal release is distinct for each lobe of the pituitary. The simplest arrangement is that for the posterior pituitary. Neuronal cells of the supraoptic and paraventricular nuclei of the hypothalamus send axons through the infundibular stalk to this posterior pituitary lobe, where they end adjacent to perivascular spaces. The pituitary "hormones" oxytocin and vasopressin are synthesized in the soma of these neurons and transported to the axon terminals together with a carrier protein called neurophysin.

Suitable stimulation of these neurons causes release of vasopressin and oxytocin from the nerve terminals in the posterior pituitary and from there

TABLE 10.3
Pituitary Hormones and Their Hypothalamic Controlling Factors

Brain area	Pituitary hormone	Target organ	Hypothalamic peptides[a]	
			Releasing factor	Inhibiting factor
Posterior lobe	Oxytocin	Mammary glands, uterus	—	—
	Vasopressin	Kidneys, vascular smooth muscle	—	—
Intermediate lobe	Melanocyte-stimulating hormone (MSH)	Melanocytes	MSH-RF	*MSH-IF or MIF*
Anterior lobe	Luteinizing hormone (LH, gonadotropin-releasing hormone)	Gonads	*LHRH or GnRH*	
	Follicle-stimulating hormone (FSH)	Gonads	FSH-RF	GIH
	Prolactin	Mammary glands	PRF	PIF
	Adrenocorticotropin (ACTH)	Adrenal cortex	CRH	
	Thyrotropin (TSH)	Thyroid gland	TRF	
	Growth hormone (somatotropin)	Many tissues	GRH	Somatostatin (*SRIF, SOM, GIH*)

[a] Structures are established for italicized peptides; existence of others is strongly suggested by bioassay results. Known structures are:
TRF, a tripeptide: *Pyro*-Glu-His-Pro-*NH₂*. Activity depends on glutamate cyclization.
MIF, a tripeptide: Pro-Leu-Gly-*NH₂*. This may be only one of several possible MIF(s). This is a C-terminal portion of oxytocin from which it may be derived.
LHRH, a decapeptide: *Pyro*-Glu-His-Trp-Ser-Tyr-Gly-Leu-Arg-Pro-Gly-*NH₂*.
GIH, a tetradecapeptide: *H*-Ala-Gly-Cys-Lys-Asn-Phe-Phe-Try-Lys-Thr-Phe-Thr-Ser-Cys*OH*.

into the blood. This simple explanation is complicated somewhat by the fact that oxytocin, vasopressin, and neurophysin are all released into the CSF and into the hypophyseal-portal circulation (Barker 1977).

The intermediate lobe of the pituitary is innervated by nonmyelinated nerve fibers from the hypothalamus, but control of the release of MSH from this intermediate lobe is still not well understood. It has been shown, however, that hypothalamic extracts contain both a melanocyte-stimulating-hormone releasing factor (MSH-RF) and a release-inhibiting factor (MSH-IF or MIF). The biological role of the pituitary hormone MSH itself in mammals is not known, although it appears to be important in amphibians, reptiles, and fish as a source of adaptive color changes of the skin.

The mechanism of hypothalamic control of hormonal release from the anterior pituitary lobe is more complex. Release is under the control of peptides which are themselves released from axon terminals in the median eminence which is pituitary tissue. They pass from this site into the blood of the pituitary portal system to reach the anterior lobe. Several releasing and inhibiting factors in the median eminence have so far been characterized (Table 10.3), although only a few of these have been chemically identified.

The cell bodies of the neurons which synthesize these various peptides in the median eminence are not within the median eminence itself but appear to be either in other hypothalamic nuclei (particularly the arcuate, preoptic, suprachiasmatic, periventricular, and ventromedial nuclei) or in more anterior areas of the brain, particularly the septal area. Initial information on the probable sites of origin of the neurons producing the various releasing-factor peptides came from bioassays of the median eminence following selective discrete lesions of hypothalamic nuclei. More recently, the application of radioimmunoassay and immunohistochemical techniques to a study of those releasing-factor peptides whose structure is known has led to major advances in their localization. For example, only about one-quarter of the TRH in brain is localized in the hypothalamus. It is also found in the brainstem, thalamus, and septal area (Brownstein 1977). Somatostatin has been localized by immunocytochemistry to a subpopulation of primary sensory neurons distinct from those containing substance P, the amygdala, gut, and endocrine pancreas (Hokfelt *et al.* 1976a). LHRH appears to occur in a preoptic-amygdaloid pathway (Leonardelli and Poulain 1977).

There is also strong evidence for other—so far unisolated and unidentified—releasing factors, particularly a growth-hormone-releasing hormone (GRH), a corticotropin-releasing hormone (CRH), a prolactin-releasing factor (PRF), a prolactin-inhibiting factor (PIF) and possibly an FSH-releasing factor (although this may be identical with LHRH). One may expect that the isolation and identification of any of these materials will lead to intensive investigation of their possible neurotransmitter roles.

The activity of these releasing-factor peptidergic neurons in the hypothalamus is in turn under complex and poorly understood neuronal control. There is abundant physiological and anatomical evidence for neural input to the hypothalamus from many CNS areas, particularly the midbrain reticular formation and limbic structures. The hypothalamus contains high concentrations of various neurotransmitters such as noradrenaline, dopamine, serotonin, acetylcholine, and histamine, all of which may play some role in stimulating or inhibiting these peptide-producing hypothalamic neurons (McCann and Ojeda 1976; Rose and Ganong 1976).

Feedback loops help regulate hypothalamic activity and thus pituitary hormone release. For example, increased blood levels of corticosteroids will cause a decrease in the release of adrenocorticotrophic hormone (ACTH) from the anterior pituitary providing a long feedback loop. Pituitary hormones apparently also act directly on the hypothalamus, forming a short feedback loop.

Whatever the complex mechanisms of control may be, it is clear that the release of various pituitary hormones is connected with stimuli associated with their function. Thus, for example, oxytocin is a milk-ejecting and uterine-contracting peptide which is associated with reproductive function. It is released in response to suckling, coitus, and parturition as well as milk ejection. The antidiuretic and presser activities of vasopressin, a vascular smooth-muscle contractor, help to maintain constant blood volume and, in turn, changes in osmotic pressure or in blood volume lead to the release of vasopressin from the pituitary. Pain and emotional disturbance may also precipitate release.

The picture is much more complex than the simple account given here. Several anterior pituitary hormones such as ACTH and prolactin have now been found in brain. In addition, there are other peptides such as gastrin, a series of kinins, and others that have not even been mentioned. For further information, the reader is referred to an excellent monograph "Peptides in Neurobiology" (Gainer 1977).

10.9 Summary

The promising peptides vary greatly in size, the strength of evidence supporting a neurotransmitter role, and their probable physiological function. The smallest is the dipeptide carnosine, which seems to be localized in primary olfactory neurons, where it may be the transmitter. The largest is β-lipotropin (91 amino acids), which is concentrated in the pituitary and hypothalamus and which has no known role, except possibly to produce other peptides with opiate activity.

Substance P is the one best established as a neurotransmitter. There is extremely strong anatomical, biochemical, and physiological evidence that this undecapeptide is the transmitter agent for about 20% of the primary sensory afferent cells in the dorsal root ganglion of the spinal cord. Although the modalities are not established, pain has been suggested as one. Substance P has a widespread distribution in brain and therefore undoubtedly serves other pathways. It plays a role in the nigrostriatal and habenulointerpeduncular complexes. It was originally extracted from intestine, (where it appears to be associated with Auerbach's plexus) and its structure determined after isolation from the hypothalamus (where its role is yet to be established).

Methionine enkephalin and leucine enkephalin are two pentapeptides which bind to the same receptor sites in brain as morphine and other opiates. Met-enk, the most common of the two in rat brain, has been identified in nerve terminals in many areas by immunohistochemistry. It is particularly concentrated in the substantia gelatinosa of the spinal cord, central gray, hypothalamus, and globus pallidus. It has transient analgesic properties when administered intraventricularly, and depresses selective neurons when applied iontophoretically. Presumably, "enkephalinergic" pathways exist in brain.

Three endorphins, fragments of the 91-amino-acid molecule β-lipotropin, have been isolated from the pituitary hypothalamus. β-Endorphin, the most active, has also been shown to exist in specific neuronal tracts in brain. It produces catalepsy when administered intraventricularly and also binds to opiate receptors centrally and peripherally.

Neurotensin, a tetradecapeptide, has been shown by immunohistochemistry to be localized to specific neuronal pathways different from the enkephalins and β-endorphin. Angiotensin II, somatostatin, and thyrotropin-releasing factor (TRF) all have wide distribution in brain.

The hypothalamus and median eminence contain a variety of peptides which control or make up the secretions of the pituitary (Table 10.3). The posterior pituitary secretes the nonapeptides oxytocin and vasopressin, which have milk-ejecting and antidiuretic functions, respectively. They are manufactured in neurons of the paraventricular and supraoptic nuclei and are transported down axons of the supraopticoneurohypophyseal tract. The intermediate lobe of the pituitary manufactures MSH, which is under the control of MSH-releasing factor and the tripeptide MSH-inhibiting factor (MIF). These latter two peptides are extractable from the median eminence. The anterior lobe of the pituitary manufactures six hormones: LH, FSH, prolactin, TSH, ACTH, and growth hormone. These, in turn, are controlled by several peptides which again can be extracted from the median eminence. They include TRF (a tripeptide); LHRH (a decapeptide which may be identical with FSH-RF); and somatostatin (GIH, a tetradecapeptide) (Table 10.2). These median eminence peptides would not qualify strictly as neurotransmitters in this area since they do not innervate pituitary cells directly, but are released from axonal nerve endings in the median eminence (which is actually part of the pituitary) into the capillary loops of the portal circulation. Somatostatin and LHRH, however, seem also to be associated with more conventional neuronal pathways within the brain which may well represent a pattern for the others.

11

Putative Transmitters and Transmitter Candidates

In the last few years, many materials have been proposed as neurotransmitters and await rigorous evaluation of their qualifications. In this chapter we will discuss briefly some of the amino acids and amines which are possible candidates, just as in Chapter 10 we considered peptides which are presently being studied from this point of view.

Of the compounds to be discussed here, histamine seems to be the best established candidate for the role of a neurotransmitter in the central nervous system of mammals. The CNS activities of taurine are of sufficiently wide interest to be the subject of a book (Huxtable and Barbeau 1975), but the evidence so far does not demonstrate neurotransmitter action as one of them. Many other amines, amino acids, and similar molecules have been suggested as possible neurotransmitters, usually on the basis of their presence in brain plus a demonstration of activity in iontophoretic or behavioral experiments (Table 11.6). In each case the evidence is fragmentary, and thus there are major opportunities for significant experiments based on applying the criteria set out in the Introduction to Part II. Of the various materials suggested, piperidine, proline, N-acetyl-L-aspartic acid, and 5-methoxytryptamine are perhaps the most likely candidates for such a role in mammalian systems, although others such as octopamine most probably play an important part in the nervous system of crustacea, mollusks, and similar species. Adenosine and its derivatives have been suggested as neurotransmitters and there has been considerable interest in the possibility of "purinergic neurons" (McIlwain 1973; Burnstock 1972). Current evidence suggests that they may instead be modulators which are released along with the specific neurotransmitters from many types of nerve endings. We will discuss histamine and taurine here at some length, and then include short sections on some of the other candidates where the available evidence cannot be summarized as succinctly as in Table 11.6.

11.1 Histamine

The history of histamine shows several close parallels with that of acetyl-choline. Like ACh it was first synthesized as a chemical curiosity (Winders and Vogt 1907). It was soon recognized as a uterine stimulant in extracts of ergot by Barger and Dale (1910) who immediately undertook intensive pharmacological studies showing that it stimulated a host of smooth muscles (Dale and Laidlaw (1910). They identified it chemically in extracts of intestinal mucosa, and with characteristic acumen drew attention to the fact that it produced signs similar to those shown by animals sensitized to foreign proteins. This observation anticipated by more than 15 years the demonstration by Lewis that histamine is liberated by injurious stimulation of skin cells including the union of antigen and antibody (Lewis and Grant 1927).

Although much of the attention since has been given to histamine in peripheral tissues, knowledge of its presence in brain traces back as far as that of other biogenic amines (Abel and Kubota 1919). In 1943 Kwiatkowski found that histamine was concentrated more in gray than white matter of brain, and Harris and his co-workers (Harris *et al.* 1952) determined that the hypothalamus contained considerably higher concentrations than any other brain region. Local synthesis of histamine in the brain from radioactive histidine was first shown by White (1959). By the early 1960s it had been postulated that the histamine might be in histaminergic neurons partly because of the central effects of antihistaminic drugs.

Interest in brain histamine was rekindled about 1970 by the development of specific and much more sensitive assays for both histamine and its synthetic enzyme, histidine decarboxylase. This has permitted a rapid accumulation of evidence, mostly of a biochemical nature, on a possible neurotransmitter role (for reviews, see Dismukes and Snyder 1974; Schwartz 1975; Calcutt 1976).

The possibility of histaminergic neurons is most strongly supported by changes in histamine and histidine decarboxylase levels following selective brain lesions. A complicating factor, however, is the knowledge that mast cells, which are nonneurological and which contain very large quantities of histamine, are associated with the blood vessels, meninges, and parenchyma of brain.

11.1.1 Anatomy and Distribution of Histamine

No good histochemical method exists for localization of histamine or for specific enzymes in the histamine system. The possible anatomy of histaminergic neurons must therefore be inferred from distribution data and changes in levels in various areas following lesions. Distribution data in the literature for histamine itself are difficult to interpret for two reasons: first, because the assay methods available until very recently were all subject to lack of specificity, and second, because of the existence of at least two pools of histamine in brain, one of which may be neuronal and one of which, in mast cells, is almost certainly nonneuronal.

The early bioassay and spectrophotofluorometric methods for histamine, although suitable for most tissues, yielded erroneous levels for brain because of the presence of high levels of spermidine (Pearce and Schanberg 1969) and other interfering materials. The availability of a sensitive and specific radiometric method (Taylor and Snyder 1972) has permitted detailed distribution data to be obtained (Table 11.1). As shown in the table, there is a wide variation from region to region, characteristic of association with specific neuronal pathways. The highest concentrations are found in certain nuclei of the hypothalamus, the pineal gland, substantia nigra, the posterior pituitary, the central gray, and the raphe nucleus.

The major difficulty now in interpretation of such distribution data concerns the role of mast cells. An important source of histamine in peripheral tissues, including the sheaths of peripheral nerves, is the mast cell, a connective tissue cell holding in its granules large amounts of heparin and histamine (together with smaller amounts of serotonin and dopamine in some animal species). The importance of the mast cell as a potential source of histamine in brain was overlooked for a long time because these cells were only rarely encountered upon histological examination of mammalian brain. Since the histamine level in brain is only about 50 ng/g, however, a relatively small number of mast cells containing histamine in amounts similar to typical peritoneal mast cells (13 ng/1000 cells) could account for all the histamine in brain. Mast cells, identified by classical staining and by histamine histofluorescence, have now been located in the mammalian brain (Dropp 1972; Ibrahim 1974). They are especially numerous along meningeal blood vessels but they are also found in the brain parenchyma. Regional variations in mast cells might, to some extent, parallel variations found in histamine content. There are a number of ways in which histamine in mast cells can be distinguished from histamine presumed to be in neurons, however (Table 11.2).

The properties refer to two identifiable pools of brain histamine, one presumed to be in mast cells, the other in nerve endings. Mast cells sediment with the nuclear component in subcellular fractionation studies. Some histamine is found with this fraction, but some is also found in the synaptosomal fraction, presumably derived from nerve endings of histaminergic cells. When the synaptosomes are subjected to hypoosmotic shock, a substantial fraction of histamine is attached to synaptic vesicles (Snyder *et al.* 1974b). In the brain of neonatal rats, at a time when very few nerve endings can be observed, histamine is almost six times higher than in adult brain. But the histidine decarboxylase during this period is very low. It does not increase until nerve endings are established. Since the histidine decarboxylase of peripheral mast cells is also low, and since mast cells are seen in abundance in neonatal brain, it is presumed that neonatal histamine is almost entirely associated with mast cells (Martres *et al.* 1975). On the other hand, it is presumed that the increase in histidine decarboxylase is associated with the development of histaminergic neurons. In adult rat brain approximately half of the histamine has shifted to a store presumably associated with histaminergic nerve endings.

TABLE 11.1
**Regional Localization of Histamine and Histidine in the Monkey Brain
and of Histamine in Human Brain**

Brain area	Monkey Histidine[a] mg per 100 mg protein	Monkey Histamine[a] ng per 100 mg protein		Human histamine[b] (ng/g tissue)
Postcentral gyrus				
Cerebral cortex	9.4	47		105
White matter	12.1	54		—
Corpus callosum	16.3	52		—
Limbic cortex				
Olfactory bulb	8.7	74		213
Hippocampus	12.1	46		109
Amygdala	16.4	69		95
Septal nuclei	16.1	127		—
Pituitary, posterior	15.8	449		
Pituitary, anterior	19.2	84		
Pineal	10.3	1465		
Hypothalamus				
Supraoptic nucleus	24.7	2059		
Paraventricular nucleus	15.6	709	Ant:	630
Median eminence	25.5	1142		
Infundibulum	41.7	521		
Preoptic nucleus	16.9	616	Middle:	1068
Ventromedial nucleus	22.8	1701		
Ventrolateral nucleus	23.2	1578		
Mammillary bodies	14.9	2619	Post:	1381
Thalamus				
Anterior nucleus	18.2	264		139
Dorsal	11.8	151		118
Extrapyramidal nuclei				
Caudate	18.4	77		91
Putamen	13.4	63		102
Midbrain				
Substantia nigra	21	271		215
Superior colliculus	9.7	71		259
Inferior colliculus	6.9	117		320
Interpeduncular nucleus	14.3	22		—
Red nucleus	23.2	161		225
Raphe nucleus	29.3	301		—
Central gray	26.6	361		—
Cerebellum and lower brainstem				
Pons	8.3	86		109
Cerebellar cortex	11.4	15		40
Dentate nucleus	8.3	68		122

[a] From Taylor *et al.* 1972.
[b] From Lipinski *et al.* 1973.

TABLE 11.2
Tentative Synopsis of Differential Properties of "Neuronal"
and "Mast Cell" Stores of Histamine in Rat Brain[a]

Property	"Neuronal" histamine	"Mast cell" histamine
Subcellular localization	Synaptosomes	Crude nuclear fraction
Appearance during development	At least 3 weeks postnatal	Before birth
Associated histidine decarboxylase activity	High	Low
Half-life	A few minutes	Several days
Effect of mast cell degranulaters, e.g., Cpd. 48/80	Not released	Released
Effect of K$^+$-induced depolarization	Released and synthesis accelerated	Not released
Effect of reserpine	Released and synthesis accelerated	Not released

[a]From Schwartz 1975.

Brain histamine can be labeled by administering a pulse of radioactive histidine parenterally. By this method it can be determined that the histamine is associated with two very different chemical pools. Some, presumably in mast cells, is difficult to label and has a turnover rate of several days. The remainder labels easily and turns over in a few minutes (Schayer and Reilly 1973). This histamine is presumably in nerve endings.

The two pools also respond differently to pharmacological stimulation. Mast cell degranulators, such as compound 48/80 and Polymixin B, induce significant release of histamine from incubated neonatal brain slices, which are high in mast cell histamine (Verdiere *et al.* 1974). On the other hand, reserpine administration or K$^+$-induced depolarization depletes a different pool of histamine, not found in neonatal brain, which is presumed to be associated with nerve endings (Taylor and Snyder 1973).

The distribution and developmental studies suggest that changes in histidine decarboxylase rather than histamine would be a more reliable way of distinguishing neuronal degeneration in histaminergic neurons following lesions, because mast cells, which have high histamine but low histidine decarboxylase, might proliferate in the lesioned area. Medial forebrain bundle lesions in the rat affect histidine decarboxylase in all regions rostral to the lesion, but not in the cerebellum, midbrain, pons, and medulla (Garbarg *et al.* 1974). These changes in histidine decarboxylase are probably not secondary to the interruption of monoaminergic ascending pathways since no change in histidine decarboxylase activity in rat telencephalon was found after treatment with either 6-hydroxydopamine or 5,6-dihydroxytryptamine. On this basis it has been supposed that there may be a histaminergic pathway emanating from the midbrain or brainstem and ascending through the lateral hypothalamic area to project in a widespread manner to the whole ipsilateral telencephalon.

Since there is almost complete disappearance of hippocampal histidine decarboxylase after various deafferentations of the hippocampus, it has been suggested that histaminergic fibers enter the hippocampus through both a dorsal route via the fimbria, fornix superior, and cingulum, and a ventral route via the amygdala (Barbin *et al.* 1976).

An ascending inhibitory histaminergic pathway to the cerebral cortex has been suggested by Sastry and Phillis (1976) since metiamide, a specific histamine antagonist, blocks the inhibition of cortical neurons induced by either stimulation of the medial forebrain bundle or iontophoretic application of histamine.

11.1.2 Chemistry of Histamine

Experiments performed in various animal species have shown that labeled histamine does not diffuse readily through the blood–brain barrier except in newborn animals, indicating that the brain depends upon local biosynthesis. Histamine formation is a simple, one-step process catalyzed by L-histidine decarboxylase (EC 4.1.1.2) (Figure 11.1).

Histidine decarboxylase is clearly a different enzyme from Dopa or 5-HTP decarboxylase (Schwartz *et al.* 1970). After the destruction of dopaminergic neurons with 6-OHDA, the Dopa decarboxylase activity drops to about 50% but the histidine decarboxylase activity is unchanged (Garbarg *et al.* 1974). Histidine decarboxylase is inhibited by α-hydrazinohistidine, brocresine, and α-methylhistidine. It is unaffected by Dopa decarboxylase blockers such as α-methyl Dopa and Ro-4-4602. Thus, administration of the former, but not the latter, agents results in a rapid loss of endogenous histamine (Schwartz 1975).

The K_m of histidine decarboxylase for histidine is reported to be about

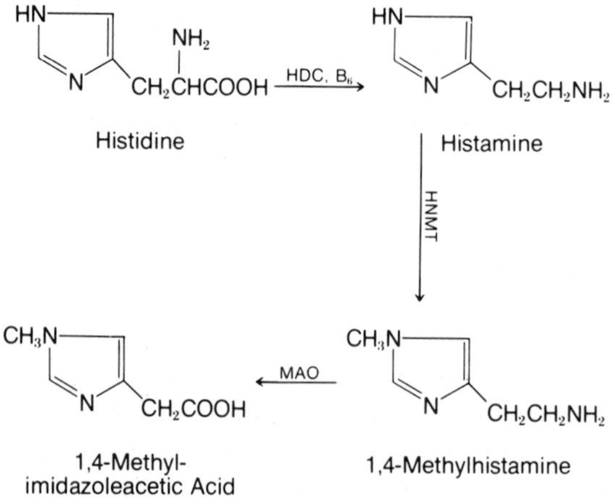

Figure 11.1 Synthesis and metabolism of histamine.

TABLE 11.3
Rat Brain Levels of Histamine, Histidine, and Related Enzymes[a]

	Histamine (ng/g)	Histidine (ng/g)	Histidine decarboxylase (nmol/h/100 mg protein)	HNMT (nmol/h/100 mg protein)
Hypothalamus	209	25.4	2.88	105
Pons/medulla	25	16.3	0.47	47
Cerebellum	26	18.7	0.33	37
Hippocampus	50	21.8	0.65	83
Corpus striatum	56	19.7	1.04	92
Thalamus/midbrain	75	13.5	1.26	91
Cerebral cortex	39	9.4	0.63	82

[a] From Taylor et al. 1972.

10^{-4} M with some variation depending on the pH (Palacios et al. 1976). This is comparable to the concentration of free histidine in brain tissue, suggesting that the enzyme may not be saturated in vivo and explaining the increase in histamine levels in brain following L-histidine loads (Schwartz et al. 1972).

Histidine decarboxylase is a pyridoxal phosphate-requiring enzyme, as are most decarboxylases, and it is therefore inhibited by compounds which are pyridoxal phosphate antagonists.

The regional distribution of histidine decarboxylase in rat brain is shown in Table 11.3 in comparison with the distributions of histamine, histidine, and histamine-N-methyltransferase (HNMT) (Taylor et al. 1972). There is not always a correlation between histamine levels and those of the synthesizing enzyme which may reflect the existence of neuronal and nonneuronal pools of the amine.

Although the levels of histidine decarboxylase and histamine are generally lower than those of other putative neurotransmitters because of the very short half-life of "neuronal" histamine, it is calculated that the number of molecules of histamine synthesized per unit time is probably greater than that of noradrenaline, dopamine, or serotonin, but may be less than that of ACh (Snyder et al. 1974b). This suggests that there is relatively little excess histidine decarboxylase activity in brain.

So far, no specific uptake process for histamine has been demonstrated. It may be, therefore, that enzymic breakdown plays a major role in the inactivation of the amine. Two major pathways of histamine catabolism are well known in peripheral tissue. These are: (1) direct oxidative deamination catalyzed by diamine oxidase and resulting in the formation of imidazoleacetic acid; and (2) methylation to 1,4-methylhistamine followed by oxidative deamination to 1,4-methylimidazoleacetic acid (Figure 11.1). The relative contributions of these pathways varies between species and also from one organ to another. It seems probable that in almost all mammalian species, however, the methylation pathway (Schayer and Reilly 1973) is the major catabolic route in brain (Van Balgooy et al. 1972). Several authors have demonstrated the presence of HNMT in brain tissue, and the product 1,4-

methylhistamine is found along with histamine in the nerve ending fraction on subcellular studies.

Although no diamine oxidase has yet been demonstrated in rat brain, imidazoleacetic acid, its ribotide and riboside have been found after the injection of radioactive histidine. N-Acetylhistamine has been reported in squid optic ganglia (Roseghini and Ramorino 1970) and stellate ganglion so that acetylation may represent yet another route of catabolism in some organs and some species.

11.1.3 Physiology and Pharmacology of Histamine

Histamine is known primarily for its actions on peripheral smooth muscle and for its relationship to allergy. Many different substances are released from damaged tissues following an allergic reaction, including histamine, ACh, serotonin, heparin, choline, adenosine, and at least one proteolytic enzyme. But it is histamine, released from mast cells, that is thought to be responsible for the serious reactions.

There are two types of histamine receptors: H_1 and H_2. The allergic reaction, which involves capillary dilatation and venous constriction, is mediated by H_1 receptors. Gastric secretion is mediated by H_2 receptors. H_1 receptors are blocked by an almost bewildering variety of conventional antihistamines, including various ethylenediamines, alkylamines, piperazines, and even certain phenothiazines. H_2 receptors are not blocked by these agents but are blocked by metiamide and butamide (Table 11.4) (Ash and Schild 1966).

These peripheral actions bear no direct relationship to the proposed existence of central histaminergic neurons, except to emphasize the general principle that many substances which have powerful effects on peripheral smooth muscle are also transmitters for central neurons associated with subcortical function (i.e., the catecholamines, ACh, serotonin, substance P). The common central and peripheral mobilization of these substances in response to emotionally stressful situations might be behind many aspects of psychosomatic medicine.

The iontophoretic application of histamine into the immediate vicinity of single neurons has been reported to alter resting potentials as well as firing rates. In a number of cases, however, the significance of these responses remains unclear since the strict conditions required for such experiments were rarely met. Histamine has been reported to depress the firing rate of many interneurons and to hyperpolarize motoneurons of the spinal cord (Phillis et al. 1968). A depressant, generally weak, action has been reported after application of histamine into the vicinity of neurons of the cuneate nucleus (Galindo et al. 1967), brainstem reticular formation, motor precruciate cortex, and cerebellar Purkinje cells (Siggins et al. 1971). In contrast, the great majority of cells in the hypothalamus, including antidromically identified supraoptic neurons (Haas et al. 1975), have been reported to be excited by histamine.

The iontophoretic effects of histamine thus fall into the same ill-defined category as those of the catecholamines and serotonin. Nevertheless, the existence of specific histamine blockers, and the effects of histamine on cAMP production, suggest the possibility of a metabotropic function.

Histamine causes large increases in cAMP formation in slices from rabbit cerebral cortex, cerebellum, or striatum (Kakiuchi and Rall 1968; Shimizu *et al.* 1969), but not to the same extent in rat slices. Thus, there are species variations. In the chick, H_2 receptor antagonists (buramide or metiamide) block the rise of cAMP but H_1 antagonists (mepyramine and diphenhydramine) do not (Nahorski *et al.* 1977). The effect is not blocked by α- or β-adrenergic blockers (Chasin *et al.* 1973), indicating the possibility of specific coupling to histaminergic receptors.

The overall central actions of histamine are less well defined than the peripheral. There has been considerable speculation that histaminergic neurons might be involved in extrapyramidal function (P. L. McGeer *et al.* 1961), in behavior (Gerald and Maickel 1972), in the schizophrenic process (Lovett-Doust 1955), in morphine tolerance (Wong and Robert 1975), in response to cold stress (Taylor and Snyder 1971), in emesis (Bhargava and Dixit 1968), in pain perception, and in the release of vasopressin (Dogterom *et al.* 1976) and prolactin (Donoso and Bannza 1976). But these speculations are largely based on drug action (Table 11.4).

Antihistamines (H_1 blockers) are well known for their sedative effects. Histamine, *N*-acetylhistamine, and *N,N*-dimethylhistamine, as might be ex-

TABLE 11.4
Drugs Affecting the Histamine System

Drug	Presumed Action
Brocresine	Competitive histidine decarboxylase inhibitor
L-Hydrazinohistidine	Competitive histidine decarboxylase inhibitor
L-3-Methylhistidine	Competitive histidine decarboxylase inhibitor
Decaborane	Histidine decarboxylase inhibitor
Thiazol-4-ylmethoxyamine	Histidine decarboxylase inhibitor
Reserpine	Releases neuronal brain histamine
Compound 48/80	Releases mast cell histamine
Polymixin B	Releases mast cell histamine
Ethanolamines (diphenhydramine)	H_1 receptor antagonist
Ethylenediamines (pyrilamine)	H_1 receptor antagonist
Alkylamines (chlorpheniramine)	H_1 receptor antagonist
Piperazines (chlorcyclizine)	H_1 receptor antagonist
Phenothiazines (promethazine)	H_1 receptor antagonist
Metiamide	H_2 receptor antagonist
Butamide	H_2 receptor antagonist
Mypyramine cimetidine	H_2 receptor antagonist
4-Methylhistamine	H_2 receptor agonist
2-Pyridylethylamine	H_2 receptor agonist
2-Methylhistamine	H_2 receptor agonist
Guanylhistamine	Weak H_2 receptor agonist
Clonidine	H_2 receptor agonist

pected, all counteract barbiturate effects in rabbits. Histamine turnover is greatly decreased in barbiturate-treated animals (Pollard *et al.* 1974). All of these facts suggest a role for histamine in arousal. Loads of L-histidine (20–48.6 g/day for 5–16 days) to normal humans and to persons with intractable narcolepsy do not promote arousal, however (Gillen *et al.* 1975). On the other hand, treatment of rats with the histidine decarboxylase inhibitor, thiazol-4-ylmethoxyamine, which is known to lower brain histamine, causes a temporary decrease in motor activity, a decline in food and water intake, and a total suppression of REM sleep (Menon *et al.* 1971). Injections of histamine either into the hypothalamus or cerebral ventricles elicit hypothermia, sleepiness, muscular weakness, salivation, increased respiratory rate, and increased water intake (Costentin *et al.* 1973). Injections into the cat supraoptic nucleus produced marked antidiuresis (Bennett and Pert 1974), while in the area postrema they elicit an emetic response (Bhargava and Dixit 1968). These results must be regarded with caution since the method of administration is nonphysiological.

Despite the rather vague nature of these findings, the conclusion that they represent a stimulation of specific histamine receptors is strengthened by the results with various pharmacological agents. Neither the central nor the peripheral effects of administered histamine are blocked by agents specific for other transmitters. Thus the GABA blocker bicuculline, the glycine blocker strychnine, the cholinergic blocker atropine, and the various catecholamine and serotonin blockers fail to prevent the histamine effects.

A large number of drugs do affect the histamine system. A few examples are shown in Table 11.4. They include H_1 and H_2 receptor agonists and antagonists, histidine decarboxylase inhibitors, histamine releasers, and HNMT inhibitors.

Polymixin B and compound 48/80 release mast cell histamine. Reserpine releases a different, presumably neuronal pool, suggesting that the mechanism of binding may be similar to that of the catecholamines and serotonin. Reserpine also increases histamine synthesis and decreases its methylation. This would seem to indicate that reserpine-released histamine does not come into contact with extraneuronal HNMT. This may be similar to the situation for the catecholamines, where there is reduced O-methylation following reserpine because the intraneuronally released amines are metabolized by a different route involving MAO.

11.1.4 Summary of Histamine Action

The evidence for histamine as a neurotransmitter in the mammalian brain is rather convincing. A significant fraction of the cerebral amine is synthesized at a high rate, seems to be held in specific neuronal tracts, can be released during depolarization, and affects specific receptors. An important forward step will be to develop more precise techniques for localizing histamine or its synthesizing enzyme histochemically.

It also appears that another fraction of the cerebral amine is held in nonneuronal cells, probably mast cells or a closely related variant. The function of such cells, as in peripheral tissues, is far from clear although they might be involved in vascular control, immune responses, or in the inflammatory process. It should be noted that the mechanism of histamine release from mast cells bears many similarities to the mechanism of release of transmitters from nerve endings. It suggests that in the process of evolution similar molecules might have been utilized in immune processes and primitive nervous systems.

11.2 Amino Acids and Their Derivatives

Many amino acids have been considered as possible neurotransmitters. Some, such as GABA and glycine, are now accepted and some of their pathways defined (Chapter 7). Others, such as glutamate and aspartate (Chapter 6), are putative transmitters in that they probably play such a role but further work of a more definitive nature is required. Still others can only be defined as transmitter candidates. Of these taurine, N-acetylaspartate, proline, and serine are the most interesting on the basis of available data. Several others are mentioned in Table 11.6.

11.2.1 Taurine

On the basis of available evidence, taurine could not be considered a neurotransmitter. Nevertheless, it has suggestive neurophysiological properties and possesses at least some of the expected chemical properties.

The structure of taurine was established by 1850, and by 1900 it was recognized as one of the most ubiquitous and abundant amino acids in the body. But since that time, only one physiological function for taurine has been firmly established. That is the conjugation of bile acids.

Roberts and Awapara independently discovered high concentrations of taurine in brain in 1950 (Chapter 7), but it was a further 10 years before Curtis and Watkins (1960) demonstrated that it had a strong inhibitory action on spinal cord neurons. Jasper and Koyama (1969) added a further step by demonstrating its increased release from the cerebral cortex of cats during arousal. Inevitably, the hypothesis was put forward that taurine is a neurotransmitter, but a strong argument against this possibility is that the rate of metabolism of taurine is far slower than that of any of the established candidates. As a result, the alternative hypothesis has been put forward that taurine is a generalized stabilizer of membrane excitability.

The highest concentrations of taurine are found in the peripheral areas of the body, particularly in the heart and muscles. The central nervous system contains relatively high levels (Table 11.5). Taurine is much more evenly distributed than some established neurotransmitters such as ACh or the catecholamines, but it does have extremely high values in the pituitary and pineal glands.

Specific details as to the possible anatomy of any taurine-dependent neurons that may exist are not available. As in the case of glycine, most of the

TABLE 11.5
Taurine Concentration in Various Brain
Areas of the Cat[a]

Brain area	μmol/g
Pituitary	8.08
Pineal	5.20
Cortex	2.10
Caudate	2.90
Thalamus	2.31
Lateral geniculate	4.30
Medial geniculate	1.70
Hypothalamus	1.81
Cerebellum	2.71
Pons	2.10
Medulla	1.70
Spinal Cord	1.70
Heart (rabbit)	10.8

[a] From Guidotti *et al.* 1972.

relevant information is concerned with the regional distribution of the amino acid and of its high-affinity uptake system. So far, autoradiographic studies of taurine uptake which might demonstrate its localization to specific nerve endings or neurons have not been done.

The subcellular distribution of taurine is not entirely clear. Only about 17% of the taurine in whole rat brain is in the synaptosomal fraction, but almost two-thirds of the synthesizing enzyme, cysteinesulfinic acid decarboxylase,* is localized to this fraction.

A high-affinity uptake system has been reported for rat brain synaptosomes (Hruska *et al.* 1976) as well as slices (Schmid *et al.* 1975). The uptake system is temperature, energy, and sodium dependent. It is unaffected by GABA, glutamic acid, glycine, and leucine, but is inhibited by β-alanine, hypotaurine, ouabain, potassium cyanide, and 2,4-dinitrophenol. The accumulated radioactive material is released rapidly from rat slices on electrical stimulation (Kaczmarek and Davison 1972; Lahdesmaki *et al.* 1975). All of this would be expected of a neurotransmitter.

Taurine is a sulfur-containing amino acid and as such is part of the sulfur pool of the body (Awapara 1975). Some taurine occurs in common foodstuffs, and it has been shown that about 1% of an exogenously administered radioactive taurine dose reaches the brain. Most of the taurine in the body, however, is probably formed from the essential amino acid methionine with a small proportion possibly coming from cystine and cysteine. Thus, the major pathway for the formation of taurine is believed to be as outlined in Figure 11.2. All three steps in the generally accepted route have been demonstrated *in vivo* and *in vitro,* but the enzymic and other requirements of the two oxidation steps (steps 1 and 3) remain obscure. The second step, the conversion of

*There is a strong probability that much of the cysteinesulfinic acid decarboxylase activity measured in brain homogenates is due to CAD.

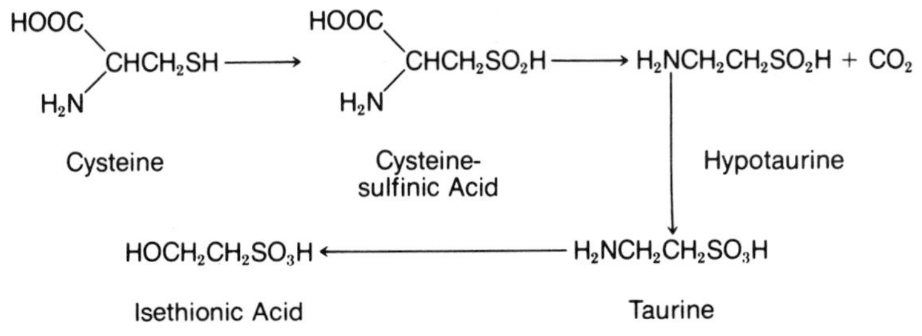

HOOC   HOOC
    \   \
     CHCH$_2$SH ⟶   CHCH$_2$SO$_2$H ⟶ H$_2$NCH$_2$CH$_2$SO$_2$H + CO$_2$
    /   /
H$_2$N   H$_2$N

Cysteine   Cysteine-   Hypotaurine
                      sulfinic Acid

HOCH$_2$CH$_2$SO$_3$H ⟵——————————— H$_2$NCH$_2$CH$_2$SO$_3$H

Isethionic Acid   Taurine

Figure 11.2 Synthesis and metabolism of taurine.

cysteinesulfinic acid to hypotaurine, is better understood and both the enzyme and hypotaurine itself have been demonstrated in brain. This decarboxylase, like most decarboxylases, reportedly requires pyridoxal phosphate (vitamin B$_6$) for its activity.

A strong argument against a transmitter role for taurine is its extremely slow turnover. Measured half-lives of taurine in rat brain are of the order of 9–16 h for "fast" decay and 40–238 h for "slow" decay (Lombardini 1975). The existence of several pools of taurine, and particularly the existence of most of the taurine in a very large, slowly exchangeable, metabolically inert pool, has been suggested to account for such observations.

It is generally agreed that much taurine is excreted unchanged or as taurine-conjugated bile acids. Some oxidation of taurine to isethionic acid probably occurs but the rates measured in rat heart (1.5 μmol/g/day) or brain (0.25 μmol/g/day), which are the only two organs where this conversion appears to occur, suggests that this oxidation is not a significant factor in taurine turnover.

Taurine, like so many other suspected neurotransmitters, was first shown to be physiologically active by microiontophoretic experiments on the spinal cord of cats (Curtis and Watkins 1960). Its action is very comparable to that of GABA or glycine, the principal features being a strong depression of unit firing or field potentials evoked by orthodromic or antidromic stimulation. The effect has a very rapid onset but is relatively prolonged, outlasting the application by several seconds. From a variety of experiments, it appears that taurine is comparable to both GABA and glycine in its effects on spinal neurons and to GABA on cortical neurons.

The depressant action of taurine on spinal, medullary, and retinal neurons is blocked by strychnine but not by bicuculline, suggesting that taurine is acting in these areas predominantly on glycine-like receptors. The depression of cortical, thalamic, and cerebellar neurons by taurine, on the other hand, is affected by both strychnine and bicuculline and in these areas taurine is generally considered to be much less inhibitory than is GABA. Taurine is also weaker than GABA in its depolarizing action on cat sympathetic ganglia and in this tissue the taurine effect is blocked by both strychnine and picrotoxin (Krnjevic and Puil 1975).

A region which has received particular attention as being one which might involve taurine-dependent neurons is the retina and optic system

(Pasantes-Morales *et al.* 1975). There is a high concentration of taurine in both the retina and the lateral geniculate body. A calcium-dependent taurine release has been obtained in the retina after both light and electrical stimulation, and taurine accumulates in the retina of light-deprived chicks. Taurine has a powerful inhibitory effect on the bioelectric activity of the retina as well as on synaptic transmission in the retinotectal pathways. In rat and mouse models of retinitis pigmentosa, it is the only amino acid found to be reduced. [35S]Taurine is transported to the optic tectum of the goldfish following intraocular injection (Ingoglia *et al.* 1976). All of these facts would be consistent with neurotransmitter action in retinal ganglion cells or other cells of the optic system.

There are few data on the effects of drugs on taurine metabolism or action. Its depressant action on central neurons, however, would naturally direct attention to its possible role in the mysterious phenomenon of epilepsy.

Van Gelder *et al.* (1972) reported that low levels of glutamic acid and taurine, in combination with high glycine concentrations, seemed to characterize the site of maximum seizure activity in the epileptic foci of man. Although this finding has been challenged, a number of studies have suggested that taurine administration blocks convulsions in various experimental epilepsy models. Results have been obtained in mice, rats, cats (Van Gelder and Courtois 1972), or baboons (Derouaux *et al.* 1973), and the models used have included seizures induced by ouabain, pentylene tetrazole, or acute or chronic cobalt-induced lesions. These results imply that taurine therapy might be effective in epilepsy. Definite, though mild anticonvulsive activity in humans has been reported (Mutani *et al.* 1974). The failure to obtain more dramatic results has been attributed to the fact that very little of the administered taurine crosses the blood–brain barrier.

In summary, taurine is a sulfur-containing amino acid which occurs in unusually high concentrations in brain. It has GABA- and glycine-like depressant actions on cerebral neurons and is taken up in synaptosomes by a high-affinity uptake system. It has been particularly studied in the optic system where it meets several criteria of a neurotransmitter, and it may produce some clinical benefit in epilepsy. Its rate of metabolism is extremely slow, which is not compatible with a neurotransmitter function. An alternative hypothesis is that it is a membrane stabilizer.

11.2.2 *N*-Acetyl-L-Aspartate (NAA)

NAA (Figure 11.3A) was reported by Tallan *et al.* in 1956 to be unique in its restricted localization to the brains of vertebrate species (birds and mammals) and in its very high concentration in that organ. In birds, the distribution appears to be neuronal (Curatolo *et al.* 1965a). In mice, the brain content is about 5.6 μmol/g (McIntosh and Cooper 1964). Similar regional distributions are reported in the cow and the horse with a gradual decrease from rostral to caudal areas (6–7 μmol/g in the cortex, 4–5 in the thalamus, 2–5 in the mesencephalon, 2–4 in the pons, 3 in the medulla, and 1–3 in the spinal cord (Curatolo *et al.* 1965b). On subcellular fractionation, NAA (like

N-acetyl-L-glutamate) appears largely in the supernatant fraction but the material in the P_2 fraction appears to be synaptosomal (Reichelt and Fonnum 1969). Since many low molecular neurotransmitters such as dopamine and GABA show a somewhat similar distribution on subcellular fractionation— due presumably to considerable leakage from the synaptosomal to the supernatant fraction—this cannot be taken as definitive evidence against (or for) a neurotransmitter role for NAA.

The function of NAA in brain is not understood. As an alternative to a neurotransmitter role, it has been suggested that NAA may interact with serotonin systems and possibly function as a carrier of the acetyl group (Berlinguet and Laliberte 1970; Gebhard and Velstra 1964).

11.2.3 Proline

This cyclic amino acid (Figure 11.3B) is present in brain at relatively low concentrations in mammalian systems (Cohen and Lajtha 1972; Perry *et al.* 1964, 1971a, 1971b), but very high concentrations have been reported in the nervous system of the crab (Evans 1973) and lobster (Gilles and Schoffeniels 1968). Its concentration in 12 areas of human brain ranged from about 0.2 μmol/g (in the corpus callosum and cortical areas) to about 0.6 μmol/g (in the cerebellum, thalamus, putamen, and globus pallidus) (Perry *et al.* 1971a). Significantly higher concentrations occur in dorsal than in ventral roots in the spinal cord (Roberts *et al.* 1973).

Proline has a weak glycine-like action on cat spinal neurons (Felix and Kunzle 1974). It is taken up by synaptosomes by a high-affinity, sodium-dependent system (Bennett *et al.* 1972; Peterson and Raghupathy 1972) which is inhibited by aminooxyacetic acid (Johnston and Balcar 1974). Transport of free L-[^{14}C]proline has been indicated in the carp optic tract (Csanyi *et al.* 1973).

11.2.4 Serine

The chief interest in serine (Figure 11.3C) is that it is interconvertible with glycine in the CNS (Figure 7.9, Section 7.6). It also has a weak glycine-like action in iontophoretic studies (Curtis *et al.* 1968). Serine concentrations are lower than those of glycine in all areas of the CNS other than the telencephalon and cerebellum (Shank and Aprison 1970; Shaw and Heine 1965). The concentration in human brain varies from about 0.6 μmol/g of tissue in the frontal cortex to approximately 1.3 μmol/g in the thalamus and amygdala (Perry *et al.* 1971a).

L-Serine does not influence the high-affinity uptake of glycine by slices of cat and rat spinal cord (Johnston and Iversen 1971; Balcar and Johnston 1973) and is itself taken up by a low-affinity system with a K_m of about 6×10^{-4} M into slices and homogenates of rat cord (Johnston and Iversen 1971; Logan and Snyder 1972). There is a blood–brain barrier to serine in mature rats (Banos *et al.* 1971) but administration of glycine raises central levels of serine (Richter and Wainer 1971). The evidence that serine is a neurotransmitter is weak, except for the possible parallelism with glycinergic neurons.

11.3 Purine and Pyrimidine Derivatives

Phillis and his co-workers (1974, 1975) have suggested that adenosine (Figure 11.3D) and possibly other purines and pyrimidines may be neurotransmitters. There are several reviews on "purinergic neurons" (Burnstock 1972; McIlwain 1973). It appears, however, that adenosine may only be released from neuronal tracts biochemically "fingerprinted" as belonging to other known neurotransmitters (Rose and Schubert 1977). Edstrom and Phil-

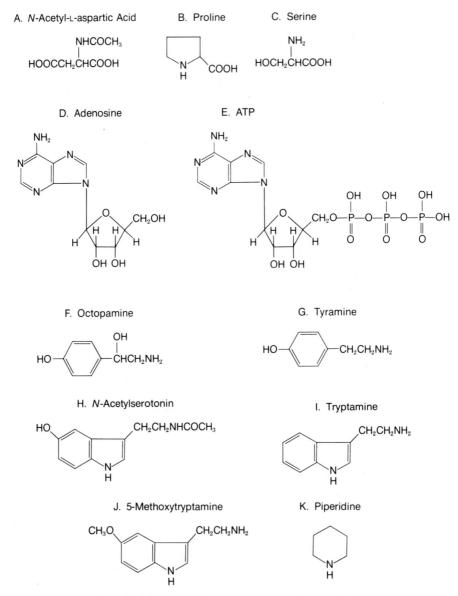

Figure 11.3 Structure of some neurotransmitter candidates.

lis (1976) indicate that the depressant action may be presynaptic and adenosine has been shown to inhibit the release of ACh (Vizi and Knoll 1976). The evidence suggests, therefore, that the purines and pyrimidines may be modulators which are released along with the specific neurotransmitters from many types of nerve endings and may act at the presynaptic rather than the postsynaptic membrane.

Adenosine triphosphate (ATP) (Figure 11.3E) has also been suggested as a neurotransmitter (Boyd 1973), particularly for primary sensory fibers, but its ubiquitous occurrence and use in many energy-requiring processes, including the vesicular binding of catecholamines (Chapter 8), renders this suggestion difficult to substantiate. It does have a high concentration in spinal roots, and release from peripheral nerve endings has been shown following antidromic stimulation (Holton and Holton 1954). Moreover, it strongly excites some cells in the cuneate nucleus when applied iontophoretically (Galindo *et al.* 1967).

Adenosine has been shown to stimulate cAMP formation by rat brain slices *in vitro* and the adenylate cyclase stimulation is selectively blocked by 2′-desoxyadenosine (Shimizu *et al.* 1975). Both anterograde and retrograde axonal transport of adenosine derivatives has been reported (Wise 1976) and label anterogradely transported in the corticothalamic projection has been visualized on postsynaptic membranes (Schubert and Kreutzberg 1974). Electrically stimulated release of adenosine derivatives can be obtained both from slices (Pull and McIlwain 1972) and from cortical synaptosomal preparations (Kuroda and McIlwain 1974). But all these data are consistent with other roles than that of neurotransmission, and more critical experiments will have to be devised before purine or pyrimidine derivatives could be considered to have this function.

11.4 Phenethylamine Derivatives

Catecholamines (Chapter 8) have paved the way for investigations of a variety of closely related phenethylamine derivatives as potential transmitters. Octopamine and tyramine are the best known, but others are, or could be, easily formed by metabolic routes similar to those used for the accepted catecholaminergic transmitters.

11.4.1 Octopamine

There is excellent evidence for a neurotransmitter role for octopamine (Figure 11.3F) in the lobster and crab as well as suggestive evidence for such a role in various other submammalian species (Robertson and Juorio 1976; Axelrod and Saavedra 1977). There is, however, no conclusive evidence as yet for a neurotransmitter role in mammalian brain although both octopamine (Axelrod and Saavedra 1974; Molinoff and Axelrod 1969; Molinoff *et al.* 1969) and its principal metabolite, *p*-hydroxymandelic acid (Karoum *et al.* 1975), have been shown to occur in the brains of mammals, including humans

(Saavedra 1974). Levels of octopamine in rat brain are in the range of only 15–40 pmol/g (Harmar and Horn 1976; Molinoff and Axelrod 1969), which is extremely low for an important neurotransmitter role. Higher levels are found in rat heart, salivary gland (0.3 nmol/g), and adrenals (3 nmol/g) (Molinoff and Axelrod 1969). Both the regional and subcellular distribution in brain parallels that of noradrenaline (Fischer *et al.* 1965; Molinoff and Axelrod 1972; Saavedra 1974) and, while a high-affinity uptake system has been demonstrated, it is probably into noradrenaline synaptosomes (Baldessarini 1971; Baldessarini and Vogt 1971, 1972; Baldessarini and Yorke 1974; Jaim-Etcheverry and Zieher 1975).

11.4.2 *p*-Tyramine

Tyramine (Figure 11.3G) is present in brain (Axelsson and Bjoerklund 1973; Phillips *et al.* 1974) in amounts only slightly higher than octopamine (40–90 pmol/g in the rat) (Suzuki and Yagi 1976; Tallman *et al.* 1976). There is slight regional heterogenity (Axelsson and Bjoerklund 1973; Phillips *et al.* 1974) with the highest levels being in the caudate (0.14–0.29 nmol/g) and hypothalamus (0.08–0.19 nmol/g) where there are high concentrations of dopaminergic and noradrenergic nerve endings, respectively. Higher concentrations are found in some peripheral tissues such as rat heart atria, salivary gland, and kidney (Tallman *et al.* 1976).

Tyramine has been reported to have a synaptosomal localization and a high-affinity uptake system, but these again have been attributed to noradrenergic nerve endings (Baldessarini and Vogt 1971). Precursors have been variously reported as L-tyrosine (David *et al.* 1974), dopamine (Boulton and Wu 1973; Brandau and Axelrod 1974), phenylalanine (Boulton *et al.* 1974), and 2-phenethylamine (Silkaitis and Mosnaim 1976), so its method of production is still in doubt. Exogenous tyramine is released from cortical slices on stimulation or from striatal slices on K^+ depolarization (Stoof *et al.* 1976).

Evidence presently available suggests that both octopamine and tyramine (as well as phenethylamine and phenylethanolamine, Table 11.6) in mammalian brain are formed, perhaps accidentally, in catecholamine neurons and do not have an important physiological role under normal conditions. The "false transmitter" action may, however, play a role in abnormal conditions, especially the tachyphylaxis seen in patients treated with an MAO inhibitor and ingesting an exogenous load of tyramine (Poch and Kopin 1966) (the "beer and cheese phenomenon," Chapter 14).

11.5 Indoles Other Than Serotonin

Fluorescent histochemical evidence for an indoleaminergic neuron other than the serotonergic neuron has been presented (Chapter 9) but the chemical structure of the indoleamine (which may be one or several) has not been established. Three possible candidates are N-acetylserotonin, 5-methoxytryptamine, and tryptamine itself (Figures 11.3H, 11.3I, and 11.3J).

11.5.1 N-Acetylserotonin (NAS)

NAS has been known for some years in the pineal gland where it is a precursor of melatonin (Chapter 9). Immunoassay techniques have indicated that NAS occurs in brain in areas which do not contain significant amounts of melatonin, such as the cerebellum, spinal tract of the trigeminal roots, and the brainstem reticular formation. It is particularly high in the cerebellum where it appears to be localized to the granule layer (Bubenik *et al.* 1974, 1976a, 1976b; Brown *et al.* 1976). These data are substantiated by studies indicating that the cerebellum contains the highest level of *N*-acetyltransferase (the synthesizing enzyme) of 15 brain regions studied (Hsu and Mandell 1975).

The distribution of NAS is not consistent with that of the indolic neurotransmitter postulated from histochemical studies and its role in brain must be considered as not established.

11.5.2 5-Methoxytryptamine

5-Methoxytryptamine occurs in brain (Green *et al.* 1973) in barely detectable amounts (less than 0.3 nmol/g in rat brain) (Green *et al.* 1975), except in the hypothalamus which has about 0.62 nmol/g. It has been suggested on the basis of microspectrophotofluorometric studies that this may be the unidentified indoleamine reported in rat mesencephalic neurons (Bjorklund *et al.* 1970, 1971a, 1971b); this indoleamine might be present only in very low concentrations in other areas of brain.

5-Methoxytryptamine mimics serotonin in iontophoretic studies (Bradley and Briggs 1974) and produces hyperactivity on injection (Green *et al.* 1975). Its formation from serotonin by brain slices has been shown (Banerjee and Snyder 1973), but some of the material found in brain may be of peripheral origin since it has been shown to cross the blood–brain barrier (Arutyunyan and Mashkovsky 1972; Green *et al.* 1975).

11.5.3 Tryptamine

Tryptamine is also found in brain in the comparably low concentrations of about 0.1–0.3 nmol/g (Sloan *et al.* 1975; Martin *et al.* 1972; Saavedra and Axelrod 1972). There is regional heterogenity in its distribution with the highest reported levels in the spinal cord (Sloan *et al.* 1975), followed by the hypothalamus and striatum (Saavedra and Axelrod 1974). In the dog, spinal cord transection did not decrease tryptamine levels below the lesion (Sloan *et al.* 1975) but did increase the levels above the lesion (Martin *et al.* 1975).

The conversion of tryptophan to tryptamine by brain tissue has been demonstrated (Snodgrass and Iversen 1974) and tryptamine levels in brain are reportedly increased seven-fold by treatment with the combination of tryptophan, iproniazid, and *p*-chlorophenylalanine (Saavedra and Axelrod 1973). It has been suggested that the formation of tryptamine may be responsible for some of the behavioral effects of tryptophan loads (Jackman and Radulovacki 1975).

In iontophoretic studies tryptamine facilitates the spinal cord reflex (Vaupel and Martin 1976; Bell *et al.* 1976); release of about 0.6 pmol per minute has been obtained in push–pull cannula studies (Martin *et al.* 1974; Knott *et al.* 1974); and there is some indication that tryptamine binds to a specific receptor (Ungar *et al.* 1976). It must be remembered, however, that tryptamine could easily be formed from tryptophan in any location where L-aromatic amino acid decarboxylase occurs. Decreased levels of tryptamine following raphe lesions or treatment with 6-OHDA have suggested that this occurs (Marsden and Curzon 1974). Like tyramine and octopamine, the appearance of tryptamine in mammalian brain could easily be fortuitous rather than of physiological importance.

11.6 Piperidine

Piperidine (Figure 11.3K) is a normal constituent of the mammalian brain (Honegger and Honegger 1960; Kase *et al.* 1969a; Perry *et al.* 1967), human CSF (Perry *et al.* 1964), and human urine (Blau 1961; Kase *et al.* 1969a; von Euler 1944) and it has a heterogeneous distribution in brain. In analyses on six regions of dog brain, the cerebellum was found to have 5 nmol/g of tissue which was some 22-fold greater than found in subcortical regions (Kataoka *et al.* 1970). Mass spectrographic analysis of small regions of waking mouse brain indicated high levels in the cervical cord (87 nmol/g) and in the olfactory bulb (24 nmol/g) (Dolezalova and Stepita-Klauco 1975). Radioactive piperidine taken up by rat brain slices *in vitro* concentrates primarily in the synaptosomal fraction (Kase *et al.* 1974).

TABLE 11.6
Miscellaneous Amines and Amino Acids Which Are Physiologically Active and/or Appear in Brain But Where Other Indicators of a Neurotransmitter Role Are Negative or Unreported

Physiologically active but not yet reported in brain:

6-Aminocaproic acid	5-Methoxy-N,N-dimethyltryptamine[a]
β-Hydroxyglutamic acid	N-Methyltryptamine[a]
N,N-Dimethyltryptamine[a]	

Found in brain but not known to be physiologically active:

1,3-Diaminopropane	Carnitine
α-Amino-n-butyric acid	Ergothionine
γ-Aminobutyrylcholine	Glycerophosphoethanolamine

Found in brain and physiologically active but no other positive indicators:

L-α-Alanine	L-Cysteinesulfinate
L-β-Alanine	L-Cysteate
L-Cystathionine	2,4-Diaminobutyrate
γ-Hydroxybutyrate	Hypotaurine
Imidazole-4-acetic acid	3-Methoxytyramine
Phenethylamine	Phenethanolamine

[a]These hallucinogenic materials have been said to be made in brain tissue, but confirmation failed.

In vitro production of piperidine from pipecolic acid (2-carboxypiperidine) in the presence of brain tissue has been demonstrated (Kase *et al.* 1967), as has the presence of tritiated piperidine in the brain tissue of rats given tritiated pipecolic acid by intraperitoneal injection (Kase *et al.* 1970). Pipecolic acid is probably formed from lysine.

Piperidine can mimic the action of ACh on synaptic sites in the brains of mammals (Tasher *et al.* 1959) and in the cerebral ganglion of mollusks (Stepita-Klauco *et al.* 1973). In snails piperidine is found in all tissues assayed, but the brain showed the highest concentration (3.3 nmol/g) (Dolezalova *et al.* 1973).

The concentration of piperidine in the CNS has been shown to increase during hibernation in mollusks and during behavioral sleep in mammals (Stepita-Klauco *et al.* 1974). Administration of piperidine is said to induce sleep (Alm 1976; Drucker-Colin and Giacobini 1975), as well as muscle relaxation, weakness, ataxia, and a general decrease in motor responsiveness (Abood *et al.* 1961; Kase *et al.* 1969b).

Piperidine is an unusual compound in that it is much more volatile than the established neurotransmitters and other transmitter candidates. While this complicates investigation of its role, it is nevertheless a material which deserves considerable attention in the future.

11.7 Miscellaneous Transmitter Candidates

Many materials other than the ones already mentioned have been investigated with a view to their potential neurotransmitter function. A number of these are listed in Table 11.6, but again the list is far from complete. Further information can be obtained from various reviews on the subject (Phillis 1970; Curtis and Johnston 1974; Krnjevic 1974).

The Integrative Aspects of Brain Function

The detailed studies of anatomy and physiology in Part I and of neurochemistry in Part II have to be given functional meaning in the most important integrative performances carried out by the brain.

Before the four chapters on these themes, however, it is necessary to have an introductory chapter (12) that gives an account of the mechanisms concerned in the embryological processes by which the brain is built. All neurons are generated in special mitosis of the germinal epithelium of the neuronal tube. There are extraordinary problems in developing hypotheses that attempt to account for the factors responsible for the migration of the neurons to their ultimate destinations and for the establishment of the correct synaptic connectivities. It has been proposed that chemical coding and sensing provide the key factors. The specific molecular configurations are as yet unknown, but they appear to be a different class from the synaptic transmitters. They are assumed to be specific macromolecules, probably polypeptides associated with the surface membranes of the neurons and the glia.

The control of movement involves an immense range of neuronal performance from the simplest spinal reflexes dealt with in Chapter 4 to the responses of progressively higher levels—the cerebellum, the basal ganglia, and ultimately the cerebral cortex. It is not yet possible to give a detailed and coherent account of all the operations of the neural machinery involved in carrying out even some simple movement. Much of our account in Chapter 13 will be impregnated with theories that go far beyond the present level of scientific investigation. But this procedure is justified because it reveals how far we are from fully understanding the control of movement, and also because it points the way to fruitful experimental investigations. As yet, our understanding is fragmentary at the best. An important feature of this chapter concerns the complex neuronal systems that comprise what is often called the extrapyramidal system. This system is of great clinical significance because its disorders are the cause of many important neurological diseases.

Consideration of the control of movement by all the skeletal musculature leads on to the problems of behavior in Chapter 14. Knowledge in this area is in an even more primitive state. Yet there is a tremendous overlap with movement as is evidenced by the commonality of effects of drugs and disease on these two integrative aspects of brain function. This is where the interface between neurology and psychiatry exists in the clinical world, and where psychology and pharmacology meet in the realm of basic science.

There is no more important function of the brain than learning and memory. Yet the essential mechanism is still obscure despite the immense experimental efforts over the last decades. The reason for this discouraging situation is that there has been no well-defined hypothesis of the neural mechanisms concerned. In Chapter 15, the attempt is made to formulate such a hypothesis. It is encouraging that the same neurobiological principles seem to underlie the mechanism used by the cerebellum in motor learning and by the cerebrum in cognitive learning. And there is even a suggestion that the basal ganglia may have an analogous mechanism. Many good experimental studies in molecular neurobiology can be developed in the testing of theoretical explanations in these structures.

It is appropriate that the final chapter of the book (Chapter 16) treats the ultimate problems of cerebral performance—perception, speech, and consciousness. Necessarily, it is possible to define only general principles in the enormous fields of perception and speech. But we believe it to be important for the molecular neurobiologist to be confronted with the problems arising in higher brain functions. Finally, in the section on consciousness there is a necessary introduction to the philosophy of the brain–mind problem. Here there is a formation of a strong dualist–interactionist hypothesis that is the principal theme of a recent book, *The Self and Its Brain* (Popper and Eccles 1977).

12

The Building of the Brain and Its Adaptive Capacity

12.1 The Building of the Brain

12.1.1 Introduction

In the last few decades there has been great progress in our understanding of the principal features of the most marvelous constructional operation—the building of brains (Gaze 1970; Jacobsen 1970; Angevine 1970; Sidman 1970; Sidman and Rakic 1973). In its organized complexity the human brain far exceeds that of other mammalian brains, and even more the brains of other animals, vertebrates and invertebrates. The foundations of neuroembryology were laid around the turn of the century by His, Ramón y Cajal, and Schaper, and many of their basic discoveries with light microscopy have survived rigorous testing with powerful new techniques such as electron microscopy, radiolabeling with autoradiography, and DNA estimations of specific cell types. It is proposed to summarize the early stages of the neuroembryological development and then to treat in more detail the two best understood developmental stories of the mammalian brain, that for the cerebral cortex and that for the cerebellar cortex.

In the earliest stage the future central nervous system can be recognized as an elongated neural plate already differentiated from the surrounding ectoderm by its thickening due to elongation of the constituent neuroepithelial cells that extend right across the plate. Soon, the rapid multiplication of these cells causes the inward longitudinal folding of the plate so that it eventually becomes the neural tube lying beneath the epidermis that has united over it. The neuroepithelial cells thus extend between an inner ventricular attachment and an outer attachment to mesodermal tissue that eventually becomes the pia mater. In the earlier stages, the neuroepithelial or germinal cells multiply in a simple clonal manner with no differentiation. Each mitosis is accompanied by a remarkable cycle of changes illustrated in Figure 12.1 in the sequential series for a single cell at approximately half-hour intervals (Sauer 1935). Beginning at the left, the cell extends across the wall of the neural tube and the nucleus migrates outward and then inward during the stage of DNA replication by synthesis (S stage). There is then a short premitotic growth phase (G2), followed by retraction of the cell to the ventricular surface for the mitotic phase (M). The mitosis is transverse so that the two daughter cells lie side by side in position to resume the extended form across the tube in phase G1.

The cycle then repeats for each of the daughter cells, the cycle time being no more than 6 h for a chick embryo at an early stage. With a mouse fetus at 10 days the cycle time is 8.5 h, while at 15 days it has lengthened to 11 h for that part of the neural tube forming the cerebral cortex. Since the mitotic cycles of the individual neuroepithelial cells are out of phase, in the postmitotic phase (G1 in Figure 12.1) there will be adjacent neuroepithelial cells stretching across the tube in phases S and G2 that can act as guides for the extension of the cytoplasmic processes of the daughter cells toward the external surface.

Soon there is an enormous population of these primitive germinal cells, and this simple clonal multiplication eventually ceases. In what is called a *differentiating mitosis*, a germinal cell divides to give rise to immature nerve cells. Or, alternatively, the differentiating mitosis may result in one immature nerve cell and one germinal cell (Rakic 1973). These immature nerve cells eventually mature as nerve cells and they never divide again. They have lost their *mitotic competency*, but largely in their nuclei they carry the genetic in-

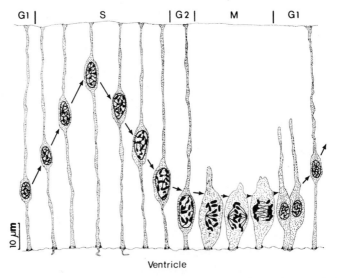

Figure 12.1 Diagram of a section of neural tube of the chick embryo, showing the intermitotic migration of the nucleus of a single neuroepithelial germinal cell at approximately half-hour intervals throughout the mitotic cycle. (Jacobson 1970.)

structions to develop into their appointed roles in the nervous system. They set about growing first an axon and then dendrites, and these sprouts seem to know where they are going, as if guided by an intelligence and a knowledge of the ultimate design of the brain they are helping to build. It is a wonderful, self-organizing, self-developing process. This biological process in constructing a brain provides one of the most challenging scientific problems confronting us: How is it controlled? Some general statements can already be made:

1. At no stage in development are the neurons of the brain connected together as a random network.
2. Only in a few special regions has there been evidence of an excess of unused and unwanted neurons which rapidly die. Elsewhere it can be assumed that there is no redundancy. No doubt there appears to be a redundancy in the enormous populations of neurons in many parts of the brain, but we may assume that this appearance derives from our deficient understanding of the modes of operation.
3. The organization of the brain is highly specific, not merely in terms of connections between particular neurons, but also in terms of the number and location of synaptic knobs upon different parts of the same cell (cf. Figure 1.3B, Section 1.2.2) and the precise distribution of synaptic knobs arising from that cell.
4. The slow changes in performance of the brain during life, in particular learned performances, are due to subtle changes in the microstructure and microfunction that we shall consider later in this chapter and in the next chapter.
5. Apart from these microchanges, the structure of the brain is precisely determined by genetic coding and the secondary instructions deriving therefrom, as well as from the medium in which the brain grows.

Fantastic challenges are raised by the problems of neurogenesis. It is important to realize that the brain is constructed in all its detail before it is used. It was postulated in the 1930s that the brain was built as a more or less randomly organized structure, and then by use it was modeled to the appropriate design. This hypothesis has been proved false by the many experimental demonstrations that the nervous system is already constructed in its detailed connectivity before it is used.

Besides forming neurons the germinal cells also undergo differentiating mitoses to form glia cells. There is conflicting evidence with respect to the details of glia generation. It is not known if there are separate lines of germinal cells for neurons and for glia, nor is the relationship of astroglial and oligodendroglial production known. It is now generally accepted that microglia do not arise from neuroendothelial cells, however, but are phagocytic cells of mesenchymal origin that invade the brain from the pia mater.

All types of glial cells differ from neurons in that they retain their mitotic competency. This generation of glial cells throughout life has one serious consequence, namely, the production of gliomas or brain tumors. Brain tumors rarely if ever have a neuronal origin. The immature glial cells or glioblasts assemble in the subventricular zone (S in Figure 12.2D) and from there may migrate laterally. Later in this chapter there will be consideration of the probable glial role in the guidance of neuronal migration and in the regenerative replacement of degenerating synaptic terminals.

12.1.2 The Building of the Cerebral Cortex

Figure 12.2 gives diagrammatically the early stages in the development of the neocortex (Sidman and Rakic 1973). Figure 12.2A shows the state of affairs depicted in Figure 12.1, but for a human fetus just less than 6 weeks of age. At 6–8 weeks a cell-sparse marginal zone develops (M in Figure 12.2B) and some of the neuroepithelial cells undergo differentiating mitoses, becoming immature nerve cells. They have permanently lost their mitotic competency and mature into fully developed nerve cells as they migrate up to form the intermediate zone (I in Figure 12.2C at 8–10 weeks) and sprout axons up into the marginal zone. The differentiating mitoses continue in zone V (Figure 12.2D at 8–10 weeks). The immature nerve cells so formed migrate more superficially to form the cortical plate (CP), that can be seen with its developing pyramidal cells in Figure 12.2E at 10–12 weeks. An additional feature in Figures 12.2D and 12.2E is the subventricular zone (S) that is composed of subventricular cells migrating up from zone V and continuing their mitotic division there. The nerve cells so formed are not constrained by the vertical alignment of the CP layer and may migrate laterally. Glia are also formed by differentiation mitoses in zone V.

An important feature is that the later differentiating neurons migrate to more superficial locations than the earlier neurons. Thus, these later neurons have a very long migration path across several millimeters of cortical thickness, whereas the earliest have a migration of only 100 μm. At the earlier stages the

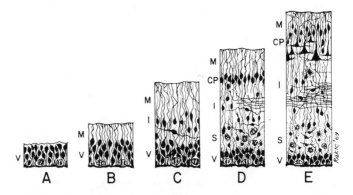

Figure 12.2 Semidiagrammatic drawing of the development of the basic embryonic zones and the cortical plate. Abbreviations: CP, cortical plate; I, intermediate zone; M, marginal zone; S, subventricular zone; V, ventricular zone. (Sidman and Rakic 1973.)

migration of the cell appears to be accomplished simply by the movement of the nucleus up the cytoplasmic cylinder of the immature nerve cell that extends right across the cerebral wall, there being a subsequent retraction of this elongated cylinder (cf. Altman and Bayer 1978). At later stages of development of the fetal monkey cerebrum, however, the thickness of the cerebral wall may be as much as 5 mm (Figure 12.3A).

A most attractive explanation has been suggested by Rakic (1972). The glioblasts generated by the neuroepithelial cells in the S layer eventually grow into glial cells that extend right across the cerebral wall as thin fibers (RF in Figure 12.3B). Possibly this great elongation in parallel fascicles is to be explained by their early origin from neuroepithelial cells spanning the thin cerebral wall at early stages and being progressively stretched as the cerebral wall thickens. The small starred rectangle in Figure 12.3B is drawn greatly enlarged in Figure 12.3C. The elongated glial filaments (RF) are shown crosshatched. One neuron (A), with its nucleus (N), is shown closely attached to a glial filament. As it migrates upward, there is a leading pole (LP) with several pseudopodia (PS) at its growing tip, one of which is already continuing along the filament. Below there is a trailing pole (TP) of the neuron that is eventually retracted. Thus, the upward migration of the neuron would seem to be guided by the glial filament, which may also aid in the movement. Two other neurons with associated glial filaments are also shown in Figure 12.3C, one (B) is moving out of the frame, the other (C) is coming out of the dense fiber mesh of the optic radiation (OR). That the glial filament provides a specific guidance is illustrated by the finding that the neuron follows all the bends of the glial filaments, and that it ignores the multitude of alternative filamentous paths (nerve fibers and blood vessels) that it crosses on its upward migration. It can be conjectured that there must be some affinity between the surfaces of the glial filament and of the neuron.

The radial guidance by glial cells can be demonstrated only at later stages of development, as in Figure 12.3. At earlier stages glia may be guiding the vertical migration of neurons, but they cannot then be distinguished from neurons, and the guidance may be shared by glia and by the vertically directed processes of already-migrated neurons. After the glia have completed their task of vertical guidance, there is retraction of their long processes seen in Figure 12.3A with their transformation into astroglia.

A special problem in this vertical guidance relates to the inside-out positioning of the migrating neurons. As indicated in Figure 12.2E, the later migrating neurons take up a position external to the earlier, which have already been transformed into pyramidal cells. It is suggested by Sidman and Rakic (1973) that the final position of the neurons is dictated by interaction with processes of other neurons and even of afferent fibers that have already grown into the cortex, probably from the thalamus. The cerebral disorganization in the reeler mutant mouse provides evidence in support of this suggestion. The vertical migrations of the immature neurons to their final places in the cerebral cortex form the developmental basis of the columnar organiza-

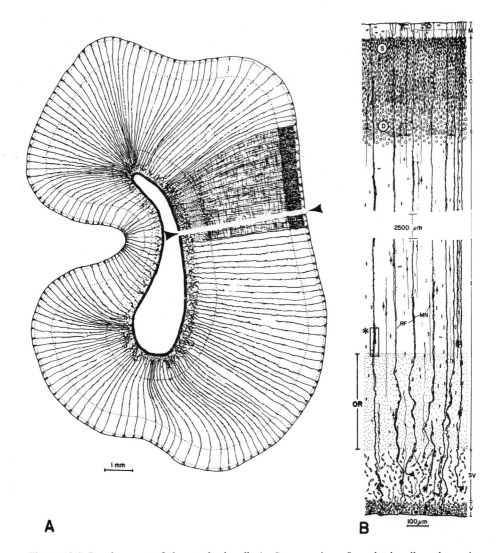

A

B

Figure 12.3 Development of the cerebral wall. A. Cross section of cerebral wall as shown in Figure 12.2E. (Rakic 1972.) B. Enlargement of the portion of cerebral wall indicated in 12.3A. The middle 2500 μm of the intermediate zone, also spanned by radial fibers, is omitted. The rectangle marked with an asterisk shows the approximate position of the three-dimensional reconstruction in 12.3C. 100-μm scale indicates the magnification. Abbreviations: C, cortical plate; D, deep cortical cells; I, intermediate zone; M, marginal layer; MN, migrating cell; RF, radial fiber; S, superficial cortical cells; SV, subventricular zone; V, ventricular zone. (Sidman and Rakic 1973.) C. Three-dimensional reconstruction of migrating neurons, based on electron micrographs of semiserial sections. The reconstruction was made at the level of the intermediate zone indicated by the rectangle and asterisk in 12.3B. The subventricular zone lies some distance below the reconstructed area whereas the cortex is more than 1000 μm above it. The lower portion of the diagram contains uniform, parallel fibers of the optic radiation (OR) and the remainder is occupied by a more variable and irregularly disposed fiber system; the border between the two systems is easily recognized. Except at the lower portion of the figure, most of these fibers are deleted from the diagram to expose the radial fibers (striped vertical shafts RF^1 to RF^6) and their

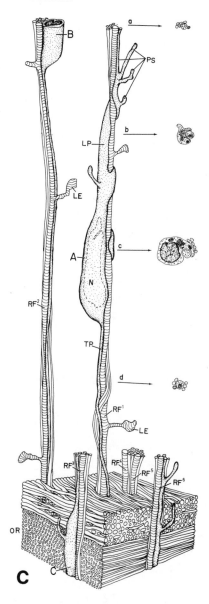

relationships to the migrating cells (A, B, and C) and to other vertical processes. The soma of
migrating cell A, with its nucleus (N) and voluminous leading process (LP), is within the reconstructed space, except for the terminal part of the attenuated trailing process and the tip of the
vertical ascending pseudopodium. Cross sections of cell A in relation to the several vertical fibers
in the fascicle are drawn at levels a to d at the right side of the figure. The perikaryon of cell B is
cut off at the top of the reconstructed space, whereas the leading process of cell C is shown just
penetrating between fibers of the optic radiation (OR) on its way across the intermediate zone.
LE, lamellate expansions; PS, pseudopodia. (Rakic 1972.)

tion of the cerebral cortex that is illustrated in Figure 15.16, Section 15.7). The horizontal arrangements of dendrites and axonal branches develop at a much later stage.

After the immature neurons have taken up their final position there is a relatively long period of dendritic development to the fully mature form. It is generally agreed that the sprouting of dendrites does not begin until the axonal process has grown. Its direction of growth seems to be determined by external factors. The early stage of development of dendrites can be seen in Figure 12.2E at 11–13 weeks, but the full structural development of the human cerebral cortex is not achieved until after birth. The initial stages of dendritic outgrowth are determined by factors internal to the neuron, but in later stages the influence of the surrounding tissues is paramount. There is need of a far more intensive and systematic study of the details of cortical development than has yet been carried out.

Histologically identifiable synapses have been detected at the surprisingly early age of 8 weeks in the human fetus, and increase very rapidly over the next few weeks (Molliner *et al.* 1973). These early synapses are located on dendrites both superficially and deep to the cortical plate (CP in Figures 12.2D, 12.2E). It is conjectured that the second phase of synaptogenesis, from 23 weeks onward (with synapses forming within the cortical plate), is due to the further development of dendrites and to the growth of afferent fibers into the cortical plate. The beautiful pictures drawn by Ramón y Cajal (1911) of the Golgi-stained neonatal cerebral cortex reveal that it is in an advanced stage of development. By contrast, the rat cerebral cortex is very immature at birth. Recognizable synapses in the rat are rare even 14 days postnatally, but are up to the adult number by Day 24 (Aghajanian and Bloom 1967).

12.1.3 The Building of the Cerebellum

The origin of the human cerebellum (see Figures 13.10, Section 13.4.1; and 13.12, Section 13.4.3) occurs in the same basic zones (V, I, and M) as for the cerebrum (Figures 12.2A, 12.2B, 12.2C). At 9–10 weeks the Purkinje cells (cf. Figure 13.12) move outward much as with the CP cells in Figure 12.2D and can be seen below the EG layer on the right side of Figure 12.4A and as the P layer in all frames of Figure 12.4B. At 10–11 weeks there is migration of SV cells out over the cerebellar surface from the rhombic lip (upper part of RL in upper Figure 12.4A) to form the external granular layer (EG in Figure 12.4A) in which there is a rapid clonal multiplication of germinal cells to give a layering 6–9 cells thick by 20–21 weeks despite the great increase in cortical area that is evident in the drawings in Figure 12.4B (Sidman and Rakic 1973). The maturation of the Purkinje cells begins in Figure 12.4B at 16 weeks and continues on to 25 weeks with profusely branching dendrites, some being from the soma, which is distinctively an embryonic feature. Meanwhile, in the external granular layer the simple clonal multiplication has been transformed by the onset of differentiating mitoses, immature granular cells being so formed in the external granule layer. By 16 weeks these granular cells are

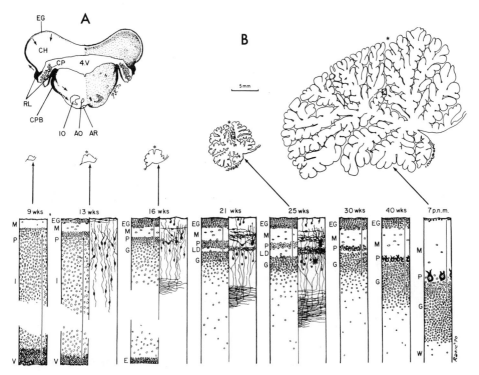

Figure 12.4 A. Transverse section of the human rhombencephalon at the end of the third lunar month of gestation. The arrows indicate the migration pathways (which are partly discussed in the text). Abbreviations: AO, accessory olive; AR, arcuate nucleus; CH, cerebellar hemisphere; CP, choroid plexus; CPB, corpus pontobulbare; EG, external granular layer; IO, inferior olive; RL, rhombic lip; 4V, fourth ventricle. B. Semidiagrammatic summary of the main morphogenetic and histogenetic events during development of the human cerebellum from the ninth fetal week to the seventh postnatal month (p.n.m.). The upper row shows outlines of the cerebellum in the midsagittal plane at 9, 13, 16, and 25 fetal weeks and at 7 postnatal months, all at the same magnification as indicated by 5-mm scale. The asterisk indicates the primary fissure. The columns in the lower row illustrate histogenesis of the cerebellar cortex in the depth of the primary fissure. At 9, 30, and 40 fetal weeks and the seventh postnatal month, only cresyl violet-stained material was available, while at 13, 16, 21, and 25 weeks both cresyl violet- (left column) and silver-stained specimens (right column) were analyzed. The arrows point in the direction of the mainstream of cell migrations. Note the spread of Purkinje cell somas from multilayer to monolayer distribution as the cerebellar surface increases in area. Abbreviations: E, ependyma; EG, external granular layer; G, granular layer; I, intermediate layer; LD, lamina dissecans; M, molecular layer; P, Purkinje layer; V, ventricular zone; W, white matter. (Sidman and Rakic 1973.)

migrating inward past the Purkinje cell bodies to form the internal granular layer (G in Figure 12.4B at 16 weeks and later). At term (Figure 12.4B at 40 weeks) the cerebellum is at an advanced stage of construction, but neurogenesis continues until 7 months postpartum (7 p.n.m.), by which time the EG has disappeared.

The monkey cerebellar development is rather earlier than the human, its maturation being almost completed by birth (Rakic 1973).

In the rat all neuronal formation from the external granular layer occurs postnatally, and Altman (1969) in particular has studied this process in more detail than has been possible with the primate cerebellum.

The valuable technique of pulse radiolabeling of developing cells depends on the fact that before the stage of each mitotic division there is a doubling of the deoxyribonucleic acid (DNA) of the cell nucleus, and thymidine is one of the four constituent purines and pyrimidines used in the building of DNA. So, if [³H]thymidine is injected a few hours prior to this DNA synthesis, it will be incorporated into the DNA, thus giving a radiolabel to the two daughter cells. Since a single injection is effective in labeling for only a few hours, the time of mitotic origin—the "birthday"—of the labeled cell is specified. Furthermore, since neurons do not again divide, they carry the radiolabel on their nuclei throughout life. By contrast, after a brief period of labeling of germinal cells, the radiolabel is halved in each mitosis and so is rapidly diluted beyond recognition. Neurons formed earlier than the injection cannot, of course, acquire a label. This radiolabeling establishes that the Purkinje cells and cerebellar nuclear cells are formed at almost the same time (the 11th to 13th embryonic days of the mouse), and, since they are formed in the same place, probably the origin is from the same germinal cells in the ependyma of the rhombic lip (Figure 12.4A; Miale and Sidman 1961). It should be mentioned that the Golgi cells of the cerebellar cortex (cf. Figures 13.12, Section 13.4.3; and 13.13, Section 13.4.4) are also formed there about about 2 days later, and like the Purkinje cells they migrate up to the cortex.

In Figure 12.5 are plotted the numbers of intensely labeled cells as observed in adults after a single pulse labeling at 0.25, 2, 6, 13, 30, or 120 days postnatally. Since nerve cells never divide, they retain their radiolabel throughout life and this is shown by the very similar numbers of intensely labeled granule cells observed at 60, 120, and 180 days postnatally in Figure 12.5A.

In Figure 12.5A the numbers of intensely labeled granule cells are greatest for the 13-day injection, which indicates that the maximum rate of generation occurs at about this time. There is even a trace of generation at 2 days, and at 6 days it is considerable, but by 30 days it has ceased. Figure 12.5B shows the differential timing in the "birthdays" of basket and stellate cells. These cells are defined by their level in the molecular layer (cf. Figure 13.12), as shown in the inset for stellate (S) and basket (B) cells. In their radiolabeling there is an overlapping period from 6 to 13 days, however.

Figure 12.6 is assembled from several pictures that Ramón y Cajal (1929) drew from his Golgi-stained preparations (cf. Figure 1.2B, Section 1.2.1) of the developing cerebellum. He brilliantly interpreted what was happening. Figure 12.6 is constructed as a perspective drawing to show sections along the folium (left frame) and across the folium (right frame). It represents the situation at about the tenth postnatal day of the rat, corresponding approximately to Figure 12.4B at 21–25 weeks. On the surface (MU) there are about

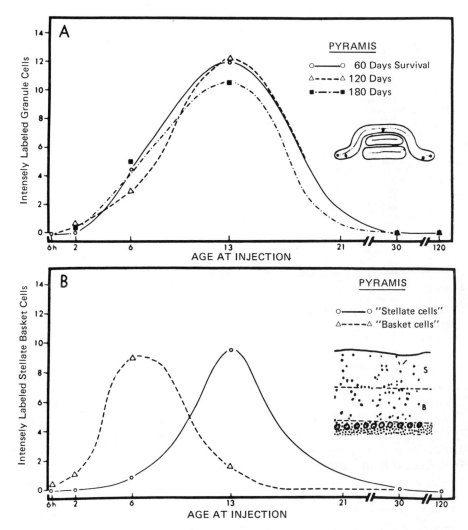

Figure 12.5 "Birthday" dating of cerebellar neurons generated in the external granule layer of rat. Mean numbers of intensely labeled neurons in pyramis region of rat cerebellar cortex in animals injected at various ages (indicated by abscissas) and examined some months later. A. Granule cells after the indicated days of survival; inset shows sampling sites. B. Presumed stellate and basket cells, as indicated in inset, after 120 days survival. (Altman 1969.)

five layers of proliferating germinal cells. In the left frame, the small irregular dark cells (a) are the immature granular cells (GrN). They can be identified because they stain differently and they are already sprouting little axons that rapidly grow out in both directions along the folium (b). These are the beginnings of the parallel fibers along the folium. They grow very rapidly and very nicely in parallel, as is well seen in the figure.

What determines the direction of the growth? The cerebellar cortex has already become folded and we suggest that the folding exercises a mechanical

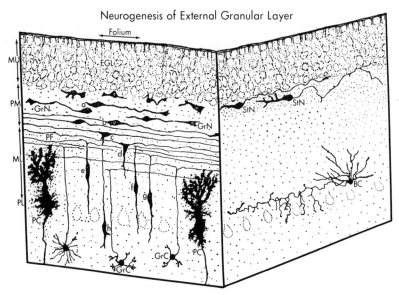

Figure 12.6 Neurogenesis of external granular layer of cerebellum, in a montage perspective diagram composed from several drawings by Ramón y Cajal in order to show the various stages of neurogenesis and morphogenesis from the cerebellar cortex, both along a folium (left) and across it (right). It could represent approximately postnatal Day 10 for the rat. (From Eccles 1970a.)

constraint on growth across the folium, thus favoring the longitudinal direction (Eccles 1970a). Once a few of these fibers have grown, the others follow because they chemically sense each other in a way we will be describing later in this chapter. So they all grow together by surface chemical identification (selective fasciculation) and by growth alignment make the dense bundles of parallel fibers that are so distinctive of the molecular layer of the adult cerebellar cortex (cf. Figure 13.10, Section 13.4.1). It is important to realize that these immature granular cells are genetically coded to grow sprouts in two directions from the cell body so that a symmetrical structure results with two oppositely directed primitive nerve fibers.

The next stage in granule cell development is seen in c, d, e, f, and g, of Figure 12.6. When the two axonal sprouts are fully grown, pressure may rise in the soma because protein manufacture by the nucleus remains very active. So a sprout goes downward from the nuclear region, finding its way between the deeper parallel fibers. After a while the nucleus follows down the sprout (cf. e, f, and g in Figure 12.6), which meanwhile has grown down below the layer formed by the Purkinje cells.

Rakic has produced convincing evidence that the vertical orientation of the downward growth of granule cells is due to guidance by the fibers of the Bergmann glia. As shown by Golgi pictures (Altman 1975), Bergmann glial filaments are already in position as early as the second postnatal day in the rat. They extend from the cell body at the Purkinje cell level up to the surface of the external granular layer. By Day 8 the glial filaments are well formed for their function of guiding the downward migration of granule cells (Bignami and Dahl 1973). In the elaborate drawing of Figure 12.7 Rakic (1973) shows the way in which the Bergmann glial filaments (BGF) act as guides in the monkey. Granule cell 2 is poised ready to descend, 3 is already on the way, 4 is half down, and 5 has completed its descent and is beginning to grow dendrites, while 6 and 7 are mature granule cells with synapses from a mossy fiber, MF. There is an exact parallel with the stages in Figure 12.6.

We can surmise that the granule cells descend below the Purkinje cell layer in order to meet the mossy fiber sprouts that are growing into that location. Once below the Purkinje cell layer the granule cell grows a lot of sprouts (i, in Figure 12.6), but eventually only a few remain. The granule cells (GrC in Figure 12.6) are now fully formed. Meanwhile, the mossy fibers (MF) come in and form synapses on the granule cell dendrites (Figure 12.7, granule cells 6 and 7). We can think that their synaptic "desire" has been consummated. It is a mutual desire. If the granule cells have been destroyed by virus or by X-rays, or in weaver mutant mice, the mossy fibers go on looking for the granule cells and grow above the Purkinje cell layer, where they never normally go (Figure 13.10, Section 13.4.1), and, apparently, finding no granule cells, they make aberrant synapses on dendrites of various cells there: Purkinje cells and basket cells. We have to postulate that all cellular elements have the instructions adequate for the normal building operation. If you fool them by distorting the situation, as, for example, by massive depletion of some elements such as the granule cells, the remaining elements do the best they can, carrying on and searching for some means of satisfying their synaptic "desire." Thus, there is displayed a quite wonderful living performance in the neurogenetic story.

We now will consider the right frame of the perspective drawing of Figure 12.6, in order to see the simultaneous happenings in the generation of cells that are oriented across the folium. At the surface are the same layers of germinal cells, and below are two stellate cells (StN) with their dendrites branching from one pole and the axon from the other. The general orientation of growth is transverse both to the length of the folium and to the parallel fibers shown in the left frame. Far down in the right frame there is a basket cell (BC) already in position. It developed much earlier, as indicated in Figure 12.5B, but it is essentially the same sort of cell as the stellate with a comparable genome; both are inhibitory and both grow axons oriented transversely to the folium. You can see that the axon of the basket cell grows sprouts that are searching for the Purkinje cells, shown faintly by dotted outlines. The stellate cells in the upper part of the right frame have still to mature before they will synapse with the Purkinje cell dendrites that eventually will grow into the upper zone of the molecular layer. In Figure 12.7, S_2 and S_3 represent more mature stellate cells while S_4 is a very immature stellate cell that has not yet formed any sprouts.

The stellate cells stay stacked in successive levels of the molecular layer according to the time of their development. This is true also for the parallel fibers that are stacked with a strict archeological layering. It can be conjec-

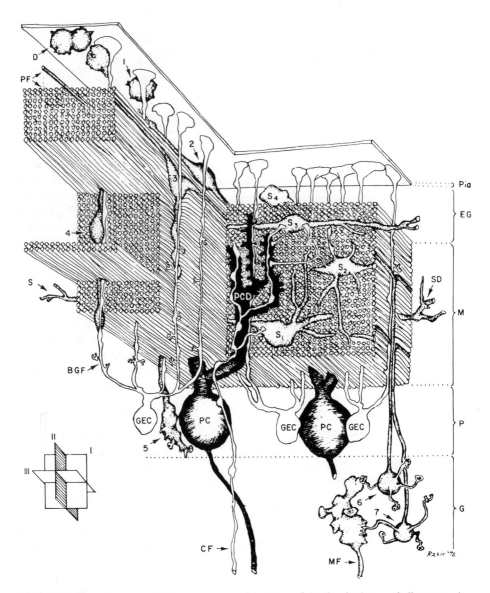

Figure 12.7 Four-dimensional (time–space) reconstruction of the developing cerebellar cortex in the rhesus monkey. The geometric figure in the left lower corner indicates the orientation of the planes: I, transverse to the folium (sagittal); II, longitudinal to the folium; III, parallel to the pial surface. On the main figure the thicknesses of the layers are drawn in their approximately true proportions for the 138-day monkey fetus, but the diameters of the cellular elements, particularly the parallel fibers, are exaggerated in order to make the reconstruction more explicit. Abbreviations: BGF, Bergmann glial fiber; CF, climbing fiber; D, dividing external granule cell; EG, external granule layer; GEC, Golgi epithelial cell (Bergmann glia); G, granular layer; M, molecular layer; MF, mossy fibers; P, Purkinje layer; PC Purkinje cell; PCD, Purkinje cell dendrite; PF, parallel fiber; S_{1-4}, stellate cells; SD, stellate cell dendrite. (Rakic 1973).

tured that the stellate cells differ from granule cells in that they have no chemical affinity with Bergmann glia, hence their inability to move vertically through the phalanx of stacked parallel fibers that are so effectively displayed in Figure 12.7. Instead, the basket and the stellate cells are seen in Figures 12.6 and 12.7 to grow transversely to the parallel fibers by what appears to be a series of avoidance reactions. Evidently, in their surface sensing the immature basket and stellate cells display a pattern of desirability very different from that of immature granule cells. Despite their apparent avoidance of parallel fibers, the stellate neurons readily accept synapses from parallel fibers. On the output side the basket and stellate cells make inhibitory synapses on Purkinje cells (Figures 13.12, Section 13.4.3; and 12.7).

There are many mysteries in the clonal mechanisms of the external granular layer. First, what is the mechanism whereby external germinal cells can form excitatory granule cells or inhibitory basket and stellate neurons? What sets the ratios of production of excitatory and inhibitory neurons? The germinal cells in Figure 12.6 make 80 times more granule cells than the combined basket and stellate cell population. Why does the whole clonal process age? Germinal cells at the onset look as if they had an everlasting life of clonal multiplication, but at about the fourth postnatal day they start differentiating into neurons, and this process occurs with increasing probability so that eventually the whole neurogenesis comes to an end. All the germinal cells have been eliminated by mitotic transformation into neurons (Figure 12.4, 7 P.N.M.) which never divide again. And why don't neurons ever divide again? These are all mysteries arising out of this early exploratory work on neurogenesis.

There are as yet no answers to these various questions, but most interesting investigations are being carried out in which some of the neurogenetic processes can be interfered with by radiation or by virus action (Altman 1976; Sidman and Rakic 1973). Mutations give all kinds of disorders and distortions of neurogenesis which lead to remarkable changes in the structure and function of the cerebellum (Rakic and Sidman 1973). Fundamental problems thus arise in what we may term molecular neurobiology.

In Figure 12.6 the two primitive Purkinje cells can be seen to have their axons passing downward. In fact, it trails behind as the body and dendrites migrate up to the cerebellar cortex, apparently in search of synaptic contacts. But the axon remains in position, ready to establish synaptic contacts with the deep nuclear cells (cf. Figure 13.12). We may call this the axon-trailing method of establishing synaptic contacts. The same axon-trailing process occurs with the granule cells that migrate, leaving their axons, the parallel fibers, in the molecular layer (Figures 12.6, 12.7).

In Figure 12.8 Altman (1972) summarizes the maturation of the cerebellar cortex of the rat by drawings of the situations at significant periods: 3, 7, 12, 15, and 21 days postnatally. Reference has already been made to the numbers of cell layers in the proliferative and premigratory zones during the successive stages of development (Figure 12.5). Figure 12.8 is particularly informative on the development of a Purkinje cell from the rudimentary form

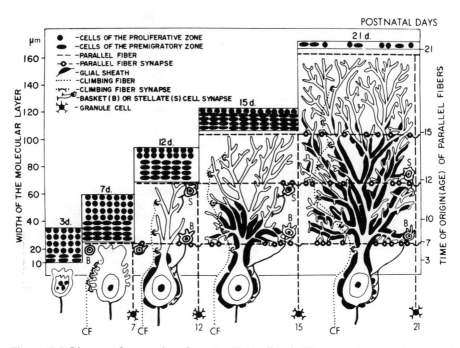

Figure 12.8 Diagram of maturation of a rat Purkinje cell and of the synaptic connections upon it. Five stages of development are shown at postnatal Days 3, 7, 12, 15, and 21. The key to the various structures is given in the inset. Climbing fibers (CF), granule cells, and basket cells (B) are shown at Days 7–21. It is to be noted that four granule cells are shown with their parallel fibers stacked at successive levels according to their "birthdays," as indicated in Figure 12.5. (Slightly modified from Altman 1972.)

at 3 days to the almost mature form, with its profusely branching dendrites, at 21 days. A puzzling transitional relationship is shown by the climbing fiber (CF) synapses on the somatic spines of the Purkinje cell at 7 days and the displacement of the climbing fiber synapses up to the dendrites at 12 days, the somatic dendrites (cf. the two Purkinje cells in Figure 12.6) having meanwhile disappeared with replacement of the climbing fiber synapses by the basket cell (B) synapses on the soma. Another interesting feature is the relatively late stage at which synapses are formed by parallel fibers on Purkinje cells. Granule cells formed at Day 7 do not give parallel fiber synapses to Purkinje cells at Day 12, but only at Day 15. Similarly 12-day and 15-day granule cells give parallel fiber synapses at Day 21.

Evidently there are many challenging problems in the Purkinje cell maturation. What is the significance of the transitory somatic spines that are so striking in the drawing by Ramón y Cajal in Figure 12.6 (left frame)? Is the displacement of the climbing fiber synapses from somatic spines related to the simultaneous development of basket cell synapses on the soma? Why are the parallel fiber synapses on Purkinje cells so delayed in formation relative to their synapses on basket and stellate cells? What is the significance of the massive development of glial coverage of the soma and main dendrites? What is the cause of the great proliferation of the Purkinje cell dendrites at 15 and

21 days though no synapses are yet formed on them by parallel fibers? Apparently, there is as yet no recognition between the parallel fibers and these newly grown dendrites. Evidence will be given later, however, that climbing fibers induce the branching of dendrites and the growth of the dendritic spines, and on these the parallel fibers make synapses. Rakic and Sidman (1973) suggest that the maturation of Purkinje cells is influenced by early contact with climbing fibers, and this is supported by the experiments by Kawaguchi *et al.* (1975) and Bradley and Berry (1976), as will be discussed later. A key role in the maturation of the Purkinje cell is played by the parallel fibers. When there is a deficiency of parallel fibers, the Purkinje cell dendrites are also deficient and deformed. It seems that the espalier-like shape of Purkinje cells (Figure 13.12) is due to the sculpturing effect by parallel fibers by what has been called an "exclusion principle" (Eccles 1970a; Altman 1976).

12.1.4 The Building of Brain Nuclei and the Hippocampus

In Figure 12.4A the lower RL line points to an intense clonal multiplication zone which gives rise to the corpus pontobulbare (CPB), from which there is migration of subventricular cells as shown by the two lower arrows. The immature neurons migrate along the two lower arrows to form deep collections of neurons such as the pontine nuclei, the lateral reticular nucleus, and the inferior olive (IO and AO). On the right side of Figure 12.4A the dots indicate neurons, and in the primitive cerebellum there can be seen a deep collection of neurons that will form the cerebellar nuclei; these arose from the dorsal rhombic lip at the same time as the Purkinje cells. In the human fetus another analogous structure to the CPB can be seen bulging into the lateral ventricle. This is the ganglionic eminence, and it is formed by rapidly mitosing subventricular cells, rather as in S of Figure 12.2E. It appears before the 13th fetal week and persists until after birth. Sidman and Rakic (1973) have shown that the immature neurons so generated are destined to form the large nuclei of the diencephalon—the thalamus, the basal ganglia, and the pulvinar of the thalamus. The hippocampus is also formed from the rapidly mitosing subventricular cells of the lateral ventricle that follow complex paths of migration eventually to build the patterned structure of the hippocampus (cf. Figure 15.14, Section 15.6). Reference should be made to Angevine (1970) for a review of the experimental work. In all examples studied, the early stages of the building are as diagrammed in Figure 12.2. Clonal multiplication of neuroendothelial cells eventually generates immature neurons that migrate to form the various nuclei.

12.2 Principles of Neuronal Recognition and Connectivity

12.2.1 Introduction

So far we have concentrated on neurogenesis in the mammalian brain, describing the genesis of the various cell types and the manner in which the immature neurons develop into the fully formed neurons. Now we come to two fundamental questions: How do these various neurons get connected together? How do they know where to go? In the 1930s it was believed, as mentioned earlier, that there was initially some kind of disorderly arrangement, with usage gradually bringing about the organization. On the contrary, however, wherever they have been investigated the neuronal connectivities were found to be established in their final form before being used. These experimental findings indicate that somehow encoded in the developing nervous system there are enormous numbers of detailed specifications for building the final structure. This requirement led Sperry to formulate this most challenging hypothesis: that the precise building of the nervous system is due to an immense variety both in the chemical coding of neurons that are seeking to make synapses and in the complementary mechanisms for specific recognition by neuronal surfaces "ripe" for synapses. We quote from his eloquent writings (Sperry 1971):

> The complicated nerve fiber circuits of the brain grow, assemble and organize themselves through the use of intricate chemical codes under genetic control. Early in development the nerve cells ... acquire individual identification tags, molecular in nature, by which they can be recognized and distinguished one from another. ... The outgrowing fibers in the developing brain are guided by a kind of probing chemical touch system. ... By selective molecular preferences nerve fibers are guided correctly into their separate channels at each of the numerous forks or decision points which they encounter as they travel through what is essentially a three-dimensional multiple Y-maze of possible channel choices.

This is a comprehensive and challenging hypothesis and it has no rivals, though, of course, detailed modifications have been suggested. We think it important to appreciate the extraordinary power and scope of this hypothesis before giving an account of some of the experimental testing to which it has been subjected. It is interesting that, so far as we know, mirror identity occurs. Symmetrical cells on the right and left sides have identical chemical coding.

Figure 12.9 illustrates Sperry's hypothesis by giving in diagrammatic form the sequences of growth of a nerve fiber that can be seen in time-lapse cinemicrography such as Pomerat so beautifully photographed in tissue cultures of nerve cells. It is easy to see that the fiber growing toward the right is sending out little searching probes. These several outgrowths at any one time are probing all the time, looking for the right kind of chemical surfaces with attractive molecular configurations. If they do not find them, they may regress or they may grow on, as in the main shaft of Figure 12.9, perhaps getting some encouragement from the general chemical environment. So it grows on, at each stage sending out searching probes in all directions in three dimensions and then it goes on following up some attraction. In A, B, and C of Figure 12.9 you can see "ghosts" of offshoots that finally failed after they had grown some distance. This diagram gives some kind of picture of what you can imagine happening to an incredible degree for the axon of every nerve cell finding its way to the neuronal surfaces on which it eventually forms synapses.

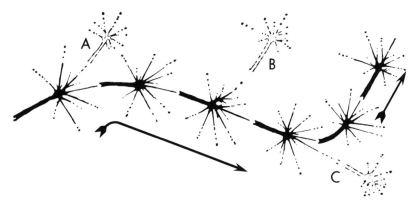

Figure 12.9 Sequential representation of steps in chemotactic guidance of a growing nerve fiber. (From Sperry 1963.)

Sperry's hypothesis has been studied in many sites, but not in the cerebrum and cerebellum, which are too complex structurally for these kinds of experiments. So we now leave the cerebrum and cerebellum, and indeed the mammalian brain, for the simpler brains of lower vertebrates. There is a tremendous amount of experimental work of high quality in this field. The books of Gaze (1970) and Jacobson (1970) are largely concerned with an appraisal of the chemical specification hypothesis in relation to searching experimental tests. It certainly has been a very fruitful hypothesis.

We will first discuss experimental evidence supporting the general hypothesis and later some speculations concerning the mechanisms and substances involved.

12.2.2 Chemical Sensing in the Visual Pathways of Fish, Amphibia, and Birds

These pathways have been very extensively investigated and have proved to be exceedingly valuable, and virtually irreplaceable, as preparations for rigorous experimental investigation. Initial diagrams of the connections from retina to brain (optic tectum) of amphibia are given in Figure 12.10A (viewed from the left side) and Figure 12.10B (viewed from ventral surface). In Figure 12.11A it can be seen that the eyes of the frog look out sideways. The optic nerves from each eye cross completely in the optic chiasma (Figure 12.11A) so that the right eye projects to the left side of the brain and left eye to the right side. Figures 12.10A and 12.10B show the pathway from the right eye going across the ventral surface of the brain to pass into the left optic tectum, the left eye and its pathway being removed as in Figure 12.10B so that the pathway from the right eye is fully revealed.

There had been sectioning of the right optic nerve some weeks previous to Figure 12.10A. Regeneration had occurred across the nerve scar and there was full functional recovery of vision in the right eye. The significance of this recovery may be appreciated by reference to Figure 12.10C, where the nor-

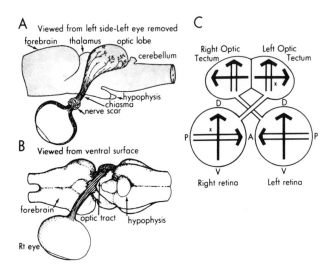

Figure 12.10 Visual pathways of the amphibian. A and B. The right eye with optic nerve is shown after section and regeneration. The left eye and optic nerve are not shown. The right optic pathway crosses at the optic chiasma to form the left optic tract that is distributed to the left optic lobe of tectum (as shown in *A* and *B*). (From Sperry 1951b.) C. Diagram of the orderly map-like projections from the right and left retinas to the left and right optic tecta. (Constructed from observations illustrated by Jacobson 1970.) P, A, D, and V on the retinae signify posterior, anterior, dorsal, and ventral.

mal spatial relations of the retinal projection for the optic tectum are shown. Thus, retinal points along the solid retinal arrow project to corresponding points on the solid tectal arrow on the contralateral side, and similarly for the open retinal and tectal arrows. Corresponding points in between have corresponding projections, for example, from point x on the retina to point x on the tectum, i.e., there is point-to-point connectivity.

One interpretation of this functional recovery is that the retinotectal projections were regrown in their detailed spatial connectivity. Alternatively, the recovery after the nerve section could be explained by a random reconstitution of the retinotectal projections, the frog then learning to use these aberrant connections in order to achieve a valid picture of its surround, much as is done by human subjects observing the world for days through inverting spectacles.

Sperry (1951b) eliminated the latter alternative by showing that the frog never learns to adapt to an eye rotation. For example, 180° rotation of both eyes without optic nerve section results in the frog's striking for flies just 180° in error. Even for a year or more there is no adjustment. The frog would quickly die if it were not fed artificially. But as soon as the eyes are rotated back to the normal position, the frog regains its old skills and suffers no disturbance from the long period of complete incoherence of movement. Therefore, learning cannot contribute to the functional recovery after optic nerve section in Figure 12.10.

We can conclude with Sperry (1951b) that correct connectivity was reestablished in the process of regeneration of the severed optic nerve in Figure 12.10A. There appeared to be an accurate recovery of the normal retinotectal topographic relations. This finding is in excellent accord with the Sperry hypothesis. There seems no other way in which the severed fibers from the optic nerve could recover their "correct" targets in the neurons of the optic tectum. And this was found to occur no matter how scrambled was the approximation of the two cut ends of the optic nerve.

Figure 12.11 illustrates a simple experimental way of changing the retinotectal connections in a frog. Normally there is a complete crossing of the optic nerves at the optic chiasma, so that the left eye projects to the right tectum and the right eye to the left (Figure 12.11A). In Figure 12.11B the chiasma is excised and the right optic nerve connected to the right optic tract so that it regenerates to the right tectum, and vice versa for the left optic nerve. After some weeks of total blindness full functional connections are reestablished, because, as already stated, there is mirror identity in the specific coding. But now, as shown in Figure 12.11C, the frog is in trouble. With a fly at X, for example, the instructions from the visual system give its location at X'; the frog strikes there and misses. When at Y, he strikes at Y'. The only success is for a fly in the midline. Thus, the behavior pattern of the frog indicates that the retinal connections to the optic tectum have regenerated in correct pattern but to the wrong side. The frog never learns to strike correctly, even for a year of errors during which he has to be fed. So here we have evidence that the retinal specification also holds, but in a mirror-image manner for the optic tectum on the same side, giving a permanently maladaptive result. As expressed by the Sperry hypothesis, there is a symmetry in the molecular labels of the retinal ganglion cells and their optic nerve fibers. Retinotectal connections are established in experiments such as illustrated in Figure 12.11 as quickly and as accurately as they are after simple severance of

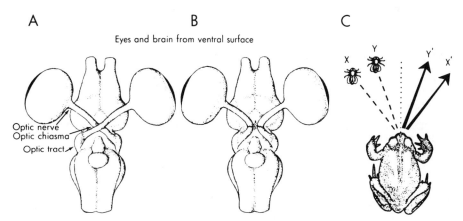

Figure 12.11 Contralateral transfer of retinal projection on the brain. A. By excising the optic chiasma and cross-uniting the two sets of optic nerve stumps as diagrammed, the central projections of the two retinas are interchanged. B. After recovery, the animals respond as if everything viewed through either eye were being seen through the opposite eye. (From Sperry 1951a.)

an optic nerve (cf. Figure 12.10A). There is no sign of any discrimination between the chemical specifications of the two sides. We therefore can ask: What guides the optic nerve fibers in their initial growth so that there is a complete decussation in the optic chiasma? The answer is not sure, but it is assumed to be largely by mechanical guidance and by the tendency of fibers of like function to grow along the same pathway by what Weiss (1960) calls selective fasciculation.

More severe tests of the Sperry hypothesis are illustrated in Figure 12.12, which displays the results of three different transplantations of both eyes. These operations take advantage of the fact that when eyes are transplanted from one side to the other, they regenerate connections to the optic tectum crossing in the usual manner by the optic chiasma. After some weeks of complete blindness the frog recovers vision, but it is disordered from the normal orientation (Figure 12.12D) in the manner indicated in Figure 12.12A to 12.12C.

Figure 12.12A represents the simplest situation. As shown in the enlarged inset of the frog head, after section of the optic nerve the right eye was rotated 180°, and similarly for the left eye. The arrows in Figure 12.12D signify the normal eye position. When the frog regained vision, it made extreme errors in striking at flies, showing that it always saw the world in reverse, and it continued indefinitely to do so. This is precisely the same behavior as was already mentioned after rotation of the eyes 180° without sectioning of the optic nerves. Hence, we can assume that after the nerve section and eye rotation, the optic nerve fibers grew into the optic tectum and reestablished their original connections to the tectal neurons regardless of the eye rotation. This result is exactly what would be predicted from the Sperry hypothesis.

Figure 12.12B illustrates a more complex situation because, as shown in the inset, at the tadpole stage the left eye was excised and, after the rotation as shown in the inset, it was substituted for the right eye and its optic nerve was sutured to the right optic nerve, and vice versa for the right eye. Again vision was regained after several weeks, but the adult frog now suffered from a visual world seen upside down, the vertical axis being reversed (see inset). It only could successfully strike at flies in the horizontal plane (the dotted line). Those above were struck at below, as in the illustration. The inset shows the actual position of the eye axes, and this systematic error again is in accord with prediction that the retinal maps are accurately reconstituted in their tectal projection, regardless of the fact that the eyes have been transposed.

Figure 12.12C shows another experiment of eye transplantation. Now after recovery the frog suffers from an anterior-posterior inversion of its visual field, as indicated in the illustration and its inset. It now only succeeds in correct striking in the vertical plane (dotted line). Again this is precisely what would be expected for a reconstruction of the retinotectal connections in accord with the original spatial relations. In all these cases the unfortunate frogs never learn to correct their errors and are kept alive by artificial feeding.

These behavioral experiments have been confirmed by many inves-

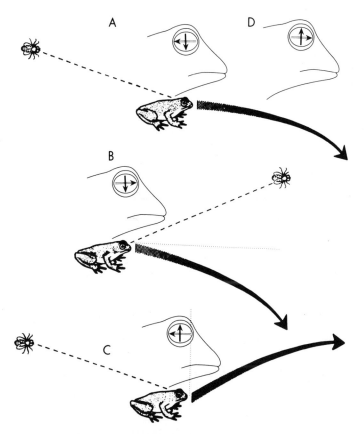

Figure 12.12 Eye rotations and transplants giving errors in striking. (From Sperry 1951a.)

tigators and have been carried out on several species of amphibia and fish with essentially the same results. Many finer variants have also been tried by Sperry, such as excision of small patches of the optic tectum after regeneration had been established (Meyer and Sperry 1974). These localized lesions resulted in a localized blind spot in the corresponding part of the visual field, just as occurs normally. Evidently after regeneration there was an orderly map-like projection of the regenerated fibers to the tectum reestablishing the normal point-to-point projection indicated in Figure 12.10C.

It is important, however, to use more rigorous testing techniques in order to establish finer details. In the goldfish, Attardi and Sperry (1963) excised various segments of the retina, then cut the optic nerve. It regenerated in the medial zone of the tectum when the dorsal part of the retina was removed (Figure 12.13A) and in the lateral zone after ablation of the ventral retina (Figure 12.13B). So histologically the fibers can be seen finding the right places in the optic tectum. Again in Figure 12.13C the posterior part of the retina regenerated selectively into the anterior part of the tectum. Of particular interest was the regeneration after excision of the posterior part of the

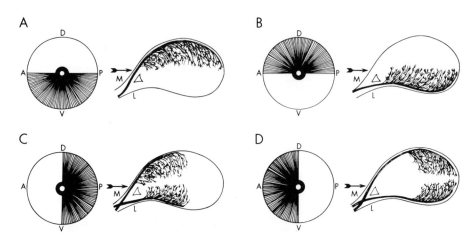

Figure 12.13 Regeneration patterns of retinal halves to optic tract and optic tectum. As in Figure 12.10, P, A, D, and V signify posterior, anterior, dorsal, and ventral on the retina. M and L signify medial and lateral aspects of the tectum. (Attardi and Sperry 1963.)

retina. The retina correctly regenerated into the posterior tectum (Figure 12.13D), and on the way grew past all of the cells of the anterior tectum despite their demand for synapses due to the fact that they, like all the tectal cells, had been denuded. Nevertheless, the anterior retinal fibers bypassed them and made endings in just the right place. We must conclude that those fibers growing out from the retina are guided by a specific sensing mechanism giving precise information on the correct neuronal surfaces on which to establish synaptic contacts. The outgrowing fibers grow in a stream over the tectal surface, plunging down when in the correct zone, as indicated in Figure 12.13D. Cowan (1971) has performed similar experiments on the regeneration of the chick retinotectal connections. Radiotracer investigations showed them to be reestablished to the correct tectal sites exactly in accord with the Sperry hypothesis.

Gaze (1970, 1974), Jacobson (1970), and Meyer and Sperry (1974) have used electrophysiological field analysis to map the relationship between retina and tectum. Microelectrode recording has been carried out from a grid of tectal sites and for each the corresponding retinal point has been determined by scanning the retina with a light spot. The retinotectal relationships shown in Figure 12.10C were mapped in this way. This technique has been used to test the retinotectal projections that were reestablished after various operative procedures as in Figures 12.11 and 12.12. In general there has been good confirmation of the regenerative pattern given by the original behavioral experiments. There were minor discrepancies. On the whole, the longer the time allowed for regeneration, the more the "correct" pattern was established. Detailed study of several regenerating processes has shown that there is a reshuffling of synapses, the "correct" synapses replacing less "correct" ones that were formed initially.

Gaze (1970, 1974) suggested that the Sperry model should be modified in that the optic nerve fibers do not grow to chemically predetermined targets, but grow out to the tectum and establish connections on a kind of sliding scale as the developing retina and tectum expand in size. Meyer and Sperry (1974) agree with Gaze that at early developmental stages there is a plasticity in the chemical specification in terms of morphogenetic fields. An extreme example of this is displayed in the experiments of Jacobson described below. At a later stage Meyer and Sperry (1974) found that the rigid specificity was displayed as in the experiments illustrated in Figures 12.10 to 12.13. They concluded that some of the discrepancies in the reported experiments on different preparations can be explained by variations in the times at which plasticity of the chemical specificity is lost.

Further investigation has raised interesting questions. Jacobson (1970) found that if you rotate the eyes through 180° at an early embryonic stage of the amphibian *Xenopus,* before any retinotectal connections have been formed, the toad develops normal vision despite the rotation. For this to occur, the rotation has to be performed before a critical embryonic age which Jacobson has been able to pinpoint within 10 h. It is of particular interest that this critical period spans the time of the differentiating mitosis of the neurons that develop into retinal ganglion cells. When there are only germinal cells in the embryonic retina, any rotation it may undergo is without influence on the specification of retinal neurons formed at subsequent mitoses. But at the time of the differentiating mitoses that results in the generation of retinal ganglion cells the situation is transformed. The immature ganglion cells then receive their chemical coding, giving them the precise instructions for the tectal location to which they must grow. This, of course, occurs before the onset of the actual fiber outgrowth from the retinal ganglion cells to the tectum. It is certainly a remarkable discovery that the complete coding of the immature retinal neurons is established in the 10 h during the last neurogenesis and that this coding is somehow given by the orientation of the eye in the head at that time.

It must be pointed out that we have no idea how the position of the eye in the head can result in the specification of the retinal ganglion cells. Nor is there any answer to the question: How does it come about that there is a complementary specification of the tectal neurons, matching that of the fibers growing out from the retina? Yoon (1975) has demonstrated that this complementary specification remains unchanged when rectangular slabs of the goldfish tectum are excised, rotated or inverted, and then reinnervated. Again, as already mentioned, mechanical guidance must play an important part in the massive control needed for ensuring a complete decussation in the optic chiasma, and in fact in the general establishment of the whole pathway from retina to tectum. Chemical specification need only operate at the last stage of the growth in insuring that the "correct" synaptic ˙connections are established.

12.2.3 Neuronal Connectivity in Mammals

Regeneration Studies.   We have chosen to deal at length with the visual pathways of amphibia and fish because they are the best exemplars of investigations on the factors determining the establishment of neuronal connections in the brain. There has been an immense range of such investigations using techniques such as limb transplantation, skin transplantation, and cross-union of various nerves to limb muscles by a technique similar to that of Figure 12.11. Many of these experiments have been carried out on mammals (Sperry 1951a). There is always the hope that operative procedures may help disabled patients by utilizing one or more nerves to reinnervate muscles deprived of their normal innervation. There is almost no specification in the innervation of muscles. Any set of motoneurons via their motor axons (cf. Figure 4.1, Section 4.1) will establish functional end plates on denervated muscle fibers and excite them effectively. In mammals, however, this innervation of some extraneous muscle does not result in any change in the central connections onto these motoneurons. Even many months after nerve cross-unions performed on kittens a few days after birth, there were very few significant changes in the monosynaptic connections onto the motoneurons (cf. Figure 4.1) which were thus given an inverted function (Eccles, 1964b; Chapter 16).

It is important to realize that after an early stage in the mammalian brain there is no regeneration of long-severed pathways. There is no trace of the incredible reconstitution of pathways observed in the amphibian or fish retinotectal pathways. Yet it can be presumed that, in the embryonic mammal, the growth of nerve fibers and the establishment of precise topographical maps such as that of motor cortex to muscle (Figure 13.2) was effected by chemical specification of the outgrowing fibers of the pyramidal tract. The tragedy for the paraplegic patient is that all this embryonic know-how is lost and the severed fibers of his pyramidal tract (cf. Figure 13.3) cannot regenerate along spinal cord pathways that were precisely traversed in the embryo. Such regeneration of the spinal cord is possible in the fully grown fish and in many amphibia. Can this embryonic know-how be recovered in the adult mammal and even in man? An encouraging answer to this question may eventuate from intensive and comprehensive investigations into the many factors that initially enter into the building of the nervous system. Undoubtedly, the Sperry hypothesis of chemical specification is the principal factor, but many other factors play an important role, such as mechanical guidance, selective fasciculation, temporal sequences, and competition for synaptic sites.

The elegant experiments of Wiesel and Hubel (1963) on kittens illustrate the complexities of the factors concerned in establishing and maintaining neuronal connectivities in the mammalian brain. It is some days after birth before kittens open their eyes; nevertheless, at birth the retina is connected with detailed topography to the visual cortex in the adult pattern, much as with the amphibian retinotectal connectivity. Moreover, the orientation sensitivity of neurons in their columnar arrangements can be demonstrated, as in Figure 16.3 (Section 16.1.3), for the adult. Here we have an excellent illustration of the establishment of precise, patterned connections from eye to brain prior to usage. We may assume that this is due in large part to the operation of chemical specification similar to that for amphibia and fish. But if this beautifully grown structure is not used for some weeks after the time the kitten normally opens its eyes, it becomes disorganized, the destruction being permanent. Use by patterned vision is essential for stabilizing this fragile organization of retinocortical connectivities that presumably is grown largely by chemical specification as in fish and amphibia.

In the last few years many remarkable investigations have established beyond all doubt that a very effective regeneration occurs in the mammalian brain, even in the adult. These regenerations differ, however, from the regenerations of pathways in lower vertebrates in that the regeneration occurs only for very short distances—perhaps for no more than 50 μm. Such regenerations have been demonstrated in many regions of the mammalian brain: the septal nuclei, the hippocampus, the red nucleus, the lateral geniculate nucleus, the superior colliculus, and the visual cortex.

This work at higher levels of the central nervous system can best be introduced by considering the fine work of Raisman and his colleagues on neurons of the septal nuclei. As illustrated in Figure 12.14, the septal nuclei provide ideal conditions for the experimental investigations (Raisman 1969; Raisman and Field 1973). The two principal inputs to these nuclei are the fimbrial pathway from the hippocampus and the medial forebrain bundle (MFB). The former input forms synapses almost exclusively on the dendrites

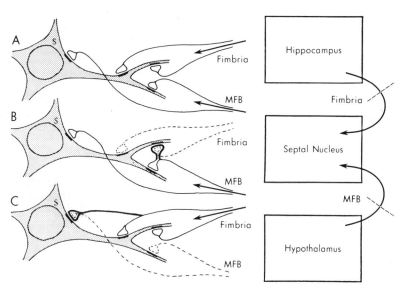

Figure 12.14 Schematic representation of synaptic regeneration of septal neurons. A. In the normal situation, afferents from the medial forebrain bundle (MFB) terminate in boutons on the cell soma (S) and on dendrites, and the fimbrial fibers are restricted in termination to the dendrites. B. Several weeks after a lesion of the fimbria, the medial forebrain bundle fiber terminals extend across from their own sites to occupy the vacated sites, thus forming double synapses. (Degenerated connections are shown with discontinuous lines, presumed plastic changes with heavy black line). C. Several weeks after a lesion of the medial forebrain bundle, the fimbrial fibers now give rise to terminals occupying somatic sites, which are presumably those vacated as a result of the former lesion. (Modified from Raisman 1969.)

of septal cells, while the latter input ends on both dendrites and somata (Figure 12.14A).

After sectioning of either pathway in adult rats (3–6 months old) within a few days there was the usual degeneration and disappearance of the synapses formed by the sectioned pathway. Electron microscopic observations revealed that the number of synapses was reduced almost to half. But some 30 days later the full population was restored. Convincing evidence was presented that this restoration was due to sprouting of the fibers of the intact pathway, the sprouts growing to form new synapses that occupied the vacated synaptic sites, as is illustrated in Figures 12.14B and 12.14C. In the electron micrographs it could be seen that some weeks after section of the MFB the fimbrial fibers occupied a considerable number of synapses on the somata of the septal neurons, as is illustrated in Figure 12.14C. Conversely, after section of the fimbrial pathway, there was evidence that the fibers of the MFB sprouted to form many new synapses, which often had the double configuration shown in Figure 12.14B. These new synapses would all be on the dendrites, which is the site of the vacated fimbrial synapses.

It appears that there has been a loss of the embryonic growth specificity, so that collaterals growing from the intact axons now heterotypically innervate synaptic sites originally reserved for and occupied by the sectioned pathway. Raisman (1969) regards this heterotypic regeneration of synapses as being functionally meaningless. Nevertheless, their regeneration is of great interest because it shows that axonal sprouting and synaptic formation can occur in the adult rat, but only at micro distances, perhaps no more than 50 μm. Raisman further suggests that this heterotypic regeneration may be aided by the chromatolytic reaction of the septal neurons that follows the inadvertent section of their axons in the initial operation (Chapter 1). Since the axons of septal neurons are in both the fimbria and the MFB, many would be sectioned in the initial operation. Two important negative findings are illustrated in Figures 12.14B and 12.14C. First, the sectioned fibers make only abortive growths and do not regenerate over the distances of millimeters along their degenerating pathways, and second, there is no evidence for development of new synaptic sites on the septal cells, there being merely heterotypic reinnervation of old sites.

Raisman and Field (1973) have illustrated the manner in which the glial cells of the septal nucleus participate in the degenerative process. As illustrated in Figure 12.15, the astroglia are specially concerned in the ingestion of the degenerating synapses, and at the same time they move in to occupy the vacated sites (as described in the figure legend). The synaptic sites can be identified by the characteristic postsynaptic membrane densities that remain after the synaptic knobs disappear. Westrum and Black (1971) have reported a similar action of astroglia on degenerating synapses in the trigeminal nucleus.

In Figure 12.15 Raisman and Field (1973) give a diagrammatic illustration of the various stages in synaptic degeneration, namely, ingestion by astroglia which then occupy the vacated synaptic sites until ejected by the new

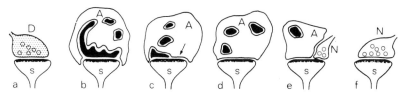

Figure 12.15 Proposed scheme for reinnervation of deafferented spine: a schematic representation of some of the structures which have been found in apposition to dendritic spines in the septum during the period of the proposed collateral reinnervation. These configurations have been arranged in sequence from a to f in order to illustrate one possible series of changes which would result in reinnervation of the deafferented spine. a. A degenerating axon terminal (D) lies in contact with the dendritic spine (s). b. The degenerating terminal, now much darker and more shrunken, is surrounded by a swollen astrocytic process (A) which is indenting its surface and has partially engulfed two detached fragments of degenerating terminal cytoplasm. c. The reactive astrocytic process has now partially displaced (arrow) the degenerating terminal from the region of the synaptic thickening. d. The displacement of the degenerating terminal from the synaptic thickening is now completed ("vacated synaptic thickening"). e. The reactive astrocytic process is partially displaced from the synaptic thickening by a nondegenerating axon terminal (N) ("shared synaptic thickening"). f. Complete reinnervation of the synaptic site by the nondegenerating axon terminal. (Raisman and Field 1970.)

ingrowing synaptic terminals. This figure is important because it leads to the formulation of questions relating to the problems of sprouting and reinnervation of the synaptic sites. We may ask two questions: How is the sprouting initiated? How is the sprout guided to the vacated synaptic sites?

Guiding to Vacated Sites. The synaptic regeneration to septal neurons gives very clear documentation of the remarkable ability of presynaptic nerve fibers to sprout and reconstitute synaptic contacts at vacated sites. It has been suggested by Raisman and Field (1973) and Watson (1974) that glial cells may provide the guidelines for the newly growing fiber in the same way as Rakic (1972, 1973) has demonstrated for glia in the initial process of neurogenesis in the monkey, both in the cerebrum (Figure 12.3B) and in the cerebellum (Figure 12.7). No theory has yet been developed to account for this close adhesion of sliding growth, but one can envisage that it is due to contacts between surfaces resembling those that guide the establishment of synaptic connections. There must be some chemical recognition depending presumably upon surface-membrane configurations.

The Primary Process. A second problem has not so far been raised, however, and that concerns the initial process, which is the triggering of an axon to produce a branch. It is one thing for the glia to guide the branch home to the vacant synaptic site. Quite another problem is raised when one asks how the branch starts in the first place. Figure 12.16 diagrammatically displays a suggestion as to how a degenerated synapse exerts this triggering role on an adjacent normal presynaptic fiber. Raisman and Field (1973) have shown by beautiful illustrations (Figure 12.15) the way in which the astroglia ingest the degenerating synaptic terminal and eventually break it up and apparently digest it. It is suggested in this theory that, because of this ingestion of the degenerating synapse, the astroglial cell develops a changed internal constitution which affects the surface contacts it makes with presynaptic fibers so that it acts as a stimulant for their growth. Thus, we would have two separate functions for glia. One is the trigger function whereby the glia through the ingested synaptic knobs stimulate the presynaptic fiber to branch, and the second is the guiding role of glia whereby the branch grows so as eventually to reach and occupy the vacated synaptic site.

In Figure 12.16 there is shown an astroglial cell ingesting the degenerating synaptic knob and in that way providing a trigger stimulus to induce a sprout from the nerve fiber and guide it to the vacant site. This concept of astroglial stimulation of nerve sprouting after ingesting degenerating synaptic knobs raises several interesting conjectures. We can imagine that the enzymes of the astroglial cell break down the protein structures of the synaptic knob, converting them into macromolecules which could be the specific molecules that trigger growth. Furthermore, it seems that the effect is relatively nonspecific because in the septal nuclei there are synaptic knobs differing in their synaptic vesicles, some having many dense core vesicles and others being free of dense cores.

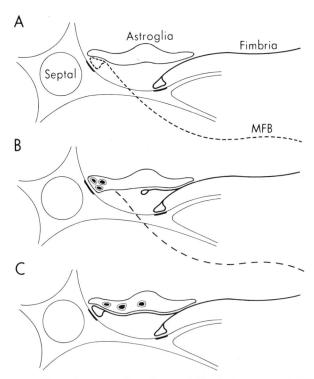

Figure 12.16 Degeneration and regeneration of synaptic knobs in septum. A. Diagram resembling that of Figure 12.14 showing degeneration of the MFB. B and C. The ingestion of the degenerated fragments of the synaptic knob. As a consequence it is proposed that the astroglial cell triggers sprouting of the fimbrial axon (B) and guides it to the vacated synaptic site (C). (Eccles 1976a.)

Adrenergic fibers are stimulated powerfully by a nerve-growth factor (NGF) discovered several years ago by Levi-Montalcini (1964). The NGF, which is highly selective in its action, will be discussed later in this chapter. NGFs for other kinds of nerve fibers have not been identified, but it is now suggested that comparable factors may be generated by astroglia digesting synaptic knobs. A test would be to have a large tissue culture of astroglia and feed the cultured glia with synaptosomes (broken-down fragments of a brain homogenate composed largely of synaptic knobs) in the hope that they would carry out the same process *in vitro* as is here postulated *in vivo*. In that case, growth factors for other nerve terminals could hopefully be extracted from the astroglia at the appropriate time after their synaptosomal meal. If such substances could be isolated, there would be most interesting possibilities of their use in aiding regenerative processes in the central nervous system (Eccles 1976a).

It has been suggested that this synaptic regeneration as disclosed in the septal nucleus has no functional meaning because, as illustrated in Figure 12.14, synapses from one type of input regenerate to occupy synaptic sites of quite different inputs, the so-called heterotypic regeneration. It must be envisaged, however, that the experimental demonstration by Raisman and his colleagues could be obtained only if there were massive degeneration. If only a small fraction of the inputs from the fimbria or from the MFB were cut, then it would be impossible to discover any regeneration that presumably would be homotypic. In other words, we have to recognize that, compared with possible naturally occurring random degenerations, the surgically induced degenerations are massive and exclusive.

Physiological Effectiveness.   We now come to some very recent work in which it has been most elegantly shown by Tsukahara and associates that not only is there synaptic regeneration as found by Raisman, but also that this synaptic regeneration may be physiologically effective. Figure 12.17 shows the synaptic connections upon a neuron of the red nucleus. These have been thoroughly investigated, both histologically (Nakamura *et al.* 1974) and also with intracellular recording (Tsukahara *et al.* 1974, 1975a). Figure 12.18B shows that the stimulation of the interpositus nucleus (IP) gives a very sharp and short EPSP, as would be expected for the somatic locations of the synapses that are illustrated in Figure 12.17A. On the contrary, stimulation of the cerebral peduncle (Ped.) gives a much more slowly rising and declining EPSP (Figures 12.18A, 12.18D), as would be expected if its monosynaptic sites were far out on the dendrites, as illustrated in Figure 12.17A. The difference in the time course of the respective EPSPs is fully accountable to the electrotonic distortions resulting from the transmission from the distal generating sites to the intracellular recording from the microelectrode in the soma of the neuron (Tsukahara *et al.* 1974, 1975b).

This preparation leads to an exquisite experimental testing of synaptic regeneration. When the interpositus nucleus is destroyed (Figure 12.17B), it is found that, after 2–3 weeks, stimulation of the peduncle results in an EPSP having features of both synapses on the soma and synapses on the dendrites,

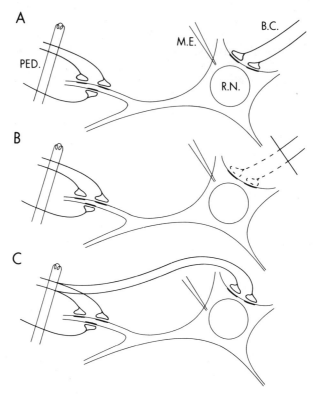

Figure 12.17 Regeneration of somatic synapses of the red nucleus cell. A. Diagram to show a red nucleus neuron (R.N.) with synapses of the brachium conjunctivum (B.C.) fibers on its soma and of the peduncle fibers (PED.) on its dendrites. M.E. is the recording microelectrode. B. Destruction of the interpositus nucleus results in degeneration of the B.C. synapses on the soma. C. Sprouts of the peduncle fibers are shown growing in to occupy the vacated synaptic sites on the soma. (Eccles 1976a.)

the intracellular EPSPs of Figure 12.18A being changed to those of Figure 12.18C. All tests carried out have shown that the terminals of the peduncular pathway ending on the dendrites have sprouted to give synapses occupying the vacant sites on the soma (Figure 12.17C), hence transforming the EPSP from that of Figure 12.18A to that of Figure 12.18C. We can conjecture that glia may again have been responsible for triggering this growth and for guiding the growing fibers to the vacant synaptic sites. Here again we have a heterotypic regeneration, just as with the septal nuclei.

Schneider (1973) has described elegant investigations on the visual pathway of very young hamsters. In the visual pathway the fibers decussate completely in the optic chiasma so that the left eye projects entirely to the right geniculate body and superior colliculus and vice versa for the right eye. The distribution of the axonal branches in these nuclei is shown in Figure 12.19A as viewed from the right side. From the optic chiasma (OCh) the fibers of the optic pathway traverse both components of the lateral geniculate body (LGv and LGd) and terminate in the superior colliculus (SC). When the right superior colliculus is removed from Day 0 to Day 5, the terminals of that part of the visual pathway projecting to the superior colliculus are severed and there is a reconstitution of connections, as indicated in Figure 12.19B. The new growths have been studied histologically and are shown by double lines. Two explanations are offered by Schneider to account for the observed distribution of the new sprouts.

1. As shown in Figure 12.19A, the superior colliculus sends fibers to the LP and LGv nuclei. These will degenerate with removal of the superior colliculus and this degeneration will induce sprouting in the optic fibers that are traversing these nuclei. It will be noted that there is no new formation of sprouts into the LGd nucleus, where there are no degenerating synapses from

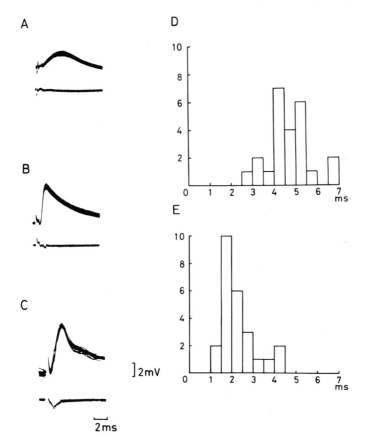

Figure 12.18 Intracellular responses of red nucleus neurons. A–C. Upper traces are intracellular responses in red nucleus (RN) neurons, while the lower traces show the corresponding field potentials recorded at a just extracellular position. A and B illustrate a Ped.-EPSP and an IP-EPSP respectively. C illustrates a PED.-EPSP after IP destruction. Time and voltage calibrations for all intra- and extracellular responses are shown at C. D and E. The histograms in D (normal cats) and E (after IP lesion) illustrate the frequency distribution (number of cells on the ordinate) of the "time-to-peak" of Ped.-EPSPs. (From Tsukahara *et al.* 1974.)

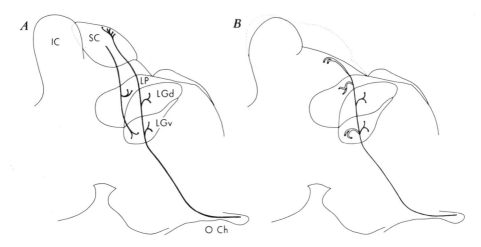

Figure 12.19 Formation of anomalous optic tract connections. A. Lateral-view reconstruction of rostral brainstem of normal adult hamster. Heavy line schematically depicts the course of a group of optic tract axons and some of their terminations. The tectothalamic pathway is shown in a similar manner. B. Similar view of brainstem of adult hamster which had undergone destruction of the superficial layers of the superior colliculus in the neonate. Anomalous optic tract connections are depicted by double lines. Abbreviations: IC, inferior colliculus; LGd, dorsal part of the lateral geniculate body; LGv, ventral part of the lateral geniculate body; LP, lateroposterior nucleus of thalamus; OCh, optic chiasm; SC, superior colliculus. (From Schneider 1973.)

the superior colliculus. This explanation is in line with the explanations for the regenerations in the septal nucleus and the red nucleus as described above.

2. Schneider suggests that there is also another factor concerned in the sprouting of the severed terminals in the optic tract projection to the superior colliculus, namely, that these fibers sprout from their severed ends and try to find neurons upon which they can make synapses. This growth gives rise to a quite remarkable new tract that crosses the midline and ends upon neurons in the intact superior colliculus on the other side. Schneider suggests that this excessive sprouting of the severed fibers is related to the pruning effect obtained with plants. Certainly, the formation of this tract cannot be attributed to the fibers growing along glia or other guidelines to vacated synaptic sites. It would seem to be a growth, guided perhaps by glia, but in itself an exploration into new neuronal territory.

It must be pointed out that these results were obtained in very young hamsters, 0–5 days of age. Experiments have not yet been done in old hamsters. In summary, this work of Schneider (1973) in part falls into line with the two previous investigations, but adds another feature of growth into neuronal territory devoid of degeneration.

Somewhat analogous is the demonstration that superior colliculi fragments transplanted from fetal to newborn rat brains develop complex internal organization and receive visual afferents from the host providing that they lie sufficiently close to host visual pathways (Lund and Hauschka 1976).

There have been many other investigations showing regeneration at the higher levels of the mammalian central nervous system. Particularly notable are the experiments of Cotman and his colleagues on the hippocampus in which degeneration by removal of the entorhinal cortex on one side leads to the new growth of connections from the intact entorhinal cortex (Steward *et al.* 1973, 1974). This work is important because it was possible to show that these new synaptic growths exhibited effective synaptic action. The work of Guillery (1972), Lund and Lund (1971), and Wall and Egger (1971) should also be mentioned.

Aging and Cell Death. Let us now turn to aging. The field is receiving increasing attention, as is warranted by the growing importance of its associated problems. The progress of medicine is leading to an ever-increasing life span, and as it does, new physiological problems are unmasked. In the case of aging, there is now an enormous cost to society of maintaining individuals whose mental and neurological capacities have slipped below the threshold necessary for independence.

Much needs to be done in defining the particular cellular and chemical losses which occur with aging, but an even more fundamental question that needs to be addressed is why these losses should occur at all. We have been concerned in earlier parts of this chapter with the building of the brain and its ability to regenerate after injury. We have been less concerned with what maintains it in its dynamic state and what factors are lost when it begins to degenerate.

We wish to approach this problem in an opposite way, by describing in more detail the catecholamine and dorsal root ganglion nerve growth factor (NGF), as an indication of the types of compounds we should be seeking to explain how a neuron may maintain sufficient growth capacity to repair itself and avoid degeneration. NGF is a protein of a molecular weight of approximately 20,000 whose structure is partially determined (Angeletti *et al.* 1973) and which can be obtained in abundance from the saliva of *adult* male mice.

The discovery and isolation of NGF is a saga in itself. Originally a factor was sought that could be responsible for attracting nerve fibers to transplanted limbs. A test system was chosen in which a rapidly growing connective tissue tumor was transplanted into a chick embryo. There was a profuse outgrowth of sensory (dorsal root) and sympathetic (catecholamine) nerve fibers from the embryo to the tumor. Next, it was shown that the tumor cells had a dramatic effect on growth of chick embryo ganglia in tissue culture. Suspecting that the stimulating material produced by the tumor cells was a nucleoprotein, Levi-Montalcini and her colleagues added a snake venom to hydrolyze the nucleic acids and neutralize the effect. Instead, growth was stimulated, suggesting that the venom itself contained the material. Since the venom came from salivary glands, it was reasoned that salivary glands from other animals might also contain a similar factor. This proved to be the case, but luck played a part in the result. The adult male mouse was chosen, a fortunate circumstance, because the saliva of female and young mice, and that

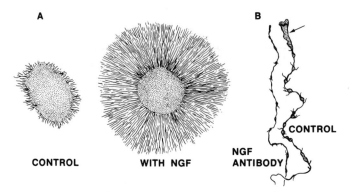

Figure 12.20 Effects of nerve growth factor (NGF) on sympathetic ganglia. A. Effect of NGF on neuron growth in 7-day-old chick sensory ganglion. Culture time, 24 h. To the left, ganglion from control medium; to the right, ganglion from a medium supplemented with NGF from adult male mouse saliva. B. Sympathetic trunks from mice. To the left is a ganglion chain, much smaller in size, from a mouse treated with rabbit anti-NGF serum. To the right is a normal sympathetic trunk. Arrow points to stellate ganglion. (Kuffler and Nichols 1976, after Levi-Montalcini 1964.)

from other species, contains much less NGF (Levi-Montalcini and Angeletti 1968).

The effect of mouse saliva containing NGF on growth of a chick embryo ganglion cultured for 24 h is shown in Figure 12.20A. The effect on immature animals *in vitro* is not just to increase the size, as shown in the figure: it also increases the numbers of cells. To prove that the effect was not just a pharmacological accident, a specific antibody to NGF was produced in rabbits. When the immune serum was injected into normal mice, they produced only a shriveled sympathetic trunk. This is shown in Figure 12.20B, where a normal mouse sympathetic ganglion chain is compared with one deprived of normal NGF as a result of reaction with the antibody (Levi-Montalcini and Cohen 1960).

NGF has been compared to insulin in some of its actions, particularly with regard to its supposed interaction with specific receptor sites on sensitive nerve cells. High-affinity NGF receptors localized on sympathetic and sensory ganglia have been characterized (Herrup *et al.* 1974; Herrup and Shooter 1975) and solubilized (Banerjee *et al.* 1976). These receptors seem not only to have a high affinity for NGF, but show a very high degree of peptide specificity. Identification of further nerve growth factors and elucidation of their mechanism of action is a major challenge in neuroscience with enormous potential clinical benefits.

Figure 12.21 Loss of cells in substantia nigra and locus coeruleus with age. A. Cell count in zona compacta of the human substantia nigra as a function of age. × refers to normal individuals; P to Parkinsonians. (P. L. McGeer *et al.* 1976.) B. Tyrosine hydroxylase (TH,nmol/100 mg protein/h), choline acetyltransferase (CAT, μmol/100 mg protein/h) and glutamic acid decarboxylase (GAD, μmol/100 mg protein/h) in human caudate nucleus as a function of age. Note sharp decrease in TH but only modest declines in CAT and GAD (P. L. McGeer *et al.* 1976). C. Cell count in human locus coeruleus as a function of age. (Data taken from Brody 1976.)

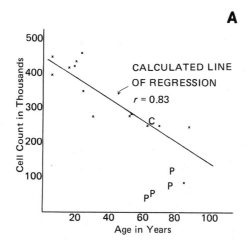

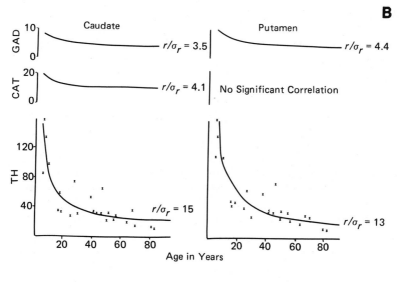

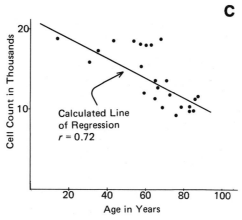

Although no correlation with NGF has been shown, catecholamine cells in the CNS are the only ones where a definite and dramatic decrease of cell numbers with age has been shown. Figure 12.21A shows the steady decline in dopaminergic cells of the substantia nigra with age in humans. At birth there are about 400,000 dopaminergic cells in each substantia nigra. By age 60, the number has dropped to about 250,000 (P. L. McGeer *et al.* 1976). In Parkinson's disease (designated by the symbol "P" in the figure), cell counts range from about 60,000 to 120,000. The symptoms of the disease are due to the loss in function of these substantia nigra cells (Chapter 13), but obviously many cells can disappear before decompensation occurs. The only stigmata may be the shuffling gait and stooped posture often seen in the very elderly.

Of even greater importance than outright cell loss may be a loss of vitality of the cells: youthful substantia nigra cells seen by the light microscope are plump with a clear cytoplasm; older cells often appear shrunken and blackened by the melanin pigment which begins to appear between the ages of 5 and 15. During this period there is a remarkable loss in tyrosine hydroxylase levels in the neostriatum where the nerve endings (Figure 8.4, Section 8.2.1) are located. Figure 12.21B shows an approximately four- to five-fold drop in tyrosine hydroxylase levels in the human caudate between the ages of 5 and 15, with a slower decline thereafter. By contrast, CAT and GAD, the neurotransmitter synthetic enzymes for cholinergic (Chapter 5) and GABAnergic (Chapter 7) neostriatal cells, only showed a mild decline (P. L. McGeer *et al.* 1976).

As described in Chapter 8, there is ordinarily a considerable excess of tyrosine hydroxylase over that required to synthesize sufficient dopamine, so that the drop may not signify anything beyond a normal physiological adjustment to termination of body growth. On the other hand, each substantia nigra cell body must serve truly remarkable numbers of boutons in the neostriatum. Bouton counts in the human caudate and putamen number about 6×10^{11} cc (T. Hattori, personal communication) which would give a total of about 150×10^{11} per caudate and putamen. It has been estimated that approximately 10% in the cat (Kemp and Powell 1971) and 16% in the rat (Hokfelt and Ungerstedt 1969) of these are dopaminergic, so that the total of such synapses would be 15×10^{11}. Assuming 300,000 dopaminergic neurons per substantia nigra, the number of dopaminergic boutons per SN cell would then be 5×10^6! This number, of course, needs verification by more exact counts of dopaminergic terminals in the human.

Brody (1976) has found a similar cell loss with noradrenergic cells of the locus coeruleus (Figure 12.21C). Here the numbers drop from about 19,000 for youths to about 10,000 for people in their 80s. The calculated lines of regression are significant in both cases, although the rate of loss is greater in the substantia nigra than in the locus coeruleus. As Brody (1976) points out, such losses are not usual, for most brainstem nuclei show no detectable decrements in cell number.

It thus seems clear that catecholamine cells require not only NGF to develop, but also undetermined factors to maintain themselves. These must be in marginal supply because of the steady decline in numbers with age.

Neuronal cell loss in other areas with aging in the absence of disease is less clear-cut. Cragg (1975) reported 15.6×10^6 neurons per cc in several areas of the human neocortex and approximately 40,000 synapses per neuron. There was no detectable difference in these counts with mental retardation or aging. Brody (1976), on the other hand, reports that granule cells are predominant in layers 2 and 4 in the newborn but are infrequently observed in 70- to 95-year-olds. Colon (1972) and Shefer (1973) both report decreased cell counts in the cortex with aging, and Scheibel and Scheibel (1975) demonstrated a progressive loss of the horizontal dendrites of pyramidal cells. Synaptic losses are reported in senescent rats by Feldman (1976) on apical dendrites of the cortex, and by Bondareff and Geinisman (1976) in the hippocampus.

It is well documented that there is an accumulation with age of pigments (lipofuscin pigments), which may be derived from incomplete degradation of damaged membranes, and an accumulation of abnormal neurofibrils in "patchy" localized regions of various parts of the brain, usually starting in the hippocampus. Samorajski and Rolsten (1973) report a fall in RNA and an alteration in the DNA:RNA ratio consistent with neuronal loss or glial proliferation in both aged human and aged monkey brain. It has therefore been postulated that the abnormal neurofibrils result from defective or inadequate translation mechanisms. It has also been hypothesized that the primary difficulty may be the accumulation of pigments which displace RNA and other normal neuronal components.

An approach which might yield information on losses of particular types of neurons or their nerve endings is to measure the neurotransmitter synthetic enzymes. This has already been described for tyrosine hydroxylase in the neostriatum. Most areas of brain tended to show mild decrements in other enzymes with age (E. G. McGeer and McGeer 1976b), but the most substantial losses were noted in the thalamus for GAD and the neocortex for CAT. More precise knowledge of GABA and ACh pathways in these areas will need to be obtained for meaningful interpretation of these data.

In the ongoing process of cell death during aging or from minor clinical lesions, we would have random degeneration of one cell or another so that on any one neuron there would be only a few degenerated synaptic sites. Under these conditions one can imagine that the homotypic regeneration would be dominant. Not only would the appropriate fibers be in close juxtaposition to the degenerative sites, but also there would be the advantage of the correct steric structures for appropriate synaptic formation, such as occurs in the initial process of neurogenesis. Thus, we have to reinterpret the experiments of Raisman, Tsukahara, and Schneider, for example, as being of great importance in the responses to the random neuronal deaths occurring throughout life. Because of this new growth, postsynaptic neurons can go on having their normal complement of synapses despite the death of a number of nerve cells that normally provide synaptic inputs. This recovery is biased strongly for homotypic synaptic connections, as has been suggested and as is indicated in Figure 12.22. Thus, the disability suffered by the nervous system as a result of neuronal death can be considered as being in many cases from loss of the normal convergence number, by which we mean the number of neurons of any one species converging synaptically upon a neuron. This loss (from 4 to 2 in Figure 12.22) would result in some coarsening of the grain of the control of neuronal responses, but no reduction in their effectiveness (Eccles 1976a). These investigations in many sites in the mammalian central nervous system certainly show that under appropriate conditions there is an effective new growth compensating for the death of neurons and their axons. It is, therefore, with optimism that one now looks at the often enigmatic and ambiguous clinical findings described during stages of recovery from lesions of the human brain.

The most remarkable account of this has recently been given by Brodal (1973), a distinguished neuroanatomist who suffered a vascular accident in his right cerebral cortex and made quite a remarkable recovery in a year or two, as has been fully described in a paper he has published. Apart from these major accidents, however, we are always continually suffering from random neuronal death and we have no way of replacing our neuronal population. This new story of synaptic regeneration shows, however, that the nervous system can go on functioning reasonably well with the disability of a less fine grain despite losses which may even halve the neuronal population. The postsynaptic sites can be recognized by virtue of their dense staining, and it is generally recognized that very few of these vacant sites can be seen in electron micrographs of the central nervous system at all ages. This indicates a general principle that vacant synaptic sites tend to be occupied, and in fact one can say that the central nervous system abhors a vacant synaptic site!

Chemical Specification in Neuronal Connectivity.   In the first chapter

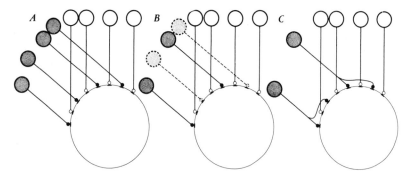

Figure 12.22 Synaptic terminals that are made on a neuron by two sets of afferent pathways. A. A "complete" set of terminals is shown. B. There is degeneration of half of the fibers of one set. C. There is occupation of the vacated synaptic sites by the remaining fibers of that set (homotypic regeneration) rather than by the fibers of the other set (heterotypic regeneration). (Eccles 1976a.)

we briefly discussed the chemical structure of the surface membrane of nerve fibers, and it was illustrated diagrammatically in Figure 1.16 (Section 1.2.5). The features of present interest are the specific proteins or other macromolecules that are shown attached to the outer surface of the membrane or partly penetrating the bimolecular leaflet of phospholipids. We can imagine that these macromolecules are responsible for the chemical labeling of the neurons. We can assume that the specificity resides in the surfaces that these macromolecules present to other neurons in the environment of that neuron or nerve fiber. This property could reside with the macromolecules as individuals, or, alternatively, with the patterned arrangement of the macromolecules as well. It will be an immense task for microneurochemistry to identify these macromolecules and to discover how they give to the surface of the neuron the chemical specifications which, as we have seen, play such a vital role in the building of the brain.

It will be our general postulate that in the differentiating neurogenesis the immature neuron receives a genetic coding that gives it the potentiality to manufacture these specific tagging macromolecules. These are most complex problems in molecular neurobiology and can only be defined in quite general terms at present (Sidman 1974). It is known, however, that genetic defects result in failures of neuronal organization such as have been described by Sidman and Rakic (1973) in the "reeler" mice, and by Brenner (1973).

A companion problem is how these macromolecules, which must be synthesized in the neuronal soma, reach their destination. They may require repetitive replacement during life and they may be involved in continual "cross talk" between neurons. To consider these aspects we must return to a discussion of axonal transport, the general features of which were described in Chapter 1. However, we shall now be concerned with the transneuronal aspect.

Axonal Transport of Materials Designed for Transneuronal Effects. Grafstein (1971) first demonstrated the possible transneuronal transport of labeled macromolecules from the injection site in the retina across the synapses in the lateral geniculate body and so to the visual cortex. This work has been dramatically confirmed by Wiesel and associates, who have shown by autoradiography that from the injected eye there is a highly specific transport to the zones of the visual cortex that receive the visual input from the injected eye. Thus, the path is selectively via the synapses of the lateral geniculate body (cf. Figure 16.2, Section 16.1.3) and not diffusely by nonspecific channels. Transneuronal transport has been demonstrated in several other sites, and Cowan (1971) and Graybiel (1975) envisage the possibility that transsynaptic transport of macromolecules may be a generalized phenomenon in the brain. Of course, we also know that proteins received by the nerve endings can be transported back to the cell body by retrograde axoplasmic flow (Figure 1.15, Section 1.2.4), creating the possibility of reciprocal effects.

A remarkable heterotopic trophic influence has been discovered by Kawaguchi et al. (1975). When, in kittens a few days old, the inferior olive on one side is killed by thermocoagulation, there is degeneration of the climbing fibers to the Purkinje cells of the opposite side. The cells deprived of climbing fiber innervation failed to grow the elaborate branching dendritic tree (cf. Figure 13.12) that provides optimum synaptic sites for the parallel fibers. Instead, the Purkinje cells had a single main dendrite running up through the molecular layer. The excitatory action of a mossy fiber input was correspondingly reduced. More recently, Bradley and Berry (1976) report quantitative investigations on Purkinje cells of rats which corroborate the findings of Kawaguchi *et al.* (1975) on kittens. The climbing fiber intput was destroyed by operation (cutting of the olivocerebellar decussation) in rats 3 days old. When examined at 30 days postpartum, the branching of the dendritic tree was found to be reduced to half, as also was the total length of the dendrites.

Thus, at the developmental level there is good evidence of a trophic influence of the climbing fiber innervation on the receptivity for the mossy fiber input to the Purkinje cell. This heterotopic influence might be regarded as a sufficient justification for the existence of the climbing fiber innervation of Purkinje cells, but this suggestion appears odd because other species of nerve cells seem to be able to grow extensive dendritic branches apparently without such a heterotopic aid. In Chapter 15 we shall see that it may be a trophic accompaniment of a plastic influence that is more significant because it probably provides the basic mechanism concerned in the synaptic modifications subserving learning.

The muscles of a limb can be classified in general as fast and slow contracting, the former being for quick movements, the latter for maintenance of posture. Several years ago it was shown that when the nerves to these muscles were severed and cross-united, the contractions of the muscles also crossed over, the fast becoming slow and the slow fast. It was postulated that this was caused by specific substances manufactured in the motoneurons that traveled down the axons and so across the neuromuscular synapse (Figure 3.1A, Section 3.2.1) to the muscle fiber. An alternative and perhaps supplementary hypothesis is that the speed of muscle contraction is modified by the frequency of discharge of impulses from the motoneurons. The hypothesis that motoneurons control the speed of muscle contractions can now be much more precisely defined because it has been shown, first by Buller *et al.* (1969) and later by Baring and Close (1971), that the cross innervation results in changes in the contractile protein, myosin, of the reinnervated muscles. The changed speed of the muscle contraction results from the change in the kinetic properties of the adenosine triphosphatase of the myosin. This finding is perhaps the best documented example of the transsynaptic production of changes in physiological and biochemical properties. It is an attractive hypothesis that these changes are produced by the transport of specific macromolecules down the motor axons and then across the neuromuscular synapses to the muscle fibers. It would be a remarkable discovery to identify such specific macromolecules.

Besides the subtle changes in the myosin and the associated changes in the speed of muscular contraction, innervation has more profound influences on muscle. These are displayed when muscle is deprived of innervation. In a day or so it changes its properties, the sensitivity to ACh spreading over the whole surface from the extremely localized site characteristic of normal muscle (Figure 3.4, Section 3.2.4). Moreover, spasmodic twitches occur in the individual muscle fibers and the muscle wastes progressively, eventually to become replaced by fibrous tissue. In part these changes are due to deprivation of the normal activation by nerve impulses, but the more severe effects are attributable to the nerve degeneration and the consequent failure of what are called the trophic influences, which we can now ascribe to the macromolecular transport along the motor axons and across to the muscle fiber via the neuromuscular synapses, as generally indicated in Figure 3.1A. We have no knowledge of these macromolecules, but the transport has been found to be also fast (cf. Figures 1.12, 1.13, and 1.14, Section 1.2.4) with an approximate speed of 360 mm/day. This measurement is derived from observations of Miledi and Slater (1970) showing that the onset of degenerative changes at the neuromuscular junction is more delayed the further the nerve section is from the muscle. The key role of macromolecular transport along the nerve fibers has now been demonstrated by the use of colchicine and vinblastin to block this transport. These substances break up the neurotubular system, but for some days do not prevent impulse propagation; nevertheless, the muscles display the excitability changes characteristic of denervated muscle.

A suggestion as to the nature of the materials required for the maintenance of synapses has been made by Geinisman *et al.* (1977). They found reduced axonal transport of glycoproteins labeled by [³H]fucose in the septal-hippocampal pathway of senescent rats. This correlated with a reduced number of synapses in the dentate gyrus of such rats (Bondareff and Geinisman 1976).

12.2.4 Summary of Neuronal Recognition and Connectivity

We can now conjecture that the whole nervous system has communication not only by impulses that signal quickly in the manner described in the preceding chapters, but also chemically by transport of specific proteins and other macromolecules. We can assume that this transport must go on in the most incredibly complex manner between the neurons of the brain, organizing all their interrelationships in a way that we still cannot even imagine, but with the life as well as the development of the neuron depending upon it. We do know that, if a cell has most of the synapses on it degenerated, much like denervated muscle fibers, it too dies. This process is called transneuronal degeneration. Apparently, a neuron does require the trophic transport from the synapses on it in order to keep alive, and this dependence goes on in turn from cell to cell. Every cell, as it were, is talking chemically to all the other cells that it is connected with and instructing them how to talk to the next ones, and so on. So this chemical manner of communication keeps the whole immense organized structure of the brain in some kind of dynamic state that we may call trophic resonance by this vast interlocking process of specific chemical communications. Here we are imagining far beyond experimental evidence, but such a vision is of the greatest value if it leads to the formulation of problems that can be experimentally attacked.

12.3 General Conclusions

In the preceding chapters we have dealt with elementary levels of brain operation, taking the brain and its constituent neurons as given. At all levels the responses of the brain have been explained as being due to transmission by nerve impulses. The input from receptor organs via synaptic action and the many sequential synaptic relays lead eventually to the discharge of motoneurons, with the ensuing muscle contractions giving movement. At the higher levels of the brain the neuronal pathways are of immense complexity, and in principle it can be conceived that these pathways provide an adequate explanation of even the most complex and subtle human performances. This claim provides the theme for discussion in Chapter 16. Yet many problems remain, even if this program were successful in identifying all brain responses as being due to the operations of what we may call its constituent neuronal machinery.

First, as treated in this chapter, there is the construction of the brain with its detailed neuronal topography and lines of communication. But neurons are not isolated entities. They develop together with their chemical sensing and recognition of interconnected neurons and there are many indications that this specific chemical communication continues throughout life. The neurons of the brain are not only linked by impulse communication but also by special chemical transmission, such as is displayed in the trophic reactions.

Second, there is plasticity in the connections and synaptic action at higher levels in the brain, as we shall see in Chapter 15. It is postulated that synaptic activity leads to their growth by its effect in causing the synthesis of RNA and so of the enzymes building proteins and other macromolecules. In electron micrographs of the brain, Sotelo and Palay (1971) have distinguished synapses in various stages of growth and dissolution, and comparable observations have been made by Barker and associates on the neuromuscular synapses. In contrast to the commonly accepted belief of a static structure, we have to think of the brain as being structurally plastic at the microlevel—some synapses being mature, others developing, others regressing.

Finally, it should be noted that the brain is composed of glial cells as well as neurons. The glia have been relatively neglected in this account of the brain which has been concentrated on the neuronal mechanisms and impulse transmission. The investigations of Kuffler (1966) and his associates (Kuffler and Nicolls 1976) have now established that glia are not concerned in impulse transmission, but function in metabolic relationship with neurons, in guiding their growth and possibly in limiting ion accumulation in the extracellular spaces. We can think of them as the "housemaids" of the brain, being particularly concerned in the transport of materials from the blood vessels to the neurons and vice versa. They fulfill an important role not only in the nutrition of the brain but also in its protection against poisons by contributing to the so-called blood–brain barrier (Figure 1.18, section 1.3).

13

Control of Movement by the Brain

13.1 Introduction

In attempting an analysis of movement and its control by the brain, it is immediately evident that there are many hierarchical levels. This was appreciated by Sherrington (1906) in his great book *The Integrative Action of the Nervous System*, where in Chapter 9, "The Physiological Position and Dominance of the Brain," he recognized the simplest reflexes with a superimposition thereon of more and more complex controls at spinal, supraspinal, cerebellar, and cerebral levels. More recently, Granit (1970) has developed the same theme in his book *The Basis of Motor Control*, which in many ways can be considered an updating of Sherrington on the basis of the great advances of knowledge since 1904 when Sherrington delivered his classic lectures. Granit, however, was more concerned with lower levels of the hierarchy, particularly with the spinal and supraspinal controls of automatic movements, and the influence of inputs from muscle receptors on these controls. Here these levels will be treated as a base on which are superimposed the controls from higher levels—cerebellar and cerebral.

In an attempt to give an account of how we can move, we must ask ourselves: How can we control our musculature to give us actions in accordance with the situations that we find ourselves in? How can we, for example, move our arm so that with our eyes shut we can smoothly put a finger on the tip of our nose? But we can think of much more complicated movements in the immense repertoire of skill that we have in games, in technology, in playing musical instruments, and, most importantly and very complexly, in speech and song and gesture, so that our whole personality can stand revealed. And it stands revealed simply because of our movements resulting from our muscular contractions, as, for example, in all facial gesture and eye movement. If one is fixed like a corpse with a mask-like face, no personality is revealed.

Our most complex muscle movements are carried out subconsciously and with consummate skill. The more subconscious we are in a golf stroke, the better it is, and the same with tennis, skiing, skating, or any other skill. In all these performances, we do not have any appreciation of the complexity of muscle contractions and joint movements. All that we are voluntarily conscious of is the general directive given by what we may call our voluntary command system. All the finesse and skill seem naturally and automatically to flow from that. Throughout life, particularly in the earlier years, we are engaged in an incessant teaching program for motor performance. As a consequence, the brain can carry out all of these remarkable tasks that we command it to do in the whole repertoire of our skill and movements in games, in techniques, in musical performance, in speech, dance, song, and so on. Dance and song are the most fantastically demanding of all. How is it that Margo Fontaine, Maria Callas, and Joan Sutherland can do what they can do? This requires an exquisite finesse which can only cause us wonder at how it is achieved.

What we will do in this chapter is describe the main influences on motor control, recognizing that much of it must be speculative at this stage of our knowledge. It was once thought that a pyramidal system, including the motor cortex and the pyramidal tract fibers descending from it to the spinal cord, controlled voluntary or higher level movements. An extrapyramidal system, including the cerebellum and basal ganglia, was thought to control lower level, or automatic movements. More recent anatomical and physiological studies, however, make it clear that the pyramidal and extrapyramidal systems are not separable.

The concept of an extrapyramidal system was introduced by the great neurologist S. A. Kinnier Wilson in 1912. Wilson never defined this system and to this date it remains a vague entity. Wilson used the term to explain the fact that in hepatolenticular degeneration (later called "Wilson's disease"), a motor deficit of major proportions occurred in the absence of damage to the corticospinal tract. Today, the term *extrapyramidal* refers to those aspects of posture and motor activity which seem to be controlled by five interconnected subcortical nuclei which can be included in another loosely defined term, *the basal ganglia*. They are: the caudate, putamen, globus pallidus, subthalamic nucleus, and substantia nigra. They are connected to some regions of the thalamus. Cerebellar function is thought of separately. Thus, at the supraspinal level we really have three systems playing upon descending pathways to anterior horn cells in the spinal cord: the cerebral cortex, the basal ganglia, and the cerebellum. The principal descending pathway is the trunk line extending from Betz cells in the motor strip down the pyramidal tract to anterior horn cells. In addition, a series of auxiliary descending reticulospinal tracts exists which is interconnected with the three great nuclear systems.

Broad hints as to the function of individual parts of this incredibly complex array of motor influences comes from diseases which affect various parts

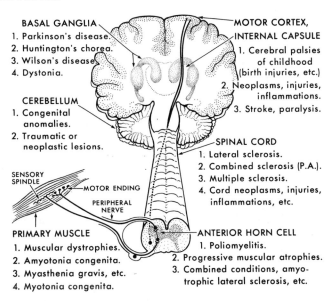

Figure 13.1 Some diseases producing motor symptoms, showing the level at which they attack the nervous system.

of the system. These are summarized in Figure 13.1 Muscular diseases (i.e., muscular dystrophy, amyotonia congenita, myasthenia gravis, etc.) or diseases of anterior horn cells (i.e., poliomyelitis, progressive muscular atrophy, amyotrophic lateral sclerosis, etc.) lead to flaccid paralysis of the limb. Spinal cord lesions (i.e., combined sclerosis of the cord from pernicious anemia, multiple sclerosis, cord neoplasms, etc.) will lead to spastic paralysis. Diseases of the motor cortex or any portion of the pyramidal tract (i.e., cerebral palsies in childhood, neoplasms, injuries, multiple sclerosis, inflammations, strokes, etc.) will also produce spastic paralysis. Lesions of the cerebellum, or diseases of the cerebellum (i.e., congenital anomalies, traumatic or neoplastic lesions), will produce tremor and uncertain movement but not paralysis. There is an unsteady gait, and a coarse tremor precipitated by movement. There is an inability to estimate the range of voluntary movement referred to as past-pointing. Lesions of the basal ganglia (i.e., Parkinson's disease, Huntington's chorea, Wilson's disease, dystonia musculorum deformans, etc.) produce tremors at rest rather than in motion. There may be difficulty initiating movement (Parkinson's disease) or an inability to control it (Huntington's chorea).

To summarize, muscle cells and peripheral nerves must be intact for there to be movement at all. There must be a pyramidal tract if voluntary movement is to occur. The subcortical motor nuclei are essential for initiating or controlling voluntary movement while the cerebellum is essential for its smooth execution. We shall now discuss each of these influences in turn, remembering that they all work in concert in normal circumstances.

13.2 Motor Control from the Spinal Cord and Brainstem

As already described (cf. Figure 3.1, Section 3.2.1) all movements are brought about by contractions induced in muscles by impulses that are discharged by motoneurons. The very simple reciprocal arrangement represented in Figure 4.1A (Section 4.1) can have functional meaning. When you are standing with slightly bent knees, your weight is stretching the knee extensor muscle (E) and the AS stretch receptors are firing into the spinal cord, exciting the knee extensor motoneurons to fire impulses so that the extensor muscle contracts and holds your weight. If this muscle contraction is inadequate the knee gives a little, stretching the extensor muscle more with more firing from its AS receptors (Figure 2.3A, Section 2.1), giving an increased reflex discharge to the muscle, which in this way is nicely adjusted to give a steady posture. At the same time the reciprocal inhibitory pathway prevents the antagonist motoneurons (F) from firing to give contraction of the antagonist flexor muscles (F). Such a contraction would oppose the extensors that are engaged in the essential task of supporting weight. This description of the mode of action of the pathways in Figure 4.1A illustrates a simple reflex performance with functional meaning. Superimposed on these simple pathways of Figure 4.1A there are the γ-motoneurons with their biasing action on the annulospiral endings as already described (Figure 2.3A).

Figure 13.2 is an elemental representation of various centers in the brainstem which are concerned with the maintenance of posture and with automatic movements such as breathing, stepping, and swimming. These actions occur when both cerebrum and cerebellum have been removed. For example, the respiratory center of the brainstem functions normally in controlling all the complex movements of respiration independently of the cerebrum and the cerebellum. It operates on the basis of inputs from various peripheral and central receptors that sense the composition of the blood and the movements of the respiratory muscles. In contradistinction, only some caricature of standing and stepping is possible in the decerebrate animal when the more rostral part of the brainstem is removed along with the cerebrum. Nevertheless, the study of the decerebrate animal has been most valuable in revealing the mode of operation of the neural machinery concerned in simpler supraspinal control systems.

13.3 Motor Control from the Cerebral Cortex

13.3.1 The Motor Cortex

After this introduction it is appropriate that we begin with motor control from the motor cortex because it is the control that has been most thoroughly investigated. Figure 13.3 shows the position of the left motor cortex as a band across the surface of the cerebral hemisphere. It lies just anterior to the central fissure (the fissure of Rolando), and many of its constituent nerve cells

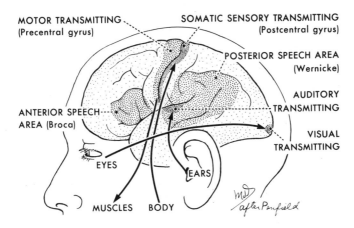

Figure 13.2 The motor-transmitting areas of the cerebral cortex. The approximate location of the motor-transmitting areas is shown in the precentral gyrus, while the symmetric sensory receiving areas are in the similar location in the postcentral gyrus.

are pyramidal cells (cf. Figure 1.2B, Section 1.2) whose axons are in the nerve fibers running down the pyramidal tract (Wiesendanger 1969). The motor cortex is a tremendously important structure but it is not the prime initiator of a movement, such as a voluntary bending of your finger. It is only the final relay station of what has been going on in widely dispersed areas in your cerebral cortex, subcortical nuclei, and cerebellum, as will be partially illustrated in Figure 13.4. The pyramidal cells of the motor cortex with their axons passing down the pyramidal tract are important because they provide a direct channel out from the brain to the motoneurons (Figure 13.3) that in turn cause the muscle contractions as described in the earlier chapters and as illustrated in Figure 3.1C.

When brief stimulating currents are passed through electrodes placed on the surface of the motor cortex, there are contractions of localized groups of muscles. In this way it was discovered that all the various parts of the body can be shown on a striplike map that is represented by the homunculus in Figure 13.3. This was done first with monkeys and anthropoid apes, but the map has now been completely established for humans, particularly by Penfield and his associates (Penfield and Jasper 1954), because during brain surgery it is often important to determine part of the motor map, using for this purpose the conventional stimulating technique. In the inset of Figure 13.3, along the strip of motor cortex are marked the general areas for toes, foot, leg, thigh, body, shoulder, arm, hand, fingers and thumb, neck, head, face, etc., starting in the midline and progressing downward over the outer surface. You will note that there is a large representation for hand, fingers, and thumb, and an even larger area for face and tongue. The motor cortex is not uniformly spread in proportion to muscle size—far from it. The muscles controlling the thumb have a large representation, but then we use them in so many skilled actions; and even more important are the areas for movement of tongue, lip, and

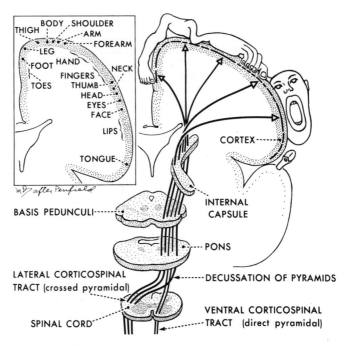

Figure 13.3 The homunculus of the motor strip, localizing the functions of large pyramidal cells. The descending tracts through the internal capsule and brainstem into the spinal cord are also shown. The tracts mostly decussate to descend in the dorsolateral column of the spinal cord on the opposte side. (Adapted from Netter 1953.)

larynx that are used in all the subtleties of expression in talking and singing. It is skill and finesse that is reflected in the representation.

Figure 13.3 shows diagrammatically the course of the pyramidal tract from the motor cortex to the spinal cord. After descending through the brainstem and giving off many branches (Phillips 1973; Porter 1973), the pyramidal tracts cross or decussate in the medulla and so course down the spinal cord to terminate at various levels, with primates making strong monosynaptic connections on motoneurons. This very direct connection of the motor cortex with motoneurons is of the greatest importance in ensuring that the cerebral cortex, via the motor cortex, can very effectively and quickly bring about the desired movement. Nevertheless, two fundamental problems remain: How can one's willing of a muscle movement set in train neural events that lead to the discharge of pyramidal cells? How do the cerebellum and basal ganglia contribute to the finesse and skill of movement? The first question will be considered scientifically in this chapter and philosophically in Chapter 16, and the second in the latter part of this chapter.

13.3.2 Cerebral Cortex Controlling Motor Cortex

The first question has been investigated physiologically by searching for signs of neuronal activity in the cerebrum before the discharge down the pyramidal tract. In an initial investigation by Grey Walter (1967), the subject was trained to perform a movement after a double stimulus sequence: a conditioning, then a later indicative stimulus. An expectancy wave was observed as a negativity over the cerebral cortex before the indicative stimulus. Essentially, however, this wave is produced by the conditioned expectancy of the indicative stimulus and not by the voluntary command for movement. The problem is to have a movement executed by the subject entirely on his or her own volition, and yet to have accurate timing in order to average the very small potentials recorded from the surface of the skull.

This problem has been solved by Kornhuber and his associates (Deecke *et al.* 1969; Kornhuber 1974), who use the onset of action potentials of the muscle involved in the movement to trigger a reverse computation of the potentials up to 2 s before the onset of the movement. The movement illustrated in Figure 13.4A was a rapid flexion of the right index finger, but many other movements of limbs have been investigated with similar results. Even the movements involved in vocalization have been investigated (Figure 13.4B). The subject initiates these movements "at will" at irregular intervals of many seconds, extreme care being taken to exclude all triggering stimuli. In this way it was possible to average 250 records of the potentials evoked at each of several sites over the surface of the skull, as shown in Figure 13.4A for the three upper traces. The slowly rising negative potential, called the readiness potential, was observed as a negative wave with unipolar recording over a

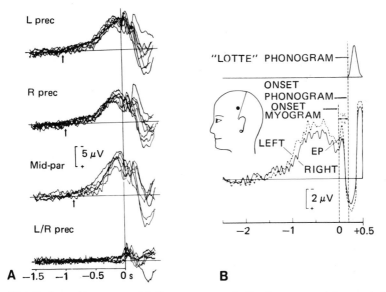

Figure 13.4 Scalp readiness potentials for movement. A. Readiness potentials recorded at indicated sites from the scalp in response to voluntary movements of finger. Zero time is at the onset of the movement, the preceding potentials being derived by backward computation, with averaging of 250 responses. L prec., left precentral; R prec., right precentral; Mid-par, midparietal; L/R prec, recording left precentral against right precentral. (Kornhuber 1974). B. As in A, but for articulation of word "Lotte," zero time being at onset of myogram's readiness potential at left recording site (dot), which was larger than at right. (Grozinger *et al.* 1975.)

wide area of the cerebral surface (recorded by scalp leads against an indifferent lead), but there were small positive potentials of similar time course over the most anterior and basal regions of the cerebrum. Usually the readiness potential began (arrows) about 0.8 s before the onset of the muscle action potentials, and led on to sharper potentials, positive then negative, beginning at about 0.09 s before the movement. In Figure 13.4B the readiness potential began almost 1.5 s before the articulatory movements. In the lowest trace of A there was bipolar leading from symmetrical zones over the motor cortex, that on the left being over the area concerned in the movement of the right index finger (cf. Figure 13.3). There was no detectable asymmetry until a sharp negativity developed at 0.05 s before the onset of the muscle action potentials. We can assume that the readiness potential was generated by complex patterns of neuronal discharges that originally were symmetrically distributed in the frontal and parietal lobes. Eventually, at only 0.05 s before the muscle response, there was concentration of the neuronal activity onto pyramidal cells of the motor cortex. The time of 0.05 s is just adequate for transmission from the pyramidal cell discharge to motoneurons and so to muscle action potentials (cf. Figure 13.3).

These experiments at least provide a partial answer to the question: What is happening in the brain at a time when a willed action is in process of being carried out? It can be presumed that during the readiness potential there is a developing specificity of the patterned impulse discharges in neurons so that eventually the pyramidal cells are activated in the correct motor cortical areas (Figure 13.3) to bring about the required movement. The readiness potential can be regarded as the neuronal consequence of the voluntary command. The surprising features of the readiness potential are its wide extent and its gradual buildup. Apparently, at the stage of willing a movement, the influence of the voluntary command is distributed widely onto the patterns of neuronal operation.

13.3.3 Discharge of Motor Pyramidal Cells

We will now study in more detail how the firing of the pyramidal cells in the motor cortex is related to a movement initiated from the cortex. For this purpose it is necessary to have a microelectrode implanted in the motor cortex so that impulse discharges from individual pyramidal cells are recorded. This requirement precludes human experimentation, and we will make special reference to the experiments performed by Evarts (1967) on monkeys. In an initial operation the monkey in Figure 13.5A has an electrode implanted in his motor cortex in the right place for recording pyramidal cells concerned in an action he has been trained to do. During an experimental run he is seated comfortably in a cage and he has to move the control bar from one stop to the other, shown in detail in Figure 13.5A, backward and forward in a time that must be between 0.4 and 0.7 s or else he is not rewarded by grape juice. As a variant, the amount of load on the movement can be changed.

In Figure 13.5B you can see the traces of the movement and the firing of

A   B

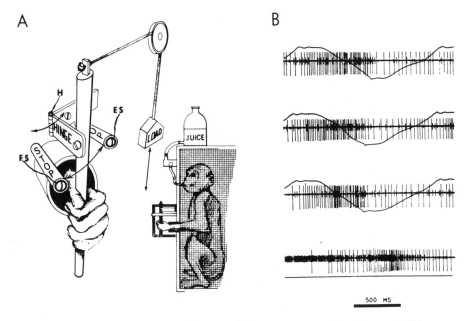

Figure 13.5 Pyramidal cell discharges during movement. (From Evarts 1967.)

the pyramidal cell which clearly is related to the downward movement that he is doing. Actually, there are two pyramidal cells: one is very close to the electrode and hence gives large action potentials, and the other is further away and gives smaller responses. We can assume that the size is only a matter of proximity and not of actual cell size. The activities of the two cells are nicely correlated with the downward movement. There is no doubt that these particular pyramidal cells are concerned with a particular movement, the flexion and not the extension. But if, as in the lowest trace, the monkey is carrying out some quite different movement, like moving his shoulder, the two units are no longer correlated in their discharges but go fast and slow in quite unrelated ways. The two pyramidal cells are only correlated when they are carrying out the specific actions corresponding to their locations on the motor map. Doubtless there are many more pyramidal cells in that location also firing in effective relation to the movement.

This is a very important result, showing just what one would expect in the functional performance of the motor cortex when it is carrying out some specific action. There is a group or colony of pyramidal cells in close proximity whose job is to keep on firing and stopping and firing and stopping during a rhythmically repeated movement. And there would be other colonies specifically related to other movements, and so on for the whole extent of the motor cortex of Figure 13.3. It should be noticed in Figure 13.5B that the two cells are not silent during the reverse phase of the movement, but only slowed. They are, as it were, modulated in their frequency. As mentioned earlier, there is a general tendency for all cells to be firing all the time. Their responses are graded in frequency, coding intensity of action as frequency. Figure 13.5 is a beautiful example of this coding.

13.3.4 Arrangement of Pyramidal Cells in Colonies

The arrangement of pyramidal cells in colonies having similar actions has long been postulated by Phillips (1966) because it has been recognized that the detailed cortical map defined by stimulation (Figure 13.3) could only thus be explained. Nevertheless, the geometry of these colonies was unknown (cf. Brooks 1969). It has now been investigated by Asanuma and Rosén (1972) and by Rosén and Asanuma (1972), using microstimulation of the monkey's motor cortex as illustrated in Figure 13.6. The microelectrode was inserted in seven numbered tracks, and various sites noted at which one or another thumb movement was evoked by weak electrical stimulation. The effective sites are noted and are indicated by the symbols. It will be seen that, as revealed by this stimulation, the actual cells doing one action or another tend to lie in columns whose boundaries are indicated by the curved lines, ortho-

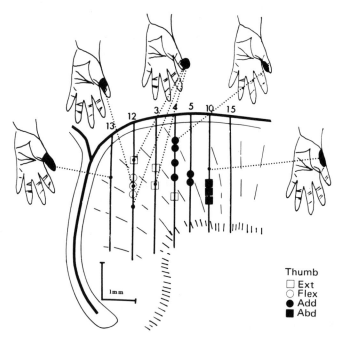

Figure 13.6 Correlation of cortical cell locations and movement: reconstruction of electrode tracks and cell locations in a section through the right motor cortex of a monkey. Electrode penetrations are indicated by numbers and passed through efferent zones projecting to various thumb muscles. The peripheral motor effects on the thumb produced by intracortical micro-stimulations of 5-mA strength at various locations are indicated by symbols, as identified in lower right corner. Positions of cells encountered along the track are indicated by dots, and dotted lines join these points to the figurines of the monkey's hand on which are indicated the receptive fields as determined by adequate stimulation. (Asanuma and Rosén 1972; Rosén and Asanuma 1972.)

gonal to the surface of the cortex. In Figure 13.6 this arrangement is particularly well shown for the movements of thumb extension (open squares) and adduction (solid circles) that are produced along more than one track. There would be many hundreds of pyramical tract cells in one such column, all presumably having the same action. It is all part of the motor map of Figure 13.3.

The other interesting point about Figure 13.6 is that when stimulation is applied to the skin of the thumb at the various indicated sites in the figurines, you find out that usually this afferent input excites pyramidal cells just in the area that is the departure point for the excitation that moves the thumb to that site of skin stimulation. This means that the movement is in the sense of making closer contact with the stimulated site for the purpose of exploration. Possibly it also accounts for the grasp reflex. For example, an object placed in a baby's hand is grasped so strongly that the baby has difficulty in letting it go. These examples illustrate the simplest way in which the cerebral motor cortex could be activated, namely, by afferent pathways coming to the cerebral cortex via three synaptic relays, and projecting to those pyramidal cells in the motor cortex which are the departure points for the contraction. It is a kind of self-stimulating circuit arrangement with a positive feedback control.

We can now ask the question: Why are the motor cells in columns all doing much the same thing? The reason for this is revealed when the synaptic connectives are studied in detail (Szentágothai 1969). Figure 13.7 shows two pyramidal cells in a column with their dendritic spines. If this section belonged to the motor area of the cerebral cortex, the axons would be going down to become pyramidal tract fibers. There would be hundreds of such pyramidal cells in any one of the columns of Figure 13.6. What is the reason for this apparent redundancy? Let us look at the pathway drawn in Figure 13.7 for a specific afferent (spec. aff.) which is an afferent fiber from the thalamus that could be conveying information from the thumb in Figure 13.6. It branches and synaptically excites a special type of interneuron (S_1) which gives off an axon that powerfully excites the pyramidal cell dendrites by a multiple synaptic arrangement that Szentágothai has named the cartridge type of synapse. Thus, in this column there is a kind of amplifying effect. The inputs by specific afferents are amplified by these powerful cells that make the cartridge type of synapses on the pyramidal cells. So the column gives an economy of arrangement, because the input is used to excite many pyramidal cells; hence, there is a large output—many pyramidal cells fire together. That is the way to achieve effective action. One cell firing alone is ineffective. Experimental investigation of motor cortex stimulation suggests that it may be necessary to have as many as 100 pyramidal cells firing impulses in order to evoke a movement. There has to be convergence of many pyramidal tract fibers onto the motoneurons in order to excite them to fire impulses out to the muscle.

The other operational significance of the columnar arrangement can be appreciated by reference to the other interneurons in the column (Figure 13.7). Some are excitatory (S_1, S_5) so that all manner of excitatory circuits are

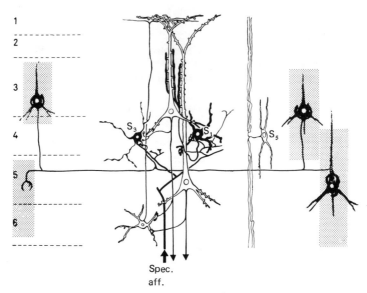

Figure 13.7 Neocortical neurons: semidiagrammatic drawing of some cell types of the cerebral cortex with interconnections (as discussed in the text). Two pyramidal cells are seen centrally in lamina 3 and 5. The specific afferent fiber is seen to excite a stellate interneuron S_1 (crosshatched) whose axon establishes cartridge-type synapses on the apical dendrites. The specific afferent fiber also excites a basket-type stellate interneuron S_3 that gives inhibition to pyramidal cells in adjacent columns, as indicated by shading. Another interneuron is shown in lamina 6 with ascending axon, and S_5 is an interneuron also probably concerned in vertical spread of excitation through the whole depth of the cortex. (Modified from Szentágothai 1969.)

activated with self-reexciting loops, so building up by positive feedback a strong excitation in the column. At the same time the inhibitory neurons (S_3) of the column have an inhibitory action on the cells of adjacent columns, three of which are indicated, along with the shaded areas signifying inhibition. And so the column is self-excitatory with built-in amplifying circuits, but it also operates for its dominance by inhibiting the surrounding columns, an action that may be on columns having the opposite action on the thumb: extension as opposed to flexion, for example.

It is a general principle of operation of the nervous system that when groups of cells are activated, they try to sharpen their effectiveness by inhibiting the other groups in the surround. But, of course, these other groups are doing the same. There is reciprocal inhibition with a continual fight for dominance. The result is that the movement occurring at a particular time is an integrated response and is one that is brought about by the most strongly excited pathways.

13.3.5 Alpha and Gamma Motoneurons and the Gamma Loop

We will now refer again to an earlier illustration, Figure 3.1 (Section 3.2.1), which shows a motoneuron. In primates the pyramidal tract fibers directly excite such a motoneuron to fire impulses and so to make the muscle contract. In Figure 3.1B there is a transverse section of a muscle nerve containing only motor nerve fibers, all afferent fibers having been degenerated by dorsal root ganglia excision some 3 weeks earlier. It can readily be seen that besides the large fibers that are concerned in the muscle contraction, there are also small fibers. In fact, there are two quite distinct populations. The small motor fibers provide the motor innervation of the muscle fibers in the muscle spindles which can be seen in the diagram of Figure 13.8 to be in parallel with the muscle fibers responsible for the contraction, the so-called extrafusal fibers. The small motor nerve fibers come from small motoneurons lying interspersed with the large motoneurons for the same muscle. The large motoneurons are called alpha (α) and the small ones gamma (γ), and their axons (γ-fibers) exclusively innervate the muscle spindles which will be lying in parallel to the extrafusal fibers that are exclusively α-innervated. This α-innervation of extrafusal fibers was the theme of much of Chapter 3. The enormous literature on muscle spindles is excellently reviewed by Matthews (1972).

In the simplified diagram of Figure 13.8 it can be seen that an annulospiral ending around a muscle fiber of the spindle discharges along a group Ia afferent fiber, as already described in Chapter 4 (Figure 4.1A). The several intrafusal fibers bundled together in a spindle form two distinct species, and there are also secondary endings with smaller afferent fibers (group 2) than the large Ia afferent fibers of the annulospiral ending. Nevertheless, for our present purpose the simplified diagram of Figure 13.8 is adequate. It is the Ia

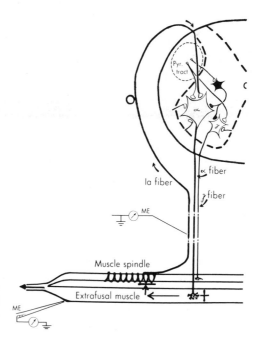

Figure 13.8 Alpha and gamma innervation: nerve pathways to and from the spinal cord showing the essential features of α- and γ-motoneuron action and interaction.

fiber that gives the monosynaptic innervation of motoneurons that was the theme of much of Chapter 4.

If you pull on the tendon of the muscle, as shown by the arrow to the left, you excite the spindle to discharge impulses up the Ia fiber (cf. Figure 2.3A, Section 2.1). If the intrafusal fibers are excited to contract by γ-motor impulses, there is powerful excitation of the annulospiral endings (as illustrated in c to b to a in Figure 2.3A), and so there is an intensification of the monosynaptic activation of the α-motoneurons. If, on the other hand, a motoneuron discharge causes the extrafusal muscle fibers to contract, tension is taken off the muscle spindle that is in parallel with it and the annulospiral ending will discharge less or not at all. But, if you fire the γ-motoneurons at the same time as the α, the muscle spindle will contract and so will not be slackened. This arrangement gives a nice servomechanism performance. The more the α-motoneurons are firing in response in part to the Ia input, the stronger the extrafusal contraction and the less the Ia activation of the α-motoneurons by the so-called γ-loop. But the action of that loop can be biased over a wide range of levels by the discharge of γ-motoneurons. There is thus an adjustable servoloop control of muscle contraction in accord with the biasing by γ-motoneuron discharge (cf. Granit 1970).

13.3.6 Pyramidal Tract Innervation of Alpha and Gamma Motoneurons

In Figure 13.8 it is shown that the fast pyramidal tract fibers monosynaptically excite both α- and γ-motoneurons. There also would be inhibitory action via interneurons. When performing a voluntary movement, both α- and γ-motoneurons are caused to discharge by the pyramidal tract. The next point of the story is a very challenging theory by Merton (cf. Granit 1970), who proposed that the usual sequence was that γ-motoneurons were first caused to discharge and then the resulting spindle contraction activated a powerful Ia discharge that reflexively caused the α-motoneurons to discharge with the resultant muscular contraction. That complex sequence involving loop operation has some advantages in that the servomechanism is working before the contraction starts. We will now describe a rigorous testing of this theory and its refutation.

It has now been shown in beautiful experiments by Hagbarth and Vallbo that in voluntary movement the pyramidal tract impulses excite both α- and γ-motoneurons (coactivation), the whole α-γ complex being put into action in an approximately synchronous manner. The α-motoneurons are excited to discharge impulses and so bring about the muscle contraction. At about the same time the γ-motoneurons discharge, thus exciting the muscle spindles and setting the γ-loop in operation. Because of the time involved in traversing the γ-loop, the α-motoneuron discharge so generated by the Ia impulses would not occur until after the muscle had started to contract. It thus occurs at just the right time for the onset of the servomechanism control.

How did they make this discovery? They carried out a very elegant study first on themselves and then on volunteers. They moved a finger or wrist voluntarily, and recorded both from the contracting muscle and from single nerve fibers in the nerve to that muscle (cf. the upper ME in Figure 13.8), using for this purpose a fine wire electrode, insulated except at its tip, which was about 10 μm in diameter. They probed this wire into their muscle nerves by an accurate insertion technique. Then they fiddled around with the wire tip in the nerve. With luck and some courage, they often succeeded in having this wire electrode record from a single Ia fiber, which could be identified by quite reliable tests.

Figure 13.9A shows such an experiment (Vallbo 1971). In the lowest trace of A are records of two voluntary movements of the wrist, which are very quick, as may be seen by reference to the half-second scale. Above are shown the action potentials from the surface of the contracting muscles (EMG) and the impulses in a single Ia fiber of the nerve (Ia) to that muscle. It can be seen that the impulses in the fiber begin a little later than the muscle action poten-

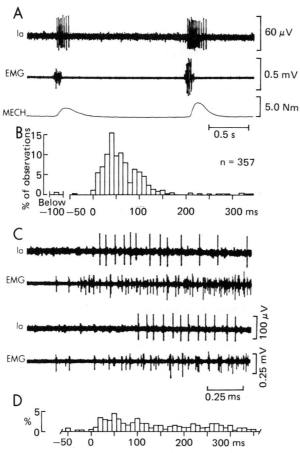

Figure 13.9 Voluntary electromyograms, mechanograms, and Ia discharges. A. Two wrist flexions. B. The histogram shows that the onset of the electromyogram precedes the Ia discharge, with zero time signaling simultaneity. C and D. Slower wrist movements (as described in the text). (From Vallbo 1971.)

tials for both of the contractions. Vallbo has carried out this experiment several hundred times with 16 Ia fibers and assembled the results in graphical form (Figure 13.9B). This display makes it quite clear that the Ia fibers are almost always activated later than the muscle contraction. In fact, after making allowance for the three factors below, the time differential corresponds closely to what would be expected for a coactivation of the α- and γ-motoneurons by the pyramidal-tract discharge, and even by the same fast pyramidal-tract fibers. First, since the γ-motor fibers conduct impulses more slowly than the α-, the muscle spindle contraction would begin later than the extrafusal contraction. There is the additional time for, second, the transduction from muscle-spindle contraction to Ia impulse discharge, and, third, the conduction time of Ia impulses to the recording point.

Figure 13.9C illustrates the same type of recording when there is a slow, prolonged movement of the wrist. In both series the Ia unit does not fire until well after the muscle action potentials have started. It is important to notice that the unit continues to fire throughout the whole duration of the contraction. In Figure 13.9D there is assemblage of the total number of observations (many hundreds) on the 27 Ia fibers so investigated. In these slow, gentle movements it is also certain that the α-motoneurons almost always initiate the muscle contraction and that the γ-loop servomechanism comes in at a later time, which is explicable by coactivation of α- and γ-motoneurons when the same allowance is made as for Figure 13.9A and 13.9B.

These experiments have shown quite clearly that voluntary movements are not initiated by a prior activation of the γ-motoneurons with follow-up by γ-loop activation of the α-motoneurons. On the contrary, they show that there is a close follow-up of the γ-loop control after the initial α- contraction. In considering the meaning of this physiological arrangement, it is important to recognize that in most voluntary movements the load opposing the movement can only be approximately anticipated, hence the necessity for adjustment up or down with the utmost expedition. The servomechanism operating via the γ-loop is thus of importance in giving an automatic adjustment at the spinal level, though controls at a higher level are of great importance, as will be discussed below.

13.3.7 Projection of Ia Fibers to the Cerebral Cortex

Recent experimental evidence has been important in revealing that despite the inability of the group Ia afferent fibers to provide a specific sensation of muscle length or tension, they nevertheless provide essential information to the cerebral cortex on the progress of movements that it initiates. It has been customary to think that the information about the movements of the limb with respect to position and load were signaled by such receptors as those in the joints and fascia and perhaps even skin, and that the muscle receptors such as the annulospiral endings with their Ia fibers did not contribute any information to the cerebral cortex that could be used for determining limb position and movement. It has been recognized for several years by Oscarsson, Landgren, and associates that group Ia fibers project to the cerebral cortex via several synaptic relays, but only recently has it been shown by Matthews (1972) that they give important information to the cerebral cortex with respect to the movement and position of the limb. In fact, their principal function may be to signal to the cerebral cortex the progress of the movement that it has just previously programmed. This is not to deny the importance of the other sensory inputs from the limb to the cortex, but merely to recognize this additional information input to the cortex.

After a penetrating analysis of these problems Granit (1972) concludes that "coactivated spindles, which are the only end organs reflecting demand and execution, play a most essential role in our judgments about muscular exertion, difficult though it be to formulate the proprioceptive experience in the way we describe things seen and heard." Furthermore, Granit has shown how various illusions of muscular movements—the kinesthetic illusions (for example, those illusions occurring in relation to postural aftercontractions or the muscle contractions produced by vibratory stimulation of its tendon)—can be explained. The suggested explanation is that the illusion results from a perceived mismatch between the willed movement and the feedback to the cortex of group Ia afferent information. Only the unexpected is perceived with the vividness to give an illusion. On this basis Granit has been able to explain many kinesthetic illusions that are described in the classical literature.

It is important to realize that much information to the cerebral cortex from receptor organs does not immediately give a special sensation, as with hearing or vision, but rather something so much less vivid that it is not ordinarily recognized. For example, the vestibular receptors in the inner ear give us a feeling of the rightness of the situation. The world is as it should be. You move your head around and the world remains fixed. If this vestibular sense goes wrong, you experience vertigo in which the world is moving when your head is fixed. This demonstrates that the vestibular receptors do in fact signal to consciousness. It is the same with the Ia receptors of muscle. Quite a lot of the Ia input to the cerebrum is not appreciated normally, but when it is disordered, when there is some mismatch as in the illusions, then you know that it is getting through to consciousness.

13.4 Motor Control by the Cerebellum [1]

13.4.1 Introduction

Figure 13.10A shows the cerebellum lying posteriorly and below the cerebrum in a brain viewed from the left side, and in Figure 13.10B the cerebellum is seen from the top after the cerebrum has been removed. Three main components are seen in medial lateral sequence, the vermis (V), the pars intermedia (PI), and the hemispheres (H). It will be noticed that the cerebel-

[1]General references: Eccles et al. 1967; Eccles 1973.

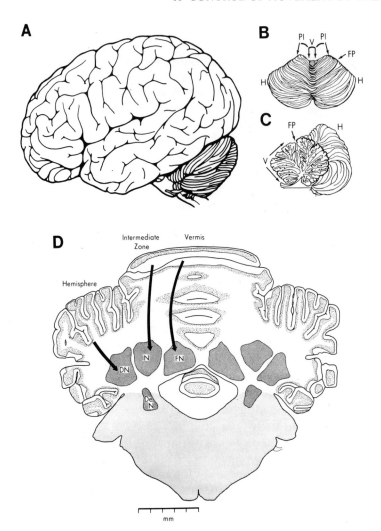

Figure 13.10 Parts of the human cerebrum and cerebellum. A. Cerebrum and cerebellum in position. B and C. The cerebellum is seen on the same scale from the dorsal aspect and after a sagittal section in the midline. D. Drawing of transverse section through the cerebellum and the cerebellar nuclei together with the lines of projection from the Vermis, Intermediate Zone, and Hemisphere to the respective nuclei. FN, fastigial nucleus; IN, interpositus nucleus; DN, dentate nucleus. Deiters nucleus (DeN) is also shown. (Drawing kindly provided by P. Scheid.)

lum is built up from a multitude of transverse folia. These are shown cut across in the sagittal section of Figure 13.10C, which displays how the complex folded arrangement maximizes the area of cerebellar cortex that can be packed in the limited area below the cerebral hemispheres. Note also the fissura prima (FP), the deep fissure that separates the anterior from the posterior lobe.

In Figure 1.1 (Section 1.1) it can be seen that as the brain grew the cerebellum grew commensurately through all the vertebrate evolution. We have to think of it as a part of the brain designed to function as a computer in handling all the complex inputs from receptors or from other parts of the brain. It is a computer which was built, originally, in relationship to the swimming of fish, using data from receptors in their lateral line organs and their vestibular mechanisms. But in the evolutionary process it turned out that it could be used for a wide variety of data computations, particularly in all the complex mechanisms involved in skilled movements, for example, in bird flight. A simple clinical test is the familiar placing of the finger on the tip of the nose. It has long been known that the cerebellum is concerned very importantly indeed in the control of all complex and subtle movements, as in the playing of musical instruments, for example. This relationship of the cerebellum to motor control was already appreciated in the last century as a result of surgical lesions in animals and clinical lesions in man. Clumsiness and all kinds of badly disordered movements resulted from such lesions. With the great development of the cerebellum in the evolution of humans (cf. Figure 1.1) came all our motor skills. We are accustomed to think of these skills as being particularly exemplified by tool manufacture and usage that eventually developed into our technology and civilization. But an even more important motor skill was in being able to speak. It was this development of the cerebellum with the cerebrum, together as a linked evolutionary process, that gave the human species its immense superiority in survival.

13.4.2 Evidence from Cerebellar Lesions

The best studies ever made on human cerebellar lesions were done in the early 1920s by Gordon Holmes on patients from the First World War who had had the cerebellum on one side destroyed by gunshot wounds with the other side normal and thus available as a control (Holmes 1939). In Figure 13.11 is a beautiful example of a simple technique that he developed. He placed a little electric bulb on the finger of his subject and caused it to flash at 25 times a second. There were two columns of three red lights and the subject was instructed to point with his outstretched arm and as quickly as possible to the succession of red lights from one and the other column. Meanwhile the movement, as revealed by the flashing light, was recorded by a fixed camera. In Figure 13.11A you can see that on the normal side there was a smooth and accurate movement from side to side to each red light in turn. On the contrary, on the side with the cerebellar lesion, Figure 13.11B, the subject gave an irregular and clumsy performance, as can be seen in tracings as he approached one after another of the target points, particularly on the two lowest where he failed badly with irregular tremor.

So, quite clearly, each side of the cerebellum is concerned in the smooth and reliable control of movement of the arm on that side. Another disability suffered by these cerebellar patients was that they could not smoothly carry out any movement involving several joints of the arm. They had to do it joint by joint. This disability is called decomposition of movement. One of Holmes's

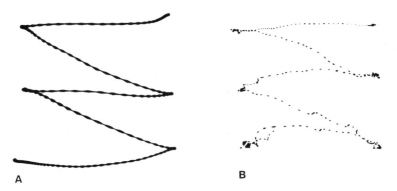

A   B

Figure 13.11 Normal (A) and dysmetric (B) movements recorded as described in the text, the range of movement of the finger being about 75 cm. (From Holmes 1939.)

patients described this very well, saying, "The movements of my left hand are done subconsciously, but I have to think out each movement of my right arm. I come to a dead stop in turning and have to think before I start again." This shows you how much we are spared this mental concentration by the cerebellum. What you do with ordinary movements is to give a general command—such as "place finger on nose," or "write signature," or "pick up glass"!—and the whole motor performance goes automatically. For example, you do not have to spell out your name letter by letter when you are writing your signature; if you did, the bank manager would not recognize it. You just give the general command from the cerebrum and let the subconscious motor circuits take over in order to give the fine characteristic details.

13.4.3 Neuronal Structure

In attempting to understand how the cerebellum carries out its amazing action as a software computer, we are fortunate to have a precise picture of its neuronal arrangement, which we owe in the first place to the genius of Ramón y Cajal (1911) (Fox *et al.* 1966; Eccles 1973; Eccles *et al.* 1967). It is a surprisingly uniform, stereotyped structure that is laid out, as you might expect for a computer, with geometrical precision. It is a rectangular laminated lattice as illustrated in Figure 13.12, where a small fragment of a folium is shown in a perspective drawing. The principal neurons are the Purkinje cells (PC), which provide the only output lines, their axons ending in the cerebellar nuclei (CN), i.e., these axons convey all the computational messages from the cerebellar cortex. The information that flows into the cerebellar cortex is entirely by two types of afferent fibers: the climbing fibers (CF) that twine around the Purkinje cell dendrites, and the mossy fibers (MF) that branch enormously and synapse on the little granule cells (GrC) in the granular layer (GrL) whose axons pass up to the molecular layer (MoL) to bifurcate and form the parallel fibers (PF) that run along the folium for about 5 mm. Thus, they are orthogonal to the espalier-like dendritic trees of the Purkinje (PC), basket (BC), and stellate cells (SC) with which they make numerous synapses, the so-called

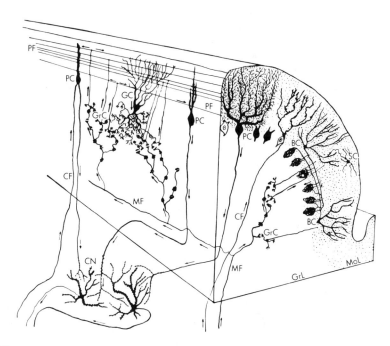

Figure 13.12 Schematic drawing of a segment of a cerebellar follium. (Fox 1962.)

crossing-over synapses. The basket cell axons are also transverse, traveling about 0.6 mm in either direction to form synapses on the Purkinje cell somata. Finally, there are the Golgi cells (GC), one being shown in Figure 13.12, which also receive synapses from the parallel fibers and have profusely branched axons that end on the granule cell dendrites.

The neuronal numbers are enormous. There are, for example, about 30,000 million granule cells, 30 million Purkinje cells, and 200 million basket and stellate cells in the human cerebellum. Each Purkinje cell receives about 100,000 parallel fiber synapses on its dendritic spines, which is its mossy fiber input. In contrast, it receives only one climbing fiber, but this makes a massive series of synapses on the dendrites of the Purkinje cell.

13.4.4 Neuronal Functions

In recent years (1963 onward) there has been an intensive investigation on the modes of action of the many synaptic connections in the cerebellar cortex; the results are summarized in very simplified form in Figure 13.13, in order to indicate the excitatory or inhibitory function of the various elements, which are drawn merely as single units (Eccles *et al.* 1967; Eccles 1973). All inhibitory neurons are shown in black. First, in Figure 13.13A the climbing fiber (CF) has been shown to make an enormously powerful excitatory synapse with the Purkinje cell (PC) that fires several times to a single climbing fiber impulse. By contrast, the mossy fiber (MF) is very diversified, each fiber branching so as to contribute excitation (Figure 13.13C) to about 400 granule

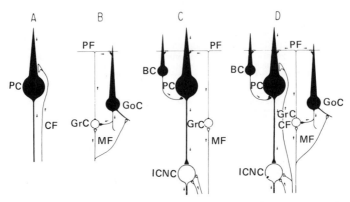

Figure 13.13 The synaptic connections of the cerebellar cortex. The component circuits of A, B, and C are assembled together in D. Arrows show lines of operation. Inhibitory cells are shown in black. PC, Purkinje cell; CF, climbing fiber; GrC, granule cell; PF, parallel fiber, GoC, Golgi cell; MF, mossy fiber; BC, basket cell; ICNC, intracerebellar nuclear cell.

cells, and 100,000 granule cells via their parallel fibers (PF) contribute to the excitation of each Purkinje cell. The basket cells (BC) are also excited by parallel fibers (Figure 13.13C) and their axons go transversely to end as inhibitory synapses on the Purkinje cell somata, which are in this way encased in a basket-like structure (cf. Figure 13.12). The Golgi cells (GoC) are excited by parallel fibers (Figure 13.13B) and also by mossy fibers, and their inhibitory synapses on granule cells complete a simple negative feedback loop (cf. Figure 4.17A, Section 4.7.4). Figure 13.13D is an ensemble of Figures 13.13A, 13.13B, and 13.13C.

We now have some basic information on which to build ideas of how the neuronal machinery of the cerebellar cortex can perform a computation. Two lines come in: one, the mossy fiber, is dispersed and powerful, and it gives both excitation and inhibition; and the other, the climbing fiber, is a direct monosynaptic in-and-out excitation. Furthermore, the sole output from the cerebellar cortex is by Purkinje cells, which, as indicated in Figure 13.13D, inhibit the intracerebellar nuclear cells (ICNC). In fact, all neurons of the cortex are inhibitory except the granule cells. Nowhere else in the brain do we know of such dominance of inhibition.

It has been suggested that this dominance is of great value in the computer-like operation of the cerebellar cortex. Since all inputs are transmuted into inhibitory action within at most two synaptic relays, there can be no prolonged chattering in chains of excitatory neurons such as would occur by the Purkinje cell axon collaterals if these were excitatory. As a consequence, within 0.4 s after some computation that area of the cerebellar cortex is "clean," ready for the next computation. This automatic "cleansing" is very important in giving reliable performance during quick movements. We can regard the arrangements to have the sole output via the inhibitory Purkinje cells as being a very clever piece of evolutionary design (Eccles *et al.* 1967).

In Chapter 15 there will be full reference to a hypothesis of learning according to which, as in a perceptron, the climbing fiber acts as a teacher in respect to the potency of the mossy fiber/parallel fiber synapses on a Purkinje cell. The dual innervation of the Purkinje cells thus is a further clever piece of evolutionary design (Eccles 1977).

So far we have discussed the excitatory and inhibitory actions in the cerebellar cortex as if they were all superimposed without any pattern, though already in Figure 13.12 we have seen the design as a rectangular lattice. The essential pattern is idealized in Figures 13.14A and 13.14B, where there is a

Patterns of inhibitory action on Purkyně cells

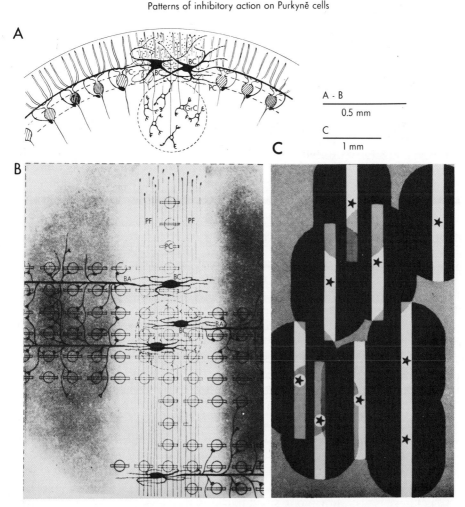

Figure 13.14 Patterns of inhibitory action on Purkinje cells: illustration of the concept of the higher order integrative unit of the mossy afferent–parallel fiber neuronal chain. Neuron matrix of folium is seen in transverse section (A) and from the surface (B), with special reference to the basket cells (BC) and the inhibitory surround produced by their axons (BA). (From Eccles *et al.* 1967.)

transverse section of a folium (13.14A) much as seen to the right of Figure 13.12, and (13.14B) the same folium seen from above as a sort of plan. It is assumed in Figure 13.14A that an input of mossy fiber impulses excites a compact group of granule cells (GrC). In Figure 13.14B their axons run as parallel fibers (PF) about 5 mm along the folium, exciting the Purkinje cells (PC), which are shown diagrammatically by a circle superimposed on a rectangle. Shown in black are the basket cells (BC) whose axons (BA) give Purkinje cell inhibitions on each side, as symbolized by the dark shading, just as in the cerebral cortex (Figure 13.7). So we have here, powerfully developed, an inhibitor surround on either side of the beam of excited Purkinje cells. Thus, a mossy fiber input excites a beam of parallel fibers that gives Purkinje cell excitation on-beam and inhibition off-beam on either side. But that, of course, is only for a single mossy fiber input focused, as it would be, by the negative feedback of the Golgi cells.

In normal operation of the cerebellum there would be multitudes of mossy fiber inputs resulting in antagonistic excitatory and inhibitory actions. The battle of excitatory and inhibitory action on Purkinje cells is fought all the time from moment to moment in every part of the cerebellum. That is the essence of this computational operation (Eccles 1969b, 1973). No excitatory input is just allowed to have unchallenged excitatory action on the Purkinje cells. It has to fight its way through against the opposed inhibitory action. The situation is thus seen to be comparable with that illustrated in the cerebral cortex in Figure 13.7.

13.4.5 Cerebrocerebellar Pathways[2]

Closed Loop via the Pars Intermedia of the Cerebellar Cortex. Figure 13.15 is a very greatly simplified diagram to show how the cerebrum and the cerebellum are linked together. All the neuronal pathways drawn in Figure 13.15 are securely based on anatomical and physiological investigations. Simplification is achieved by having the individual species of cells reduced to just one example. Each species shown, however, has been recorded as individual cells, and the connectivities shown have all been checked by the timing of their discharges in each of the pathways. This is a fairly complete diagram so far as it goes, but, of course, it misses all the operational features deriving from the enormous number of cells in parallel with the wealth of convergence and divergence that has been already mentioned in part. Thus, Figure 13.15 should be regarded merely as a skeleton drawing.

Let us start the operative sequences of Figure 13.15 by the firing of a large pyramidal cell (L.Pyr.C.) of the motor cortex. These cells are, of course, the principal cells of origin of the pyramidal tract (PT) (cf. Figures 13.3, 13.5, 13.7). There is also shown one small pyramidal cell (S.Pyr.C.). Axons of these cells form small fibers of the pyramidal tract, but it is not known how effective these fibers are in the spinal cord. For this reason only large pyramidal fibers

[2]General references: Evarts and Thach 1969; Allen and Tsukahara 1974; Eccles 1973.

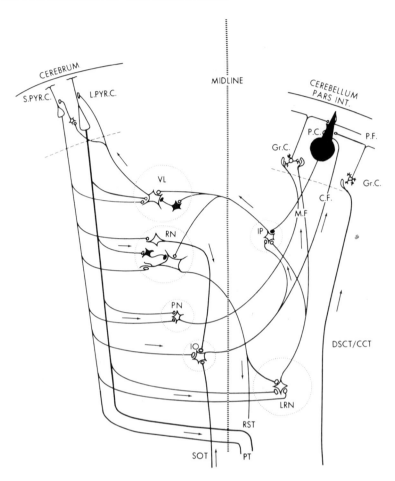

Figure 13.15 Pathways linking the sensorimotor areas of the cerebrum with the pars intermedia of the cerebellum.

are shown in Figure 13.8, sending collaterals to both α- and γ-motoneurons, and so being responsible for the action potentials of muscle fibers and of Ia fibers in Figure 13.9. In Figure 13.15 the large pyramidal tract fiber sends off branches that, after synaptic relay in the nuclei pontis (PN) and the lateral reticular nucleus (LRN), give mossy fiber (MF) inputs to the cerebellar cortex (pars intermedia). Thus, impulses fired down the pyramidal tract in order to begin a movement (as we have seen in the monkey in Figure 13.5 making to-and-fro movements) will at the same time go to the cerebellar cortex on the opposite side from the cerebral cortex and on the same side as the movement. So there is an extremely fast and reliable input from the cerebrum to the cerebellum. The cerebrum cannot begin instituting any action without the cerebellum knowing about it immediately. There is little doubt that the cerebrum is the command center, but all instructions it fires to the motor machinery of the spinal cord are immediately fired into all the computational machinery of the cerebellar cortex via these two mossy fiber pathways, one via the nuclei pontis (PN) and the other via the lateral reticular nucleus (LRN). In addition, impulses fired from the small pyramidal cells go to these two nuclei, presumably aiding in their responses. More importantly, only small pyramidal cell discharges go to the inferior olive (IO), which is the exclusive source for the climbing fiber (CF) input to the cerebellar cortex (Desclin 1974).

After the stage of interaction in the cerebellar cortex, as in Figure 13.14, for example, there is the final stage of the return circuit to the motor cortex of the cerebrum. In Figure 13.15 the Purkinje cell inhibits the nuclear cell in the interpositus nucleus (IP), which is also excited by collaterals from mossy and climbing fibers as shown for the LRN and IO pathways respectively. Here, then, is a further site for computation in the clash of excitatory and inhibitory actions on the IP cells. Thence the pathway is very fast and direct, there being only one synaptic relay in the ventrolateral thalamus (VL) on the way to the pyramidal cells of the motor cortex. Another pathway for action is also shown in Figure 13.15, namely, from IP to RN (the red nucleus) and so via the rubrospinal tract (RST) to the motoneurons of the spinal cord.

We can thus appreciate the authority of the cerebellar influence on the course of all movements that are initiated by the motor cortex. There is a very rapid and complete signaling to the cerebellar cortex of the whole array of impulse discharges down the pyramidal tract. We can assume that the input is computed in the cerebellar cortex with utilization of its memory stores (cf. Chapter 15), and after a further computation in the cerebellar nuclei (IP) it is returned to the same motor area of the cerebrum, much as occurred from the periphery in Figure 13.6. In the cat the circuit time for this complete loop would be less than a hundredth of a second. With man it would be longer, about a fiftieth of a second. With respect to the motor cortex this system operates in a closed-loop manner.

Open-Loop System in the Cerebellar Hemispheres. The cerebellar hemispheres comprise almost 90% of the human cerebellum (cf. Figure 13.12B), the principal cerebrocerebellar circuits being shown in Figure 13.16. In contrast to the pars intermedia, the cerebellar hemispheres receive most of their cerebral inputs from extensive areas of the cortex, such as the motor association cortex that is anterior to the motor cortex in Figure 13.2, and less from the motor cortex via collaterals of pyramidal tract fibers (dotted lines in Figure 13.16). This distinctive circuitry is well developed in primates and is preeminent in humans, where widespread zones of one cerebral hemisphere provide 20 million fibers passing to lower levels, as against only 500,000 pyramidal tract fibers (Allen and Tsukahara 1974). In Figure 13.16 impulses discharged from the pyramidal tract cells of the association cortex pass to the contralateral cerebellar hemisphere via relays in the pontine nuclei (PN) and the inferior olive (IO). After computation in the cerebellar hemisphere, the return circuit is via the VL thalamus to the motor cortex and so down the pyramidal tract (PT) to effect the movement. Thus, the cerebrocerebellar circuit is essentially an open-loop system.

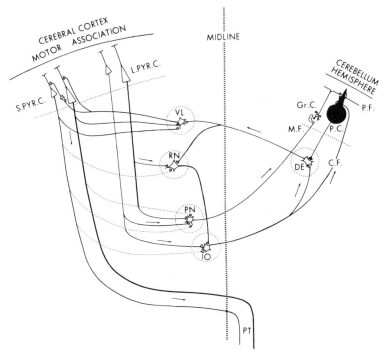

Figure 13.16 Cerebrocerebellar pathways linking association and motor cortices with the cerebellar hemisphere. DE is nucleus dentatus. Other symbols are as in Figure 13.15. In the red nucleus only a small cell is shown. (Allen and Tsukhara 1974.)

Dynamic Operation of the Cerebrocerebellar Circuits. The circuit operations are best appreciated by the very simplified diagrams of Figure 13.17, in which lines of communication are shown by arrows that substitute for all the synaptic connectivities of Figures 13.15 and 13.16. Figure 13.17A illustrates the mode of operation of the pars intermedia that has already been considered in relation to Figure 13.15, but it adds the input–output components of spinal centers and the evolving movement.

When pyramidal cells of the motor cortex (area 4) are firing impulses down the pyramidal tract (PT) in order to bring about a voluntary movement (a motor command), the patterns of this discharge in all details are transmitted to the cerebellum (pars intermedia) by virtue of the collateral branches of the pyramidal tract fibers (Eccles 1969b; Allen and Tsukahara 1974). Computation occurs in the cerebellar cortex (PI) and the resulting output is returned to the motor cortex so that there is an ongoing "comment" from the cerebellum within 10–20 ms of every motor command. We may regard this "comment" as being in the nature of an ongoing correction continuously provided by the cerebellum and being immediately incorporated in the modified motor commands issued by the motor cortex. Figures 13.15 and 13.17A also illustrate a longer feedback loop that operates through the same region of the

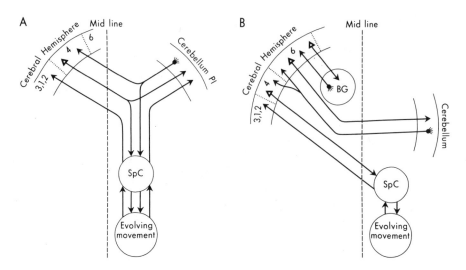

Figure 13.17 Cerebrocerebellar circuits in motor control are shown simplified by omission of the synaptic connectivities. A. The circuits from pyramidal cell Δ in motor cortex (4) via pyramidal tract to spinal cord, and so the evolving movement, and with collateral to the pars intermedia (PI) of the cerebellum. The Purkinje cell in PI communicates (via synaptic relays) back to the motor cortex and also down the spinal cord to the spinal centers (SpC). Also shown is the projection from spinal centers to PI and to the somesthetic area (3,1,2). B. The circuits are shown from the cerebrum (principally area 6) to the hemisphere of the cerebellum. The return circuit from the Purkinje cell is back to areas 4 and 6. From area 4, there is the projection down the spinal cord by the pyramidal tract as in A, and the return circuit from the evolving movement via the spinal centers to areas 3,1,2. Additionally there is shown the circuit from area 6 to the basal ganglia (BG) and the return to the cerebrum.

cerebellum. When the motor command brings about a movement, this evolving movement excites a wide variety of peripheral receptors, in muscles, skin, joints, etc., and these signal back to the same regions of the cerebellar cortex (upgoing arrow) that were concerned in the more direct loop. A computation of the two sets of input forms the basis of the cerebellar response. Thus, the motor command centers are provided with an ongoing cerebellar comment synthesized from these two loops. In addition, the pars intermedia has a more direct path for influencing the spinal centers via the red nucleus (RN) and the rubrospinal tract (RST), as indicated in Figure 13.15 and by the downward arrow in Figure 13.17A.

In summary, we can regard the pars intermedia of the cerebellum as acting like the controlling system on a target-finding missile. It does not give a single message for correction of a movement that is off-target. Instead, it provides sequences of correcting messages, thus giving a continuously updating control by closed dynamic loops.

In the primate there is no effective peripheral input to the cerebellar hemispheres such as that for the pars intermedia (compare Figure 13.17A with 13.17B) (Allen and Tsukahara 1974). Hence, the only feedback information into this open-loop system is via the sensory pathways to the cerebral cortex from peripheral receptors in muscle, skin, joints, etc., that project to the somesthetic areas, 3, 1, 2. Because of these special features of connectivity Allen and Tsukahara (1974) have proposed that the cerebellar hemisphere is concerned in the planning of a movement rather than in its actual execution and correction by follow-up control. Its function is largely anticipatory, based upon learning and previous experience, and also upon preliminary, highly digested sensory information transmitted from some of the association cortex.

Experimental investigation of unitary neuronal responses at all stages of the circuits suggests that in principle they give a satisfactory explanation of the performance of motor control (Eccles 1973; Allen and Tsukahara 1974). But much more experimental investigation is required, particularly on the mapping of related cerebral and cerebellar areas. We have to envisage that there is some sort of topographic design of the cerebellar hemispheres and pars intermedia somewhat on the lines of the cortical motor map of Figure 13.3. We know, however, that the maps are congruent only in part. The task of the cerebellum is to blend into a harmonious whole the movements of different joints of a limb, for example, so it would be anticipated that there would be convergence from these diverse motor areas to specific integrational areas of cerebellar cortex. Tests of this conjecture have yet to be carried out on the primate cerebellum.

Figures 13.18A and 13.18B shows examples of the background "or resting" discharges of single Purkinje and nuclear cells, with the characteristic irregularity of the rhythmic responses. The dots mark the unique responses of the Purkinje cell evoked by climbing fiber impulses, the so-called complex spikes.

Figures 13.18C to 13.18F (Thach 1968) take us back to experiments on monkeys resembling that of Figure 13.5 on the discharge of pyramidal tract cells during alternating movements. In Figures 13.18D and 13.18F the monkey is performing the same rhythmic to-and-fro movements, which are

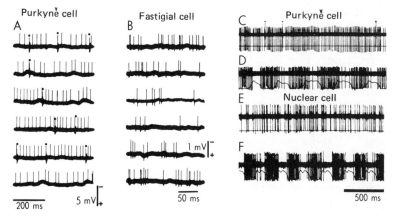

Figure 13.18 Responses of Purkinje and nuclear cells. A and B. The normal irregular firing of these cells. C–F. Records from the monkey cerebellum to illustrate the firing of both Purkinje and nuclear cells in phase with movements as illustrated in Figure 13.4A. (From Thach 1968.)

indicated by the superimposed traces. The Purkinje cell is seen to exhibit a rhythmic discharge that is modulated in accord with the movement, which contrasts with the fairly steady firing at rest in Figure 13.18A and 13.18C. This is precisely the effect that would be expected from the diagram of Figure 13.15, and the rhythmic PT cell discharges of Figure 13.5B. It shows that the Purkinje cell was related to that movement and it is explicable by the postulate that the Purkinje cell was being driven by the PT discharge. Similarly, the nuclear cell in Figure 13.18F showed the rhythm of the movement in contrast to the fairly steady discharge at rest in 13.18E. In accordance with the inhibitory action of Purkinje cells on nuclear cells (Figures 13.13 and 13.14) it would be expected that the rhythmic discharge of the nuclear cell would be 180° out of phase with respect to the Purkinje cells projecting onto it. But we do not yet have the experimental techniques to test out this simple idea. We can conclude that Figure 13.18, in general, illustrates the correctness of one aspect of the cerebrocerebellar connections illustrated in Figure 13.15.

13.4.6 Cerebellospinal Connectivities[3]

These connections have been far more intensively studied than the cerebrocerebellar connections so far considered. They are specially concerned in walking, standing, reacting, balancing, and in all the postural adjustments that follow active movements—the stabilizing positions attained thereby. Essentially, it can be seen from Figure 13.19 that the same general circuits from the spinal cord act on the cerebellar cortex as those diagrammed in Figures 13.15 and 13.17A. There are, first, the two main tracts up the spinal cord that end as mossy fibers (MF), the fast dorsal spinocerebellar tract (DSCT) and the slower tract (bVFRT) up to the lateral reticular nucleus (LRN). Second, there are the slow tracts to the inferior olive (IO) and so to the cerebellum as climbing fibers (CF). The inhibitory outputs of the Purkinje cells go initially to the fastigial (FN) or the Deiters (DN) nucleus and secondarily, via FN, either to DN or to the reticular nucleus (ReN) and so down the spinal cord to motoneurons via the vestibulospinal (VST) or reticulospinal (ReST) tracts, respectively.

The important feature of these connections of the cerebellar vermis is that there is only one loop in the dynamic control system. The evolving movement results in the discharge of various kinds of receptors that project via the various ascending pathways to the cerebellar cortex, thus modifying the Purkinje cell output to the cerebellar nuclei and so via the ReST and VST to the motoneurons. In this way the cerebellum is able to control posture and movements by the simplified version of dynamic-loop operation. We have studied intensively most of the stages in the neuronal system illustrated in Figure 13.19 and find that the various species of neurons show the responses that would be expected for inputs from receptor organs of skin and muscle via the DSCT and bVFRT. It should be mentioned that Figure 13.19 is a diagram

[3]General references: Eccles 1969b, 1973.

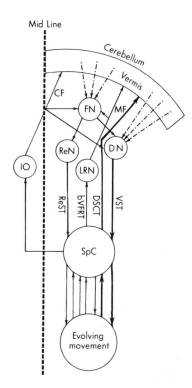

Figure 13.19 Pathways linking the cerebellar vermis with the spinal centers and so to the evolving movement.

for the hindlimb. For the forelimb the cuneocerebellar tract (CCT) substitutes for the DSCT.

In Figures 13.13C and 13.13D the intracellular nuclear cell (ICNC) is shown with excitatory inputs that provide the background against which is pitted the inhibitory action of the Purkinje cell. In Figure 13.19 this excitatory background to the fastigial nucleus (FN) is shown to be provided by collaterals from two pathways to the cerebellum, climbing fibers from the inferior olive (IO) and mossy fibers from the lateral reticular nucleus (LRN). It was found experimentally that the fast mossy fiber pathway (DSCT) had almost no excitatory action on the fastigial nucleus. Figure 13.15 shows a similar state of affairs for the collateral excitation of the interpositus nucleus (IP). These arrangements have been fully substantiated by stimulating the IO and LRN nuclei and observing the monosynaptic excitation of the cerebellar nuclei, FN and IP. They are also substantiated by anatomical studies employing radiotracer and horseradish peroxidase techniques.

13.4.7 General Comments on the Cerebellum

An intensive study of about 1000 Purkinje and 1000 nuclear cells has revealed a wide variety of responses and a surprising individuality. This individuality of response results from the built-in connectivities in the pathways to these cells that are only shown in skeletal form in Figures 13.15 and 13.19. We can assume that these connectivities are in part a result of the initial genetic coding and in part of superimposed "learned" connections that we will discuss in Chapter 15. The responses are all explicable in terms of these connectivities together with the basic integrated performance of the cerebellar cortex as illustrated in Figures 13.13 and 13.14.

Our samplings of Purkinje cells by several microelectrode insertions into the cerebellum in any one experiment give but fragmentary glimpses of the immense number of colonies constituted by similarly behaving neurons (cf. Figures 13.6, 13.7). We had earlier made a comprehensive study of the responses evoked in single Purkinje cells by afferent volleys from many nerves, however, both cutaneous and muscular, and had found remarkable examples of convergence. In order to preserve and develop the integration of information occurring in colonies having somewhat similar inputs, it is postulated that such colonies of Purkinje cells would tend to project onto a common target of nuclear neurons. This arrangement would give averaging of inputs from many Purkinje cells, reducing noise, as already described, and it would give opportunity for further integration of different subsets of the total input to the cerebellum. If there were randomized projection, there would be a loss of all specificity of information. It is the diversified convergence that enhances the pattern-generating capability of the neuronal machinery of the cerebellar cortex. This postulate of organized projection from Purkinje cells to the nuclear cells has already been investigated by a systematic investigation of both nuclear cell responses and the neurons to which these nuclear cells project.

The relative simplicity of the neuronal design of the cerebellum together with its well-defined action in control of movement provide a most enticing challenge, both experimentally and theoretically. It is our belief that the cerebellum will be the first part of the brain to be "understood" in its total performance. For this reason we have given a rather detailed account of the linkage between structure and function in specific cerebellar performances, and of theoretical developments from these investigations. Much of this theory is speculative, going beyond what has been scientifically demonstrated. But this is the essence of scientific advance, to have theory leading and guiding experiments which essentially are tests of theoretical predictions.

Perhaps the most puzzling feature of the cerebellar design is the existence of the two quite different inputs, mossy and climbing fibers, that convey much the same information. These two inputs have been preserved from the most primitive cerebella. We now have a theory that accounts satisfactorily for this linked input. There will be a further reference to this problem in Chapter 15.

Since, as we have seen, there is good reason to believe that the cerebellum functions as a special type of computer (Eccles 1973), there have been many attempts at computer modeling of the cerebellum. It fails, we think, because it is premature. We do not yet have sufficient "hard data" as a basis for computer modeling. There has to be much more experimental investigation using the recordings of single cells as is illustrated in Figure 13.18. In such investigations we are, as it were, listening in to the communications of data as coded in the cell discharges, in an attempt to understand these codes. So gradually, piece by piece, we are building a coherent story of cerebellar performance. In this way there will be a gradual winning of the "hard data" for effective computer modeling. But there is still much to accomplish and all too few experiments are at an adequate level.

A remarkable and apparently unique feature of the cerebellar cortex is that its output is entirely by the inhibitory Purkinje cells (cf. Figure 13.13, Section 13.4.4). At first it might be thought that this is poor neuronal design. How can information be effectively conveyed in this negative manner? But it has to be remembered that this inhibitory action is exerted on nuclear cells that have a strong background discharge (Figures 13.18B, 13.18E) and that most of the inputs to the cerebellum give excitatory collaterals to the nuclear cells, as indicated in Figures 13.15, 13.16, and 13.19. A good analogy is to the sculpture of stone. You have a block of stone and achieve form by "taking away" stone by chiseling. Similarly, the Purkinje cell output of the cerebellum achieves form in the nuclear cell discharges by taking away from the background discharges by inhibition. It is a challenging thought to think of this unique design for cerebellar output. As already suggested, the remarkable inhibitory bias in cerebellar design may be related to its computer function.

13.5 Motor Control by the Basal Ganglia

13.5.1 Introduction

We now turn to the last group of nuclei exerting control over motor function. These are the subcortical masses loosely referred to as the basal ganglia. We shall include in the definition of the basal ganglia the caudate nucleus, the putamen, the globus pallidus, the subthalamic nucleus, and the substantia nigra. Although not directly concerned in our story of motor function, two other structures of the basal ganglia, the nucleus accumbens and the amygdala, should also be mentioned because they play a prominent role in the associated behavioral phenomena which will be discussed in Chapter 14.

Information as to the physiological functioning of the basal ganglia is still extremely deficient. It cannot be said, for example, how these nuclei eventually affect peripheral musculature, or how postural signals from muscle spindles influence their internal activity. Too little is known about the inflow and outflow routes. There must be integration with the major motor systems of the cerebellum and cortex, but how this is achieved is still a matter of speculation.

Much of the information regarding the function of the basal ganglia has come from investigations that are very different from the precise neuroanatomy and elegant neurophysiology upon which knowledge of cerebellar function has been built. Instead, knowledge has been built up largely as a result of motor deficits produced by lesions, the effects of various drugs, and the consequences of certain diseases.

Pathological signs associated with disorders of the basal ganglia vary between two extremes: akinesia-rigidity and chorea-athetosis. Circumstances which will result in the appearance of these opposite signs are summarized in Table 13.1. Lesions, drugs, chemical poisons, altered neurotransmitter concentration, and a number of diseases of unknown etiology all are capable of producing the classical signs. The two extremes can be thought of as a postural rigidity on the one hand leading to poverty of movement, and a postural hyperactivity on the other leading to incessant, involuntary activity. The system is like a teeter-totter, with normal activity occurring only when these two opposing forces are in balance (Figure 13.25). Tremors at rest are frequently seen when the system is out of balance, indicating that rhythmic activity is involved in normal functioning and is probably a feature of the integration of this system with cerebellar, cortical and peripheral motor loops.

Table 13.1 shows that there are many causes of akinesia-rigidity: in Parkinson's disease; with certain types of poisoning; with reduced dopamine concentration or blockade of dopaminergic receptors; with excess ACh; or with lesions to the substantia nigra.

Parkinson's disease is, of course, the classic example of the akinetic-rigid syndrome. Although the disease is most easily spotted because of the telltale pill-rolling tremor, this is the least of the problems confronting the unfortunate victim. It is the difficulty in executing willed movements that is the overwhelming curse of his existence. In advanced stages such simple tasks as rising from a chair can require the kind of concentration that goes with

TABLE 13.1
Functional Output of the Basal Ganglia

Overactivity (Chorea-hypotonia)	Underactivity (Akinesia-rigidity)
Diseases	
Huntington's chorea	Parkinson's disease
Sydenham's chorea	Manganese poisoning
	Carbon monoxide poisoning
	Carbon disulfide poisoning
Neurotransmitter functional activity	
Excess dopamine	Excess ACh
Lowered ACh	Lowered dopamine
Drug action	
Cholinergic blockade (atropine, artane, etc.)	Dopaminergic blockade (butyrophenones, phenothiazines)
L-Dopa	Dopamine depletion (rauwolfia alkaloids)
Lesions	
Caudate	Substantia nigra
Putamen	

executing a battle plan. Even maintaining a fixed position can be fraught with danger. A patient thrown off balance falls like a log before compensatory reflexes can be brought into play. Life becomes an endless series of struggles to force obstinate and rigid muscles to perform. The inadequacy of reflexes that do not ordinarily come to notice are a perpetual worry. The Parkinsonian patient would never agree with the classical physiological notion that extrapyramidal function is somehow concerned with "background" or "automatic" movement. It is very much the here and now as he formulates and rejects plans for setting his limbs in motion, seeking some compromise that will permit him to start climbing the stairs or whatever other action he had in mind.

The victim of chorea has quite the opposite problem. Head, neck, and limbs wander endlessly, and no act of will can bid them cease. Postural adjustment is never achieved. In athetosis there is a more regular oscillation between normal postural extremes while in chorea the motion is seemingly more random. There is no problem in initiating movement. Rather, the difficulty is in holding it in check. Every motion is overdone.

Chorea-athetosis occurs in Huntington's and Sydenham's chorea, with excess dopamine, with cholinergic blockade, and with lesions of the caudate and putamen. While the true picture is much more complicated than indicated in the table, it nevertheless serves as a starting point to show that definite correlations exist between neurotransmitter amines, subcortical motor nuclei, and disease states of unknown etiology.

Knowledge regarding these correlations is currently growing at a rapid rate. Some firm foundations have been laid, which should form the basis for many new and effective treatments for extrapyramidal diseases as well as indicating which experimental approaches may lead to new breakthroughs.

The door to discovery was opened by the unlikely tool of clinical psychiatry. It was noted soon after the introduction of the first antipsychotic drugs, reserpine and chlorpromazine, that a prominent side effect was a Parkinsonian-like rigidity. This clinical oddity was soon given neurochemical substance by the discovery by Carlsson (1959) that the corpus striatum contained extraordinarily high quantities of dopamine and that dopamine was depleted by reserpine. One year later Ehringer and Hornykiewicz (1960) published their classic paper showing a decreased concentration of dopamine in the striatum in Parkinson's disease. In subsequent papers they correlated this finding with the known dropout of nigral cells in Parkinson's disease to postulate a dopaminergic connection between the nigra and the striatum. Such a connection was verified by combined lesion and biochemical studies by Poirier and Sourkes (1965) about the same time as the elegant Swedish techniques unequivocally demonstrated the dopaminergic nigrostriatal tract by histochemistry.

It was several years before Cotzias et al. (1967) clearly demonstrated the beneficial effects of dopamine replacement therapy in Parkinson's disease by using large oral doses of L-Dopa, but that practical event completed a cycle involving only one facet of extrapyramidal function.

New investigational techniques soon revealed there were ACh- and GABA-containing cells in the caudate and putamen as well as GABA-containing cells in the globus pallidus (Chapters 5 and 7). Thus, when reports first began to appear indicating that there was a deficiency of GABAnergic and cholinergic cells in the basal ganglia of Huntington's chorea patients (Perry *et al.* 1973; Bird and Iversen 1974; P. L. McGeer *et al.* 1973), the motor deficit of Huntington's chorea became explainable on the basis of a loss of these cells.

The information we can obtain by correlating the motor deficits in disease states such as Parkinson's disease and Huntington's chorea with basal ganglia neuronal loss, however, gives only a glimmering of the physiological action of the neuronal machinery which exists in this part of the brain. A better indication is given by considering the intricate anatomical relationships that have so far been uncovered.

13.5.2 Anatomical Interconnections of the Basal Ganglia

A simplified way of regarding the organization of the basal ganglia is to divide it into input, processing, and output neuronal arrangements. These arrangements are indicated in Figure 13.20A to 13.20C. A number of the pathways have only recently been described as a result of employing new and sophisticated neuroanatomical techniques. It can be conjectured that the ground has not been completely tilled, and that considerable modification and improvement can be anticipated in the future. (For reviews, see Kemp and Powell 1971; Mehler and Nauta 1974; Carpenter 1975).

The principal processing areas of the basal ganglia are the caudate and putamen. These twin nuclei, divided in primates by the internal capsule but not so divided in smaller rodents like the rat, are enormous assemblages of neurons of uncertain internal organization. Together they are referred to as the neostriatum. Afferent input comes from the cortex, the thalamus, and the midbrain (Figure 13.20A).

The cortical projection upon the neostriatum is massive. Nearly all cortical regions participate in this cortical-striate pathway which is organized more or less topographically (Kemp and Powell 1971). The more frontal regions project to the caudate, the pre- and postcentral gyri project to the putamen, and so on. Some cortical fibers cross over to project to the contralateral side. The thalamus has a heavy, but less massive input from the intralamellar nuclei including the centrum medianum (Powell and Cowan 1956; Mehler and Nauta 1974). Midbrain input comes from the zona compacta of the substantia nigra (Figure 8.3, Section 8.2.1) and the rostral raphe (Figure 9.1, Section 9.2). The nigrostriatal dopaminergic tract (Anden *et al.* 1966), which is damaged in Parkinson's disease, has already been extensively discussed (Chapter 8). The serotonergic pathway from the rostral raphe region to the neostriatum and SN has also been discussed (Chapter 9), but the functional contribution of this pathway is not yet known.

Processing systems within the basal ganglia are shown in Figure 13.20B.

Figure 13.20 Pathways of the basal ganglia. A. Input circuits to the basal ganglia. B. Processing circuits within the basal ganglia. C. Output circuits of the basal ganglia. (1) Corticostriatal, (2) Nigrostriatal, (3) Thalamostriatal, (4) Raphestriatal, (5) Striatopallidal, (6) Striatonigral, (7) Pallidal-subthalamic-pallidal, (8) Pallido-nigral, (9) Nigrothalamic, (10) Nigrobrainstem, (11) Pallidothalamic, (12) Pallidohabenula, (13) Pallidotegmental. CX, cerebral cortex; C, caudate nucleus; P, putamen; GPE, external globus pallidus; GPI, internal globus pallidus; S, subthalamic nucleus; SNC, substantia nigra, pars compacta; SNR, substantia nigra, pars reticulata; VL, ventral lateral thalamic nucleus; VA, ventral anterior thalamic nucleus; I, intralaminar thalamic nucleus; CM, centrum medianum thalamus; SC, superior colliculus; RF, reticular formation; CG, central gray; R, raphe system; A, aqueduct of Sylvius.

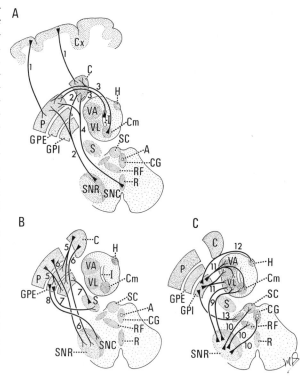

The striped appearance implied in the Latin term *corpus striatum* is due to bundles of striatal fugal fibers which converge radially upon the globus pallidus. The majority terminate in the globus pallidus but some course through the external and internal pallidal segments, "comb" through the cerebral peduncle, and terminate chiefly in the pars reticulata of the substantia nigra. Thus, the principal output of the caudate and putamen is to the globus pallidus and to a much lesser extent the pars reticulata of the substantia nigra. Considerable convergence is involved because the globus pallidus is much smaller than the neostriatum, and the substantia nigra pars reticulata is much smaller than the globus pallidus. The globus pallidus itself (the paleostriatum) is divided into two segments, the external and internal divisions. The internal division is equivalent to the entopeduncular nucleus in smaller mammals. The external globus pallidus projects primarily to the subthalamic nucleus, a small body embryologically related to the thalamus, but also to the pars compacta of the substantia nigra (P. L. McGeer *et al.* 1974a; Hattori *et al.* 1975; Grofova 1975; Bunney and Aghajanian 1976). The subthalamic nucleus projects back to the internal segment of the globus pallidus.

Figure 13.20C shows the main outflow circuits from the basal ganglia. The most important of these is from the internal segment of the globus pallidus to the thalamus. While the processing systems within the basal ganglia are highly convergent from the caudate and putamen to the internal globus pallidus and substantia nigra, the output from these structures again becomes highly divergent. The fibers from the internal globus pallidus follow one of two routes: they traverse the internal capsule as the fasciculus lenticularis or loop around the internal capsule as the ansa lenticularis. Terminations are in the ventrolateral (VL) and ventroanterior (VA) nuclei of the thalamus, and to a lesser extent in the centrum medianum. There is also a pallidotegmental tract terminating in the pedunculopontine nucleus (Carpenter 1975).

It is in the thalamus that there are such extensive overlaps between input from the basal ganglia and the cerebellum.

Lesions of VL of the thalamus are successful in removing Parkinsonian tremor and rigidity, presumably because they remove incompatible information being supplied to VL from the cerebellum and globus pallidus which is being relayed to area 4 of the cortex. Such lesions, unfortunately, are ineffective at curing akinesia and have largely been replaced by more effective chemotherapeutic regimens, particularly L-Dopa therapy. The internal segment of the globus pallidus also projects to the lateral habenula (Nauta 1974). This connection is believed to be more related to behavioral rather than motor performance.

The substantia nigra has two main efferent projections emanating from different lamina of the pars reticulata. The intermediate lamina, lying just beneath the zona compacta, projects mainly to the thalamus (Faull and Mehler 1976). It distributes partly to the VL and VA thalamic nuclei, but mainly to centrum medianum. The extent to which these nigral thalamic fibers overlap with cerebellar projections to the thalamus is still being evaluated. The most ventral cells of the pars reticulata project to the superior colliculus (Graybiel and Sciascia 1975; Faull and Mehler 1976), and to a wide medial zone of the ipsilateral midbrain tegmentum including the anterior parabrachial, intrabrachial regions (Domesick 1976), and central gray (Hopkins and Niessen 1976). These latter pathways will no doubt receive considerable attention in the future to consolidate the long-held belief that the basal ganglia must have projections directly to descending motor pathways exclusive of the pyramidal tract.

13.5.3 Physiology of the Basal Ganglia

Unfortunately, very little is known about how this complicated anatomical machinery operates. Repeated electrophysiological studies have demonstrated that most cells in the caudate and putamen are quiescent under normal circumstances, as if they were held in check by tonic inhibition. Pallidal cells, on the other hand, are normally very active, discharging at rates of 60–70 per second. Beyond this, the picture is confusing. Stimulation of the various input and output pathways has produced variable, and so far uninterpretable results. Low-frequency stimulation of the centrum medianum–parafascicular complex of the medial thalamus elicits excitatory postsynaptic potentials of long duration in caudate target neurons (Purpura 1975). Occasionally EPSP–IPSP sequences are seen, particularly in caudate cells exhibiting a higher rate of spontaneous activity than those showing pure EPSPs. Cortical input is also thought to be mainly excitatory (Spencer 1976), but Buchwald *et al.* (1973) have observed EPSP–IPSP sequences in caudate neurons receiving input not only from cortical, but also from thalamic, peripheral, and brainstem stimulation. The meaning of such EPSP–IPSP sequences remains a mystery but it does suggest that complicated intracaudatal reactions may be involved.

A similar confusion exists with respect to nigrocaudatal projections. Some studies have reported inhibitory effects (Connor 1970), others excitatory-inhibitory sequences (Buchwald *et al.* 1975; Feltz and Albe-Fessard 1972), and still others purely excitatory effects (Kitai *et al.* 1975). Again, the proposal has been put forward that more than one nigrocaudatal pathway exists (Brawley *et al.* 1976).

Low-frequency stimulation of the caudate will produce EPSP–IPSP sequences or pure inhibitory ones in the pallidal neurons (Malliani and Purpura 1967). The inhibitory action is profound and long lasting. Yoshida *et al.* (1971) recorded only monosynaptic IPSPs in the substantia nigra following caudate stimulation in pentobarbital-anesthetized cats. A more sophisticated study of Feger and Ohye (1975) in unanesthetized monkeys has established a dual pattern of excitation–inhibition or pure inhibition following caudate stimulation. Approximately half the nigral cells examined exhibited one or the other pattern. Thus, this more sophisticated study seems to substantiate the existence of two caudatonigral pathways.

Several patterns of synaptic activity have been demonstrated in thalamic neurons during pallidal stimulation. Desiragu and Purpura (1969) described short-latency monosynaptic EPSPs, medium-latency IPSPs, and long-latency EPSP–IPSP sequences.

To summarize, cortical and thalamic afferents exert largely monosynaptic excitatory actions on caudate and putamen target cells, but long-latency inhibitory and dual excitatory-inhibitory actions are also observed. All three patterns have also been observed with stimulation of the nigrostriatal input.

The striatopallidal and striatonigral projections which represent the main long pathways of internal processing produce monosynaptic excitation, monosynaptic inhibition, and dual excitatory-inhibitory actions on target cells in the pallidum and substantia nigra.

The efferent projections to the thalamus (pallidal and nigrothalamic projections) also exert monosynaptic excitation, short-latency inhibition, and dual excitatory and inhibitory effects on thalamic neurons.

All of these physiological studies provide little insight into the dynamic loop activity which is implied by the sophisticated neuroanatomy of the basal ganglia and which is evident from the effects of drugs and disease. Fruitful results await the neurophysiologists of the future who tackle this region of the brain. An indication of directions that might be taken come from the experiments of York and Lentz (1976), who briefly stimulated the substantia nigra in rats and found in the caudate a prolonged enhancement of response to peripheral stimulation. This result may have great meaning in terms of the conjunction theory of motor learning postulated in Chapter 15.

13.5.4 Biochemistry of the Basal Ganglia

Through the combined application of biochemical and neuroanatomical techniques, it has been possible to assign transmitters to some of the pathways shown in Figure 13.20 and even to suggest their interconnections. These are shown in Figure 13.21, although it must be remembered that some are speculative at this time.

The most significant interrelationship from a physiological point of view is the one connecting dopaminergic nerve endings in the neostriatrum with cholinergic interneurons. The existence of the nigrostriatal dopaminergic tract has been discussed in Chapter 8 and cholinergic interneurons in Chapter 5. The connection between these two has been demonstrated through the combined use of 6-OHDA administered intrathecally to cause degeneration of dopaminergic nerve endings, and immunohistochemical staining to localize cholinergic neuronal processes. Figure 13.22 shows degenerating dopaminergic nerve endings making contact with dendrites and dendritic spines of CAT-containing cells in the neostriatum. These electron micrographs provide positive proof of the direct relationship of dopaminergic and cholinergic structures in the basal ganglia which had long been hypothesized on pharmacological grounds (Hattori *et al.* 1976b).

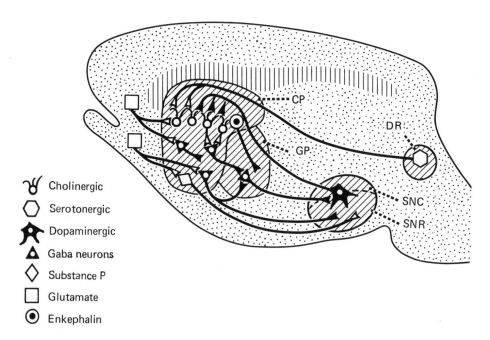

- 𝒴 Cholinergic
- ⬡ Serotonergic
- ★ Dopaminergic
- ▲ Gaba neurons
- ◇ Substance P
- ☐ Glutamate
- ◉ Enkephalin

Figure 13.21 Biochemical interconnections of the basal ganglia. CP, caudate–putamen; GP, globus pallidus; SN, substantia nigra; DR, dorsal raphe.

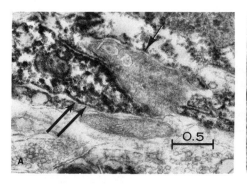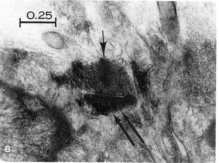

Figure 13.22 Dopaminergic-cholinergic linkage in the neostriatum. A. Guinea pig neostriatal dendrite (double arrow) stained for CAT. Degenerating nerve ending (single arrow) is making contact. Degeneration was induced by 6-OHDA administration. B. Same conditions as *A*. Vesicles in degenerating nerve ending (single arrow) can just be detected. Calibration bar 0.5 μm in *A* and 0.25 μm in *B*. (From Hattori *et al.* 1976b.)

A similar direct connection has been demonstrated between descending GABA pathways from the pallidum, and to a lesser extent from the neostriatum, with dopaminergic cell bodies in the substantia nigra. In these experiments, the cell bodies were made to degenerate by the same technique of intrathecal administration of 6-OHDA. Nerve endings of descending pathways to the nigra were identified by anterograde axoplasmic flow following the injection of [³H]leucine into either the neostriatum or pallidum (Hattori *et al.* 1975). As described in Chapter 1, the labeled leucine is synthesized in the cell somata of the injection site into labeled protein. This is transported to the nerve endings, and is detected by characteristic traces or grains on electron microscopic radioautograms.

Following pallidal injections in rats, the ratio of grains making contact with degenerating, as opposed to normal, dendrites in the substantia nigra was much higher than following caudate–putamen injections. The implication of this study is that pallidal afferents are much more likely to make contact with dopaminergic dendrites than are the neostriatal afferents. This is consistent with the finding that the striatonigral tract ends primarily in the pars reticulata while the pallidonigral tract preferentially terminates in the compacta (Hattori *et al.* 1975).

These experiments do not, of course, establish that the descending pathways are GABA-containing. Some GABA transport has been demonstrated from the pallidum to degenerating dendrites of the substantia nigra, however (P. L. McGeer *et al.* 1974c).

Figure 13.23 illustrates these results. 6-OHDA had been administered to the rat to distinguish, through degeneration, which of the substantia nigra dendrites were dopaminergic. The site of injection was the globus pallidus. The figure illustrates a labeled protein molecule in a nigral nerve ending making contact with a degenerating dopaminergic dendrite.

It has been established that GABA interneurons exist in the neostriatum

(Chapter 7). These are indicated in Figure 13.21, although the interconnections are not yet known. The corticostriatal pathway in Figure 13.21 is shown as a glutamate pathway. Spencer (1976) has postulated that the corticostriatal pathway is glutamatergic on the grounds that it is excitatory and that the effects of stimulation of the pathway are blocked by iontophoretic application of diethyl glutamate. Such evidence is not strong, since diethyl glutamate is a nonspecific blocking agent. Decortication of rats, however, brings about a sharp reduction of glutamate uptake in the synaptosomal fraction of neostriatal homogenates (Chapter 6; P. L. McGeer *et al.* 1977; Divac *et al.* 1977). These are somewhat more suggestive data but are not completely conclusive.

Immunohistochemical (Hokfelt *et al.* 1975) and lesion (Kanazawa *et al.* 1977a; Mroz *et al.* 1977) evidence suggests that there is a substance-P-containing descending striatonigral pathway (Chapter 10). Such a pathway would help to explain some of the EPSP–IPSP neurophysiological sequences which evidently occur in addition to the IPSP pattern attributed to a GABA pathway. This too is indicated in Figure 13.21. The serotonin pathway from the midbrain to the striatum (Figures 9.1 to 9.3, Section 9.2) is shown, as is the enkephalin pathway from the neostriatum to the globus pallidus (Figure 10.4, Section 10.3.1).

The interconnections shown in Figure 13.21 help to explain some of the observations in Parkinson's disease and Huntington's chorea which are not revealed by normal pathological examination. It had been known since the

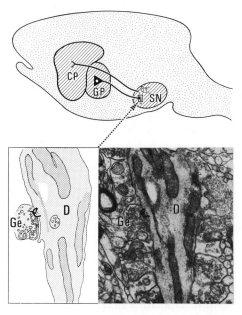

Figure 13.23 GABA–dopamine linkages in the substantia nigra. *Top:* diagrammatic representation of the pallidonigral pathway showing termination of globus pallidus (GP) axons on dendrites of dopaminergic cells which lie in the pars compacta of the substantia nigra. *Bottom left:* enlargement of the insert area showing a terminal of a globus pallidus neuron (Ge) making contact with a degenerating dendrite (D) of a dopaminergic neuron. *Bottom right:* actual electron microscopic autoradiograph of the area diagrammed on left. Degeneration of the dopaminergic dendrite, indicated by the darkened color, was brought about by administration of the neurotoxic agent 6-hydroxydopamine. The dopaminergic dendrite (D) is receiving a nerve ending (Ge) which contains labeled protein as established by the silver grain. [^{3}H]Leucine had been injected into the GP, resulting in the synthesis of labeled protein which was then transported to the substantia nigra by the process of axoplasmic flow. Magnification of electron micrograph × 30,000. (From Hattori *et al.* 1975.)

time of Tretiakoff (1919) that there was a loss of pigmented cells in the substantia nigra in Parkinson's disease. Similarly, it had been known for many years that there was a loss of cells in the caudate, putamen, and globus pallidus in Huntington's chorea. But until dopamine was measured by Hornykiewicz, no neostriatal pathology was suspected in Parkinson's disease, and until GABA was measured by Perry *et al.* (1973) no nigral pathology was suspected in Huntington's chorea.

Figure 13.24 illustrates the contrast between morbid and biochemical pathology in these two conditions. Figure 13.24B shows the cellular content of the substantia nigra as determined by standard pathological staining techniques for normal, Huntington's chorea, and Parkinsonian cases. Figure 13.24A shows the same comparisons for the caudate. In Figure 13.24B, the loss of neurons in the Parkinsonian case is very evident, but the Huntington's chorea case appears normal. But the biochemical results in the figure show that in Huntington's chorea there are very large decreases of glutamic acid decarboxylase even though there is no evidence of cellular loss. Tyrosine hydroxylase is much decreased in the Parkinsonian case, as would be expected.

In Figure 13.24A, the Huntington's chorea case shows the typical severe neuronal loss with replacement by smaller glial cells. The Parkinsonian case appears normal. Nevertheless, the Parkinsonian case shows a profound loss of tyrosine hydroxylase, while the Huntington's chorea case shows only a loss of choline acetyltransferase and glutamic acid decarboxylase. The explanation has already been provided in Chapters 7 and 8. It is nerve endings that contain glutamic acid decarboxylase in the nigra and dopamine in the neostriatum. The cellular processes that were elegantly displayed in Figures 1.3 and 1.4 (Section 1.2.2) are, of course, not revealed by the simple cellular staining techniques used in pathology. Thus, if only the processes are damaged or disappear, evidence may never be found. The fine structure of neurons revealed in the many electron micrographs of this and earlier chapters can only be obtained following the use of delicate perfusion techniques at the time of sacrifice. Such structural details are never obtained on human postmortem material. Thus, it cannot be assumed that current pathological techniques are adequate for revealing the true state of affairs in many human diseases.

While Figure 13.24 illustrates the advantage of biochemical studies as an adjunct to normal pathological investigation, Figure 13.21 suggests approaches to be taken in the treatment of Parkinson's disease and Huntington's chorea. This requires some appreciation of the pharmacology of the basal ganglia.

13.5.5 Pharmacology of the Basal Ganglia

The most fundamental principle of the pharmacology of the basal ganglia is the antagonism between dopaminergic and cholinergic tone. It had been

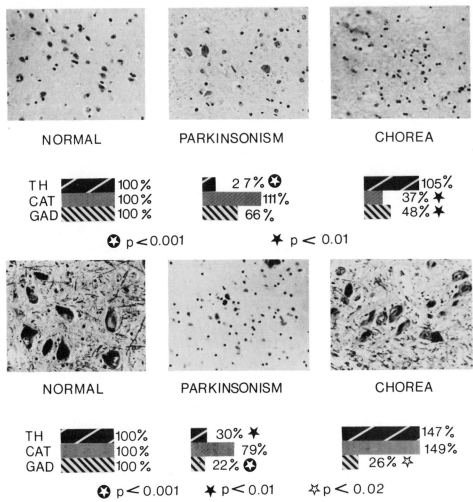

Figure 13.24 Tissue and enzyme damage in Parkinsonism and Huntington's chorea. A. Histological and enzyme changes in the caudate nucleus of patients with Parkinson's disease and Huntington's chorea. In chorea there is extensive replacement of neurons with glial cells (small nuclei) while larger neurons can be seen in normals and Parkinsonian cases. Note the sharp losses in tyrosine hydroxylase (TH) in Parkinsonism, however, and choline acetyltransferase (CAT) and glutamic acid decarboxylase (GAD) in Huntington's chorea. TH is normal in chorea. Thus, morbid pathology failed to detect the loss of dopaminergic processes in Parkinson's disease and also failed to indicate their presence in Huntington's chorea. B. Histological and enzyme changes in the substantia nigra in Parkinson's disease and Huntington's chorea. Note the presence of normal neurons in Huntington's chorea, but their absence and replacement with smaller glial cells in Parkinsonism. Enzyme changes show that there is a sharp reduction in TH and GAD in Parkinson's disease, and GAD in Huntington's chorea. Thus, the morbid pathological techniques failed to indicate the loss of GAD-containing processes in the substantia nigra in Huntington's chorea. (Tissue sections provided by E. Zis, National Institutes of Health; biochemical data from P. L. McGeer and McGeer 1976.)

known from the time of the great French physician Charcot (1825–1893) that Parkinsonian patients responded to the anticholinergic belladonna derivatives. But experimental Parkinsonism was not produced until the dopaminergic depletor reserpine was introduced into clinical medicine in the 1950s.

We suggested many years ago (P. L. McGeer *et al.* 1961; P. L. McGeer 1963) that some kind of balance must exist in the basal ganglia between dopamine and ACh. It can be represented as a teeter-totter (Figure 13.25). Excess cholinergic balance on the teeter-totter, produced by depleted or blocked dopamine, results in Parkinsonian-like akinesia and rigidity. Excess dopaminergic balance, produced by depleted or blocked ACh, results in choreiform manifestations.

This balance theory has only recently been given a sound neuroanatomical and neurophysiological foundation. The linkup of dopaminergic nerve endings in the neostriatum with cholinergic interneurons (Figure 13.22) (Hattori *et al.* 1976b) has already been described.

Drugs that block or deplete dopamine cause large decreases in steady-state striatal ACh levels while drugs that enhance dopamine cause substantial increases (P. L. McGeer *et al.* 1974d; Sethy and Van Woert 1974; Agid *et al.* 1975). This would be expected if dopaminergic nerve endings were exerting a tonic inhibition on cholinergic interneurons.

Support for this hypothesis comes from two sources. First, iontophoretically applied dopamine has an inhibitory action on caudate neurons (Bloom *et al.* 1965; Connor 1970; McLennan and York 1967). Second, dopamine blockers enhance the release of striatal ACh as measured by permanently implanted cannulae (Stadler *et al.* 1973). The effect can be reversed by stimulating the receptor sites with apomorphine. A weakness in this theory is the fact that *in vivo* studies of the nigrostriatal tract have failed to verify its inhibitory nature. It is possible that the excitatory-inhibitory sequences that have been observed give a net inhibitory effect or that the measurements are contaminated by interference with other pathways. Whatever the explanation, all available evidence substantiates an antagonism between the actions of ACh and dopamine in the basal ganglia.

The role of GABA is less well defined. Presumably it always acts in an inhibitory fashion (Chapter 7), but there are both interneurons and long pathways to consider. γ-Butyrolactone (GBL), a presumed stimulator of GABA receptor sites, when administered parenterally depresses the activity of dopaminergic neurons by specifically blocking the firing of unit cells in the substantia nigra. As a consequence GBL selectively increases brain dopamine (Gessa *et al.* 1968) while decreasing that of its metabolite dihydroxyphenylacetic acid (Walters and Roth 1972). This shutting down of dopaminergic neurons is what would be anticipated from an inhibitory GABA input.

Figure 13.25 Dopaminergic-cholinergic "teeter-totter" in extrapyramidal function. AC, acetylcholine; DA, dopamine.

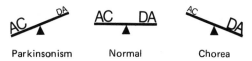

Parkinsonism   Normal   Chorea

We can now return to Figure 13.21 to consider modes of therapy in Parkinson's disease and Huntington's chorea. Therapy in Parkinson's disease must, of course, be aimed primarily at compensating for the deficit in the nigrostriatal pathway. This can be done by administering L-Dopa so that there is more dopamine in the dopaminergic nerve ending, by stimulating the dopamine receptor sites with materials such as apomorphine or bromocriptine, and by blocking by anticholinergic agents the cholinergic cells which are released from dopamine inhibition. Although the possibility has not yet been exploited, there might be some benefit to be gained by blocking GABA receptor sites on dopaminergic substantia nigra neurons. This might be difficult to achieve without seriously affecting other systems. In Huntington's chorea, the objectives would be the opposite. Blocking the dopamine receptor sites with agents such as haloperidol is the current method of choice. Giving cholinomimetics or GABAmimetics should also prove useful (Walters *et al.* 1972). In this respect deanol, lioresal, and γ-hydroxybutyrate have all received clinical trials.

None of these treatments, of course, addresses itself to the causation of either disease. That remains a mystery. Discovery of the cause is by far the most important objective of future research.

In this respect it is interesting that toxic agents have now been found which will duplicate the biochemical findings in both these diseases when locally injected into the brain. 6-OHDA will reproduce the changes of Parkinson's disease when injected into the substantia nigra (Ungerstedt 1971), while kainic acid or other glutamate analogs will do so when injected into the neostriatum (Coyle and Schwartz 1976; E. G. McGeer and McGeer 1976a).

Ungerstedt was also able to show that denervation of the striatum by injection of 6-OHDA bilaterally into the substantia nigra of rats produced a long-lasting adipsia and aphagia. There was hypoactivity and difficulty in initiating movement, with concomitant loss of exploratory behavior and curiosity. Such a result would, of course, be expected since such lesions would induce the akinetic state of Parkinsonism. But these are the same signs as those seen in the lateral hypothalamic syndrome. There is a vast literature in the field of physiological psychology on this syndrome which is produced by lesioning the lateral hypothalamus. The effects have always been thought to be related to consummatory centers in this region. The data are now all suspect in that the nigrostriatal dopaminergic pathway passes through that same region of the brain.

Ungerstedt (1971) has also devised a simple rotometer test to evaluate in rats the overall effect of unilateral damage to the nigrostriatal tract and the consequent supersensitivity of striatal receptors. Lesioned animals wearing a simple harness attached to a microswitch are placed in a bowl. Their rotations ipsilateral or contralateral to the lesioned side are automatically recorded. In animals where 6-OHDA was administered unilaterally into the substantia nigra, rotation was toward the side of the lesion, indicating a dominance of the unoperated side. This was enhanced by the administration of amphetamine, a dopamine releaser, which gave even more power to the unaffected side. The effect was abolished by inhibiting tyrosine hydroxylase, thus lowering dopamine levels on the intact side. But when the dopamine stimulator apomorphine was administered, the rotation was away from the side of the lesion. This effect indicated that the dopamine receptor sites were not only intact on the lesioned side, but that they were supersensitive as a result of being deprived of normal stimulation. Haloperidol, which blocks the receptor sites, strongly inhibited the supersensitivity. Supersensitivity is thought to be a factor in the choreaform activity seen in some Parkinsonian patients treated with high doses of L-Dopa.

As far as chorea is concerned, local injection of small doses of kainic acid, or higher doses of less powerful glutamate analogs such as homocysteic acid or glutamate itself, into the neostriatum will destroy cholinergic and GABA cells in that structure (E. G. McGeer and McGeer 1976a). This is evidenced by disappearance of neurons histologically, by decreases of the synthetic enzymes choline acetyltransferase and glutamic acid decarboxylase, by decreases of choline uptake and decreased binding of the cholinergic receptor agonist QNB (Coyle and Schwartz 1976). As in Huntington's chorea, tyrosine hydroxylase was relatively unaffected, indicating preservation of dopaminergic nerve endings. A theory put forward by Olney *et al.* (1975) to account for actions of kainic acid and glutamate analogs on other regions of brain is that they attach to glutamate receptors. Since glutamate is excitatory, the analogs may permanently open up ionic channels so that the cell is unable to maintain an appropriate concentration of intracellular materials. Although such a theory is highly speculative, there is an undoubted value in having an animal model of Huntington's chorea.

13.5.6 Summary of the Basal Ganglia

An appropriate way to consider the manner in which the basal ganglia fit into the overall motor scheme is to compare and contrast it with the cerebellum. They are both concerned with the initiation and smooth execution of movement. The basal ganglia have a more powerful role to play in initiation and gross control, while the cerebellum is responsible for fine tuning and the close matching of actual performance with cerebral command. "Static" from the basal ganglia produces a tremor at rest, while that from the cerebellum produces a tremor with movement.

In both the cerebellum and the basal ganglia there are two distinct subdivisions, the first of which receives a great majority of afferent fibers, mainly from the cortex, and which is densely cellular. This projects upon the second subdivision, which is composed of relatively fewer neurons and which sends afferents to other centers in a divergent fashion. In the cerebellum the divisions are the cortex and the deep nuclei respectively, while in the basal ganglia they are the neostriatum and globus pallidus.

There is not, of course, the same kind of laminated structure in the neostriatum as in the cerebellar cortex, but a considerable order may emerge when more is understood about the cellular arrangement. Both are obviously designed for extensive internal computation, with cells containing the highest concentration of dendritic spines in the nervous system, and with inter-neurons having axons making numerous boutons *en passant*. Each sends a major output from its deeper segment to the ventrolateral thalamus for playback upon area 4 of the cerebral cortex, and each sends a minor output to the midline nuclei of the thalamus. Thus, there is not only a similarity in design of the two systems; there is a conjunction of the principal loops to the cortex.

Having noted the similarities between the basal ganglia and the cerebel-lum, attention should be drawn to two key differences. The first of these is that the neurons of the basal ganglia utilize different neurotransmitters. Dopamine and ACh are major transmitters in the basal ganglia but appear to play no role in the cerebellum. GABA is the only neurotransmitter so far known to be shared by the two systems, although future research may uncover others.

The second is in the matter of behavior. The cerebellum appears to be purely a motor structure. But there is an important overlap between motor performance and behavior which seems to reside in the basal ganglia. This is evidenced by the close correlation between the antipsychotic activity of drugs and their tendency to produce a Parkinsonian-like state. It is also indicated by the mental symptoms seen in Huntington's chorea, a disease marked by striatal neuron loss. This behavioral overlap will be further discussed in Chap-ter 14.

13.6 Synthesis of Various Neuronal Mechanisms Concerned with the Control of Voluntary Movement

We can now recapitulate the discussion by synthesizing, to the extent possible, the contribution of cortex, basal ganglia, and cerebellum to volun-tary movement. In order to do so, a proposed sequence of events is illustrated in Figure 13.26. Initially, there is the conception of a movement which results in willing an action to take place. This achieves expression in patterns of excitation in the association cortex, which are recognized as the readiness potential in diffuse scalp recording (Figure 13.4). These play down to the basal ganglia and cerebellum, starting in train the open- and closed-dynamic-loop sequences that establish and guide the movement.

Loops playing through the basal ganglia are required for the initiation of the movement. This is established by the fact that in Parkinson's disease there is failure or delay in executing a willed action. At the same time the open loop of the basal ganglia is set in train, so is the open loop of the cerebellum to commence the initial programming of the movement. These open loops pro-

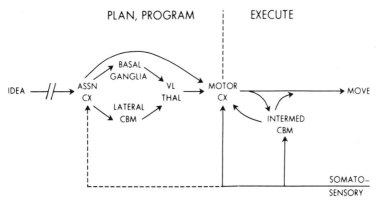

Figure 13.26 Pathways for planning, control, and execution of voluntary movement: proposed sequence of events in initiating movement.

ject back to the motor cortex via VL thalamus. In addition they also project back to the association cortex (Figure 13.17) with the opportunity of further dynamic loop circuitry. The synthesis of all these loop inputs with the ongoing activities of the association cortex provides what we may call preprogrammed information for the motor cortex. As a consequence of this preprogramming the appropriate discharges are generated down the pyramidal tract as the final motor command for bringing about the desired movement. This story has recently been given a penetrating and vivid expression by Mountcastle (1975).

At the stage of motor discharge, by the two closed loops illustrated in Figure 13.17A, the pars intermedia of the cerebellum makes an important contribution by updating the movement that is based upon the sensory description of the limb position and velocity upon which the intended movement is to be superimposed. This closed-loop operation is a kind of short-range planning as opposed to the long-range planning of the association cortex and the cerebellar hemispheres. Certainly, both cerebellar zones must cooperate in the performance of every skilled movement (Allen and Tsukahara 1974).

In learning the movement, we first execute it very slowly because it cannot yet be adequately preprogrammed. Instead, it is performed largely by intense cerebral concentration with playdown through the open loops as well as with the constant updating via the pars intermedia of the cerebellum. With practice and the consequent motor learning (cf. Chapter 14), a greater amount of the movement can be preprogrammed in the basal ganglia and cerebellar hemispheres and the movement can be executed more rapidly. With very rapid movements, e.g., piano playing or typing, we rely entirely on preprogramming by the circuits to the left of Figure 13.26 because there is no time for on-target correction by the pars intermedia once a fast movement has begun.

We may thus conjecture that trained movements are largely preprogrammed, whereas exploratory movements, which constitute an important fraction of our movement repertoire, are imperfectly preprogrammed, being provisional and subject to continuous revision.

The role of the basal ganglia can be envisaged by considering the consequences of the misfiring of its neurons. Units in the putamen and globus pallidus are known to discharge prior to the onset of movement (DeLong 1974). With damage to the neostriatum, as in Huntington's chorea, or with excess dopamine from L-Dopa therapy, preprogrammed circuits play back in random manner to area 4 of the cortex in the absence of cortical command. The result is the dancing movement from which chorea gains its name. The "volume control" of voluntary movement is turned too high and every movement is overdone. With damage to the nigrostriatal tract, as in Parkinson's disease, preprogrammed circuits can be played only with intense concentration. Even then the "volume control" is too low and the movement is poverty stricken. Background "static" produces a tremor at rest, which can sometimes be cured by lesions of VL thalamus. The same poverty of movement can be produced in monkeys by local cooling of the globus pallidus (Hore and Vilis 1976).

The role of the cerebellum, presumably the pars intermedia, in untrained exploratory movements is attested to by the clumsiness and slowness with which they are performed when, after cerebellectomy, the cerebrum and basal ganglia have to function in the absence of cerebellar cooperation in both preprogramming and updating (cf. Figure 13.11). If only the cerebellar hemisphere, with the circuits of Figure 13.17B, is put out of action, a tremor often results because the movement is so poorly preprogrammed that the pars intermedia can only ineffectively perform its normal function, which is updating a movement that is already a good guess. Presumably, the pars intermedia carries an immense store of information coded in its specific neuronal connectivities so that, in response to any pattern of pyramidal tract input, computation by the integrational machinery leads to an output to the cerebellum that appropriately corrects its pyramidal tract discharges.

Brooks (1975) reviews almost a decade of investigation on the effect of localized cooling in the region of the cerebellar nuclei on trained movements of monkeys. Cooling of the dentate nucleus (DN in Figure 13.10D) results in slower, poorly controlled, even tremulous movements (dysmetria), which indicates failure of preprogramming. By contrast, cooling of the intermediate nucleus (IN in Figure 13.10D) seemed merely to slow movements.

Let us now try to visualize what would be happening in the cerebro-basal ganglia–cerebellar circuits during some skilled action, for example, a golf stroke or a stroke in billiards. There will initially be a motor command with a preprogramming of the movement by circuits shown to the left of Figure 13.26. This preprogramming mobilizes all the learned skills and leads on to the pyramidal tract discharges with the consequent report of this discharge in detail to the pars intermedia of the cerebellum and the basal ganglia. At the next stage there will be a report back to the motor cortex of corrective information in accord with its learned skills. There will be a consequent modification of the pyramidal discharge which also reaches the pars intermedia and basal ganglia for further revision. But, meanwhile, the pre-programming circuits will also be in continual action, for the movement is not just some staccato action but a smooth, highly integrated performance with a duration of hundreds of milliseconds. Thus, we have to envisage that in the carrying out of a skilled movement, there is an immense integration of neuronal activities in interacting dynamic loops.

It is important to realize that no part of the cerebellar cortex or basal ganglia knows in general what other parts are doing. All the integration of input is going on in little subsets of machinery, as indicated for the cerebellar cortex in Figure 13.17. Each small zone of the cerebellar cortex is working with some subsets of the total information input, from a limb, for example, and the integration from the whole limb is accomplished in a piecemeal manner (cf. Eccles 1969b). It will be recognized that one could not integrate all the complex input coming from the movement of a limb in a single center. As would be expected, we find that the total information is broken down into subsets so that there are many centers for many different kinds of integration. These computing centers, if we may so call them, are not sharply demarcated from each other. On the contrary, they are diffusely organized with much overlap and replication. But all this diversity of operation is integrated in the end in the smooth control of the evolving movement. It will be appreciated that it is this end result that is important, not all the confusing diversity of operation that may appear in the impulse discharges in single units at the various stages of the neuronal circuits. In summary, we can regard the basal ganglia and cerebellum as acting like the gross and fine control system on a target-finding missile. They act similarly in that initially they aid in programming the movement but they do not give a single message for correction of a movement that is off-target. Instead, they provide continuous sequences of corrections.

Here it is important to recognize the role of the cerebral cortex in evaluating the movement and guiding the learning experience that goes with each execution. Athletes refer to this process as concentration and recognize the critical part it plays in their performance. The read-out of the preprogrammed motion must be perfect. The tiniest failure in execution must be carefully analyzed and the read-out improved in the next trial. Champion golfers, for example, go through this process hundreds of times each day, calculating the risk involved in "laying off" such practice for even the briefest of periods.

When we come to the question of the reliability of basal ganglia and cerebellar action in control of movement, we have to consider not only the operational character of the continuous dynamic-loop controls, but most importantly the enormous numbers of cells that are involved, in contrast to the single units of Figure 13.15. At all stages of the synaptic relays there is a convergence of many fibers onto the cell next in sequence, with the sole exception of the single climbing fiber input onto a Purkinje cell. Thus, there is averaging of many individual inputs so that the unreliabilities of the units are minimized. It may be questioned if this immense redundancy is worthwhile. The answer is that if reliability can be secured by having hundreds of units in parallel instead of one, then it certainly is a biological advantage. Moreover, we can assume that the multiplication of units gives in addition more subtlety and flexibility of performance in a way that we still do not fully understand. Anyway, this is what has happened in the evolutionary process and the marvelous control of movements exhibited by birds and mammals shows that in basal ganglia and cerebellar design it has paid off handsomely to go into numbers to give reliability, despite the counterweight of all the inherent complexity. We still cannot account for the relative numbers of the cerebellar cells, particularly for the enormous numbers of the granule cells—an estimated 30,000 million in the human cerebellum. It would seem that in order to secure reliability there is this prodigious neural cost. But in reality the neural cost is low because the metabolism of single cells is so minute. The total brain consumes only 20 W for a total of tens of thousands of millions of nerve cells. Hence, in evolutionary terms the animal could afford metabolically the prodigious neural cost, as numerically evaluated.

14

Basic Behavioral Patterns

14.1 Introduction

Our discussion so far on the integrated action of the brain has dealt only with the control of the musculoskeletal system. An attempt was made to account for how a willed movement could be smoothly executed. In giving this account, no effort was made to deal with what may have motivated the movement.

Why are there times during the day when we will our muscles to prepare a pleasant meal? Why at other times do we seek a comfortable bed, or wish to play a game of golf? Why does our heart pound if someone shouts at us in a dark alley? Why are there butterflies in our stomach before an examination or an important athletic contest? Why does our blood pressure rise under provocation? To say that we wish to eat, sleep, or get some exercise, or that we are afraid, anxious, or angry, is the most superficial kind of answer. It merely says that we have *feelings* that penetrate our conciousness, *motivate* our behavior, and produce noticeable changes in the functioning of our internal organs.

Feelings and motivation are *emotional* characteristics of the nervous system and, like all other aspects of brain function, must have an anatomical base and a physiological mode of expression. To date relatively little is known about these aspects of brain function, but the practical value of such knowledge would be difficult to overestimate.

Disturbances in these feelings and motivations, both major and minor, occupy a large proportion of the total time and effort of physicians. It is estimated that over half of all prescriptions written in North America are for minor tranquilizers. Beds occupied by the mentally ill, even though reduced to less than half their former total by the introduction of the major tranquilizers, nevertheless represent a significant proportion of all hospital beds. So it is understandable why probing the emotional aspects of the nervous system should receive so much fundamental research attention.

In this chapter we shall discuss some basic behavioral patterns, and where

465

possible, point out anatomical and biochemical correlations. Although we are concerned with the brain, it will be necessary for us to make some reference to visceral innervation and circulating hormones because they play such an important part in the overall behavioral response. Attention will be concentrated, however, on the operating characteristics of those parts of the brain that have been identified with behavior. We shall also discuss the psychoactive drugs, drawing together much of the information that was presented in scattered fashion under individual transmitters in Part II. Finally, we shall use the information developed to discuss from a biochemical point of view the mystery of mental illness. Throughout the chapter, the recurring theme will be the participation of certain metabotropic transmitters, particularly the catecholamines and serotonin. One of the amazing characteristics of the brain is that the relatively few neurons using these transmitters (probably less than .01% of the total in the human) should have such a profound influence over so many others.

14.2 Visceral Innervation and Feelings

Feelings are deeply intertwined with the demands of our internal organs. It was the great French physiologist Claude Bernard (1813–1878) who first focused attention on the homeostatic mechanisms designed to maintain the constancy of the "milieu interieur." The autonomic nervous system plays a vital role in maintaining this milieu through innervation of the viscera. Autonomic functions include the movement of the heart and gastrointestinal tract; the control of the smooth muscles of the blood vessels, the urinary bladder, the bronchi, and the skin; and the activities of various glands.

Stimulation of the musculoskeletal system in response to voluntary command must also, through simultaneous action of the autonomic system, produce coincident changes in the internal environment in ways which will preserve the fitness of this internal environment to meet the challenges placed upon it. For example, the decision to sprint to the corner to catch a bus requires increased heart rate and output of blood to the muscles, increased oxygen intake from the lungs, increased blood sugar from the liver, dissipation of the heat formed by sweating, and inhibition of digestion, micturation, or defecation. Thus, the autonomic system must be brought into play to adjust the functions of the viscera to the advantage of the organism as a whole.

We shall not be concerned with any of the details of the autonomic nervous system. They are part of classical neuroanatomy and neurophysiology and can be obtained from any number of excellent textbooks on the subject (Truex and Carpenter 1969; Goodman and Gilman 1970). We shall only refer to some of the broader principles to indicate how autonomic function may be integrated with central emotional perception and consequent behavior. Cannon (1929) described how the sympathetic nervous system prepares the body for "fight or flight," while the parasympathetic system is geared for rest and recuperation.

The sympathetic system mobilizes the existing reserves of the body quickly during emergencies or emotional crises. The eyes dilate, respiration is deepened, and the rate and force of the heart is increased. The blood vessels of the viscera and skin are constricted, providing priority flow to the voluntary muscles, heart, and brain. Peaceful activities are slowed down or stopped entirely. Appetite disappears, peristalsis and alimentary secretion are inhibited, and the urinary and rectal outlets are blocked by contraction of their sphincters.

The parasympathetic system performs the great service of eliminating wastes, building up reserves, and fortifying the body against times of need and stress. The sacral division rids the body of its intestinal and urinary wastes by controlling bladder and bowel function. The cranial division narrows the pupils (to shield the retinas from excessive light), slows the heart rate (to give the cardiac muscles longer periods for rest and invigoration), and ensures proper digestion (by providing for the flow of saliva and gastric juice and by supplying the necessary muscular tone for the contraction of the alimentary canal).

The efferent nerves of the autonomic nervous system arise from groups of cells extending from the midbrain caudally to the sacral region of the spinal cord. After issuing from the brain or cord, the fibers make connections with nerve cells situated either in a ganglion or in the innervated organ itself, whether it be muscular or glandular. As described in Chapter 5, ACh is the preganglionic transmitter for both parasympathetic and sympathetic divisions, and the main postganglionic transmitter for the parasympathetic system. Substance P (Chapter 10) may serve some fibers in the intestine, however, and serotonin (Chapter 9) serves argentaffin cells which also promote gut mobility. Noradrenaline is the main postganglionic sympathetic transmitter, while adrenaline is the principal catecholamine released from the adrenal medulla (Chapter 8).

Despite only limited voluntary controls over the viscera, we are nevertheless very conscious of their presence. They give us strong *sensations* or *feelings* and demand behavioral response. The unmyelinated postganglionic fibers, at 0.3–1.3 μm diameter, are classified as C fibers (0.7–2 m/s conduction rate), and response times are approximately 10-fold slower than with skeletal muscle. Thus, visceral changes have a more prolonged time course once initiated. That may be one reason why "feelings" leave a persistent impression on consciousness.

There is only limited voluntary control over autonomic function. There is almost no capacity to control heart rate, blood pressure, or digestion. Breathing, and bowel and bladder function, can only be controlled within limits.

There is a system of visceral afferent fibers which are not included in the classical autonomic system definition of Langley (1921) but which accompany visceral motor fibers and form the afferent arm of most visceral reflexes. It should be noted that visceral afferent neurons, like their motor counterparts, are interweaved with the musculoskeletal system so that neither acts independently of the other. Throughout most neural levels there is intermingling and association of visceral and somatic neurons. Visceral reflexes may be initiated by impulses passing through somatic afferent fibers coming from any type of external receptor, and, conversely, visceral changes may give rise to active somatic movement.

Such afferent pathways are not the only source of "feelings" in the brain. Input also comes from circulating hormones produced by the endocrine glands. These materials readily cross the blood–brain barrier where they may influence susceptible neurons. The sex hormones are particularly striking examples. The dramatic behavioral changes following castration of domestic animals has been exploited since the earliest days of animal husbandry. Over 2000 years ago, Aristotle made record of the fact that "the ovaries of sows are excised with the view of quenching their sexual appetites." But the conversion of a bland capon into an aggressive rooster by the transplantation of testicular tissue was not known until Berthold's description in 1849. And it was not until the availability of estrogen that veterinarians could promote coitus in "sterile" females by a simple injection. Some authors even state that administration of stilbestrol may evoke nymphomania in cows.

There are myriad papers on the relationship between sexual behavior and the circulating level of sex hormones. But how is this achieved? The uptake of [^{3}H]estradiol by rat neurons illustrated in Figure 14.1 gives one indication (Pfaff and Keiner 1973). The two sagittal sections of rat brain demonstrate the highly preferential uptake of sex hormone by the hypothalamus, septum, and amygdala, all areas prominently involved in sexual response. Presumably, the estrogen not only modulates the excitability of complex behavioral pathways, but also influences the appropriate release of the peptidergic hypothalamic hormones described in Chapter 10. Estrogens do not stimulate cyclic AMP production, so the receptor mechanism is different from dopamine and some other neurotransmitters.

Thyroid hormones as well as steroid hormones also have a profound effect on behavior. Hyperthyroidism is associated with nervousness, irritability, and general emotional instability, while hypothyroidism produces apathy and depression. Cushing's syndrome, which is caused by an overproduction of steroid hormones, is very frequently associated with frank psychosis. Addison's disease, which involves a failure to produce such hormones, results in anorexia, apathy, irritability, and depression. The appropriate hormones from these endocrine glands are all taken up by neurons, presumably through selective binding processes, and must thereby influence neuronal processes in a similar fashion to the sex hormones (McEwen 1976).

It is only natural to anticipate that the integration of autonomic function at higher levels should take place in brain areas where feelings associated with these nerve and chemical afferents can be detected and interpreted. Similarly, it could be anticipated that input from the memory bank, or other intrinsic brain sources, playing into these same parts of brain would result in motor reactions to the viscera and endocrine glands at the same time these "feelings" were registered upon the conscious mind. Where are these areas of brain that respond to feelings and coordinate behavioral response?

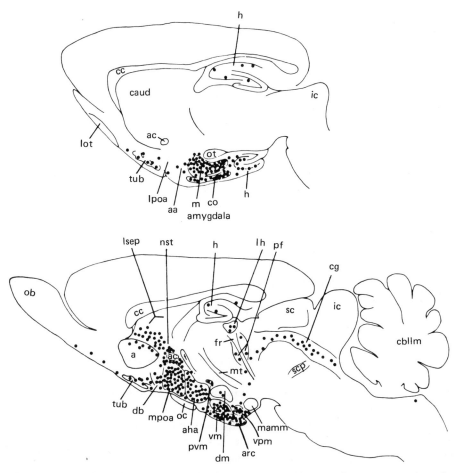

Figure 14.1 Uptake pattern of estradiol by rat brain: distribution of estrogen-concentrating neurons in the brain of the female rat represented schematically in two sagittal sections. Estradiol-concentrating neurons are represented by black dots. Most labeled neurons are found in the amygdala (upper section) and hypothalamic, preoptic, and septal areas (lower section). Labeled cells are also found in the olfactory tubercle. (From Pfaff and Keiner 1973; details of legend in original text.)

14.3 Central Coordination of Behavior

Hess was the first investigator to undertake a systematic exploration of the central integration of behavior. He studied the nature of the integration by electrically stimulating various regions of the brain in conscious, unanesthetized cats in order to observe their reactions. These experiments commenced in the 1930s and continued for many years. Hess recognized that such stimulation led to automatic responses amazingly similar to spontaneous behavior. Observations were generally recorded with moving pictures and then the area was usually lesioned with diathermy through the same electrodes in order to localize the site of electrode placement postmortem. After lesioning but before sacrifice, behavior opposite to the effects of stimulation was often obtained.

Hess found that lesions in the caudal region of the hypothalamus produced functional defects opposite to lesions in the rostral region. He conceived of an ergotropic system designed to integrate sympathetic and somatomotor activities to produce behavioral patterns preparing the body for positive action. The overall effects of ergotropic predominance would be arousal, increased sympathetic activity, enhanced skeletal muscle tone, and an activated psychic state, characterized by outgoing, exploratory behavior. He also conceived of a trophotropic system designed to integrate parasympathetic and somatomotor activities to produce rest and recuperation. With trophotropic predominance, there would be drowsiness and sleep, increased parasympathetic activity, decreased skeletal muscle tone and activity, withdrawal from external stimuli, increased appetite, enhanced digestive function and elimination of bodily wastes. The systems would be in rhythmic opposition resulting in times of trophotropic predominance (sleep, relaxation, or apathy) or ergotropic predominance (alertness, exploration, or aggressive activity) (Hess 1964).

He concluded that the integrating center for ergotropic function was the posterior hypothalamus, while that for trophotropic function was the anterior hypothalamus. Hess as well as many other investigators, however, recorded highly similar behavioral sequences following stimulation or ablation of a wide variety of subcortical structures, so that ascribing the integration of such broad functions exclusively to the hypothalamus must be regarded with caution.

The types of behavior he induced ran the emotional range from happiness to rage and included many which were associated with vicarious sensations from the viscera. A desire to eat, or a desire to drink, were created by brain stimulation in the absence of any physiological need. Similarly, a desire for sex was created in the absence of appropriate stimulation, as was an urgency to urinate or defecate.

Table 14.1 summarizes some of the dominant behavioral patterns that various investigators have observed after stimulation or ablation of areas in the limbic system, diencephalon, and brainstem. These regions seem to be the ones most critically involved. It by no means covers all of the phenomena observed, and does not attempt to deal with the complicated effects seen when more than one of these areas is stimulated simultaneously. The reader is referred to detailed reviews for more thorough descriptions (Hess 1964; Mogensen and Huang 1973; MacLean 1958).

Apart from the hypothalamus, the structures producing the most dramatic behavioral changes following stimulation or ablation are the amygdala and septum. Kluver and Bucy (1939) first noted that bilateral surgical excisions of the amygdala and its overlying pyriform cortex brought about drastic alterations in behavior. Monkeys, whose normally preferred diet is fruit, will eat raw meat or fish after ablation of this region, and wild animals seem to lose

TABLE 14.1
Behavioral Responses Observed in Selected Areas of the Brain

Brain area	Response following:	
	Stimulation	Ablation
Cingulate gyrus	Tameness or aggression	Tameness, lack of anxiety
Septum	Defecation, micturation, tameness, hypersexuality	Irritability and attack
Amygdala	Aggression	Tameness, hyper-sexuality
Thalamus		
Paramedian	Somnolence, relaxation	
Ventrolateral	Laughter	Apathy
Midline	Attack	
Anterior		Docility
Dorsomedial		Rage
Hypothalamus		
Lateral	Eating and drinking	Adipsia and aphagia
Medial	Fight	Hyperphagia
Dorsomedial	Flight	
Anterior		Trophotropic deficits
Posterior		Ergotropic deficits
Midbrain		
Central gray	Fight and flight	Hyperphagia

their sense of fear and to become tame. Animals may also show bizarre hypersexuality, as observed, for example, by Schreiner and Kling (1953). In their experiments, lesioned cats attempted to copulate with chickens.

Stimulation of the amygdala will produce behaviors that may be considered opposite. Mark and Ervin (1970) investigated a series of patients with epileptiform abnormalities of the temporal lobe whose problem was uncontrollable violence. In one of these patients, amygdaloid stimulation produced feelings ranging from elated and floating, through warm and pleasant, to unpleasant, according to the exact location of the electrodes. Stimulation of the medial part of the amygdala was dangerous, provoking him to lose control. Another patient with episodes of violence had this medial region stimulated while she was playing a guitar during a therapeutic session with a psychiatrist. As seizure activity took hold, she swung her guitar at the astonished psychiatrist, narrowly missing his head and smashing the instrument against the wall. Relief of episodes of rage was obtained in both these patients and several others, not by psychotherapy, but by surgical extirpation of the amygdala.

The septal and amygdala areas seem to have a partially reciprocal relation. Lesions in the septal area will produce the same effects as stimulation of the medial amygdala, i.e., irritability and a tendency to attack upon the slightest provocation. Stimulation, on the other hand, may produce penile erection and other evidence of sexual arousal. It also may stimulate the desire to defecate. Thus, the suggestion has been made that one of the consequences of amygdala extirpation is to permit septal dominance to emerge.

Only mild effects are produced with stimulation of the cingulate gyrus. Both tameness and aggression have been reported. The cerebral cortex may play an inhibitory role. Bard and Mountcastle (1948) reported that removal of the neocortex produces placidity unless the amygdala is subsequently lesioned, in which case ferocity is observed as in classical "sham rage."

Since the thalamus is the main sensory relay station of the brain, it would be reasonable to expect that stimulation or ablation of appropriate thalamic areas would greatly modify behavioral responses. Destruction of the dorsomedial nucleus or stimulation of midline nuclei increases irritability and attack reactions, while destruction of the anterior nucleus produces the opposite. These ergotropic patterns are balanced by trophotropic ones in other areas. Stimulation in the paramedian fields of the thalamus in the region of the massa intermedia (the so-called somnogenous zone in the diencaphalon) produces a decreased response to external stimuli that would normally provide a response, such as the strength of noise required to disturb a cat from sleep. During stereotaxic operations for Parkinsonism Hassler and Reichert (1961) observed that stimulation of the inner margin of the ventral oral nucleus of the thalamus frequently produced smiles or laughter in awake patients, which could not be suppressed even when the surgeon instructed that it be done prior to stimulating the area. Everything seemed to be funny to the patient or some amusing situation was recalled.

There is little doubt that the most prominent area for observing behavioral responses following stimulation or lesioning is the hypothalamus. In the cat, stimulation of the most medial sites results in considerable sympathetic arousal, with hissing, flattened ears, wide eyes, and a tendency to strike out at random with unsheathed claws. More laterally and ventrally, the animal stalks quietly, but will attack a rat with lethal effect (fight). In this same perifornicular region, stimulation will also stimulate feeding and drinking behavior, while these are inhibited by medial stimulation. Thus, the concept has emerged that facilitatory systems for ingestive behavior are represented in the lateral hypothalamus, while inhibitory ones are in the ventromedial hypothalamus. Lesions of this ventromedial hypothalamus will cause rats or cats to eat incessantly until they become disgustingly obese. The role of the hypothalamus in feeding and drinking has been the subject of many reviews (see Epstein 1971; Fitzsimons 1972; Mogensen and Huang 1973).

Lesioning of the central gray area of the midbrain will blunt rage reactions produced by hypothalamic stimulation, while midbrain stimulation will facilitate such attack. It is not known whether ascending or descending pathways are involved, or both.

To summarize the results of these diverse experiments, it can be said that complex behavior can be triggered by external stimulation of widely separated subcortical areas of the brain.

Olds was responsible for an exciting, yet remarkably simple advance over the external electrical stimulating techniques of Hess. He had the insight to perceive that animals, given an opportunity, would stimulate their own brains if the electrodes were correctly placed. He initiated the concept of pleasure and punishment centers in the brain. The key discovery came in 1953. Olds and Milner (1954) were investigating subcortical centers in rats in the hope of finding areas which, upon stimulation, would provoke the rat to perform some act to avoid the stimulus. To their surprise, they found that in certain brain areas the rats seemed to enjoy the stimulation. To confirm this impression, they placed a rat in a box with a pedal which the rat could press each time he wished to deliver a brief pulse of current to his brain. With this arrangement, and with electrodes placed in appropriate areas, the rat would stay at the pedal continuously for hours, stimulating his brain thousands of times. As Olds describes a typical rat he had prepared:

> He begins to search and pursue, eagerly after his very first stimulation. . . . Sometimes while the animal is self-stimulating, the circuit is cut off by the experimenter, so that stepping on the pedal will not produce any brain shock; in this case an animal that is self-stimulating will give a series of forceful ("frustrated-looking") responses and turn away finally from the pedal to groom himself for sleep; but he will go back from time to time and press the pedal (as though to make sure he is not missing anything).

The term "reward center" was devised to describe those areas of the brain where such self-stimulation could be elicited.

The avoidance effect, which had been originally sought, was also eventually found. In such brain areas, termed "punishment centers," the rat would never self-stimulate but would work to avoid getting stimulated.

With his simple experimental arrangement, Olds was able to map the rat brain for reward and punishment centers and to establish a hierarchy among the reward areas, based on how seriously the animal would seek such stimulation. He concluded that about 35% of the cells of the rat brain lay in reward centers, about 5% in punishment centers, and approximately 60% were neutral. Neocortical areas, i.e., the "newest" part of the brain developmentally, made up the majority of the neutral areas. The paleocortex, or cingulate gyrus, showed rates of approximately 200 stimulations per hour. This was increased up to 3000 per hour in certain subcortical nuclei, particularly in the septal area, and, in "lower" regions yet, even higher rates were found. Thus, rats would self-stimulate up to 5000 times per hour in the hypothalamus and up to 7000 times per hour in certain areas of the brainstem connected with the medial forebrain bundle. Neither self-stimulation, nor avoidance of stimulation, by itself gives insight into the nature of the sensation being experienced by the animal. Different experiments were devised to provide such information. Rats were taught to run a maze for the privilege of self-stimulation. At times they would run faster for electrical stimulation than for food. Furthermore, with electrodes in certain hypothalamic areas, the rats would run faster for an electrical stimulus even when they were starved. Given the choice between food or electrical stimulation at the end of the maze, the rats would choose electrical stimulation.

In other experiments, a rat was obliged to run across a grid which gave him successively more unpleasant shocks in order to reach the pedal for stimulation. The shock delivered by the grid could be raised to the point where the rat would forego the pleasure of brain stimulation and no longer cross the box. But the animal would accept roughly twice as powerful electrical shocks for self-stimulation as he would for food, even if he had been starved. The implication, then, is that electrical stimulation in these areas satisfied the equivalent of hunger centers in the brain, and that the electrical stimulation was more pleasant than the sensation which the brain received following food intake (Olds 1962).

Rats in another series were castrated after they had been taught to self-stimulate. In animals with electrodes correctly placed, the rate of stimulation dropped after castration, but increased when sex hormone was given to the rat. In these areas, which are distinct from the "hunger centers," it seemed as though the stimulation was providing sexual satisfaction.

Self-stimulation is obtained in virtually all areas of the limbic system, and to a lesser extent in areas of the extrapyramidal system. The medial forebrain bundle is the most preferred region of all.

Following Olds's discovery of self-stimulation centers in the brain of rats, it was not long before comparable studies were being undertaken in animals with more advanced brains. Cats, monkeys, dolphins, and humans were all studied.

Opportunities for making observations in conscious humans during subcortical electrical stimulation have occurred with patients suffering from epilepsy, brain tumors, and mental illness. The pattern previously described for animals has so far been generally corroborated. Reward centers have been found in the human associated with intense, but nonspecific feelings of well-being, with pleasant sensations ascribed to distinct parts of the body, and with sexual arousal. Punishment areas have also been found, the stimulation of which can elicit terror, anger, or pain (Delgado 1976; Sem-Jacobsen 1976). The implication of all these studies is that much learning and decision making, which presumably take place chiefly in the cortex, are directed toward stimulating pleasure centers and away from stimulating punishment centers. In other words, the cells which are "neutral" for reward or punishment are marshalled to direct the animals' behavior in such a way as to provide a sensory input which is "just right" for pleasure and punishment cells. That many of the functions associated with pleasure and punishment should have survival value for self and species is, no doubt, of more than casual consequence to our own existence.

Figure 14.2 indicates a number of the better known connections of the midbrain, diencephalon, and limbic system. These may provide some insights into the functional interrelationships suggested by self-stimulation and the other behavioral phenomena described in Table 14.1.

A key pathway is the medial forebrain bundle (MFB). It is shown in Figure 14.2 extending from the midbrain to the septal area with prominent connections along the way to the hypothalamus. It serves as the main two-way-traffic highway between the brainstem, diencephalon, and limbic lobe. The septum

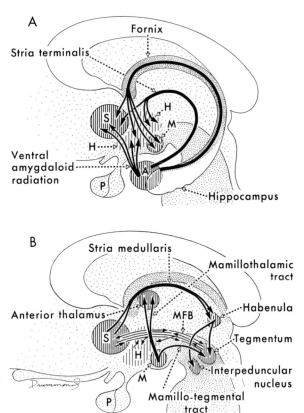

Figure 14.2 Some anatomical interconnections between the limbic system and brainstem. Abbreviations: H, hypothalamus; S, septal area; M, mammillary body; A, amygdala; P, pituitary.

has prominent connections with the amygdala via the stria terminalis and the diagonal band of Broca. The septum also gives and receives fibers from the hippocampus via the fornix. The majority of fibers in the fornix, however, are traveling from the hippocampus to the mammillary bodies with some being given off to other hypothalamic areas. From there, the mammillothalamic tract takes fibers to the anterior thalamus, from there to the cingulate gyrus, and then back to the hippocampus. The septum also sends fibers via the stria medullaris to the medial habenula, which sends its projection to the midbrain, particularly the interpeduncular nucleus. The hypothalamus also receives fibers from the amygdala and thalamus.

Close conjunctions are formed with the basal ganglia at a number of locations. The nucleus accumbens forms an intermediate zone between the head of the caudate and the septal area. The tail of the caudate merges with the amygdala. Fibers of the mesolimbic and nigrostriatal dopaminergic tracts intermingle and lie adjacent to medial forebrain bundle fibers, and the internal globus pallidus sends fibers to the lateral habenula. This close relationship between behavior and extrapyramidal function has already been mentioned in Chapters 8 and 13.

Figure 14.3 summarizes, in schematic form, the diverse behavioral relationships discussed in this situation. Suppose a man in a strange building

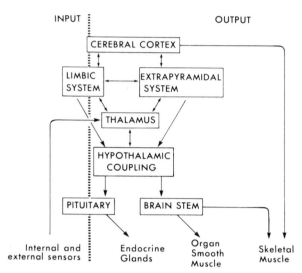

Figure 14.3 Schematic diagram of proposed input and output systems for behavioral responses.

smells smoke and hears a fire alarm. The external stimuli are immediately accompanied by internal signals connoting the "feeling" of fear, with consequent visceral sensations. These external and internal inputs are relayed, as shown on the left side of the figure, from the thalamus to the cerebral cortex, limbic system, extrapyramidal system, and hypothalamus. Coordination of response commences within milliseconds as motor systems come into play. These are shown on the right-hand side of the figure. The cerebral cortex gives a voluntary command to run for the exit, and the express motor pathways described in Chapter 13 start the musculoskeletal system into motion. The limbic system plays the danger signal into the hypothalamus to start the internal, involuntary forces working. Mechanisms in the hypothalamus couple these signals to the midbrain and spinal cord for sympathetic action and to the pituitary for endocrine response. Upon stimulation of these more slowly conducting fibers, sympathetic nerve endings and the adrenal gland release noradrenaline and adrenaline. The heart pounds, cardiac output increases, blood flow to the muscles is enhanced, and blood glucose rises. Hypothalamic releasing factors also stimulate the pituitary to release ACTH, which circulates to the adrenal cortex to produce the stress reaction. This endocrine response will carry on for hours after the emergency has been answered, as part of the body's long-term adaption to stress.

14.4 Central Amines and Behavior

Up to this point we have considered only anatomical correlates of behavior. Are there chemical correlates as well? Do the behavioral pathways (Figure 14.2) have transmitters that are relatively specific to them and not used by other pathways with different functions? Are there drugs which specifically interact with these transmitters? And do they bring about chemical modification of behavior? The answer to all these questions is a partial yes. But the evidence is at best fragmentary. Only the barest beginning has been made in exploring behavioral pathways. Most of the transmitters for the pathways shown in Figure 14.2 are not known, and the mechanism of action of many major psychoactive agents is still doubtful.

The individual chapters of Part II made some references to behavioral pathways and drugs which interact with them. The catecholamines and serotonin are the transmitters most involved in our behavioral story, no doubt because they have their richest innervation in the anatomical areas displayed in Figure 14.2. For example, serotonin (Figures 9.1, 9.2, 9.3, Section 9.2), adrenaline (Figure 8.18, Section 8.4.1), and possibly noradrenaline (Figures 8.14, Section 8.3; 8.15, Section 8.3.1) send rostral projections through the medial forebrain bundle (Figure 14.2B) to serve the hypothalamus, septal area, and structures beyond. Just laterally, dopaminergic (Figure 8.3, Section 8.2.1) fibers project to the cortex, accumbens, neostriatum, and amygdala. Dopaminergic fibers also project from the arcuate nucleus of the hypothalamus to the infundibulum. It would appear as if the overwhelming majority of all central pathways involving the catecholamines and serotonin are involved in behavioral and extrapyramidal circuits. Peripherally, of course, noradrenaline and adrenaline are the effectors of the sympathetic system. Dopamine modifies transmission in sympathetic ganglia (Figures 8.9; 15.9, Section 15.4). Serotonin is concentrated in argentaffin cells in the gut, and while release is not directly stimulated by nerve fibers, it profoundly affects intestinal motility.

ACh is also involved in behavioral and extrapyramidal circuits, although it has much broader general distribution. The septal-hippocampal, and habenulointerpeduncular pathways are cholinergic (Table 5.4, Section 5.2) and some amygdala pathways may be as well (Chapter 5). There is a rich supply of cholinergic striatal interneurons (Figure 5.4, Section 5.2). Peripherally, ACh is the effector substance at all autonomic ganglia and at most parasympathetic nerve endings.

Can we relate any of these amines and their pathways directly with behavioral response? Elmadjian (1959) was the first to do so, but only with the peripheral sympathetic system. He measured urinary noradrenaline and adrenaline in athletes and psychiatric patients under various conditions. His experiments demonstrated that the "fight" sympathetic response was primarily associated with noradrenaline excretion while the "flight" response was primarily associated with adrenaline excretion.

Elmadjian took pre- and postgame urine samples of professional hockey players. Typically, there was about a six-fold increase in noradrenaline excretion, but some exceptions were noted. Two players who sat on the bench and were worried about their injuries had no increase in noradrenaline, but an appreciable increase in adrenaline. One player became involved in a fight and was ejected from the game. He showed a 9-fold increase in noradrenaline excretion and a 20-fold increase in adrenaline secretion.

A comparable result was obtained with measurements from boxers, and basketball and baseball players. Noradrenaline excretion was high whenever aggressive play was undertaken without anxiety being associated with the contest. In the latter case, there was an elevated excretion of adrenaline.

Psychiatric patients followed the same general pattern. Those with a passive, self-effacing emotional display had normal levels of urinary noradrenaline, but one subject showing periodic bursts of excitement with expressions of fear and guilt showed high adrenaline excretion.

One patient and his psychiatrist were followed in a series of psychotherapeutic interviews. In one session, the therapist was severely criticized by the patient and showed a sharp increase both in urinary adrenaline and noradrenaline. In the following session, the psychiatrist severely criticized the patient and in other ways showed considerable emotional expression. On this occasion, it was the patient who showed the elevated adrenaline and noradrenaline excretion.

Other studies by Goodall (1962), involving astronauts in training, have confirmed the general conclusion that noradrenaline excretion is associated with active, aggressive action while adrenaline excretion is associated with apprehension and anxiety.

The physiological response to intravenous infusion of these two materials is also somewhat different. Noradrenaline causes an increase in systolic blood pressure due to vasoconstriction but no change in diastolic pressure. Adrenaline causes a greatly increased stroke volume of the heart, an increased systolic but decreased diastolic pressure, a mobilization of blood glucose, and a net vasodilation of blood vessels. Adrenaline clearly prepares the body for the more severe environmental stress of the "flight" reaction.

These correlations between the different emotional states associated with voluntary activity and response of the autonomic nervous system, as evidenced by excretion of some of its neurotransmitters, is an indirect indication of central integration. What direct evidence can be found?

Even before the neurotransmitter status of noradrenaline and serotonin had been established in the CNS, Brodie et al. (1959) reasoned that many drugs affecting the brain probably did so by modifying chemical transmission rather than by interfering with such universal processes as intermediary metabolism. They suggested that the ergotropic system of Hess used noradrenaline as its neurohormone, and considered that at least one component of the reticular activating system of Magoun (1958) must be noradrenergic. They further proposed that serotonin was the mediator of the trophotropic system. They cited as evidence for their ergotropic-trophotropic theory the concentration of these two amines within the brainstem, diencephalon, and limbic system, and the effects on behavior of such agents as Dopa, amphetamine, LSD, reserpine, chlorpromazine, and some monoamine oxidase inhibitors. They acknowledged that their evidence was fragmentary and circumstantial, but advanced their theory as a guide to new experiments.

Almost a decade later, Stein (1968) suggested that noradrenaline was the principal mediator for drive reward. He argued that the systemic administration of drugs which influence catecholamine metabolism had the most pronounced effects of any drugs on rates of self-stimulation. For example, substances that release catecholamines, such as amphetamine, α-methyl-m-tyrosine, or tetrabenazine in combination with an MAO inhibitor, facilitate self-stimulation. Conversely, drugs such as reserpine that deplete the brain of catecholamines, or drugs such as chlorpromazine that block receptor sites, suppress self-stimulation. Furthermore, stimulation is enhanced by MAO inhibitors or by tricyclic antidepressants. These drug data, taken in conjunction with the fact that the most rapid rates for self-stimulation are found in the region of the medial forebrain bundle, persuaded Stein that noradrenaline represented the chemical substrate for reward.

While there is little doubt that noradrenergic systems are important, there is evidence to refute this limited view. Self-stimulation is found in many areas of the brain which lack significant noradrenergic innervation. Structures such as the substantia nigra and striatum, which are interconnected by a dopaminergic neuron rather than a noradrenergic path, consistently support stimulation, although the rate is considerably lower than in the medial forebrain bundle. Two drugs which suppress self-stimulation (pimozide and haloperidol) do so regardless of whether electrodes are placed in dopaminergic or noradrenergic projections (Fibiger et al. 1976). Dopamine, it must always be remembered, is required to initiate movement (Chapter 13), and it is not possible to distinguish in the experiments of Stein between the effects of drugs on reward itself as opposed to the motor control needed to press the stimulating bar.

It is not even certain that the self-stimulation in the medial forebrain bundle is primarily governed by noradrenaline. Clavier and Routtenberg (1976) obtained good self-stimulation in the medial forebrain bundle even after bilateral lesions of the locus coeruleus which destroyed the major ascending noradrenergic pathway. Further destruction of the ventral noradrenergic system also failed to attenuate the stimulation even though lesions of the medial forebrain bundle itself severely reduced the rate. It seems clear from the data available that reward may be associated with various amine pathways (and possibly with still unknown neurotransmitters).

Serotonin is associated with more passive, introverted activities related to comfort and relaxation. It is particularly involved with sleep (Chapter 9). Peripherally, it may play some part in digestion.

Dopamine, of course, is associated with the initiation and execution of movement (Chapters 8 and 13). It also has an important governing influence on the endocrine system, particularly in promoting the release of growth hormone and inhibiting the release of prolactin (Chapters 8 and 10).

Other amines are suspected of playing a role in behavior but the evidence is much slimmer and depends mainly on their location within the brain. These amines include ACh (Table 5.3, Section 5.2), GABA (Table 7.2, Section 7.2.7; note the rich concentration in the extrapyramidal system as well as in certain areas of the limbic forebrain and hypothalamus), and histamine (Table 11.2, Section 11.1.1; note the localization in the hypothalamus).

By far the most convincing evidence as to the possible role of the aromatic amines in behavior has come from clinical observations on drug effects combined with basic studies on their molecular mode of action. The subtlety of action of any drug in a behavioral sense can never be appreciated through animal experiments. This is why detailed clinical observation is so important to establish correlations between behavior and the anatomy and chemistry of the brain. For these reasons, we shall give brief descriptions of the behavioral effects of major psychoactive drugs, correlating, wherever possible, their spectrum of action with neurotransmitter amines.

14.5 Psychoactive Drugs

A classification of the most widely used psychopharmacological agents is outlined in Table 14.2 and the structures of representatives are shown in Figure 14.4. Detailed pharmacological and clinical information can be found in many useful textbooks on the subject (Goodman and Gilman 1970; Cecil-Loeb 1971; Efron 1968, 1970; Clark and del Guidia 1970).

14.5.1 Neuroleptics or Major Tranquilizers

Reserpine. Reserpine (Figure 14.4A) was the first of the neuroleptics and is still an extremely valuable research tool. It is no longer widely used in psychiatry because of its side effects and because other, more useful agents have been discovered for the same clinical purposes. When an animal is given a large dose of reserpine, it is broadly distributed in the body without preferential uptake by CNS. The immediate effect is the loss of monoamines from storage granules in the brain and the periphery. Reserpine apparently acts on the membrane of the intraneuronal storage vesicle in a process that is Mg^{2+} and ATP dependent. Serotonin, dopamine, noradrenaline, and adrenaline are all similarly depleted. The clinical effects of reserpine must be interpreted, therefore, in terms of the loss of these neurotransmitters. These clinical effects may, in turn, give clues as to the function of brain pathways affected.

Reserpine has a marked calming effect. It produces indifference to environmental stimuli and a tendency to sleep, but arousal is nevertheless possible. The effects of a parenteral dose commence about an hour after administration and persist for 1–2 days. Nasal congestion is common, as well as excessive salivation. There is orthostatic hypotension, occasional nausea, and a risk of gastric ulcers on repeated dosages. In the initial phases, as amines are being released, there are paradoxical effects. These include excitement and hypertension. There is also a characteristic flushing and diarrhea.

Reserpine was the first truly effective chemotherapeutic agent for schizophrenics. It has clearly an antipsychotic action but unfortunately it also has one serious side effect. It produces mental depression. This is of sufficient intensity to cause suicide even in non-mental patients. In one study, for example, 26% of hypertensive patients receiving reserpine became depressed. Parkinsonian-like movement disorders are produced with high doses. It reduces sex drive, interferes with the menstrual cycle, and reduces self-

TABLE 14.2

Classification and Properties of Some Psychoactive Drugs

Category	Typical examples	Broad pharmacological effects	Clinical
Neuroleptics, or major tranquilizers	Phenothiazines Butyrophenones Thioxanthenes Reserpine Benzoquinolizines	Decreased sympathetic activity; decreased motor activity; decreased conditioned avoidance reflex; no anesthesia	Antipsychotic effect; tranquilization; indifference to environment; in high doses, extrapyramidal symptoms produced; calming action
Anxiolytic, or minor tranquilizers	Diazepoxides	Mild sedation, muscle relaxation	Tranquilization, but without any antipsychotic effect; sedation
	Barbiturates	As for other minor tranquilizers; also, there is some blockade of parasympathetic functions; in higher doses, ataxia and anesthesia	Tranquilization; hypnotic effect
Antidepressants	Imipramine Chlorimipramine Desipramine Amitriptyline MAO inhibitors	Reduced autonomic activity, especially of cholinergic systems; central sympathetic activation	Antidepressant action; stimulation
Stimulants	Amphetamine	Increased motor activity; alerting behavior; reduction of appetite and sleeping time	Decreased fatigue; anxiety already present may be increased; if psychotic symptoms are already present, they may be increased
Hallucinogenic agents	Lysergide Mescaline Psilocybin	Increased motor activity; variable autonomic effects, catatonia	Marked action on affect, perception, thought processes; psychotic disturbance may be aggravated.

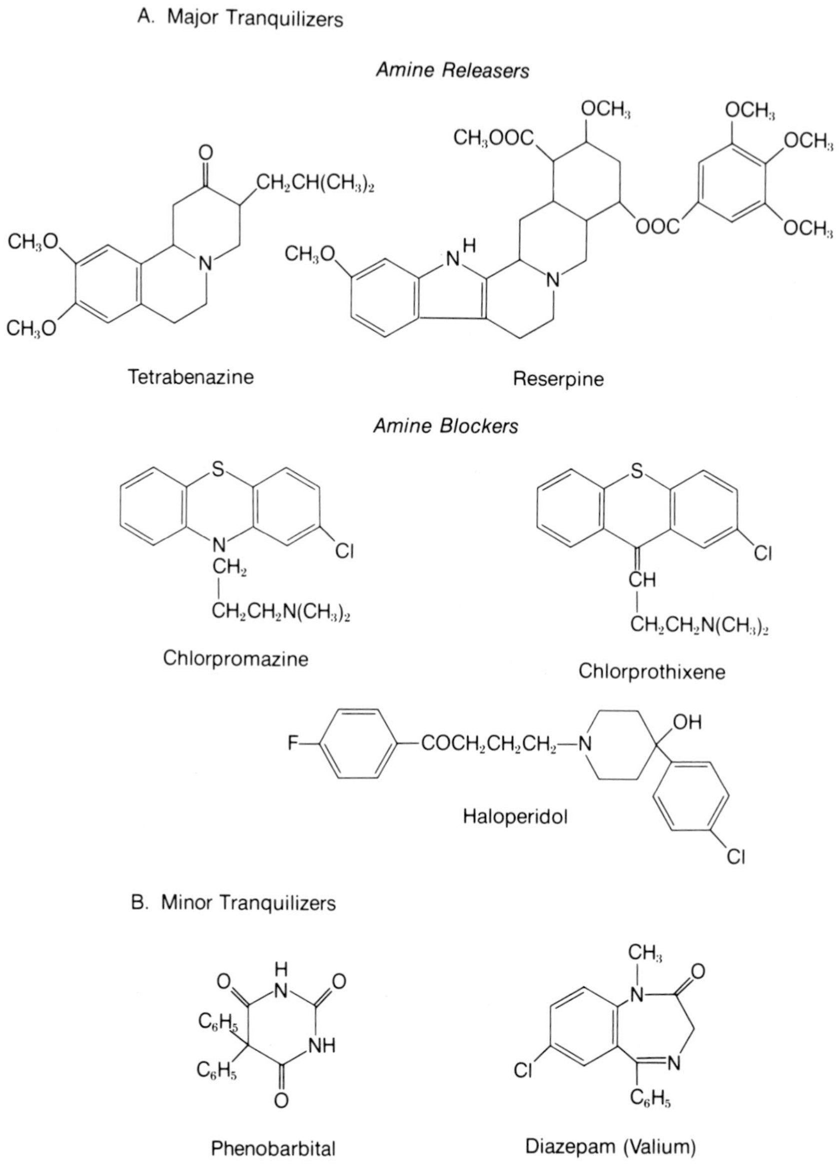
A. Major Tranquilizers
Amine Releasers
O
$CH_2CH(CH_3)_2$
CH_3O
CH_3O
N
Tetrabenazine
OCH_3
CH_3OOC
OCH_3
OCH_3
OCH_3
OOC
CH_3O
H
N
N
Reserpine
Amine Blockers
S
N
Cl
CH_2
$CH_2CH_2N(CH_3)_2$
Chlorpromazine
S
Cl
CH
$CH_2CH_2N(CH_3)_2$
Chlorprothixene
F
$COCH_2CH_2CH_2$—N
OH
Cl
Haloperidol
B. Minor Tranquilizers
H
O
N
O
C_6H_5
NH
C_6H_5
O
Phenobarbital
CH_3
N
O
Cl
N
C_6H_5
Diazepam (Valium)

Figure 14.4 Structures of some psychoactive agents.

C. Antidepressants

Monoamine Oxidase Inhibitors

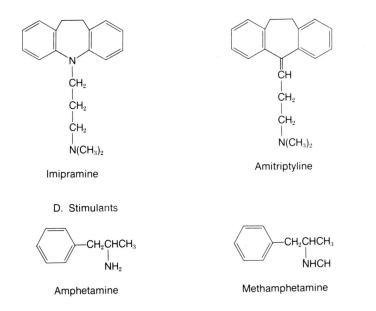

Iproniazid

Pargyline

Deprenyl

Tranylcypromine

Clorgyline

Tricyclic Uptake Inhibitors

Imipramine

Amitriptyline

D. Stimulants

Amphetamine

Methamphetamine

Figure 14.4 *(Continued)*

E. Hallucinogens

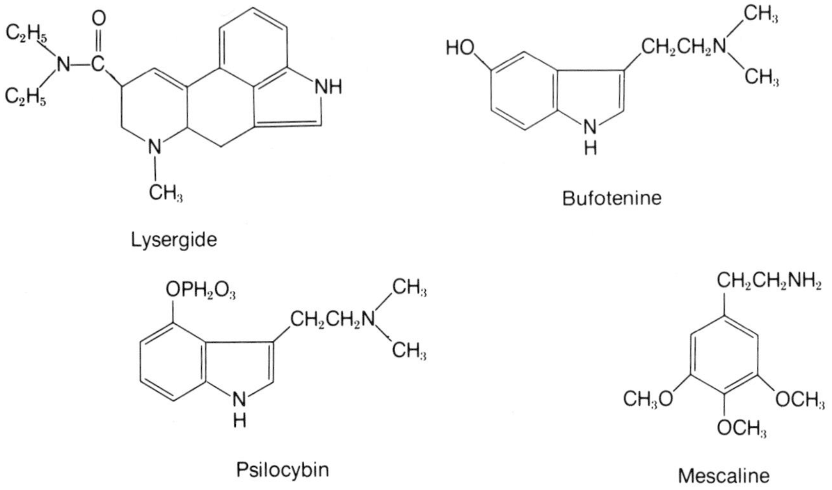

Lysergide

Bufotenine

Psilocybin

Mescaline

Figure 14.4 (*Continued*)

stimulation in animals. Reserpine has a prolonged action, apparently because it binds irreversibly to storage vesicles. Fluorescent histochemical studies have demonstrated that amines will begin to reappear within 12 h after the administration of a single dose. But the recovery of fluorescence commences in the cell body and only gradually extends to the nerve endings, so that full restoration of function lags behind. This would be in keeping with the formation and axonal transport of new binding proteins for the amines to replace those inactivated by the drug.

It seems fairly certain that the Parkinsonian and endocrine side effects of reserpine are due to depletion of dopamine, and the antihypertensive effects are due to peripheral depletion of noradrenaline. But other actions, particularly that of tranquilization and depression, are still somewhat conjectural.

Dopa will temporarily reverse the sedation caused by reserpine in some species, but 5-hydroxytryptophan will not. This has led to the conclusion that a catecholaminergic rather than a serotonergic system is involved in the tranquilization. Dopamine is the most likely candidate, since the reversal takes place before significant amounts of noradrenaline are formed.

Further evidence that noradrenaline is not responsible is that 3,4-dihydroxyphenylserine, a synthetic β-hydroxy derivative of Dopa which upon decarboxylation produces noradrenaline directly, has proved to be relatively ineffective at reversing the reserpine tranquilization. It crosses the blood–brain barrier poorly, however, and may not have comparable access to Dopa in reaching neurons.

The flushing and diarrhea experienced on initial dosing are probably due to serotonin release, because highly similar effects are seen in the carcinoid syndrome where rapid release of large stores of serotonin takes place. It has been speculated that reserpine-induced depression is due to serotonin depletion largely because PCPA, which is a tryptophan hydroxylase inhibitor, and thus a selective serotonin depletor, also produces depression in susceptible individuals.

Reserpine does not completely extinguish any of the many behavioral patterns it influences. Thus, none of them is absolutely dependent on the amines it depletes. Rather, the results argue for a modulatory effect of the amines on pathways whose primary function may be governed by other neurotransmitters.

In addition to the rauwolfia alkaloids, some benzoquinolizines have a reserpine-like action, both pharmacologically and on the storage of monoamines. The best known of these is tetrabenazine (Figure 14.4A) which behaves qualitatively like reserpine although it has a much shorter time of action.

Phenothiazines.   Phenothiazines were introduced into medicine some 40 years ago for the treatment of intestinal worms. They then were utilized as antihistaminics and basal anesthetics. It was not until 1952 that Delay and Deniker (1952) of France first reported on the psychiatric effects of the most famous phenothiazine, chlorpromazine (Figure 14.4A). They were convinced their results did not represent any simple relief of anxiety, but reflected a genuine antipsychotic action. For some years skeptical psychiatrists thought chlorpromazine no more effective than placebo, but by the mid-1960s over 50 million mental patients had been treated with a variety of phenothiazines and more than 10,000 articles had testified to their usefulness.

Delay and Deniker gave this classic description of the behavioral effects of chlorpromazine, upon which their term "neuroleptic" was based:

> He is usually aware of improvement induced by the treatment but does not show euphoria. The apparent indifference or the slowing of responses to external stimuli, the diminution of initiative and of anxiety without a change in the state of waking and consciousness or of intellectual faculties constitute the psychological syndrome attributable to the drug.

The effects of phenothiazines are qualitatively different from the sedative action of barbiturates. There is little ataxia and uncoordination with phenothiazines and the patient may be easily aroused.

The fundamental difference, however, can best be illustrated by a simple screening test developed by Courvoisier and colleagues (1953) to separate the two. A phenothiazine such as chlorpromazine impairs the ability of an animal to respond to an auditory warning signal (conditioned response) that an electric shock is coming, but not to its ability to run up a pole when it arrives (unconditioned response). Barbiturates, which make the animal sluggish, impair both escape and avoidance equally. High doses of most phenothiazines (and lower doses of those containing a piperazine moiety in the side chain) will, however, produce Parkinsonian-like side effects which can impair motor performance.

Phenothiazines have a prominent antiemetic effect which has been of important clinical benefit in such conditions as gastroenteritis, uremia, motion sickness, the nausea of pregnancy, carcinomatosis, and so on. They also strongly influence the endocrine system. Like reserpine, they block ovulation and interfere with menstrual cycles. They induce lactation, decrease testicular weight, and suppress growth. The latter can be overcome by injections of growth hormone.

Peripherally, they have a hypotensive effect and block the actions of injected adrenaline and noradrenaline. Antihistaminic effects seem to vary inversely and anti-Dopa effects proportionally with antipsychotic potency.

Phenothiazines are remarkably safe. Dosage may vary between 5 and 5000 mg/day. Unlike narcotic analgesics, barbiturates, and alcohol, there are no addicting qualities to phenothiazines. How are these remarkable clinical effects related to our known neurotransmitter pathways?

Phenothiazines block dopamine, noradrenaline, and serotonin receptor sites and therefore interfere with all the central and peripheral pathways described in Chapters 8 and 9. But many of the physiological actions are, to a large extent, explicable on the basis of dopamine blockade, including the antipsychotic, antiemetic, extrapyramidal, and endocrine effects. Since phenothiazines prevent apomorphine-induced emesis, the antinauseant effect is probably due to prevention of dopamine stimulation of the chemoreceptor trigger zone. The Parkinsonian side effects are undoubtedly due to blockade of the nigrostriatal dopaminergic tract. Effects on the endocrine system, particularly growth suppression and induction of lactation, could be due to interruption of hypothalamic dopamine since dopamine promotes the release of growth hormone and inhibits the release of prolactin. The most important action of phenothiazines, that of counteracting psychosis, could also be due to dopamine blockade, since there is such excellent correlation between antipsychotic potency and ability to prevent dopamine stimulation of brain adenylate cyclase (Chapter 8). On the other hand, the peripheral vascular effects are probably due to noradrenaline blockade.

Butyrophenones. Haloperidol (Figure 14.4A), the most important butyrophenone, was introduced in Europe in 1958 and the United States in 1967. Although butyrophenones are highly important clinically, they do not provide a significantly different spectrum from some phenothiazines. There is also a considerable degree of structural similarity between butyrophenones and piperazine-substituted phenothiazines (Figure 14.4).

Butyrophenones are virtually devoid of peripheral autonomic effects, but centrally they have been shown to block both dopamine and noradrenaline. They are potent antipsychotic and antiemetic agents, and tend to produce Parkinsonian side effects.

The only novel suggestion about the mechanism of action of butyrophenones as opposed to phenothiazines is that they may mimic GABA and block glutamate (Janssen 1967). Haloperidol has been superior to most phenothiazines in the treatment of Huntington's chorea (Chapter 13), which helps to lend credence to that suggestion. Although it blocks dopamine-sensitive adenylate cyclase, its activity in this model system is below what would be anticipated from its clinical potency (Chapter 8).

Thioxanthenes. This family represents only a minor structural variation from the phenothiazines (Figure 14.4A), the nitrogen in the central ring being replaced by a carbon atom. Chlorprothixene (Figure 14.4A) is the main clinical representative, but others are in use as well. The spectrum of action of the thioxanthenes is almost identical to phenothiazines and to butyrophenones. They are potent blockers of dopamine-sensitive adenylate cyclase (Table 8.3, Section 8.2.4). Support for the theory that blockade of this second messenger system for dopamine is the reason for antipsychotic action is the fact that only the cis isomers are active both clinically and against the cyclase. Trans isomers are inactive in both spheres (Iversen 1975).

14.5.2 Minor Tranquilizers or Anxiolytics

These are the most widely sold of all prescription drugs. There are only two major groups: the barbiturates and the benzodiazepines. At one time, barbiturates such as phenobarbital (Figure 14.4B) dominated the market, but it is now fashionable to prescribe a benzodiazepine for tension, anxiety, or any of a variety of minor complaints involving physical or mental discomfort. There is little solid evidence that any of these drugs interact with the behavioral monoamines which are the primary concern of this chapter. Rather, they seem to exert a nonspecific depression on all neurons. This is of some value in the treatment of epilepsy and alcoholic withdrawal, but there is the disadvantage of drug dependence with both these classes. The benzodiazepines—chlordiazepoxide, diazepam (Figure 14.4B), and oxazepam—have special muscle-relaxant qualities which has suggested a possible interaction with glycine and GABA (Chapter 7).

14.5.3 Antidepressants

MAO Inhibitors. Until the late 1950s, the only pharmacological treatment for depression was stimulant drugs. Psychotherapy and electroshock treatment were therefore used almost exclusively. But the euphoriant properties of iproniazid (Figure 14.4C) were noted when it was being tested alongside the antitubercular agent isoniazid. Iproniazid was discontinued on these grounds as a treatment for tuberculosis, but not before Zeller had noted that, unlike isoniazid, it inhibited MAO. Kline and colleagues (Loomer *et al.* 1957) introduced iproniazid into psychiatry, reasoning that its euphoria-inducing properties would be valuable in the treatment of depression. They coined the term "psychic energizer" to describe its action.

The properties of MAO itself have been described in Chapters 8 and 9 as well as the properties of some of the wide variety of agents that have been found to inhibit it. The first MAO inhibitors to be used clinically were all hydrazine derivatives. They turned out to be the most dangerous of all drugs used for treating mental illness because of their toxicity to liver, brain, and the cardiovascular system.

Since the recovery rate from depression is high, these dangerous side effects had to be weighed against the seriousness of the disease itself. MAO participates in a wide variety of actions that have nothing to do with the catecholamines and serotonin, and some of the side effects were not anticipated. One particularly bizarre example involved the ingestion of tasty cheese such as Camembert and Stilton. These produced violent headaches, hypertensive crises, and, in a few instances, cerebral hemorrhage and death in depressed individuals on MAO inhibitors.

Asatoor and colleagues (1963) discovered that the high content of tyramine in the sharp cheeses was escaping destruction by liver MAO and acting as a false transmitter for noradrenaline. The released noradrenaline was causing a lethal increase in blood pressure. Other foods containing aromatic amines soon became added to the danger list.

The vulnerability of the body to exogenous foods or to overdosing by hydrazine-type MAO inhibitors is not easily corrected since these compounds bind irreversibly to MAO. New MAO must be synthesized in the cell body and transported to the nerve endings before the effect of the inhibitors is counteracted.

The nonhydrazine MAO inhibitors such as tranylcypromine (Figure 14.4C) and pargyline (Figure 14.4C) are less toxic than the irreversibly inhibiting hydrazines. Nevertheless, all MAO inhibitors are used sparingly, and a number have been removed from the market because of toxicity.

Mood elevation is the principal psychic effect of the MAO inhibitors. It is often not easy to establish this fact because the clinical response takes several weeks, and depression is not a stable disorder. It is not known why the response is so slow. In animals, behavioral changes are similarly difficult to see even when amine levels are sharply altered. MAO inhibitors must be combined with amine precursors such as Dopa or 5-HTP, or a releaser such as reserpine, for excitement and agitation to be noted readily. In humans, the effects of overdosage include insomnia, agitation, hallucinations, hyperpyrexia, and convulsions. In some cases treatment may bring about manic or hypomanic reactions.

Tissue levels of dopamine, noradrenaline, adrenaline, and serotonin are all elevated by MAO inhibitors in animals. Measurements in humans at clinically effective doses have not been done. The antidepressant effect could be due to enhancement of any one or any combination of these amines. It could also be due to some other, as yet undetermined, action of these MAO inhibitors. Combinations of MAO inhibitors with precursors would suggest, however, that the hyperpyrexia is a serotonin effect while the agitation and insomnia are catecholamine effects.

A paradoxical effect of MAO inhibitors on blood pressure has never been adequately explained. Instead of the expected hypertensive effect due to enhanced noradrenaline, a hypotensive effect is often observed. Pargyline, which is mildly preferential for type B MAO inhibition, consistently produces hypotension, and this is its principal clinical application.

The recent availability of more narrow-spectrum MAO inhibitors, such as clorgyline (Figure 14.4C) and deprenyl (Figure 14.4C), may cast new light on the problem of correlating physiological effects with amine metabolism.

Despite the theoretical importance of MAO inhibitors, they have been eclipsed clinically by the safer, and probably more effective, tricyclic antidepressants.

Tricyclic Antidepressants. The tricyclic antidepressants include dibenzazepines such as imipramine (Figure 14.4C) and dibenzocycloheptadienes such as amitryptyline (Figure 14.4C). Their structure is basically similar to the phenothiazines except that the sulfur atom of the central ring is replaced by an ethylene bridge. The result is to bend the molecule out of its planar shape and to alter substantially its spectrum of action.

The first tricyclic to be investigated clinically, imipramine, was found not to be effective in schizophrenia as anticipated, but to be remarkably beneficial in depression (Kuhn 1958). From that point on, many cogeners were synthesized and tested and much research undertaken to establish their mechanism of action. They soon became a particularly attractive class of drugs, largely because of the drawbacks of alternative modes of therapy. MAO inhibitors carry several risks, while ECT is not appealing in the long term for a condition as subject to relapse as depression. Tricyclics are relatively safe and can be administered on a long-term basis. They can also be used safely in conjunctions with phenothiazines for schizophrenics who are subject to depressive episodes.

The principal effect of the tricyclics is not to inhibit MAO. It is to interfere with the "pump" systems that cause reuptake of the monoamines, particularly noradrenaline and serotonin, from the synaptic cleft back into the nerve ending.

Unlike the MAO inhibitors, imipramine will not usually cause euphoria in normal individuals. Rather, it tends to produce fatigue, plus annoying side effects such as dry mouth, palpitations, and urinary retention. The way in which it counteracts depression in mental patients is not really understood. Since it seems to impair cognitive processes, some have suggested that it blunts depressive ideation. It can, however, produce euphoria, insomnia, hallucinations, delusions, and even manic excitement in some psychiatric patients, so that it has the potentiality to overcompensate for depression.

Administration of imipramine concurrently with, or shortly after, treatment with MAO inhibitors has resulted in hyperpyrexia, convulsions, coma, and death. Even if the tricyclics are given some days after cessation of MAO inhibitor treatment, untoward reactions can occur. Thus, the two classes have an adjuvant effect, as would be suspected. Although the combination is dangerous, it has been used occasionally under very carefully controlled conditions with good effect.

The overlap in spectrum of action between tricyclics and MAO inhibitors focuses attention on the catecholamines and serotonin as being the molecules primarily related to the antidepressant effect. The evidence particularly points to serotonin and noradrenaline, because the dopamine pump is unaffected by clinically effective tricyclics. Instead, these agents inhibit the serotonin and noradrenaline pumps. Imipramine, chlorimipramine, and amitryptyline preferentially inhibit reuptake of serotonin, while desipramine, nortryptyline, and protryptyline are more effective against noradrenaline. There is overlap between the two groups, and all are effective clinically. Benztropine, which strongly inhibits dopamine reuptake but not that of noradrenaline or serotonin, has no antidepressant action. Instead, it counteracts drug-induced Parkinsonism, a property not shared by the tricyclics.

14.5.4 Stimulants

Amphetamine (Figure 14.4D) and similar compounds are stimulants but not antidepressants. In normals they produce wakefulness, alertness, a decreased sense of fatigue, increased ability to concentrate or perform physical tasks, euphoria, and increased initiative and confidence. They can also produce headache, palpitation, dizziness, agitation, and, with the famous amphetamine "payback" principle, an overwhelming fatigue as the effect wears off. They suppress appetite, and were at one time used to encourage weight reduction.

The clinical use of these agents is now largely restricted to narcolepsey and the treatment of hyperkinetic children. Illicit use is, however, common. The improvement in athletic performance has been widely capitalized on by professional athletes with little evidence of serious harm. The same cannot be said for intravenous use by soft-drug addicts. The slogan "speed kills" can equally refer to driving or to "speed," the street name for amphetamine derivatives and the inevitable outcome of its promiscuous parenteral use for thrills. The drug produces an instant feeling of well-being upon intravenous administration, but at high doses this is always replaced by a paranoia strikingly similar to the endogenous disease. The amphetamine psychosis is said to mimic schizophrenia more closely than any other form of drug-induced or toxic psychosis. It is this latter action which is of greatest behavioral interest because it may provide a decisive clue to the etiology of schizophrenia, the most severe of the mental illnesses.

The effects of amphetamine, methamphetamine (Figure 14.4D), and similar agents can probably be attributed exclusively to their interaction with catecholamines. This can take place in a number of ways. First of all, amphetamine acts as a releaser in the same way as mentioned previously for tyramine. It penetrates the nerve ending and releases the endogenous amine, producing an enhanced physiological result.

The extent to which the pharmacological effects of amphetamine reflect changes in dopaminergic systems, as opposed to noradrenergic or even adrenergic systems, is still controversial. The classical idea that the central stimulant and anorexic actions are mediated by noradrenaline whereas the stereotyped behaviors are mediated by dopamine has been challenged, with turnover rates suggesting that most of the effects may be due to dopamine (Chapter 8).

A second important action of amphetamine is as a competitive inhibitor of MAO. Amphetamine, being a monoamine, is destroyed by MAO and it competes successfully with catecholamines for this enzyme.

Amphetamine also inhibits reuptake of catecholamines, especially dopamine. Again, this appears to be a competitive phenomenon. In this manner it acts similarly to cocaine or the tricyclic antidepressants. Amphetamine may also have some small influence at catecholamine receptor sites, but in view of the "payback" principle, which is reminiscent of tachyphylaxis, it is unlikely that this contributes in a significant way to its initial positive activity.

Whether the effects of amphetamine-like stimulants are due to dopamine or noradrenaline, they produce bizarre behavioral disturbances on overdosage. Thus, they lie on the borderline with the next category of psychoactive drugs, the hallucinogenic agents.

14.5.5 Hallucinogenic Agents[1]

From a theoretical point of view, drugs which disturb normal behavior are as important as those which correct abnormal behavior. They are a major social problem instead of a significant benefit, but they may nevertheless be capable of yielding equivalent insights into the workings of the behavioral areas of brain.

Hallucinogenic agents, also called psychotogens, disturb the normal sensorium and thus affect mood. The terms are somewhat inappropriate since they seldom cause either full-blown hallucinations or frank psychosis. They are a motley group of compounds whose mechanisms of action are not understood. We shall be concerned in this chapter with three main groups, which are thought to interact with the behavioral amines. These are: indole derivatives, which may interact with serotonin; phenylalkylamine derivatives, which may interact with catecholamines; and glycolic acid esters, which may interact with ACh.

Indolic Hallucinogens. A number of these have structures strikingly similar to serotonin. The two closest are bufotenine (N,N-dimethyl-5-hydroxytryptamine, Figure 14.4E), and psilocybin (N,N-dimethyl-4-phosphoryltryptamine, Figure 14.4E). The active principal of the latter compound is psilocin, the 4-hydroxy analog of bufotenine, which is produced when the phosphoryl group is cleaved. Bufotenine is found both in plants and in the poison gland of the toad. Psilocybin is obtained from certain mushrooms found in the southern highlands of Mexico. N-Substituted tryptamines such as dimethyl- and diethyl-N-tryptamine are also hallucinogenic.

All these compounds produce disturbances in sensation and mood. They also produce thought disturbances. These latter effects have been pursued in relation to schizophrenia. Some investigators, for example, claim that bufotenine is a constituent of the urines of schizophrenic patients, normal subjects, or both. The implication is that schizophrenia may have as its fundamental cause the production of this hallucinogenic compound. Such results have not been generally corroborated, however. Although proof has been hard to obtain, it is thought that these materials interact with serotonin recep-

[1]General reference: Efron 1970.

tor sites with the N-methyl substituents making them poorer substrates than serotonin for MAO.

By far the most dramatic hallucinogenic compound of all that have been discovered is D-lysergic acid diethylamide (Figure 14.4E). LSD is the most powerful known psychotomimetic compound. Its discovery by Hofmann in 1943 was an accidental one. He had been synthesizing a series of ergot derivatives with a view to finding agents having oxytocic action. One day he was unaccountably seized with peculiar sensations while at work. He went home to experience a period of exaggerated imagination coupled with, in his words, "fantastic pictures of extraordinary plasticity and intensive color." He retained enough insight to suspect that the effect might have come from the new compound LSD. He returned after a few days to ingest experimentally 250 μg, a dose he believed to be extremely minute. This is now known to be 5–10 times the dose of LSD required to cause disturbances in susceptible subjects, and the effect on poor Hofmann was spectacular. From then on there was no doubt that LSD was an extremely potent psychotogen (Hofmann 1963).

The social upheaval instigated by this compound is hard to credit even today. It not only revived theories of a toxic chemical etiology for schizophrenia, but it sparked a search for other psychotomimetic agents. An astonishing number of these were uncovered, including new synthetic compounds, as well as a variety of natural products that had been hidden in the legends and religions of primitive peoples. After a lag of almost 20 years, the use of LSD broke the bounds of scientific control and became the focus of a drug craze. Seldom has an agent unleashed such an unpredictable chain of events.

LSD is generally considered to be either a serotonin agonist or antagonist. It has a powerful blocking action on peripheral smooth muscle preparations which respond to serotonin. Centrally, there is considerably evidence that it stimulates rather than blocks some serotonin receptor sites (Chapter 9).

Phenylalkylamine Derivatives.   The most famous representative of this group is mescaline or 3,4,5,-trimethoxyphenylethylamine (Figure 14.4E). This constituent of the peyote button is famous for its use by Aldous Huxley and in religious rites of North American Indians. It produces visual hallucinations of a vivid and striking quality. There may be the appearance of brightly colored lights, vivid geometric patterns, animals, and people, all with a distortion of space and time. Several similar compounds such as 2,5-dimethoxy-4-methylamphetamine (DOM) and 2,5-dimethoxy-4-ethylamphetamine (DOET) are also hallucinogenic.

Exactly how mescaline and the amphetamine-like derivatives exert their halluconogenic effects is not known. Their resemblance to the catecholamines, however, definitely suggests that it is through some interaction with them. In fact, the close structural resemblance to dopamine has led to the suggestion that in schizophrenia, methylation of the ring hydroxyl groups of dopamine might lead to an endogenous toxin of structural similarity to mescaline. Unfortunately, there has been no solid evidence that significant amounts of this material are made in the human body, under normal or abnormal circumstances.

Glycolic Acid Esters. A series of compounds studied by Abood and Biel (1962), which are esters of a heterocyclic imino alcohol and a glycolic acid, proved to produce in humans delerium, hallucinations, and many abnormalities associated with psychotic-like behavior. Their effects are unpleasant so that they have not become a problem on the illicit market. They have in common a strong anticholinergic action. The best known representative is Ditran. Physostigmine has a temporary antidotal effect. Abnormal behavior is also noted with high doses of much milder anticholinergics such as atropine and hyoscine. Thus, ACh receptors are implicated in behavioral abnormalities along with serotonin and catecholamine receptors.

14.5.6 Summary

To summarize the results of this section on drugs, it can be said that the main interfaces of all these materials of widely differing structure are the catecholamines and serotonin. It seems evident that perturbations in the metabolism of these amines can result in thought disorder and shifting of mood, as well as disturbances in movement, appetite, sleep, and functioning of the endocrine glands.

Yet is is equally clear that none of these functions is exclusively dependent on any of the aromatic amines. Therefore, it must be concluded that they somehow modulate the functions.

Dopamine seems to be involved in modulation of movement, certain endocrine functions, and possibly also thought. Noradrenaline seems to be involved in alertness and exploratory behavior for food, etc. Serotonin, on the other hand, seems to be concerned with sleep, relaxation, and digestive function.

14.6 The Mystery of Mental Illness

We have now considered some anatomical pathways concerned with behavior, and the neurotransmitters that govern their actions. We have classified psychopharmacological agents, drawing attention to the interaction of many of them with these same neurotransmitters. This leads to a consideration of the problems of mental illness, and whether the information developed so far leads to insights as to the etiology of this group of illnesses.

Mental disorders, by definition, are those in which physical damage to the brain cannot be detected postmortem by the usual pathological techniques. Nevertheless, it is clearly unacceptable to a large body of scientists that a deranged mind could exist in the presence of a normal brain. It is this conviction which spurs biological research into mental illness. It is strengthened not only by the effects of psychoactive drugs already described, but by the knowledge that there is a genetic or familial nature to the major mental illnesses.

The genetic nature of schizophrenia seems to have been proven beyond question by the elegant studies of Kety and co-workers (Kety 1974). Using the detailed birth records of the Danes, they were able to assemble statistically significant groups of offspring raised in foster homes in order to separate genetic from environmental factors.

Children with a schizophrenic genealogy raised by nonschizophrenic families had approximately the same attack rate as children from similar genealogy raised by their own families. On the other hand, children with a clear genealogy raised by schizophrenic families did not experience a higher attack rate than the general population. This result confirmed the classic study of Kallman (1938) which had been attacked on the basis that the environment in schizophrenic is different than that of normal families. Kallman found that the incidence of schizophrenia in the general population was 0.85%. The concurrence rate for parent–offspring was 9%; that for full siblings 14%; that for two-egg twins 15%; and that for one-egg twins 86%.

Such impressive statistics have not been accumulated for manic-depressive psychosis, but nevertheless the strong familial tendency has been consistently noted in epidemiological studies. Thus, while it is recognized that environmental factors can precipitate a psychotic illness, there can be little doubt that a constitutional vulnerability exists which must someday be revealed by appropriate physiological studies.

Another basis for belief in the theory of constitutional vulnerability is that psychoactive drugs frequently produce a different response in mental patients than in normals. For example, S. Gershon et al. (1975) administered L-Dopa to psychiatric inpatients. In the nonpsychotic group, the behavioral effects were quite similar in quality and range to those seen in neurological patients such as Parkinsonians. When L-Dopa was administered to 10 schizophrenic patients, however, a clear deterioration occurred in every instance. Therefore, L-Dopa can induce paranoid psychosis in the occasional nonpsychotic patient but dramatically aggravates psychopathology in schizophrenics. Similarly, when L-Dopa was administered to manic-depressive patients, there was a tendency to precipitate hypomanic behavioral episodes which were strikingly similar to the natural episodes.

Amphetamine and methylphenidate are similar to L-Dopa in that they cause dramatic exacerbation of symptoms in schizophrenics, but only produce a reaction in normals at very high doses.

Thus, in predisposed individuals, psychotic episodes can be precipitated by alterations in catecholamine metabolism. The effect is not restricted to catecholamines, however. In a well-known study conducted in 1961 at the U.S. National Institute of Health, Pollin and colleagues administered various amino acids together with an MAO inhibitor to schizophrenic patients (Pollin et al. 1961). Most of the amino acids brought about no changes, but L-tryptophan caused a feeling of euphoria, while methionine produced anxiety, a flood of associations, increased hallucinatory activity, and periods of disorientation.

In view of these general observations, it is not surprising that most of the presently popular biochemical theories of mental illness suggest disturbances of either catecholamine or serotonin metabolism.

14.6.1 Theories of Schizophrenia

The Dopamine Hypothesis. The most prominent current theory of schizophrenia suggests that there is an abnormality in dopamine metabolism in which excess dopamine reaches critical receptor sites, probably in the mesolimbic and/or mesocortical dopaminergic pathways. The main evidence is the following:

1. Excessive systemic doses of CNS stimulants, such as amphetamine, which are primarily false transmitters for catecholamines, will produce a drug-induced psychosis which is extremely difficult to distinguish from the disease itself (Snyder *et al.* 1974a).
2. In about 15% of cases receiving high doses of L-Dopa for other therapeutic purposes, a psychotic reaction is produced (S. Gershon *et al.* 1975).
3. Antipsychotic drugs either block dopaminergic receptor sites, as evidenced by inhibition of dopamine-stimulated adenyl cyclase, or deplete dopamine from its stores (Iversen 1975). Propranolol (Atsmon *et al.* 1972), a noradrenergic blocker (Table 8.4), seems to be the only exception.

Against this theory may be cited the fact that no abnormality has yet been detected in dopamine metabolism in schizophrenia. Tyrosine hydroxylase levels are not elevated postmortem (P. L. McGeer and McGeer 1978) and homovanillic acid levels in the urine and CSF are not increased premortem. Tyrosine hydroxylase and dopamine levels are also not severely altered in Huntington's chorea, a neurological disorder which produces behavioral changes which may be strikingly similar to schizophrenia in the early stages (Bird and Iversen 1974; P. L. McGeer *et al.* 1973).

It has been reported that platelet MAO is low in schizophrenic patients, and that this is genetically determined (Wyatt and Murphy 1975). This might imply that all MAO is low in schizophrenia, and that enhanced dopaminergic activity would result. But platelet MAO has different characteristics from both types A and B tissue MAO, and studies on brain MAO have failed to turn up differences between schizophrenics and normals.

The Noradrenaline Hypothesis. The suggestion was made by Wise and Stein (Stein 1968) that drive reward was mediated by noradrenaline. They further hypothesized that schizophrenia resulted from a deficit in the reward mechanism caused by abnormal endogenous production of 6-OHDA and subsequent destruction of noradrenergic pathways (Stein and Wise 1971). Apart from one doubtful report of reduced dopamine-β-hydroxylase levels in schizophrenia (Wise *et al.* 1975), this remains a highly speculative theory.

Further negative results regarding this theory come from studies of serum dopamine β-hydroxylase where no difference was found between schizophrenics and normals (Wyatt and Murphy 1975). The fact that noradrenaline blockers such as the phenothiazines and propranolol (Atsmon *et al.* 1972) are effective in treating schizophrenia is also contrary to this hypothesis.

The Serotonin Hypothesis.   Woolley and Shaw (1954) put forward the original postulate that some form of defective serotonin metabolism might be at the root of mental illness. Much of the data that have been accumulated since have been an extension of this evidence indicating an interaction of serotonin with agents known to produce behavioral changes. For example, LSD is a powerful blocker of some peripheral serotonin receptor sites and may stimulate or block central sites. Psilocybin, bufotenine, and some other hallucinogenic agents are close structural relatives of serotonin.

There are impressive overlaps between the LSD syndrome and schizophrenia. They include hallucinations, delusions, disturbances in mood and affect, and thinking disorders: however, LSD-induced hallucinations tend to be visual rather than auditory, and the individual usually retains an insight into his condition which is totally lacking in schizophrenia.

Based on the LSD model, many studies have been undertaken to determine whether there is abnormal serotonin metabolism in schizophrenia. Conflicting results are reported but, in any case, the fact that over 95% of serotonin metabolism takes place outside the CNS makes it difficult to rule out selective alterations in brain.

The recent finding that $(-)$-propranolol prevents the behavioral effects of excess serotonin or the hallucinogen 5-methoxy-N,N-dimethyltryptamine (Green and Grahame-Smith 1976, Chapter 9) and is also effective in treating schizophrenia (Atsmon *et al.* 1972) has renewed interest in the original Woolley hypothesis. It is thought that this is a counterbalancing effect on serotonin with the primary action at catecholamine receptors.

Since tryptophan is a precursor of serotonin, the bizarre behavioral effects of tryptophan combined with an MAO inhibitor have heightened speculation that the formation of an N-methylated hallucinogenic serotonin derivative could underlie schizophrenia. Workers have actually reported the presence of N,N-dimethyltryptamine, bufotenine (N,N-dimethylserotonin), and 5-methoxydimethyltryptamine in untreated schizophrenics and in schizophrenics receiving various precursor loads. The amounts are miniscule, however, even compared with hallucinogenic doses, and are so low that verification of their presence has been difficult (Rosengarten and Friedhoff 1976).

Despite the speculative excitement aroused by the effects of hallucinogenic agents, no firm evidence has so far been produced indicating defective serotonin metabolism, or the presence of significant quantities of any hallucinogenic metabolite of serotonin in schizophrenia.

Hypotheses Involving ACh and GABA.   Huntington's chorea involves a loss of ACh and GABA cells in the basal ganglia (Chapter 13). It also results in a behavioral syndrome reminiscent of the early stages of schizophrenia. Both cell types interact with dopamine pathways (Chapter 13). It is logical to hypothesize, therefore, that an abnormality in one or both of these transmitters could underlie schizophrenia. In the case of ACh, it would be a deficiency in cells postsynaptic to ascending dopamine pathways. Dopamine blockers, by releasing tonic dopamine inhibition, would boost the activity of cholinergic cells. Atropine, hyoscine, Ditran-type cholinergic blockers, or cell destruction through Huntington's chorea would produce a cholinergic deficit and psychosis.

In the case of GABA, there would be a deficiency of GABA inhibition on dopaminergic cells of the midbrain and substantia nigra. The resulting disinhibition would result in overactivity of dopaminergic cells and would produce an amphetamine-like psychosis. This would be blocked by the antipsychotic phenothiazines. Unfortunately, measurements on postmortem schizophrenic brain have not yet shown evidence of abnormalities to support these attractive hypotheses. The only abnormality so far reported is increased levels of choline acetyltransferase in the basal ganglia and certain limbic areas of the brain (E. G. McGeer and McGeer 1977).

Other Theories. Numerous other biochemical theories of schizophrenia have appeared over the years but most have been abandoned for lack of supportive evidence. The adrenochrome hypothesis, that envisaged a cyclization of adrenaline to an indolic hallucinogen, was the first of these. It was followed by the taraxein hypothesis, that imagined a serum protein which altered brain function being produced by schizophrenics. The methylation hypothesis for dopamine supposed that dimethoxyphenylethylamine was produced in schizophrenia and that it might have mescaline-like properties. The nicotinic acid theory of treatment, which escalated into megavitamin therapy, was initially based on the notion that nicotinic acid, by virtue of its conversion to a methylated product, would divert the schizophrenic brain from abnormal methylation. This expanded into the somewhat vague theory that only massive vitamin doses could compensate for unknown metabolic weaknesses. A number of hypotheses involving endocrine abnormalities have also surfaced.

In all cases, independent laboratories have tried and failed to confirm the original supportive experimental evidence, and so they remain as historical curiosities and a testimony to the difficulty of experimentation in this field.

14.6.2 Affective Psychoses

The affective psychoses may be bipolar (manic-depressive) or unipolar (depressive only). Again, theories as to the causation have primarily involved the catecholamines and serotonin because of the effectiveness of the psychoactive drugs in treating these illnesses. But endocrine and other functions have been implicated as well in attempts to find a physiological explanation.

The Noradrenaline Hypothesis. Noradrenaline has been preferred over dopamine in catecholamine theories of affective disorders because amphetamines, which are primarily dopamine releasers, are relatively ineffective at treating the depressive aspects, while tricyclics, which primarily enhance noradrenaline action, are effective. Benztropine, a selective dopamine reuptake inhibitor as well as an anticholinergic, is not useful in treating depression. There have been some reports of low CSF homovanillic acid in depression but others have found no alteration. Urinary homovanillic acid is not representative of central dopamine metabolism.

Equally controversial results have been obtained with noradrenaline metabolites. As discussed in Chapter 8, 3-methoxy-4-hydroxyphenylglycol (MHPG) is the principal noradrenaline metabolite in brain, but not the periphery. In the periphery it is converted to vanillylmandelic acid (VMA). Between 25% and 60% of urinary MHPG is from brain, while nearly all VMA is from the periphery. Thus, some distinction can be made between central and peripheral noradrenaline metabolism. It has been reported that daily excretion of MHPG is low in patients with endogenous depressions as compared with controls, that it increases during recovery, and that some patients with bipolar affective psychosis have substantially higher levels during the manic phase. This finding could not be confirmed (Goodwin and Post 1975), however. Equally controversial reports have been obtained with respect to CSF MHPG. There is considerable scatter to all the data so far obtained which would have the effect of obscuring small differences. Nevertheless, it must be concluded that large differences do not exist, which is evidence against a catecholamine hypothesis based simply upon abnormal production of the neurotransmitter.

A more subtle hypothesis might fit in with the action of lithium salts. This monovalent ion is considerably smaller in ionic radius than sodium, and has the unusual property of being highly effective in the treatment of manic psychosis. Obviously, lithium could affect a tremendous variety of cellular activities because of its universal distribution, but it appears to have a somewhat selective action in reducing the availability of noradrenaline at receptor sites. In acute experiments dopamine and serotonin are not so affected (Sourkes 1976).

The Serotonin Hypothesis.   Support for the hypothesis that deficient serotonin production is connected with depression comes from drug studies. Both PCPA and reserpine are noted for inducing depression in susceptible individuals. On the other hand, in patients being treated with 5-HTP for certain types of epilepsy, euphoria is one of the symptoms of overdose. Isolated reports of successful treatment of depression with both tryptophan and 5-HTP have appeared, but several groups have published negative clinical results with these agents after careful trials.

Three groups of workers have reported a decrease in concentration of either serotonin or 5-HIAA in the brains of suicide victims. Five out of eight groups that studied 5-HIAA in the CSF drawn from the spinal canal of depressed patients found lower than normal levels while the other three found no change. One of the groups that found a substantially reduced 5-HIAA noted that there was not a return to normal upon recovery of the patients, raising the question as to whether the levels were low before the illness commenced. Another weakness is that 5-HIAA in the lumbar CSF may reflect spinal cord 5-HT metabolism rather than brain itself (Goodwin and Post 1975).

In summary, there is suggestive evidence of serotonin involvement in depression, but a substantial body of negative data has also been accumulated. This would seem to diminish the likelihood that deranged serotonin metabolism can be established as the fundamental cause of depression.

14.7 Summary

In this chapter we have dealt with the anatomy, physiology, neurochemistry, and neuropharmacology of mood and behavior as they are understood today. We have touched on theories of mental illness, recognizing that this important clinical problem is unsolved from an etiological point of view despite impressive gains in chemotherapy. Many behavioral patterns have their origin in "feelings" penetrating the conscious brain. These feelings are deeply intertwined with demands of our internal organs and with the instinctive need for survival of self and species. Thus, the peripheral sympathetic motor system prepares the body for fight or flight while the parasympathetic section prevails over rest and recuperation. The fibers which carry these motor impulses are not like the rapid express lines that run to the musculoskeletal system. Rather, they are thin, nonmyelinated fibers which conduct slowly and are reflected in the more tonic response of the smooth muscle they innervate. Feelings are not subject to the rapid shifts that characterize the voluntary motor system.

Stimulation of appropriate subcortical centers in the brain will precipitate complex behavioral patterns which, without knowledge that such stimulation had taken place, might be mistaken for voluntary but inappropriate action. For example, stimulation of areas in the limbic system, diencephalon, or brainstem can provoke evasive or aggressive reactions in the absence of any external provocation. Sexual initiative, feeding, or drinking in the absence of appropriate stimulus or need may be obtained. Ablation of selected areas in these same regions can bring about similar results and, in some cases, reciprocal effects can be noted.

It is apparent from these stimulation and ablation studies that selected pathways in the brain, operating at the subcortical level, are responsible for the task of coordinating psychological and somatotropic action to prepare the body externally and internally to meet appropriate life situations.

Given the option of self-stimulation of selected brain areas, animals will seek or avoid initiating the electric shock. On this basis, reward and punishment centers have been defined on the simple intuitive basis that the brain and body serve these two great masters. Reward centers are spread throughout the limbic, diencephalic, and, to a lesser extent, the extrapyramidal systems. Punishment zones are confined to a much smaller region in the periventricular area of the midbrain and diencephalon.

The areas involved, both central and peripheral, are richly served by pathways using the transmitters dopamine, noradrenaline, adrenaline, and serotonin. ACh and GABA are also involved to a lesser degree. The use of selective metabolic inhibitors of these neurotransmitters has permitted some sorting out of the probable functions that are modulated. For example, dopamine is most concerned with extrapyramidal movement and coordination of some thought processes. Noradrenaline is particularly concerned with some ergotropic functions of brain and seems to modulate self-stimulation. Serotonin, on the other hand, is concerned with sleep, possibly digestive function, and increased temperature and comfort. None of these functions seems to depend absolutely upon any of the amines. Rather, they seem to influence the functions in a way that is not properly understood at the cellular level.

Much of the information regarding the individual roles of these amines has come from the discovery of psychoactive drugs. These drugs not only interact with the amines to bring about behavioral modification, but many of them are of tremendous practical benefit in the treatment of mental illness. Thus, antipsychotic agents such as the phenothiazines block dopaminergic receptor sites while reserpine depletes the amine. Reserpine also depletes noradrenaline, adrenaline, and serotonin. MAO inhibitors are psychic energizers which elevate levels of all four amines, while trycyclic antidepressants inhibit the reuptake, particularly of noradrenaline and serotonin.

Out of these drug studies have emerged a number of theories with regard to schizophrenia and affective psychosis. The theories provide an attractive basis on which to base further investigation, but as yet they offer no firm conclusions as to the causation of any form of mental illness.

The amines we have discussed in this chapter are both modulators of central neurotransmission and profoundly affect the action of peripheral smooth muscle and the endocrine glands; thus, they can be said to provide links between the psyche and the soma. Psychosomatic medicine is an important branch which has in the past relied more upon art than science but may in the future find a firmer physiological base.

This chapter on instinctive behavior makes no attempt to deal with two features of higher brain activity. One is the process of learning and memory, which will be discussed in Chapter 15. The other is the phenomenon of consciousness itself and the realm of original, creative thought. That will be the concern of Chapter 16.

15

Neuronal Mechanisms Involved in Learning and Memory

15.1 Introduction

Since this book is devoted to molecular neurobiology, attention will be focused on the structural and functional changes in the brain that form the basis of learning. Learning is essentially a process of storage in the brain and memory is the retrieval from the storage or, in modern terminology, from the data banks in the brain. There are two quite distinct types of learning and memory, though in many situations they may be employed in conjunction. First, there is motor learning and memory, which is the learning of all skilled movements. The repertoire is immense: the playing of all muscial instruments; the playing of all games; the learning of all arts and crafts, and of all technologies. Furthermore, there are all the expressive movements as in speech, dance, song, and writing. Second, there is what we may call cognitive learning and memory. At the simplest level there is the ability to recall some perceptual experience, but all levels can be involved, e.g., the remembrance of faces, names, scenes, events, pictures, musical themes. Then at a higher level there is the learning of language, and of stories, and of the contents of disciplines from the simplest technologies to the most refined academic studies in the humanities and the sciences.

It is a familiar observation that there may be an enduring cognitive memory of some single highly emotional experience. On the other hand, motor memories require reinforcement by continual practice if they are to be retained at a high level of skill. Quite distinct parts of the brain are concerned in these two types of memory. Nevertheless, it appears likely that the same kind of neural mechanism is concerned. We shall start with a brief resume of the various types of neural mechanism that have been proposed, and of the related experimental evidence.

15.2 Structural and Functional Changes Possibly Related to Learning and Memory

There have been theories of long-term memory based upon a supposed analogy with genetic or immunological memory. For example, it has been conjectured that memories are encoded in specific macromolecules, in particular RNA (Hydén 1965, 1967), or that it is analogous to immunological memory (Szilard 1964). These theories fail for various reasons (cf. Eccles 1970b; Szentágothai 1971) and need not here be further discussed. A brief account will now be given of the evidence for the *growth theory of learning* in the central nervous system.

In general terms, following Sherrington (1940), Adrian (1947), Hebb (1949), Lashley (1950), Kandel and Spencer (1968), and Szentágothai (1971), it is supposed that long-term memories are somehow encoded in the neuronal connectivities of the brain. We are thus led to conjecture that the structural basis of memory lies in modifications of synapses (cf. Eccles 1970b, Szentágothai 1971). In mammals there is no evidence for growth or change of major neuronal pathways in the brain after their initial formation. It is not possible to construct or reconstruct major brain pathways at such a gross level. But it should be possible to secure the necessary changes in neuronal connectivity by means of microstructural changes in synapses. For example, they may be hypertrophied or they may bud additional synapses, or, alternatively, they may regress. Since it would be expected that the increased synaptic efficacy would arise because of a strong conditioning synaptic activation, experiments such as those illustrated in Figure 15.1 have been carried out on many types of synapses.

Figure 15.1B is remarkable in showing that repetitive stimulation results in a large increase (up to six times) in the excitatory postsynaptic potentials, EPSPs, monosynaptically produced in an α-motoneuron by pyramidal tract fibers (cf. Figure 13.3, Section 13.3.1). By contrast, in Figure 15.1A the EPSPs generated monosynaptically in that same motoneuron by Ia fibers from muscle spindles (cf. Figure 4.2, Section 4.2) were not potentiated. Evidently, the pyramidal tract synapses display an extreme range of modifiability by what we may call *frequency potentiation*. The synaptic mechanism involved in this potentiation is not understood, but most probably it is due to an equivalent increase in emission of the synaptic transmitter substance. Many types of synapses at the higher levels of the brain have this ability to build up operationally during intense activation.

The series of Figures 15.1C and 15.1D give another example for synapses in a primitive part of the cerebrum, the hippocampus (cf. Figure 15.10, 15.14). The hippocampus is of particular interest because it is believed to be important in the laying down of memory traces, as will be described below.

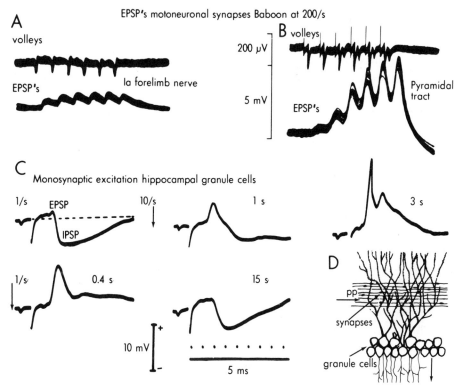

Figure 15.1 Frequency potentiation of excitatory synapses. A, B. The lower traces are monosynaptic EPSPs of the same motoneuron of the cervical enlargement of the baboon spinal cord, there being in each case six stimuli at 200 per second to the Ia afferent pathway in A and to the pyramidal tract in B. (Landgren *et al.* 1962). C. Shows frequency potentiation of monosynaptic EPSPs of hippocampal granule cells when the frequency of stimulation of the perforant pathway (pp of D) was increased from 1 to 10 per second, and its decline on return to 1 per second. (From Bliss and Lømo 1973.)

Figure 15.1D shows the excitatory synapses from the perforating pathway (pp) onto the dendrites of the granule cells. In Figure 15.1C the intracellular record from a granule cell during the initial series at 1 per second stimulation of pp showed a very small initial EPSP followed by a large IPSP. With the stimulus frequency raised to 10 per second, already within 1 s there was a large potentiation of the EPSP that counteracted to some extent the IPSP. After 3 s of this stimulation the very large EPSP completely submerged the IPSP and is seen to generate an impulse discharge from the cell. On again slowing the stimulation to 1 per second, the frequency potentiation had already considerably declined at 0.4 s and had disappeared in 15 s. It is attractive to think that synapses responding so enthusiastically during and for some seconds after moderate activation (posttetanic potentiation) could be the *modifiable synapses* responsible for the phenomena of learning and memory.

Figure 15.2 shows a much more enduring kind of posttetanic potentiation in those same synapses of the hippocampus. A very mild stimulation of 20 per second for 15 s (300 pulses) was applied at the first arrow. The plotted points show that there was only a small transient potentiation. But with successive repetitions of this mild stimulation about every half hour, there was a progressive increase in the potentiation so that after the fifth there was an enormous potentiation of the impulse discharge from the granule cells. Actual records are given in the insets, where three test responses may be compared with the three controls below that are given by the other side. The plotted measurements are of the sharp downward extracellular spikes marked by the arrows in these test responses. This large potentiation continued for 3 h. This amazing effect was observed in many such experiments, potentiations being fully maintained even for 10 h in acute experiments (Bliss and Lømo 1973). In chronic experiments with implanted electrodes a similar potentiation was observed even for several weeks after conditioning by six brief trains of stimulation, 15 per second for 15 s (Bliss and Gardner-Medwin 1973). The potentiation was built up as in Figure 15.2, and declined to about half by 12 h, but it declined little further at 1 day, 6 days, and 16 weeks thereafter. We can conclude that in these experiments it has been demonstrated that the spine synapses on the dendrites of the hippocampal granule cells are modifiable to a high degree and exhibit a prolonged potentiation that could be the physiological expression of the memory process.

Physiological experiments have thus indicated that the *modifiable synapses* which could be responsible for memory are excitatory and are specially prominent at the higher levels of the brain. In the cerebral cortex the great majority of excitatory synapses on pyramidal cells are on their dendritic spines, as illustrated in Figure 1.2 (Section 1.2.1). There is also much evidence by Val-

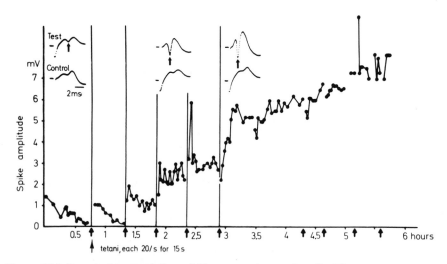

Figure 15.2 Posttetanic potentiation of hippocampal granule cells. The measurements were made on the extracellular recording of the positive spike shown in the specimen records by arrows, and it may be taken to be a measure of the number of granule cells firing impulses in the zone sampled by the recording electrode. (Bliss and Lømo 1973.)

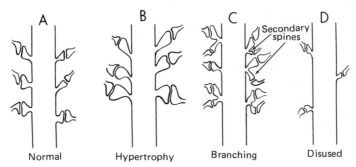

Figure 15.3 Plasticity of dendritic spine synapses. The drawings are designed to convey the plastic changes in spine synapses that are postulated to occur with growth in B and C and with regression in D.

verde (1968) and others that these spine synapses regress during disuse (cf. Eccles, 1970b). Hence, it is postulated that these spine synapses on the dendrites of such neurons as the pyramidal cells of the cerebral cortex and the hippocampus, the granule cells of the hippocampus, and the Purkinje cells of the cerebellum are the modifiable synapses concerned in learning. These would be the synapses displaying the indefinitely prolonged potentiation illustrated in Figure 15.2. One can imagine that the superior performance by these synapses was indefinitely prolonged because a growth process had developed in the dendritic spines giving a structural change which could have great endurance. There is as yet no convincing demonstration of this growth in electron micrographs,[1] but there is much circumstantial evidence. The conjectured changes are diagrammatically shown in Figure 15.3, where Figure 15.3A represents the normal state and 15.3B and 15.3C the hypertrophied states. An alternative to the synaptic spine hypertrophy of Figure 15.3B is shown in 15.3C, being an increase in synaptic potency secured by branching.

We are on much surer histological grounds in showing the effects of disuse in causing a regression and depletion of spine synapses (Figure 15.3D), because this has been beautifully demonstrated by Valverde (1967) on the dendrites of the pyramidal cells in the visual cortex of mice raised in visual deprivation; and indeed similar demonstrations have been made with other spine synapses, and even with excitatory synapses in the spinal cord (Szentágothai 1971). So it can be assumed that normal usage results in the maintenance of the dendritic spine synapses at the normal level depicted in Figure 15.3A.

It can be concluded that the excitatory spine synapses are probably the modifiable synapses concerned in memory, but much more rigid experimental investigation with systematic electron microscopic examination is urgently needed to test this hypothesis. It is surprising that there has as yet[1] been no such systematic study of synapses in the hippocampus under conditions that would be expected to show synaptic hypertrophy.

[1]*Note added in proof:* This has now been demonstrated (Fifkova and van Harreveld. 1977. *J. Neurocyt.* **6,** 211).

If synaptic growth is required for learning, there must be an increase in brain metabolism of a special kind with the manufacture of proteins and other macromolecules required for the membranes and the chemical transmission mechanism. Presumably, in the synaptic growth theory of learning it must be assumed that RNA is responsible for the protein synthesis required for growth. This assumed growth, however, would not be the highly specific chemical phenomenon conjectured in Hydén's (1967) molecular theory of learning, where the coding of memories is attributed to specific macromolecules, each memory being associated with unique macromolecules. Instead, the specificities would be encoded in the structure and in the synaptic connections of the nerve cells, which are arranged in the unimaginably complex pattern that has already been formed in development. From then onward, all that seems to be required for the functional reorganization that is assumed to be the neuronal substrate of memory is merely the microgrowth of synaptic connections already in existence, as indicated in Figures 15.3B and 15.3C, e.g., of the spine synapses on pyramidal cells and Purkinje cells (Eccles 1966, 1970a, 1972). The flow of impulses from receptor organs into the nervous system (cf. Figures 2.2, 2.3A, Section 2.1) will result in the activation of specific spatiotemporal patterns of neurons linked by sequential impulse discharges. The synapses so activated will grow to an increased effectiveness and even sprout branches to form secondary synapses; hence, the more a particular spatiotemporal pattern of impulses is replayed in the cortex, the more effective become its synapses relative to others. And by virtue of this synaptic efficacy, later similar sensory inputs will tend to traverse these same neuronal pathways and so evoke the same responses, both motor and cognitive, as the original input.

In neurochemistry and neuropharmacology there have now been many fine studies by Barondes (1969, 1970), Agranoff (1967, 1969), and others, which reveal that long-term learning (beyond 3 h) does not occur when either cerebral protein synthesis or RNA synthesis is greatly depressed by poisoning of the specific enzymes by cyclohexamide or puromycin. It is conjectured that in the process of learning, synaptic activation of neurons leads first to specific RNA synthesis and this in turn to protein synthesis, and so finally to unique structural and functional changes involved in the synaptic growth and the coding of the memory. Unfortunately, the crucial step is not yet understood, namely, how synaptic activation can trigger the appropriate molecular response. It is known (Barondes 1970), however, that critical synthesis in the brain is in action during the learning process and in some circumstances can be effective in laying down the memory traces within minutes. Apparently long-term memory can be established only if there is an intact protein-synthesizing capacity, an appropriate "state of arousal," and availability of the information in a short-term memory store (Barondes (1970). Experiments performed by Agranoff (1976) illustrate the principles. He injected puromycin, which blocks protein synthesis by about 80%, directly into the brains of conscious goldfish. The goldfish were trained in a shuttle box to swim over a barrier upon a light signal in order to avoid a punishing electrical shock administered through the water. If puromycin was given immediately after the training session, the effects of training were abolished when response was measured several days later. The training was not abolished, however, when that same amount of puromycin was injected into the fish after they had been first returned to their home tanks for about 1 h. When puromycin was injected before training, there was also no effect on subsequent performance. If puromycin was injected at various times after training, there was a slow gradient of abolition of the training effect. Thus, puromycin does not immediately obliterate memory, but rather interferes with some ongoing process that is initiated by the training process. Acetoxycycloheximide and cycloheximide, both inhibitors of protein synthesis, produce the same effect.

These experiments parallel those of a number of other workers indicating that inhibition of protein synthesis abolishes the consolidation of memory. They do not affect short-term memory.

Whether or not a critical protein or series of proteins is task specific for memory is a matter of conjecture. If enzyme induction is critical to some aspect of the formation of long-term memory, a block at the transcriptional level of RNA formation would be expected to produce a similar effect. Such agents are often toxic, making the interpretation of results difficult. Nevertheless, inhibitors of nucleic acid synthesis such as 8-azaguanine, actinomycin D, d-amanitin, and camptothecin, all interfere with the consolidation of memory.

If RNA synthesis is truly involved in the memory consolidation process, then increased turnover would be anticipated during the period when this is taking place. Glassman (1969) has reported increased [³H] uridine incorporation into macromolecular material, presumably RNA, in the diencephalon of mice during the consolidation period. Although such a change could be due to responses other than just the learning procedure, restrained control animals subjected to the same amount of training experience but not permitted to execute the trained response did not show labeling changes. Overtrained animals permitted to make the trained response also failed to show increased labeling. It appears, then, as if some changes in RNA metabolism take place in the training situation. A difficulty is that concomitant alterations in protein labeling have not been found in these situations. Such changes should be detectable since RNA subserves protein synthesis (Agranoff 1976). As the theory presented in this chapter makes clear, however, the enhanced protein synthesis need only be associated with those selective synapses involved in the learning pathways.

15.3 Motor Learning and Memory

In the preceding section an account was given of the growth theory of learning and forgetting according to which excess usage of synapses leads to hypertrophy and enhanced function, whereas disuse leads to regression and diminished function. This theory can be criticized because it is now recognized that almost all cells at the highest levels of the nervous system are discharging impulses continuously. One can imagine, therefore, that there would be overall hypertrophy of all synapses under such conditions of continued activation, and hence but little possibility of any selectivity in the hypertrophic change. Evidently, frequent synaptic excitation alone could hardly provide a satisfactory explanation of synaptic changes involved in learning. Such "learned" synapses would be too ubiquitous! This criticism of the simple "growth theory" of learning can perhaps be contained by the recent suggestion of Szentágothai (1968) and Marr (1969) that synaptic learning is a dual and dynamically linked happening, namely, that activation of a special type of synapse provides instructions for the growth of other activated synapses on the same dendrite. This may be called the *conjunction theory of learning.*

It was originally suggested that the unique operation of climbing fibers on the Purkinje dendrites of the cerebellum (Figure 13.12, Section 13.4.3) was to give "growth instructions" to the spine synapses that at about the same time were activated by the parallel fibres. Otherwise, activation of the spine synapses did not result in growth and the long-term memory coded thereby. So far it has proved technically impossible to test this hypothesis in a simple direct experiment. Fortunately, Ito and associates in Tokyo (Ito *et al.* 1974a, 1974b; Ito and Mijishita 1975; Ito 1975) have addressed themselves to the same problem, but by a more indirect method that is reliable for many hours, and have carried out a convincing series of experiments of the greatest importance.

There is a very ancient part of the cerebellum that controls the eye movements when the head is turned. One can think that the function of this part of the cerebellum is to maintain, as far as possible, a fixed position of the visual image on the retina. The movements of the head are sensed by the semicircular canals of the vestibular system. For our present purpose we have only to consider the horizontal canal because, in the experiments of Ito, the

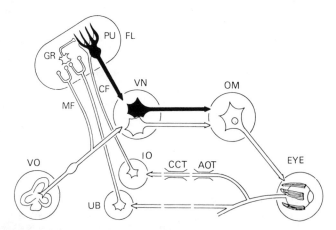

Figure 15.4 Construction of the flocculovestibulo-ocular system. OM, oculomotor neuron; IO, inferior olive; CCT, central tegmental tract; AOT, accessory optic tract; VO, vestibular organ; FL, flocculus; VN, second-order vestibular neurons. (Modified from Ito 1975.)

head of the rabbit was rotated about a vertical axis. The pathway diagram of Figure 15.4 shows the direct projection of the vestibular nerve to the vestibular nucleus (VN) that in turn projects directly to the oculomotor nucleus (OM) for the eye muscles. By itself this pathway can achieve some stabilization of the retinal image, but the control is much more regular in operation when it is aided by the pathway through the cerebellar flocculus. In Figure 15.4 there are two mossy fiber (MF) inputs, one directly from the vestibular nerve, and the other a feedback from the visual system, which probably acts as a controlling device on the vestibulo-ocular reflex, improving the stability of the visual field during head movements. So far we have neglected the role of the climbing fiber (CF) input, which is predominantly activated by the visual input via the pathway indicated to the inferior olive via the accessory optic tract (Figure 15.4, AOT; Maekawa and Simpson 1973).

In Figure 15.5 the head of the rabbit is subjected to a sine wave horizontal rotation with a standard amplitude of 10° and a frequency of 0.15 Hz. The plotted left eye movements (Figure 15.5A) show that the compensation for the head rotation was much better when there was a fixed vertical strip of light

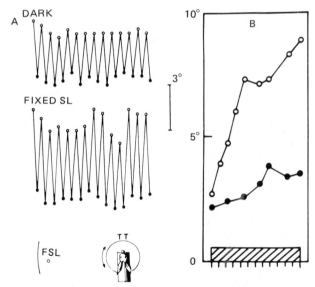

Figure 15.5 Horizontal eye movement induced by sinusoidal head rotation and its modification during presentation of slit lights in the visual field. A. Open circles indicate the most nasal, and close circles the most temporal, positions of the cornea mark on the left eye during each cycle of 10° head rotation. Plottings on the top indicate the eye movement in darkness and provide the control for the immediately succeeding measurement with slit light (SL) to the left eye, as plotted below. Diagrams at the bottom indicate dorsal views of the turntable (TT) mounting the rabbit, rotating-screen, fixed-slit light (FSL). (Ito *et al.* 1974a.) B. Changes of rabbit's horizontal vestibulo-ocular reflex during sustained head rotation with the fixed-slit light presented. Normal rabbit. Ordinates, mean angular amplitudes of the horizontal eye movement during 10° head rotation; abscissae, time; O, eye movements measured with the fixed-slit light shown; ●, that measured in temporary darkness. (Ito *et al.* 1974a.)

than in darkness, but it was well below the 10° requirement for complete compensation. It was of great interest that when the head rotation was continued for many hours (Figure 15.5B), there was a progressive increase in the compensation when tested with illuminated slit (open circles). After several hours it became almost perfect for eliminating retinal image slip. Evidently, there had been learning of the adaptive change, and this was very much less with the continual rotation when the measurements were carried out in darkness (filled circles).

In an alternative experiment, the vertical light was moved synchronously with the head rotation, and in the same direction but with an amplitude of 20°. Thus the influence from the visual input was opposed to the vestibulo-ocular reflex arising from the 10° head rotation. As would be expected, the visual input acted to decrease the eye movements below the amplitude observed in darkness. When the rotation was continued for hours there was a progressive decrease in the vestibulo-ocular reflex, the eye movements continuing to be less than when tested in darkness, i.e., again there was an enhancement of the visual influence with repetition over hours. All these adaptive changes had disappeared by 24 h.

These observations provide clear evidence of a learning process that over some hours tends to improve the stability of the retinal image. Two experiments define the neural pathways concerned in this learning. In the first experiment, after ablation of the flocculus there was no learning. Two possible explanations remain. One is that it could be by the mossy fiber return circuit to the flocculus from the visual system (cf. Figure 15.4). That was disproved by the second experiment which showed that learning did not occur, or rarely in a minor late form, when the inferior olive was destroyed (Figure 15.6). Thus, we have good evidence that the CF input to the cerebellar Purkinje cells of the flocculus (Figure 15.4) is effective in causing a learned cerebellar modification that gives improvement in the stabilization of the retinal image over and above that achieved by the vestibulo-ocular reflex.

It is important to find out if this improvement is due to the more effective responses of Purkinje cells to the MF input from the vestibular receptors, or, alternatively, if it could be due to a more prominent role of the CF input to Purkinje cells. Recording of the responses of individual Purkinje cells has established that the simple spikes generated by the MF input to Purkinje cells have a rhythmically modulated discharge at the frequency of the head rotation (Ghelarducci et al. 1975). With learning with a fixed-slit light there is an increased tendency to be 180° out of phase so that their discharge giving an inhibitory action on cells of the vestibular nucleus (Figure 15.4) would account for the cerebellar influence on the vestibulo-ocular reflex. Thus, there is good empirical evidence that the climbing fiber input to the Purkinje cell has been effective in causing a response to the mossy fiber/parallel fiber input that is adaptive in that it minimizes retinal image slip. It can be recognized as a learned movement. A remarkable feature is that in Figure 15.5 about 1300 trials were required for the good adaptive response. This experimental investigation by Ito is still at an early stage, however. As yet there is no evidence

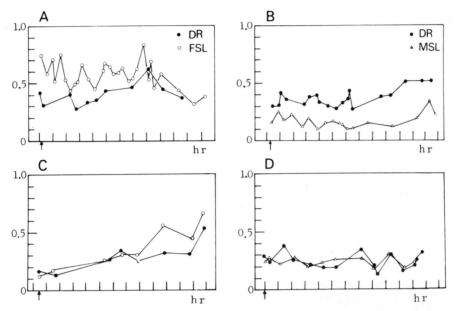

Figure 15.6 Effects of the continuous application of the combined vestibular and visual stimulation. Rabbit with inferior olive ablated. Ordinates, gain of the horizontal vestibulo-ocular reflex measured with 5° head rotation at 0.1 Hz. Open circles in A and C plot the gain under influences of FSL as in Figure 15.5B, and open triangles in B and D that under MSL. Closed circles in A–D indicate the gain measured in temporary darkness. Upward arrows on the abscissae mark the moment of onset of the continuous rotation. (Ito and Mijishita 1975.)

about the essential parameter, which is the time relationship for effective conjunction of the CF and MF inputs. This remains enigmatic, yet it is the key component of the selection theory. The slow sine wave rotations of Figure 15.5 do not give the sharp timing that would be required for a measurement of the conjunction time.

Gonshor and Melville Jones (1976a, 1976b) investigated the plastic changes in the human vestibulo-ocular reflexes (VOR) by a technique resembling that of Ito in that the head of the subject was given a horizontal sinusoidal rotation at 0.17 Hz, but the amplitude (60°) was much larger. When the subjects had a horizontal visual inversion by dove prisms that were worn continuously for 1 week, there was a progressive reduction in the amplitude of the vestibulo-ocular reflex (VOR). With the visual inversion the retinal image slip would be in the opposite direction during the ordinary day-to-day visual inputs, and hence the VOR would be in the inverse phase for reducing the retinal image slip that consequently would be accentuated. Hence, the observed diminution of the VOR to about 25% indicates a response to the inverted visual inputs that was adaptive in that it effectively reduced the retinal image slip. When normal vision was restored, the VOR recovered with the same slow time course of days.

It can be conjectured that, as with Ito's experiments on rabbits, these plastic changes for reducing retinal image slip were the result of CF inputs into the flocculus with the consequent conjunction–potentiation of the appro-priate MF–parallel fiber synapses on Purkinje cells. Supporting this conjec-ture are the recent experiments of Robinson (1976) showing that, in cats subjected to horizontal visual inversion, adaptive changes in the VOR did not occur after ablation of the flocculonodular lobe.

15.4 The Instruction–Selection Theory of Learning in the Cerebellum

Figure 15.7 displays very well the extreme localization of the CF input, (C fibers) usually only to one Purkinje cell in an area (Armstrong *et al.* 1973), whereas there is a great dispersion of the input by a single mossy fiber to the Purkinje cells of several square millimeters (Eccles *et al.* 1967; Mugnaini 1976). Of course, convergence of several MF fibers onto individual granule

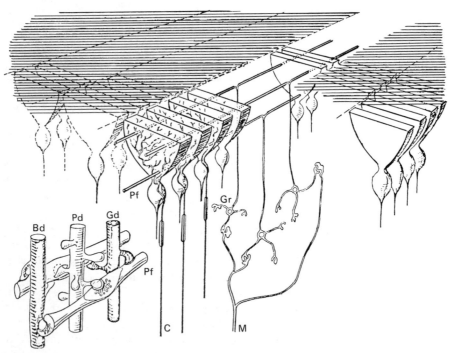

Figure 15.7 Stereo diagram of the cerebellar cortex showing five rows of Purkinje cells with their dendrites arranged in parallel arrays. The climbing fiber inputs (C) to three Purkinje cells are drawn, but all will have a similar input. The mossy fiber (M) is shown branching to give synapses to three granule cells (Gr), but normally it would supply over 400. The granule cell axons are shown bifurcating in the molecular layer to form parallel fibers (Pf). Instead of the three drawn traversing the Purkinje cell dendritic trees there would be 400,000, but only one-half to one-quarter would make synapses on any one Purkinje cell. The inset diagram shows in detail the crossing-over synapses made on spines by parallel fibers (Pf) on Purkinje cell dendrites (Pd), basket cell dendrites (Bd), and Golgi cell dendrites (Gd). (Modified from Szentágothai 1972.)

cells is required for evoking their discharges along parallel fibers. It has always been a serious problem how the cerebellum can be effective in such a refined way when there is this broad distribution of the principal input by the MFs. The conjectured role of the CF input in plastic modification of MF effectiveness gives an entirely new vision of cerebellar performance. It links with the trophic influences reported by Hámori, Kawaguchi, and Bradley and Berry (Chapter 12), and provides a mechanism for greatly sharpening the influence of the MF input on the cerebellar cortex.

But it is not enough simply to postulate this potential influence of single CF impulses in sharpening the response to an MF input. It is required to show that the CF inputs themselves have a degree of refinement that matches their potentiality for refining the MF responses by selecting a class of the MFs that represents the context of the CF input and that can effectively operate in the absence of this CF input, as postulated by Marr (1969).

There is good progress in this field, particularly by Oscarsson and associates (Oscarsson 1969; Larson *et al.* 1969; Ekerot and Larson 1973; Oscarsson 1973, 1976) in relation to a different part of the cerebellum from that investigated by Ito and associates, namely, the anterior lobe (vermis plus pars intermedia) that is oriented specially to the limbs and body. They have described narrow parasagittal strips of the cerebellum, about 0.5 mm across (Figure 15.8A) that have specific spino-olivocerebellar pathways and CF responses of the Purkinje cells. These would cut across the parallel fibers of the MF input that run transversely for about 6 mm, as shown very precisely by Mugnaini (1976). Even at this relatively crude level of display, the CF input in these parasagittal strips would potentially be able to cause plastic changes in an MF input according to the many different parasagittal strips of the CF input that its parallel fibers would traverse. But now a much finer grain for the CF input has been recognized in an elaborate somatotopic organization of microstrips (Figure 15.8C). Together these microstrips form a systematic somatotopic organization of the entire ipsilateral body surface. The grain of this is remarkably fine: for example, in some cerebellar folia the digits of the forepaw project to fine overlapping microstrips with their maxima 100 μm apart. According to the conjunction hypothesis, the MF input itself with its parallel fiber relay would be unaltered, but the synaptic effectiveness of its parallel fibers would be selected in accordance with the CF territory which it traverses, provided that there is this rather precise conjunction in time between its activation and the CF activation. It then has been selected to form the MF context for that particular CF input (Marr 1969).

This hypothesis of learning brings it into relationship with Jerne's (1967) selection theory of immunity. He revolutionized immunology by his hypothesis that production of immune bodies was not on the basis of instruction, namely, that the antigen induced the antibody formation by some instructional process of chemical complementarity. On the contrary, all the antibodies are already in existence as unitary characters of the vast variety of lymphocytes. The action of the antigen is *to select from* this vast array those lymphocytes with the correct antibodies and then to call forth an immense

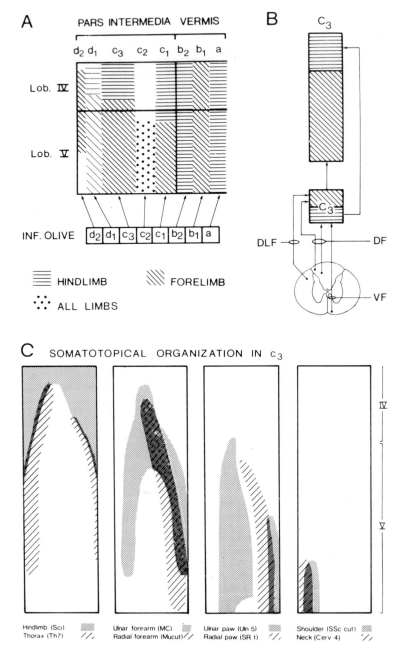

Figure 15.8 Organization of sagittal zones in lobules IV and V of cerebellar anterior lobe. A. Each zone represents the CF projection from a compartment in the inferior olive. The zones and compartments are labeled a, b_1, b_2, etc. The somatotopical organization of the zones is indicated by different hatchings and stippling (see key). Zones indicated by sparse hatching (d_1) and stippling (c_2) are activated after relatively long latencies. B. Organization of the SOCP projections to the c_3 zone. The c_3 olivary compartment (square) consists of two subdivisions which project to the rostral and caudal parts of the c_3 zone (rectangle) and relay information related to the hindlimb

clonal multiplication of these specific lymphocytes that in turn shed their antibodies into the bloodstream. Jerne suggested that the learning processes in the brain similarly might be by selection rather than by instruction, and this selection is in line with the present hypothesis. There are immense numbers of synapses formed by the MF-parallel fiber pathway onto each Purkinje cell, at least 100,000 (Chapter 13). It is proposed that in learning there is no instruction to form new synapses. The CF input to a Purkinje cell instructs it to increase the potency of those parallel fiber synapses that *by the conjunction in time* are selected from the existing multitude of synapses. This potentiation is the plastic influence of the CF input. It is highly selective. Perhaps no more than a few thousand of the 100,000 parallel fiber synapses on that cell are selected to form the MF context of its CF input (cf. Marr 1969). Because of the operative features of this conjunction theory of learning, it is appropriate to call it the instruction–selection theory of learning (cf. Eccles 1977).

We would agree with Marr's (1969) conjecture that the inferior olive sets the precision of the cerebellum by its instructional role in selecting from the multitude of potentialities provided by the widespreading mossy fiber input, even when this is generated in the most discriminating manner, e.g., taps to one toe pad. It has to be realized also that such a mossy fiber input is not selected for one Purkinje cell or for one set of Purkinje cells. Rather, one must think that an MF input is selected in order to be the context of a CF input on a Purkinje cell. Moreover, some of these same mossy fibers may participate in the contexts of other CF inputs, being selected again by conjunction in space and time. It is implicit in Marr's theory that any one mossy fiber can participate in the contexts of several climbing fibers. Thus, the set of 8000 Purkinje cells that potentially can be excited by one mossy fiber (Eccles 1977) give it that many chances of being potentiated by conjunction with a CF input to the 8000 Purkinje dendritic trees which it contacts via parallel fibers to the spine synapses.

Figure 15.7 gives but a skeletal drawing of the relationship of parallel fibers to dendritic trees. These numbers of Purkinje cells must be viewed in relation to the output pathway from the cerebellum, where there is a great reduction in numbers. Any one nuclear cell averages the inhibitory synaptic actions generated on it by the axonal terminals of the 200 Purkinje cells that converge onto it. It must be assumed that this projection to nuclear cells is organized and not randomized, and this assumption is supported by studies of

---

and forelimb, respectively (see key). The spino-olivary path ascending through the dorsal funiculus (DF) projects to both olivary regions, whereas the path ascending through the dorsolateral funiculus (DLF) projects only to the "forelimb region" and the path ascending through the ventral funiculus (VF) only to the "hindlimb region." C. Somatotopical organization of the DF-SOCP projection to the c_3 zone. (Based on unpublished observations made by Ekerot and Larson.) The four rectangles represent the c_3 zone in lobules IV and V. Stippling and hatching show the areas activated through the DF-SOCP on stimulation of the nerves indicated (see key). Abbreviations of nerves: Sci, sciatic; Th7, Th7 dorsal root; MC, medial cutaneous; Mucut, musculocutaneous; Uln 5, ulnar branch to fifth digit; SR 1, superficial radial branch to first digit; SSc cut, cutaneous branch of suprascapular; Cerv 4, fourth cervical nerve. (Oscarsson 1976.)

the responses of single nuclear cells (cf. Eccles 1973). Evidently, we are far from being able to model anything but the simplest level of cerebellar computation.

The widespread distribution of MF inputs via parallel fibers constitutes a relatively undifferentiated pool available by conjunction potentiation for selection as a context of a wide range of CF inputs. Any one parallel fiber synapses on about 100 Purkinje cells and hence has that number of chances to be selected as a CF context. Only a small fraction of these chances would normally be realized. It is a common finding that many different MF inputs from peripheral nerves converge on a single Purkinje cell (Eccles *et al.* 1971b, 1972; Leicht and Schmidt 1978). They can be considered as potential components of the MF context of the climbing fiber of that cell.

We have given an account of the general principles involved in the instruction–selection theory of learning for the cerebellum. There is already a wealth of detailed anatomical knowledge with precise study of the climbing fiber and mossy fiber pathways. There is evidence of trophic interaction (Chapter 12) and the studies of Ito and associates provide the first experiments on mammals where there has been an analytical investigation of the neural machinery involved in learning. Important corroborative evidence for the role of the cerebellar flocculus in the adaptive changes of the vestibulo-ocular reflex (Gonshor and Melville Jones 1976b) has recently been reported by Robinson (1976), but the possible role of the climbing fibers has not yet been investigated. Much more experimental evidence is urgently needed on this extremely important problem of motor learning in the cerebellum. There is a convincing story on the cerebellar neural machinery involved in motor learning related to the vestibulo-ocular reflex (VOR), but for motor learning of limb and body actions it is not possible to give an account to match that for the VOR because the systems are much more complex.

Fundamental problems in neurochemistry arise in relation to the proposed conjunction–potentiation theory. Synaptic changes are postulated, but as yet the metabolic processes responsible for these changes are poorly understood. A recent discovery by Libet, Kobayski, and Tanaka (1975) is of great interest because it is interpreted to be a model for a synaptic memory process that depends on conjunction between two species of synapses on the same sympathetic ganglion cell (Figure 15.9). They state:

> A heterosynaptic interaction takes place between two types of synaptic inputs to the same neurone; the memory trace is initiated by a brief (dopaminergic) input in one synaptic line, while "read-out" of the memory consists simply in the enhanced ability of the postsynaptic unit to produce its specific response to another (cholinergic) synaptic input. This arrangement provides for a "learned" change in the response to one input as a result of an "experience" previously carried by way of the other input.

This model is based on carefully controlled responses of sympathetic ganglion cells, which display a doubling of response to acetyl-β-methylcholine for many hours after a brief exposure to dopamine. A brief perfusion of the isolated superior cervical ganglion of the rabbit with dopamine will bring about an

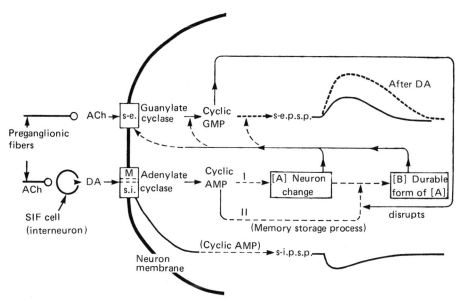

Figure 15.9 Neurotransmitter effects on superior cervical ganglion. Schema of pathways for synaptic inputs and intracellular mediators involved in the responses to dopamine (DA) and acetylcholine (ACh) for a principal neuron (or ganglion cell) in the rabbit's superior cervical ganglion. The s-e.p.s.p. is a slow depolarizing change in the membrane potential in response to ACh's acting at a muscarinic type of membrane receptor (s-e); it is represented both before and after (dashed line) its enhancement by a modulatory action of DA. The s-i.p.s.p. is a slow hyperpolarizing change in membrane potential produced by the action of DA. DA is released by a special type of interneuron (the small, intensely fluorescent, SIF, cell as seen with the Falck–Hillarp technique), in response to preganglionic impulses or to an applied muscarinic agonist of ACh. The two postsynaptic actions of DA are designated as M for the modulatory one and s-i, for the hyperpolarizing response or s-i.p.s.p.; the actions could be mediated by a single or by two separate membrane receptors. Persisting enhancement of the s-e.p.s.p. response after exposure to DA would be somehow dependent on neuronal change (A) and its more durable form (B). (Libet *et al.* 1975.)

enhancement of the ganglion's response to the muscarinic agonist methacholine (MCh) for several hours or as long as the ganglion survives the experiment. Dibutyryl cAMP is able to substitute for dopamine, indicating that the modulatory change is probably due to dopamine-sensitive adenylate cyclase producing increased cAMP. A temporary addition of the soluble dibutyryl-cGMP abolishes the enhancement, but after a delay of 4 min, the counteracting effect progressively diminishes.

These experiments suggest that after the initial modulating change induced by dopamine, a "storage" process develops which leads to a more durable form of modulatory change. Cyclic GMP is able to disrupt this initial memory storage process (II in Figure 15.9), but not the enduring change.

The "read-in" is by dopamine, which induces a "storage" process through enduring change, with the "read-out" being by ACh stimulation. The cGMP effect, with its striking time dependency, is likened to the well-known disrupt-ability feature of the memory storage process. It is evident that this discovery is of great significance in relation to the conjunction theory of learning.

Kandal *et al.* (1975) have also produced evidence that cAMP is concerned in a prolonged facilitation process in an *Aplysia* ganglion, but have located this effect in the presynaptic terminals, suggesting that serotonergic synapses on these terminals cause a mobilization of cAMP that in turn results in an increased liberation of transmitter. If the serotonergic synapses acted post-synaptically and not presynaptically, the potentiating system would closely parallel that proposed by Libet *et al.* (1975).

15.5 Cognitive Learning and Memory

It will be recognized that two distinct problems are involved in cognitive learning and memory—storage and retrieval—or, in relation to the present theme—learning and remembering. It is proposed to deal with these problems at two levels.

First, they will be considered as a problem of neurobiology, namely, the structural and functional changes which form the neural basis of memory. We are all familiar with short-term memory of a few seconds' duration. For example, when we look up a telephone number we have to keep it "in mind" by continual rehearsal else it is lost beyond recall. A distraction during this rehearsal interrupts the rehearsal and the short-term memory is lost. There is evidence that longer memories are of two kinds: intermediate, up to an hour or so; and truly long term—for days, months, years. The necessity for keeping the remembered information "in mind" indicates that there is no special storage problem in short-term memory. As long as the information is carried encoded in the dynamic operations of the neuronal circuitry, it is available for read-out. The short-term memory can be plausibly explained by its coding in specific patterns of continued neuronal activity. On the other hand, with memories enduring for minutes to years, it has to be discovered how the neuronal connectivities are changed so that there is stabilized some tendency for replay of the spatiotemporal patterns of neuronal activity that occurred in the initial experience, and that have meanwhile subsided.

Second, the role of the self-conscious mind has to be considered. It will be proposed in Chapter 16 that a conscious experience arises when the self-conscious mind enters into an effective relationship with certain activated modules, "open" modules, in the cerebral cortex. In the willed recall of a memory the self-conscious mind must again be in relationship to a pattern of modular responses resembling the original responses evoked by the event to be remembered, so that there is a reading out of approximately the same experience. We have to consider how the self-conscious mind is concerned in calling forth the neuronal events that give the remembered experience on demand, as it were. Furthermore, the self-conscious mind acts as an arbiter or assessor with respect to the correctness or relevance of the memory that is delivered on demand. For example, the name or number may be recognized as incorrect by the self-conscious mind and a further recall process may be instituted, and so on. Thus, the recall of a memory involves two distinct processes in the self-conscious mind: first, that of recall from the "data banks" in the brain; second, the recognition memory that judges its correctness.

There are patients who suffer from a tragic loss of all but short-term memories. In the most severe and complete cases there is an almost complete failure to establish new memories, a clinical condition known as the amnestic syndrome or Korsakoff syndrome (Milner 1966, 1972; Victor *et al.* 1971; Kornhuber 1973). The unfortunate victim is disoriented in time and place. Gaps in memory are covered up by confabulation: a bedridden patient, for example, may insist he has just taken a walk in the garden.

Many of these cases are not useful for our purpose of trying to discover the actual neural mechanism by which new memories are laid down for semipermanent storage because the patients suffer from general cerebral degeneration, often from alcoholism (Victor *et al.* 1971). The mammillary bodies are characteristically involved, however. The mediodorsal and anterior thalamus plus the terminal portion of the fornices are also frequently affected. More exact information can be obtained by considering cases where the lesion was sharply defined because it was due to an operative excision.

In particular we shall begin with the case H.M., which has been the most extensively investigated. The operation was a bilateral excision of the hippocampus, as indicated in Figure 15.10, that was performed to cure epileptic seizures initiated bilaterally in that area. Several such operations had been performed before the tragic results of such an excision were known. In the inset of Figure 15.10 the 6-cm anteroposterior extent of the hippocampus is shown together with the levels of the four sections A, B, C, and D. In order to illustrate the extent of the ablation, it is shown on the left side only for comparison with the intact right side. But, of course, this complete excision of the hippocampus and the adjacent part of the hippocampal gyrus was carried out on both sides. Since that time this man has had an extremely severe loss of ability to lay down memory traces. There is an almost complete failure of memory for all happenings and experiences after the lesion, i.e., he has an anterograde amnesia. He lives entirely with the short-term memories of a few seconds' duration and the memories retained from before the operation.

He can keep current events in mind so long as he is not distracted. Distraction completely eliminates all trace of what he had been doing only a few seconds before. There are cited many remarkable examples of his failure to remember as soon as he is distracted. The only way in which this patient can hold onto new information is by constant verbal rehearsal. Forgetting occurs as soon as this rehearsal is prevented by some new activity claiming his attention. Recently Marlen-Wilson and Teuber (1975) have shown by a testing procedure of prompting that a minimal storage of information even occurs for experiences after the operation, but it is of no use to the patient.

There are three other recorded cases where a comparable severe anterograde amnesia resulted from destruction of both hippocampi (Milner 1966). There was almost no recovery even after 11 years. The variable retro-

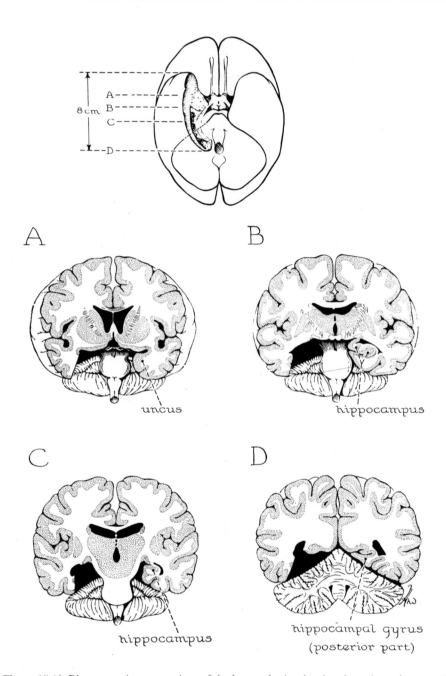

Figure 15.10 Diagrammatic cross sections of the human brain, showing the estimated extent of removal in Scoville's medial temporal ablation in the case discussed in the text. The anterior-posterior extent of the hippocampus is shown in the upper drawing with A, B, C, and D indicating the level of the transverse sections below. For illustrative purposes the removal is shown on the left side only, but the removal was made on both sides in a single operation. (Milner 1972.)

grade amnesia, i.e., the memory of events preceding the hippocampal destruction, however, showed a continued recovery. There are two other reported cases where unilateral hippocampectomy resulted in a comparable anterograde amnesia, but there was evidence that the surviving hippocampus was severely damaged. We can conclude that the severe anterograde amnesia only occurs with grave bilateral hippocampal deficiency.

It is important to recognize that the hippocampus is not the seat of the memory traces. Memories from before the hippocampectomy are well retained and recalled. The hippocampus is merely the instrument responsible for the laying down of the memory trace or engram, which presumably is very largely located in the cerebral cortex in the appropriate areas. There is no obvious impairment of intellect or personality in these subjects despite the acute failure of memory. In fact, they live either in the immediate present or with remembered experiences from before the time of the operation. There is one small relieving feature, namely, that they still have ability to learn motor acts, which indicates the distinctiveness of the neuronal mechanisms concerned in motor memory, as described above. Thus, the subject can build up skills in motor performances such as drawing a line in the narrow space between the line drawings of one five-pointed star and another surrounding star using only the guidance provided by the view in a mirror of his hand and the two stars; but he has no memory of how he learned the skill!

Partial amnestic syndromes have been observed in patients with a variety of lesions in structures related to the hippocampus: the cingulate gyrus, the fornix, the anterior and mediodorsal nuclei of the thalamus (Victor *et al.* 1971), and the prefrontal lobe. We are now in a position to consider the neural pathways concerned in the laying down of memory traces in the neocortex.

15.6 Neural Pathways Involved in the Laying Down of Cognitive Memories According to the Instruction-Selection Theory

The theory here proposed is developed from Kornhuber's theory (1973), which is illustrated in Figure 15.11. The sensory association areas play a key role, being, first, on the input pathway to the limbic system and frontal cortex, and second, in an intimate two-way relationship to the frontal cortex that receives a "selection input" from the limbic system. It is to be noted that the hippocampus is given a dominant role in the two limbic circuits. One circuit is the so-called Papez loop: hippocampus, mammillary body, anterior thalamic nucleus, cingulate gyrus, parahippocampus, hippocampus, (cf. Figure 15.8). The other circuit is of special interest because it leads from the association cortices to the hippocampus via the cingulate gyrus and thence via the mediodorsal (MD) thalamus to the prefrontal lobe. Kornhuber (1973) conjectures that with special neurons of the sensory association areas: "... the synapses of afferents coming (directly or indirectly) from the limbic system are essential for forming long-term memory, while other synapses on the

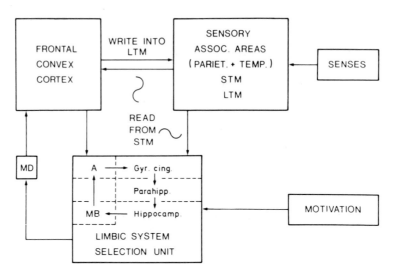

Figure 15.11 Scheme of anatomical structures involved in selection of information between short-term memory (STM) and long-term memory (LTM). MB, mammillary body; A, anterior thalamic nucleus; MD, mediodorsal thalamic nucleus. (Kornhuber 1973.)

same neurons are essential for information processing and for recall." He even conjectures that "long-term memory could involve coincidence of thalamic and cortico-cortical afferents at a given cortical neuron or cell column." These theoretical developments by Kornhuber provide the basis for the further developments here described.

A general schema of the connectivities linking sensory association areas, the limbic system, and the prefrontal lobe is illustrated in Figure 15.12 for the monkey brain (Kornhuber 1973). This diagram shows very well the inputs from frontal, temporal, and parietal cortical association areas to the hippocampus via the cingulate gyrus. The output from the hippocampus is shown via the mediodorsal thalamus to the convexity of the frontal lobe. The reciprocal connections of the frontal convexity to the parietal and temporal lobes are in accord with the diagram in Figure 15.11. Actually, both these figures greatly simplify the connectivities, particularly in the limbic system, as appears later when a more detailed presentation will be attempted. For the present we may note the important operation of motivation in Figure 15.11 and also the labeling of the limbic system as a selection unit. According to the present theory the hippocampal output does indeed act to select, but this is done in the association cortex.

A more detailed map of connectivities of the limbic-oriented pathways is given in Figure 15.13, but this too is very crude and special features of the hippocampal region must be further detailed in order to gain even a simple understanding of how it could play the key role of selection in the laying down of memories in the neocortex. Figure 15.13 is devoted to the pathways in both directions from neocortex to hippocampus (HI). First, pathways to the hippocampus are shown relaying in the hippocampal gyrus (HG) or a special

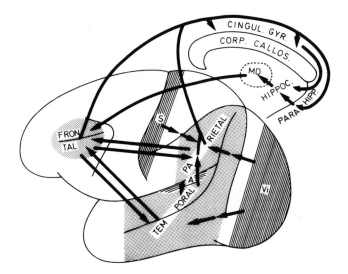

Figure 15.12 Scheme of pathways in the monkey brain involved in the flow of information from primary sensory areas via the sensory association areas of the temporal and parietal lobe and the cortex of the frontal convexity to the limbic system and then the loop back via the mediodorsal nucleus of the thalamus (MD) and the frontal cortex to the temporal and parietal areas for long-term storage. Primary sensory areas: Vi, visual; A, auditory; S, somatosensory; the vestibular area is in the lower part of S. (Kornhuber 1973.) (Based, in part, on data of Pandya and Kuypers 1969, and Pandya *et al.* 1971.)

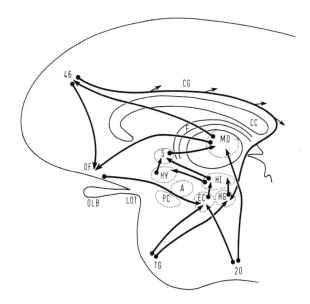

Figure 15.13 Schematic drawing to show connectivities from the neocortex to and from the mediodorsal thalamus (MD). OF is orbital surface of prefrontal cortex; TG, temporal pole; HG, gyrus hippocampi; HI, hippocampus; S, septum; F, fornix; CC, corpus callosum; OLB, olfactory bulb; LOT, lateral olfactory tract; PC, pyriform cortex; EC, entorhinal cortex; A, amygdala; HY, hypothalamus; CG, cingulate gyrus.

zone of it called the entorhinal cortex (EC). In addition to the pathway from area 46 via the cingulate gyrus (CG) shown in Figure 15.12, there are also pathways from temporal areas 20 and TG and from the orbital zone of the prefrontal lobe (OF).

On the output side, the hippocampus is seen projecting to the septal nucleus (S) where it converges with the hypothalamic (HY) input, even onto the same neurons (Raisman 1969). From there the path is via the MD thalamus as in Figures 15.11 and 15.12, but to area OF as well as 46, and probably even more widely to the prefrontal areas.

Before we embark on a detailed consideration of the proposed mode of selective action of the hippocampal output on the immensely complex neuronal connectivities in the association cortex (cf. Figure 15.12), we should inquire into the neuronal circuitry of the hippocampus in order to see if it is built so as to work in a highly selective manner with respect to the inputs it receives from the neocortex. Recent investigations by Andersen and associates have shown to an amazing degree that the hippocampus is indeed organized in a series of narrow transverse lamellae (cf. Figure 15.14A) which function independently through all the complex connectivity. This functional discrete-

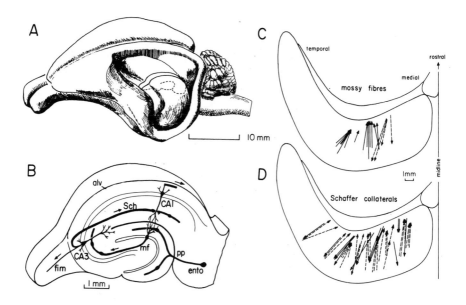

Figure 15.14 Anatomy of the hippocampal formation. A. Lateral view of the rabbit brain with the parietal and temporal neocortex removed to expose the hippocampal formation. B. The indicated section of the hippocampal formation is enlarged to show the main neuronal elements of the structure: alv, alveus; ento, entorhinal area; mf, mossy fibers; pp, perforant path; Sch, Schaffer collaterals. C, D. Comparison of direction of mossy fibers, Schaffer collaterals. Each arrow represents a full experiment with recording or stimulation from a series of tracks. The fully drawn lines indicate experiments with impulses running in an orthodromic direction whereas the broken lines indicate experiments with antidromic activation. C gives the direction of mossy fibers and D that of the Schaffer collaterals. (Andersen *et al.* 1971.)

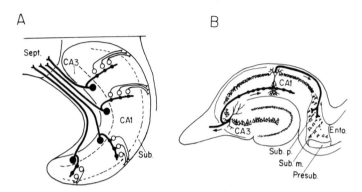

Figure 15.15 Distribution of CA3 and CA1 axons. A. Diagram of a dorsal view of the hippocampal formation showing the distribution and orientation of axons of CA3 cells (large filled circles) and CA1 cells (open circles). The broken lines represent the CA3–CA1 border and the border between CA1 and the subiculum. Small filled circles denote Schaffer synapses. B. Semidiagrammatic transverse section of the hippocampal formation showing an approximation of the histological relationship of CA3 and CA1 cells and axons. Axonal thickness grossly exaggerated. (Andersen *et al.* 1973.)

ness of the lamellae is first demonstrated from the entorhinal cortex (ento) via the perforating pathways (pp) in Figure 15.14B to the granule cells of the fascia dentata (Lømo 1971). Thence, as shown in Figure 15.14B, the pathway is via mossy fibers (mf) to CA3 pyramidal cells, whose axon collaterals (Schaffer collaterals, Sch) act as powerful excitants of the CA1 pyramidal cells (Anderson *et al.* 1971). As shown in Figure 15.14C the mossy fiber–CA3 connection is in narrow transverse lamellae, as also is the connection of Schaffer collaterals to CA1 (Figure 15.14D). Finally, as beautifully illustrated in Figure 15.15 (Andersen *et al.* 1973), the output of the hippocampus is by two independent channels, each conveying the lamellar specificity of the hippocampal processing. The CA3 output goes rostrally by the fimbria to the septal nuclei and the CA1 output caudally to the subiculum, whence it goes to the mammillary bodies and so participates in the Papez circuit of Figure 15.11. The CA3 output goes via the septal nucleus to the MD thalamus (cf. Figure 15.11, 15.12, 15.13) and so is of our immediate concern.

It can be concluded that the hippocampus has a functional subdivision from input to output in a large, as yet undefined, number of lamellae; and this discrimination is maintained in the output line of the CA3 pyramidal cells by a strict segregation of the CA3 axons according to location in the fimbria, the more rostral being medial and the more caudal, lateral (as shown in Figure 15.15). It can be presumed that this segregation leads on to a segregation in the septal nucleus. Andersen, Bliss, and Skrede (1971) sum up their findings: "A point source of entorhinal activity projects its impulses through the four-membered pathway along a slice, or lamella, of hippocampal tissue oriented normally to the alvear surface and nearly sagittally in the dorsal part of the hippocampal formation."

15.7 Structural Features of the Neocortex of Special Significance for the Instruction–Selection Learning Hypothesis

According to Szentágothai (1971) the same general structural principles obtain for all regions of the neocortex, though in contrast to the cerebellum there is much diversity in detail, as witness the more than 50 Brodmann areas so defined. In trying to identify the features of cortical structure that could be related to the storage and retrieval of memories, our interest must center on the association areas of the cortex and particularly on the prefrontal areas. Unfortunately, such areas have not yet been subjected to a rigorous histological study, so we have to build up our theoretical story on the basis of the geometrical and architectural descriptions given by Szentágothai (1969, 1970a,b, 1971, 1972, 1974, 1975) for a standard region of the neocortex. Furthermore, the account will be restricted to the special features that form the basis of the instruction–selection hypothesis of learning that is being formulated.

Any area of neocortex has two main categories of afferent fibers. As illustrated in Figure 15.16 the specific afferents (Spec. aff.) from the specific

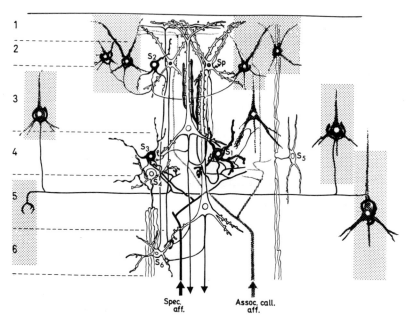

Figure 15.16 Interconnections of some cell types in the cerebral cortex. Two pyramidal cells are seen centrally in lamina 3 and 5. The specific afferent fiber (Spec. aff.) is seen to excite a stellate interneuron S_1 (crosshatched) whose axon establishes cartridge-type synapses on the apical dendrites. The specific afferent fiber also excites a basket-type stellate interneuron, S_3, that gives inhibition to pyramidal cells in adjacent columns, as indicated by shading. Another interneuron (S_6) is shown in lamina 6 with ascending axon, and S_5 is an interneuron also probably concerned in vertical spread of excitation through the whole depth of the cortex. Sp are stellate pyramidal cells and S_2 are the short-axon inhibitory cells in lamina 2. The afferents formed by association and callosal fibers (Assoc. call. aff.) are shown ascending to branch in lamina 1. (Szentágothai 1969.)

nuclei of the thalamus branch profusely in giving excitatory synapses to varieties of stellate cells in laminae 3, 4, and 5 and also to the basal dendrites of
pyramidal cells. Szentágothai (1972) states that the fibers from nonspecific
thalamic nuclei have similar endings. One species of stellate cell (S_1 in Figure
15.16) is of particular importance because the subdivisions of its axon run
perpendicularly along the apical dendrites of two or three pyramidal cells to
form hundreds of synapses on the dendritic spines in a unique assemblage
that is invested by a glial sheath. The whole ensemble is called a cartridge
(Szentágothai 1969). It is further illustrated in Figure 15.17, where the
apical dendrites of two pyramidal cells are invested. There are many
similarities to the synapses of a single climbing fiber on a Purkinje cell of the
cerebellum (Figures 13.12, Section 13.4.3; 15.7), there being likewise only one
cartridge synapse on any one apical dendrite. Actually, the fibers of the cartridge have been referred to as climbing fibers (Colonnier 1966; Szentágothai
1972), such fibers being first described by Ramón y Cajal (1911).

The other category is from other regions of the neocortex by association
fibers from that cerebral hemisphere and by commissural fibers from the
contralateral hemisphere. In Figure 15.16 this class of afferent is named
"Assoc.-call. aff." It is seen to send some branches to lamina 4, but the main
distribution is to lamina 1 and the superficial part of lamina 2 (Heimer *et al.*
1967; Szentágothai 1969). As depicted in Figure 15.17 the COM and ASS
fibers bifurcate to form long horizontal fibers in lamina 1 or 2, much as do the
axons of granule cells in forming the parallel fibers of the molecular layer of
the cerebellar cortex (Eccles *et al.* 1967). According to a rough estimate the
ASS fibers are at least six times more numerous than the COM fibers, which
number about 100 million for one cerebral hemisphere. The lengths of these
fibers (ASS and COM) are about the same as parallel fibers, 2.5 mm in each
direction, and similarly they form crossing-over synapses with the spines of the
terminal dendrites of pyramidal cells and of stellate pyramidal cells, just as do
the parallel fibers with cerebellar Purkinje cells.

Much more quantitative investigation is needed on the connectivities in
lamina 1 and the superficial zone of lamina 2, but there appear to be considerable differences from those for the parallel fiber/Purkinje cell relationships
in the cerebellar cortex. One difference is that the horizontal fibers can be in
any direction, which is in striking contrast to the strictly parallel array in the
cerebellar cortex. Reference should be made to Eccles (1978) for a quantitative comparison between the cerebellar and cerebral connectivities.

The other element forming horizontal fibers in lamina 1 (Szentágothai
1965) are the Martinotti cells of lamina 6 (S_6 Figure 15.16) whose axons
ascend to bifurcate as shown in Figure 15.17, MA. Presumably these synapses
also would be available for selection by conjunction, and this would give the
possibility for learning in the neuronal circuits originating in Martinotti cells
in a zone extending radially for 2.5 mm from the cartridge assemblage providing the conjunction. The ASS fibers in Figure 15.17 would tend to be from
more remote zones of that hemisphere, while the COM are from the other
hemisphere, largely from a symmetrical zone (Jones and Powell 1969b). The

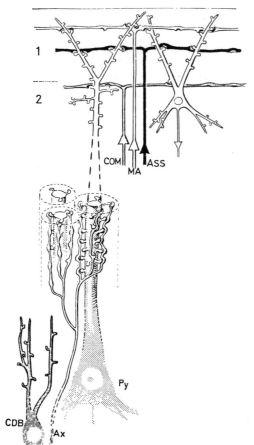

Figure 15.17 Simplified diagram of connectivities in the neocortex. In laminae 1 and 2 there are shown horizontal fibers arising as bifurcating axons of commissural (COM) and association (ASS) fibers and also of Martinotti cells (MA). The horizontal fibers make synapses with the apical dendrites of a pyramidal and a stellate-pyramidal cell. Deeper there is shown a spiny stellate cell (CDB) making cartridge synapses with the shafts of apical dendrites of pyramidal cells. (Szentágothai 1970a.)

ASS and COM inputs presumably can also act indirectly via local cortical circuits activating Martinotti cells, however. Marr's (1970) conjunction theory was exclusively for Martinotti cells.

The contribution of the commissural input to the horizontal fibers can be related to the finding that it is possible to lay down memories for the special functions of the hemisphere on the side of a unilateral hippocampectomy. This ability is, however, gravely deteriorated, which would correspond to the relatively small population of horizontal fibers arising from the commissural input.

It is of great interest that after commissurotomy each hemisphere can learn and remember its own particular tasks, but with a considerably diminished competence: the left hemisphere for verbal and numerical tasks; the right hemisphere for spatial, musical, and pictorial tasks. As already indicated, the respective left and right hippocampi are concerned in this memory storage. Each would work alone because the hippocampal commissure is also sectioned in the commissurotomy.

As already mentioned, Marr (1970) attempted to formulate a conjunction theory of learning for the neocortex that was analogous to his cerebellar conjunction theory (1969). This theory was inadequate in that it made no reference to the role of the hippocampus, however, being restricted to local neocortical circuits. Also, on analogy with the granule cells of the cerebellar cortex, and on the basis of earlier publications of Szentágothai (cf. 1965), the Martinotti cells were assumed to be the sole origin of the crossing-over synapses in lamina 1 (cf. Figure 15.17). More recent work (cf. Heimer *et al.* 1967; Szentágothai 1969), however, has established that the commissural and association fibers make large contributions to the horizontal fiber system of lamina 1 (cf. Figures 15.16 and 15.17). These contributions are of particular importance because they provide a direct input from a wide zone of the ipsilateral neocortex (cf. Figure 15.12) and of a symmetrical zone on the other side. Presumably, as stated above, the Martinotti cells are important in providing a localized mnemonic system. It would be of great value to have some approximate estimate of the fractions of the horizontal fibers arising from the three inputs to lamina 1 in the prefrontal lobe.

15.8 Formulation of the Instruction–Selection Hypothesis for Cognitive Learning

In general there are remarkable similarites between the dual-input systems to the Purkinje cells of the cerebellum on the one hand and to the pyramidal cells of the neocortex on the other. There is much experimental evidence supporting the instruction–selection role of the climbing fiber on the parallel fiber input to the Purkinje cell (Section 15.3 above, and Eccles 1978), which has been developed on analogy with the selection theory of immunity by Jerne (1967). The question arises: Does the dual-input system to the pyramidal cells of the neocortex function similarly in learning, and can this be linked with the role of the hippocampus?

We must now return to the general circuit diagram of Kornhuber (Figure 15.11). Two pathways are shown converging on the frontal convex cortex—that directly from the sensory association areas, and that indirectly, via a detour through the limbic system and the MD thalamus. In the frontal cortex we would propose that the indirect input would be via nonspecific thalamic afferents from the MD thalamus that would excite the spiny stellate cells forming the cartridge type of synapse (CDB Figure 15.17), while the direct input would be the association fibers that terminate as horizontal fibers in laminae 1 and 2 and that are particularly well shown in Figure 15.17. On analogy with the cerebellum, it is proposed that the cartridge-type synapse on a pyramidal cell acts similarly to the climbing fibers in selecting from the total input of horizontal fibers on the apical dendrites of that same pyramidal cell. This selection would be dependent on conjunction of the two inputs in some specific time relationship, as yet undefined, and would result in an enduring potentiation of the selected synapses on the apical dendrite. Just as with parallel fibers, it is assumed that several association, commissural, and Martinotti fibers would be selected from the 2000 as forming the context of the cartridge synaptic activity on that pyramidal cell (Marr 1970; Eccles 1978). Thus, activity of the cartridge system is the instruction that selects for potentiation those horizontal fiber synapses that are activated in the appropriate temporal conjunction. As indicated in Figure 15.17, Szentágothai (1972) proposes that a single cartridge system comprises the apical dendrites of about three pyramidal cells, which thus form a unitary selection system. For further quantitative consideration, see Eccles (1978).

Possibly the Papez circuit (cf. Figure 15.11) functions to provide the reverberatory activation of the hippocampus with its CA3 output through the septal nucleus to the MD thalamus. In Figure 15.18 part of Figure 15.11 is redrawn, incorporating the detail of Figure 15.15. Figure 15.18 raises the question: Is the lamella specificity of CA1 activation preserved through the Papez circuit so that it is fed back into the entorhinal cortex with uncontaminated specificity?

The diagrammatic representation in Figure 15.18 gives deep meaning to the fundamental design feature discovered by Andersen et al. (1973), namely, that the CA3 and CA1 pyramidal cells of the hippocampus are sharply discriminated by their distinctive projections (Figure 15.15). One of the synaptic links in the circuits of Figure 15.18, the entorhinal cortex to the granule cells of the fascia dentata (Fasc. Dent.), exhibits remarkable responses to repetitive stimulation, which would make it function very effectively in a reverberating loop such as that proposed for the Papez circuit in Figure 15.11. There is very large potentiation during repetitive stimulation at 10 per second (Figure 15.1C), and with repeated short episodes there is a progressive buildup of a potentiation that is maintained for hours (Figure 15.2; Bliss and Lømo 1973) and even for days (Bliss and Gardner-Medwin 1973). Thus, this synaptic transmission would operate with greatly increased potency during reverberating circuit action. As shown in Figure 15.14B, this potentiation would also be on the circuit from CA3 neurons to the prefrontal lobe and so be of importance in causing a progressive buildup in the activation of the cartridge synapses.

It is interesting that motivation comes into Kornhuber's circuit diagram (Figure 15.11). This implies attention or interest in the experiences that are coded in the neuronal activities of the association cortex and that are to be stored. It implies a process of mind–brain interaction. We are all cognizant that we do not store memories of no interest to us and to which we pay no attention. It is a familiar statement that a single sharp experience is remembered for a lifetime, but it overlooks the fact that the intense emotional involvement is reexperienced incessantly immediately after the original, highly charged emotional experience. Evidently, there has been a long series of "replays" of the patterns of cortical activity associated with the original experience, and this activity would particularly involve the limbic system as indicated by the strong emotional overtones. Thus, there must be built into the

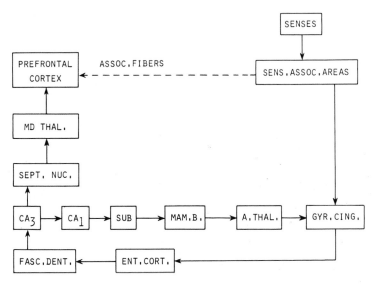

Figure 15.18 Scheme of structures participating in the circuits involved in cerebral learning. Figure 15.11 is redrawn to show the two circuits emanating from the CA3 and CA1 hippocampal pyramidal cells as in Figure 15.15.

neuronal machinery of the cortex the propensity for the reverberating circuit activity which would cause the synaptic potentiation giving the memory.

15.9 Recall of a Memory

At one stage of consideration the recall of a memory can be explained by the replay of neocortical circuits that were consolidated by this learning process. Thus, the replayed circuits would closely resemble those giving the original experience; hence, the remembered experience is recognized as genuine. But there are many problems overlooked in this simple story. No special difficulty would attend an explanation of a memory triggered by some related present experience or a sequence of memories in some train of thought. But we recognize that we can at will attempt to recall a memory. Evidently we are now involved with the mind–brain problem (cf. Chapter 16). Even more challenging is the experience that we can assess the validity of a recalled memory, recognizing, for example, that the telephone number or ZIP Code or name is almost correct, so that a new demand is made on the memory data banks for a deliverance that can be recognized as correct. Thus, we have to entertain the idea of two kinds of long-term memory: a data bank memory that is stored in the neocortex; and a recognition memory that is in the conscious mind.

In the further development of our hypothesis of long-term conscious memory we would propose that the self-conscious mind would enter into this transaction between the modules of the liaison brain (cf. Chapter 16) and the hippocampus in two ways: first, by keeping up the modular activity by the general action of interest or attention so that the hippocampal circuit would be continuously reinforced; second, in a more concentrated manner by probing into the appropriate modules to read out their storage and if necessary to reinforce it or modify it by direct action on the modules concerned. All of these proposed actions are from the self-conscious mind to those modules that have the special property of being "open" to it. By its direct action on "open" modules, however, the self-conscious mind can exercise an indirect action on those "closed" modules to which the "open" modules project (Chapter 16). Evidence for this supportive action by "closed" modules has been presented by Sperry (1974). He finds that there is a pronounced memory impairment of the left cerebral hemisphere after commissurotomy. This would be expected if the "closed" modules of the minor hemisphere were indirectly active in the memory storage and retrieval, as here proposed. Sperry (1974) makes a comparable comment: "Any storage, encoding, or retrieval process dependent normally on integration between symbolic functions in the left hemisphere and spatio-perceptual mechanisms in the right would also be disrupted by commissurotomy."

In retrieval of a memory we have further to conjecture (cf. Chapter 16) that the self-conscious mind is continuously searching to recover memories, e.g., words, phrases, sentences, ideas, events, pictures, melodies, by active scanning through the modular array, and that, by its action on the preferred "open" modules, it tries to evoke the full neural patterned operation that it can read out as a memory. Largely this could be by a trial-and-error process. We are all familiar with the ease or difficulty of recall of one or another memory, and of the strategies we discover in order to recover memories of names that for some unknown reason are refractory to recall. We can imagine that our self-conscious mind is under a continual challenge to recall the desired memory by discovering the appropriate entry into module operation that would by development give the appropriate patterned array of modules.

Penfield and Perot (1963) gave a most illuminating account of the experiential responses evoked in 53 patients by stimulation of the cerebral hemispheres during operations performed under local anesthesia. These responses differed from those produced by stimulation of the primary sensory areas, which were merely flashes of light or touches, and paraesthesia (Chapter 16), in that the patients had experiences that resembled dreams, the so-called dreamy states. During the continued gentle electrical stimulation of sites on the exposed surface of their brains, the patients reported experiences that they often recognized as being recalls of long-forgotten memories. As Penfield states, it is as if a past stream of consciousness is recovered during this electrical stimulation. The most common experiences were visual or auditory, but there were also many cases of combined visual and auditory. The recall of music and song provided very striking experiences for both the patient and the neurosurgeon. All of these results were obtained from brains of patients with a history of epileptic seizures.

It can be concluded that the stimulation acts as a mode of recall of past experiences. We may regard this as an instrumental means for recovery of memories. It can be suggested that the storage of these memories is likely to be in cerebral areas close to the effective stimulation sites. It is important to recognize, however, that the experiential recall is evoked from areas in the region of the disordered cerebral function that is displayed by the epileptic seizures. Conceivably, the effective sites are abnormal zones that are thereby able to act by association pathways to the much wider areas of the cerebral cortex which are the actual storage sites for memories.

15.10 Durations of Conscious Memory

An analysis of the durations of the various processes involved in memory provides evidence for three distinct memory processes (cf. McGaugh 1969). We have already presented evidence for the short-term memory, usually lasting a few seconds, that can be attributed to the continual activity in neural circuit that holds the memory in a dynamic pattern of circulating impulses. The patients with bilateral hippocampectomy have almost no other memory. Second, there is the long-term memory that endures for days to years. According to the conjunction theory of learning, this memory (or memory trace) is encoded in the increased efficacy of synapses that have been activated in conjunction during and after the original episode that is being remembered. In the present context of conscious memory it can be conjectured that this synaptic growth would occur in multitudes of synapses in patterned array in the modules strongly reacting in response to the original episode that sets in train the operation of the circuits diagrammed in Figure 15.18. As a consequence of this synaptic growth, the self-conscious mind would be able to develop strategies for causing the replay of modules in a pattern resembling that of the original episode, hence the memory experience. Moreover, this replay would be accompanied by a renewed reverberatory activity through the hippocampus resembling the original, with a consequent strengthening of the memory trace.

We are confronted, however, with the urgent problem of filling in the temporal gap between the short-term memory of seconds and the hours required for the synaptic growth of long-term memory. Barondes (1970) reviews the experiments testing for the time course of action of substances, cycloheximide for example, that prevent protein synthesis in the brain, which at the same time is unable to learn. The approximate time of about 30 min to 3 h seems to be required for the synaptic growth giving long-term memory. McGaugh (1969) has proposed an intermediate-term memory to bridge the gap of seconds to hours between the end of short-term and the full development of synaptic growth giving the long-term memory. We would propose that, as indicated in Figure 15.19, the posttetanic potentiation described earlier (Figures 15.1, 15.2) is exactly fitted for bridging this gap. It would be induced by the repetitive synaptic actions of short-term memory and would immediately follow those actions, being restricted to the activated synapses

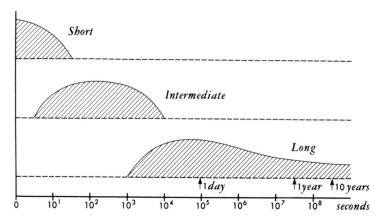

Figure 15.19 Diagrammatic representation of the durations of the three memories described in the text. Note the logarithmic time scale and the conjectured rise and fall of the memories with time.

and being graded in accord with their action. In Figure 15.2 posttetanic potentiation enduring for hours followed quite mild repetitive stimulation of hippocampal synapses. As that physiological process of synaptic potentiation declines, the metabolically induced synaptic growth supervenes to provide an enduring basis for the strategic read-out by the self-conscious mind.

15.11 Conclusions

In conclusion, it can be stated that this formulation of the instruction–selection hypothesis is restricted to the elemental features of the proposed mnemonic machinery. At a further stage of its development the neglected features of neuronal structure and function can be introduced. In particular there has been a neglect of the inhibitory machinery, both in the hippocampus and in the neocortex, and of all the neocortical excitatory circuits involving stellate cells to stellate cells and to pyramidal cells. Furthermore, no reference has been made to the fundamental arrangements of neocortical neurons into columns or modules (cf. Szentágothai 1975). This arrangement is a special feature of laminae 3, 4, 5, and 6, whereas in lamina 2 the short-axon inhibitory neurons would give a more restricted functional zoning (Szentágothai 1969, 1975). Finally, the role of the axon collaterals from pyramidal cells is overlooked. Presumably this diffuse wide-spreading influence (Scheibel and Scheibel 1970; Szentágothai 1975) can have no more significance than the provision of a local excitatory background that could be of importance in establishing and maintaining a "state of alertness" in that region of the neocortex.

Figure 15.20 is designed to show the parallel design features of the instruction–selection hypotheses of learning for the cerebellum and the cere-

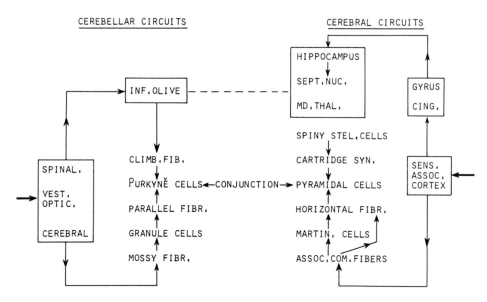

Figure 15.20 Drawings of cerebellar and cerebral circuits that are postulated for the instruction–selection hypothesis of learning. They are arranged in order to display the similar features. Conjunction occurs with the two distinctive inputs to the Purkinje cells and the pyramidal cells respectively.

brum. The respective inputs are shown projecting in from the left and the right. Each divides into two circuits: those projecting upward go through the inferior olive and the hippocampus and act in an instruction–selection capacity on the dendritic trees of the Purkinje cells and pyramidal cells respectively; those projecting downward go through mossy fibers and association–commissural fibers respectively, and their synapses on the dendritic trees are selected by the process of conjunction. The design features of the two learning systems are not identical, but have much in common, despite the differences in circuit arrangements. Of particular significance are the similarities in the design of the conjunction zones: climbing fibers or cartridge synapses on the one hand, and crossing-over synapses on dendritic spines on the other.

16

Perception, Speech, and Consciousness

16.1 Perception

16.1.1 Introduction

There are certain principles relating to the neural events that lead to perceptions of the various sensory experiences (Adrian 1947; Mountcastle 1975). Touch and vision have been most thoroughly investigated, but there is good reason to believe that all other sensory experiences are dependent upon similar neuronal mechanisms. Of necessity the crucial experimental investigation of sensory experiences must be carried out on conscious human subjects, but both the design and interpretation of these experiments are dependent on the wonderful successes that have attended investigations on animal, and particularly monkey, sensory systems in the last few decades. The powerful techniques designed for precision and selectivity of stimulation have been matched by microelectrode recording from single neurons. But just as importantly there has been the success in defining the neural pathways from receptor organs to cerebral cortex and within the cerebral cortex by precise anatomical investigations.

There is a large variety of these receptor organs with built-in properties that enable them to encode in a highly selective manner some environmental change into a discharge of nerve impulses. In general it can be stated that intensity of stimulus is encoded as frequency of discharge of impulses (cf. Chapter 2). In this way there are transmitted to the higher levels of the central nervous system signals from receptor organs that result in the conscious experiences of vision, hearing, and touch, for example. An introduction to the problem of conscious perception is best given in relation to cutaneous sensing. In the skin are receptor organs specialized for converting some mechanical stimulus such as a touch or tap into impulse discharges in nerve fibers (Figure 2.2A, Section 2.1).

The pathways from receptor organs to the brain are never direct. There are always synaptic linkages from neuron to neuron at each of several relay stations. Each of these stages gives the opportunity for modifying the coding of the "messages" from the sensory receptors. Even the simplest stimuli such as a flash of light or a tap on the skin are signaled to the appropriate primary receiving area of the cerebral cortex in the form of a code of nerve impulses in various temporal sequences and in many fibers in parallel.

Our special interest is focused on the neural events that are necessary for giving a conscious experience. It is now generally agreed that a conscious experience does not light up as soon as impulses in some sensory pathway reach the primary sensory areas in the cerebral hemisphere. In response to some brief peripheral stimulus the initial response is a sharp potential change, the evoked response in the appropriate primary cortical area. Immediately afterward there is a change in the background frequency of firing of numerous neurons in this area—an increase or a decrease, or some complex temporal sequence thereof. Our present problem is to gain some insight into the neural events that have a necessary relationship to the conscious experience.

16.1.2 Cutaneous Perception (Somesthesis)

Figure 16.1 is a diagram of the simplest pathway from receptor organs in the skin up to the cerebral cortex. For example, a touch on the skin causes a receptor to fire impulses. These travel up the dorsal columns of the spinal cord (the cuneate tract for the hand and arm) and then, after a synaptic relay in the cuneate nucleus and another one in the thalamus, the pathway reaches the cerebral cortex. There are only two synapses on the way and, you might say, why have any at all? Why not have a direct line? The point is that each one of these relays gives an opportunity for an inhibitory action that sharpens the neuronal signals by eliminating all the weaker excitatory actions, such as would occur when the skin touches an ill-defined edge. In this way a much more sharply defined signal eventually comes up to the cortex and there again there would be the same inhibitory sculpturing of the signal by modular interaction (cf. Figure 15.16). As a consequence, touch stimuli can be more precisely located and evaluated. In fact, because of this inhibition a strong cutaneous stimulus is often surrounded by a cutaneous area that has reduced sensitivity.

Also shown in Figure 16.1 are the pathways down from the cerebral cortex to both of these relays on the cutaneous pathway. In this way, by exerting presynaptic and postsynaptic inhibition, the cerebral cortex is able to block these synapses and so protect itself from being bothered by cutaneous stimuli that can be neglected. This is, of course, what happens when you are very intensely occupied, for example, in carrying out some action or in experiencing or in thinking. Under such situations you can be oblivious even of

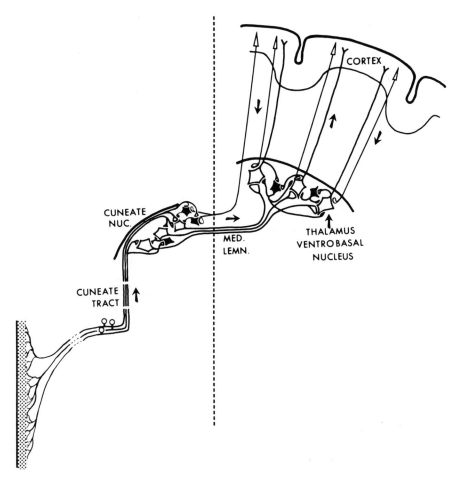

Figure 16.1 Pathway to the sensorimotor cortex for cutaneous fibers from the forelimb. Note the inhibitory cells shown in black in both the cuneate nucleus and the ventrobasal nucleus of the thalamus. The inhibitory pathway in the cuneate nucleus is of the feed-forward type and in the thalamus it is the feedback type. Also shown is one presynaptic inhibitory pathway to an excitatory synapse of a cuneate tract fiber. Efferent pathways from the sensorimotor cortex are shown exciting the thalamocortical relay cells and exciting both postsynaptic and presynaptic inhibitory neurons in the cuneate nucleus.

severe stimulation. For example, in the heat of combat severe injuries may be ignored. At a less severe level it has long been a practice to give counterirritation to relieve pain. Presumably in this way inhibitory suppression of the pain pathway to the brain is produced. Thus, we can account for the apparent anesthesias of hypnosis or of yoga or of acupuncture by the cerebral pathways' inhibition of the cutaneous pathways to the brain. In all these cases, discharges from the cerebral cortex down the pyramidal tract and other pathways will exert an inhibitory blockage at the relays in the spinocortical pathways such as those diagrammed in Figure 16.1. This ability of the cerebral cortex is important because it is undesirable to have all receptor organ discharges from your body pouring into your brain all the time. The design pattern of successive synaptic relays, each with various central and peripheral inhibitory inputs, gives opportunity for turning off inputs according to the exigencies of situations.

Conventional studies on animals and man have defined the area of cortex that is primarily involved in responding to cutaneous sense, the somesthetic area. As shown in Figures 13.2 and 13.3 (Section 13.3.1), the principal area is laid out as a long strip map in the postcentral gyrus with locations matching those of the motor cortex. All areas of the body surface from the extreme caudal to the extreme rostral lie in linear sequence along the postcentral gyrus from its dorsomedial end over the convex surface of the cerebral hemisphere. There is also a subsidiary somesthetic area which is partly concerned in the perception of pain. It will be seen in Figure 13.3 that cortical areas are apportioned in relationship to the fineness of discrimination of the cutaneous areas and not to their relative areas. This map has been explored in detail by two main procedures: recording in nonhuman primates of the cortical responses evoked by exploratory stimulation applied systematically to the whole surface of the body, limbs, neck, and head; and electrical stimulation of the sensory cortex in conscious human subjects who report the skin areas to which the evoked sensations are referred (Penfield and Jasper 1954).

Usually the subjects report abnormal sensory experiences—paresthesia such as tingling, numbness, "pins and needles"—though there are also reports of normal sensations—touch, tap, and pressure. The paresthesias are plausibly explained by the outrage that the applied stimulation perpetrates on the highly organized neuronal machinery of the cerebral cortex. Even the weakest electrical stimulus will excite in a manner dependent on the relationship of the immense neuronal assemblage to the applied electrical current. As a consequence, there will be "a neuronal shock wave" having little resemblance to the pattern of neuronal activation generated by a natural input from the receptor organs; hence the paresthesia, just as when the ulnar nerve at the elbow joint is bumped—the so-called funny-bone.

The thalamocortical fibers in Figure 16.1 would have the distribution in the cortex illustrated for Spec. aff. in Figure 15.16 (Section 15.7), exciting many stellate cells—the excitatory that amplify and distribute the message in that column or module, and the inhibitory that inhibit adjacent modules. Figure 15.16 is a greatly simplified picture of the real state of affairs. An important design feature of the cerebral cortex is that the whole neuronal machinery, activated by many similar fibers of some specific cutaneous input, is arranged in a column that is vertically oriented to the cortical surface. So, as a first approximation, we can say that all these neurons in Figure 15.16 are engaged in an integrated operation. The same kind of cutaneous sense (modality) from the cutaneous area provides the input to a column, where there is mutual help in neuronal activation. You might ask, what is the point of this columnar arrangement? Isn't it enough for each afferent fiber to register its own ongoing activity in isolation? The point is that impulses in one fiber and discharges from one neuron or a few neurons will be virtually ineffective at the next stage of the synaptic relay. There must be many parallel lines because the input from one afferent fiber is lost in all the incessant "noise" of background firing, which is the actual state of the neurons of a "resting" cerebral cortex. But, if you have 100 or even perhaps 20 fibers coming in with approximately synchronized bursts of impulses, the whole ensemble of cells in the column would be stirred up in a highly significant manner. This in turn will result in a spreading of meaningful signals widely and selectively in the cerebral cortex that could lead eventually to a conscious experience.

It is important to realize that there is a considerable "incubation period" between the arrival of impulses at the primary sensory area in Figure 16.1 and the experience of a conscious sensation (Popper and Eccles 1977, Chapter E2). By accurately timing experiments with stimulation of the somesthetic cortex of conscious subjects, Libet (1973) has shown that the "incubation period" is as long as 0.5 s for weak cortical stimuli presented in a repetitive train of shocks at about 50 per second. Evidently there is opportunity for a great elaboration of neuronal activity in complex spatiotemporal patterns before there is a conscious experience.

16.1.3 Visual Perception

Highly complicated and exquisitely designed structures are involved in all steps of the visual pathways. The optical system of the human eye gives an image on the retina which is a sheet of closely packed receptors, some 10^7 cones and 10^8 rods, that feed into the complexly organized neuronal systems of the retina. Thus, the first stage in visual perception is a radical fragmentation of the retinal picture into the independent responses of a myriad of punctate elements, the rods and cones. In some quite mysterious way the retinal picture appears in conscious perception, but nowhere in the brain can there be found neurons that respond specifically to some fragment of the retinal image or of the perceived picture. The neuronal machinery of the visual system of the brain has been shown to accomplish a very inadequate reconstitution that can be traced in many sequences (cf. Kuffler 1973).

The initial stage of reconstitution of the picture occurs in the complex nervous system of the retina. As a consequence of this retinal synthetic mechanism the output in the million or so nerve fibers in each optic nerve is not a simple translation of the retinal image into a corresponding pattern of impulse discharges that travel to the primary visual center of the brain (Figure 13.2, Section 13.3.1). Already in the nervous system of the retina there has begun the abstraction from the richly patterned retinal mosaic into elements of pattern, which we may call features, and this abstraction continues in the many successive stages that have now been recognized in the visual centers of the brain (Popper and Eccles 1977, Figures E1-E7H and 8B).

The complex interactions in the retinal nervous system eventually are expressed by the retinal ganglion cells that discharge impulses along the optic nerve fibers and so to the brain. These cells respond particularly to spatial and temporal changes of luminosity of the retinal image by two neuronal subsystems signaling brightness and darkness, respectively. The brightness contrasts of the retinal image are converted into contoured outlines by several neuronal stages of information processing. One type of ganglion cell is excited by a spot of light applied to the retina over it and is inhibited by light on the surrounding retina. The other type gives the reverse response, inhibition by light shone into the center and excitation by the surround. The combined responses of these two neuronal subsystems result in a contoured abstraction of the retinal image in the visual cortex. Hence, what the eye tells the brain by the million fibers of the optic nerve is an abstraction of brightness and color contrasts.

As illustrated in Figure 16.2, the optic nerves from each eye meet in the optic chiasma where there is a partial crossing. The hemiretinas of both eyes (nasal of right and temporal of left) that receive the image from the right visual field have their optic nerve projections rearranged in the chiasma so that they coalesce to form the pathway to the left visual cortex, and vice versa for the left visual field projecting to the right visual cortex. Thus, with the exception of a narrow vertical (meridional) strip of the visual field that is directly in the line of vision, the visual imagery of the right and left fields

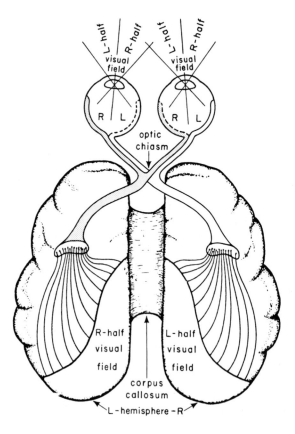

Figure 16.2 Diagram of visual pathways showing the L-half and R-half visual fields with the retinal images and the partial crossing in the optic chiasma so that the R-half of the visual field of each eye goes to left visual cortex, after relay in the lateral geniculate body (LGB), and correspondingly for left visual field to right visual cortex.

comes to the left and to the right visual cortices, respectively, to form an ordered map much as with the cortical map for cutaneous sensation. There is, of course, topographic distortion. The fine visual sensing in the center of the visual field results from a much more amplified cortical projection area than for the retina concerned with peripheral vision.

Microelectrodes can be used to record the impulses discharged from single nerve cells, for example, as has been done with great success by Hubel and Wiesel (1962) in the primary visual cortex. We use one aspect of their work to illustrate the very subtle and well-controlled experiments that are being carried out in many laboratories. In Figure 16.3A there is a single cell firing impulses, having been "found" by a microelectrode which has been inserted into the primary visual cortex of the cat. The track of insertion is shown for example in Figure 16.3B as the sloping line with short transverse lines indicating the locations of many nerve cells along that track. With the microelectrode it is possible to record extracellularly the impulse discharges of a single cell if it is positioned carefully (cf. Figure 13.5, Section 13.3.3). The cell has a slow background discharge (upper trace of Figure 16.3A), but if the retina is swept with a band of light, as illustrated in the diagram to the left, there is an intense discharge of that cell when light sweeps across a certain zone of the retina and there is immediate cessation of the discharge as the light band leaves the zone (lowest trace of Figure 16.3A). If you rotate the direction of sweep, the cell discharges just a little, as in the middle trace. Finally, if the sweep is at right angles to the most favorable direction, it has no effect whatever (uppermost trace). It is a sign that this particular cell is most sensitive for movements of the light strip in one orientation and is quite

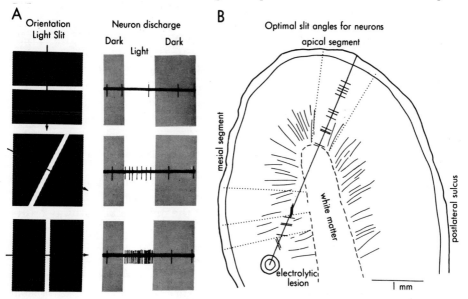

Figure 16.3 Orientational responses of neurons in primary visual cortex of cat. (From Hubel and Wiesel 1962.)

insensitive for movements at right angles thereto. As illustrated by the direction of the lines across the microelectrode track in Figure 16.3B, all cells along that track have the same orientational sensitivity. This is found when the track runs along a column of cells that is orthogonal to the surface, as in the upper group of 12 cells. In Figure 16.3B, however, the track continued on across the central white matter and then proceeded to pass through three groups of cells with quite different orientation sensitivities. Evidently, the track was crossing several columns with different orientational sensitivities in the way illustrated by the dotted sectors. This columnar arrangement for mutual reinforcement of similar receiving cells has already been illustrated (Figures 13.6, Section 13.3.4; and 15.16, Section 15.7).

It has been demonstrated by Jung, Bishop, and others that retinal zones adjacent to those giving excitation have an inhibitory action on the firing of a neuron such as that of Figure 16.3A. For example, if the cell in Figure 16.3A had a relatively high background discharge, this discharge would be depressed by illumination of adjacent areas of the retina. This can be explained by the diagram of Figure 15.16, where excitation of one area of the cerebral cortex results in inhibition of adjacent areas, much as with basket cell inhibition in the cerebellum in Figure 13.14 (Section 13.4.4). This observation has great interest because it explains the so-called Mach bands of perceptual physiology. At the edge between fields of uniform bright and dull illumination there is perceived to be a narrow light-dark zone, the Mach band, which is explicable by this lateral inhibition.

The activation of cells in the primary visual cortex does not in itself give one a sensation. This is merely the first coded response of the visual cortex. Before one can experience a flash of light, neuronal discharges must be induced in extremely complicated pathways after many successive levels of neuronal transmission. As shown by Hubel and Wiesel, each one of these cells with its simple orientational sensing (Figure 16.3B) relays into interacting systems of cells which will respond specifically to angles of different degrees between the two light bands, and that also may be specified as to the lengths and widths of the effective light bands. As the record is moved further from the primary visual cortex out into the surrounding areas of the visual cortex, these more synthetic responses are observed for the so-called complex and hypercomplex cells. It is a further stage of the processing that goes on in the brain.

In addition to the first visual system that is indicated in Figure 16.2 there is a second visual system, the residuum from an earlier visual system (Schneider 1969). As indicated in Figure 16.4 the optic nerve pathway bifurcates, one projecting to the lateral geniculate nucleus as in Figure 16.2, the other to the superior colliculus on the same side. There are then many pathways, the simplest being indicated in Figure 16.4. The pulvinar is the principal further projection site, both ipsilateral and contralateral, and from thence there is projection to the association visual cortices on both sides. In normal subjects there is no evidence for the projection of the left visual field of the right eye to the contralateral visual cortex, as indicated in Figure 16.4. When

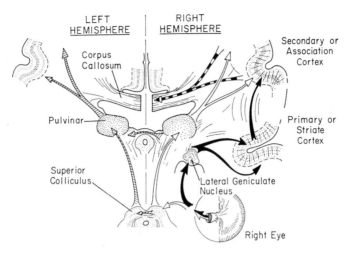

Figure 16.4 Anatomical pathways of a second visual system whereby, despite commissurotomy, objects far out in the visual field can be projected to the visual cortices of both hemispheres via the superior colliculus and the pulvinar. The first visual system (cf. Figure 16.2) is shown in black from the right eye to lateral geniculate to striate (visual) cortex. (Trevarthen and Sperry 1973.)

the corpus callosum is sectioned, however, preventing information transfer from the ipsilateral visual cortex (the black–white arrows), information from the left peripheral visual field of the right eye can be received by the conscious subject that is in liaison only with the left hemisphere. Trevarthen and Sperry (1973) suggest that the decussating pathways of the second visual system are responsible. This system does not give sharp visual information, but may be important in signaling brightness and some kinds of pattern.

16.1.4 Auditory Perception

There is a highly specialized transduction mechanism in the cochlea where, by a beautifully designed resonance mechanism, there is a frequency analysis of the complex patterns of sound waves and conversion into the discharges of neurons that project into the brain. After several synaptic relays the coded information reaches the primary auditory area (Heschl's gyrus) in the superior temporal gyrus (cf. Figures 13.2, 16.5). The right cochlea projects mostly to the left primary auditory area, and vice versa for the left cochlea. There is a linear somatotopic distribution, the highest auditory frequencies being most medial in Heschl's gyrus (Figure 16.5), and the lowest most lateral. There is apparently no attempt at even a fragmentary reconstitution of the initiating stimulus such as occurs in the visual centers. It remains quite mysterious how a sequence of tones gives rise to a new synthesis, a melody. Nevertheless, there are parallels between the connections in cascade (Popper and Eccles 1977, Figure E1–7I to L) and those for somesthesis and vision.

As in the visual system, there is a crossed connection from the ear to the primary auditory sensory area, i.e., to Heschl's gyrus in the temporal lobe

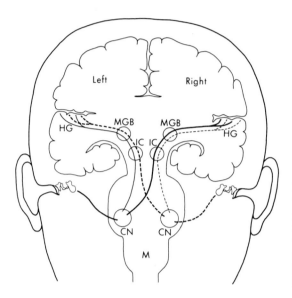

Figure 16.5 Schematic drawing of the auditory pathways to Heschl's gyri (HG) on each side, showing the dominance of the crossed connections. CN, cochlear nucleus; IC, inferior colliculus; MGB, medial geniculate body; M, medulla oblongata.

(Figure 16.4). The situation differs from that in vision, however, where the connections of one visual field to the respective primary visual cortex are entirely crossed (Figure 16.2). There exist also ipsilateral connections from one ear to Heschl's gyrus of the same side. This ipsilateral connection is much weaker than the contralateral, however; furthermore, the ipsilateral pathways are suppressed by the contralateral during dichotic presentation (Milner *et al.* 1968; Sparks and Geschwind 1968), presumably by inhibition in the cerebral cortex. Thus, we would attribute the right ear advantage to the fact that the right ear has a more direct access to that hemisphere in which the encoded auditory input is decoded into recognizable words, vis., to the left hemisphere, the speech hemisphere.

16.1.5 Olfactory Perception

In most lower mammals olfaction (smell) is the dominant sensory input into the forebrain, but in the evolution of primates to man olfaction became subordinated to vision and hearing, and even to somesthesis, particularly when this became vital in manual skills. Chemical sensing in the olfactory mucosa is by receptor cells that are specialized neurons with axons that pass to the olfactory bulb where there is a processing of information by a complex nervous system much as in the retina (Figure 4.13, Section 4.6). From the olfactory bulb (OLB) the lateral olfactory tract (LOT) passes to the brain (cf. Figure 15.13) where it has a complex distribution. The principal termination is in the pyriform cortex (PC), a primitive cerebral cortex. Thence there are connections (not shown) to many structures of the limbic lobe. Connection to the neocortex (orbitofrontal cortex) is effected only after several relays in the limbic system and is in part via the MD thalamus. Thus, the olfactory connections are quite different from the somesthetic, visual, and auditory systems, where the connections are first to the neocortex and after several relays reach the limbic system (Figure 15.12, Section 15.6).

16.1.6 Emotional Coloring of Conscious Perceptions

It is a common experience that the conscious perception derived from some sensory input is greatly modified by emotions, feelings, and appetitive drives. For example, when one is hungry the sight of food gives an experience deeply colored by an appetitive drive! Nauta (1971) conjectures that the state of the organism's internal milieu (hunger, thirst, sex, fear, rage, pleasure) is signaled to the prefrontal lobes from the hypothalamus, the septal nuclei, and various components of the limbic system such as the hippocampus and the amygdala. The pathways would be mainly through the MD thalamus to the prefrontal lobes (Figures 15.12, 15.13). Thus, by their projections to the prefrontal lobes the hypothalamus and the limbic system modify and color with emotion the conscious perceptions derived from sensory inputs and superimpose on them motivational drives. No other part of the neocortex has this intimate relationship with the hypothalamus.

Figure 15.12 shows for the somesthetic, visual, and auditory system the many projections to the prefrontal lobes from the primary sensory and the related association areas of the parietal, occipital, and temporal lobes. Simultaneously, these areas project to the limbic system, and in Figure 15.13 there are also projections from the prefrontal lobe (areas 46 and OF) to the limbic system. Thus, there are pathways for complicated circuitry from the various sensory inputs to the limbic system and back to the prefrontal lobe, with further circuits from that lobe to the limbic system and back again (Nauta 1971). From the connectivities of Figure 15.13 it can be seen that the prefrontal and limbic systems are in reciprocal relationship and have the potentiality for continuously looping interaction. Thus, by means of the prefrontal cortex the subject may be able to exercise a controlling influence on the emotions generated by the limbic system. An additional sensory input (olfaction) comes directly into the limbic system for cross-modal transfer to the other senses and thus contributes to the richness and variety of the perceptual experience. For example, the neocortical sensory systems via areas 46, OF, 20, and TG project to the hypothalamus (Figure 15.13), the entorhinal cortex, and the hippocampal gyrus, and so to the hippocampus, to the septal nuclei, and to the MD thalamus, while, after relay in the pyriform cortex and amygdala, the olfactory input also goes to the hypothalamus, septal nuclei, and the MD thalamus. Thus, the MD nucleus is the receiving station for all inputs and in turn it projects widely to the cortex of the prefrontal lobe. Thus, one can think of the prefrontal cortex as being the area where all emotive information is synthesized with somesthetic, visual, and auditory to give conscious experiences to the subject and guidance to appropriate behavior.

16.2 The Language Centers of the Human Brain[1]

16.2.1 Introduction

The representation of language in the cerebral cortex has been investigated by four methods: (1) the study of linguistic disorders arising from cerebral lesions; (2) the effects of stimulation of the exposed brain of conscious subjects and of the transient aphasias resulting from the exposure; (3) the effects of intracarotid injections of sodium amytal (a neural depressant); and (4) the dichotic listening tests.

16.2.2 Aphasia

For over a century disorders of speech (aphasia) have been associated with lesions of the left cerebral hemisphere (Figures 13.2, 16.6). There was first the motor aphasia described by Broca in 1861 as arising from lesions of an area that we now call the anterior speech center of Broca. The patient had lost the ability to speak although he could understand spoken language. Broca's area lies just in front of the cortical areas controlling the speech muscles; nevertheless, motor aphasia is due not to paralysis of the vocal musculature, but to disorders in their usage. Much more important, however, is the large speech area lying more posteriorly in the left hemisphere. On the basis of evidence from lesions, it was originally thought by Wernicke in 1974 to be only in the superior temporal convolution, but now it is recognized as having a much more extensive representation on the parietotemporal lobes (Figure 16.6). This posterior speech center of Wernicke is specially associated with the ideational aspect of speech. The aphasia is characterized by failure to understand speech—either written or spoken. Although the patient could speak with normal speed and rhythm, his speech was remarkably devoid of content, being a kind of nonsense jargon. In the great majority of patients, lesions anywhere in the right hemisphere do not result in serious disorders of speech.

Aphasia itself has been subjected to most detailed and diverse descriptions and classifications. Areas specialized for reading and writing, for example, have been recognized by the alexia or agraphia resulting from their destruction. It is essential to recognize the incredible complexity of the encoding and decoding in speech.

As an illustration we can consider the neural events concerned in some simple linguistic performance. For example, in reading aloud, black marks on white paper are projected from the retina to the brain, in the encoded form of impulse frequencies in the optic nerve fibers, and so eventually to the primary visual cortex (cf. Figures 13.2, 16.2). The next stage is the transmission of the encoded visual information to the visual association areas (Figure 16.6), where there is a further stage of reconstitution of the visual image. As described

[1]General references: Penfield and Roberts 1959; Geschwind 1970; Popper and Eccles 1977, Chapter E.

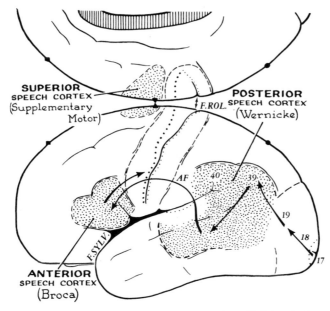

Figure 16.6 Left hemisphere showing speech areas. Redrawing of left hemisphere as in Figure 13.2 (Section 13.3.1), but with supramarginal and angular gyri labeled as Brodmann areas 39 and 40. Also shown by arrows is the pathway from area 17 to 18, 19, to 39 (angular gyrus), to the Wernicke speech area, thence by the arcuate fasciculus (AF) to Broca's area (cf. Geschwind 1972.)

above, this reconstitution is still most inadequate. Neurons specifically respond to simple geometrical forms, the so-called feature recognition neurons. At the next stage, however, lesions of the posterior part of Wernicke's area (the angular gyrus, 39 in Figure 16.6) result in dyslexia, suggesting that the relay from the visual association neurons provides information that is converted into word patterns and that these in turn are interpreted as meaningful sentences in the semantic process of conscious recognition of meaning that occurs in the Wernicke area.

The further stage in the process of reading aloud is via the arcuate fasciculus (Figure 16.6) to the motor speech area (Broca's area). Lesions of the arcuate fasciculus result in conduction aphasia (Geschwind 1970). There is comprehension of spoken language but a gross defect in its repetition and in normal speaking. At the terminal stage, appropriate patterns of neural activity in Broca's area lead to the motor areas for vocalization and so to the coordinated contraction of the speech muscles that give spoken language. A similarly complex chain of encoding and decoding is involved in writing out language that is heard.

As a general summary, it can be stated that great difficulties arise in a sharp classification of aphasias because of the irregular destructive action of clinical lesions. For our present purpose it is not necessary to go into all of the detailed disputation between the various experts on the many types of aphasia or on the causative cerebral lesions. The remarkable discovery is that the enormous proportion of aphasics have lesions in their left cerebral hemisphere. Only very rarely is a right cerebral lesion associated with aphasia. There was originally a general belief that right-handed patients had their speech centers on the left side and vice versa for the left-handed patients. This has proved to be untrue. The great majority of the left-handed subjects also have their speech centers in the left cerebral hemisphere (Penfield and Roberts 1959).

16.2.3 Experiments on Exposed Brains

In the hands of Penfield and his associates, stimulation of the cerebral cortex has been responsible for quite remarkable discoveries relating to the localization of speech centers. Stimulation of the motor areas in either hemisphere (Figure 13.2) innervating structures concerned in sound production such as tongue and larynx cause the patients to produce a variety of calls and cries (vocalization), but not recognizable words. These are the motor areas of voice control and are bilateral. Only rarely does a similar stimulation in animals give vocal responses. On the other hand, stimulation of the speech areas (Figure 16.6) results in an interference or an arrest of speech. For example, if the subject is engaged in some speech performance such as the counting of numbers, his voice may be slurred or distorted, or the same number may be repeated. Often the application of the gentle stimulating current to the speech areas causes a cessation of speech which is resumed as soon as the stimulation stops, or there is a temporary inability to name objects during the stimulation. One can imagine that the stimulus has caused a widespread interference with the specific spatiotemporal patterns of neuronal activity that are responsible for speech. In this way Penfield and his associates have been able to delimit the two speech areas that have been recognized from clinical studies of aphasia, namely, the anterior and posterior speech areas, and also a subsidiary third area (Figure 16.6).

Inadvertent sequelae to operative procedures have been important in demonstrating the cerebral hemisphere that is responsible for speech—whether it is the right or left hemisphere of the subject. It has been observed that after a cerebral operation involving exposure of one cerebral hemisphere, a transient aphasia often develops some days after the operation and continues for two to three weeks. This is attributed to the neuroparalytic edema resulting from brain exposure. A systematic study of the neuroparalytic aphasia of patients by Penfield and Roberts (1959) showed that it developed in over 70% of patients with a left hemisphere operation regardless of whether they were right- or left-handed. By contrast, with operations on the right hemisphere aphasia was very rare. These observations indicated the very strong dominance of speech representation (98%) in the left hemisphere. Other investigators using various techniques are in general agreement with these results, the dominance being about 95%, but left-handed patients have right hemisphere representation of speech more frequently, though still not as frequently as left hemisphere representation (Piercy 1967).

16.2.4 Intracarotid Injections of Sodium Amytal

Another method of determining speech representation and relating it to handedness was developed by Wada with the injection of sodium amytal into the common or internal carotid arteries of subjects in which it was important preoperatively to identify the speech hemisphere. There was likewise the overwhelming dominance of the left hemisphere representation of speech for right-handed subjects and a considerable dominance also for the left-handed subjects. Speech in very young children is bilateral, the left hemisphere gradually assuming dominance over the first few years of life. By 4 to 5 years of age, speech has become fully lateralized (Kimura 1967). Damage to the left hemisphere in infancy may result in the development of speech areas in the right hemisphere (Milner 1974). There appears to be considerable neural plasticity at this early age.

16.2.5 Dichotic Listening Test

Through headphones, the subject receives simultaneously two different auditory stimuli, one to the right ear, the other to the left ear. The test was first tried on word recognition. Three pairs of digits (say 2, 5; then 3, 4; then 9, 7) were presented dichotically to normal subjects in rapid succession, after which the subject was asked to report in any order as many of the digits as he could. It was surprising to find that those digits presented to the right ear were more accurately reported than those to the left ear, although it was shown that there were no differences in the respective sensory auditory channels. The asymmetry in normal auditory recognition of words is explained by the dominance of crossed transmission in the neural pathways for hearing (cf. Figure 16.5). At the level of word recognition the left hemisphere displays its linguistic superiority, but it has been shown that the auditory areas of both hemispheres perform equally well at the earlier stage of auditory pattern analysis.

16.2.6 Anatomical Substrates of Speech Mechanisms

The unique association of speech and consciousness with the dominant (left) hemisphere gives rise to the question: Is there some special anatomical structure in the dominant hemisphere that is not matched in the minor hemisphere? In general, the two hemispheres have been regarded as being mirror images at a crude anatomical level, but recently it has been discovered that in about 65% of human brains there is hypertrophy of a part, the planum temporale, of the left superior temporal gyrus in the region of the posterior speech area of Wernicke. Wada *et al.* (1975) found the left-right asymmetries of the planum temporale not only in infants who died at birth, but also in a 29-week-old fetus. Thus, the speech localization appears to be genetically determined, the speech centers being built in preparation for their eventual usage after birth. On the other hand, handedness would appear to be much more flexible and to be at least in part determined by environmental habits. There will be further reference to the anatomical substrates of speech mechanisms in the next section, particularly in relation to the effects produced by global lesions of the cerebrum—commissurotomy and hemispherectomy.

16.3 Language and Self-Consciousness

16.3.1 Effects of Global Cerebral Lesions

Commissurotomy. The corpus callosum is a tract of about 200 million fibers that provides an immense commissural linkage in an approximately mirror-image manner between almost all regions of the cerebral hemispheres. The intense impulse traffic in the corpus callosum keeps the two hemispheres of the brain working together. The corpus callosum has been completely severed in about 20 human subjects who were suffering from almost incessant epileptic seizures that could not be controlled, even by heavy medication. It was surmised that seizures developed in one cerebral hemisphere and then excited the other via the corpus callosum so that the seizure rapidly became generalized. Hence, it was proposed that section of the corpus callosum would at least keep one hemisphere free from the seizures. It turns out that the operation does better than predicted. There is a remarkable diminution of seizures in both hemispheres.

Sperry and associates (Sperry 1974) have developed testing procedures in which information can be fed into one or the other hemisphere of these "split-brain" patients and in which the responses of either hemisphere can be observed independently. Essentially, the information is fed into the brain from the right or left visual fields (cf. Figure 16.7). The subject fixates a central point and the signal, for example a word, is flashed on for only 0.1 s in order to eliminate changes in the visual field by eye movements. All of the seven investigated subjects had the speech areas in the left hemisphere (cf. Figure 16.6), which was thus always the dominant hemisphere. The most remarkable discovery of these experiments was that all the neural activities in the right hemisphere (the so-called minor hemisphere) are unknown to the speaking subject who is only in liaison with the neuronal activities in the left hemisphere, the dominant hemisphere. Only through the dominant hemisphere can the subject communicate with language. Furthermore, in liaison with this hemisphere is the conscious being or self that is recognizably the same person as before the operation.

Figure 16.7 is a diagram drawn by Sperry several years ago. It is still valuable, however, as a basis of discussion of the whole split-brain story. The diagram illustrates the right and left visual fields with their highly selective projection to the crossed visual cortices, as indicated by the letters R and L (cf. Figure 16.2). Also shown in the diagram is the strictly unilateral projection of smell, and the predominantly crossed projection of hearing (cf. Figure 16.5).

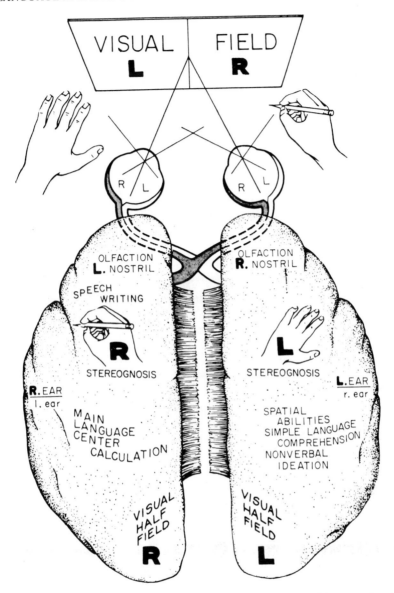

Figure 16.7 Schema showing the way in which the left and right visual fields are projected onto the right and left visual cortices, respectively, due to the partial decussation in the optic chiasma. The schema also shows other sensory inputs from right limbs to the left hemisphere and from left limbs to the right hemisphere. Similarly, hearing is largely crossed in its input, but olfaction is ipsilateral. (From Sperry 1974.)

The crossed representations of both motor and sensory innervation of the hands are indicated, and also the further finding that arithmetical calculation is predominant in the left hemisphere. Only very simple addition sums can be carried out by the right hemisphere.

We can say that the right hemisphere is a very highly developed brain except that it cannot express itself in language, so it is not able to disclose any experience of consciousness that we can recognize. Sperry postulates that there is another consciousness in the right hemisphere, but that its existence is obscured by its lack of expressive language. On the other hand, the left hemisphere has a normal linguistic performance, so it can be recognized as being associated with the prior existence of the ego or the self with all the memories of the past before the commissural section. On this view there has been split off from the talking self a nontalking self which cannot communicate by language, so it is there but mute, or aphasic. For the present we can say that we do not know if there is some inexpressible consciousness in the isolated right hemisphere just as we do not know of any consciousness animals might have. We cannot discover it in any way, but we have to realize the limitations of our testing procedure both for higher animals and for the right hemisphere after section of the corpus callosum. It is important to distinguish between the self-consciousness associated with the left hemisphere and the postulated consciousness associated with the right hemisphere.

In general, the dominant hemisphere is specialized in respect to fine imaginative details in all descriptions and reactions, i.e., it is analytic and sequential. Also, it can add and subtract and multiply and carry out other computer-like operations. But, of course, its dominance derives from its verbal and ideational abilities and its liaison to self-consciousness. Because of its deficiencies in these respects the minor hemisphere deserves its title, but in many important properties it is preeminent, particularly in respect to its spatial abilities with a strongly developed pictorial and pattern sense. For example, the minor hemisphere programming the left hand is greatly superior in all kinds of geometrical and perspective drawings. This superiority is also evidenced by the ability to assemble colored blocks so as to match a mosaic picture. The dominant hemisphere is unable to carry out even simple tasks of this kind and is almost illiterate in respect to pictorial and pattern sense, at least as displayed by its copying disability. It is an arithmetical hemisphere but not a geometrical hemisphere.

All of the very fine work with flash testing has been superseded by a new technique (Zaidel 1976a, 1976b) in which a contact lens is placed in the right eye with an optical device that limits the input into that eye to the left visual field no matter how the eye is moved. At the same time, an eye shield prevents the left eye from being used. In this way there can be up to 2 h of continuous investigation of the subject, which gives opportunity for much more sophisticated testing procedures than with the flash testing. The tests have been concerned with the ability of the right (minor) hemisphere to understand complex visual imagery, as shown by appropriate reactions with the left hand. It must be recognized that the left eye of the subject is covered and that the optical device mounted on the contact lens on the right eye allows an input to the retina only from the left visual field. Thus, there is elimination of all input to the subject from the right visual field, i.e., the subject is quite blind so far as any conscious visual experiences are concerned.

Examples of the pictorial understanding of the right hemisphere are given by testing procedures in which there is a picture, say, of a cat and below the words "cat" and "dog." The subject with the left hand can correctly point to the appropriate word. Symmetrically, if there are two pictures, a cup and a knife, and below is only one word, "cup," the subject will point to the correct object (cup) with the left hand. An even more sophisticated test of picture identification is provided by a drawing of landscapes below which are a correct and an incorrect name. For example, below the pictures were the names "summer" and "winter," and the subject was able to point to the word "winter" rather than "summer" in correct identification of the picture. Despite this intelligent performance with pictorial and verbal presentation to the minor hemisphere, this hemisphere is totally unable to complete sentences, even the simplest. The evidence from language testing of chimpanzees indicates likewise that they are unable to complete sentences, though some dubious claims have been made. This, of course, arises from the fact that neither the minor hemisphere nor the chimpanzee brain has a Wernicke area that provides the necessary semantic ability.

These more rigorous testing procedures (Zaidel 1976b) have shown that after commissurotomy the right hemisphere has access to a considerable auditory vocabulary, being able to recognize commands and to relate words presented by hearing or vision to pictorial representations. It is particularly effective in recognition of pictorial representations that occur in common experiential situations. It was also surprising that the right hemisphere responded to verbs as effectively as to action names. Response to verbal commands was not recognized by the flash technique. Despite all this display of language comprehension, the right hemisphere is extremely deficient in expression in speech or in writing, which is effectively zero. It is also incapable of understanding instructions that include many items which have to be remembered in correct order. The highly significant finding is the large difference between comprehension and expression in the performance of the right hemisphere.

Hemispherectomy. The investigations on patients subjected to commissurotomy have provided far more definitive and challenging information than have other cerebral lesions. Nevertheless, the conclusions derived from the commissurotomy investigations can be subjected to testing procedures applied to patients with global or circumscribed cerebral lesions. The most radical are the hemispherectomized patients, where either the minor or the dominant hemisphere has been radically removed in the treatment of a gross cerebral tumor. They supplement the more definitive observations on the commissurotomy patients, and are in general agreement therewith (Zaidel 1976b).

Excisions of the minor hemisphere result in symptoms that, except for the hemiplegia and the absence of far-out left visual field recognition, are not appreciably different from those described in detail by Sperry in his study of patients with commissurotomy reported above. Thus, after minor hemispherectomy consciousness is derived from neural activities in the dominant hemisphere. One of the two cases of minor hemispherectomy reported by

Gott (1973a) is remarkable because it occurred in a young woman who was a music major and an accomplished pianist. After the operation there was a tragic loss of her musical ability. She could not carry a tune, but could still repeat correctly the words of familiar songs. The left (dominant hemisphere) excisions in the adult have much more serious sequelae. In the four cases that have been reported there are traces of residual self-consciousness and some slight recovery in very primitive linguistic ability. But the patients were very difficult to study as they were almost completely aphasic. Smith (1966) reported that his patient could use expletives and simple words in a song that he used to know. He had extreme restriction in language usage. Nevertheless, the isolated minor hemisphere had more linguistic ability than occurs in the minor hemisphere of Sperry's patients, where it is overshadowed by the dominant hemisphere. One wonders how much transfer of dominance had occurred in this patient before the operation because there had been a severe lesion of the dominant hemisphere for at least two years, from age 45 to age 47 at the time of the operation.

Hillier (1954) gave a more encouraging account of dominant hemispherectomy in a boy of 14 who survived about 2 years. This boy achieved a good recovery in general performance, but linguistically was very handicapped. Again, one suspects that there was some transfer of dominant hemisphere functions before the operation, and the youthfulness of the patient could help in his recovery. Despite the rather optimistic tone of this report it can be recognized that there was a tragic linguistic disability, a sequel that is to be expected after excision of the Wernicke speech area.

A better recovery was reported by Gott (1973b) in a girl who had complete hemispherectomy (dominant) at age 10. At age 8 there had been an excision of a tumor from that hemisphere, the final operation two years later being for a recurrence in the parietal area. At age 12 the patient had a linguistic ability gravely reduced, but surpassing the cases of dominant hemispherectomy considered above. It is suggested by Gott that there was a better recovery because at the age of 7 to 8 language may already be in the process of transfer from the damaged dominant hemisphere. It was remarkable that despite the very limited speaking ability, the patient could sing well and liked to do so, usually with the correct words. Despite the grave linguistic handicap it cannot be doubted that this girl had retained a self-conscious mind after the dominant hemispherectomy.

Infants present a much more encouraging situation. There is good evidence of a remarkable plasticity, the functions of the dominant hemisphere being effectively transferred up to five years of age (Milner 1974). There is bilateral speech representation at birth. Then, over the first few years of life cerebral dominance is established with regression of the linguistic ability of the minor hemisphere. When speech is transferred to the minor hemisphere it is always defective, and there is in addition a deterioration of the normal functions of the minor hemisphere. Thus, there is a limit to the capabilities of the remaining hemisphere so that it is deficient in mediating its normal functions and in accepting the functions from the other hemisphere.

Complete removal of the dominant hemisphere gives somewhat enigmatic results. Yet, so far as tests are possible in the almost aphasic cases, there is clearly some residual self-consciousness. Certainly, years after hemispherectomy in the first 5 years of life, tests reveal that the hemisphere has taken over linguistic functions and hence assumed a "dominant" status, at least to a partial extent. Possibly a small transfer of this kind occurs even in adults because of the destruction of large areas of the dominant hemisphere in the years preceding the operation.

Discussion. Zaidel (1976b) has formulated a most interesting hypothesis. Up till the age of 4 or 5 years both hemispheres develop together in linguistic competence, but the great increase in linguistic ability and skill coming on at that age demands fine motor control in order to give well-formed speech. It is at this stage that one hemisphere, usually the left, becomes dominant in linguistic ability because of its superior neurological endowment. Meanwhile, the other hemisphere, usually the right, regresses in respect to speech, but retains its limited competence in understanding. This understanding is particularly valuable when there are gestalt concepts to be interpreted. We suggest also that the right hemisphere is important for expressiveness and rhythm in speech, particularly in song, which is well preserved after dominant hemispherectomy and lost after minor hemispherectomy. This hypothesis accounts well for the transfer of speech to the minor hemisphere when there is severe damage to the linguistic areas of the dominant hemisphere before the age of 5, and the progressive limitation of transfer at later ages.

16.3.2 Dominant and Minor Hemispheres

Figure 16.8 shows that in their properties the two hemispheres are complementary. The minor is coherent and the dominant is detailed. Fur-

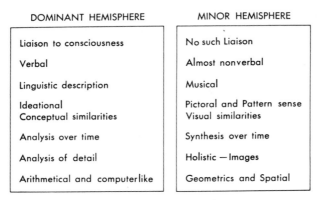

Figure 16.8 Various specific performances of the dominant and minor hemispheres as suggested by the new conceptual developments of Levy-Agresti and Sperry (1968) and Levy (1973). There are some additions to their original list.

thermore, not only is the minor hemisphere pictorial, but there is much recent evidence that it is musical. Music is essentially coherent and synthetic, being dependent on a sequential input of sounds. A coherent, synthetic, sequential imagery is made for us in some holistic manner by our musical sense. Furthermore, there is accumulating evidence by Milner (1974) that excision of the right temporal lobe does, in fact, seriously limit musical ability, as displayed in the Seashore tests.

The distinctiveness of the functions of the two hemispheres listed in Figure 16.8 is also indicated by the results of the dichotic technique that has been described above. It provides essentially a study of the subject's response to signals of a given modality that are applied so as to give competitive inputs into the two hemispheres. This new psychological technique of interhemispheric challenge has the great advantage that it can be applied to normal subjects, but on the other hand the results are not so discriminative as with studies of the effects of hemispheral lesions, both global and circumscribed.

It is an attractive hypothesis of Sperry's that the two hemispheres have complementary functions, which is an efficient arrangement because each can independently exercise its own peculiar abilities in developing and fashioning the neuronal input. Then, as illustrated in Figure 16.9, by commissural transfer the two complementary performances can be combined and integrated in the ideational, linguistic, and liaison areas.

It is our thesis that the philosophical problem of brain and mind has been transformed by these investigations of the functions of the separated dominant and minor hemispheres in the split-brain subjects. In Figure 16.9 the corpus callosum is shown as an extremely strong communication system so that all happenings in the minor hemisphere can very quickly and effectively be transmitted to the liaison brain of the dominant hemisphere, and so to the conscious self. This occurs for the contribution of the minor hemisphere for all perceptions, all experiences, all memories, and in fact for the whole content of consciousness. Removal of the immense interhemispheral communication by the commissurotomy would deplete the performance of the minor hemisphere. Normally it may have some direct liaison with the self-conscious mind as indicated by the broken arrows of Figure 16.9. The possible role of commissural support to the minor hemisphere will be referred to again later.

In Figure 16.9 the communication lines are going both ways: out to the muscles via the motor pathways; and receiving from the world, from the receptors to the sensory cortical areas. It will also be recognized that in its communication to the conscious self the minor hemisphere normally suffers no material disability relative to the dominant hemisphere from its necessity to transmit through the corpus callosum to the liaison areas of the dominant hemisphere. First, the corpus callosum is such an immense tract that in transmission through it the minor hemisphere would have no traffic problem. Second, as seen in Figure 16.9, the greater part of the dominant hemisphere presumably also has to transmit to the liaison areas.

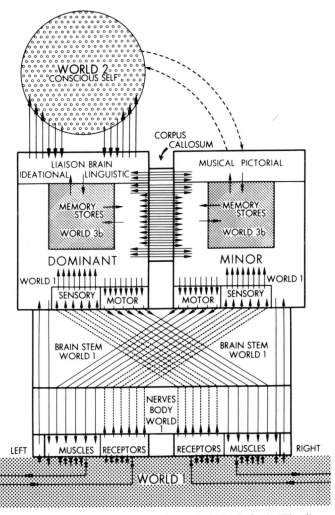

Figure 16.9 Communications to and from the brain and within the brain. The diagram shows the principal lines of communication from peripheral receptors to the sensory cortices and so to the cerebral hemispheres. Similarly, the diagram shows the output from the cerebral hemispheres via the motor cortex and so to muscles. Both these systems of pathways are largely crossed as illustrated, but minor uncrossed pathways are also shown by the vertical lines in the brainstem. The dominant left hemisphere and minor right hemisphere are labeled, together with some of the properties of these hemispheres that are found displayed in Figure 16.8. The corpus callosum is shown as a powerful cross linking of the two hemispheres, and, in addition, the diagram displays the modes of interaction between Worlds 1, 2, and 3, as described in the text, and also illustrated in Figure 16.10.

16.4 Relationship of Brain to Mind[2]

16.4.1 Introduction

In general terms there are two theories about the way in which the behavior of an animal (and a man) can be organized into the effective unity which it so obviously is.

First, there is the explanation inherent in monist materialism plus all varieties of parallelism. In current neurological theory the diverse inputs into the brain interact on the basis of all the structural and functional connectivities to give some integrated output of motor performance. The aim of the neurosciences is to provide a more and more coherent and complete account of the manner in which the total performance of an animal or man is explicable on those terms. Without making too dogmatic a claim, it can be stated that the goal of the neurosciences is to formulate a theory that can in principle provide a complete explanation of all behavior of animals and man, including man's verbal behavior. With some important reservations we share this goal in our own experimental work and believe that it is acceptable for all automatic and subconscious movements, even of the most complex kind. We believe, however, that the reductionist strategy will fail in the attempt to account for the higher levels of conscious performance of the human brain.

Second, there is the dualist-interactionist explanation which has been specially developed for the self-conscious mind and human brains. It is proposed that, superimposed upon the neural machinery in all its performance, there are at certain sites of the cerebral hemispheres (the so-called liaison areas) effective interactions with the self-conscious mind, both in receiving and in giving.

It is desirable to make brief reference to the philosophical basis of the discussion, which derives from Popper (1972). As illustrated in Figure 16.10, everything in existence and in experience is subsumed in one or another of the categories enumerated under Worlds 1, 2, and 3.

In Figure 16.10, World 1 is the world of physical objects and states. It comprises the whole cosmos of matter and energy, all of biology including human brains, and all artifacts that man has made for coding information, for example, the paper and ink of books or the material base of works of art. World 1 is the total world of the materialists. They recognize nothing else. All else is fantasy.

World 2 is the world of states of consciousness and subjective knowledge of all kinds. The totality of our perceptions comes in this world. There are several levels, as indicated in Figure 16.11, but it may be more correct to think of it as a spectrum.

The first level (outer sense) would be the ordinary perceptions provided by all our sense organs, hearing and touch and sight and smell and pain. All of these perceptions are in World 2, of course: vision with light and color; sound

[2]General reference: Popper and Eccles 1977.

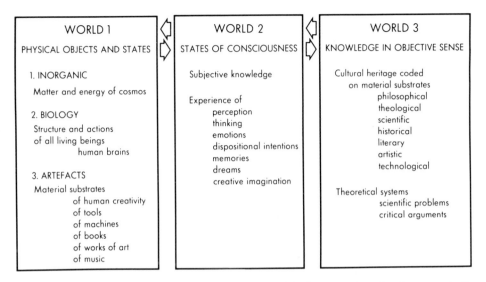

Figure 16.10 Tabular representation of the three worlds that comprise all existents and all experiences as defined by Popper (1972) and Eccles (1970b).

with music and harmony; touch with all its tactual qualities and vibration; the range of odors and tastes; and so on. These qualities do not exist in World 1, where correspondingly there are but electromagnetic waves, pressure waves in the atmosphere, material objects, and chemical substances.

In addition there is a level of inner sense, which is the world of more

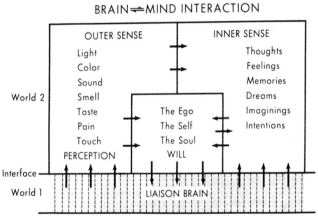

Figure 16.11 Information flow diagram for brain–mind interaction. The three components of World 2—outer sense, inner sense, and the ego or self—are diagrammed with their connectivities. Also shown are the lines of communication across the frontier between World 1 and World 2, that is, from the liaison brain to and from these World 2 components. The liaison brain has the columnar arrangement indicated (cf. Figure 15.16, Section 15.7). It must be imagined that the area of the liaison brain is enormous, with open modules probably numbering several hundred thousand and not just the 40 here depicted.

subtle perceptions. It is the world of emotions, of feelings of joy and sadness and fear and anger and so on. It includes all memories and all imaginings and planning into the future. In fact, there is a whole range of levels which could be described at length. All the subtle experiences of the human person are in this inner sensory world. It is all private to you but you can reveal it in linguistic expression, and by gestures of all levels of subtlety.

Finally, at the core of World 2 there is the self or ego, which is the basis of our unity as an experiencing being throughout our whole lifetime.

This World 2 is our primary reality. Our conscious experiences are the basis of our knowledge of World 1, which is thus a world of secondary reality, a derivative world. We are all the time, in every action we do, incessantly playing backward and forward between World 1 and World 2.

As shown in Figure 16.10, World 3 is the whole world of culture. It is the world that was created by man and that reciprocally made man. The whole of language is here. All our means of communication, all our intellectual efforts coded in books, coded in the artistic and technological treasures in the museums, coded in every artifact left by man from primitive times—this is World 3, right up to the present time. It is the world of civilization and culture. Education is the means whereby each human being is brought into relation with World 3. In this manner he becomes immersed in it throughout life, participating in the heritage of mankind and so becoming fully human. World 3 is the world that uniquely relates to man. It is the world which is completely unknown to animals. They are blind to all of World 3.

Furthermore, it is proposed that there is interaction between these worlds. There is reciprocal interaction between Worlds 1 and 2, and between Worlds 2 and 3—in part via the mediation of World 1. The objective knowledge of World 3 (the man-made world of culture) is encoded on various objects of World 1—books, pictures, structures, machines—and can be perceived only when projected to the brain by the appropriate receptor organs and afferent pathways. Reciprocally, the World 2 of conscious experience can bring about changes in World 1, first in the brain, then in muscular contractions; in that way World 2 is able to act extensively on World 1. This is the operation in voluntary movement that has been considered in Chapter 13. We may diagram the interactions of this trialist-interactionist hypothesis as: World 1 $\rightleftharpoons$ World 2 and World 3 $\rightleftharpoons$ World 1 $\rightleftharpoons$ World 2, where World 2 $\rightarrow$ World 1 contains the problem of voluntary action and World 1 $\rightarrow$ World 2 the problem of conscious perception.

16.4.2 Statement of Dualist-Interactionist Hypothesis

On this background we can now give a more complete statement of the strong dualist hypothesis. The self-conscious mind is actively engaged in reading out from the multitude of active centers at the highest level of brain activity, namely, the liaison areas of the dominant cerebral hemisphere (Figures 16.6, 16.9, 16.11). The self-conscious mind selects from these centers according to attention, and from moment to moment integrates its selection to

give unity even to the most transient experience. Furthermore, the self-conscious mind acts upon these neural centers, modifying the dynamic spatiotemporal patterns of the neural events. Thus, the self-conscious mind exercises a superior interpretative and controlling role upon the neural events.

A key component of the hypothesis is that the unity of conscious experience is provided by the self-conscious mind and not by the neural machinery of the liaison areas of the cerebral hemisphere. Hitherto it has been impossible to develop any neurophysiological theory that explains how a diversity of brain events comes to be synthesized so that there is a unified conscious experience of a global or gestalt character. The brain events remain disparate, being essentially the individual actions of countless neurons that are built into complex circuits (cf. Figure 15.16), and so participate in the spatiotemporal patterns of activity. Our present hypothesis regards the neuronal machinery as a multiplex of radiating and receiving structures (modules). *The experienced unity comes not from a neurophysiological synthesis, but from the proposed integrating character of the self-conscious mind.* We conjecture that in the first place the *raison d'être* of the self-conscious mind was to give this unity of the self in all its conscious experiences and actions.

In developing the statement of the hypothesis we propose that in the liaison areas of the cerebral hemisphere some sensory input here and there causes an immense ongoing dynamic pattern of neural activity. The different sensory modalities project from the primary sensory areas indicated in Figure 13.2 (Section 13.3.1) to common areas, the polymodal areas. In these areas most varied and wideranging information is being processed in the unitary components, the modules of the cerebral cortex (Figure 15.16). We may ask how this is to be selected from and put together to give the unity and the relative simplicity of our conscious experience from moment to moment. As an answer to this question, it is proposed that the self-conscious mind plays through the whole liaison brain in a selective and unifying manner. An analogy is provided by a searchlight. Perhaps a better analogy would be some multiple scanning and probing device that reads out from and selects from the immense and diverse patterns of activity in the cerebral cortex and integrates these selected components, so organizing them into the unity of conscious experience. Figure 16.11 illustrates a fragment of a momentary "fix" by the self-conscious mind. Thus, we conjecture that the self-conscious mind is scanning the modular activities in the liaison areas of the cerebral cortex, as may be appreciated from the very inadequate diagram in Figure 16.11. From moment to moment it is selecting modules according to its interest, the phenomenon of attention, and is itself integrating from all this diversity to give the unified conscious experience.

This hypothesis may be considered to be just an elaborated version of parallelism—a kind of selective parallelism. This is a mistake, however. It differs radically in that *the selectional and integrational functions are conjectured to be attributes of the self-conscious mind,* which is thus given an active and dominant role. There is a complete contrast with the passivity postulated in parallelism (cf. Feigl 1967). Furthermore, the active role of the self-conscious mind is extended in our hypothesis to effect changes in the neuronal events. Thus, not only does it read out selectively from the ongoing activities of the neuronal machinery, but it also modifies these activities. For example, when following up a line of thought or trying to recapture a memory, it is proposed that the self-conscious mind is actively engaged in searching and in probing through specially selected zones of the neural machinery and so is able to deflect and mold the dynamic patterned activities in accord with its desire or interest. A special aspect of this intervention of the self-conscious mind upon the operations of the neural machinery is exhibited in its ability to bring about movements in accord with some voluntary desired action, what we may call a motor command. We have seen that the readiness potential (Figure 13.4, Section 13.32) is a sign that this command brings about changes in the activity of the neuronal machinery.

16.4.3 Summary and Conclusions

We can now briefly consider the implications of the strong dualist hypothesis that we have formulated. Its central component is that primacy is given to the self-conscious mind. It is proposed that the self-conscious mind is actively engaged in searching for brain events that are of its present interest, the operation of attention; but it also is the integrating agent, building the unity of conscious experience from all the diversity of the brain events. Even more importantly, it is given the role of actively modifying the brain events according to its interest or desire. The scanning operation by which it searches can be envisaged as having an active role in selection. Sperry (1968) has made a similar proposal.

> Conscious phenomena in this scheme are conceived to interact with and to largely govern the physiochemical and physiological aspects of the brain process. It obviously works the other way round as well, and thus a mutual interaction is conceived between the physiological and the mental properties. Even so, the present interpretation would tend to restore mind to its old prestigious position over matter, in the sense that the mental phenomena are seen to transcend the phenomena of physiology and biochemistry.

It has been suggested here that this interaction of the self-conscious mind and the brain is dependent on the arrangement of the cerebral neurons in the modules that are defined by anatomical and physiological studies. It is proposed that each module has an intense and subtle inner dynamic life based upon the collective interaction of its many thousands of constituent neurons. These components of the physical world (World 1) come in this way to be momentary constituents of a fundamental interface (cf. Figure 16.11), being "open" to two-way influences from another world, the self-conscious mind of World 2. Not all of the modules of the cerebral hemispheres are "open" in this way. After the commissurotomy operation the self-conscious mind is in liaison only with the dominant hemisphere, and it is proposed that the liaison area is further restricted to the linguistic areas in the widest sense, to the polymodal sensory areas, particularly of the prefrontal lobe, and to the ideational areas whereby the self-conscious mind communicates nonverbally, for example, pictorially and musically. We propose that the self-conscious mind can read out at will from the modules of that great area of neuronal activation in the dominant hemisphere. From moment to moment only a minute fraction is so sampled, and much that is read is held only for seconds in the short-term memory. Thus, the greater part of our conscious experiences are ephemeral. Concentration on special deliverances of the self-conscious mind, however, can start neuronal processes of storage that are the basis of intermediate and long-term memories. We would conjecture that the self-conscious mind is actively engaged in the process of laying down this memory storage and in retrieving from it.

It can be claimed that the strong dualist-interactionist hypothesis that has been developed has the recommendation of its great explanatory power. In principle it at least gives explanations of the whole range of problems relating to brain–mind interaction (cf. Popper and Eccles 1977). It also aids in understanding some aspects of memory and illusion and creative imagination.

Finally, it can be claimed that the hypothesis belongs to science because it is based on empirical data and it is objectively testable. It must be emphasized that, just as with other scientific theories of great explanatory power, the present hypothesis has to be subjected to empirical testing. It is claimed, however, that it is not refuted by any existing knowledge. It can be predicted optimistically that there will be a long period of remodeling and development, but not an irretrievable falsification.

Epilogue

We have told our story. Our objective in this monograph has been to build concepts of how the incredibly complex machinery of the mammalian brain operates. Topics of increasing subtlety have been treated as the book progressed, and we hope that the foundations laid at each stage are adequate for the next. It must be appreciated that we have not attempted to deal with every aspect of mammalian neurobiology. To do so would have been too ambitious for a single book and beyond the expertise of authors whose own horizons are limited. Rather, we have tried to select some of the areas that are fundamental to all branches of mammalian neurobiology, and to explain in simple terms the underlying principles. To do so, we have selected for description only a few of the many critical experiments that have led to their establishment. Where possible, these principles have been presented in terms of the underlying hypotheses that have been formulated and the investigations that were undertaken to test them.

As these pages have attempted to illustrate, the whole brain, with its amazing powers of integration, is immeasurably greater than the sum of the parts. That is why a broad knowledge is more important in neurobiology than in any branch of science. It is not always easy to obtain and assimilate the background information necessary to approach the now-massive literature on neurobiology. We hope the book will assist in this endeavor.

Necessarily, much of the theme has been concerned with the conventional methods of signaling in the nervous system: impulse generation, impulse transmission along pathways, and synaptic action at terminals. But it is clear that the neurons must participate as well in other methods of signaling that are barely touched upon in the book. First of all, there is the immense problem of the building of the brain with correct establishment of all the lines of communication and their detailed topographic relations. We have seen in Chapter 12 that this turns out to be a problem of genetic instruction that achieves expression as specific molecular labels on individual neurons. The neurons are born, in a sense, with the postal addresses of certain other neurons. With the help of glia, their axon terminals are delivered on time, thence to begin the process of giving and receiving impulses.

But the brain, once built, is not a fixed, unchanging machine. At the higher levels of the brain, modifiability is the essence of its performance, as is evidenced by learning and memory. In Chapter 15, we have seen that this could be explained by microstructural changes. It is postulated that synaptic activity leads to terminal growth by its effect in causing the synthesis of RNA and, through this, synthesis of the enzymes building proteins and other macromolecules. There are many indications that not only this synaptic growth but neuronal life itself depends upon specific interneuronal chemical signals of a type not yet understood or explored.

Our hypothesis of memory and learning implies that activity is as important to the brain as exercise is to the muscles. The DNA template is there from the earliest embryonic age at which a cell develops neuronal competency. It can, at least in theory, provide transcribable information from that moment throughout the life of the cell. But whether or not transcription occurs may depend upon receipt of appropriate signals from sources external to that cell. Although we do not know the nature of the signals, we are aware of the consequences of their disappearance. A neuron deprived of its nerve endings rapidly undergoes retrograde degeneration. It may attempt to sprout, as does a sectioned nigrostriatal neuron, but if appropriate postsynaptic contacts are not made, it will give up and degenerate. It needs some signal from its postsynaptic neuron to survive. Similarly, a neuron deprived of its presynaptic input can also wither and die by the process referred to as transneuronal degeneration.

What manner of signals are involved in these life-sustaining communications, without which the competency of the DNA template is of no value? What signals are involved in the plastic processes of cognitive and motor learning, which can only be sustained through reinforcement? By what means do they alter so dramatically subsequent responses to the familiar types of impulse transmission? They emanate from teaching cells, but what are the methods of instruction (Chapter 15)?

Is it possible that ways of substituting for these signals, or ways of altering their strength, can be found? Do macromolecules exist in the CNS with properties along the lines of the catecholamine nerve-growth factor (Chapter 8) that can alter the structure and function of the brain during life, and help to sustain its machinery in the face of disease and degenerative processes?

We are only at the beginning of our understanding of signaling in the nervous system. In each chapter of this book, great gaps in our knowledge are evident. We have, however, felt a particular satisfaction in being able to present this book to students who represent a new generation of neuroscientists. We feel that no other field of science today offers as much challenge to young investigators as does the brain. There is virtually an infinitude of problems. Of course, some fields may show signs of drying up, but it is only necessary to lift one's eyes to see new and almost never-ending vistas for research.

Glossary

α-adrenergic receptor: A site where the action of noradrenaline is blocked or mimicked by certain drugs which are known as α-adrenergic antagonists or agonists. See Table 8.5, Section 8.3.3.

β-adrenergic receptor: A site where the action of noradrenaline is blocked or mimicked by different drugs which are known as β-adrenergic antagonists or agonists. See Table 8.5, Section 8.3.3.

afferent: Axons conveying impulses *toward* the specified area.

agonist: A substance that mimics the action of a transmitter at the receptor.

alveus: The thin sheet of pyramidal cell axons that covers the ventricular surface of the hippocampus.

amacrine cells: Special inhibitory cells in the retina.

amnestic syndrome: Loss of all memory except short-term memory.

anaerobic: Absence of oxygen.

anion: A negative ion such as Cl^-.

anode: Positive pole to which anions are attracted.

anoxia: Reduction of oxygen below physiological levels.

antagonist: A substance that tends to nullify the action of a neurotransmitter at the receptor.

anterograde transneuronal degeneration: Degeneration of a neuron subsequent to destruction of afferents.

antibody: A circulating protein of the immunoglobulin class which interacts only with the antigen that induced its synthesis or (in cross reactions) with molecules closely related to that antigen.

antidromic: Impulses going from axon terminals toward the neuronal soma, i.e., opposite to the usual (orthodromic) direction.

antigen: Any substance (usually a protein) which is capable of inducing the formation of antibodies and of reacting specifically with the antibodies so induced.

antimitotic: Inhibiting cell division.

aphasia: Inability to express oneself properly through speech, or loss of verbal comprehension.

archicortex: The ancient part of the cerebral cortex including the hippocampus.

569

argentaffin cells: Type of cell in the gastrointestinal tract easily stained with silver salts and containing high concentrations of serotonin.

ataxia: Motor incoordination of an irregular type during purposive movement.

atony: Lack of normal strength or tone.

atrophy: A wasting away.

autoradiograph: A photographic picture made by exposing photographic emulsion to the radiation emitted from radioactive substances contained in tissue or other material.

axolemma: A delicate membrane between an axon and the surrounding myelin sheath.

axon collaterals: Branches from the main axon.

axoplasm: Intracellular fluid within an axon.

bipolar cell: Neuron with two major processes arising from the cell body.

cable transmission: Transmission of a potential charge along a nerve fiber in the manner of an electric cable and not as an impulse.

carcinoid syndrome: Symptoms characteristic of persons having a tumor of the argentaffin cells of the gastrointestinal tract.

carotid body: A nucleus at the bifurcation of each carotid artery with receptors sensitive to changes in oxygen tension in the blood. Impulses arising in these receptors help in reflex control of respiration.

cathode: The negative pole to which positive ions are attracted.

cation: A positively charged ion such as Na^+, K^+ or Ca^{2+}.

caudal: Toward the tail or posterior end (in humans, inferior).

causalgia: A burning pain, often accompanied by trophic skin changes, due to injury of a peripheral nerve.

central nervous system (CNS): Brain and spinal cord.

cerebellar nuclei: Collections of nerve cells deep in the cerebellum that mediate the output from the cerebellar Purkinje cells; see Figures 13.10, 13.13.

cerebral ventricles: Cavities (lumen) in the brain filled with cerebrospinal fluid (CSF).

cerebrospinal fluid (CSF): Clear liquid filling the ventricles of the brain and spaces between meninges, arachnoid, and pia.

chemical specification: Specific chemical structures on the surface of nerve fibers and nerve cells that control the development of connectivities.

cholinergic: Neurons releasing acetylcholine as the transmitter.

cholinoceptive: Applies to neurons or other structures possessing receptors for acetylcholine.

cholinomimetic: An agent that mimics the action of acetylcholine at cholinergic receptors (acetylcholine agonist).

choroid plexus: Folded processes rich in blood vessels that project into ventricles of the brain and secrete cerebrospinal fluid.

chorea: Involuntary jerky movements of skeletal muscles.

chromaffin cells: Cells which stain strongly with chromium salts due to catecholamine content. Such cells are found in the adrenal gland, along sympathetic nerves, and in certain organs.

chromatography, paper, column or gas: Separation technique in which the materials to be separated are carried through an adsorbent (or stationary phase) by a liquid or gas as the moving phase.

clone: The assemblage of cells made by successive subdivisions of an initial parent cell.

cognitive experience: A conscious thought or experience.

complex and hypercomplex cells: Names given to cells in the visual cortex surrounding the primary visual cortex that exhibit sensitivity to special spatial properties of the retinal stimulation.

conditioned reflex: A reflex response which is developed in special training procedures.

conductance (g): Reciprocal of electrical resistance and thus a measure of the ability of a circuit to conduct electricity; in excitable cells, a useful measure of permeability for an ion or ions.

contralateral: Relating to the opposite side of the body.

coronal: Vertical section through the skull at right angles to the front–back (sagittal) axis, i.e., a plane parallel to the face.

corpus callosum: The tract of nerve fibers connecting the cerebral hemispheres.

CSF: See cerebrospinal fluid.

cytoplasm: The protoplasm of a cell exclusive of the nucleus.

dalton: A unit of mass, being $\frac{1}{16}$th the mass of the oxygen atom or approximately 1.65×10^{-24}g.

deafferentation: The elimination or interruption of afferent nerves.

decussation: The intercrossing of similar structures.

depolarization: Reduction of membrane potential from the resting value toward zero.

desmosome: A thickening that forms the site of attachment between cells and consists of local differentiations of the opposing cell membranes.

dielectric coefficient: Measure of the insulating capacity of a substance.

diencephalon: The "in-between brain" consisting principally of the hypothalamus and thalamus.

differentiating mitosis: Cell divisions in which the daughter cells have an effectively different genetic coding from the parents.

distal: Away from the center of the body.

dorsal: Pertaining to or situated near the back of an animal.

dysgenesis: Defective development.

dyslexia: Reading disability.

dyspnea: Difficult or labored breathing.

ectoderm, ectodermal: The outer, investing cellular membrane of multicellular animals; applies especially to the outer germ layer of embryos.

edema, cerebral: The presence of abnormally large amounts of fluid in the intercellular spaces of the brain.

efferent: Axons conducting impulses away from the named nucleus or area.

electrochemical gradient: The value representing the voltage across a membrane for a particular charged particle, being compounded of the electrochemical potential for that particle and the voltage gradient across the membrane.

EC enzyme classification: In this book, the numbers assigned to various enzymes, e.g., monoamine oxidase (EC 1.4.3.4) are those taken from *Enzyme Nomenclature* (1972), Elsevier Scientific Publishing Co., Amsterdam.

electrochemical potential: Voltage difference between two solutions insulated from each other. It is derived by the Nernst equation from the logarithmic relationship of the concentrations of the charged particles on either side of the insulating membrane.

Electrophorus: A genus of cyprinoid fishes which includes electric eels.

end plate: Postsynaptic area of vertebrate skeletal muscle fiber.

endogenous: Arising from within the body.

endothelial cells: Layer of cells lining blood vessels and other closed cavities.

engram: A patterned arrangement of neuronal connections via synapses that have been given increased effectiveness in learning.

ependyma: Layer of cells lining cerebral ventricles and central canal of spinal cord.

equilibrium potential: Membrane potential predicted from Nernst equation on the basis of concentration differences for an ion to which the membrane is permeable.

ergastoplasm: Granular endoplasmic reticulum.

exocrine glands: A gland which discharges its secretion through a duct opening on an internal or external surface of the body, as a lacrimal gland.

exocytosis: The discharge from a cell of particles that are too large to diffuse through the wall.

exogenous: Originating outside the organism.

expectancy wave: A potential change recorded from the surface of the brain and which arises in anticipation of a conditioned movement.

extensor muscle: A muscle which acts to straighten a joint.

external germinal layer: A layer of cells (germinal cells) formed on the surface of the developing cerebellum and which generates many cell types of the cerebellum.

extrafusal: Muscle fibers making up the mass of a skeletal muscle (i.e., not within the sensory muscle spindles).

extrasynaptic: The surface membrane that is not covered by synapses.

extravesicular: Lying outside of the vesicles.

facilitation: Increased effectiveness of synaptic transmission by successive presynaptic impulses.

fasciculus: A bundle of nerve fibers.

fenestrated: Pierced with one or more openings.

festinating: An involuntary tendency to take short accelerating steps in walking.

fissure: Any cleft or groove, normal or otherwise; especially a deep fold in the cerebral cortex which involves the entire thickness of the brain wall.

folium: Leaflike subdivisions of the cerebellar cortex.

γ-loop: The operative path employing γ-fibers and γ-motoneurons in movement control: γ-motoneurons, muscle spindles, Ia fibers, α-motoneurons, extrafusal muscle contraction.

γ-motoneurons: The small motoneurons innervating intrafusal muscle.

ganglion: Collection of neurons which send and receive impulses. Many in the autonomic system relay synaptically to the viscera.

gemmules: Bud-like excrescences of nerve process, e.g., such as found in the olfactory bulb or Purkinje cells of the cerebellum.

genome: The operative DNA of a cell which expresses genetic characteristics.

genu: A general term used to designate any anatomical structure bent like the knee, e.g., the most rostral part of the corpus collosum.

germinal cells: Embryonic cells.

Golgi apparatus: A secretory organelle in the neuronal cytoplasm easily recognized in electron microscopic studies.

Golgi type II neuron: A short-axoned neuron with an axon that arborizes in dendrite-like fashion in the neighborhood of the cell body.

gonadotrophins: Applies to hormones which stimulate the gonads.

heterotypic regeneration: Regeneration in which the vacated synaptic sites are occupied by pre-synaptic fibers different from the original fibers.

hippocampus: A special part of the cerebral cortex (see archicortex).

histones: Simple proteins derived from cell nuclei which contain many basic groups and are thought to be involved in suppressing transcription.

horizontal cells: Special inhibitory cells in the retina.

hydrophilic: A material that attracts water.

hypertonic: Solutions which have greater osmotic pressure than normal extracellular fluids when bathing body cells.

hypothermia: A state of low temperature of the body.

hypotonic: Solutions which have less osmotic pressure than normal extracellular fluids.

hypoxia: Low oxygen content or tension; deficiency of oxygen in the inspired air.

infundibulum: The funnel-shaped stalk of the pituitary gland.

inhibition: Effect of one neuron upon another tending to prevent it from initiating impulses: *postsynaptic inhibition* is mediated through a permeability change in the postsynaptic cell, holding the membrane potential away from threshold; *presynaptic inhibition* is mediated by an inhibitory fiber upon an excitatory terminal, reducing the release of transmitter; *electrical inhibition* is mediated by currents in presynaptic fibers that hyperpolarize the postsynaptic cell and do not involve the secretion of a chemical transmitter.

interneuron: Short-axoned excitatory or inhibitory nerve cells in the central nervous system; see Golgi type II neurons.

internode: Myelinated portion of a nerve axon lying between two nodes of Ranvier (see below).

interstitial: Pertaining to spaces between essential parts of organs or tissues.

intrafusal fibers: Fibers within sensory motor spindles.

iontophoresis: Application of ions by passing current through a micropipette; used for applying charged molecules with a high degree of temporal and spatial resolution.

ipsilateral: On the same side of the body.

isotonic: Physiological solutions which have the same osmotic pressure as normal extracellular fluids.

karyoplasm: The nucleoplasm, or protoplasm, of the nucleus of a cell.

K_i: An equilibrium constant expressing the concentration of inhibiter required to reduce the product of a reaction by 50%.

kinases: See protein kinases.

latency: The time between the initiation of a physiological or chemical sequence and a measurable result.

ligand: A molecule that donates the necessary electrons to form coordinate covalent bonds with metallic ions, as oxygen is bound to the central iron atom of hemoglobin.

ligature: The use of a substance, such as catgut, cotton, silk, or wire, to tie a vessel or strangulate a part.

limbic lobe: A part of the brain thought to be associated with emotion. It includes the archicortex (hippocampal formation and dentate gyrus), paleocortex (pyriform cortex of the anterior parahippocampal gyrus), and mesocortex (cingulate gyrus).

lipofuscin: Material accumulating in neurons with age. It is composed of fatty pigment and is presumed to be insoluble refuse.

medullation: Myelination, the formation of a laminated wrapping (myelin) around nerve fibers.

meninges: The three membranes enveloping the brain and spinal cord. From without inward they are the dura mater, arachnoid mater, and pia mater.

mesodermal: Derived from the middle germinal layer of the embryo. Most internal parts of the body, including all muscle, are mesodermal in origin.

microwave: An electromagnetic wave of very high frequency and of wavelength in the centimeter range.

miosis: Contraction of the pupil.

mitosis: Reproductive cell division.

molar: Concentration of a solution in moles per liter.

morphogenetic: Producing growth; producing form or shape.

motoneuron (motor neuron): A neuron that innervates muscle fibers.

multiple sclerosis: A nervous disease resulting from interruptions of neural transmission to and from the brain as a consequence of degeneration of the myelin sheaths of the nerve fibers.

muscarinic: Resembling acetylocholine or muscarine in actions at specific receptors.

muscle spindles: Encapsulated bundles of fine muscle fibers specially related to muscle stretch receptors.

myasthenia gravis: A neurological disease associated with partial paralysis due to defect in nerve-muscle transmission and with extreme fatigability.

myelin: Fused membranes of Schwann cells or glial cells forming a high-resistance sheath around an axon.

myoneural junction: The junction pertaining to both muscle and nerve; said of nerve terminations in muscles.

myosin: The contractile protein of muscle fibers.

neocortex: The most recently developed part of the cerebral cortex composing the cerebral hemispheres.

nerve fiber: An axon (the principal branch from a nerve cell) that may extend for long distances.

neuroblastoma: A tumor arising from neuroblasts.

neuroblasts: Special cells that are the precursors of nerve cells, developing directly into them.

neuroepithelium: Epithelial cells receiving terminations of sensory nerves such as the specific cells of the taste buds, olfactory mucosa, and the cochlear and vestibular apparatus.

neurofibrils: Fine fibers running along the interior of nerve axons.

neurogenesis: The process whereby nerve cells are made by the sequence of germinal cells to neuroblasts to fully fashioned nerve cells.

neuroleptic: Denoting a neuropharmacologic agent that has antipsychotic action principally affecting psychomotor activity, and is generally without hypnotic effects, as a tranquilizer; it may produce extrapyramidal syndrome.

neuroglia: Nonneuron satellite cells associated with neurons. In the mammalian central nervous system the main groupings are astocytes and oligodendrocytes; in peripheral nerves the satellite cells are called Schwann cells.

neuromuscular junctions: Junctions of the muscles and nerves.

neuron (nerve cell): The biological unit of the brain and of the remainder of the nervous system; see Figures 1.2, 1.3, and 1.4A (Section 1.2).

neuron theory: The theory that the nervous system is composed of individual neurons biologically independent but informationally communicating by synapses.

neuropil: Network of axons, dendrites, and synapses.

neurosecretory cells: Secreting nerve cells.

neurotoxic: Poisonous or destructive to nerve tissue.

neurotubules: Very fine tubules running along the interior of nerve axons.

Nissl bodies: Intracytoplasmic, basophilic masses that ramify loosely through the cytoplasm, also known as rough endoplasmic reticulum. See Figures 1.8 and 1.10 (Section 1.2).

nucleolus: Dense, basophilic structure within the nucleus containing RNA as well as DNA. Females have a small nucleolar satellite known as a Barr body.

nucleus: Large basophilic mass, usually centrally placed within the cell body. Contains the DNA which provides the genetic instruction for the cell.

orthodromic: Wave of neuronal excitation passing in the physiological direction from cell body to axonal ending.

orthostatic hypotension: Sudden drop in blood pressure upon assuming a vertical posture due to gravitational pooling of blood.

parenchyma: A general term to designate the functional elements of an organ, as distinguished from its framework or stroma.

perikaryon: The body of the neuron.

perinuclear: Situated or occurring around a nucleus.

phagocytes: Glia or other cells that ingest and destroy waste or foreign materials including cellular debris.

phospholipids: Molecules composed of fatty acid chains and a phosphorylated polar end forming the basic structure of cell membranes.

pia mater: The innermost of the three membranes (meninges) covering the brain and spinal cord.

pinocytosis: The imbibition of liquids by cells, especially the phenomenon in which minute incuppings or invaginations are formed in the surface of cells, which close to form fluid-filled vacuoles.

polysomal: Structure possessing many bodies, as in a complex of individual ribosomes.

positive feedback control: The control by which the output is fed back to give intensification of input.

posterior: At or toward the hind end of the body; in a tailward or caudal direction (in humans, behind or dorsal rather than below).

postsynaptic membrane: The nerve cell or other receptor cell membrane immediately related to the synapse formed by presynaptic fibers ending on it.

postsynaptic potentiation: The increased synaptic action that follows intensive synaptic stimulation.

presynaptic fibers: The terminal branches of nerve fibers that end as synaptic knobs.

prolactin: One of the hormones of the anterior pituitary gland that stimulates and sustains lactation in postpartum mammals, the mammary gland having been prepared by other hormones, including estrogens, progesterone, growth hormone, and corticosteroids.

proliferative zone: That part of the developing cerebellar cortex in which the germinal cells are dividing.

prostaglandins: A group of naturally occurring, chemically related, long-chain hydroxy fatty acids that stimulate contractility of the uterine and other smooth muscle and have the ability to lower blood pressure and to affect the action of certain hormones.

prosthetic group: A chemical group on an enzyme molecule essential for its activity.

protein kinases: Any protein that catalyzes the transfer of a phosphate group from ATP to form a phosphoprotein.

punctate: Resembling or marked with points or dots.

pyriform cortex: Refers to the anterior parahippocampal gyrus. See limbic lobe.

rami: A branch; a general term for a smaller structure given off by a larger one, or into which the larger structure, such as a blood vessel or nerve, divides.

resting potential: The steady electrical potential across the membrane in the quiescent state.

reticular activating system: A system emanating from the reticular formation of the brainstem which, when stimulated, has a powerful recruiting influence upon the cortex and other neuronal systems.

retrograde axonal transport: The transport of substances along nerve axons toward the cell body in contrast to the usual transport away from the cell body.

rostral: Toward the rostrum or nose; thus, in the anterior direction in the CNS.

rough endoplasmic reticulum: See Nissl bodies.

sacral: Situated near or pertaining to the sacrum (the triangular base just below the lumbar vertebrae).

sagittal section: A section taken parallel to the anterior-posterior plane.

sarcoplasm: The interfibrillary matter of the striated muscles; the substance in which the fibrillae of the muscle fiber are embedded.

secondary endings: The receptor endings on the muscle spindle in addition to the primary endings or annulospiral endings.

selective fasciculation: The growth of nerve fibers along pioneering paths of earlier similar fibers so that nerve bundles or fascicles are built.

sensorium: The seat of sensation, located in the brain; the term is often used to designate the condition of a subject relative to his consciousness or mental clarity.

sensory modality: The distinguishing names of all the various sensations arising from diverse inputs with their specific receptor organs.

servoloop control: Automatic control by feedback pathways such as given by γ-loops.

servomechanism: Mechanism designed for feedback-control operation.

stereotypy: The persistent repetition of senseless acts or words.

stretch receptor: Stretch-sensitive nerve endings embedded in skeletal muscle (muscle spindle) and tendon (Golgi tendon organ) which play a part in position sense and postural reflexes.

synaptic plasticity: The property of synapses whereby they are changed in functional efficiency, probably by virtue of changes in size.

syncitium: A multinucleate mass of protoplasm produced by the merging of cells.

tachyphylaxis: Sudden diminution in response to repeated injections of the same material.

tegmentum: That part of the midbrain dorsal to the substantia nigra and ventral to the cerebral aqueduct.

tetraploid: An individual or cell having four sets of chromosomes.

tonic: Characterized by continuous tension or producing and restoring normal tone.

transcription: The process by which genetic information contained in DNA produces a complementary sequence of bases in an RNA chain.

transducer: Device for converting one form of energy into another.

trophic influences: Actions from one part of a cell to another, or from one cell to another, which are concerned with the growth, maintenance, and metabolism of the cell.

trophic transport: The postulated mechanism of transport of specific macromolecules within a cell and between cells that is concerned in exerting trophic influences.

ventral: Toward the under or belly side of the body.

ventricles: See cerebral ventricles.

vertigo: An illusion of movement.

voltage clamp: Technique for displacing membrane potential abruptly to a desired value and keeping the potential constant while measuring currents across the cell membrane; devised by Cole and Marmont.

white matter: Part of the central nervous system appearing white; consisting largely of myelinated fiber tracts.

References

Abel, J. J., and Kubota, S. 1919. On the presence of histamine (β-imidazolylethylamine) in the hypophysis cerebri and other tissues of the body and its occurrence among the hydrolytic decomposition products of proteins. *J. Pharmacol. Exp. Ther.* **13,** 243–300.

Abood, L. G., and Biel, J. H. 1962. Anticholinergic psychotomimetic agents. *Int. Rev. Neurobiol.* **4,** 217–273.

Abood, L. G., Rinaldi, R., and Eagleston, V. 1961. Distribution of piperidine in the brain and its possible significance in behavior. *Nature* **191,** 201–202.

Adrian, E. D. 1947. *The Physical Background of Perception.* The Clarendon Press, Oxford.

Aghajanian, G. K., and Bloom, F. E. 1967. The formation of synaptic junctions in developing rat brain. A quantitative electron microscopic study. *Brain Res.* **6,** 716–727.

Aghajanian, G. K., and Haigler, H. J. 1973. Direct and indirect actions of L.S.D., serotonin and related compounds on serotonin-containing neurons. In: *Serotonin and Behavior,* edited by J. Barchas and E. Usdin. Academic Press, New York and London, pp. 263–266.

Aghajanian, G. K., Foot, W. E., and Sheard, M. H. 1973. Direct and indirect actions of L.S.D. In: *Serotonin and Behavior,* edited by J. Barchas and E. Usdin. Academic Press, New York, pp. 263–266.

Agid, Y., Guyenet, P., Glowinski, J., Beaujouan, J. C., and Javoy, F. 1975. Inhibitory influence of the nigrostriatal dopamine system on the striatal cholinergic neurons in the rat. *Brain Res.* **86,** 488–492.

Agranoff, B. W. 1967. Agents that block memory. In: *The Neurosciences: A Study Program,* edited by G. C. Quarton, T. Melnechuk, and F. O. Schmitt. Rockefeller University Press, New York, pp. 756–764.

Agranoff, B. W. 1969. Protein synthesis and memory formation. In: *The Future of the Brain Sciences,* edited by S. Bogoch. Plenum Press, New York, pp. 341–353.

Agranoff, B. W. 1976. Learning and memory: Approaches to correlating behavioral and biochemical events. In: *Basic Neurochemistry,* edited by G. S. Siegel, R. W. Albers, R. Katzman, and B. W. Agranoff. Little, Brown and Co., Boston, pp. 765–786.

Akert, K. 1973. Dynamic aspects of synaptic ultrastructure. *Brain Res.* **49,** 511–518.

Akert, K., and Peper, K., 1975. Ultrastructure of chemical synapses: A comparison between presynaptic membrane complexes of the motor end plate and the synaptic junction of the central nervous system. In: *Golgi Centennial Symposium Proceedings,* edited by M. Santini. Raven Press, New York, pp. 521–527.

Akert, K., Moor, K., Pfenninger, K., and Sandri, C. 1969. Contributions of new impregnation methods and freeze etching to the problems of synaptic fine structure. In: *Progress in Brain Research,* vol. 31, edited by K. Akert and P. G. Waser. Elsevier Publ. Co., Amsterdam, pp. 223–240.

Akert, K., Peper, K., and Sandri, C. 1975. Structural organization of motor end plate and central synapses. In: *Cholinergic Mechanisms,* edited by P. G. Waser. Raven Press, New York, pp. 43–57.

Alberici, M., de Lores Arnaiz, G. R., and De Robertis, E. 1969. Glutamic acid decarboxylase inhibition and ultrastructural changes by the convulsant drug allylglycine. *Biochem. Pharmacol.* **18,** 137–143.

Allen, G. I., and Tsukahara, N. 1974. Cerebrocerebellar communication systems. *Physiol. Rev.* **54,** 957–1006.

Allen, G. I., Eccles, J. C., Nicoll, R. A., Oshima, T., and Rubia, F. J. 1977. The ionic mechanisms concerned in generating the IPSP of hippocampal pyramidal cells. *Proc. R. Soc. Lond. [Biol.]* **198,** 363–384.

Alm, B. 1976. Piperidine: Effects on locomotor activity and brain monoamine turnover. *Psychopharmacology* **50,** 301–304.

Altman, J. 1969. Autoradiographic and histological studies of postnatal neurogenesis. III. Dating the time of production and onset of differentiation of cerebellar microneurons in rats. *J. Comp. Neurol.* **136,** 269–294.

Altman, J. 1972. Postnatal development of the cerebellar cortex in the rat. II. Phases in the maturation of Purkinje cells and of the molecular layer. *J. Comp. Neurol.* **145,** 399–464.

Altman, J. 1975. Postnatal development of the cerebellar cortex in the rat. IV. Spatial organization of bipolar cells, parallel fibers and glial pallisades. *J. Comp. Neurol.* **163,** 427–448.

Altman, J. 1976. Experimental reorganization of the cerebellar cortex. VII. Effects of late x-radiation schedules that interfere with cell acquisition after stellate cells are formed. *J. Comp. Neurol.* **165,** 65–76.

Altman, J., and Bayer, S. A. 1978. Prenatal development of the cerebral cortex in the rat: I. Histogenesis and cytogenesis. *J. Comp. Neurol.* **179,** 23–48.

Amin, A. H., Crawford, T. B., and Gaddum, J. H. 1954. The distribution of substance P and 5-hydroxytryptamine in the central nervous system of the dog. *J. Physiol.* **126,** 596–618.

Anden, N. E., and Fuxe, K. 1971. A new dopamine-β-hydroxylase inhibitor: Effects on the noradrenaline concentration after L-Dopa in the spinal cord. *Brit. J. Pharmacol.* **43,** 747–756.

Anden, N. E., Fuxe, K., Hamberger, B., and Hokfelt, T. 1966. A quantitative study of the nigro-neostriatal dopamine neuron system in the rat. *Acta Physiol. Scand.* **67,** 306–312.

Andersen, P., and Lømo, T. 1966. Mode of activation of hippocampal pyramidal cells by excitatory synapses on dendrites. *Exp. Brain Res.* **2,** 247–260.

Andersen, P., Eccles, J. C., and Løyning, Y. 1964. Location of postsynaptic inhibitory synapses on hippocampal pyramids. *J. Neurophysiol.* **27,** 592–607.

Andersen, P., Holmqvist, B., and Voorhoeve, P. E. 1966a. Excitatory synapses on hippocampal apical dendrites activated by entorhinal stimulation. *Acta Physiol. Scand.* **66,** 461–472.

Andersen, P., Holmqvist, B., and Voorhoeve, P. E. 1966b. Entorhinal activation of dentate granule cells. *Acta Physiol. Scand.* **66,** 448–460.

Andersen, P., Bliss, T. V. P., and Skrede, K. K. 1971. Lamellar organization of hippocampal excitatory pathways. *Exp. Brain Res.* **13,** 222–238.

Andersen, P., Bland, B. H., and Dudar, J. D. 1973. Organization of the hippocampal output. *Exp. Brain Res.* **17,** 152–168.

Angeletti, P. U., Angeletti, R. H., Frazier, W. A., and Bradshaw, R. A. 1973. Nerve growth factor. In: *Proteins of the Nervous System,* edited by D. J. Schneider, R. A. Hogue-Angeletti, R. A. Bradshaw, A. Grasso, and B. W. Moore. Raven Press, New York, pp. 133–155.

Angevine, J. B., Jr. 1970. Critical cellular events in the shaping of neural centers. In: *The Neurosciences: Second Study Program,* edited by F. O. Schmitt. Rockefeller University Press, New York, pp. 62–72.

Aprison, M. H., and Werman, R. 1965. The distribution of glycine in cat spinal cord and roots. *Life Sci.* **4,** 2075–2083.

Aprison, M. H., Davidoff, R. A., and Werman, R. 1970. Glycine: Its metabolic and possible roles in nervous tissue. In: *Handbook of Neurochemistry,* vol. 3, edited by A. Lajtha. Plenum Press, New York, pp. 381–397.

Aprison, M. H., Tachiki, K. H., Smith, J. E., Lane, J. D., and McBride, W. J. 1974. In: *Advances in*

Biochemical Pharmacology, edited by E. Costa, G. L. Gessa, and M. Sandler. Raven Press, New York, pp. 31–41.

Araki, T., Ito, M., and Oscarsson, O. 1961. Anion permeability of the synaptic and non-synaptic motoneuron membrane. *J. Physiol.* **159,** 410–435.

Armstrong, D. M., Harvey, R. J., and Schild, R. F. 1969. The distribution on the surface of the cat cerebellum of some responses evoked via climbing fibre afferents. *J. Physiol.* **203,** 34P.

Armstrong, D. M., Harvey, R. J., and Schild, R. F. 1973. The spatial organization of climbing fibre branching in the cat cerebellum. *Exp. Brain Res.* **18,** 40–58.

Armstrong, M. D., McMillan, A., and Shaw, K. N. F. 1957. 3-Methoxy-4-hydroxy-d-mandelic acid, a urinary metabolite of norepinephrine. *Biochem. Biophys. Acta* **25,** 422–425.

Arregni, A., Logan, W. J., Bennett, J. P., and Snyder, S. H. 1972. Specific glycine-accumulating synaptosomes in the spinal cord of rats. *Proc. Natl. Acad. Sci. USA* **69,** 3485–3489.

Arutyunyan, G. S., and Mashkovsky, M. D. 1972. Comparative pharmacological investigation of tryptamine derivatives substituted by a methoxy-group in positions 4, 5, 6 and 7 of the indole ring. *Russ. Pharmac. Tox.* **35,** 2–7.

Asanuma, H., and Rosén, I. 1972. Topographical organization of cortical efferent zones projecting to distal forelimb muscles in the monkey. *Exp. Brain Res.* **14,** 243–256.

Asatoor, A. M., Levi, A. J., and Milne, M. D. 1963. Tranylcypromine and cheese. *Lancet* **ii,** 733–734.

Ascher, P. 1973. Excitatory effects of dopamine in molluscan neurons. In: *Frontiers in Catecholamine Research,* edited by E. Usdin and S. H. Snyder, Pergamon Press, New York, pp. 667–672.

Ash, A. S. F., and Schild, H. O. 1966. Receptors mediating some actions of histamine. *Brit. J. Pharmacol.* **27,** 427–443.

Atsmon, A., Blum, I., Steiner, M., Latz, A., and Wijsenbeek, H. 1972. Further studies with propranolol in psychotic patients. *Psychopharmacol.* **27,** 249–254.

Attardi, D. G., and Sperry, R. W. 1963. Preferential selection of central pathways by regenerating optic fibres. *Exp. Neurol.* **7,** 46–64.

Awapara, J. 1975. Metabolism of taurine in the animal. In: *Taurine,* edited by R. Huxtable and A. Barbeau. Raven Press, New York, pp. 1–19.

Awapara, J., Landua, A. J., Fuerst, R., and Seale, B. 1950. Free γ-aminobutyric acid in brain. *J. Biol. Chem.* **187,** 35–39.

Axellsson, S., and Bjoerklund, A. 1973. Identification of the para isomer of tyramine in the rat brain. *Life Sci.* **13,** 1411–1419.

Axelrod, J. 1972. Dopamine-β-hydroxylase: Regulation of its synthesis and release from nerve terminals. *Pharmacol. Rev.* **24,** 233–243.

Axelrod, J., and Saavedra, J. M. 1974. Octopamine, phenylethanolamine, phenylethylamine and tryptamine in the brain. *Ciba Found. Symp.* **22,** 51–66.

Axelrod, J., and Saavedra, J. M. 1977. Octopamine. *Nature* **265,** 501–504.

Axelrod, J., and Tomchek, R. 1958. Enzymatic *O*-methylation of epinephrine and other catechols. *J. Biol. Chem.* **233,** 702–705.

Axelrod, J., and Wurtman, R. J. 1968. Photic and neural control of indoleamine metabolism in the rat pineal gland. *Adv. Pharmacol.* **6A,** 157–166.

Bacq, Z. M. 1934. La pharmacologie du système nerveux autonome, et particulièrement du sympatique, d'après la théorie neurohumorale. *Ann. Physiol. Physiochim. Biol.* **10,** 467–528.

Baker, P. F., Hodgkin, A. L., and Shaw, T. I. 1962. Replacement of the axoplasm of giant nerve fibres with artificial solutions. *J. Physiol.* **164,** 330–354.

Baker, P. F., Hodgkin, A. L., and Ridgway, E. P. 1970. Two phases of calcium entry during the action potential in giant axons of Loligo. *J. Physiol.* **208,** 80P–82P.

Balcar, V. J., and Johnston, G. A. R. 1973. High affinity uptake of transmitters: Studies on the uptake of L-aspartate, GABA, L-glutamate and glycine in cat spinal cord. *J. Neurochem.* **20,** 529–539.

Balcar, V. J., Borg, J., and Mandel, P. 1977. High affinity uptake of L-glutamate and L-aspartate by glial cells. *J. Neurochem.* **28,** 87–93.

Baldessarini, R. J. 1971. Release of aromatic amines from brain tissues of the rat *in vitro. J. Neurochem.* **18,** 2509–2518.

Baldessarini, R. J., and Vogt, M. 1971. The uptake and subcellular distribution of aromatic amines in the brain of the rat. *J. Neurochem.* **18,** 2519–2533.

Baldessarini, R. J., and Vogt, M. 1972. Regional release of aromatic amines from tissues of the rat brain *in vitro. J. Neurochem.* **19,** 755–761.

Baldessarini, R. J., and Yorke, C. 1974. Uptake and release of possible false transmitter amino acids by rat brain tissue. *J. Neurochem.* **23,** 839–848.

Baldissera, F., and Gustafsson, B. 1974a. Firing behavior of a neurone model based on the afterhyperpolarization,conductance time course. First interval firing. *Acta Physiol. Scand.* **91,** 528–544.

Baldissera, F., and Gustafsson, B. 1974b. Firing behavior of a neurone model based on the afterhyperpolarization, conductance time course and algebraical summation. Adaptation and steady state firing. *Acta Physiol. Scand.* **92,** 27–47.

Banerjee, S. P., and Snyder, S. H. 1973. Methyltetrahydrofolic acid mediates N- and O-methylation of biogenic amines. *Science* **182,** 74–75.

Banerjee, S. P., Cuatrecacas, P. P., and Snyder, S. H. 1976. Solubilization of nerve growth factor receptors in rabbit superior cervical ganglia. *J. Biol. Chem.* **251,** 5680–5685.

Banos, G., Daniel, P. M., Moorhouse, S. R., and Pratt, O. E. 1971. The entry of amino acids into the brain of the rat during the postnatal period. *J. Physiol.* **213,** 45P–46P.

Barber, R. P., Vaughn, J. E., Saito, K., McLaughlin, B. J., and Roberts, E. 1978. GABAnergic terminals are presynaptic to primary afferent terminals in the substantia gelatinosa of the rat spinal cord. *Brain Res.* **141,** 35–55.

Barbin, G., Garbarg, M., Schwartz, J. C., and Storm-Mathisen, J. 1976. Histamine-synthesizing afferents to the hippocampal region. *J. Neurochem.* **26,** 259–264.

Bard, P., and Mountcastle, V. B. 1948. Some forebrain mechanisms involved in expression of rage with special reference to suppression of angry behavior. *Res. Publ. Assoc. Res. Nerv. & Ment. Dis.* **27,** 362–404.

Barger, G., and Dale, H. H. 1910. The presence in ergot and physiological activity of β-imidozolylethylamine. *J. Physiol.* **40,** 38–40.

Barker, P. 1977. Physiological roles of peptides in the nervous system. In: *Peptides in Neurobiology,* edited by H. Gainer. Raven Press, New York, pp. 295–332.

Baring, M., and Close, R. I. 1971. The transformation of myosin in cross-innervated rat muscles. *J. Physiol.* **213,** 455–477.

Barondes, S. H. 1969. The mirror focus and long-term memory storage. In: *Basic Mechanisms of the Epilepsies,* edited by H. H. Jasper, A. A. Ward, and A. Pope. Little, Brown and Co., Boston, pp. 371–374.

Barondes, S. H. 1970. Multiple steps in the biology of memory. In: *The Neurosciences,* vol. 2, edited by F. O. Schmitt. Rockefeller University Press, New York, pp. 272–278.

Battistin, L. Grynbaum, A., and Lajtha, A. 1969. Distribution and uptake of amino acids in various regions of the rat brain *in vitro. J. Neurochem.* **16,** 1459–1468.

Baumgarten, H. G., and Lachenmayer, L. 1972. Chemically induced degeneration of indoleamine-containing nerve terminals in rat brain. *Brain Res.* **38,** 228–232.

Baxter, C. F. 1976. Some recent advances in studies of GABA metabolism and compartmentation. In: *GABA in Nervous System Function,* edited by E. Roberts, T. N. Chase, and D. B. Tower. Raven Press, New York, pp. 61–87.

Bazemore, A. W., Elliott, K. A. C., and Florey, E. 1957. Isolation of Factor I. *J. Neurochem.* **1,** 334–339.

Beart, P. M. 1976. The autoradiographic localization of L-[³H] glutamate in synaptosomal preparations. *Brain Res.* **103,** 350–355.

Bell, J. A., Martin, W. R., Sloan, J. W., and Buchwald, W. F. 1976. The effect of L-tryptophan on spinal cord C-fiber reflexes. *J. Pharmacol. Exp. Ther.* **196,** 373–379.

Bennett, C. T., and Pert, A. 1974. Antidiuresis produced by injections of histamine into the cat supraoptic nucleus. *Brain Res.* **78,** 151–156.

Bennett, J. P., Jr., Logan, W. J., and Snyder, S. H. 1972. Amino acid neurotransmitter candidates—sodium dependent high affinity uptake by unique synaptosomal fractions. *Science* **178,** 997–999.

Bennett, J. P., Jr., Arregui, A., and Snyder, S. H. 1976. Angiotensin II as a possible mammalian central neurotransmitter: Synaptic neurochemistry in normal mammalian and Huntington chorea brain tissue. Neuroscience Meeting, Abstract #1102.

Benuck, M., and Marks, N. 1975. Enzymatic inactivation of substance P by a partially purified enzyme from rat brain. Biochem. Biophys. Res. Commun. 65, 153-160.

Berlinguet, L., and Laliberte, M. 1970. Biosynthesis of N-acetyl-L-aspartic acid in vivo and in brain homogenates. Can. J. Biochem. 48, 207-211.

Bertler, A., Falck, B., Hillarp, N. A., Rosengren, E., and Torp, A. 1959. Dopamine and chromaffin cells. Acta Physiol. Scand. 47, 251-258.

Bezanilla, F., and Armstrong, C. M. 1975. Kinetic properties and inactivation of the gating currents of sodium channels in squid axon. Phil. Trans. R. Soc. Lond. 270B, 449-458.

Bhargava, K. P., and Dixit, K. 1968. Role of the chemoreceptor trigger zone in histamine induced emesis. J. Pharmacol. 34, 508-513.

Bignami, A., and Dahl, D. 1973. Differentiation of astrocytes in the cerebellar cortex and the pyramidal tracts of the newborn rat. An immunofluorescence study with antibodies to a protein specific to astrocytes. Brain Res. 49, 393-402.

Bird, E. D., and Iversen, L. L. 1974. Huntington's chorea: Post mortem measurement of glutamic acid decarboxylase, choline acetyltransferase and dopamine in basal ganglia. Brain 97, 457-472.

Birk, Y., and Li, C. H. 1964. β-Lipotropin. J. Biol. Chem. 239, 1048-1052.

Birkmayer, W., and Hornykiewicz, O. 1961. The effect of (3,4-dihydroxyphenyl)-L-alanine (= Dopa) on akinesia in Parkinson's disease. Wien. Klin. Wschr. 73, 787-788.

Birks, R., and MacIntosh, F. C. 1961. Acetylcholine metabolism of a sympathetic ganglion. Can. J. Biochem. Physiol. 39, 788-827.

Birks, R., Huxley, H. E., and Katz, B. 1960. The fine structure of the neuromuscular junction of the frog. J. Physiol. 150, 134-144.

Bjorklund, A., Falck, B., and Steveni, U. 1970. On the possible existence of a new intraneuronal monoamine in the spinal cord of the rat. J. Pharmacol. Exp. Ther. 175, 525-532.

Bjorklund, A., Falck, B., and Steveni, U. 1971a. Classification of monoamine neurones in the rat mesencephalon: Distribution of a new monoamine neurone system. Brain Res. 32, 1-17.

Bjorklund, A., Falck, B., and Steveni, U. 1971b. Microspectrofluorimetric characteristics of monoamines in the central nervous system: evidence for a new neuronal monoamine-like compound. Brain Res. 32, 269-285.

Bjorklund, A., Cegrell, L., Falck, B., Ritzen, M., and Rosengren, E. 1970. Dopamine containing cells in sympathetic ganglia. Acta Physiol. Scand. 78, 334-338.

Bjorklund, A., Baumgarten, H. G., and Nobin, A. 1974. Chemical lesioning of central monoamine axons by means of 5,6-dihydroxytryptamine and 5,7-dihydroxytryptamine. Adv. Biochem. Pharmacol. 10, 13-33.

Blankenship, J. E., and Kuno, M. 1968. Analysis of spontaneous subthreshold activity in spinal motoneurons of the cat. J. Neurophysiol. 31, 195-209.

Blaschko, H. 1939. The specific action of L-Dopa decarboxylase. J. Physiol. 96, 50P-51P.

Blau, K. 1961. Chromatographic methods for the study of amines from biological material. Biochem. J. 80, 193-200.

Blinzinger, K., and Kreutzberg, G., 1968. Displacement of synaptic terminals from regenerating motoneurons by microglial cells. Z. Zellforsch. Mikrosk. Anat. 85, 145-157.

Bliss, T. V. P., and Gardner-Medwin, A. R. 1973. Long-lasting potentiation of synaptic transmission in the dentate area of the unanesthetized rabbit following stimulation of the perforant path. J. Physiol. 232, 357-374.

Bliss, T. V. P., and Lømo, T. 1973. Long-lasting potentiation of synaptic transmission in the dentate area of the anesthetized rabbit following stimulation of the perforant path. J. Physiol. 232, 331-356.

Bloom, F. E., and Iversen, L. L. 1971. Localizing [³H]-GABA in nerve terminals of rat cerebral cortex by electron microscopic autoradiography. Nature, 229, 628-630.

Bloom, F. E., Oliver, A. P., and Salmoiraghi, G. C. 1963. The responsiveness of individual hypothalamic neurons to microelectrophoretically administered endogenous amines. *Int. J. Neuropharmacol.* **2**, 181-183.

Bloom, F. E., Costa, E., and Salmoiraghi, G. C. 1965. Anesthesia and the responsiveness of individual neurons of the caudate nucleus of the cat to acetylcholine, norepinephrine and dopamine administered by microelectrophoresis. *J. Pharmacol. Exp. Ther.* **150**, 244-252.

Bloom, F. E., Hoffer, B. J., Siggins, G. R., Barker, J. L., and Nicoll, R. A. 1972. Effect of serotonin on central neurons: Microiontophoretic administration. *Fed. Proc.* **31**, 97-106.

Bloom, F., Segal, D., Ling, N., and Guillemin, R. 1976. Endorphins: Profound behavioral effects in rats suggest new etiological factors in mental illness. *Science* **194**, 630-632.

Bloom, F., Rossier, J., Battenberg, E., Vargo, T., Minik, S., Ling, N., and Guillemin, R. 1977. Regional distribution of β-endorphin and enkephalin in rat brain. *Neuroscience Abst.* **7**, #907.

Bodian, D. 1967. Neurons, circuits and neuroglia. In: *The Neurosciences: First Study Program*, edited by G. C. Quarton, T. Melnechuk, and F. O. Schmitt. Rockefeller University Press, New York, pp. 6-24.

Bogdanski, D. F., Weissbach, H., and Udenfriend, S. 1957. The distribution of serotonin, 5-hydroxytryptophan decarboxylase and monoamine oxidase in brain. *J. Neurochem.* **1**, 227-228.

Bolme, P., Fuxe, K., Hokfelt, T., and Goldstein, M. 1977. Studies on the role of dopamine in cardiovascular and respiratory control: Central versus peripheral mechanisms. In: *Advances in Biochemical Psychopharmacology*, vol. 16, edited by E. Costa and G. L. Gessa. Raven Press, New York, pp. 281-290.

Bondareff, W., and Geinisman, Y. 1976. Loss of synapses in the dentate gyrus of the senescent rat. *Am. J. Anat.* **145**, 129-136.

Bondy, S. C., and Purdy, J. L. 1977. Putative neurotransmitters of the avian visual pathway. *Brain Res.* **119**, 417-426.

Boulton, A. A., and Wu, P. W. 1973. Biosynthesis of cerebral phenolic amines. II. *In vivo*, regional formation of p-tyramine and octopamine from tyrosine and dopamine. *Can. J. Biochem. Physiol.* **51**, 428-435.

Boulton, A. A., Dyck, L. E., and Durden, D. A. 1974. Hydroxylation of β-phenylethylamine in the rat. *Life Sci.* **15**, 1673-1684.

Boyd, H. J. 1973. Structural analysis of ATP suggests neurotransmitter role. *J. Theor. Biol.* **42**, 49-53.

Bradford, H. F., and Richards, C. D. 1976. Specific release of endogenous glutamate from piriform cortex stimulated *in vitro*. *Brain Res.* **105**, 168-172.

Bradley, P. B. 1973. Excitatory effects of catecholamines in the central nervous system. In: *Frontiers in Catecholamine Research*, edited by E. Usdin and S. H. Snyder. Pergamon Press, New York, pp. 653-656.

Bradley, P., and Berry, M. 1976. The effects of reduced climbing and parallel fibre input on Purkinje cell dendritic growth. *Brain Res.* **109**, 133-151.

Bradley, P. B., and Briggs, I. 1974. Further studies on the mode of action of psychotomimetic drugs: Antagonism of the excitatory actions of 5-hydroxytryptamine by methylated derivatives of tryptamine. *Brit. J. Pharmacol.* **50**, 345-354.

Brandau, K., and Axelrod, J. 1974. Ring dehydroxylation and N-methylation of noradrenaline and dopamine in the intact rat brain. *Biochem. Pharmacol.* suppl. pt. 1, 83-85.

Brawley, P., Doyle, J., and Marmar, C. 1976. Is nigroneostriatal facilitation dopaminergic or not? Neuroscience Meeting, Abstract #79.

Brenneman, A. R., and Kaufman, S. 1964. The role of tetrahydropteridines in the enzymic conversion of tyrosine to dopa. *Biochem. Biophys. Res. Comm.* **17**, 177-183.

Brenner, S. 1973. The genetics of behaviour. *Brit. Med. Bull.* **29**, 269-271.

Brightman, M. W., and Reese, T. S. 1975. Membrane specialization of ependymal cells and astrocytes. In: *The Nervous System*, vol. 1, *The Basic Neurosciences*, edited by D. B. Tower and R. O. Brady. Raven Press, New York, pp. 267-277.

Brock, L. G., Coombs, J. S., and Eccles, J. C. 1952. The recording of potentials from motoneurons with an intracellular electrode. *J. Physiol.* **117**, 431-460.

Brodal, A. 1973. Self-observations and neuro-anatomical considerations after a stroke. *Brain* **96,** 675–694.

Brodie, B. B., Spector, S., and Shore, P. A. 1959. Interaction of drugs with norepinephrine in the brain. *Pharmacol. Rev.* **11,** 548–564.

Brody, H. 1976. An examination of cerebral cortex and brain stem aging. In: *Aging,* vol. 3, edited by R. D. Terry and S. Gershon. Raven Press, New York, pp. 177–182.

Brooks, V. B. 1969. Information processing in the motorsensory cortex. In: *Information Processing in the Nervous System,* edited by K. N. Leibovic. Springer-Verlag, New York, pp. 231–243.

Brooks, V. B. 1975. Roles of cerebellum and basal ganglia in initiation and control of movements. *Can. J. Neurol. Sci.* **2,** 265–277.

Brown, G. M., Pang, S. F., Grota, L. J., Chambers, J. W., and Rodman, R. L. 1976. Activities of N-acetylserotonin and melatonin in biological tissues. Neuroscience Meeting, Abstract #949.

Brownstein, M. J. 1977. Biologically active peptides in mammalian central nervous system. In: *Peptides in Neurobiology,* edited by H. Gainer. Raven Press, New York, pp. 145–163.

Brownstein, M. J., Mroz, E. A., Kizer, J. S., Palkovits, M., and Leeman, S. E. 1976. Regional distribution of substance P in the brain of the rat. *Brain Res.* **116,** 299–305.

Bubenik, G. A., Brown, G. M., Uhlir, I., and Grota, L. J. 1974. Immunohistochemical localization of N-acetylserotonin in pineal gland, retina and cerebellum. *Brain Res.* **81,** 233–242.

Bubenik, G. A., Brown, G. M., and Grota, L. J. 1976a. Immunohistochemical investigations of N-acetylserotonin in the rat cerebellum after p-chlorophenylalanine treatment. *Experientia* **32,** 579–581.

Bubenik, G. A., Brown, G. M., and Grota, L. J. 1976b. Immunohistochemical localization of melatonin in the rat Harderian gland. *J. Histochem. Cytochem.* **24,** 1173–1177.

Buchwald, N. A., Price, D. D., Vernon, L., and Hull, G. D. 1973. Caudate intracellular response to thalamic and cortical inputs. *Exp. Neurol.* **38,** 311–323.

Buchwald, N. A., Hull, C. D., and Levine, M. S. 1975. Neurophysiological and anatomical interrelationships of basal ganglia. In: *Brain Mechanisms in Mental Retardation,* edited by N. A. Buchwald and M. A. B. Brazier. Academic Press, New York, pp. 187–203.

Buller, A. J., Mommaerts, W. F. H. M., and Seraydarian, K. 1969. Enzymic properties of myosin in fast and slow twitch muscles of the cat following cross-innervation. *J. Physiol.* **205,** 581–597.

Bunge, M. B., Bunge, R. P., and Ris, H. 1961. Ultrastructural study of remyelination in an experimental lesion in adult cat spinal cord. *J. Biophys. Biochem. Cytol.* **10,** 67–94.

Bunney, B. S., and Aghajanian, G. K. 1976. The precise localization of nigral afferents in the rat as determined by a retrograde tracing technique. *Brain Res.* **117,** 423–435.

Bunney, B. S., Walters, H. R., Roth, R. H., and Aghajanian, G. K. 1973. Dopaminergic neurons: Effect of antisychotic drugs and amphetamine on single cell activity, *J. Pharm. Exp. Ther.* **185,** 560–571.

Burgen, A. S. V., Hiley, C. R., and Young, J. M. 1974. The binding of [³H]propylbenzilylcholine mustard by longitudinal muscle strips from guinea pig small intestine. *Brit. J. Pharmacol.* **34,** 99–115.

Burke, R. E. 1967. Composite nature of the monosynaptic excitatory postsynaptic potential. *J. Neurophysiol.* **30,** 1114–1137.

Burnstock, G. 1972. Purinergic nerves. *Pharmacol. Rev.* **24,** 509–581.

Butcher, L. L., and Bilezikjian, L. 1975. Acetylcholinesterase containing neurons in the neostriatum and substantia nigra revealed after punctate intracerebral injection of Di-isopropyl fluorophosphate. *Eur. J. Pharmacol.* **34,** 115.

Calcutt, C. R. 1976. Mini-review: The role of histamine in the brain. *Gen. Pharmacol.* **7,** 15–25.

Calne, D. B. 1973. *Parkinsonism.* Edward Arnold Ltd., London.

Calvin, W. H., and Schwindt, P. C. 1972. Steps in production of motoneuron spikes during firing. *J. Neurophysiol.* **35,** 297–310.

Cannon, W. B. 1929. *Bodily Changes in Pain, Hunger, Fear and Rage.* D. Appleton and Company, New York.

Cannon, W. B., and Uridil, J. E. 1921. Studies on the conditions of activity in endocrine glands. VIII. Some effects on the denervated heart of stimulating the nerves of the liver. *Am. J. Physiol.* **58,** 353–354.

Carlsson, A. 1959. Occurrence, distribution and physiological role of catecholamines in the nervous system. *Pharmacol. Rev.* **11,** 300–304.

Carlsson, A. 1974. Measurement of monoamine synthesis and turnover with special reference to 5-hydroxytryptamine. *Adv. Biochem. Psychopharmacol.* **10,** 75–81.

Carlsson, A., and Lindquist, M. 1963. Effect of chlorpromazine or haloperidol on formation of 3-methoxytyramine and normetanephrine in mouse brain. *Acta Pharmacol. Toxicol. (Kbh)* **20,** 140–144.

Carlsson, A., Lindquist, M., Magnusson, T., and Atack, C. 1973. Effect of acute transection on the synthesis and turnover of 5HT in the rat spinal cord. *Naunyn-Schm. Arch. Pharmacol.* **277,** 1–12.

Carmon, A., Lavy, S., Gordon, H., and Portnoy, Z. 1975. Hemispheric differences in rCBF during verbal and non-verbal tasks. In: *Brain Work: The Coupling of Function, Metabolism and Blood Flow in the Brain,* edited by D. H. Ingvar and N. A. Larssen. Munksgaard, Copenhagen, pp. 414–423.

Carpenter, M. B. 1975. Anatomical organization of the corpus striatum and related nuclei. In: *The Basal Ganglia,* Res. Publ. Assoc. Res. Nerv. & Ment. Dis., vol. 55, edited by M. D. Yahr. Raven Press, New York, pp. 1–36.

Carraway, R. E., and Leeman, S. E. 1975. The amino acid sequence of a hypothalamic peptide, neurotensin. *J. Biol. Chem.* **250,** 1907–1911.

Cecil-Loeb Textbook of Medicine, 1971. Edited by P. B. Beeson and W. McDermott. W. B. Saunders Co., Philadelphia.

Cedar, H., and Schwartz, J. H. 1972. Cyclic adenosine monophosphate in the nervous system of Aplysia californica. II. Effect of serotonin and dopamine. *J. Gen. Physiol.* **60,** 570–587.

Celesia, G. G., and Jasper, H. H. 1966. Acetylcholine released from cerebral cortex in relation to state of activation. *Neurology* **16,** 1053–1064.

Chandler, W. K., and Meeves, H. 1965. Voltage clamp experiments on internally perfused giant axons. *J. Physiol.* **180,** 788–820.

Chang, M. M., and Leeman, S. E. 1970. Isolation of a sialogic peptide from bovine hypothalamus tissue and its characterization as substance P. *J. Biol. Chem.* **245,** 4784–4790.

Chao, L. P. 1976. Subunits of choline acetyltransferase. *J. Neurochem.* **25,** 261–266.

Chao, L. P., and Wolfgram, F. 1973. Purification and some properties of choline acetyltransferase from bovine brain. *J. Neurochem.* **20,** 1075–1081.

Chasin, M., Mamrak, F., Samaniego, S., and Hess, S. M. 1973. Characteristics of the catecholamine and histamine receptor sites mediating accumulation of cyclic adenosine-3' 5'monophosphate in guinea pig brain. *J. Neurochem.* **21,** 1415–1427.

Cheney, D. L., LaFevre, H. F., and Racagni, G. 1975. Choline acetyltransferase activity and mass fragmentographic measurement of acetylcholine in specific nuclei and tracts of rat brain. *Neuropharmacology* **14,** 801–809.

Christenson, J. G., Dairman, W. D., and Udenfriend, S. 1972. On the identity of dopa decarboxylase and 5-hydroxytryptophan decarboxylase. *Proc. Natl. Acad. Sci. USA* **69,** 343–347.

Cicero, T. J., Sharpe, L. G., Robins, E. P., and Grote, S. S. 1972. Regional distribution of tyrosine hydroxylase in rat brain. *J. Neurochem.* **19,** 2241–2243.

Clark, W. G., and del Guidia, J. 1970. *Principles of Psychopharmacology.* Academic Press, New York.

Clarke, G., and Straughan, D. W. 1977. Evaluation of the selectivity of antagonists of glutamate and acetylcholine applied microiontophoretically onto cortical neurons. *Neuropharmacol.* **16,** 391–398.

Clavier, R. M., and Routtenberg, A. 1976. Brainstem self-stimulation alleviated by lesions of medial forebrain bundles but not by lesions of locus coeruleus or caudal ventral norepinephrine bundle. *Brain Res.* **101,** 251–271.

Clement-Cormier, Y. C., Parrish, R. G., Petzold, G. L., Kebabian, J. W., and Greengard, P. 1975. Characterization of a dopamine-sensitive adenylate cyclase in the rat caudate nucleus. *J. Neurochem.* **25,** 143–150.

Cohen, S. R., and Lajtha, A. 1972. Amino acid transport. In: *Handbook of Neurochemistry,* vol. 7, edited by A. Lajtha. Plenum Press, New York, pp. 543–572.

Collier, H. O. J. 1972. Pharmacological mechanisms of drug dependence. In: *Pharmacology and The Future of Man,* vol. 1, edited by G. H. Acheson. Karger, Basel, pp. 65–76.

Colon, E. J. 1972. The elderly brain. *Psychiatr. Neurol. Neurochir. (Amst.)* **75,** 261–270.

Colonnier, M. L. 1966. The structural design of the neocortex. In: *Brain and Conscious Experience,* edited by J. C. Eccles. Springer-Verlag, New York, pp. 1–23.

Connett, R. J., and Kirschner, N. 1970. Purification and properties of bovine phenylethanola-mine-*N*-methyltransferase. *J. Biol. Chem.* **245,** 329–334.

Connor, J. D. 1970. Caudate nucleus neurons: Correlation of the effects of substantia nigra stimulation with iontophoretic dopamine. *J. Physiol.* **208,** 691–703.

Conradi, S. 1969. Ultrastructure of dorsal root boutons on lumbosacral motoneurons of the adult cat, as revealed by dorsal root section. *Acta Physiol. Scand.* **332,** Supp. 85–115.

Coombs, J. S., Eccles, J. C., and Fatt, P. 1955a. The electrical properties of the motoneuron membrane. *J. Physiol.* **130,** 291–325.

Coombs, J. S., Eccles, J. C., and Fatt, P. 1955b. The specific ionic conductances and the ionic movements across the motoneuronal membrane that produce the inhibitory post-synaptic potential. *J. Physiol.* **130,** 326–373.

Coombs, J. S., Eccles, J. C., and Fatt, P. 1955c. Excitatory synaptic action in motoneurons. *J. Physiol.* **130,** 374–395.

Coombs, J. S., Curtis, D. R., and Eccles, J. C. 1957a. The interpretation of spike potentials of motoneurons. *J. Physiol.* **139,** 198–231.

Coombs, J. S., Curtis, D. R., and Eccles, J. C. 1957b. The generation of impulses in motoneurons. *J. Physiol.* **139,** 232–249.

Cooper, J. R., Bloom, F. E., and Roth, R. H. 1974. *The Biochemical Basis of Neuropharmacology,* 2nd edition. Oxford University Press, New York, p. 127.

Costa, E. 1970. Simple neuronal models to estimate turnover rate of noradrenergic transmitters *in vivo.* In: *Advances in Biochemical Psychopharmacology,* vol. 2, edited by E. Costa and E. Giacobini. Raven Press, New York, pp. 169–216.

Costa, E., and Gessa, G. L. Editors. 1977. *Nonstriatal Dopaminergic Neurons,* vol. 16, *Advances in Biochemical Psychopharmacology,* Raven Press, New York.

Costa, E., and Sandler, M. Editors. 1972. *The Monoamine Oxidases: New Vistas,* vol. 5, *Advances in Biochemical Psychopharmacology.* Raven Press, New York.

Costa, E., Guidotti, A., Mao, C. C., and Suria, A. 1975. New concepts on the mechanism of action of benzodiazepines. *Life Sci.* **17,** 167–186.

Costentin, J., Boulu, P., and Schwartz, J. C. 1973. Pharmacological studies on the role of his-tamine in thermoregulation. *Agents & Actions* **3,** 177.

Cotzias, G. C., Van Woert, M. H., and Schiffer, L. M. 1967. Aromatic amino acids and modifica-tion of Parkinsonism. *New Engl. J. Med.* **276,** 374–379.

Cotzias, G. C., Papavasilion, P. S., and Gellene, R. 1969. Modification of Parkinsonism—chronic treatment with L-Dopa. *New Engl. J. Med.* **280,** 337–345.

Courvoisier, S., Fournel, J., Ducrot, R., Kolsky, M., and Koetschef, P. 1953. Proprietes phar-macodynamiques du chlorhydrate de chloro-3-(dimethylamino-3'-propyl)-10-phenothiazine (4560RP). *Archs. Int. Pharmacodyn. Ther.* **92,** 305–361.

Couteaux, R. 1958. Morphological and cytochemical observations on the postsynaptic membrane at motor end-plates and ganglionic synapses. *Exp. Cell Res. Suppl.* **5,** 294–322.

Cowan, W. M. 1970. Anterograde and retrograde transneuronal degeneration in the central and peripheral nervous system. In: *Contemporary Research Methods in Neuroanatomy,* edited by W. H. Nauta and S. O. E. Ebbesson. Springer-Verlag, Berlin, Heidelberg, New York, pp. 217–251.

Cowan, W. M. 1971. Studies on the development of the avian visual system. In: *Cellular Aspects of Neural Growth and Differentiation,* edited by D. S. Pearse. University of California Press, Los Angeles, pp. 177–218.

Coyle, J. T., and Axelrod, J. 1972. Tyrosine hydroxylase in rat brain: Developmental characteris-tics. *J. Neurochem.* **19,** 1117–1123.

Coyle, J. T., and Snyder, S. H. 1969. Antiparkinsonian drugs: Inhibition of dopamine uptake in the corpus striatum as a possible mechanism of action. *Science* **166,** 899–901.

Coyle, J. T., and Schwarcz, R. 1976. Lesions of striatal neurons with kainic acid provides a model for Huntington's chorea. *Nature* **263**, 244–246.

Cragg, B. G. 1975. The density of synapses and neurons in normal, mentally defective and aging human brains. *Brain* **98**, 81–90.

Crawford, J. M. 1963. The effect upon mice of intraventricular injection of excitant and depressant amino acids. *Biochem. Pharmacol.* **12**, 1443–1444.

Creveling, C. R., Borchardt, R. T., and Isersky, C. 1973. Immunological characterization of catechol-O-methyltransferase. In: *Frontiers in Catecholamine Research*, edited by E. Usdin and S. H. Snyder. Pergamon Press, New York, pp. 117–120.

Creed, R. S., Denny-Brown, D., Eccles, J. C., Liddell, E. G. T., and Sherrington, C. S. 1932. *Reflex Activity in the Spinal Cord*. Oxford University Press, London.

Creutzfeldt, O. D., 1975. Neurophysiological correlates of different functional states of the brain. In: *Brain Work: The Coupling of Function, Metabolism and Blood Flow in the Brain*, edited by D. H. Ingvar and N. A. Larssen. Munksgaard, Copenhagen, pp. 21–46.

Csanyi, V., Gervai, J., and Lajtha, A. 1973. Axoplasmic transport of free amino acids. *Brain Res.* **56**, 271–284.

Cuenod, M., Felix, D., Henke, H., Hunt, S., Kuzle, H., Le Fort, D., Reubi, J. C., Schenker, T., and Streit, P. 1976. Glycinergic pathway in the pigeon optic lobe and retrograde migration of material taken up as an amino acid. Neuroscience Meeting, Abstract #111.

Curatolo, A., d'Arcangelo, P., Lino, A., and Brancati, A. 1965a. Distribution of N-acetylaspartic and N-acetylaspartyl-glutamic acids in the neuroaxis of birds. *Boll. Soc. Ital. Biol. Sper.* **41**, 591–593.

Curatolo, A., d'Arcangelo, P., Lino, A., and Brancati, A. 1965b. Distribution of N-acetylaspartic and N-acetylaspartylglutamic acids in nervous tissue. *J. Neurochem.* **12**, 339–342.

Curtis, D. R., and Eccles, J. C. 1959. The time courses of excitatory and inhibitory synaptic actions. *J. Physiol.* **145**, 529–546.

Curtis, D. R., and Johnston, G. A. R. 1974. Amino acid transmitters in the mammalian central nervous system. *Ergeb. Physiol.* **69**, 97–188.

Curtis, D. R., and Watkins, J. C. 1960. The excitation and depression of spinal neurons by structurally related amino acids. *J. Neurochem.* **6**, 117–141.

Curtis, D. R., Phillis, J. W., and Watkins, J. C. 1961. Actions of amino acids on the isolated hemisected spinal cord of the toad. *Brit. J. Pharmacol.* **16**, 262–283.

Curtis, D. R., Hosli, L., and Johnston, G. A. R. 1967. Inhibition of spinal neurons by glycine. *Nature* **215**, 1502–1503.

Curtis, D. R., Hosli, L., and Johnston, G. A. R. 1968. A pharmacological study of the depression of spinal neurons by glycine and related amino acids. *Exp. Brain Res.* **6**, 1–18.

Curtis, D. R., Duggan, A. W., Felix, D., and Johnston, G. A. R. 1970. Bicuculline and GABA receptors. *Nature* **228**, 676–677.

Curtis, D. R., Felix, D., Game, C. J. A., and McCullock, R. M. 1973. Tetanus toxin and the synaptic release of GABA. *Brain Res.* **51**, 362–385.

Dahlstrom, A., and Fuxe, K. 1964a. Evidence for the existence of monoamine-containing neurons in the central nervous system. *Acta Physiol. Scand.* **62**, Suppl. 232.

Dahlstrom, A., and Fuxe, K. 1964b. A method for the demonstration of monoamine-containing fibers in the central nervous system. *Acta Physiol. Scand.* **60**, 293–295.

Dairman, W., Christenson, J. C., and Udenfriend, S. 1972. Changes in tyrosine hydroxylase and dopa decarboxylase induced by pharmacological agents. *Pharmacol. Rev.* **24**, 269–289.

Dale, H. H. 1914. The action of certain esters and ethers of choline, and their relation to muscarine. *J. Pharmacol.* **6**, 147–190.

Dale, H. H. 1935. Pharmacology and nerve endings. *Proc. R. Soc. Med.* **28**, 319–332.

Dale, H. H. 1937. Transmission of nervous effects by acetylcholine. *Harvey Lectures* **32**, 229–245.

Dale, H. H. 1938. Acetylcholine as a chemical transmitter substance of the effects of nerve impulses. *The William Henry Welch Lectures, 1937. J. Mt. Sinai Hosp.* **4**, 401–429.

Dale, H. H., and Dudley, H. W. 1929. The presence of histamine and acetylcholine in the spleen of the ox and the horse. *J. Physiol.* **68**, 97–123.

Dale, H. H., and Laidlaw, P. P. 1910. The physiological action of β-imidazolylethylamine. *J. Physiol. London* **41**, 318–344.

Dale, H. H., Feldberg, W., and Vogt, M. 1936. Release of acetylcholine at voluntary motor nerve endings. *J. Physiol.* **86,** 353-380.

David, J., Dairman, W., and Udenfriend, S. 1974. Decarboxylation to tyramine: A major route of tyrosine metabolism in mammals. *Proc. Natl. Acad. Sci. USA* **71,** 1771-1775.

Davidoff, R. A., Graham, L. T., Jr., Shank, R. P., Werman, R., and Aprison, M. H. 1967. Changes in amino acid concentrations associated with loss of spinal interneurons. *J. Neurochem.* **14,** 1025-1031.

Davis, J. N., Carlsson, A., MacMillan, V., and Siesjo, B. K. 1974. Brain tryptophan hydroxylation: Dependence on arterial oxygen tension. *Science* **182,** 72-73.

Davson, H. 1976. The blood-brain barrier. *J. Physiol.* **255,** 1-28.

Davson, H., and Danielli, J. F. 1943. *The Permeability of Natural Membranes.* Cambridge University Press, Cambridge, p. 361.

de Boilley, D., and Sorel, L. 1969. Premiers resultats en clinique d'un nouvel anti-epileptique: Di-*n*-propylacetate de sodium (DPA), specialise sous la marque Depakine. *Acta Neurol. Belg.* **69,** 909-913.

Deecke, L., Scheid, P., and Kornhuber, H. H. 1969. Distribution of readiness potential, pre-motion positivity, and motor potential of the human cerebral cortex preceding voluntary finger movements. *Exp. Brain Res.* **7,** 158-168.

Delay, J., and Deniker, P. 1952. Trente-huit cas de psychoses traiteés par la cure prolongée et continuée de 4560 R.P. Les Congrès des Alet Neurol de Langue Fr. In: *Compte rendu du Congres.* Masson et Cie, Paris.

del Castillo, J., and Katz, B. 1954. Quantal components of the end-plate potential. *J. Physiol.* **124,** 560-573.

del Castillo, J., and Katz, B. 1956. Biophysical aspects of neuromuscular transmission. *Prog. Biophys.* **6,** 121-170.

Delgado, J. M. P. 1976. New orientations in brain stimulation in man. In: *Brain-Stimulation Reward,* edited by A. Wanquier and E. T. Rolls. Elsevier Press, Amsterdam, pp. 481-503.

DeLong, M. R. 1974. Motor functions of the basal ganglia: Single-unit activity during movement. In: *The Neurosciences: Third Study Program,* edited by F. O. Schmitt and F. G. Worden. MIT Press, Cambridge, Mass., pp. 319-325.

De Robertis, E. 1971. Molecular biology of synaptic receptors. *Science* **171,** 963-971.

De Robertis, E. 1975. *Modern Pharmacology-Toxicology,* vol. 4, *Synaptic Receptors.* Marcel Dekker, Inc., New York.

De Robertis, E., and Bennett, H. S. 1955. Some features of the submicroscopic morphology of synapses in frog and earthworm. *J. Biophys. Biochem. Cytol.* **1,** 47-58.

De Robertis, E., and de Plazas, S. F. 1976. Differentiation of L-aspartate and L-glutamate high affinity binding sites in a protein fraction isolated from rat cerebral cortex. *Nature* **260,** 347-349.

De Robertis, E., de Iraldi, A. P., De Lores Arnaiz, G. R., and Salganicoff, L. 1962. Cholinergic and non-cholinergic nerve endings in rat brain. I. Isolation and subcellular distribution of acetyl-choline and acetylcholinesterase. *J. Neurochem.* **9,** 23-35.

De Robertis, E., De Lores Arnaiz, G. R., Salganicoff, L., Peregrino de Iraldi, A., and Zieher, L. M. 1963. Isolation of synaptic vesicles and structural organization of the acetylcholine system within brain nerve endings. *J. Neurochem.* **10,** 225-235.

Derouaux, M., Puil, E., and Naquet, R. 1973. Antiepileptic effect of taurine in photosensitive epilepsy. *EEG Clin. Neurophysiol.* **34,** 770.

Desclin, J. C. 1974. Histological evidence supporting the inferior olive as the major source of cerebellar climbing fibers in the rat. *Brain Res.* **77,** 365-384.

Desiraju, T., and Purpura, D. P. 1969. Synaptic convergences of cerebellar and lenticular projections to the thalamus. *Brain Res.* **15,** 544-547.

Devanandan, M. S., Eccles, R. M., and Westerman, R. A. 1965. Single motor units of mammalian muscle. *J. Physiol.* **178,** 359-367.

Dismukes, K., and Snyder, S. H. 1974. Dynamics of brain histamine. In: *Second Canadian Conference on Parkinson's Disease,* edited by F. H. McDowell and A. Barbeau. Raven Press, New York. *Adv. Neurol.* **5,** 101-109.

Divac, I., Fonnum, F., and Storm-Mathisen, J. 1977. High affinity uptake of glutamate in terminals of cortico striatal axons. *Nature* **266,** 377–378.

Dixon, W. E. 1906. Vagus inhibition. *Brit. Med. J.* **(ii),** 1807.

Dodge, F. A., and Rahamimoff, R. 1967. Cooperative action of calcium ions in transmitter release at the neuromuscular junction. *J. Physiol.* **193,** 419–432.

Dodge, F. A., Jr., Miledi, R., and Rahamimoff, R. 1969. Strontium and quantal release of transmitter at the neuromuscular junction. *J. Physiol.* **200,** 267–283.

Dogterom, J., van Wimersma Greidanus, T. B., and De Wied, D. 1976. Histamine as an extremely potent releaser of vasopressin in the rat. *Experientia* **32,** 659–660.

Dolezalova, H., and Stepita-Klauco, M. 1975. Regional distribution of piperidine in the brain of waking mice. Neuroscience Meeting, Abstract #1132.

Dolezalova, H., Giacobini, E., Seiler, N., and Schneider, H. H. 1973. Evidence for piperidine as a physiological constituent of the invertebrate nervous system. Am. Soc. Neurochem. Meeting.

Domesick, V. B. 1976. Projections of the nucleus of the diagonal band of Broca in the rat. *Anat. Rec.* **184,** 391–392.

Domino, E. F., and Wilson, A. E. 1972. Psychotropic drug influences on brain acetylcholine utilization. *Psychopharmacol.* **25,** 291–298.

Donoso, A. O., and Bannza, A. M. 1976. Acute effects of histamine on plasma prolactin and luteinizing hormone levels in male rats. *J. Neurol. Transm.* **39,** 95–101.

Dowling, J. E., and Boycott, B. B. 1966. Neural connections of the retina: Fine structure of the inner plexiform layer. *Cold Spring Harbor Symp. Quant. Biol.* **30,** 393–402.

Downing, A. C., Gerard, R. W., and Hill, A. V. 1926. The heat production of nerve. *Proc. R. Soc. London [Biol]* **100,** 223–251.

Dropp, J. J. 1972. Mast cells in the central nervous system of several rodents. *Anat. Rec.* **174,** 227–238.

Droz, B. 1975. Synaptic machinery and axoplasmic transport: maintenance of neuronal connectivity. In: *The Nervous System,* edited by D. B. Tower and R. O. Brady, vol. 1. *The Basic Neurosciences.* Raven Press, New York, pp. 111–128.

Droz, B., Koenig, H. L., and Di Giamberardino, L. 1974. Axonal transport of presynaptic macromolecules. In: *Actualites Neurophysiologiques,* vol. 10, edited by A. M. Monnier. Masson, Paris, pp. 236–260.

Drucker-Colin, R. R., and Giacobini, E. 1975. Sleep-inducing effect of piperidine. *Brain Res.* **88,** 186–189.

Drummond, G. I., and Ma, Y. 1973. Metabolism and function of cyclic AMP in nerve. *Prog. Neurobiol.* **2,** 119–176.

Duffy, M. J., Mulhall, D., and Powell, D. 1975. Subcellular distribution of substance P in bovine hypothalamus and substantia nigra. *J. Neurochem.* **25,** 305–307.

Eccles, J. C. 1953. *The Neurophysiological Basis of Mind.* Clarendon Press, Oxford.

Eccles, J. C. 1957. *The Physiology of Nerve Cells.* John Hopkins Press, Baltimore.

Eccles, J. C. 1960. The properties of dendrites. In: *Structure and Function of the Cerebral Cortex,* edited by D. B. Tower and J. P. Schadé. Elsevier Press, Amsterdam, pp. 192–202.

Eccles, J. C. 1961. Membrane time constants of cat motoneurons and time courses of synaptic action. *Exp. Neurol.* **4,** 1–22.

Eccles, J. C. 1964a. Excitatory responses of spinal neurons. In: *Progress in Brain Research,* vol. 12, *Physiology of Spinal Neurons,* edited by J. C. Eccles and J. P. Schadé. Elsevier, Amsterdam, pp. 1–34.

Eccles, J. C. 1964b. *The Physiology of Synapses.* Springer-Verlag, Heidelberg.

Eccles, J. C. 1964c. Presynaptic inhibition in the spinal cord. In: *Progress in Brain Research,* vol. 12, *Physiology of Spinal Neurons,* edited by J. C. Eccles and J. P. Schadé. Elsevier, Amsterdam, pp. 65–91.

Eccles, J. C. 1964d. Ionic mechanism of postsynaptic inhibition. In: *Les Prix Nobel* 1963, Nobel Fdn., Stockholm, pp. 261–283.

Eccles, J. C. 1966. Concious experience and memory. In: *Brain and Conscious Experience,* edited by J. C. Eccles. Springer-Verlag, Heidelberg, pp. 314–344.

Eccles, J. C. 1969a. *The Inhibitory Pathways of the Central Nervous System* (Sherrington Lecture). Liverpool University Press, Liverpool.

Eccles, J. C. 1969b. The dynamic loop hypothesis of movement control. In: *Information Processing in the Nervous System,* edited by K. N. Leibovic. Springer-Verlag, New York, 245-269.

Eccles, J. C. 1970a. Neurogenesis and morphogenesis in the cerebellar cortex. *Proc. Nat. Acad. Sci. USA* **66,** 294-301.

Eccles, J. C. 1970b. *Facing Reality: Philosophical Adventures by a Brain Scientist.* Springer-Verlag, New York.

Eccles, J. C. 1972. Possible synaptic mechanism subserving learning. In: *Brain and Human Behaviour,* edited by A. G. Karczmar and J. C. Eccles. Springer-Verlag, Heidelberg, pp. 39-61.

Eccles, J. C. 1973. The cerebellum as a computer: Patterns in space and time. *J. Physiol.* **229,** 1-32.

Eccles, J. C. 1976a. The plasticity of the mammalian central nervous system with special reference to new growths in response to lesions. *Naturwissenschaften,* **63,** 8-15.

Eccles, J. C. 1976b. From electrical to chemical transmission in the central nervous system. *Notes and Records, Roy. Soc.* **30,** 219-230.

Eccles, J. C. 1977. An instruction-selection theory of learning in the cerebellar cortex. *Brain Res.* **127,** 327-352.

Eccles, J. C. 1978. An instruction-selection hypothesis of cerebral learning. In: *Cerebral Correlates of Conscious Experience,* edited by P. A. Buser and A. Rougeul-Buser. Elsevier, Amsterdam, pp. 155-175.

Eccles, J. C., and Sherrington, C. S. 1930. Numbers and contraction values of individual motor-units examined in some muscles of the limb. *Proc. R. Soc. Lond. [Biol.]* **106,** 326-357.

Eccles, J. C., Fatt, P., and Koketsu, K. 1954. Cholinergic and inhibitory synapses in a pathway from motor-axon collaterals to motoneurons. *J. Physiol.* **126,** 524-562.

Eccles, J. C., Eccles, R. M., and Lundberg, A. 1957. Synaptic actions on motoneurons in relation to the two components of the group I muscle afferent volley. *J. Physiol.* **136,** 527-546.

Eccles, J. C., Eccles, R. M., and Lundberg, A. 1958a. The action potentials of the alpha motoneurones supplying fast and slow muscles. *J. Physiol.* **142,** 275-291.

Eccles, J. C., Libet, B., and Young, R. R. 1958b. The behavior of chromatolyzed motoneurons studied by intracellular recording. *J. Physiol.* **143,** 11-40.

Eccles, J. C., Hubbard, J. I., and Oscarsson, O. 1961a. Intracellular recording from cells of the ventral spinocerebellar tract. *J. Physiol.* **158,** 486-516.

Eccles, J. C., Eccles, R. M., Iggo, A., and Lundberg, A. 1961b. Electrophysiological investigations on Renshaw cells. *J. Physiol.* **159,** 461-478.

Eccles, J. C., Schmidt, R. F., and Willis, W. D. 1963. Inhibition of discharges into the dorsal and ventral spinocerebellar tracts. *J. Neurophysiol.* **26,** 635-645.

Eccles, J. C., Llinas, R., and Sasaki, K. 1966. The excitatory synaptic action of climbing fibers on the Purkinje cells of the cerebellum. *J. Physiol.* **182,** 268-296.

Eccles, J. C., Ito, M., and Szentágothai, J. 1967. *The Cerebellum as a Neuronal Machine.* Springer-Verlag, Heidelberg.

Eccles, J. C., Faber, D. S., and Taborikova, H. 1971a. The action of parallel fiber volley on the antidromic invasion of Purkinje cells of cat cerebellum. *Brain Res.* **25,** 335-356.

Eccles, J. C., Faber, D. S., Murphy, J. T., Sabah, N. H., and Taborikova, H. 1971b. Afferent volleys in limb nerves influencing impulse discharges in cerebellar cortex. I. In mossy fibers and granule cells. *Exp. Brain Res.* **13,** 15-35.

Eccles, J. C., Sabah, N. H., Schmidt, R. F., and Taborikova, H. 1972. Cutaneous mechanoreceptors influencing impulse discharges in cerebellar cortex. II. In Purkinje cells by mossy fiber input. *Exp. Brain Res.* **15,** 261-277.

Eccles, J. C., Nicoll, R. A., Oshima, T., and Rubia, F. J. 1977. The anionic permeability of the inhibitory postsynaptic membrane of hippocampal pyramidal cells. *Proc. R. Soc. Lond. [Biol.]* **198,** 345-361.

Edstrom, J. P., and Phillis, J. W. 1976. Effect of adenosine 5-monophosphate on the membrane potential of cortical neurons in the rat. Neuroscience Meeting, Abstract #1083.

Efron, D. H., editor. 1968. Psychopharmacology, a review of progress, 1957-1967. U.S. Public Health Service Bulletin #1836.

Efron, D. H. 1970. *Psychotomimetic Drugs.* Raven Press, New York.

Ehinger, B. 1977. Synaptic connections of the dopaminergic retinal neurons. In: *Advances in*

Biochemical Psychopharmacology, vol. 16, edited by E. Costa and G. L. Gessa. Raven Press, New York, pp. 299-306.

Ehrenpreis, T., and Pernow, B. 1952. On the occurrence of substance P in the recto-sigmoid in Hirschsprung's disease. *Acta Physiol. Scand.* **27**, 380-388.

Ehringer, H., and Hornykiewicz, O. 1960. Verteilung von Noradrenalin und Dopamin (3-Hydroxytryamin) im Gehirn des Menschen und ihr Verhalten bei Erkrankungen des extrapyramidalen Systems. *Klin. Wochenschr* **38**, 1236-1239.

Eide, E., Jurna, I., and Lundberg, A. 1968. Conductance measurements from motoneurons during presynaptic inhibition. In: *Structure and Function of Inhibitory Neuronal Mechanisms*, edited by C. von Euler, S. Skoglund, and U. Soderberg, Pergamon Press, Oxford, pp. 215-219.

Ekerot, C. F., and Larson, B. 1973. Correlation between sagittal projection zones of climbing and mossy fiber paths in cat cerebellar anterior lobe. *Brain Res.* **164**, 446-450.

Ekerot, C. F., and Oscarsson, O. 1975. Inhibitory spinal paths to the lateral reticular nucleus. *Brain Res.* **99**, 157-161.

Elde, R., Hokfelt, T., Johannson, O., and Terenius, L. 1976. Immunohistochemical studies using antibodies to leucine-enkephalin: Initial observations on the nervous system of the rat. *Neuroscience* **1**, 349-355.

Elliott, T. R. 1904. On the action of adrenaline. *J. Physiol.* **31**, XXP.

Elmadjian, F. 1959. Excretion and metabolism of epinephrine. *Pharmacol. Rev.* **11**, 409-415.

Enna, S. J., Kuhar, M. J., and Snyder, S. H. 1975. Regional distribution of postsynaptic receptor binding for γ-aminobutyric acid (GABA) in monkey brain. *Brain Res.* **93**, 168-174.

Epstein, A. N. 1971. The lateral hypothalamic syndrome: Its implications for the physiological psychology of hunger and thirst. In: *Progress in Physiological Psychology*, vol. 4, edited by J. M. Sprague. Academic Press, New York, pp. 263-317.

Eranko, O. 1955. Histochemistry of noradrenaline in the adrenal medulla of rats and mice. *Endocrinology* **57**, 363-368.

Erspamer, V., and Anastasi, A. 1962. Structure and pharmacological actions of eledoisin, the active undecapeptide of the posterior salivary glands of Eledone. *Experentia* **18**, 58-59.

Erspamer, V., and Asero, B. 1952. Identification of enteramine, the specific hormone of the enterochromaffin cell system, as 5-hydroxytryptamine. *Nature* **169**, 800-801.

Erspamer, V., Anastasi, A., Bertaccini, G., and Cei, J. M. 1964. Structure and pharmacological activity of physalaemin, the main active polypeptide of the skin of *Physayaemus fuscumaculatus*. *Experentia* **20**, 489-490.

Evans, P. D. 1973. Amino acid distribution in the nervous system of the crab. *J. Neurochem.* **21**, 11-17.

Evans, W. O. 1968. The psychopharmacology of the normal human: Trends in research strategy. In: *Psychopharmacology*, edited by D. H. Efron. U.S. Government Printing Office, Washington, pp. 1003-1011.

Evarts, E. V. 1961. Effects of sleep and waking on activity of single units in the unrestrained cat. In: *The Nature of Sleep*, edited by G. E. W. Wolstenholme and M. O'Connor. J. & A. Churchill Ltd., London, pp. 171-182.

Evarts, E. V. 1967. Representation of movements and muscles by pyramidal tract neurons of the precentral motor cortex. In: *Neurophysiological Basis of Normal and Abnormal Motor Activity*, edited by M. D. Yahr and D. P. Purpura. Raven Press, New York, pp. 215-253.

Evarts, E. V., and Thach, W. T. 1969. Motor mechanisms of the CNS, cerebrocerebellar interrelations. *Ann. Rev. Physiol.* **31**, 451-498.

Fahn, S., and Cote, L. J. 1968a. Regional and subcellular distribution of tyrosine hydroxylase. *Neurology* **18**, 293-4.

Fahn, S., and Cote, L. 1968b. Regional distribution of γ-aminobutyric acid (GABA) in brain of rhesus monkey. *J. Neurochem.* **15**, 209-213.

Falck, B., Hillarp, N. A., Thieme, G., and Torp, A. 1962. Fluorescence of catecholamines and related compounds condensed with formaldehyde. *J. Histochem. Cytochem.* **10**, 348-354.

Farley, I. J., and Hornykiewicz, O. 1977. Noradrenaline distribution in subcortical areas of human brain. *Brain Res.* **126**, 53-62.

Fatt, P., and Katz, B. 1951. An analysis of the end-plate potential recorded with an intra-cellular electrode. *J. Physiol.* **115**, 320–370.

Fatt, P., and Katz, B. 1952. Spontaneous subthreshold activity at motor nerve endings. *J. Physiol.* **117**, 109–128.

Faull, R. L. M., and Mehler, W. R. 1976. Subdivision of the ventral tier nuclei in the rat thalamus based on their afferent fiber connections. *Anat. Rec.* **184**, 400.

Feger, J., and Ohye, C. 1975. The unitary activity of the substantia nigra following stimulation of the striatum in the awake monkey. *Brain Res.* **89**, 155–159.

Feigl, H. 1967. *The "Mental" and the "Physical."* University of Minnesota Press, Minneapolis, p. 179.

Feldberg, W., and Gaddum, J. H. 1934. The chemical transmitter at synapses in a sympathetic ganglion. *J. Physiol.* **81**, 305–319.

Feldberg, W., and Krayer, O. 1933. Des Auftreten eines acetylcholinartigen Stoffes im Herzvenenblut bei Reizung der Nervi vagi. *Arch. f. exper. Path.* **172**, 170–193.

Feldman, M. L. 1976. Aging changes in the morphology of cortical dendrites. In: *Aging*, vol. 3, edited by R. D. Terry and S. Gershon. Raven Press, New York, pp. 211–228.

Feldstein, A., Hoagland, H., Freeman, H., and Williamson, O. 1967. Effect of ethanol ingestion on serotonin-C^{14} metabolism in man. *Life Sci.* **6**, 53–61.

Felix, D., and Kunzle, H. 1974. Iontophoretic and autoradiographic studies on the role of proline in synaptic transmission. *Pflugers Arch.* **350**, 135–144.

Feltz, P., and Albe-Fessard, D. 1972. The study of an ascending nigro-caudate pathway. *Electroencephalogr. Clin. Neurophysiol.* **33**, 179–193.

Fenstermacher, J. D. 1975. Mechanisms of ion distribution between blood and brain. In: *The Nervous System*, vol. 1, *The Basic Neurosciences*, edited by D. B. Tower and R. O. Brady. Raven Press, New York, pp. 299–311.

Fernstrom, J. D., and Wurtman, R. J. 1972. Brain serotonin: Physiological regulation by plasma neutral amino acids. *Science* **173**, 149–152.

Fernstrom, J. D., and Wurtman, R. J. 1974. Control of the brain serotonin levels by diet. *Adv. Biochem. Psychopharm.* **11**, 133–142.

Ferriero, D., and Margolis, F. L. 1975. Denervation in the primary olfactory pathway of mice. II. Effects on carnosine and other amine compounds. *Brain Res.* **94**, 75–86.

Fertuck, H. C., and Salpeter, M. M. 1974. Localization of acetylcholine receptor by I-labeled α-bungarotoxin binding at mouse motor endplates. *Proc. Natl. Acad. Sci. USA*, **71**, 1376–1378.

Fibiger, H. C., and McGeer, E. G. 1973. Increased axoplasmic transport of [^{3}H]dopamine in nigro-neostriatal neurons after reserpine. *Life Sci.* **13**, 1565–1571.

Fibiger, H. C., Pudritz, R. E., McGeer, P. L., and McGeer, E. G. 1972. Axonal transport in nigro-striatal neurons. *Nature* **237**, 177–179.

Fibiger, H. C., Carter, D. A., and Phillips, A. G. 1976. Decreased intracranial self-stimulation after neuroleptics or 6-hydroxydopamine: Evidence for mediation by motor deficits rather than reduced reward. *Psychopharmacol.* **47**, 21–27.

Fischer, J. E., Horst, W. D., and Kopin, I. J. 1965. β-Hydroxylated sympathomimetic amines as false neurotransmitters. *Brit. J. Pharmacol. Chemother.* **24**, 477–484.

Fischer-Ferraro, C., Nahmod, V. E., Goldstein, D. J., and Finkelman, S. 1971. Angiotensin and renin in dog brain. *J. Exp. Med.* **133**, 353–361.

Fisher, S. K., and Davies, W. E. 1976. GABA and its related enzymes in the lower auditory system of the guinea pig. *J. Neurochem.* **27**, 1145–1155.

Fisher, G. H., Humphries, J., Folkers, K., Pernow, B., and Bowers, C. Y. 1974. Synthesis and some biological activities of substance P. *J. Med. Chem.* **17**, 843–846.

Fitzsimons, J. T. 1972. Thirst. *Physiol. Rev.* **52**, 468–561.

Florey, E. 1953. Uber einen nervosen Hemmungsfaktor in Gehirn and Ruckenmark. *Naturwissenschaften* **40**, 295–296.

Fonnum, F. 1970. Topographical and subcellular localization of choline acetyltransferase in the rat hippocampal region. *J. Neurochem.* **17**, 1029–1037.

Fonnum, F., Storm-Mathisen, J., and Walberg, F. 1970. Glutamate decarboxylase in inhibitory neurons. A study of the enzyme in Purkinje cell axons and boutons in the cat. *Brain Res.* **20**, 259–275.

Fonnum, F. Grofova, I., Rinvik, E., Storm-Mathison, J., and Waldberg, F. 1974. Origin and distribution of glutamate decarboxylase in the substantia nigra of the cat. *Brain Res.* **71**, 77–92.

Fonnum, F. 1978. Comments on localization of neurotransmitters in the basal ganglia. In: *Amino Acids as Chemical Transmitters,* edited by F. Fonnum. Plenum Press, New York, pp. 143–153.

Fox, C. A. 1962. The structure of the cerebellar cortex. In: *Correlative Anatomy of the Nervous System,* edited by E. C. Crosby, T. H. Humphrey, and E. W. Lauer. The MacMillan Co., New York, pp. 193–198.

Fox, C. A., Hillman, D. E., Siegesmund, K. A., and Dutta, C. R. 1966. The primate cerebellar cortex: A Golgi and electron microscope study. In: *Progress in Brain Research,* vol. 25, edited by C. A. Fox and R. S. Snider, Elsevier Press, Amsterdam, pp. 174–225.

Fox, C. F. 1972. The structure of cell membranes. *Sci. Amer.* **226**, 31–38.

Frank, K., and Fuortes, M. G. F. 1957. Presynaptic and postsynaptic inhibition of monosynaptic reflexes. *Fed. Proc.* **16**, 39–40.

Frederickson, R. C. A., and Norris, F. H. 1976. Enkephalin-induced depression of single neurons in brain areas with opiate receptors-antagonism by naloxone. *Science* **194**, 440–442.

Friedman, S., and Kaufman, S. 1965. 3,4-Dihydroxyphenylethylamine β-hydroxylase: Physical properties, copper content and the role of copper in the catalytic activity. *J. Biol. Chem.* **240**, 4763–4773.

Funk, C. 1911. Syntheses of dl-3:4-dihydroxyphenylalanine. *J. Chem. Soc.* **99**, 554–557.

Fuxe, K., and Johnsson, G. 1974. Further mapping of central 5-hydroxytryptamine neurons: Studies with the neurotoxic dihydroxytryptamines. In: *Advances in Biochemical Pharmacology,* vol. 10, edited by E. Costa, G. L. Gessa, and M. Sandler. Raven Press, New York, pp. 1–12.

Fuxe, K., Holmstedt, B., and Johnsson, G. 1972. Effects of 5-methoxy-*N,N*-dimethyltryptamine on central monoamine neurons. *Europ. J. Pharmacol.* **19**, 25–34.

Gage, P. W., and Armstrong, C. M. 1968. Miniature end-plate currents in voltage-clamped muscle fibers. *Nature* **218**, 363–365.

Gage, P. W., and McBurney, R. N. 1972. Miniature end-plate currents and potentials generated by quanta of acetylcholine in glycerol-treated toad sartorius fibers. *J. Physiol.* **226**, 79–94.

Gage, P. W., and Moore, J. W. 1969. Synaptic current at the squid giant synapse. *Science* **160**, 510–512.

Gainer, H. 1977. *Peptides in Neurobiology.* Raven Press, New York.

Gal, E. M. 1972. 5-Hydroxytryptamine-*O*-sulphate: An alternative route of serotonin inactivation in brain. *Brain Res.* **44**, 309–312.

Gal, E. M. 1974. Tryptophan 5-hydroxylase: Function and control. *Adv. Biochem. Psychopharmacol.* **11**, 1–11.

Gal, E. M., and Sherman, A. D. 1976. Biopterin II: Evidence for cerebral synthesis of 7,8-dihydrobiopterin *in vivo* and *in vitro. Neurochem. Res.* **1**, 627–640.

Gal, E. M., Yang, S. N., and Moses, F. 1975. Cerebral mono-oxygenases: Biochemical and immunological studies. *Abst. Int. Soc. Neurochem.* **5**, 78.

Galindo, A., Krnjevic, K., and Schwartz, S. 1967. Microiontophoretic studies on neurons in the cuneate nucleus. *J. Physiol. (Lond.)* **192**, 359–377.

Gallagher, J. P., Higashi, H., and Nishi, S. 1975. The ionic requirements for the production of the GABA depolarization at cat primary afferent neurons. *Fed. Proc.* **34**, 418.

Ganten, D., Granger, P., Hayduk, K., Brecht, H. M., Barbeau, A., Boucher, R., and Genest, J. 1971. Angiotensin-forming enzyme in brain tissue. *Science* **173**, 64–65.

Garbarg, M., Barbin, G., Bischoff, S., Pollard, H., and Schwartz, J. C. 1974. Evidence for a specific decarboxylase involved in histamine synthesis in an ascending pathway in rat brain. *Agents & Actions* **4**, 181.

Gaze, R. M. 1970. *The Formation of Nerve Connections.* Academic Press, New York.

Gaze, R. M. 1974. Neuronal specificity. *Br. Med. Bull.* **30**,(2).

Gebhard, O., and Velstra, H. 1964. *N*-Acetylaspartic acid. Experiments on biosynthesis and function. *J. Neurochem.* **11**, 613–617.

Geinisman, Y., Bondariff, W., and Telser, A. 1977. Transport of [³H]glucose labeled glycoproteins in the septohippocampal pathway of young adult and genescent rats. *Brain Res.* **125**, 182–186.

Gerald, M. C., and Maickel, R. P. 1972. Studies on the possible role of brain histamine in behavior. *Brit. J. Pharmacol.* **44**, 462–471.

Gerschenfeld, H. M., and Paupardin-Tritsch, D. 1974. Ionic mechanisms and receptor properties underlying the responses of molluscan neurons to 5-hydroxytryptamine. *J. Physiol.* **243**, 427–456.

Gershon, M. D., Dreyfus, C. F., Pickel, V. M., Joh, T. H., and Reis, D. J. 1975. Serotonergic neurons in the mammalian peripheral nervous system. *Neuroscience Abst.* **1**, 372.

Gershon, S., Angrist, B., and Shopsin, B. 1975. Drugs, diagnosis and disease. In: *Biology of the Major Psychoses,* edited by D. X. Freedman. Raven Press, New York, pp. 85–96.

Geschwind, N. 1970. The organization of language and the brain. *Science* **170**, 940–944.

Geschwind, N. 1972. Language and the brain. *Sci. Am.* **226**, 76–83.

Gessa, G. L., Craba, F., Vargin, L., and Spano, P. F. 1968. Selective increase of brain dopamine induced by γ-hydroxybutyrate: Study of the mechanism of action. *J. Neurochem.* **15**, 377–381.

Ghelarducci, B., Ito, M., and Yagi, N. 1975. Impulse discharges from flocculus Purkinje cells of alert rabbits during visual stimulation combined with horizontal head rotation. *Brain Res.* **87**, 66–72.

Gibson, G. E., and Blass, J. P. 1976. Impaired synthesis of acetylcholine in brain accompanying mild hypoxia and hypoglycemia. *J. Neurochem.* **27**, 37–42.

Gilles, R., and Schoffeniels, E. 1968. Fixation de $^{14}CO_2$ par les acides amines de la chaine nerveuse ventrale du crustace Humarus vulgaris. *M. Edw. Arch. Int. Physiol. Biochim.* **76**, 441–451.

Gillin, J. C., Fram, D. H., Wyatt, R. J., Henkin, R. I., and Snyder, F. 1975. L-Histidine: Failure to affect the sleep-waking cycle in man. *Psychopharmacol.* **40**, 305–311.

Glassman, E. 1969. Some considerations of the effects of short term learning on the incorporation of uridine into RNA and polysomes of mouse brain. In: *The Future of the Brain Sciences,* edited by S. Bogoch. Plenum Press, New York, pp. 281–291.

Glover, V., Sandler, M., Owen, F., and Riley, G. J. 1977. Dopamine is a monoamine oxidase B substrate in man. *Nature* **265**, 80–81.

Glowinski, J. 1975. Neuroleptics and dopamine neurons. In: *Biology of the Major Psychoses: A Comparative Analysis,* Res. Publ. Assoc. Nerv. & Ment. Dis., vol. 54, edited by D. X. Freedman. Raven Press, New York, pp. 233–246.

Goldberg, M. E., Satama, A. I., and Blum, S. W. 1971. Inhibition of choline acetyltransferase and hexobaritone-metabolizing enzymes by naphthyl vinyl pyridine and analogues. *J. Pharm. Pharmacol.* **23**, 384–385.

Goldstein, A. 1976. Opioid peptides (endorphins) in pituitary and brain. *Science* **193**, 1081–1086.

Goldstein, A., Lowney, L. I., and·Pal, B. K. 1971. Stereospecific and nonspecific interactions of the morphine cogenor levorphanol in subcellular fractions of mouse brain. *Proc. Natl. Acad. Sci. USA* **68**, 1742–1747.

Goldstein, D. J., Diaz, S., Finkielman, S., Nahmod, V. E., and Fischer-Farro, C. 1972. Angiotensinase activity in rat and dog brain. *J. Neurochem.* **19**, 2451–2453.

Goldstein, M., Fuxe, K., and Hokfelt, T. 1972. Characterization and tissue localization of catecholamine synthesizing enzymes. *Pharmacol. Rev.* **24**, 293–309.

Gonshor, A., and Melvill Jones, G. 1976a. Short-term adaptive changes in the human vestibulo-ocular reflex arc. *J. Physiol.* **256**, 361–379.

Gonshor, A., and Melvill Jones, G. 1976b. Extreme vestibulo-ocular adaptation induced by prolonged optical reversal of vision. *J. Physiol.* **256**, 381–414.

Goodall, M. 1951. Studies of adrenaline and noradrenaline in mammalian heart and suprarenals. *Acta Physiol. Scand.* **24**, Suppl. 85.

Goodall, M. 1962. Sympathoadrenal response to gravitational stress. *J. Clin. Invest.* **41**, 197–202.

Goodman, L. S., and Gilman, A. 1970. *The Pharmacological Basis of Therapeutics,* 4th edition. The MacMillan Co., New York.

Goodwin, F, K., and Post, R. M. 1975. Studies of amine metabolites in affective illness and in schizophrenia. In: *Biology of the Major Psychoses: A Comparative Analysis,* Res. Publ. Assoc. Nerv. & Ment. Dis., vol. 54, edited by D. X. Freedman. Raven Press, New York, pp. 299–332.

Gott, P. S. 1973a. Cognitive abilities following right and left hemispherectomy. *Cortex* **9**, 266–274.

Gott, P. S. 1973b. Language after dominant hemispherectomy. *J. Neurol., Neurosurg. and Psychiat.* **36**, 1082–1088.

Grafstein, B. 1971. Transneuronal transfer of radioactivity in the central nervous system. *Science* **172**, 177–179.

Graham, L. T. 1972. Intraretinal distribution of GABA content and GAD activity. *Brain Res.* **36**, 476–479.

Graham, L. T., Shank, R. P., Werman, R., and Aprison, M. H. 1967. Distribution of some synaptic transmitter suspects in cat spinal cord: Glutamic acid, aspartic acid, γ-aminobutyric acid, glycine and glutamine. *J. Neurochem.* **14**, 465–472.

Grahame-Smith, D. G. 1973. Does the total turnover of the brain 5HT reflect the functional activity of 5HT in brain? In: *Serotonin and Behavior,* edited by J. Barchas and E. Usdin. Academic Press, New York, pp. 5–7.

Granit, R. 1970. *The Basis of Motor Control.* Academic Press, New York.

Granit, R. 1972. Constant errors in the execution and appreciation of movement. *Brain* **95**, 649–660.

Granit, R. Kernell, D., and Shortess, G. K. 1963. Quantitative aspects of repetitive firing of mammalian motoneurones, as caused by injected currents. J. Physiol. **168**, 911–931.

Gray, E. G. 1959. Axo-somatic and axo-dendritic synapses of the cerebral cortex: An electron microscope study. *J. Anat., Lond.* **93**, 420–423.

Gray, E. G. 1962. A morphological basis for pre-synaptic inhibition? *Nature* **193**, 82–83.

Gray, E. G. 1964. Tissue of the central nervous system. In: *Electron Microscopic Anatomy.* Academic Press, New York, pp. 369–417.

Gray, E. G. 1970. The fine structure of nerve. *Comp. Biochem. Physiol.* **36**, 419–448.

Graybiel, A. M. 1975. Wallerian degeneration and anterograde tracer methods. In: *The Use of Axonal Transport for Studies of Neuronal Connectivity,* edited by W. M. Cowan and M. Cuenod. Elsevier, Amsterdam, pp. 173–216.

Graybiel, A. M., and Sciascia, T. R. 1975. Origin and distribution of nigrotectal fibers in the cat. *Neurosci. Abst.* **1**, 174.

Green, A. R., and Grahame-Smith, D. G. 1975. 5-Hydroxytryptamine and other indoles in the central nervous system. In: *Biochemistry of Biogenic Amines,* edited by L. L. Iversen, S. D. Iversen, and S. H. Snyder. Plenum Press, New York, pp. 169–246.

Green, A. R., and Grahame-Smith, D. G. 1976. Propranolol inhibits behavioural responses of rats to increased 5-hydroxytryptamine in the central nervous system. *Nature* **262**, 594–597.

Green, A. R., and Youdim, M. B. H. 1975. Effects of monoamine oxidase inhibition by clorgyline, deprenil or tranylcypromine on 5-hydroxytryptamine concentrations in rat brain and hyperactivity following subsequent tryptophan administration. *Brit. J. Pharmacol.* **55**, 415–422.

Green, A. R., Koslow, S. H., and Costa, E. 1973. Identification and quantitation of a new indolealkylamine in rat hypothalamus. *Brain Res.* **51**, 371–374.

Green, A. R., Hughes, J. P., and Tordoft, A. F. C. 1975. The concentration of 5-methoxytryptamine in rat brain and its effects on behaviour following its peripheral injection. *Neuropharmacol.* **14**, 601–606.

Greengard, P. 1976. Possible role for cyclic nucleotides and phosphorylated membrane proteins in postsynaptic actions of neurotransmitters. *Nature* **260**, 101–108.

Grofova, I. 1975. The identification of striatal and pallidal neurons projecting to substantia nigra. An experimental study by means of retrograde axonal transport of horseradish peroxidase. *Brain Res.* **91**, 286–291.

Grözinger, B., Kornhuber, H. H., and Kriebel, J. 1975. Methodological problems in the investigation of cerebral potentials preceding speech: Determining the onset and suppressing artifacts caused by speech. *Neuropsychologia* **13**, 263–270.

Guidotti, A., Badiani, G., and Pepeu, G. 1972. Taurine distribution in cat brain. *J. Neurochem.* **19**, 431–435.

Guillemin, R., Ling, N., and Burgus, R. 1976. Endorphines, peptides d'origine hypothalaminque et neurophyphphysaire a activité morphinomimetique. *C. R. Acad. Ser. D.* **282**, 783–785.

Guillery, R. W. 1970. Light- and electron-microscopical studies of normal and degenerating axons. In: *Contemporary Research Methods in Neuroanatomy*, edited by W. J. H. Nauta and S. O. E. Ebbesson. Springer-Verlag, Heidelberg, pp. 77–105.

Guillery, R. W. 1972. Experiments to determine whether retinogeniculate axons can form translaminar collateral sprouts in the dorsal lateral geniculate nucleus of the cat. *J. Comp. Neurol.* **146,** 407–419.

Gunne, L. M. 1962. Relative adrenaline content in brain tissue. *Acta Physiol. Scand.* **56,** 324–333.

Gustafsson, B. 1974. Afterhyperpolarization and the control of repetitive firing in spinal neurons of the cat. *Acta Physiol. Scand.* **416,** suppl. 1–47.

Guyenet, P., Lefresne, P., Rossier, J., Beaujouan, J. C., and Glowinski, J. 1973. Inhibition by hemicholinium-3 of [^{14}C]acetylcholine synthesis and [^{3}H]choline high affinity uptake in rat striatal synaptosomes. *Mol. Pharmacol.* **9,** 630–639.

Gyermek, L. 1966. Drugs which antagonize 5-hydroxytryptamine and related indolealkylamines. In: *Handbook of Experimental Pharmacology*, vol. 19, edited by V. Erspamer. Springer-Verlag, Berlin, pp. 471–514.

Haas, H. L., Wolf, P., and Nussbaumer, J. C. 1975. Histamine: Action on supraoptic and other hypothalamic neurons of the cat. *Brain Res.* **88,** 166–170.

Haggendal, J. 1973. Regulation of catecholamine release. In: *Frontiers in Catecholamine Research*, edited by E. Usdin and S. Snyder. Pergamon Press, London, pp. 531–535.

Hagiwara, S., and Tasaki, K. 1958. A study of the mechanism of impulse transmission across the giant synapse of the squid. *J. Physiol.* **143,** 114–137.

Halasz, N., Hokfelt, T., Ljungdahl, A., Johansson, O., and Goldstein, M. 1977. Dopamine neurons in the olfactory bulb. In: *Advances in Biochemical Psychopharmacology*, vol. 16, edited by E. Costa and G. L. Gessa. Raven Press, New York, pp. 169–177.

Hamlin, K. E., and Fisher, F. E. 1951. The synthesis of 5-hydroxytryptamine. *J. Am. Chem. Soc.* **73,** 5007–5008.

Hamlyn, L. H. 1962. An electron microscope study of pyramidal neurons in the Ammon's Horn of the rabbit. *J. Anat.* **97,** 189–201.

Hámori, J., Pasik, T., Pasik, P., and Szentágothai, J. 1974. Triadic synaptic arrangements and their possible significance in the lateral geniculate nucleus of the monkey. *Brain Res.* **80,** 379–393.

Hanin, I., and Costa, E. 1976. Approaches used to estimate brain acetylcholine turnover rate *in vivo;* effects of drugs on brain acetylcholine turnover rate. In: *Biology of Cholinergic Function*, edited by A. M. Goldberg and I. Hanin. Raven Press, New York, pp. 355–378.

Harbison, R. D., Dwivedi, C., and Rama Sastry, B. V. 1974. Development of sperm choline acetyltransferase activity: Studies on mechanisms of antifertility activity of trimethyl phosphate. *Pharmacol.* **16,** 307.

Harding, J., and Margolis, F. L. 1976. Denervation in the primary olfactory pathway of mice. III. Effects on enzymes of carnosine metabolism. *Brain Res.* **110,** 351–360.

Hare, M. L. C. 1928. Tyramine oxidase, I. A new enzyme system in liver. *Biochem. J.* **22,** 968–970.

Harmar, A. J., and Horn, A. S. 1976. Octopamine in mammalian brain: Rapid postmortem increase and effect of drugs. *J. Neurochem.* **26,** 987–993.

Harris, G. W., Jacobson, D., and Kahlson, G. 1952. The occurrence of histamine in the cerebral regions related to the hypophysis. In: *CIBA Foundation Colloquia on Endocrinology*, vol. 4, edited by G. E. W. Wolstenholme, assisted by M. P. Cameron. The Blakiston Company, New York, pp. 186–194.

Hartline, H. K. 1934. Intensity and duration in the excitation of single photoreceptors. *J. Cell Comp. Physiol.* **5,** 229–247.

Hartman, B. K., and Udenfriend, S. 1972. The application of immunological techniques to the study of enzymes regulating catecholamine synthesis. *Pharmacol. Rev.* **24,** 311–330.

Harvey, J. A., and Yunger, L. M. 1973. Relationship between telencephalic content of serotonin and pain sensitivity. In: *Serotonin and Behavior*, edited by J. Barchas and E. Usdin. Academic Press, New York, pp. 179–189.

Harvey, J. A., Schofield, C. N., Graham, L. T., Jr., and Aprison, M. H. 1975. Putative transmitters in denervated olfactory cortex. *J. Neurochem.* **24,** 445–449.

Hassler, R., and Reichert, T. 1961. Wirkungen der Reizungen und Koagulationen in den stammganglien bei stereotaktischen hirnoperation. *Nervenarzt* **32,** 97–109.

Hattori, T., McGeer, P. L., Fibiger, H. C., and McGeer, E. G. 1973. On the source of GABA-containing terminals in the substantia nigra. Electron microscopic autoradiographic and biochemical studies. *Brain Res.* **54,** 103–114.

Hattori, T., Fibiger, H. C., and McGeer, P. L. 1975. Demonstration of a pallido-nigral projection innervating dopaminergic neurons. *J. Comp. Neurol.* **162,** 487–504.

Hattori, T., McGeer, P. L., and McGeer, E. G. 1976a. Synaptic morphology in the neostriatum of the rat: Possible serotonergic synapse. *Neurochem. Res.* **1,** 451–467.

Hattori, T., Singh, V. K., McGeer, E. G., and McGeer, P. L. 1976b. Immunohistochemical localization of choline acetyltransferase containing neostriatal neurons and their relationship with dopaminergic synapses. *Brain Res.* **102,** 164–173.

Hattori, T., McGeer, E. G., Singh, V. K., and McGeer, P. L. 1977. Cholinergic synapse of the interpeduncular nucleus. *Exp. Neurol.* **55,** 102–111.

Haubrich, D. R. 1976. Choline acetyltransferase and its inhibitors. In: *Biology of Cholinergic Function,* edited by A. M. Goldberg and I. Hanin. Raven Press, New York, pp. 239–268.

Hayashi, T. 1952. A physiological study of epileptic seizures following cortical stimulation in animals and its application of human clinic. *Jap. J. Physiol.* **3,** 46–64.

Hebb, C. O., and Waites, G. 1956. Choline acetylase in antero and retrograde degeneration of cholinergic nerve. *J. Physiol.* **132,** 667–671.

Hebb, D. O. 1949. *The Organization of Behavior.* John Wiley & Sons, New York.

Heilbronn, E. 1975. Biochemistry of cholinergic receptors. In: *Cholinergic Mechanisms,* edited by P. G. Waser. Raven Press, New York, pp. 343–364.

Heilbronn, E., and Mattson, C. 1974. The nicotinic cholinergic receptor protein: Improved purification method, preliminary amino acid composition and observed auto-immuno response. *J. Neurochem.* **22,** 315–317.

Heimer, L., Ebner, F. F., and Nauta, W. J. H. 1967. A note on the termination of commissural fibers in the neocortex. *Brain Res.* **5,** 171–177.

Henke, H., Schenker, T. M., and Cuenod, M. 1976a. Uptake of neurotransmitter candidates by pigeon optic tectum. *J. Neurochem.* **26,** 125–130.

Henke, H., Schenker, T. M., and Cuenod, M. 1976b. Effects of retinal ablation on uptake of glutamate, glycine, GABA, proline and choline in pigeon tectum. *J. Neurochem.* **26,** 131–134.

Henry, J. L. 1976. Effects of substance P on functionally identified units in cat spinal cord. *Brain Res.* **114,** 435–451.

Henry, J. L., Krnjevic, K., and Morris, M. E. 1975. Substance P and spinal neurons. *Can. J. Physiol. Pharmacol.* **53,** 423–432.

Herrup, K., and Shooter, E. M. 1975. Properties of the beta-nerve growth factor receptor in development. *J. Cell Biol.* **67,** 118–125.

Herrup, K., Stickgold, R. E., and Shooter, E. M. 1974. The role of the nerve growth factor in the development of sensory and sympathetic ganglia. *Ann. N.Y. Acad. Sci.* **228,** 381–392.

Herz, A., and Zieglgansberger, W. 1968. The influence of microelectrophoretically applied biogenic amines, cholinomimetic and procaine on synaptic excitation in the corpus striatum. *Int. J. Neuropharmacol.* **7,** 221–230.

Hess, W. R. 1964. *The Biology of Mind.* University of Chicago Press, Chicago.

Hill, A. V. 1960. The heat production of nerve. In: *Molecular Biology,* edited by D. Nachmansohn. Academic Press, New York, pp. 153–162.

Hille, B. 1976. Gating in sodium channels of nerve. *Ann. Rev. Physiol.* **38,** 139–142.

Hillier, W. F., Jr. 1954. Total left cerebral hemispherectomy for malignant glioma. *Neurol.* **4,** 718–721.

Ho, B. T. 1972. Monoamine oxidase inhibitors. *J. Pharmacol. Sci.* **61,** 821–837.

Hodgkin, A. L. 1964. *The Conduction of the Nervous Impulse.* Liverpool University Press, Liverpool.

Hodgkin, A. L., and Huxley, A. F. 1952. A quantitative description of membrane current and its application to conduction and excitation in nerve. *J. Physiol.* **117,** 500–544.

Hoffer, B. J., Jiggins, G. R., Oliver, A. P., and Bloom, F. E. 1973. Activation of the pathway from the locus coereleus to rat cerebellar Purkinje neurons: Pharmacological evidence of noradrenergic central inhibition. *J. Pharmacol. Exp. Ther.* **184,** 553–569.

Hofman, A. 1963. Psychotomimetic substances. *Indian J. Pharm.* **25,** 245–256.

Hokfelt, T., and Ljungdahl, A. 1972. Autoradiographic identification of cerebral and cerebellar cortical neurons accumulating labeled gamma-aminobutyric acid (^{3}H-GABA). *Exp. Brain Res.* **14,** 331-353.

Hokfelt, T. H., and Ungerstedt, U. 1969. Electron and fluorescence microscopical studies on the nucleus caudatus putamen of the rat after unilateral lesions of ascending nigro neo-striatal dopamine neurons. *Acta Physiol. Scand.* **76,** 415-426.

Hokfelt, T., Fuxe, K., and Goldstein, M. 1973. Immunohistochemical localization of aromatic L-amino acid decarboxylase (Dopa decarboxylase) in central dopamine and 5-hydroxytrypta-mine nerve cell bodies of the rat. *Brain Res.* **53,** 175-180.

Hokfelt, T., Fuxe, K., Goldstein, M., and Johansson, O. 1974. Immunohistochemical evidence for the existence of adrenaline neurons in the rat brain. *Brain Res.* **66,** 235-251.

Hokfelt, T., Kellerth, J. O., Nilsson, G., and Pernow, B. 1975. Substance P localization in the central nervous system and in some primary sensory neurons. *Science* **190,** 889-890.

Hokfelt, T., Elde, R. V., Johansson, O., Luft, R., Nilsson, G., and Arimura, A. 1976a. Immunohis-tochemical evidence for separate populations of somatostatin-containing and substance P-containing primary afferent neurons in the rat. *Neuroscience* **1,** 131-136.

Hokfelt, T., Meyerson, R., and Nilsson, G. 1976b. Immunohistochemical evidence for substance P-containing nerve endings in the human cortex. *Brain Res.* **104,** 181-186.

Holmes, G. 1939. The cerebellum of man. *Brain* **62,** 11-30.

Holton, P. 1960. Substance P concentration in degenerating nerves. In: *Polypeptides Which Affect Smooth Muscle and Blood Vessels,* edited by M. Schachter. Pergamon Press, New York, pp. 192-194.

Holton, F. A., and Holton, P. 1954. The capillary dilator substances in dry powders of spinal roots; a possible role of adenosine triphosphate in chemical transmission from nerve endings. *J. Physiol.* **126,** 124-140.

Holtz, P. 1939. Dopadecarboxylase. *Naturwissenschaften* **27,** 724.

Holtz, P., Credner, K., and Kroneberg, G. 1947. Uber des sympathicomimetische pressorische Princip des Harn ("Urosympathin"). *Arch. exp. Path. Pharmak.* **204,** 228-243.

Honegger, C. G., and Honegger, R. 1960. Volatile amines in brain. *Nature* **185,** 530-532.

Hong, J. S., Costa, E., and Yang, H. Y. T. 1976. Effects of habenular lesions on the substance P content of various brain regions. *Brain Res.* **118,** 523-525.

Hong, J., Yang, H. Y. T., Recagni, G., and Costa, E. 1977a. Projection of substance P containing neurons from neostriatum to substantia nigra. *Brain Res.* **122,** 541-544.

Hong, J. S., Yang, H. Y. T., Fratta, W., and Costa, E. 1977b. Determination of methionine enkephalin in discrete regions of rat brain. *Brain Res.* **134,** 383-386.

Hopkins, D. A., and Niessen, L. W. 1976. Substantia nigra projections to the reticular formation, superior colliculus and central gray in the rat. *Neuroscience Letters* **2,** 253-259.

Hore, J., and Vilis, T. 1976. Initiation of monkey arm movements during globus pallidus cooling. *Neuroscience Abst.* **#86,** 63.

Horn, A. S., Baumgarten, H. G., and Schlossberger, H. G. 1973. Inhibition of uptake of 5-hydroxytryptamine, noradrenaline and dopamine into rat brain homogenates by various hydroxylated tryptamines. *J. Neurochem.* **21,** 232-236.

Horn, A. S., Cuello, A. C., and Miller, R. J. 1974. Dopamine in the mesolimbic system of the rat brain: Endogenous levels and the effects of drugs on the uptake mechanism and stimulation of adenylate cyclase activity. *J. Neurochem.* **22,** 265-270.

Hosli, L., Tebecis, A. K., and Schonwetter, H. P. 1971. A comparison of the effects of monoamines on neurons of the bulbar reticular formation. *Brain Res.* **25,** 357-370.

Hruska, R. E., Bressler, R., and Yamamura, H. I. 1976. The characterization of the sodium-dependent high-affinity uptake of taurine into rat brain synaptosomes. *Neuroscience Abst.* **#834.**

Hsu, L. L., and Mandell, A. J. 1975. Studies of extrapineal N-acetylation of amines in rat brain. *Fed. Proc.* **34,** 2444.

Hubbard, J. I. 1970. Mechanism of transmitter release. *Prog. Biophys. Mol. Biol.* **21,** 33-124.

Hubbard, J. I. 1973. Microphysiology of vertebrate neuromuscular transmission. *Physiol. Rev.* **53,** 674-723.

Hubel, D. H., and Wiesel, T. N. 1962. Receptive fields, binocular interaction and functional architecture in the cat's visual cortex. *J. Physiol.* **160**, 106–154.

Hudson, D. B., Valcana, T., Bean G., and Timiras, P. S. 1976. Glutamic acid: A strong candidate as the neurotransmitter of the cerebellar granule cells. *Neurochem. Res.* **1**, 73–81.

Hughes, J., Smith, J. W., Kosterliz, H. W., Fothergill, L. A., Morgan B. A., and Morris, H. R. 1975a. Identification of methionine-enkephalin structure. *Nature* **258**, 577–579.

Hughes, J., Smith, T., Morgan, B., and Fothergill, L. 1975b. Purification and properties of enkephalin—the possible endogenous ligand for the morphine receptor. *Life Sci.* **16**, 1753–1758.

Hunt, R., and Taveau, R. de M. 1906. On the physiological action of certain choline derivatives and new methods for detecting choline. *Brit. Med. J.* **ii**, 1788–1791.

Huxley, A. F. 1959. Ion movements during nerve activity. *Ann. New York Acad. Sci.* **81**, 221–246.

Huxley, A. F., and Stämpfli, R. 1949. Evidence for saltatory conduction in peripheral myelinated nerve fibers. *J. Physiol.* **108**, 315–339.

Huxtable, R., and Barbeau, A. 1975, editors. *Taurine.* Raven Press, New York.

Hydén, H. 1965. Activation of nuclear RNA in neurons and glia in learning. In: *Anatomy of Memory,* edited by D. P. Kimble. Science and Behavior Books, Inc., Palo Alto, Calif., pp. 178–239.

Hydén, H. 1967. Biochemical changes accompanying learning. In: *The Neurosciences,* edited by G. C. Quarton, T. Melnechuk, and F. O. Schmitt. Rockefeller University Press, New York, pp. 765–771.

Ibrahim, M. Z. M. 1974. The mast cells of the mammalian central nervous system. Part I. Morphology, distribution and histochemistry. *J. Neurol. Sci.* **21**, 431–478.

Ingoglia, N. A., and Dole, V. P. 1970. Localization of D- and L-methadone after intraventricular injection into rat brains. *J. Pharmacol. Exp. Ther.* **175**, 84–87.

Ingoglia, N. A., Sturma, J. A., Lindquist, T. D., and Gaull, G. E. 1976. Axonal migration of taurine in the goldfish visual system. *Brain Res.* **115**, 535–539.

Ingvar, D. H. 1975. Patterns of brain activity revealed by measurements of regional cerebral blood flow. In: *Brain Work: The Coupling of Function, Metabolism and Blood Flow in the Brain,* edited by D. H. Ingvar and N. A. Larssen. Munksgaard, Copenhagen, pp. 397–413.

Ito, M. 1975. Cerebellar learning control of vestibulo-ocular mechanisms. In: *Mechanisms in Transmission of Signals for Conscious Behavior,* edited by T. Desiraju. Amsterdam, Elsevier, pp. 1–22.

Ito, M., and Mijishita, Y. 1975. The effects of chronic destruction of the inferior olive upon visual modification of the horizontal vestibulo-ocular reflex of rabbits. *Proc. Japan Acad.* **51**, 716–720.

Ito, M., and Oshima, T. 1964. The extrusion of sodium from cat spinal motoneurons. *Proc. R. Soc. Lond. [Biol.]* **161**, 109–131.

Ito, M., and Yoshida, M. 1964. The cerebellar-evoked monosynaptic inhibition of Deiters neurons. *Experientia* **20**, 515–516.

Ito, M., Kostyuk, P. G., and Oshima, T. 1962. Further study on anion permeability of inhibitory post-synaptic membrane of cat motoneurons. *J. Physiol.* **164**, 150–156.

Ito, M., Udo, M., and Mano, N. 1970. Long inhibitory and excitatory pathways converging onto cat reticular and Deiters' neurons and their relevance to reticulofugal axons. *J. Neurophysiol.* **33**, 210–226.

Ito, M., Shiida, T., Yagi, N., and Yamamoto, M. 1974a. The cerebellar modification of rabbits' horizontal vestibulo-ocular reflex induced by sustained head rotation combined with visual stimulation. *Proc. Japan Acad.* **50**, 85–89.

Ito, M., Shiida, T., Yagi, N., and Yamamoto, M. 1974b. Visual influence on rabbit horizontal vestibulo-ocular reflex presumably effected via the cerebellar flocculus. *Brain Res.* **65**, 170–174.

Iversen, L. L. 1967. *The Uptake and Storage of Noradrenaline in Sympathetic Nerves.* Cambridge University Press, Cambridge.

Iversen, L. L. 1973. Neuronal and extraneuronal catecholamine uptake mechanisms. In: *Frontiers in Catecholamine Research,* edited by E. Usdin and S. H. Snyder. Pergamon Press, New York, pp. 403-407.

Iversen, L. L. 1975. Dopamine receptors in the brain. *Science* **188,** 1084-1089.

Iversen, L. L., and Bloom, F. E. 1972. Studies of the uptake of ³H-Gaba and ³H-glycine in slices and homogenates of rat brain and spinal cord by electron microscopic autoradiography. *Brain Res.* **41,** 131-143.

Iversen, L. L., and Glowinski, J. 1966. Regional studies of catecholamines in various brain regions. *J. Neurochem.* **13,** 671-682.

Iversen, L. L., and Kelly, J. S. 1975. Uptake and metabolism of gammaaminobutyric acid by neurons and glial cells. *Biochem. Pharmacol.* **24,** 933-938.

Iversen, L. L., and Schon, F. E. 1973. The use of autoradiographic techniques for the identification and mapping of transmitter specific neurons in CNS. In: *New Concepts in Neurotransmitter Mechanisms,* edited by A. J. Mandell. Plenum Press, New York, pp. 153-193.

Iversen, L. L., Jessell, T., and Kanazawa, I. 1976. Release and metabolism of substance P in rat hypothalamus. *Nature* **264,** 81-83.

Jack, J. J. B., Noble, D., and Tsien, R. W. 1975. *Electric Current Flow in Excitable Cells.* Clarendon Press, Oxford.

Jackman, H., and Radulovacki, M. 1975. Equal concentrations of ³H-5HT and ³H-tryptamine after injection of ³H-tryptophan into rat hippocampus. *Neuroscience Abst.* #391.

Jacobson, M. 1970. *Developmental Neurobiology.* Holt, Rinehard & Winston, Inc., New York.

Jaim-Etcheverry, G., and Zieher, L. M. 1975. Octopamine probably coexists with noradrenaline and serotonin in vesicles of pineal adrenergic nerves. *J. Neurochem.* **25,** 915-917.

Jankowska, E., and Lindstrom, S. 1971. Morphological identification of Renshaw cells. *Acta Physiol. Scand.* **81,** 428-430.

Jansen, J. K. S., and Matthews, P. B. C. 1962. The central control of the dynamic response of muscle spindles. *J. Physiol.* **161,** 357-378.

Janssen, P. A. J. 1967. The pharmacology of haloperidol. *Int. J. Neuropsychiat.* **3,** 510-518.

Jasper, H. H., and Koyama, I. 1969. Rate of release of amino acids from the cerebral cortex of the cat as affected by brain stem and thalamic stimulation. *Can. J. Physiol. Pharmacol.* **47,** 889-905.

Jenden, D. J. 1975. Chemical methods of acetylcholine and choline analysis. In: *Cholinergic Mechanisms,* edited by P. G. Waser. Raven Press, New York, pp. 87-98.

Jerne, N. K. 1967. Antibodies and learning: Selection versus instruction. In: *The Neurosciences,* edited by G. C. Quarton, T. Melnechuk, and F. O. Schmitt. Rockefeller University Press, New York, pp. 200-205.

Joh, T. H., and Goldstein, M. 1973. Isolation and characterization of multiple forms of phenylethanolamine-N-methyl transferase. *Mol. Pharmacol.* **9,** 117-129.

Joh, T. H., Shikima, T., Pickel, V. M., and Reis, O. J. 1975. Brain tryptophan hydroxylase: Purification of antibodies to, and cellular and ultrastructural localization in serotonergic neurons of rat midbrain. *Proc. Natl. Acad. Sci. USA* **72,** 3575-3579.

Johnson, J. L. 1977. Glutamic acid as a synaptic transmitter in the dorsal sensory neuron: reconsiderations. *Life Sci.* **20,** 1637-1644.

Johnsson, G., Malmfors, T., and Sachs, C. 1975. *6-Hydroxydopamine as a Denervation Tool in Catecholamine Research.* North Holland Publishing Co., Amsterdam.

Johnston, G. A. R. 1973. Convulsions induced in 10-day-old rats by intraperitoneal injection of monosodium glutamate and related excitant amino acids. *Biochem. Pharmacol.* **22,** 137-140.

Johnston, G. A. R. 1976a. Glutamate and aspartate as transmitters in the spinal cord. *Adv. Biochem. Psychopharmacol.* **15,** 175-184.

Johnston, G. A. R. 1976b. Physiologic pharmacology of GABA and its antagonists in the vertebrate nervous system. In: *GABA in Nervous System Function,* edited by E. Roberts, T. N. Chase, and D. B. Tower. Raven Press, New York, pp. 395-412.

Johnston, G. A. R., and Balcar, V. B. 1974. Aminooxyacetic acid; a relatively non-specific inhibitor of uptake of amino acids and amines by brain and spinal cord. *J. Neurochem,* **22,** 609-610.

Johnston, G. A. R., and Iversen, L. L. 1971. Glycine uptake in rat central nervous system slices and homogenates: Evidence for different uptake systems in spinal cord and cerebral cortex. *J. Neurochem.* **18,** 1951-1961.

Johnston, J. P. 1968. Some observations upon a new inhibitor of monoamine oxidase in brain tissue. *Biochem. Pharmacol.* **17,** 1285–1297.

Jones, E. G., and Powell, T. P. S. 1969a. Electron microscopy of synaptic glomeruli in the thalamic relay nuclei of the cat. *Proc. R. Soc. Lond. [Biol.]* **172,** 153–171.

Jones, E. G., and Powell, T. P. S. 1969b. Connections of the somatic sensory cortex of the rhesus monkey. II. Contralateral cortical connections. *Brain* **92,** 717–730.

Jordan, L. M., Frederickson, R. C. A., Phillis, J. W., and Lake, N. 1972. Microelectrophoresis of 5-hydroxytryptamine: A clarification of its action on cerebral cortical neurons. *Brain Res.* **40,** 552–558.

Jouvet, M. 1973. Serotonin and sleep in the cat. In: *Serotonin and Behavior,* edited by J. Barchas and E. Usdin. Academic Press, New York, pp. 385–400.

Jouvet, M. 1975. Cholinergic mechanisms and sleep. In: *Cholinergic Mechanisms,* edited by P. G. Waser. Raven Press, New York, pp. 455–476.

Jung, M. J., Lippert, B., Metcalf, B. W., Schechter, P. J., Bohlen, P., and Sjoerdsma, A. 1977. The effect of 4-amino hex-5-ynoic acid γ-acetylenic GABA, γ-ethynyl GABA catalytic inhibitor of GABA transaminase, on brain GABA metabolism *in vivo. J. Neurochem.* **28,** 717–724.

Kaczmarek, L. K., and Davison, A. N. 1972. Uptake and release of taurine from rat brain slices. *J. Neurochem.* **19,** 2355–2362.

Kakiuchi, S., and Rall, T. S. 1968. Studies on adenosine 3′,5′ phosphate in rabbit cerebral cortex. *Mol. Pharmacol.* **4,** 367–378.

Kallman, F. J. 1938. *The Genetics of Schizophrenia.* Augustin, New York.

Kanazawa, I., and Jessell, T. 1976. Postmortem changes and regional distribution of substance P in the rat and mouse nervous system. *Brain Res.* **117,** 362–367.

Kanazawa, I., Emson, P. C., and Cuello, A. C. 1977a. Evidence for the existence of substance P-containing fibers in striato-nigral and pallido-nigral pathways in rat brain. *Brain Res.* **119,** 447–453.

Kanazawa, I., Bird, E., O'Connell, R., and Powell, D. 1977b. Evidence for a decrease in substance P content of substantia nigra in Huntington's chorea. *Brain Res.* **120,** 387–392.

Kandel, E. R., and Spencer, W. A. 1968. Cellular neurophysiological approaches in the study of learning. *Physiol. Rev.* **48,** 65–134.

Kandel, E. R., Spencer, W. A., and Brinley, F. J. 1961. Electrophysiology of hippocampal neurons. 1. Sequential invasion and synaptic organization. *J. Neurophysiol.* **24,** 225–265.

Kandel, E. R., Brunelli, M., Byrne, J., and Castellucci, V. 1975. A common presynaptic locus for the synaptic changes underlying short-term habituation and sensitization of the gill-withdrawal reflex in Aplysia. *Cold Spring Harbor Symp. Quant. Biol.* **40,** 465–482.

Karczmar, A. G. 1975. Cholinergic influences on behavior. In: *Cholinergic Mechanisms,* edited by P. G. Waser, Raven Press, New York, pp. 501–530.

Karlsson, A., Fonnum, F., Malthe-Sorenssen, D., and Storm-Mathisen, J. 1974. Effect of the convulsive agent 3-mercaptopropionic acid on the levels of GABA, other amino acids and glutamate decarboxylase in different regions of the rat brain. *Biochem. Pharmacol.* **23,** 3052–3061.

Karoum, F., Gillin, J. C., and Wyatt, R. J. 1975. Mass fragmentographic determination of some acidic and alcoholic metabolites of biogenic amines in rat brain. *J. Neurochem.* **25,** 653–658.

Kase, Y., Kataoka, M., and Miyata, T. 1967. *In vitro* production of piperidine from pipecolic acid in the presence of brain tissue. *Life Sci.* **6,** 2427–2431.

Kase, Y., Kataoka, M., and Miyata, T. 1969a. An improved method for the determination of micro amounts of piperidine in living materials. *Jap. J. Pharmacol.* **19,** 354–362.

Kase, Y., Miyata, T., Kamikawa, Y., and Kataoka, M. 1969b. Pharmacological studies on alicyclic amines. II. Central actions of piperidine, pyrrolidine and piperazine. *Jap. J. Pharmacol.* **19,** 300–314.

Kase, Y., Okano, Y., Yamanishi, Y., Kataoka, M., Kitahara, K., and Miyata, T. 1970. *In vivo* production of piperidine from pipecolic acid in the rat. *Life Sci.* **9,** 1381–1387.

Kase, Y., Yonehara, N., Okano, Y., Miyata, T., and Takahama, K. 1974. Subcellular localization of ^{3}H-piperidine in the rat brain. *Life Sci.* **15,** 1197–1202.

Kastin, A. J., Plotnikoff, N. P., Schally, A. V., and Sandman, C. A. 1976. Endocrine and CNS effects of hypothalamic peptides and MSH. In: *Reviews of Neuroscience,* vol. 2, edited by S. Ehrenpreis and I. J. Kopin. Raven Press, New York, pp. 111–148.

Kataoka, M., Kase, Y., Miyata, T., and Kawahito, E. 1970. Piperidine in the cerebellum of the dog. *J. Neurochem.* **17,** 291–292.

Kataoka, K., Nakamura, Y., and Hassler, R. 1973. Habenulo-interpeduncular tract: A possible cholinergic neuron in rat brain. *Brain Res.* **62,** 264–267.

Katz, B. 1958. Microphysiology of the neuromuscular junction. A physiological "quantum of action" at the myoneural junction. *Bull. Johns Hopkins Hosp.* **102,** 275–295.

Katz, B. 1962. The transmission of impulses from nerve to nerve and the subcellular unit of synaptic action. *Proc. R. Soc. Lond. [Biol.]* **155,** 455–479.

Katz, B. 1966. *Nerve, Muscle and Synapse.* McGraw-Hill Book Co., New York.

Katz, B. 1969. *The Release of Neural Transmitter Substances.* Liverpool University Press, Liverpool.

Katz, B., and Miledi, R. 1963. A study of spontaneous minature potentials in spinal motoneurones. *J. Physiol.* **168,** 389–422.

Katz, B., and Miledi, R. 1965a. Propagation of electric activity in motor nerve terminals. *Proc. R. Soc. Lond. [Biol.]* **161,** 453–482.

Katz, B., and Miledi, R. 1965b. Measurement of synaptic delay and time course of acetylcholine release at neuromuscular junction. *Proc. R. Soc. Lond. [Biol.]* **161,** 483–495.

Katz, B., and Miledi, R. 1965c. The effect of calcium on acetylcholine release from motor nerve terminals. *Proc. R. Soc. Lond. [Biol.]* **161,** 496–503.

Katz, B., and Miledi, R. 1966. Input-output relation of a single synapse. *Nature* **212,** 1242–1245.

Katz, B., and Miledi, R. 1967a. Tetrodotoxin and neuromuscular transmission. *Proc. R. Soc. Lond. [Biol.]* **167,** 8–22.

Katz, B., and Miledi, R. 1967b. Release of acetylcholine from nerve endings by graded electric pulses. *Proc. R. Soc. Lond. [Biol.]* **167,** 23–38.

Katz, B., and Miledi, R. 1967c. The timing of calcium action during neuromuscular transmission. *J. Physiol.* **189,** 535–544.

Katz, B., and Miledi, R. 1968a. The role of calcium in neuromuscular facilitation. *J. Physiol.* **195,** 481–492.

Katz, B., and Miledi, R. 1968b. The effect of local blockage of motor nerve terminals. *J. Physiol.* **199,** 729–741.

Katz, B., and Miledi, R. 1969a. Tetrodotoxin-resistant electric activity in presynaptic terminals. *J. Physiol.* **203,** 459–487.

Katz, B., and Miledi, R. 1969b. Spontaneous and evoked activity of motor nerve endings in calcium Ringer. *J. Physiol.* **203,** 689–706.

Katz, B., and Miledi, R. 1972. The statistical nature of the acetylcholine potential and its molecular components. *J. Physiol.* **224,** 665–699.

Katz, B., and Miledi, R. 1973. The binding of acetylcholine to receptors and its removal from the synaptic cleft. *J. Physiol.* **231,** 549–574.

Kawaguchi, S., Yamamoto, Y., Mizuno, N., and Iwahori, N. 1975. The role of climbing fibers in the development of Purkinje cell dendrites. *Neuroscience Letters* **1,** 301–304.

Kebabian, J. W., Petzold, G. L., and Greengard, P. 1972. Dopamine-sensitive adenyl cyclase in caudate nucleus of rat brain and its similarity to the "dopamine receptor." *Proc. Natl. Acad. Sci. USA* **69,** 2145–2149.

Kebabian, J. W., Steiner, A. L., and Greengard, P. 1975. Muscarinic cholinergic regulation of cyclic guanosine 3′,5′-monophosphate in autonomic ganglia: Possible role in synaptic transmission. *J. Pharmacol. Exp. Ther.* **193,** 474–488.

Kehoe, J. 1972. Ionic mechanisms of a two-component cholinergic inhibition in Aplysia neurones. *J. Physiol.* **225,** 85–114.

Kehr, W., Carlsson, A., and Lindqvist, N. 1975. Biochemical aspects of dopamine agonists. In: *Advances in Neurology,* vol. 9, edited by D. B. Calne, T. N. Chase, and A. Barbeau. Raven Press, New York, pp. 185–195.

Kemp, J. M., and Powell, T. P. S. 1971. The structure of the caudate nucleus of the cat: Light and electron microscopy. *Philos. Trans. R. Soc. Lond. [Biol. Sci.]* **262,** 383–401.

Kety, S. S. 1961. Sleep and the energy metabolism of the brain. In: *The Nature of Sleep,* Ciba

Found. Symp. edited by G. E. W. Wolstenholme and M. O'Connor. J. & A. Churchill Ltd., London, pp. 375-385.

Kety, S. S. 1974. From rationalization to reason. *Am. J. Psychiat.* **131**, 957-963.

Keynes, R. D. 1976. Organization of ionic channels in nerve membranes. In: *The Nervous System,* vol. 1, *The Basic Neurosciences,* edited by D. B. Tower and R. O. Brady. Raven Press, New York, pp. 165-175.

Keynes, R. D., and Rojas, E. 1974. Kinetics and steady-state properties of the charged system controlling sodium conductance in the squid giant axon. *J. Physiol.* **239**, 393-434.

Killam, K. F., and Bain, J. A. 1957. Convulsant hydrazides. I. *In vitro* and *in vivo* inhibition of vitamine B_6 enzymes by convulsant hydrazides. *J. Pharmacol. Exp. Ther.* **119**, 255-262.

Kimura, D. 1967. Functional asymmetry of the brain in dichotic listening. *Cortex* **3**, 163-178.

Kinscherf, D. A., Change, M. M., Rubin, E. H., Schneider, D. P., and Ferrendelli, J. A. 1976. Comparison of the effects of depolarizing agents and neurotransmitters on regional CNS cyclic GMP levels in various animals. *J. Neurochem.* **26**, 527-530.

Kitai, S. J., Wagner, A., Precht, W., and Ohne, T. 1975. Nigro-caudate and caudate-nigral relationship: an electrophysiological study. *Brain Res.* **85**, 44-48.

Klee, W. A. 1977. Endogenous opiate peptides. In: *Peptides in Neurobiology,* edited by H. Gainer. Raven Press, New York, pp. 375-393.

Klein, D. C., and Yuwiler, A. 1973. β-Adrenergic regulation of indole metabolism in the pineal gland. In: *Frontiers in Catecholamine Research,* edited by E. Usdin and S. H. Snyder. Pergamon Press, New York, pp. 321-325.

Kluver, H., and Bucy, P. 1939. Preliminary analysis of functions of the temporal lobes in monkeys. *Arch. Neurol. Psych.* **42**, 979-1000.

Knott, P. J., Marsden, C. A., and Curzon, G. 1974. Comparative studies of brain 5-hydroxytryptamine and tryptamine. *Adv. Biochem. Psychopharmacol.* **11**, 109-114.

Kobayashi, R. M., Brown, M., and Vale, W. 1977. Regional distribution of neurotensin and somatostatin in rat brain. *Brain Res.* **126**, 584-588.

Koe, B. K., and Weissman, A. 1966. p-Chlorophenylalanine: A specific depletor of brain serotonin. *J. Pharmacol. Exp. Ther.* **154**, 499-516.

Koelle, G. B. 1970. Anticholinesterase agents. In: *The Pharmacological Basis of Therapeutics,* 4th edition, edited by L. S. Goodman and A. Gilman. The MacMillan Co., London, pp. 442-465.

Koike, H., Okada, Y., Oshima, T., and Takashi, K. 1968. Accommodative behavior of cat pyramidal tract cells investigated with intracellular injection of currents. *Exp. Brain Res.* **5**, 173-188.

Kojima, T., Saito, K., and Kakimi, S. 1975. *An Electron Microscopic Atlas of Neurons.* University of Tokyo Press, Tokyo.

Konishi, S., and Otsuka, M. 1974. The effects of substance P and other peptides on spinal neurons of the frog. *Brain Res.* **65**, 397-410.

Kornhuber, H. H. 1973. Neural control of input into long-term memory: Limbic system and amnestic syndrome in man. In: *Memory and Transfer of Information,* edited by H. P. Zippel. Plenum Press, New York, pp. 1-22.

Kornhuber, H. H. 1974. Cerebral cortex, cerebellum and basal ganglia: An introduction to their motor function. In: *The Neurosciences: Third Study Program,* edited by F. O. Schmitt and F. G. Worden, MIT Press, Cambridge, Mass., pp. 267-280.

Kravitz, E. A., Kuffler, S. W., and Potter, D. D. 1963. Gamma-aminobutyric acid and other blocking compounds in Crustacea. III. Their relative concentrations in separated motor and inhibitory axons. *J. Neurophysiol.* **26**, 739-751.

Kravitz, E. A., Slater, C. R., Takahashi, K., Bownds, M. D., and Grossfeld, R. N. 1970. Excitatory transmission in invertebrates—glutamate as a potential neuromuscular transmitted compound. In: *Excitatory Synaptic Mechanisms,* edited by P. Andersen and J. K. S. Jansen. Universitetsforlaget, Oslo, pp. 85-93.

Krnjevic, K. 1974. Chemical nature of synaptic transmission in vertebrates. *Physiol. Rev.* **54**, 418-540.

Krnjevic, K., and Miledi, R. 1958. Acetylcholine in mammalian neuromuscular transmission. *Nature* **182**, 805-806.

Krnjevic, K., and Morris, M. E. 1974. An excitatory action of substance P on cuneate neurons. *Can. J. Physiol. Pharmacol.* **52**, 736-744.

Krnjevic, K., and Phillis, J. W. 1963. Iontophoretic studies on neurones in the mammalian cerebral cortex. *J. Physiol.* **165**, 274–304.

Krnjevic, K., and Puil, E. 1975. Electrophysiological studies on actions of taurine. In: *Taurine*, edited by R. Huxtable and A. Barbeau. Raven Press, New York, pp. 179–190.

Krnjevic, K., and Schwartz, S. 1966. Is γ-aminobutyric acid an inhibitory transmitter? *Nature* **211**, 1372–1374.

Kuczenski, R. J., and Mandell, A. J. 1972. Regulatory properties of soluble and particulate rat brain tyrosine hydroxylase. *J. Biol. Chem.* **247**, 3114–3122.

Kuffler, S. W. 1942. Electric potential changes at an isolated nerve-muscle junction. *J. Neurophysiol.* **5**, 18–26.

Kuffler, S. W. 1966. Physiological properties of vertebrate and invertebrate neuroglial cells and the movement of substances through the nervous system. *Proc. R. Soc. Lond.* [*Biol.*] **168**, 1–21.

Kuffler, S. W. 1973. The single-cell approach in the visual system and the study of receptive fields. *Invest. Ophthalmol.* **12**, 794–813.

Kuffler, S. W., and Edwards, C. 1958. Mechanism of gamma-aminobutyric acid (GABA) and its relation to synaptic inhibition. *J. Neurophysiol.* **21**, 589–610.

Kuffler, S. W., and Nicholls, J. G. 1976. *From Neuron to Brain.* Sinauer Associates, Sunderland, Mass.

Kuffler, S. W., and Yoshikami, D. 1975a. The distribution of acetylcholine sensitivity at the postsynaptic membrane of vertebrate skeletal twitch muscles: Iontophoretic mapping in the micron range. *J. Physiol.* **244**, 703–730.

Kuffler, S. W., and Yoshikami, D. 1975b. The number of transmitter molecules in a quantum: An estimate from iontophoretic application of acetylcholine at the neuronmuscular synapse. *J. Physiol.* **251**, 465–482.

Kuhar, M. J. 1973. Neurotransmitter uptake: A tool in identifying neurotransmitter-specific pathways. *Life Sci.* **13**, 1623–1634.

Kuhar, M. J., Pert, C. B., and Snyder, S. H. 1973. Regional distribution of opiate receptor binding in monkey and human brain. *Nature* **245**, 447–450.

Kuhn, R. 1958. The treatment of depressive states with G22355 (imipramine hydrochloride). *Am. J. Psychiat.* **115**, 459–464.

Kuno, M. 1964. Quantal components of excitatory synaptic potentials in spinal motoneurones. *J. Physiol.* **175**, 81–99.

Kuno, M. 1969. Transmitter action and multiple discharge. In: *Basic Mechanisms of the Epilepsies*, edited by H. H. Jasper, A. A. Ward, and A. Pope. Little, Brown and Co., Boston, pp. 130–135.

Kuriyama, K., Sisken, B., Haber, B., and Roberts, E. 1968. The γ-aminobutyric acid system in rabbit retina. *Brain Res.* **9**, 165–168.

Kuroda, Y., and McIlwain, H. 1974. Uptake and release of [^{14}C]adenine derivatives at beds of mammalian cortical synaptosomes in a superfusion system. *J. Neurochem.* **22**, 691–699.

Kwaitkowski, H. 1943. Histamine in nervous tissue. *J. Physiol.* **102**, 32–41.

Kwok, R. H. M. 1968. Chinese restaurant syndrome. *N. Engl. J. Med.* **278**, 796.

Lahdesmaki, P., Pasula, M., and Oja, S. S. 1975. Effect of electrical stimulation and chlorpromazine on the uptake and release of taurine, gamma-aminobutyric acid and glutamic acid in mouse brain synaptosomes. *J. Neurochem.* **25**, 675–680.

Lakshmanan, J., and Padmanaban, G. 1974. Effect of α-oxalyl-L-α,β-diaminopropionic acid on glutamate uptake by synaptosomes. *Nature* **249**, 469–471.

Landgren, S., Phillips, C. G., and Porter, R. 1962. Cortical fields of origin of the monosynaptic pyramidal pathways to some alpha motoneurones of the baboon's hand and forearm. *J. Physiol.* **161**, 112–125.

Lane, A. C., Rance, M. J., and Walter, D. S. 1977. Subcellular localization of leucine-enkephalin-hydrolysing activity in rat brain. *Nature* **269**, 75–76.

Langley, J. N. 1921. *The Autonomic Nervous System*, vol. 1. W. Heffer and Sons, Cambridge.

Larson, B., Miller, S., and Oscarsson, O. 1969. Termination and functional organization of the dorso-lateral spino-olivocerebellar path. *J. Physiol.* **203**, 611–640.

Lashley, K. S. 1950. In search of the engram. *Symp. Soc. Exp. Biol.* **4**, 454–482.

LaVail, J. H. 1975. The retrograde transport method. *Fed. Proc.* **34**, 1618–1624.

Lee, T. P., Kuo, J. F., and Greengard, P. 1972. Role of muscarinic cholinergic receptors in regulation of guanosine 3′,5′-cyclic monophosphate content in mammalian brain, heart muscle and intestinal smooth muscle. *Proc. Natl. Acad. Sci. USA* **69**, 3287–3291.

Leeman, S. E., and Mroz, E. A. 1974. Substance P. *Life Sci.* **15**, 2033–2044.

Leicht, R., and Schmidt, R. F. 1978. Somatotropic studies on the vermal cortex of the cerebellar anterior lobe of unanesthetized cats. *Exp. Brain Res.* **27**, 479–490.

Lembeck, F. 1953a. 5-Hydroxytryptamine in a carcinoid tumor. *Nature* **172**, 910–911.

Lembeck, F. 1953b. Zur Frage der zentralen Ubetragung afferenter Impulse. *Arch. Exper. Path. u. Pharmakol.* **219**, 197–213.

Leonardelli, J., and Poulain, P. 1977. About a ventral preopticoamygdaloid pathway in the guinea pig. *Brain Res.* **124**, 538–543.

Lerner, A. B., Chase, J. D., and Heinzelman, R. V. 1959. Structure of melatonin. *J. Am. Chem. Soc.* **81**, 6084–6085.

Levi-Montalcini, R. 1964. Growth control of nerve cells by protein factor and its antiserum. *Science* **143**, 105–110.

Levi-Montalcini, R., and Angeletti, P. U. 1968. Nerve growth factor. *Physiol. Rev.* **48**, 534–569.

Levi-Montalcini, R., and Cohen, S. 1960. Effects of the extracts of the mouse submaxillary salivary glands on the sympathetic system of mammals. *Ann. N.Y. Acad. Sci.* **85**, 324–341.

Levy, J. 1973. Lateral specialization of the human brain: Behavioral manifestations and possible evolutionary basis. In: *The Biology of Behavior,* edited by J. A. Kiger, Jr., Oregon State University Press, Corvallis, Oregon, pp. 159–180.

Levy-Agresti, J., and Sperry, R. W. 1968. Differential perceptual capacities in major and minor hemispheres. *Proc. Natl. Acad. Sci. USA* **61**, 1151.

Lew, J. Y., Matsumoto, Y., Pearson, J., Goldstein, M., Hokfelt, T., and Fuxe, K. 1977. Localization and characterization of phenylethanolamine-n-methyltransferase in the brain of various mammalian species. *Brain Res.* **119**, 199–210.

Lewis, P. R., Shute, C. C. D., and Silver, A. 1967. Confirmation from choline acetylase of a massive cholinergic innervation to the rat hippocampus. *J. Physiol.* **191**, 215–224.

Lewis, T., and Grant, R. T. 1927. *The Blood Vessels of the Human Skin and their Responses.* Shaw & Sons, London.

Libet, B. 1973. Electrical stimulation of cortex in human subjects, and conscious sensory aspects. In: *Handbook of Sensory Physiology,* vol. 2, edited by A. Iggo. Springer-Verlag, New York, pp. 743–790.

Libet, B. Kobayski, H., and Tanaka, T. 1975. Synaptic coupling with the production and storage of a neuronal memory trace. *Nature* **258**, 155–157.

Liley, A. W. 1956. The quantal components of the mammalian end-plate potential. *J. Physiol.* **133**, 571–587.

Lindvall, O., and Bjorklund, A. 1974. The organization of the ascending catecholamine neuron system in the rat brain. *Acta Physiol. Scand.* **412**, suppl. 1–48.

Lipinski, J. F., Schaumberg, H. H., and Baldessarini, R. J. 1973. Regional distribution of histamine in the human brain. *Brain Res.* **52**, 403–408.

Ljungdahl, A., and Hokfelt, T. 1973. Autoradiographic uptake patterns of [³H]-GABA and [³H]glycine in central nervous tissues with special reference to the cat spinal cord. *Brain Res.* **62**, 587–595.

Lloyd, D. P. C. 1946. Facilitation and inhibition of spinal motoneurons. *J. Neurophysiol.* **9**, 421–438.

Lloyd, K. G. 1975. Special chemistry of the basal ganglia. I. Monoamines. *Pharmacol. Therap.* **1**, 49–61.

Loewi, O. 1933. Problems connected with the principle of humoral transmission of nervous impulses. *Proc. R. Soc. Lond. [Biol.]* **118**, 299–316.

Loewi, O. 1960. An autobiographic sketch. *Persp. Biol. Med.* **4**, 3–25.

Logan, W. J., and Snyder, S. H. 1971. Unique high affinity uptake systems for glycine, glutamic and aspartic acids in central nervous tissue of the rat. *Nature* **234**, 297–299.

Logan, W. J., and Snyder, S. H. 1972. High affinity uptake systems for glycine, glutamic and aspartic acids in synaptosomes of rat central nervous tissue. *Brain Res.* **42**, 413–431.

Lombard, J. P., Gilbert, J. G., and Donofrio, A. F. 1955. The effects of glutamic acid upon the intelligence, social maturity and adjustment of a group of mentally retarded children. *Am. J. Ment. Def.* **60**, 127-132.

Lombardini, J. B. 1975. Regional and subcellular studies on taurine in the rat central nervous system. In: *Taurine,* edited by R. Huxtable and A. Barbeau. Raven Press, New York, pp. 311-326.

Lømo, T. 1971. Patterns of activation in a monosynaptic cortical pathway: The perforant path input to the dentate area of the hippocampal formation. *Exp. Brain Res.* **12**, 18-45.

Loomer, H. P., Saunders, J. C., and Kline, N. S. 1957. A clinical and pharmacodynamic evaluation of iproniazid as a psychic energizer. *Psychiat. Res. Rep. Wash.* #8, pp. 129-141.

Lovett-Doust, J. 1955. Histamine and schizophrenia. *AMA Arch. Neurol. Psych.* **74**, 137-143.

Lund, R. D. 1969. Synaptic patterns of the superficial layer of the superior colliculus of the rat. *J. Comp. Neurol.* **135**, 179-208.

Lund, R. D., and Hauschka, S. D. 1976. Transplanted neural tissue develops connections with host rat brain. *Science* **193**, 582-584.

Lund, R. D., and Lund, J. S. 1971. Synaptic adjustment after deafferentation of the superior colliculus of the rat. *Science* **171**, 804-807.

Lund-Karlsen, R., and Fonnum, F. 1978. Evidence for glutamate as a transmitter in corticofugal fibers in the dorsal lateral geniculate body and the superior colliculus in rats. *Brain Res.* (in press).

Lux, H. D. 1971. Ammonium and chloride extrusion: Hyperpolarizing synaptic inhibition in spinal motoneurones. *Science* **173**, 555-557.

Lux, H. D., Loracher, C., and Neher, E. 1970. Action of ammonium on postsynaptic inhibition of cat spinal motoneurones. *Exp. Brain Res.* **11**, 431-447.

Lynch, G. S., Lucas, P. A., and Deadwyler, S. A. 1972. The demonstration of acetylcholinesterase containing neurons within the caudate nucleus of the rat. *Brain Res.* **45**, 617-621.

McBride, W. J., Dayle, E., and Aprison, M. H. 1973. Interconversion of glycine and serine in a synaptosomal fraction isolated from the spinal cord, medulla oblongata, telencephalon and cerebellum of the rat. *J. Neurobiol.* **4**, 557-566.

McBride, W. J., Aprison, M. H., and Kusano, K. 1976. Contents of several amino acids in the cerebellum, brain stem and cerebrum of the "staggerer," "weaver" and "nervous" neurologically mutant mice. *J. Neurochem.* **26**, 867-870.

McCann, S. M., and Ojeda, S. R. 1976. Synaptic transmitters involved in the release of hypothalamic releasing and inhibiting hormones. In: *Reviews of Neuroscience,* vol. 2, edited by S. Ehrenpreis and I. J. Kopin. Raven Press, New York, pp. 91-109.

McCann, S. M., Fawcett, C. P., and Krulich, L. 1974. Hypothalamic hypophysial releasing and inhibiting hormones. In: *Endocrine Physiology,* vol. 5, edited by S. M. McCann. MTP Press, Lancaster, Pa., pp. 31-65.

McDowell, F. H., and Barbeau, A., editors. 1964. *Second Canadian-American Conference on Parkinson's Disease,* vol. 5, *Advances in Neurology.* Raven Press, New York.

McEwen, B. S. 1976. Endocrine effects on the brain and their relationships to behavior. In: *Basic Neurochemistry* 2nd edition, edited by G. J. Siegel, R. W. Albers, R. Katzman, and B. W. Agranoff. Little Brown and Co., Boston, pp. 737-764.

McGaugh, J. L. 1969. Facilitation of memory storage processes. In: *The Future of the Brain Sciences,* edited by S. Bogoch. Plenum Press, New York, pp. 355-370.

McGeer, E. G., and McGeer, P. L. 1973. Amino acid hydroxylase inhibitors. In: *Metabolic Inhibitors,* edited by R. M. Hochster, M. Kates, and J. H. Quastel. Academic Press, New York, pp. 45-101.

McGeer, E. G., and McGeer, P. L. 1976a. Duplication of biochemical changes of Huntington's chorea by intrastriatal injections of glutamic and kainic acids. *Nature* **263**, 517-519.

McGeer, E. G., and McGeer, P. L. 1976b. Neurotransmitter metabolism in the aging brain. In: *Neurobiology of Aging,* vol. 3, edited by R. D. Terry and S. Gershon. Raven Press, New York, pp. 389-404.

McGeer, E. G., and McGeer, P. L. 1977. Possible changes in striatal and limbic cholinergic systems in schizophrenia. *Arch. Gen. Psychiat.* **34**, 1319-1323.

McGeer, E. G., McGeer, P. L., and McLennan, H. 1961. The inhibitory action of 3-hydroxytyramine, γ-aminobutyric acid (GABA) and some other compounds towards the crayfish stretch receptor neuron. *J. Neurochem.* **8**, 36–49.

McGeer, E. G., Ling, G. M., and McGeer, P. L. 1963. Conversion of tyrosine to catecholamines by cat brain *in vivo. Biochem. Biophys. Res. Comm.* **13**, 291–296.

McGeer, E. G., Gibson, S., and McGeer, P. L. 1967. Some characteristics of brain tyrosine hydroxylase. *Can. J. Biochem.* **45**, 1557–1563.

McGeer, E. G., Peters, D. A. V., and McGeer, P. L. 1968. Inhibition of rat brain tryptophan hydroxylase by 6-halotryptophans. *Life Sci.* **7**, 605–616.

McGeer, E. G., Wada, J. A., Terao, A., and Jung, E. 1969. Amine synthesis in various brain regions with caudate or septal lesions. *Exp. Neurol.* **24**, 277–284.

McGeer, E. G., Searl, K., and Fibiger, H. C. 1974. Chemical specificity of dopamine transport in the nigro-neostriatal projection. *J. Neurochem.* **24**, 283–288.

McGeer, E. G., Hattori, T., and McGeer, P. L. 1975. Electron microscopic localization of labeled norepinephrine transported in nigro-striatal neurons. Brain Res. **86**, 478–482.

McGeer, E. G., Innanen, V. T., and McGeer, P. L. 1976. Evidence on the cellular localization of adenyl cyclase in the neostriatum. *Brain Res.* **118**, 356–358.

McGeer, P. L. 1963. Central amines and extrapyramidal functions. *J. Neuropsych.* **4**, 247–250.

McGeer, P. L., and McGeer, E. G. 1964. Formation of adrenaline by brain tissue. *Biochem. Biophys. Res. Commun.* **17**, 502–507.

McGeer, P. L., and McGeer, E. G. 1975. Evidence for glutamic acid decarboxylase containing interneurons in the neostriatum. *Brain Res.* **91**, 331–335.

McGeer, P. L., and McGeer, E. G. 1976. Enzyme associated with the metabolism of catecholamines, acetylcholine and GABA in human controls and patients with Parkinson's disease and Huntington's chorea. *J. Neurochem.* **26**, 65–76.

McGeer, P. L., Boulding, J. C., Gibson, W. C., and Foulkes, R. G. 1961. Drug-induced extrapyramidal reactions. *JAMA* **177**, 665–670.

McGeer, P. L., McGeer, E. G., and Wada, J. A. 1963. Central aromatic amines and behaviour. *Arch. Neurol.* **9**, 81–89.

MeGeer, P. L., Bagchi, S. P., and McGeer, E. G. 1965. Subcellular localization of tyrosine hydroxylase in beef caudate nucleus. *Life Sci.* **4**, 1859–1867.

McGeer, P. L., McGeer, E. G., Fibiger, H. C., and Wickson, V. 1971. Neostriatal choline acetylase and acetylcholinesterase following selective brain lesions. *Brain Res.* **35**, 308–314.

McGeer, P. L., McGeer, E. G., and Fibiger, H. C. 1973. Choline acetylase and glutamic acid decarboxylase in Huntington's chorea. *Neurol.* **23**, 912–917.

McGeer, P. L., Fibiger, H. C., Maler, L., Hattori, T., and McGeer, E. G. 1974a. Evidence for descending pallido-nigral GABA-containing neurons. In: *Advances in Neurology*, vol. 5, edited by F. H. McDowell and A. Barbeau. Raven Press, New York, pp. 153–160.

McGeer, P. L., McGeer, E. G., Singh, V. K., and Chase, W. H. 1974b. Choline acetyltransferase localization in the central nervous system by immunohistochemistry. *Brain Res.* **81**, 373–379.

McGeer, P. L., McGeer, E. G., Fibiger, H. C., Hattori, T., Singh, V. K., and Maler, L. 1974c. Biochemical neuroanatomy of the basal ganglia in neurohumoral coding of brain function. In: *Advances in Behavioral Biology*, vol. 10, edited by R. D. Myers and R. R. Drucker-Colin, Plenum Press, New York, pp. 27–48.

McGeer, P. L., Grewaal, D. S., and McGeer, E. G. 1974d. Influence of noncholinergic drugs on rat striatal acetylcholine levels. *Brain Res.* **80**, 211–217.

McGeer, P. L., Hattori, T., and McGeer, E. G. 1975. Chemical and autoradiographic analysis of γ-aminobutyric acid transport in Purkinje cells of the cerebellum. *Exp. Neurol.* **47**, 26–41.

McGeer, P. L., McGeer, E. G., and Suzuki, J. S. 1976. Aging and extrapyramidal function. *Arch. Neurol.* **34**, 33–35.

McGeer, P. L., McGeer, E. G., Scherer, U., and Singh, K. 1977. A glutamatergic cortico-striatal path? *Brain Res.* **128**, 369–373.

McIlwain, H. 1973. Adenosine in neurohumoral and regulatory roles in the brain. In: *Central Nervous System: Studies on Metabolic Regulation and Function*, edited by E. Genazzani and H. Herken. Springer, Berlin, pp. 3–11.

McIntosh, C. H. S., and Plumer, D. T. 1973. Multiple forms of acetylcholinesterase from pig brain. *Biochem. J.* **133,** 655–665.

McIntosh, J. C., and Cooper, J. R. 1964. Function of N-acetylaspartic acid in the brain: Effect of certain drugs. *Nature* **203,** 658.

McKhann, G. M., Albers, R. W., Sokoloff, L., Michelsen, O., and Tower, D. 1960. The quantitative significance of the gamma-aminobutyric acid pathway in cerebral oxidative metabolism. In: *Inhibition in the Nervous System and Gamma-Aminobutyric Acid,* edited by E. Roberts *et al.* Pergamon Press, New York, pp. 169–181.

McLaughlin, B. J., Wood, J. G., Saito, K., Barber, R., Vaughn, J. E., Roberts, E., and Wu, J. Y. 1974. The fine structural localization of glutamate decarboxylase in synaptic terminals of rodent cerebellum. *Brain Res.* **76,** 377–391.

MacLean, P. D. 1958. Contrasting functions of limbic and neocortical systems of the brain and their relevance to psycho-physiological aspects of medicine. *Am. J. Med.* **25,** 611–626.

McLennan, H., and York, D. H. 1967. The action of dopamine on neurons of the caudate nucleus. *J. Physiol. (Lond.)* **189,** 393–402.

Maekawa, K., and Simpson, J. I. 1973. Climbing fiber responses evoked in the vestibulocerebellum of rabbit from visual system. *J. Neurophysiol.* **36,** 649–666.

Maeno, T. 1966. Analysis of sodium and potassium conductances in the procaine end-plate potential. *J. Physiol.* **183,** 592–606.

Magoun, H. W. 1958. *The Waking Brain.* Charles C Thomas, Springfield, Ill.

Main, A. R. 1976. Structure and inhibitors of cholinesterase. In: *Biology of Cholinergic Function,* edited by A. M. Goldberg and I. Hanin. Raven Press, New York, pp. 269–377.

Maitre, L., Waldmeier, P. C., Baumann, P. A., and Stachelin, M. 1974. Effect of Maprotiline, a new antidepressant drug on serotonin uptake. In: *Advances in Biochemical Psychopharmacology* vol. 10, edited by E. Costa, G. L. Gessa, and M. Sandler. Raven Press, New York, pp. 297–309.

Malliani, A., and Purpura, D. P. 1967. Intracellular studies of the corpus striatum. II. Patterns of synaptic activities in lenticular and entopeduncular neurons. *Brain Res.* **6,** 341–354.

Marchbanks, R. 1975. Biochemistry of cholinergic neurons. In: *Handbook of Psychopharmacology,* vol. 3, *Biochemistry of Biogenic Amines,* edited by L. L. Iversen, S. D. Iversen, and S. H. Snyder. Plenum Press, New York, pp. 247–326.

Margolis, F. L. 1974. Carnosine in the primary olfactory pathway. *Science* **184,** 909–911.

Margolis, F. L., Roberts, N., Ferriero, D., and Feldman, J. 1974. Denervation in the primary olfactory pathway of mice: Biochemical and morphological effects. *Brain Res.* **81,** 469–483.

Mark, V. H., and Ervin, F. R. 1970. *Violence and the Brain.* Harper and Row, New York.

Marks, N., and Stern, F. 1974. Enzymatic mechanisms for the inactivation of luteinizing hormone-releasing factor (LH-RH). *Biochem. Biophys. Res. Commun.* **61,** 1458–1463.

Marr, D. 1969. A theory of cerebellar cortex. *J. Physiol.* **202,** 437–470.

Marr, D. 1970. A theory for cerebral neocortex. *Proc. R. Soc. Lond. [Biol.]* **176,** 161–234.

Marsden, C. A., and Curzon, G. 1974. Effects of lesions and drugs on brain tryptamine. *J. Neurochem.* **23,** 1171–1176.

Marshall, J., and Voaden, M. 1974. An investigation of the cells incorporating [³H]-GABA and [³H]glycine in the isolated retina of the rat. *Exp. Eye Res.* **18,** 367–370.

Marslen-Wilson, W. D., and Teuber, H. L. 1975. Memory for remote events in anterograde amnesia: Recognition of public figures from news photographs. *Neurophysologia* **13,** 353–364.

Martin, W. R., Sloan, J. W., Christian, S. T., and Clements, T. H. 1972. Brain levels of tryptamine. *Psychopharmacol.* **24,** 331–346.

Martin, W. R., Sloan, J. W., Buchwald, W. F., and Bridges, S. R. 1974. The demonstration of tryptamine in regional perfusates of rat brain. *Psychopharmacol.* **37,** 189–198.

Martin, W. R., Sloan, J. W., Buchwald, W. F., and Clements, T. H. 1975. Neurochemical evidence for tryptaminergic ascending and descending pathways in the spinal cord of the dog. *Psychopharmacol.* **43,** 131–134.

Martres, M. P., Baudry, M., and Schwartz, J. C. 1975. Histamine synthesis in the developing rat brain: Evidence for a multiple compartmentation. *Brain Res.* **83,** 261–275.

Matsui, Y., and Deguchi, T. 1977. Effects of GABAculine, a new potent inhibitor of gamma-aminobutyrate transaminase, on the brain gamma-aminobutyrate content and convulsions in mice. *Life Sci.* **20**, 1291–1296.

Matthews, P. B. C. 1972. *Mammalian Muscle Receptors and Their Central Actions.* Williams and Wilkins, Baltimore.

Mehler, W. R., and Nauta, W. J. 1974. Connections of the basal ganglia and of the cerebellum. *Confin. Neurol.* **36**, 205–222.

Mendell, L. M., and Henneman, E. 1971. Terminals of single Ia fibers: Location, density and distribution within a pool of 300 homogeneous motoneurons. *J. Neurophysiol.* **34**, 171–187.

Menon, M. K., Clark, W. G., and Aures, D. 1971. Effect of thiazol-4-ylmethoxyamine, a new inhibitor of histamine biosynthesis on brain histamine, monoamine levels and behavior. *Life Sci.* **10**, 1097–1109.

Meyer, R. L., and Sperry, R. W. 1974. Explanatory models for neuroplasticity in retinotectal connections. In: *Plasticity and Recovery of Function in the Central Nervous System,* edited by D. G. Stein, J. I. Rosen, and N. Butters, Academic Press, New York, pp. 45–63.

Miale, I. L., and Sidman, R. L. 1961. An autoradiographic analysis of histogenesis in the mouse cerebellum. *Exp. Neurol.* **4**, 277–296.

Miledi, R. 1967. Spontaneous synaptic potentials and quantal release of transmitter in the stellate ganglion of the squid. *J. Physiol.* **192**, 379–406.

Miledi, R. 1969. Transmitter action in the giant synapse of the squid. *Nature* **223**, 1284–1286.

Miledi, R. 1973. Transmitter release induced by injection of calcium ions into nerve terminals. *Proc. R. Soc. Lond. [Biol.]* **183**, 421–425.

Miledi, R., and Slater, C. R. 1966. The action of calcium on neuronal synapses in the squid. *J. Physiol.* **184**, 473–498.

Miledi, R., and Slater, C. R. 1970. On the degeneration of rat neuromuscular junctions after nerve section. *J. Physiol.* **207**, 507–528.

Milner, B. 1966. Amnesia following operation on the temporal lobes. In: *Amnesia,* edited by C. W. M. Whitty and O. L. Zangwill. Butterworths, London, pp. 109–133.

Milner, B. 1972. Disorders of learning and memory after temporal-lobe lesions in man. *Clin. Neurosurg.* **19**, 421–446.

Milner, B. 1974. Hemispheric specialization: Scope and limits. In: *The Neurosciences: Third Study Program,* edited by F. O. Schmitt and F. G. Worden. MIT Press, Cambridge, Mass., pp. 75–89.

Milner, B., Taylor, L., and Sperry, R. W. 1968. Lateralized suppression of dichotically presented digits after commissural section. *Science* **161**, 184–186.

Minnich, J. L. Donaldson, J., and Barbeau, A. 1972. Angiotensin-induced decrease in brain region norepinephrine. 5th Intl. Cong. of Pharmacology.

Mishra, R. K., Makman, M. H., Ahn, H. S., Drorkin, B., Horowitz, S. G., Keehn, E., and Demirjian, C. 1976. Differences in primate brain regions in relative potency for antagonism of dopamine stimulated adenylate cyclases by neuroleptic drugs and possible implications for localization of antipsychotic activity. *Neuroscience Abst.* #1134.

Miyata, Y., and Otsuka, M. 1972. Distribution of γ-aminobutyric acid in cat spinal cord and the alteration produced by local ischemia. *J. Neurochem.* **19**, 1833–1834.

Mogensen, G. T., and Huang, Y. H. 1973. The neurobiology of motivated behavior. *Prog. Neurobiol.* **1**, 53–83.

Molinoff, P., and Axelrod, J. 1969. Octopamine: Normal occurrence in sympathetic nerves of the rat. *Science* **164**, 428–429.

Molinoff, P. B., and Axelrod, J. 1972. Distribution and turnover of octopamine in tissues. *J. Neurochem.* **19**, 157–163.

Molinoff, P. B., Landsberg, L., and Axelrod, J. 1969. An enzymatic assay for octopamine and other β-hydroxylated phenylethylamines. *J. Pharmacol. Exp. Ther.* **170**, 253–261.

Molliver, M. E., Kostović, I., and Van der Loos, H. 1973. The development of synapses in cerebral cortex of human fetus. *Brain Res.* **50**, 403–407.

Morest, D. K. 1971. Dendrodendritic synapses of cells that have axons. The fine structure of the Golgi type II cell in the medial geniculate body of the cat. *Z. Anat. Entwickl.-Gesch.* **133**, 216–246.

Morgan, R., Vrbova, G., and Wolstencroft, J. H. 1972. Correlation between the retinal input to lateral geniculate neurones and their relative response to glutamate and aspartate. *J. Physiol.* **224,** 41P–42P.

Mountcastle, V. B. 1966. The neural replication of sensory events in the somatic afferent system. In: *Brain and Conscious Experience,* edited by J. C. Eccles. Springer-Verlag, New York, pp. 85–115.

Mountcastle, V. B. 1975. The view from within: Pathways to the study of perception. *Johns Hopkins Med. J.* **136,** 109–131.

Mroz, E. A., Brownstein, M. J. and Leeman, S. E. 1977. Evidence for substance P in the striato-nigral tract. *Brain Res.* **125,** 305–311.

Mugnaini, E., 1976. Organization of cerebellar cortex. In: *Afferent and Intrinsic Organization of Laminated Structures in the Brain,* edited by O. Creutzfeldt. Springer-Verlag, Heidelberg.

Murphree, H. B., and Domino, E. F. 1968. Questions and comments in antipsychotic agents. In: *Psychopharmacology, a review of progress 1957–1967,* edited by D. H. Efron. U.S. Public Health Service Bulletin #1836.

Mushahawa, I. K., and Koeppe, R. E. 1971. The toxicity of monosodium glutamate in young rats. *Biochim. Biophys. Acta (Amst.)* **244,** 318–321.

Mutani, R., Bergamini, L., Fariello, R., and Delsedime, M. 1974. Effects of taurine on cortical epileptic foci. *Brain Res.* **70,** 170–173.

Myers, R. D. 1973. The role of hypothalamic serotonin in thermoregulation. In: *Serotonin and Behavior,* edited by J. Barchas and E. Usdin. Academic Press, New York, pp. 293–302.

Nadler, J. V., Vaca, K. W., White, W. F., Lynch, G. S., and Cotman, C. W. 1976. Aspartate and glutamate as possible transmitters of excitatory hippocampal afferents. *Nature* **260,** 538–540.

Nagatsu, T., Levitt, M., and Udenfriend, S. 1964. Tyrosine hydroxylase. The initial step in norepinephrine biosynthesis. *J. Biol. Chem.* **239,** 2910–2917.

Nagatsu, T., Hidaka, H., Kuzaya, H., and Takeya, K. 1970. Inhibition of dopamine-beta-hydroxylase by fuscaric acid (5-butylpicolinic acid) *in vitro* and *in vivo. Biochem. Pharmacol.* **19,** 35–44.

Nahorski, S. R., Rogers, K. J., and Smith, B. M. 1974. Histamine H_2 receptors and cyclic AMP in brain. *Life Sci.* **15,** 1887–1894.

Nahorski, S. R., Rogers, K. J., and Smith, B. M. 1977. Stimulation of cyclic adenosine 3′,5′-monophosphate in chick cerebral hemisphere slices: Effects of H_1 and H_2 histaminergic agonists and antagonists. *Brain Res.* **126,** 387–390.

Nakamura, Y., Mizuno, N., Konishi, A., and Sato, M. 1974. Synaptic reorganization of the red nucleus after chronic deafferentation from cerebellorubral fibers: An electron microscope study in the cat. *Brain Res.* **82,** 298–301.

Nauta, W. J. H. 1971. The problem of the frontal lobe: A reinterpretation. *J. Psychiat. Res.* **8,** 167–187.

Nauta, W. J. H. 1974. Evidence of a pallidohabenular pathway in the cat. *J. Comp. Neurol.* **156,** 19–28.

Neal, M. J. 1971. The uptake of [^{14}C]glycine by slices of mammalian spinal cord. *J. Physiol.* **215,** 103–117.

Neckers, L. M., and Meek, J. L. 1976. Measurement of 5HT turnover rate in discrete nuclei of rat brain. *Life Sci.* **19,** 1579–1584.

Neff, N. H., and Goridis, C. 1972. Neuronal monoamine oxidase: specific enzyme types and their rates of formation. In: *Monoamine Oxidase, New Vistas,* edited by E. Costa and M. Sandler. *Advances in Biochemical Psychopharmacology,* vol. 5, Raven Press, New York, pp. 307–323.

Neff, N. H., Yang, H. T., Goridis, C., and Bialek, D. 1974. The metabolism of indolealkylamines by type A and B monoamine oxidase of brain. In: *Serotonin-New Vistas,* edited by E. Costa, G. L. Gessa, and M. Sandler. *Advances in Biochemical Psychopharmacology,* vol. 11, Raven Press, New York, pp. 51–58.

Netter, F. H. 1953. *The Ciba Collection of Medical Illustrations,* vol. 1, *Nervous System.* Case-Hoyt Corp., Rochester, New York.

Nicoll, R. A. 1969. Inhibitory mechanisms in the rabbit olfactory bulb: Dendro-dendritic mechanisms. *Brain Res.* **14,** 157–172.

Nicoll, R. A. 1975. The action of presumed blockers of chloride transport on synaptic and amino acid responses in the frog spinal cord. *Neurosci. Abst.* **1,** 377.

Nilsson, C., Hokfelt, T., and Pernow, B. 1974. Distribution of substance P-like immunoreactivity in the rat central nervous system as revealed by immunohistochemistry. *Med. Biol.* **52,** 424–427.

Nishi, S., Minota, S., and Karczmar, A. G. 1974. Primary afferent neurones: The ionic mechanism of GABA-mediated depolarization. *Neuropharmacol.* **13,** 215–219.

Noell, W. K. 1959. The visual cell: Electric and metabolic manifestations of its life processes. *Am. J. Ophthalmol.* **48,** 347–370.

Nonner, W., Rojas, E., and Stämpfli, R. 1975. Gating currents in the node of Rannier: Voltage and time dependence. *Phil. Trans. R. Soc. Lond. [Biol.]* **270,** 483–492.

Norberg, K. A., and Hamberger, B. 1964. The sympathetic adrenergic neuron. *Acta Physiol. Scand.* **84,** 54–64.

Nyback, H. 1972. Effect of brain lesions and chlorpromazine on accumulation and disappearance of catecholamines formed *in vivo* from ¹⁴C-tyrosine. *Acta Physiol. Scand.* **84,** 54–64.

Obata, K., and Takeda, K. 1969. Release of GABA into the fourth ventricle induced by stimulation of the cat cerebellum. *J. Neurochem.* **16,** 1043–1047.

Obata, K., Ito, M., Ochi, R., and Sato, N. 1967. Pharmacological properties of the postsynaptic inhibition of Purkinje cell axons and the action of γ-aminobutyric acid on Deiter's neurons. *Exp. Brain Res.* **4,** 43–57.

Obata, K., Takeda, K., and Shinozaki, H. 1970. Further studies on pharmacological properties of the cerebellar-induced inhibition of Deiter's neurones. *Exp. Brain. Res.* **11,** 327–342.

Ochs, S. 1972a. The dependence of fast transport in mammalian nerve fibres on metabolism. *Acta Neuropathol. Suppl.* **5,** 86–96.

Ochs, S. 1972b. Fast transport of materials in mammalian nerve fibers. *Science* **176,** 252–260.

Ochs, S. 1974. Systems of material transport in nerve fibers (axoplasmic transport) related to nerve function and trophic control. *Ann. N.Y. Acad. Sci.* **228,** 202–223.

Okada, Y., and Shimada, C. 1975. Intrahippocampal distribution of γ-aminobutyric acid (GABA) in the guinea pig. In: *GABA and Nervous System Function,* edited by E. Roberts, T. N. Chase, and D. B. Tower. Raven Press, New York, pp. 223–228.

Okada, Y., Nitsch-Hassler, C., Kim, J. S., Bak, I. J., and Hassler, R. 1971. The role of γ-aminobutyric acid (GABA) in the extrapyramidal motor system. *Exp. Brain Res.* **13,** 514–518.

Okamoto, S. 1951. Epileptogenic action of glutamate directly applied into the brain of animals and inhibitory effect of proteins and tissue emulsions on its action. *J. Physiol. Soc. Japan.* **13,** 555–562.

Oldendorf, W. H. 1975. Permeability of the blood-brain barrier. In: *The Nervous System,* vol. 1, *The Basic Neurosciences,* edited by D. B. Tower and R. O. Brady. Raven Press, New York, pp. 279–289.

Olds, J. 1962. Hypothalamic substrates of reward. *Physiol. Rev.* **42,** 554–604.

Olds, J., and Milner, P. 1954. Positive reinforcement produced by electrical stimulation of septal area and other regions in rat brain. *J. Comp. Physiol.* **47,** 419–427.

Oliver, G., and Schafer, E. A. 1895. The physiological effects of extracts of the suprarenal capsules. *J. Physiol.* **18,** 230–276.

Olney, J. W. 1976. Brain damage and oral intake of certain amino acids. *Adv. Exp. Med. Biol.* **69,** 497–506.

Olney, J. W., Sharpe, L. G., and de Gubareff, T. 1975. Excitotoxic amino acids. *Neuroscience Abst.* #572.

Osborne, N. N., Wu, P. H., and Neuhoff, V. 1974. Free amino acids and related compounds in the dorsal root ganglia and spinal cord of the rat as determined by the microdansylation procedure. *Brain Res.* **74,** 175–181.

Oscarsson, O. 1969. Termination and functional organization of the dorsal spino-olivocerebellar path. *J. Physiol.* **200,** 129–149.

Oscarsson, O. 1973. Functional organization of spinocerebellar paths. In: *Handbook of Sensory Physiology,* vol. 2, *Somato-sensory System,* edited by A. Iggo. Springer, Berlin, pp. 339–380.

Oscarsson, O. 1976. Spatial distribution of climbing and mossy fibre inputs into the cerebellar cortex. In: *Afferent and Intrinsic Organization of Laminated Structures in the Brain,* edited by O. Creutzfeldt. Springer-Verlag, Heidelberg.

Oshima, T. 1969. Studies of pyramidal tract cells. In: *Basic Mechanisms of the Epilepsies,* edited by H. H. Jasper, A. A. Ward, and A. Pope. Little, Brown and Co., Boston, pp. 253–261.

Otsuka, M., Iversen, L. L., Hall, Z. W., and Kravitz, E. A. 1966. Release of gamma-aminobutyric and from inhibitory nerves of lobster. *Proc. Natl. Acad. Sci. USA,* **56,** 1110–1115.

Otsuka, M., Konishi, S., and Takahashi, T. 1972a. The presence of motoneuron-depolarizing peptide in bovine dorsal roots of spinal nerves. *Proc. Japan. Acad.* **48,** 342–346.

Otsuka, M., Konishi, S., and Takahashi, T. 1972b. A further study of the motoneuron-depolarizing peptide extracted from sorsal roots of bovine spinal nerves. *Proc. Japan. Acad.* **48,** 747–752.

Palacios, J. M., Mengod, G., Picatoste, F., Grau, M., and Blanco, I. 1976. Properties of rat brain histidine decarboxylase. *J. Neurochem.* **27,** 1455–1460.

Palay, S. L., and Chan-Palay, V. 1972. The structural heterogeneity of central nervous tissue. In: *Metabolic Compartmentation in the Brain,* edited by R. Balazs and J. E. Cremer. MacMillan Press, New York, pp. 187–207.

Pandya, D. N., and Kuypers, H. G. 1969. Cortico-cortical connections in the rhesus monkey. *Brain Res.* **13,** 13–36.

Pandya, D. N., Dye, P., and Butters, N. 1971. Efferent cortico-cortical projections of the prefrontal cortex in the rhesus monkey. *Brain Res.* **31,** 35–46.

Pappas, C. D. 1975. Ultrastructural basis of synaptic transmission. In: *The Nervous System,* vol. 1, *The Basic Neurosciences,* edited by D. B. Tower and R. O. Brady. Raven Press, New York, pp. 19–30.

Pasantes-Morales, H., Salceda, R., and Lopez Colombe, A. M. 1975. The role of taurine in retina: Factors affecting its release. In: *Taurine,* edited by R. Huxtable and A. Barbeau. Raven Press, New York, pp. 191–200.

Patrick, J., and Lindstrom, J. 1973. Autoimmune response to acetylcholine receptor. *Science* **180,** 871–872.

Pearce, L. A., and Schanberg, S. M. 1969. Histamine and spermidine content in brain during development. *Science* **166,** 1301–1303.

Penfield, W., and Jasper, H. 1954. *Epilepsy and the Functional Anatomy of the Human Brain.* Little, Brown and Co., Boston.

Penfield, W., and Perot, P. 1963. The brain's record of auditory and visual experience. *Brain* **86,** 596–696.

Penfield, W., and Roberts, L. 1959. *Speech and Brain-Mechanisms,* Princeton University Press, Princeton, N.J.

Peper, K., and McMahan, U. J. 1972. Distribution of acetylcholine receptors in the vicinity of nerve terminals on skeletal muscle of the frog. *Proc. R. Soc. Lond. [Biol.]* **181,** 431–440.

Pernow, B. 1953. Studies of substance P. Purification, occurrence and biological actions. *Acta Physiol. Scand.* **29,** Suppl. 105, 1–90.

Perry, T. L., Hansen, S., and Jenkins, L. C. 1964. Amine content of normal human cerebrospinal fluid. *J. Neurochem.* **11,** 49–53.

Perry, T. L., Hansen, S., and MacDougall, L. 1967. Amines of human whole brain. *J. Neurochem.* **14,** 775–782.

Perry, T. L., Berry, K., Diamond, S., and Mok, C. 1971a. Regional distribution of amino acids in human brain obtained at autopsy. *J. Neurochem.* **18,** 513–519.

Perry, T. L., Hansen, S., Berry, K., Mok, C., and Lesk, D. 1971b. Free amino acids and related compounds in biopsies of human brain. *J. Neurochem.* **18,** 521–528.

Perry, T. L., Hansen, S., and Kloster, M. 1973. Huntington's chorea, deficiency of γ-aminobutyric acid in brain. *N. Engl. J. Med.* **288,** 337–342.

Persson, T. 1970. *Catecholamine Turnover in Central Neurons System.* Scandinavian University Books, Goteberg.

Pert, C. B., and Snyder, S. H. 1973a. Properties of opiate receptor binding in rat brain. *Proc. Natl. Acad. Sci. USA* **70,** 2243–2247.

Pert, C. B., and Snyder, S. H. 1973b. Opiate receptor: Demonstration in nervous tissue. *Science* **179,** 1011–1014.

Pert, C. B., Kuhar, M. J., and Snyder, S. H. 1975. Autoradiographic localization of the opiate receptor in rat brain. *Life Sci.* **16,** 1849-1853.

Pert, C. B., Bowie, D. L., Pert, A., Morell, J. L., and Gross, E. 1977. Agonist-antagonist properties of N-allyl-[D-ala]²-met-enkephalin. *Nature* **269,** 73-75.

Peters, A., Palay, S. L., and Webster, H. de F. 1976. *The Fine Structure of the Nervous System.* Saunders, Philadelphia.

Peters, D. A. V., McGeer, P. L., and McGeer, E. G. 1968. The distribution of tryptophan hydroxylase in cat brain. *J. Neurochem.* **15,** 1431-1435.

Peterson, N. A., and Raghupathy, E. 1972. Characteristics of amino acid accumulation by synaptosomal particles isolated from rat brain. *J. Neurochem.* **19,** 1423-1438.

Pfaff, D. W., and Keiner, M. 1973. Atlas of estradiol-concentrating cells in the central nervous system of the rat. *J. Comp. Neurol.* **151,** 121-158.

Phillips, C. G. 1966. Changing concepts of the precentral motor area. In: *Brain and Conscious Experience,* edited by J. C. Eccles, Springer-Verlag, New York, pp. 389-421.

Phillips, C. G. 1973. Cortical localization and "sensorimotor processes" at the "middle level" in primates. *Proc. Roy. Soc. Med.* **66,** 987-1002.

Phillips, S. R., Durden, D. A., and Boulton, A. A. 1974. Identification and distribution of p-tyramine in the rat. *Can. J. Biochem.* **52,** 366-373.

Phillis, J. W. 1970. *The Pharmacology of Synapses.* Pergamon Press, New York.

Phillis, J. W., and Kostopoulos, G. K. 1975. Adenosine as a putative transmitter in the cerebral cortex. Studies with potentiators and antagonists. *Life Sci.* **17,** 1085-1094.

Phillis, J. W., and Limbacher, J. J. 1974. Substance P effect on cerebral cortical Betz cells. *Brain Res.* **69,** 158-163.

Phillis, J. W., and Tebecis, A. K. 1967. The responsiveness of thalamic neurons to iontophoretically applied monoamines. *J. Physiol.* **192,** 715-745.

Phillis, J. W., Tebecis, A. K., and York, D. H. 1968. Depression of spinal motoneurones by noradrenaline, 5-hydroxytryptamine and histamine. *Eur. J. Pharmacol.* **4,** 471-475.

Phillis, J. W., Kostopoulos, G. K., and Limbacher, J. J. 1974. Depression of corticospinal cells by various purines and pyrmidines. *Can. J. Physiol. Pharmacol.* **52,** 1226-1229.

Phillis, J. W., Kostopoulos, G. K., and Limbacher, J. J. 1975. A potent depressant action of adenine derivatives on cerebral cortical neurones. *Eur. J. Pharmacol.* **30,** 125-129.

Pickel, V. M., Tong, H. J., and Reis, D. J. 1975. Ultrastructural localization of tyrosine hydroxylase in noradrenergic neurons of brain. *Proc. Nat. Acad. Sci. USA* **72,** 659-663.

Pickel, V. M., Tong, H. J., and Reis, D. J. 1976. Monoamine synthesizing enzymes in central dopaminergic, noradrenergic and serotonergic neurons. *J. Histochem. Cytochem.* **24,** 792-806.

Pickel, V. M., Reis, D. J., and Leeman, S. E. 1977. Ultrastructural localization of substance P in neurons of rat spinal cord. *Brain Res.* **122,** 534-540.

Piercy, M. 1967. Studies of the neurological basis of intellectual function. *Modern Trends in Neurology* **4,** 106-124.

Pletscher, A., Shore, P. A., and Brodie, B. B. 1955. Serotonin release as a possible mechanism of reserpine action. *Science* **122,** 284-285.

Pletscher, A., Gey, K. F., and Burkard, W. P. 1966. Inhibitors of monoamine oxidase and decarboxylase of aromatic amino acids. In: *Handbook of Experimental Pharmacology,* vol. 19, edited by V. Erspamer. Springer, Berlin, pp. 593-735.

Poch, G. R., and Kopin, I. J. 1966. The role of octopamine in tachyphylaxis to tyramine. *Biochem. Pharm.* **15,** 210-212.

Poirier, L. J., and Sourkes, T. L. 1965. Influences of the substantia nigra on catecholamine content of the striatum. *Brain* **88,** 181-192.

Pollard, H., Bischoff, S., and Schwartz, J. C. 1974. Turnover of histamine in rat brain and its decrease under barbiturate anesthesia. *J. Pharmacol. Exp. Ther.* **190,** 88-90.

Pollin, W., Cardon, P. V., and Kety, S. S. 1961. Effect of amino acid feeding in schizophrenic patients treated with iproniazid. *Science* **133,** 104-105.

Popper, K. R. 1972. *Objective Knowledge: An Evolutionary Approach.* The Clarendon Press, Oxford.

Popper, K. R., and Eccles, J. C. 1977. *The Self and Its Brain.* Springer-Verlag, Heidelberg.

Poritsky, R. 1969. Two and three dimensional ultrastructure of boutons and glial cells on the motoneuronal surface in the cat spinal cord. *J. Comp. Neur.* **135**, 423–452.

Porter, R. 1973. Functions of the mammalian cerebral cortex in movement. *Progr. Neurobiol.* **1**, 1–51.

Powell, T. P. S., and Cowan, W. M. 1956. A study of thalamo-striate relations in the monkey. *Brain* **79**, 364–390.

Powell, D., Leeman, S. E., Tregear, G. W., Niall, H. D., and Potts, J. T., Jr. 1973. Radioimmunoassay for substance P. *Nature* **241**, 252–254.

Precht, W., Schwindt, P. C., and Baker, R. 1973. Removal of vestibular commissural inhibition by antagonists of GABA and glycine. *Brain Res.* **62**, 222–226.

Price, J. C., Waelsch, H., and Putnam, T. 1943. DL-Glutamic acid hydrochloride in treatment of petit mal and psychomotor seizures. *JAMA* **122**, 1153–1156.

Pull, I., and McIlwain, H. 1972. Adenine derivatives as neurohumoral agents in the brain. The quantities liberated on excitation of superfused cerebral tissues. *Biochem. J.* **130**, 975–981.

Purpura, D. P. 1975. Physiological organization of the basal ganglia. In: *The Basal Ganglia*, vol. 55, *Res. Publ. Ass. Res. Nerv. & Ment. Dis.*, edited by M. D. Yahr. Raven Press, New York, pp. 91–114.

Purpura, D. P., Girado, M., Smith, T. G., Gallan, D. A., and Grundfest, H. 1959. Structure-activity determinents of pharmacological effects of amino acids and related compounds on central synapses. *J. Neurochem.* **3**, 238–266.

Quastel, J. H., Tennenbaum, M., and Wheatley, A. H. M. 1936. Choline ester formation in, and choline esterase activities of, tissues *in vitro*. *Biochem. J.* **30**, 1668–1681.

Quay, W. B. 1965. Retinal and pineal hydroxyindole-*O*-methyl transferase activity in vertebrates. *Life Sci.* **4**, 983–991.

Racagni, G., Cheney, D. L., Trabucci, M., Wang, C., and Costa, E. 1974. Measurement of acetyl-choline turnover rate in discrete areas of rat brain. *Life Sci.* **15**, 1961–1975.

Raisman, G. 1969. Neuronal plasticity in the septal nuclei of the adult rat. *Brain Res.* **14**, 25–48.

Raisman, G., and Field, P. M. 1973. A quantitative investigation of the development of collateral reinnervation after partial deafferentation of the septal nuclei. *Brain Res.* **50**, 241–264.

Rakic, P. 1971. Neuron–glia relationship during granule cell migration in developing cerebellar cortex. A Golgi and electron microscopic study in *Macacus rhesus*. *J. Comp. Neurol.* **141**, 283–312.

Rakic, P. 1972. Mode of cell migration to the superficial layers of fetal monkey neocortex. *J. Comp. Neurol.* **145**, 61–84.

Rakic, P. 1973. Kinetics of proliferation and latency between final cell division and onset of differentiation of cerebellar, stellate and basket neurons. *J. Comp. Neurol.* **147**, 523–540.

Rakic, P., and Sidman, R. L. 1973. Sequence of developmental abnormalities leading to granule cell deficit in cerebellar cortex of weaver mutant mice. *J. Comp. Neurol.* **152**, 103–132.

Rall, T. W. 1972. Role of adenosine 3′,5′-monophosphate (cyclic AMP) in actions of catecholamines. *Pharmacol. Rev.* **24**, 399–410.

Rall, W. 1970. Dendritic neuron theory and dendrodendritic synapses in a simple cortical system. In: *The Neurosciences*, vol. 2, edited by F. O. Schmitt. Rockefeller University Press, New York, pp. 552–565.

Rall, W., Shepherd, G. M., Reese, T. S., and Brightman, M. W. 1966. Dendrodendritic synaptic pathway for inhibition in the olfactory bulb. *Exp. Neurol.* **14**, 44–56.

Ralston, H. J., and Herman, M. M. 1969. The fine structure of neurons and synapses in the ventrobasal thalamus of the cat. *Brain Res.* **14**, 77–97.

Ramón y Cajal, S. 1895. *Les nouvelles idées sur la structure du système nerveux chez l'homme et chez les vertébrés*. Reinwald, Paris.

Ramón y Cajal, S. 1909. *Histologie due système nerveux de l'homme et des vertébrés*, vol. 1. Maloine, Paris.

Ramón y Cajal, S. 1911. *Histologie due système nerveux de l'homme et des vertébrés*, vol. 2. Maloine, Paris.

Ramón y Cajal, S. 1929. *Studies on Vertebrate Neurogenesis* (translated by L. Guth, 1959). Charles C Thomas, Springfield, Ill.

Ramón y Cajal, S. 1934. Les preuves objectives de l'unité anatomique des cellules nerveuses. *Trab. Lab. Invest. Biol. Univ. Madr.* **29**, 1–137.

Rapport, M. M. 1949. Serum vasoconstrictor (serotonin) V. The presence of creatinine in the complex: A proposed study of the vasoconstrictor principle. *J. Biol. Chem.* **180,** 961–969.

Rapport, M. M., Green, A. A., and Page, I. H. 1948. Crystalline serotonin. *Science* **108,** 329–330.

Rasminsky, M., and Sears, T. A. 1971. Internodal conduction in normal and demyelinated mammalian single nerve fibers. *Proc. Physiol. Soc.* **217,** 66P–67P.

Reed, L., and Cox, D. J. 1966. Macromolecular organization of enzyme systems. *Ann. Rev. Biochem.* **35,** 57–84.

Reichelt, K. L., and Fonnum, F. 1969. Subcellular localization of *N*-acetylaspartyl-glutamate, *N*-acetylglutamate and glutathione in brain. *J. Neurochem.* **16,** 1409–1416.

Reiffenstein, R. J., and Neal, M. J. 1974. Uptake, storage, and release of γ-aminobutyric acid in normal and chronically denervated cat cerebral cortex. *Can. J. Physiol. Pharmacol.* **52,** 286–296.

Reis, D. J., Ross, R. A., and Pickel, V. M. 1975. Some differences in tyrosine hydroxylase in central noradrenergic and dopaminergic neurons. In: *Chemical Tools in Catecholamine Research II,* edited by O. Almgren, A. Carlsson, and J. Engel. North Holland Publishing Company, Amsterdam, pp. 53–60.

Ribak, C. E., Vaughn, J. E., Saito, K., Barber, R., and Roberts, E. 1977a. Glutamate decarboxylase localization in neurons of the olfactory bulb. *Brain Res.* **126,** 1–18.

Ribak, C. E., Vaughn, J. E., Saito, K., Barber, R., and Roberts, E. 1977b. Immunocytochemical localization of glutamate decarboxylase in rat substantia nigra. *Brain Res.* **116,** 287–298.

Richter, D. 1937. Adrenaline and amine oxidase. *Biochem. J.* **31,** 2022–2028.

Richter, J. J., and Wainer, A. 1971. Evidence for separate systems for the transport of neutral and basic amino acids across the blood brain barrier. *J. Neurochem.* **18,** 613–620.

Rizzoli, A. A. 1968. Distribution of glutamic acid, aspartic acid, γ-aminobutyric acid and glycine in six areas of cat spinal cord after transection. *Brain Res.* **11,** 11–18.

Roberts, E., editor. 1960. *Inhibition in the Nervous System and Gamma-Aminobutyric Acid.* Pergamon Press, New York.

Roberts, E., and Frankel, S. 1950. γ-Aminobutyric acid in brain; its formation from glutamic acid. *J. Biol. Chem.* **187,** 55–63.

Roberts, E., Chase, T. N., and Tower, D. B., editors. 1976. *GABA in Nervous System Function.* Raven Press, New York.

Roberts, M. H. G., and Straughan, D. W. 1967. Excitation and depression of cortical neurones by 5-hydroxytryptamine. *J. Physiol.* **193,** 269–294.

Roberts, P. J., and Keen, P. 1973. Uptake of [^{14}C]glutamate into dorsal and ventral roots of spinal cord nerves of the cat. *Brain Res.* **57,** 234–238.

Roberts, P. J., and Keen, P. 1974. Effects of dorsal root section in amino acids of rat spinal cord. *Brain Res.* **74,** 333–337.

Roberts, P. J., Keen, P., and Mitchell, J. F. 1973. The distribution and axonal transport of free amino acids and related compounds in dorsal sensory neuron of the rat, as determined by the dansyl reaction. *J. Neurochem.* **21,** 199–210.

Robertson, H. A., and Juorio, A. V. 1976. Octopamine and some related noncatecholic amines in invertebrate nervous system. *Int. Rev. Neurobiol.* **19,** 173–224.

Robertson, J. D. 1958. Structural alterations in nerve fibers produced by hypotonic and hypertonic solutions. *J. Biophys. Biochem. Cytol.* **4,** 349–364.

Robinson, D. A. 1976. Adaptive gain control of the vestibulo-ocular reflex by the cerebellum. *J. Neurophysiol.* **39,** 954–969.

Rojas, E., and Keynes, R. D. 1975. On the relation between displacement currents and activation of the sodium conductance in the squid giant neuron. *Phil. Trans. R. Soc. Lond. [Biol.]* **270,** 459–482.

Rose, G., and Schubert, P. 1977. Release and transfer of [^{3}H]adenosine derivatives in the cholinergic septal system. *Brain Res.* **121,** 353–357.

Roseghini, M., and Ramorino, L. M. 1970. Serotonin, histamine and *N*-acetylhistamine in the nervous system of *Dosidicus gigas.* *J. Neurochem.* **17,** 489–492.

Rosén, I., and Asanuma, H. 1972. Peripheral afferent inputs to the forelimb area of monkey motor cortex: Input–output relations. *Exp. Brain Res.* **14,** 257–273.

Rosengarten, H., and Freidhoff, A. J. 1976. A review of recent studies of the biosynthesis and excretion of hallucinogens formed by methylation of neurotransmitters or related substances. *Schizophrenia Bull.* **2**, 90–105.

Roskoski, R., Lim, C. T., and Roskoski, L. M. 1975. Human brain and placental choline acetyltransferase: Purification and properties. *Biochem.* **14**, 5105–5110.

Roth, R. H., and Suhr, Y. 1970. Mechanism of the γ-hydroxybutyrate induced increase in brain dopamine and its relationship to sleep. *Biochem. Pharmacol.* **19**, 3001–3012.

Ryall, R. W. 1962. The subcellular distribution of substance P and 5-HT in brain. *Biochem. Phrmacol.* **11**, 1233–1237.

Saavedra, J. M. 1974. Enzymatic-isotopic method for octopamine at the picogram level. *Anal. Biochem.* **59**, 629–633.

Saavedra, J. M., and Axelrod, J. 1972. A specific and sensitive enzymatic assay for tryptamine in tissues. *J. Pharmacol. Exp. Ther.* **182**, 363–369.

Saavedra, J. M., and Axelrod, J. 1973. Effect of drugs on the tryptamine content of rat tissues. *J. Pharmacol. Exp. Ther.* **185**, 523–529.

Saavedra, J. M., and Axelrod, J. 1974. Brain tryptamine and the effects of drugs. In: *Advances in Biochemical Psychopharmacology*, vol. 10, edited by E. Costa, G. L. Gessa, and M. Sandler. Raven Press, New York, pp. 135–139.

Saelens, J. K., Simke, J. P., Allen, M. P., and Conroy, C. A. 1973. Some of the dynamics of choline and acetylcholine metabolism in rat brain. *Arch. Int. Pharmacodyn. Ther.* **209**, 250–258.

Saito, K., Barber, R., Wu, J., Matsuda, T., Roberts, E., and Vaughn, J. E. 1974a. Immunohistochemical localization of glutamate decarboxylase in rat cerebellum. *Proc. Natl. Acad. Sci. USA* **71**, 269–277.

Saito, K., Schousboe, A., Wu, J. Y., and Roberts, E. 1974b. Some immunochemical properties of and species specificity of GABA-α-ketoglutarate transaminase from mouse brain. *Brain Res.* **65**, 287–296.

Samorajski, T., and Rolsten, C. 1973. Age and regional differences in the chemical composition of brains of mice, monkeys and humans. *Progr. Brain Res.* **40**, 251–265.

Sanders-Bush, E., Gallager, D. A., and Sulser, F. 1974. On the mechanism of brain 5-hydroxytryptamine depletion by p-chloroamphetamine and related drugs and the specificity of their reaction. In: *Advances in Biochemical Psychopharmacology*, vol. 10, edited by E. Costa, G. L. Gessa, and M. Sandler. Raven Press, New York, pp. 185–194.

Sar, M., Stumpf, W. B., Miller, R. J., Chang, K. J., and Cuatrecasas, P. 1977. Immunohistochemical localization of enkephalins in rat brain. *Neuroscience Abst.* **7**, #966.

Sastry, B. S. R., and Phillis, J. W. 1976. Evidence for an ascending inhibitory histaminergic pathway to the cerebral cortex. *Can. J. Physiol. Pharmacol.* **54**, 782–786.

Satinsky, D. 1967. Pharmacological responsiveness to lateral geniculate nucleus neurons. *Int. J. Neuropharmacol.* **6**, 387–397.

Sauer, F. C. 1935. Mitosis in the neural tube. *J. Comp. Neurol.* **62**, 377–405.

Schayer, R. W., and Reilly, M. A. 1973. Formation and fate of histamine in rat and mouse brain. *J. Pharmacol. Exp. Ther.* **184**, 33–40.

Scheibel, M. E., and Scheibel, A. B. 1970. Elementary processes in selected thalamic and cortical subsystems—the structural substrates. In: *The Neurosciences: Second Study Program*, edited by F. O. Schmitt. Rockefeller University Press, New York, pp. 443–457.

Scheibel, M. E., and Scheibel, A. B. 1975. Structural changes in the aging brain. In: *Neurobiology of Aging*, vol. 1, *Clinical, Morphological and Neurochemical Aspects of the CNS*, edited by H. Brody, D. Harmon, and J. M. Ordy. Raven Press, New York, pp. 11–37.

Schenker, C., Mroz, E. A., and Leeman, S. E. 1976. Release of substance P from isolated nerve endings. *Nature* **264**, 790–792.

Schiller, P. H. 1970. The discharge characteristics of single units in the oculomotor and abducens nuclei of the unanesthetized monkey. *Exp. Brain Res.* **10**, 347–362.

Schmid, R., Sieghart, W., and Karobath, M. 1975. Taurine uptake in synaptosomal fractions of rat cerebral cortex. *J. Neurochem.* **25**, 5–9.

Schmidt, J., and Raftery, M. 1973. Purification of acetylcholine receptors from *Torpedo california* electroplax by affinity chromatography. *Biochem.* **12**, 852–856.

Schmidt, M. J., Thornberry, J. F., and Molloy, B. B. 1977. Effects of kainate and other glutamate analogues on cyclic nucleotide accumulation in slices of rat cerebellum *Brain Res.* **121,** 182–189.

Schmidt, R. F. 1971. Presynaptic inhibition in the vertebrate central nervous system. *Ergb. Physiol.* **63,** 21–108.

Schneider, G. E. 1969. Two visual systems. *Science* **163,** 895–902.

Schneider, G. E. 1973. Early lesions of superior colliculus: Factors affecting the formation of abnormal retinal projections. *Brain Behav. Evol.* **8,** 73–109.

Schousboe, A., Wu, J. Y., and Roberts, E. 1973. Purification and characterization of the 4-aminobutyrate-2-ketoglutarate transaminase from mouse brain. *Biochem.* **12,** 2868–2873.

Schreiner, L., and Kling, A. 1953. Behavioral changes following rhinencephalic injuries in cat. *J. Neurophysiol.* **16,** 643–659.

Schubert, P., and Kreutzberg, G. W. 1974. Axonal transport of adenosine and uridine derivatives and transfer to postsynaptic membranes. *Brain Res.* **76,** 526–530.

Schwartz, J. C. 1975. Histamine as a transmitter in brain. *Life Sci.* **17,** 503–518.

Schwartz, J. C., Lampart, C., and Rose, C. 1970. Properties and regional distribution of histidine decarboxylase in rat brain. *J. Neurochem.* **17,** 1527–1534.

Schwartz, J. C., Lampart, C., and Rose, C. 1972. Histamine formation in rat brain *in vivo:* Effects of histidine loads. *J. Neurochem.* **19,** 801–810.

Schwindt, P. C., and Calvin, W. H. 1972. Membrane-potential trajectories between spikes underlying motoneurone firing rate. *J. Neurophysiol.* **35,** 311–325.

Segal, D. S., Knapp, S., Kuczenski, R. T., and Mandell, A. J. 1971. Effects of long-term reserpine treatment on brain tyrosine hydroxylase activity and behavioral activity. *Science* **173,** 847–849.

Segawa, T., Nakara, Y., Nakamura, K., Yajima, H., and Kitagawa, K. 1976. Substance P in the central nervous system of rabbits: Uptake system differs from putative transmitters. *Jap. J. Pharmacol.* **26,** 757–760.

Sem-Jacobsen, C. W. 1976. Electrical stimulation and self-stimulation in man with chronic implanted electrodes. In: *Brain-stimulation Reward,* edited by A. Wanquier and E. T. Rolls. Elsevier Press, Amsterdam, pp. 505–526.

Sethy, V. H., and Van Woert, M. H. 1974. Modification of striatal acetylcholine concentration by dopamine receptor agonists and antagonists. *Res. Comm. Chem. Pathol. Pharmacol.* **8,** 13–28.

Shank, R. P., and Aprison, M. H. 1970. The metabolism *in vivo* of glycine and serine in eight areas of the rat central nervous system. *J. Neurochem.* **17,** 1461–1475.

Shaw, R. K., and Heine, J. D. 1965. Ninhydrin positive substance present in different areas of normal rat brain. *J. Neurochem.* **12,** 151–155.

Shefer, V. F. 1973. Absolute number of neurons and thickness of cerebral cortex during aging, senile and vascular dementia in Pick's and Alzheiner's disease. *Neurosci. Behav. Physiol.* **6,** 319–324.

Shepherd, G. M. 1974. *The Synaptic Organization of the Brain.* Oxford University Press, London.

Sherrington, C. S. 1906. *Integrative Action of the Nervous System.* Yale University Press, New Haven, Conn.

Sherrington, C. S. 1929. Some functional problems attaching to convergence. *Proc. Roy. Soc. B* **105,** 332–362.

Sherrington, C. S. 1940. *Man on his Nature.* Cambridge University Press, Cambridge.

Shillito, E. E. 1970. The effect of p-chlorophenylalanine on social interaction of male rats. *Brit. J. Pharmacol.* **38,** 305–315.

Shimizu, H., Daly, J. W., and Creveling, C. R. 1969. A radioisotopic method for measuring the formation of 3′,5′-cyclic monophosphate in incubated slices of brain. *J. Neurochem.* **16,** 1609–1619.

Shimizu, H., Ichishita, H., and Odagiri, H. 1974. Stimulated formation of cyclic adenosine-3′,5′-monophosphate by aspartate and glutamate in cerebral cortical slices of guinea pig. *J. Biol. Chem.* **249,** 5955–5962.

Shimuzu, H., Ichishita, H., and Umeda, I. 1975. Inhibition of glutamate-elicited accumulation of adenosine cyclic 3′,5′-monophosphate in brain slices by α,ω-diaminocarboxylic acids. *Mol. Pharmacol.* **11,** 866–873.

Sholl, D. A. 1956. *The Organization of the Cerebral Cortex,* John Wiley and Sons, Inc., New York.

Shute, C. C., and Lewis, P. R. 1963. Cholinesterase-containing systems of the brain of the rat. *Nature* **199,** 1160–1164.

Sidman, R. L. 1970. Cell proliferation, migration and interaction in the developing mammalian central nervous system. In: *The Neurosciences,* vol. 2, edited by F. O. Schmitt. Rockefeller University Press, New York, pp. 100–107.

Sidman, R. L. 1974. Cell–cell recognition in the developing central nervous system. In: *The Neurosciences,* vol. 3, edited by F. O. Schmitt and F. G. Wordens. MIT Press, Cambridge, Mass., pp. 743–758.

Sidman, R. L., and Rakic, P. 1973. Neuronal migration, with special reference to developing human brain: A review. *Brain Res.* **62,** 1–35.

Siggins, G. R., Hoffer, B. J., and Bloom, F. E. 1971. Studies on norepinephrine-containing afferents to Purkinje cells of rat cerebellum. III. Evidence for mediation of norepinephrine effects by cyclic 3′,5′-adenosine monophosphate. *Brain Res.* **25,** 535–553.

Silkaitis, R. P., and Mosnaim, A. D. 1976. Pathways linking 1-phenylalanine and 2-phenylalanine with p-tyrosine in rabbit brain. *Brain Res.* **114,** 105–115.

Simantov, R., and Snyder, S. H. 1976. Isolation and structure identification of a morphine-like peptide "enkephalin" in bovine brain. *Life Sci.* **18,** 781–788.

Simantov, R., Kuhar, M. J., Uhl, G. R., and Snyder, S. H. 1977. Opioid peptide enkephalin: Immunohistochemical mapping in rat central nervous system. *Proc. Natl. Acad. Sci. USA* **74,** 2167–2171.

Simon, E. J., Hiller, J. M., and Edelman, I. 1973. Stereospecific binding of the potent narcotic analgesic [³H]etorphine to rat brain homogenates. *Proc. Natl. Acad. Sci. USA* **70,** 1947–1949.

Simon, J. R., Contrera, J. F., and Kuhar, M. J. 1976. Binding of [³H]kainic acid, an analogue of L-glutamate, to brain membranes. *J. Neurochem.* **26,** 141–147.

Sims, K. L., David, G. A., and Bloom, F. E. 1973. Activities of 3,4-dehydroxy-L-phenylalanine 5-hydroxy-L-tryptophan decarboxylases in rat brain: Assay characteristics and distribution. *J. Neurochem.* **20,** 449–464.

Singh, V. K., and McGeer, P. L. 1974a. Antibody production to choline acetyltransferase purified from human brain. *Life Sci.* **15,** 901–913.

Singh, V. K., and McGeer, P. L. 1974b. Cross immunity of antibodies to choline acetyltransferase in various vertebrate species. *Brain Res.* **82,** 356–359.

Singh, V. K., and McGeer, P. L. 1977. Studies on choline acetyltransferase purified from human brain. *Neurochem. Res.* **2,** 281–291.

Sloan, J. W., Martin, W. R., Clements, T. H., Buchwald, W. F., and Bridges, S. R. 1975. Factors influencing brain and tissue levels of tryptamine: Species, drugs and lesions. *J. Neurochem.* **24,** 523–532.

Sloper, J. J. 1971. Dendro-dendritic synapses in the primate motor cortex. *Brain Res.* **34,** 186–192.

Smith, A. J. 1966. Speech and other functions after left (dominant) hemispherectomy. *J. Neurol. Neurosurg. Psychiat.* **29,** 467–471.

Smith, C. M. 1972. The release of acetylcholine from rat hippocampus. *Brit. J. Pharmacol.* **45,** 172P.

Smith, S. M., Brown, H. O., Toman, J. E. P., and Goodman, L. S. 1947. The lack of cerebral effects of D-tubocurarine. *Anesthesiol.* **8,** 1–14.

Snodgrass, S. R., and Iversen, L. L. 1974. Formation and release of ³H-tryptamine from ³H-tryptophan in rat spinal cord slices. In: *Advances in Biochemical Psychopharmacology,* vol. 10, edited by E. Costa, G. L. Gessa, and M. Sandler. Raven Press, New York, pp. 141–150.

Snyder, S. H., Shaskan, E. G., and Kuhar, M. J. 1973. Serotonin uptake systems in brain tissue. In: *Serotonin and Behavior,* edited by J. Barchas and E. Usdin. Academic Press, New York, pp. 97–108.

Snyder, S. H., Banerjee, S. P., Yamamura, H. I., and Greenberg, D. 1974a. Drugs, neurotransmitters and schizophrenia. *Science* **184,** 1243–1253.

Snyder, S. H., Brown, B., and Kuhar, M. J. 1974b. The subsynaptosomal localization of histamine, histidine decarboxylase and histamine methyltransferase in rat hypothalamus. *J. Neurochem.* **23,** 37–45.

Sotelo, C., and Palay, S. L. 1971. Altered axons and axon terminals in the lateral vestibular nucleus of the rat. *Lab. Invest.* **25,** 653-671.

Sourkes, T. R. 1976. Psychopharmacology and biochemical theories of mental illness. In: *Basic Neurochemistry,* 2nd edition, edited by G. J. Siegel *et al.* Little, Brown and Co., Boston, pp. 705-736.

Spano, P. F., and Neff, N. H. 1972. Metabolic fate of caudate nucleus dopamine. *Brain Res.* **42,** 139-145.

Sparks, R., and Geschwind, N. 1968. Dichotic listening in man after section of neocortical commissures. *Cortex* **4,** 3-16.

Spencer, H. J. 1976. Antagonism of cortical excitation of striatal neurons by glutamic acid diethyl ester: Evidence for glutamic acid as an excitatory transmitter in the rat striatum. *Brain Res.* **102,** 91-101.

Sperry, R. W. 1951a. Mechanisms of neural maturation. In: *Handbook of Experimental Psychology,* edited by S. S. Stevens. John Wiley and Sons, New York, pp. 236-280.

Sperry, R. W. 1951b. Regulative factors in the orderly growth of neural circuits. *Growth* **10,** 63-87.

Sperry, R. W. 1963. Chemoaffinity in the orderly growth of nerve fiber patterns and connections. *Proc. Natl. Acad. Sci. USA* **50,** 703-710.

Sperry, R. W. 1968. Hemisphere deconnection and unity of conscious awareness. *Am. Psychol.* **23,** 723-733.

Sperry, R. W. 1971. How a developing brain gets itself properly wired for adaptive function. In: *Biopsychology of Development,* edited by E. Tobach, R. Avonson, and E. Shaw. Academic Press, New York, pp. 27-44.

Sperry, R. W. 1974. Lateral specialization in the surgically separated hemispheres. In: *The Neurosciences: Third Study Program,* edited by F. O. Schmitt and F. G. Worden. MIT Press, Cambridge, Mass., pp. 5-19.

Squires, R. F., and Lassen, J. B. 1975. The inhibition of A and B forms of MAO in the production of a characteristic behavioral syndrome in rats after L-tryptophan loading. *Psychopharmacol.* **41,** 145-151.

Stadler, H., Lloyd, K. G., Gadea-Ciria, M., and Bartholini, G. 1973. Enhanced striatal acetylcholine release by chlorpromazine and its reversal by apomorphine. *Brain Res.* **55,** 476-480.

Stavinoha, W. B., and Weintraub, S. J. 1974. Choline content of rat brain. *Science* **183,** 964-965.

Stedman, E., and Stedman, E. 1939. The mechanism of biological synthesis of acetylcholine. *Biochem. J.* **33,** 811-921.

Stein, L. 1968. The chemistry of reward and punishment. In: *Psychopharmacology: A Review of Progress 1957-1967,* edited by D. H. Efron. U.S. Public Health Service Publ. #1836, pp. 105-123.

Stein, L., and Wise, C. 1971. Possible etiology of schizophrenia: Progressive damage to the noradrenergic reward system by 6-hydroxydopamine. *Science* **171,** 1032-1036.

Stepita-Klauco, M., Dolezalova, H., and Giacobini, E. 1973. The action of piperidine on cholinoceptive neurons of the snail. *Brain Res.* **63,** 141-152.

Stepita-Klauco, M., Dolezalova, H., and Fairweather, R. 1974. Piperidine increase in the brain of dormant mice. *Science* **183,** 536-537.

Sternberger, L. A., Hardy, P. H., Cuculis, J. J., and Myer, H. G. 1970. The unlabelled antibody enzyme method by immunohistochemistry: Preparation and properties of soluble antigen-antibody complex (horseradish peroxidase-antiperoxidase) and its use in identification of spirochetes. *J. Histochem. Cytochem.* **18,** 315-333.

Steward, O., Cotman, C. W., and Lynch, G. S. 1973. Reestablishment of electrophysiologically functional entorhinal cortical input to the dentate gyrus deafferented by ipsilateral entorhinal lesions: Innervation by the contralateral entorhinal cortex. *Exp. Brain Res.* **18,** 396-414.

Steward, O., Cotman, C. W., and Lynch, G. S. 1974. Growth of a new fiber projection in the brain of adult rats: Reinnervation of the dentate gyrus by the contralateral entorhinal cortex following ipsilateral entorhinal lesions. *Exp. Brain Res.* **20,** 45-66.

Stewart, C. N., Coursin, D. B., and Bhagava, H. N. 1972. Electroencephalographic study of L-glutamate induced seizures in rats. *Toxicol. Appl. Pharmacol.* **23,** 635-639.

Stewart, R. M., Martuza, R. L., Baldessarini, R. J., and Kornblith, P. L. 1976. Glutamate accumulation by human gliomas and meningiomas in tissue culture. *Brain Res.* **118**, 441–452.

Stolz, F. 1904. Ueber Adrenaline und Alkylaminoacetobrenzcatechin. *Ber. dtsch. Chem. Ges.* **37**, 4149–4154.

Stoff, J. C., Liem, A. L., and Mulder, A. M. 1976. Release and receptor stimulus properties of p-tyramine in rat brain. *Arch. Int. Pharmacodyn. Ther.* **220**, 62–71.

Storm-Mathisen, J. 1970. Quantitative histochemistry of AcChE in rat hippocampus. *J. Neurochem.* **17**, 739–750.

Storm-Mathisen, J. 1977. Glutamic acid and excitatory nerve endings: Reduction of glutamic acid uptake after axotomy. *Brain Res.* **120**, 379–386.

Storm-Mathisen, J. 1978. Localization of putative transmitters in the hippocampal formation. In: *Functions of the Septo-hippocampal Formation,* edited by XXX. *Ciba Found. Symp.,* in press.

Storm-Mathisen, J., and Fonnum, F. 1971. Quantitative histochemistry of glutamate decarboxylase in the rat hippocampal region. *J. Neurochem.* **18**, 1105–1111.

Storm-Mathisen, J., and Fonnum, F. 1972. Localization of transmitter candidates in the hippocampal region. *Progr. Brain Res.* **36**, 40–57.

Straschell, M., and Perwein, J. 1969. The inhibition of retinal ganglion cells by catecholamines and γ-aminobutyric acid. *Pfluegers Arch.* **312**, 45–54.

Stryer, L. 1975. *Biochemistry.* Freeman & Co., San Francisco.

Studer, R. O., Trzeciak, A., and Lergier, W. 1973. Isolierung und Aminosauresequenz von Substance P aus Pferdedarm. *Helv. Chim. Acta* **56**, 860–866.

Suzuki, O., and Yagi, K. 1976. A fluorometric assay for beta-phenylethylamine in rat brain. *Anal. Biochem.* **75**, 192–200.

Swanson, L. W., and Hartman, B. K. 1975. The central adrenergic system. An immunofluorescence study of the location of cell bodies and their efferent connections in the rat utilizing dopamine-beta-hydroxylase as a marker. *J. Comp. Neurol.* **163**, 467–505.

Szentágothai, J. 1965. The synapses of short local neurons in the cerebral cortex. *Symp. Biol. Hung.* **5**, 251–276.

Szentágothai, J. 1968. Structure-functional considerations of the cerebellar neuron network. *I.E.E.E. Proc.* **56**, 960–968.

Szentágothai, J. 1969. Architecture of the cerebral cortex. In: *Basic Mechanisms of the Epilepsies,* edited by H. H. Jasper, A. A. Ward, and A. Pope. Little, Brown & Co., Boston, pp. 13–28.

Szentágothai, J. 1970a. Les circuits neuronaux de l'écorce cérébrale. *Bull. Acad. Roy. Med. Belg. vii,* 10, 475–492.

Szentágothai, J. 1970b. Glomerular synapses, complex synaptic arrangements, and their operational significance. In: *The Neurosciences: Second Study Program,* edited by F. O. Schmitt. Rockefeller University Press, New York, pp. 427–443.

Szentágothai, J. 1971. Memory functions and the structural organization of the brain. In: *Biology of Memory,* edited by G. Adam. *Symp. Biol. Hung.* **10**, 21–25.

Szentágothai, J. 1972. The basic neuronal circuit of the neocortex. In: *Synchronization of EEG Activity in Epilepsies,* edited by H. Petsche and M. A. B. Brazier. Springer-Verlag, Vienna, pp. 9–24.

Szentágothai, J. 1974. A structural overview. In: *Conceptual Models of Neural Organization,* edited by J. Szentágothai and M. Arbib. *Neurosci. Res. Prog. Bull.* **12**, 354–410.

Szentágothai, J. 1975. The "module concept" in cerebral cortex architecture. *Brain Res.* **95**, 475–496.

Szerb, J. C., Malik, H., and Hunter, E. G. 1970. Relationship between acetylcholine content and release in the cat's cerebral cortex. *Can. J. Physiol. Pharmacol.* **48**, 780–790.

Szilard, L. 1964. On memory and recall. *Proc. Natl. Acad. Sci. USA* **51**, 1092–1099.

Tachibana, M., and Kuriyama, K. 1974. Gamma-aminobutyric acid in the lower auditory pathway of the gunea pig. *Brain Res.* **69**, 370–374.

Tafuri, W. L., Maria, T. A., Pitella, E., Bogliolo, L., Hial. W., and Diniz, C. R. 1974. An electron microscopic study of Auerbach's plexus and determination of substance P of the colon in Hirschsprung's disease. *Virchow's Arch. Pathol. Anat.* **362**, 41–50.

Takahashi, T., and Otsuka, M. 1975. Regional distribution of substance P in the spinal cord and nerve roots of the cat and the effect of dorsal root section. *Brain Res.* **87**, 1–11.

Takahashi, T., Konishi, S., Powell, D., Leeman, S. E., and Otsuka, M. 1974. Identification of the motoneuron-depolarizing peptide in bovine dorsal roots as hypothalamic substance P. *Brain Res.* **73,** 59–69.

Takeuchi, A., and Takeuchi, N. 1959. Active phase of frog's end-plate potential. *J. Neurophysiol.* **22,** 395–411.

Takeuchi, A., and Takeuchi, N. 1960. On the permeability of the end-plate membrane during the action of transmitter. *J. Physiol.* **154,** 52–67.

Takeuchi, A., and Takeuchi, N. 1962. Electrical changes in pre- and postsynaptic axons of the giant synapse of *Loligo. J. Gen. Physiol.* **45,** 1181–1193.

Tallan, H. H., Moore, S., and Stein, W. H. 1956. N-Acetyl-L-aspartic acid in brain. *J. Biol. Chem.* **219,** 257–264.

Tallman, J. F., Saavedra, J. M., and Axelrod, J. 1976. A sensitive enzymatic-isotopic method for the analysis of tyramine in brain and other tissues. *J. Neurochem.* **27,** 465–469.

Tasaki, I. 1939. The electro-saltatory transmission of the nerve impulse and the effect of narcosis upon the nerve fiber. *Am. J. Physiol.* **127,** 211–227.

Tasher, D. C., Abood, L. G., Gibbs, F. A., and Gibbs, E. L. 1959. Introduction of a new type of psychotropic drug: Cyclopentimine. *J. Neuropsychiat.* **1,** 266–273.

Taxt, T., Storm-Mathisen, J., Fonnum, F., and Iversen, L. L. 1977. Glutamate/aspartate in three defined systems of excitatory nerve endings in the hippocampal formation. *Proc. Int. Soc. Neurochem.* **6,** 647.

Taylor, K. M., and Snyder, S. H. 1971. Brain histamine: Rapid apparent turnover altered by restraint and cold stress. *Science* **172,** 1037–1039.

Taylor, K. M., and Snyder, S. H. 1972. Isotopic microassay of histamine in brain tissue. *J. Neurochem.* **19,** 1343–1358.

Taylor, K. M., and Snyder, S. H. 1973. The release of histamine from tissue slices of rat hypothalamus. *J. Neurochem.* **21,** 1215–1223.

Taylor, K. M., Gfeller, E., and Snyder, S. H. 1972. Regional localization of histamine and histidine in the brain of the rhesus monkey. *Brain Res.* **41,** 171–179.

Terenius, L., and Wahlstrom, A. 1974. Inhibitor(s) of narcotic receptor binding in brain extracts and cerebrospinal fluid. *Acta. Pharmacol. Toxicol.* **35,** suppl. #87, 55.

Terenius, L., and Wahlstrom, A. 1975a. Search for an endogenous ligand for the opiate receptor. *Acta Physiol. Scand.* **94,** 74–81.

Terenius, L., and Wahlstrom, A. 1975b. Morphine-like ligand for opiate receptor in human CSF. *Life Sci.* **16,** 1759–1764.

Terzuolo, C. A., and Araki, T. 1961. An analysis of intra- versus extracellular potential changes associated with activity of single spinal motoneurones. *Ann. N.Y. Acad. Sci.* **94,** 547–558.

Teschemacher, H., Opheim, K. E., Cox, B. M., and Goldstein, A. 1975. A peptide-like substance from pituitary that acts like morphine. *Life. Sci.* **16,** 1771–1776.

Thach, W. T. 1969. Discharge of Purkinje and cerebellar nuclear neurons during rapidly alternating arm movements in the monkey. *J. Neurophysiol.* **31,** 785–797.

Thierry, A. M., Stinus, L., Blanc, G., and Glowinski, J. 1973. Some evidence for the existence of dopaminergic neurons in the rat cortex. *Brain Res.* **50,** 230–234.

Tipton, K. F. 1968. The purification of pig brain mitochondrial monoamine oxidase. *Eur. J. Biochem.* **4,** 103–107.

Tong, J. H., and Kaufman, S. 1975. Tryptophan hydroxylase: Purification and some properties of the enzyme from rabbit hindbrain. *J. Biol. Chem.* **250,** 4152–4158.

Trautwein, W., and Dudel, J. 1958. Zum Mechanismus der Membranwirkung des Acetylcholin an der Herzmuskerfaser. *Arch. ges. Physiol.* **266,** 324–334.

Trebecis, A. K. 1973. Transmitters and reticulospinal neurons. *Exp. Neurol.* **40,** 297–308.

Tretiakoff, C. 1919. Contribution à l'étude de l'anatomie pathologique du locus niger de Soemmering avec quelques déductions relatives à la pathogenie des troubles du tonus musculaire de la maladie de Parkinson. Thèse de Paris.

Trevarthen, C. B., and Sperry, R. W. 1973. Perceptual unity of the ambient visual field in human commissurotomy patients. *Brain* **96,** 547–570.

Treux, R. C., and Carpenter, M. B. 1969. *Human Neuroanatomy,* 6th edition. Williams and Wilkins, Baltimore.

Tsukahara, N., Hultborn, H., and Murakami, F. 1974. Sprouting of cortico-rubral synapses in red nucleus neurons after destruction of the nucleus interpositus of the cerebellum. *Experientia* **30,** 57-58.

Tsukahara, N., Hultborn, H., Murakami, F., and Fujito, Y. 1975a. Electrophysiological study of formation of new synapses and collateral sprouting in red nucleus neurons after partial denervation. *J. Neurophysiol.* **38,** 1359-1372.

Tsukahara, N., Murakami, F., and Hultborn, H. 1975b. Electrical constants of neurons in the red nucleus. *Exp. Brain Res.* **23,** 49-64.

Turnbull, M. J., and Slater, P. 1970. Further studies on homocarnosine induced hypothermia. *Life Sci.* **9,** 83-89.

Turnbull, M. J., Slater, P., and Briggs, I. 1972. An investigation of the pharmacological properties of homocarnosine. *Arch. Int. Pharmacodyn. Ther.* **196,** 127-132.

Twarog, B. M., and Page, I. H. 1953. Serotonin content of some mammalian tissues and urine and a method for its determination. *Am. J. Physiol.* **175,** 157-161.

Uchizono, K. 1967. Synaptic organization of the Purkinje cells in the cerebellum of the cat. *Exp. Brain Res.* **4,** 97-113.

Udenfriend, S. 1966. Tyrosine hydroxylase. *Pharmacol. Rev.* **18,** 43-51.

Udenfriend, S., Zaltzman-Nirenberg, P., and Nagatsu, T. 1965. Inhibitors of purified beef adrenal tyrosine hydroxylase. *Biochem. Pharmacol.* **14,** 837-845.

Uhl, G. R., Kuhar, M. J., and Snyder, S. H. 1977. Neurotensin immunohistochemical localization in rat central nervous system. *Proc. Natl. Acad. Sci. USA* **74,** 4059-4063.

Umrath, K., and Fuerst, G. 1976. Identification of the transmitter substance of sensory nerves as gamma-aminobutyryl-L-histidine. *Zool. Jahrn. (Abt. Zool. Physiol.)* **80,** 274-295.

Ungar, F., Hittri, A., and Alivisatos, S. G. A. 1976. Drug antagonism and reversibility of the binding of indoleamines in brain. *Eur. J. Pharmacol.* **36,** 115-125.

Ungerstedt, U. 1971. Stereotaxic mapping of the monoamine pathways in the rat brain. *Acta Physiol. Scand. Suppl.* #367, 1-48.

Vallbo, Å. B. 1971. Muscle spindle response at the onset of isometric voluntary contractions in man. Time difference between fusimotor and skeletomotor affects. *J. Physiol.* **318,** 405-431.

Valverde, F. 1967. Apical dendritic spines of the visual cortex and light deprivation in the mouse. *Exp. Brain Res.* **3,** 337-352.

Valverde, F. 1968. Structural changes in the area striata of the mouse after enucleation. *Exp. Brain Res.* **5,** 274-292.

Van Balgooy, J. N. A., Marshall, F. D., and Roberts, E. 1972. Metabolism of intracerebrally administered histidine, histamine and imidazoleacetic acid in mice and frogs. *J. Neurochem.* **19,** 2341-2353.

Van den Berg, C. J., Matheson, D. F., Ronda, G., Reijnierse, G. L. A., Blokhuis, G. G. D., Kroon, M. C., Clarke, D. D., and Barbinkel, D. 1975. A model of glutamate metabolism in brain: A biochemical analysis of heterogeneous structure. In: *Metabolic Compartmentation and Neurotransmission,* edited by S. Berl, D. D. Clarke, and D. Schneider. Plenum Press, New York, pp. 515-543.

Van Gelder, N. M. 1969. The action *in vivo* of a structural analogue of GABA: Hydrazinopropionic acid. *J. Neurochem.* **16,** 1355-1360.

Van Gelder, N. M., and Courtois, A. 1972. Close correlation between changing content of specific amino acids in epileptogenic cortex of cats and severity of epilepsy. *Brain Res.* **43,** 477-484.

Van Gelder, N. M., Sherwin, A. L., and Rasmussen, T. 1972. Amino acid content of epileptogenic human brain: Focal versus surrounding regions. *Brain Res.* **40,** 385-393.

Vaupel, D. B., and Martin, W. R. 1976. Actions of methoxamine and tryptamine and their interactions with cyproheptadine and phenoxybenzamine on cat spinal cord segmental reflexes. *J. Pharmacol. Exp. Ther.* **196,** 87-96.

Verdiere, M., Rose, C., and Schwartz, J. C. 1974. Synthesis and release of ³H-histamine in slices from rat brain. *Agents & Actions* **4,** 184-185.

Victor, M., Adams, R. D., and Collins, G. H. 1971. *The Wernicke-Korsakoff-Syndrome,* Blackwell Scientific Publications, Oxford.

Vizi, E. S., and Knoll, J. 1976. The inhibitory effect of adenosine and related nucleotides on the release of acetylcholine. *Neurosci.* **1,** 391–398.

Vogt, M. 1954. The concentration of sympathin in different parts of the central nervous system under normal conditions and after the administration of drugs. *J. Physiol.* **123,** 451–481.

von Euler, U. S. 1944. Identification of a urine base with nicotine-like action. *Nature* **154,** 17.

von Euler, U. S., and Gaddum, J. H. 1931. An unidentified depressor substance in certain tissue extracts. *J. Physiol.* **72,** 74–87.

von Euler, U. S., and Pernow, B. 1977. *Substance P,* Raven Press, New York.

Wada, J. A., Clark, R., and Hamm, A. 1975. Cerebral hemispheric asymmetry in humans. *Arch. Neurol.* **32,** 239–246.

Wall, P. D., and Egger, M. D. 1971. Formation of new connexions in adult rat brains after partial deafferentation. *Nature* **232,** 542–545.

Wallace, E. F., Krantz, M. J., and Lovenberg, W. 1973. Dopamine-β-hydroxylase: A tetrameric glycoprotein. *Proc. Natl. Acad. Sci. USA* **70,** 2253–2255.

Walter, W. G. 1967. Electrical signs of association, expectancy and decision in the human brain. *Electroenceph. Clin. Neurophysiol. Suppl.* **25,** 258–263.

Walters, J. R., and Roth, R. H. 1972. Effect of gamma-hydroxybutyrate on dopamine and dopamine metabolites in the rat striatum. *Biochem. Pharmacol.* **21,** 2111–2121.

Walters, J. R., Bunney, B. S., and Roth, R. H. 1975. Piribedil and apomorphine: pre- and post-synaptic effects on dopamine synthesis and neuronal activity. In: *Dopaminergic Mechanisms,* edited by D. B. Calne, T. N. Chase, and A. Barbeau, Raven Press, New York, pp. 273–284.

Watson, J. D., editor. 1970. *Transcription of Genetic Material,* Cold Spring Harbor Symposia on Quantitative Biology, vol. 35. Cold Spring Harbor Laboratory, Long Island, N.Y.

Watson, S. J., Akil, H., Sullivan, S., and Barchas, J. D. 1977a. Immunocytochemical localization of methionine enkephalin. *Life Sci.* **21,** 733–738.

Watson, S. J., Barchas, S. J., and Li, C. H. 1977b. β-Lipotropin: Localization of cells and axons in rat brain by immunocytochemistry. *Proc. Natl. Acad. Sci. USA* **74,** 5155–5158.

Watson, W. E. 1974. Cellular responses to axotomy and to related procedures. *Brit. Med. Bull.* **30,** 112–115.

Weiner, N. 1970. Regulation of norepinephrine biosynthesis. *Ann. Rev. Pharm.* **10,** 273–290.

Weiss, P. 1960. The concept of perpetual neuronal growth and proximodistal substance convection. In: *4th International Neurochemistry Symposium,* edited by S. S. Kety and J. Elkes. Pergamon Press, Oxford, pp. 220–242.

Weiss, P. 1970. Neuronal dynamics and neuroplasmic flow. In: *The Neurosciences, 2nd Study Program,* edited by F. O. Schmitt. Rockefeller University Press, New York, pp. 840–850.

Weiss, P., and Hiscoe, H. B. 1948. Experiments on the mechanism of nerve growth. *J. Exp. Zool.* **107,** 315–396.

Wenk, M., von Hahn, H. P., and Honegger, C. G. 1976. Partial separation of synaptosomes accumulating 4-aminobutyrate or glutamate by zonal centrifugation on a discontinuous sucrose gradient. *Hoppe Seylers Z. physiol. Chem.* **357,** 1469–1476.

Werman, R., Davidoff, R. A., and Aprison, M. H. 1968. Inhibitory effect of glycine in spinal neurons in the cat. *J. Neurophysiol.* **31,** 81–95.

Westrum, L. E., and Black, R. G. 1971. Fine structural aspects of the synaptic organization of the spinal trigeminal nucleus (pars interpolaris) of the cat. *Brain Res.* **25,** 265–287.

White, T. 1959. Formation and catabolism of histamine in brain tissue *in vitro. J. Physiol.* **149,** 34–42.

Whittaker, V. P., and Gray, E. G. 1962. The synapse: Biology and morphology. *Brit. Med. Bull.* **18,** 223–228.

Whittaker, V. P., Michaelson, I. A., and Kirkland, R. J. A. 1964. The separation of synaptic vesicles from nerve ending particles ("synaptosomes"). *Biochem. J.* **90,** 293–303.

Wiesel, T. N., and Hubel, D. H. 1963. Single-cell responses in striate cortex of kittens deprived of vision in one eye. *J. Neurophysiol.* **26,** 1003–1017.

Wiesendanger, M. 1969. The pyramidal tract. Recent investigations on its morphology and function. *Ergeb. Physiol. Biol. Chem. Exptl. Pharmakol.* **61,** 72–136.

Wilson, V. J., and Burgess, P. R. 1962. Disinhibition in the cat spinal cord. *J. Neurophysiol.* **25,** 392–404.

Wilson, V. J., Yoshida, M., and Schor, R. H. 1970. Supraspinal monosynaptic excitation and inhibition of the thoracic back motoneurons. *Exp. Brain Res.* **11,** 282–295.

Winders, A., and Vogt, W. 1907. Synthese des Imidazolylathylamins. *Ber dt. Chem. Ges.* **40,** 3691–3695.

Wise, C. D., Baden, M. M., and Stein, L. 1975. Postmortem measurements of enzymes in human brain: Evidence of a central noradrenergic deficit in schizophrenia. *J. Psychiat. Res.* **11,** 185–198.

Wise, S. P. 1976. Retrograde axonal transport of adenosine derivatives in the central nervous system. *Neuroscience Abst.* #56.

Wofsey, A. R., Kuhar, M. J., and Snyder, S. H. 1971. A unique synaptosomal fraction which accumulates glutamic and aspartic acids in brain tissue. *Proc. Natl. Acad. Sci. USA* **68,** 1102–1106.

Wong, C. L., and Robert, M. B. 1975. The possible role of histamine and H_1 and H_2 receptors in the development of morphine tolerance and physical dependence in mice. *Agents & Action* **5,** 476–483.

Wood, J. D. 1975. The role of gamma-aminobutyric acid in the mechanism of seizures. *Progr. Neurobiol.* **5,** 77–95.

Wood, J. G., McLaughlin, B. J., and Vaughn, J. E. 1976. Immunocytochemical localization of glutamate decarboxylase in electron microscopic preparations of rodent CNS. In: *GABA in Nervous System Function*, edited by E. Roberts, T. N. Chase, and D. B. Tower. Raven Press, New York, pp. 133–148.

Woolley, D. W., and Shaw, E. 1954. A biochemical and pharmacological suggestion about certain mental disorders. *Science* **119,** 587–588.

Wu, J. Y. 1976. Purification, characterization and kinetic studies of GAD and GABA-T from mouse brain. In: *GABA in Nervous System Functions*, edited by E. Roberts, T. N. Chase, and D. B. Tower. Raven Press, New York, pp. 7–60.

Wu, J. Y., Matsuda, T., and Roberts, E. 1973. Purification and characterization of glutamate decarboxylase from mouse brain. *J. Biol. Chem.* **248,** 3029–3034.

Wurtman, R. J., Axelrod, J., and Kelly, D. E. 1968. *The Pineal.* Academic Press, New York.

Wurtman, R. J., Pohorecky, L. A., and Baliga, B. S. 1972. Adrenocortical control of the biosynthesis of epinephrine and proteins in the adrenal medulla. *Pharmacol. Rev.* **24,** 411–426.

Wyatt, H. J., and Daw, N. W. 1976. Specific effects of neurotransmitter antagonists on ganglion cells in rabbit retina. *Science* **191,** 204–205.

Wyatt, R. J., and Murphy, D. L. 1975. Neurotransmitter-related enzymes in major psychiatric disorders. In: *Biology of the Major Psychoses: A Comparative Analysis*, vol. 54, Res. Pub. Assoc. Res. Nerv. & Ment. Dis., edited by D. X. Freedman. Raven Press, New York, pp. 289–298.

Wyatt, R. J., Chase, T. N., Scott, J., and Snyder, F. 1970. Effect of L-Dopa on the sleep of man. *Nature* **228,** 999–1000.

Yajima, H., and Kitagawa, K. 1973. Studies on peptides. XXXIV. Conventional synthesis of the undecapeptide amide corresponding to the entire amino acid sequence of bovine substance P. *Chem. Pharm. Bull. (Tokyo)* **21,** 682–683.

Yamamura, H. I., and Snyder, S. H. 1973. High affinity transport of choline into synaptosomes of rat brain. *J. Neurochem.* **21,** 1355–1374.

Yamamura, H. I., and Snyder, S. H. 1974. Muscarinic cholinergic binding in rat brain. *Proc. Natl. Acad. Sci. USA* **71,** 1725–1729.

Yamamura, H. I., Kuhar, M. J., Greenberg, D., and Snyder, S. H. 1974. Muscarinic cholinergic receptor binding: Regional distribution in monkey brain. *Brain Res.* **66,** 541–546.

Yarowsky, P. J., and Carpenter, D. O. 1976. Aspartate: Distinct receptors on Aplysia neurons. *Science* **192,** 807–809.

Yasunobu, K. T., Igaue, I., and Gomes, B. 1968. The purification and properties of beef liver mitochondrial monoamine oxidase. In: *Advances in Pharmacology*, vol. 6, edited by S. Garattini and P. A. Shore. Academic Press, New York, pp. 43–59.

Yoon, M. G. 1975. On topographic polarity of the optic tectum in the goldfish. In: *The Synapse, Cold Spring Harbor Symp. Quant. Biol.* **40,** 503–519.

York, D. H. 1970. Possible dopaminergic pathway from substantia nigra to putamen. *Brain Res.* **20,** 233–249.

York, D. H., and Lentz, S. 1976. Nigral modulation of peripheral inputs on cells in the striatum of rats. *Neuroscience Abst.* #100.

Yoshida, M., Rabin, A., and Anderson, M. 1971. Two types of monosynaptic inhibition of pallidal neurons produced by stimulation of the diencephalon and the substantia nigra. *Brain Res.* **30,** 235–239.

Youdim, M. B. H., Hamon, M., and Bourgoin, S. 1974. Purification of pig brainstem tryptophan hydroxylase and some of its properties. In: *Advances in Biochemical Pharmacology*, vol. 11, edited by E. Costa, G. L. Gessa, and M. Sandler. Raven Press, New York, pp. 13–17.

Youdim, M. B. H., Green, A. R., and Grahame-Smith, D. G. 1976. The role of 5-hydroxytryptamine, dopamine and MAO in the production of the hyperactivity syndrome. In: *Advances in Parkinsonism*, edited by W. Birkmayer and O. Hornykiewicz, F. Hoffmann-LaRoche Co., Basel.

Young, A. B., and Snyder, S. H. 1973. Strychnine binding associated with glycine receptors of the central nervous system. *Proc. Natl. Acad. Sci. USA* **70,** 2832–2836.

Young, A. B., Oster-Granite, M. L., Herndon, R. M., and Snyder, S. H. 1974a. Glutamic acid: Selective depletion by viral-induced granule cell loss in hamster cerebellum. *Brain Res.* **73,** 1–13.

Young, A. B., Zukin, S. R., and Snyder, S. H. 1974b. Interaction of benzodiazapines with central nervous glycine receptors: Possible mechanism of action. *Proc. Natl. Acad. Sci. USA* **71,** 2246–2250.

Zaidel, E. 1976a. Unilateral auditory language comprehension on the token test following cerebral commisurotomy and hemispherectomy. *Neuropsychol.* **15,** 1–18.

Zaidel, E. 1976b. Auditory language comprehension in the right hemisphere following cerebral commisurotomy and hemispherectomy. A comparison with child language and aphasia. In: *The Acquisition and Breakdown of Language: Parallels and Divergencies*, edited by E. Zurif and A. Caramazza. John Hopkins University Press, Baltimore.

Zieglgansberger, W., and Puil, E. A. 1973. Actions of glutamic acid on spinal neurons. *Exp. Brain Res.* **17,** 35–49.

Zimmerman, F. T., and Ross, S. 1944. Effect of glutamic acid and other amino acids on maze learning in the white rat. *Arch. Neurol. Psychiat.* **51,** 446–451.

Zimmerman, F. T., Burgemeister, B. B., and Putnam, T. J. 1946. Effects of glutamic acid on mental functioning in children and in adolescents. *Arch. Neurol. Psychiat.* **56,** 489–506.

Index